ILLUSTRATED RESIDENTIAL AND COMMERCIAL CONSTRUCTION

PETER A. MANN

Mohawk College

of Applied Arts and Technology

Hamilton, Ontario, Canada

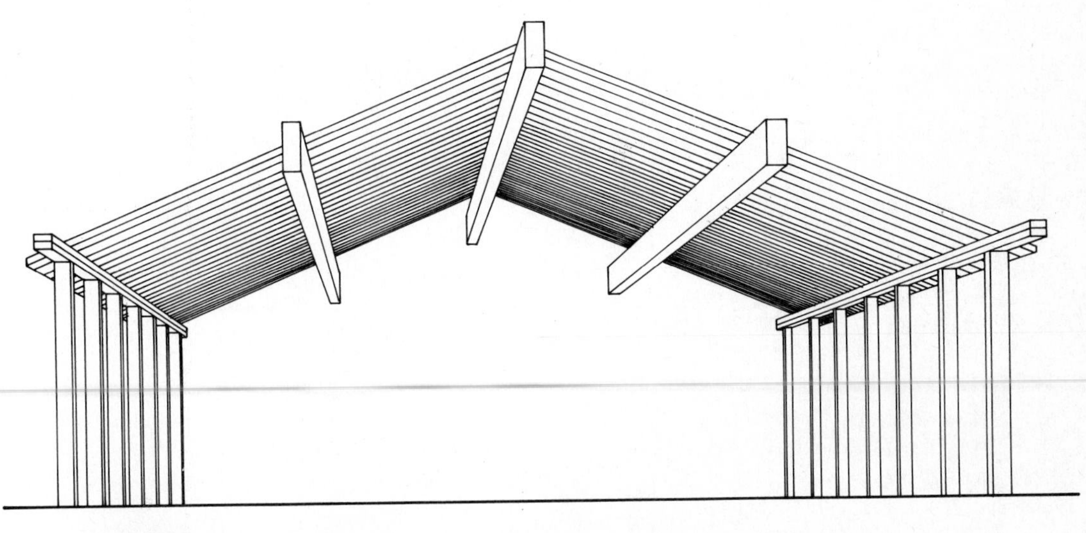

PRENTICE HALL

Englewood Cliffs, New Jersey 07632

Library of Congress Cataloging-in-Publication Data

Mann, Peter A.
Illustrated residential and commercial construction / Peter A. Mann.
p. cm.
Includes index.
ISBN 0-13-453250-3
1. Building. 2. House construction. I. Title.
TH145.M2798 1989
690--dc19 88-4135
CIP

Editorial/production supervision: Barbara Cassel
Interior design, cover design, and page layout: Peter Mann
Manufacturing buyer: Robert Anderson

A division of Simon & Schuster
Englewood Cliffs, New Jersey 07632

Printed in the United States of America

10 9 8 7 6 5 4 3 2

ISBN 0-13-453250-3

Prentice-Hall International (UK) Limited, London
Prentice-Hall of Australia Pty. Limited, Sidney
Prentice-Hall Canada Inc., Toronto
Prentice-Hall Hispanoamericana, S.A., Mexico
Prentice-Hall of India Private Limited, New Delhi
Prentice-Hall of Japan, Inc., Tokyo
Simon & Schuster, Asia Pte. Ltd., Singapore
Editora Prentice-Hall do Brasil, Ltda, Rio de Janeiro

For Denise, Jason, and Craig.

With grateful thanks to my colleagues, Horst Jung, Robert Nelson, Michael Haslam, Kenneth Mercer, and Daniel Nichols for their invaluable technical proofreading, and to Barbara Cassel of Prentice Hall for her cheerful encouragement.

CONTENTS

To assist in locating specific items of information, individual page titles are summarized on the first page of each chapter.

ILLUSTRATED RESIDENTIAL AND COMMERCIAL CONSTRUCTION describes in detail the materials and construction techniques of modern construction together with the basic principles of competent building design. All major building elements from building location and foundations to electrical and mechanical services are presented in a sequence similar to that of actual construction. In each subject chapter, comparisons are made between the wide range of options available in construction materials and installation methods.

In addition to standard construction techniques, throughout the book emphasis is placed on designs and construction methods that increase the energy efficiency of a building's structure. This aspect of the construction industry is now a major part of a modern building's function and designers and builders in all climatic regions must be well aware of the systems and methods involved. Chapter 10 is devoted entirely to this subject and details the theoretical and practical aspects of energy efficiency.

Information is presented in a somewhat unique form, with each page totally hand-drawn and lettered. A large proportion of the text is written in point form for greater information density per page. This unusual method of production has distinct advantages over standard typesetting procedures. Most significant is that each individual page or pair of facing pages have been designed as completely self-contained information units. All text is located beside or close to each illustration, eliminating illustration "lag" and making figure numbers unnecessary. Other advantages include extensive page number cross-referencing, updating of information immediately prior to publication, and the ability to arrange placement of illustrations and text to create interesting visual effects.

ILLUSTRATED RESIDENTIAL AND COMMERCIAL CONSTRUCTION is ideally suited for all students engaged in building studies, including first- and second-year architectural courses, building technologist and technician courses, apprenticeship training, and upper level high school technical courses. In addition, some of the more specialized chapters, such as Energy Conservation, Basic Structures and Passive Solar Design, and Metric Measure, will be of significant value to other interested groups. Finally, persons already involved in the construction business who require a broad-based reference book, and members of the general public planning new construction or renovation projects, will also find much of interest.

Refer to Chapter 1: Introduction, for detailed information on US Customary/metric applications, heading and page numbers, building codes, construction industry information sources, the rapid evolution of building materials, and the illustration drawing methods.

Instructors should refer to the Instructor's Manual (ISBN 0-13-453243-0) for this text, which contains additional chapter information and a complete energy-efficiency in construction course outline.

CHAPTER PAGE TITLES

GENERAL CONTENTS AND ENERGY EFFICIENCY

This introductory chapter is intended to provide an insight into the design and layout of ILLUSTRATED RESIDENTIAL AND COMMERCIAL CONSTRUCTION, hopefully allowing the reader to gain maximum benefit from the information presented in this book.

GENERAL CONTENT

In any textbook, especially those dealing with construction, choices have to be made regarding the range and type of subjects covered since it is not possible in one volume to detail adequately all aspects of the industry. As defined in the Preface, this book deals with the complete construction process for residential and light commercial buildings, describing, quantifying, and comparing industry-wide accepted and recommended construction practices. "Traditional", "conventional", and new methods are compared. Where pertinent, detailed descriptions of specific cutting, fastening, or other site work have been included, but as a general rule, this skilled trade's aspect of construction has not been emphasized. Generalized design information is also included throughout the book, and part of Chapter 3 is devoted to the principles of energy-saving passive solar design features.

ENERGY EFFICIENCY

As also defined in the Preface, a major thrust of this book is to detail practical techniques for achieving an energy-efficient building shell. In all regions the need for energy-saving (and therefore cost-saving) structures has assumed significant importance. The extent to which such techniques are employed is generally determined by the severity of the climate (hot or cold), the price range, and the intended market of the building. Considering the length of a building's useful life, the prospect of long-term rising energy costs, despite the occasional downward blip in fuel prices, is an added incentive. Throughout this book, while illustrating the standard methods of construction practice, reference is also made to additional energy-saving items. Chapter 10, however, is devoted entirely to the subject of energy conservation in both theory and practice. Most important, conservation techniques are presented as logical extensions of the traditional building process without resort to the more exotic and unnecessarily costly specialized techniques. Good conservation methods should not carry the penalty of excessive cost; there must be an acceptable payback period wherein all associated extra costs are fully recovered.

TEXT

The text has been hand-lettered to add a certain "style" to the book and to allow accurate and convenient location of the text to the illustrations. The majority of information is given in point form rather than narrative form as in most textbooks. This allows a compression and intensification of information, leaving more room for the extensive number of illustrations.

PAGE SETUP AND METRIC

PAGE SETUP

Since each page of this book is hand-drawn, advantage can be taken of page setup methods that are difficult to achieve with standard typeset production. Take the page headings, for example. Because each page is treated as a separate and complete information unit, the main heading is used to identify the page contents, while a subheading, positioned over the page number, identifies the general topic for those readers (most of us) who refer to the Index only as a last resort. Reproduced below as an example is page 7-4.

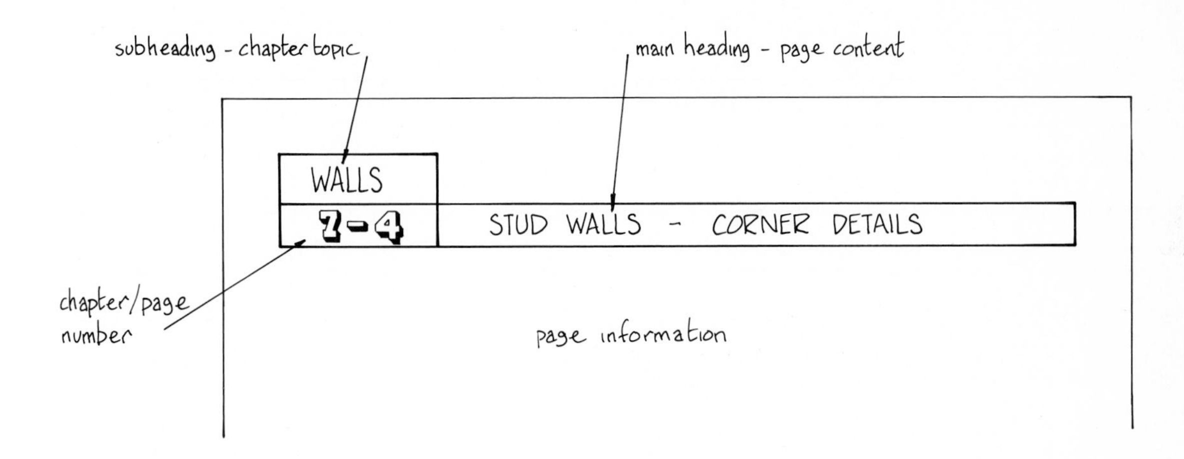

REFERENCES

Extensive use is made of cross-referencing between pages so that related information can be effectively combined. This useful tool is possible because the contents of each page was set during the initial writing process.

METRIC MEASURE

For textbook authors the use of metric measurements poses something of a problem. Officially, metric is destined to become the only system of measurement in commerce and industry, including construction. Unofficially, the progress toward metrification has been remarkably slow. In any event, both the inch-pound and metric systems must exist side by side for many years, if only for the fact that most existing buildings are built to U.S. Customary (or Imperial) measurements.

To assist the users of both systems the most effective compromise is to give all dimensions in U.S. Customary with equivalent metric sizes in parentheses. Note, however, that in almost all cases, the metric size given is "hard metric" and is not simply a "soft conversion". In cases where a soft conversion has been used, an asterisk (*) has been added (for precise details, see 2-3). In addition, Chapter 2 is devoted to metric standards as applied to all phases of the construction industry.

1-4 MATERIALS AND CODES

CONSTRUCTION MATERIALS

The building sector is probably one of the most prolific industries in its production of materials, equipment, and processes. New materials, new uses for old materials, and updated methods of installation for all areas of construction are constantly being brought forward by an amazing number of manufacturers, making it extremely difficult for designers and builders to remain current. For example, at one time there used to be little variation in modular concrete blocks (see 5-5) other than face styles and finishes. Since the initial invention of a concrete block with a high insulation value (see 10-38), there is now a wide range of competing products with varying installation methods and differing claims of performance.

The material types discussed in this book are those in general use over the whole of North America. For every "generic" type of material illustrated, there will be many variations on that theme produced by individual manufacturers.

BUILDING CODES

Throughout this book the words "refer to local codes" or similar phrases are mentioned often. All aspects of construction are covered and regulated by one or more building codes, such as ANSI (American National Standards Institute), NBC (National Building Code), UBC (Uniform Building Code), and many other national, state, provincial, regional, and local codes. The information in this book is applicable to most regions and localities in the U.S.A. and Canada, but ultimately designers and builders must be aware of and comply with the codes covering each local area.

INFORMATION SOURCES

To remain current with new materials, equipment, installation methods, and codes, designers must be constantly in touch with up-to-date information sources. Depending on the type of information required, there are a wide variety of sources available to professionals and the general public. Sources can be classified under four general headings:

Directories, Catalogs, and Libraries. There are several generalized directories and catalogs that list trade associations, companies, and government departments from whom information can be obtained. Examples are the Encyclopedia of Associations, National Trade and Professional Associations of the U.S., Directory of the Forest Products Industry, Thomas Register of American Manufacturers, and many more. A well-stocked public reference library should contain one or more of these directories.

Trade Associations. There are hundreds of trade associations whose main purpose is to disseminate information about the trade group they represent. Some are quite prolific sources and all will supply a publications list on request. In many cases information is free, although some do charge for larger items. Examples of trade associations include American Plywood Association, American Concrete Institute, Brick Institute of America, and the National Forest Products Association.

Individual Manufacturers. Of the thousands of corporate manufacturing companies, almost all are happy to supply their own promotional literature at no charge. Extensive design and product information is often available from the major manufacturers, although this information naturally tends to favor particular products. Provided that careful comparisons are made between competing claims, this type of information is very useful.

Government Agencies. For the construction industry information comes primarily from the separate departments of Energy, Commerce, Housing and Urban Development (HUD), and Agriculture. While the large volume of available information is extremely useful, the trick is knowing how to extract it from the vast governmental labyrinth. Most helpful is the U.S. Government Manual, available at libraries, which catalogs all federal departments and agencies. A more direct approach is to contact a Federal Information Center, where the staff can provide references to specific information sources.

In countries other than the United States, similar sets of information sources are readily available. In Canada, for example, in addition to sources comparable to those mentioned above, considerable research into energy-efficient structures has been undertaken and published by several agencies, including the Canadian Home Builder's Association, Ministry of Energy, Mines and Resources, National Research Council, and several provincial government departments.

ILLUSTRATIONS

Some of the more inquisitive readers, especially those not familiar with drafting techniques, may be interested in the types of illustration drawings used in this book. Aside from the standard "flat" two-dimensional orthographic views, a variety of pictorial "three-dimensional" drawing types have been used to convey information efficiently and to create interesting effects. Shown below are examples of each type, together with a brief description of the drawing method.

ORTHOGRAPHIC DRAWINGS

This flat, two-dimensional drawing system is the standard method for virtually all construction and engineering drawings. It is ideal for the clear presentation of details and dimensions, but requires the reader to interpret and mentally visualize the information in a three-dimensional, pictorial form. As shown in the example at the right, a large proportion of orthographics are drawn as sections, or cuts through parts of the structure. Since the aim of this book is to present information as visually as possible, orthographic drawings are generally used only where accurate dimensional or locational information is required. See page 10-21 for another example.

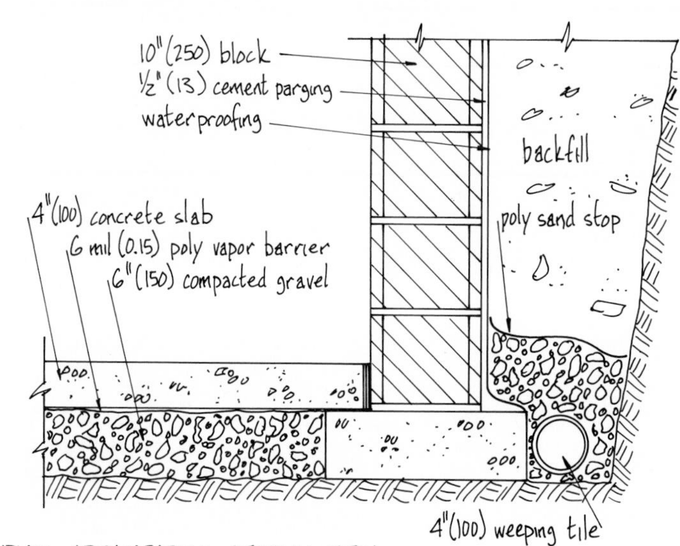

TYPICAL ORTHOGRAPHIC SECTION VIEW

OBLIQUE DRAWINGS

This is a pictorial type of drawing where one face of the object is positioned parallel to the viewer (i.e., squarely facing the viewer), while the side and top faces are drawn backward at a fixed angle, usually 45°. Since this type of view is impossible in real life, the object appears quite distorted to the eye of the viewer. The one advantage of this method is that circles, or other hard-to-draw features, can be located in the front face of the object and treated almost as an orthographic drawing. Oblique drawings are used sparingly in this book. See 3-11 (center) for an actual example.

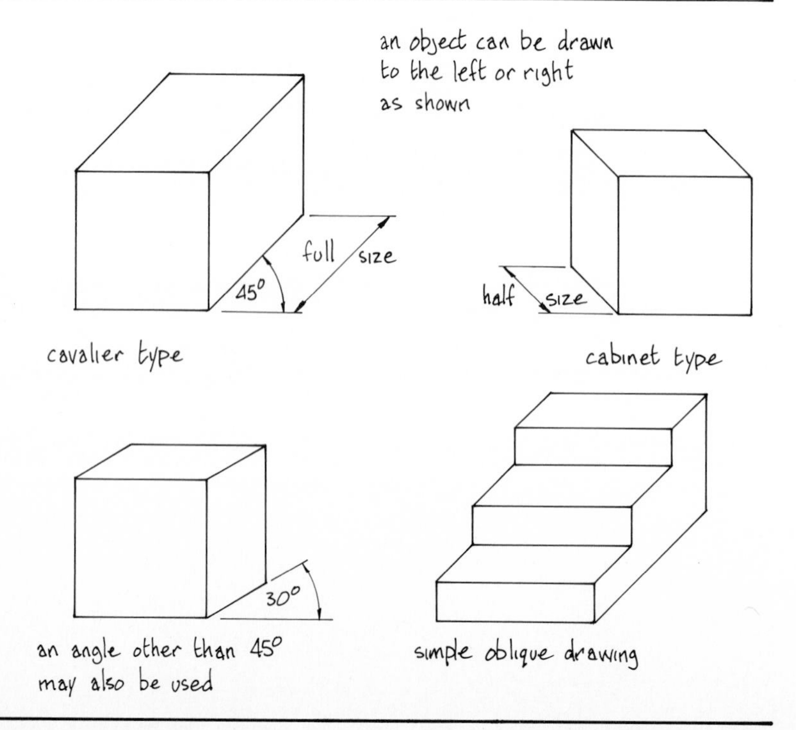

ILLUSTRATIONS

AXONOMETRIC DRAWINGS

An axonometric drawing has all three faces of the object receding from the viewer. An illustrator can choose between several variations of this method, and in this book, two types, isometric and trimetric are used:

ISOMETRIC

This is the most popular pictorial drawing, combining ease of construction with equal emphasis of all three faces. Receeding edges are drawn at 30° to the horizontal on each side of the vertical, and all dimensions are measured at full scale. Nonisometric lines (those not vertical or at 30°) such as the stair in the example at the right, are simply drawn between measured points. Isometric is used extensively in this book, and page 9-9 provides additional examples.

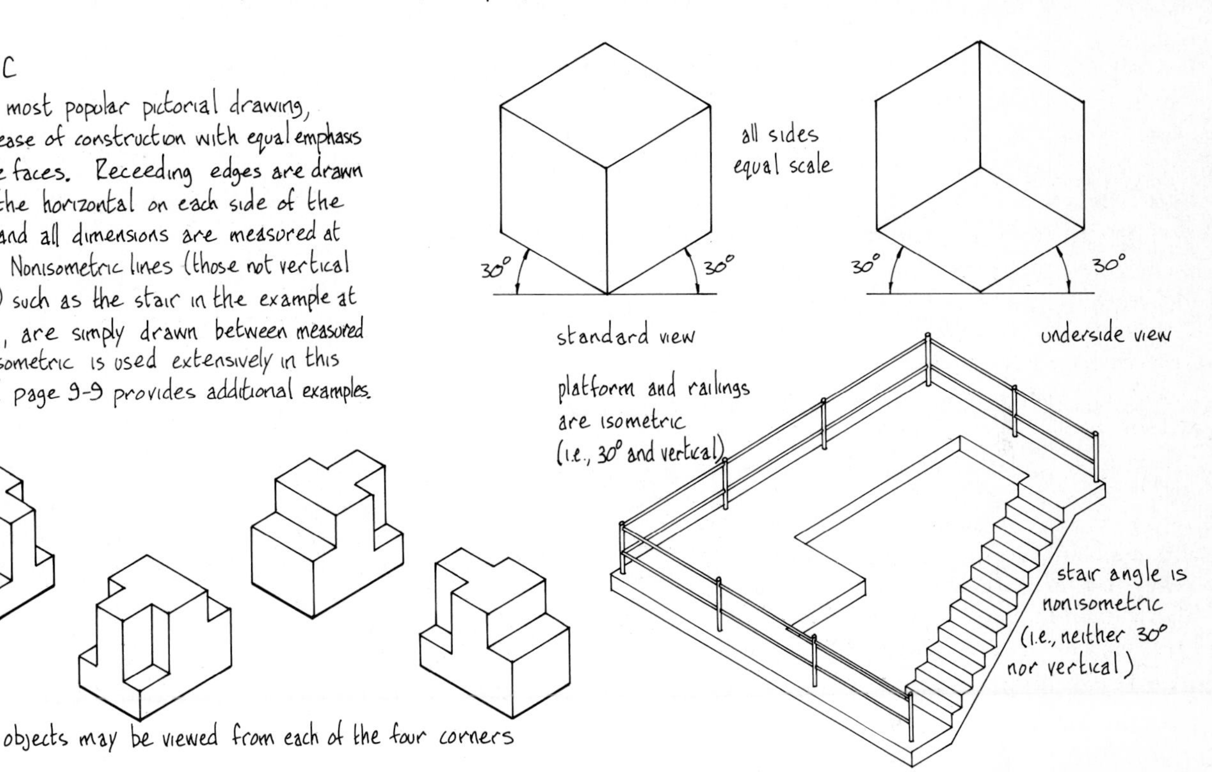

TRIMETRIC

A trimetric drawing has each face receeding from the viewer as in isometric, but the backward angles on each side of the vertical are not equal. In this book the predominant angles are 15° and 30°, these being the most convenient from a drafting point of view, but almost any pair of angles can be used. Illustrations in this mode tend to deemphasize the plan face while "softening" the whole image, and are used here as extensively as isometrics (see page 7-2 for an example).

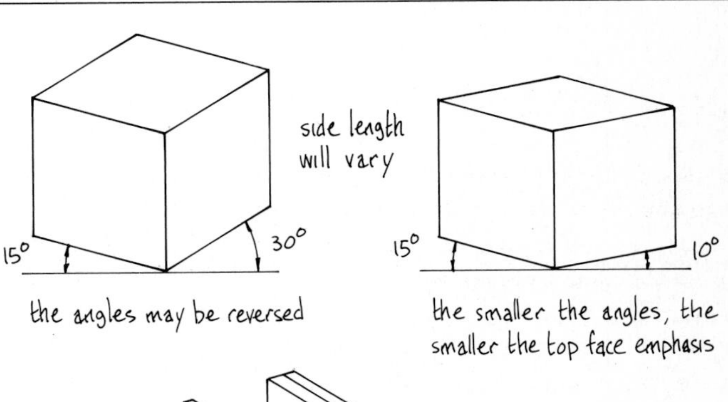

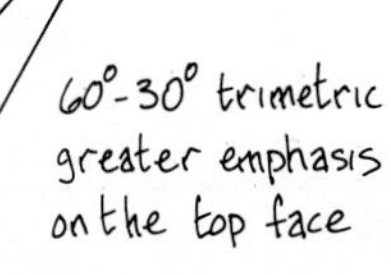

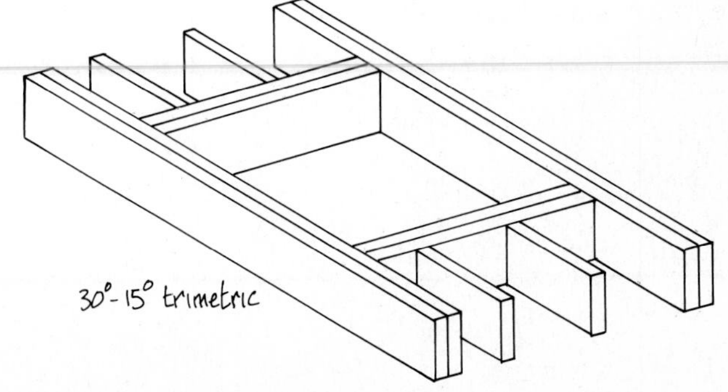

ILLUSTRATIONS

PERSPECTIVE DRAWINGS

In contrast to the preceding drawing methods where the image produced is relatively static, the perspective method produces an image which is very close to that actually seen by the eye. In addition, the point of view can be placed in almost any position, allowing the illustrator wide latitude in image presentation. Technically, the production of a proper perspective drawing is quite complex and time-consuming, requiring the setup of a plan and elevation before the perspective is drawn. In practice, however, a certain amount of creative cheating can be employed, reducing drawing time considerably.

Two methods of perspective drawing are used in this book; one-point and two-point. In one-point (or parallel) one face of the object is drawn parallel to the viewer, while the receding faces are drawn to converge and vanish at a single point on the horizon. The horizon is the viewer's eye level as seen in the far distance. In two-point (or angular) perspective, all faces of the object recede to two separate vanishing points on the horizon. Because perspectives are such important illustrative methods, they are used extensively in this book (see 3-9 and 13-13 for an example of each method).

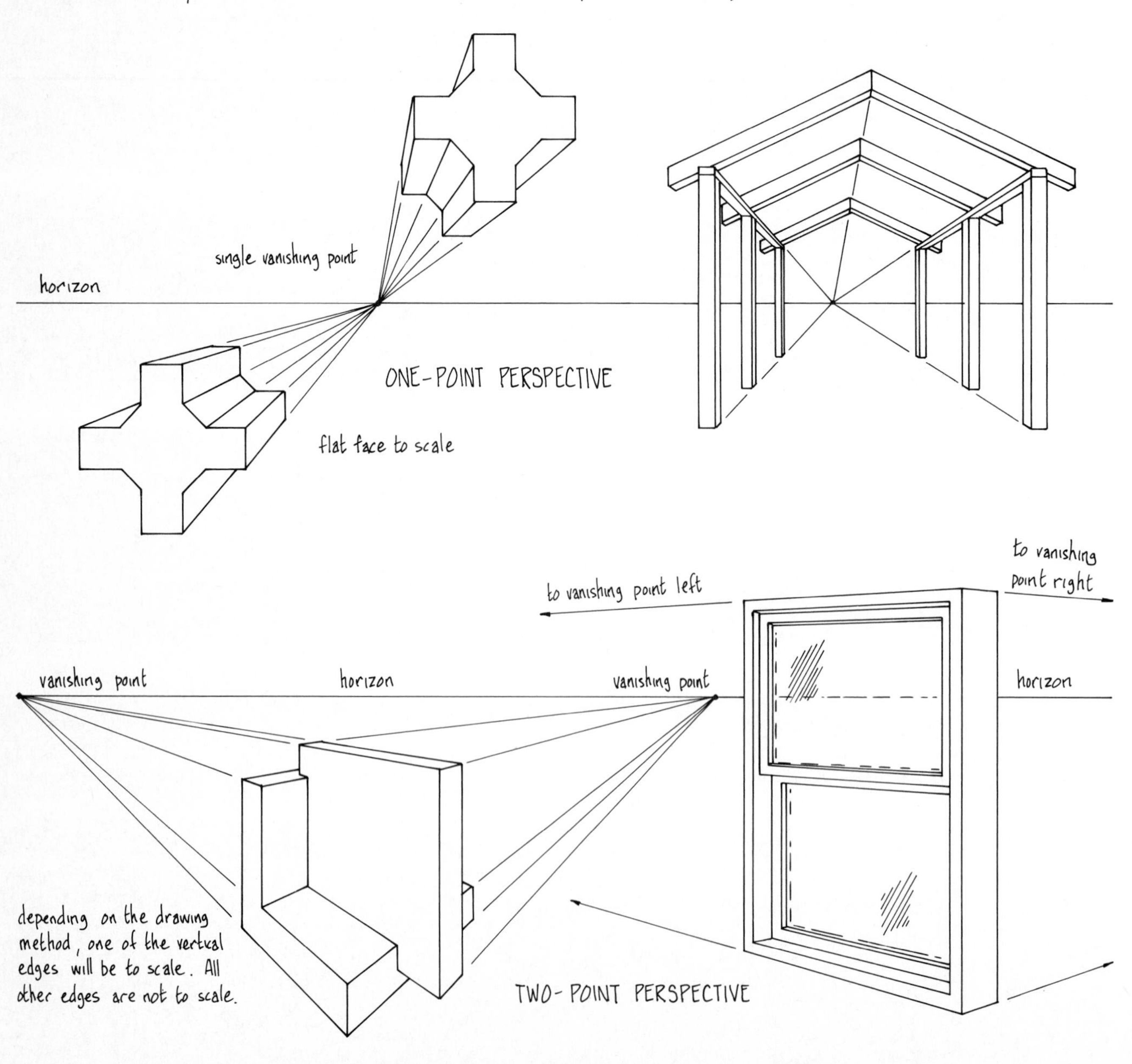

METRIC MEASURE

CHAPTER PAGE TITLES

The information contained in this chapter has been taken from the reference book "Construction Metriguide", published by Domtar Construction Materials Ltd., Montreal, Quebec, Canada, whose kind permission is gratefully acknowledged.

MEASUREMENT SYSTEMS

SYSTEMS OF MEASUREMENT

There are two systems of measurement used by the industrialized nations of the world: the inch-pound system and the metric system. Historically, most English-speaking countries use or have used some variation of the inch-pound system, while most European, South American, African, and Asian countries have used some variation of the metric system. Adding to the difficulties of trade and information transfer between countries using opposing systems are the many size differences in the measurement standards set by each country. These differences are especially common in the inch-pound system, as will be seen in the following pages.

Confusion even extends to the names given to the inch-pound system. In the United States it is correctly known as the U.S. Customary system, which differs in some respects (notably fluid measure) from the U.K. Imperial system. Canada and Australia use an Imperial system almost identical with the U.K. version. The overall system is based on a wide variety of units having roots in the sizes of various body parts or quantities that served a specific purpose at the time of their inception. Most units have little or no logical relationship with each other, making the system cumbersome to use and needlessly complex.

THE METRIC SYSTEM

In comparison, the metric system is based on clearly defined and logically interrelated units, making it easy to learn and simple to use even though there were some minor standards differences between nations. Consequently, in an effort to standardize all measurement systems, agreement was reached in 1972 on an international system of standard metric measure by the 30 nations of the International Organization for Standardization (I.S.O.). The international system is called S.I. (Système International d'Unités) and not only replaces U.S. Customary and Imperial measure but also supersedes and improves on the older metric systems. Both the United States and Canada were signatories to this agreement and the present movement toward universal metric conversion for all sectors of business and industry was begun at that time.

CURRENT TRENDS

Since 1972 the process of metric conversion has been somewhat erratic although generally progressive. Industries with extensive export opportunities have quickly converted products to compete in foreign markets. Other industries, especially construction, having little or no pressure to compete abroad, are only slowly pursuing the conversion process. It would appear that the transition phase from U.S. Customary to full metric conversion will be a long one. Current estimates are for a 50% conversion by the year 2000 (American National Metric Council).

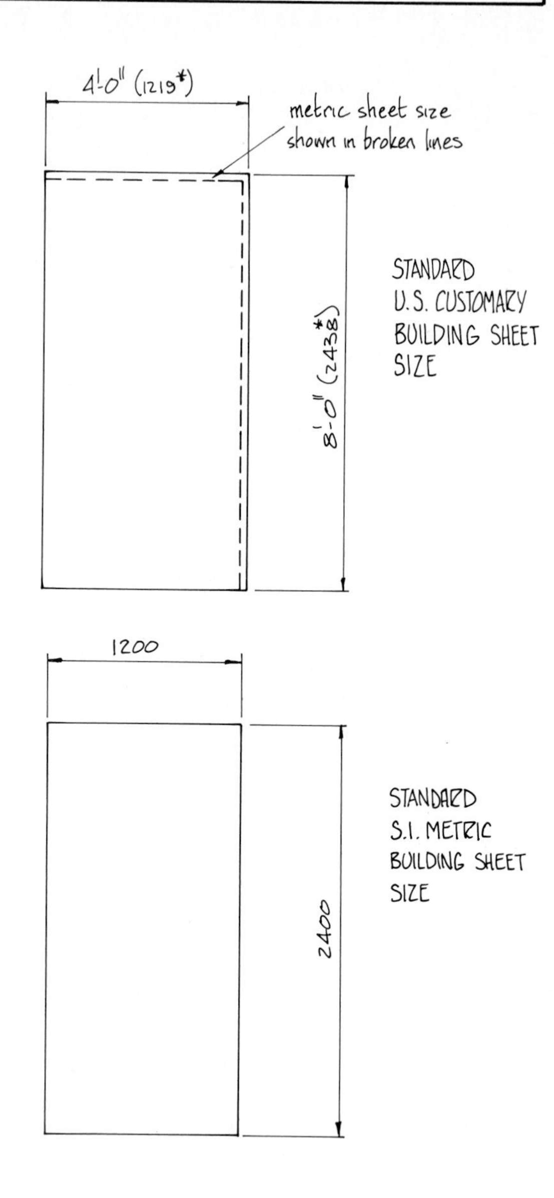

An example of hard metric conversion is show above (see also 2-7). The standard building sheet is 4'-0" x 8'-0" in Customary measure with a soft conversion size of 1219 x 2438 mm. A true metric sheet has dimensions of 1200 x 2400. The two sheets are not compatible and should not be mixed with, or substituted for, each other.

* soft conversion (see the next page).

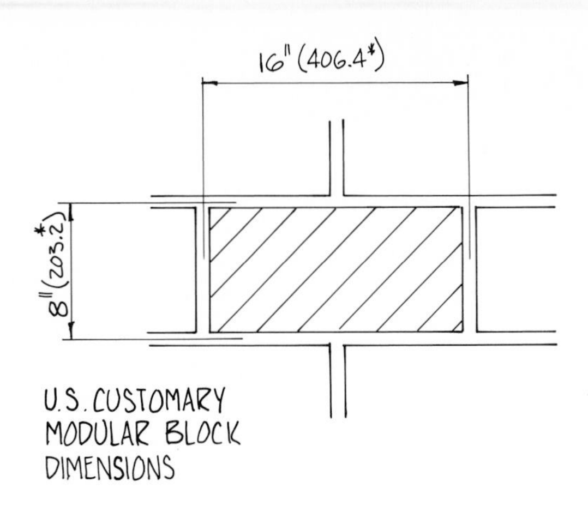

U.S. CUSTOMARY MODULAR BLOCK DIMENSIONS

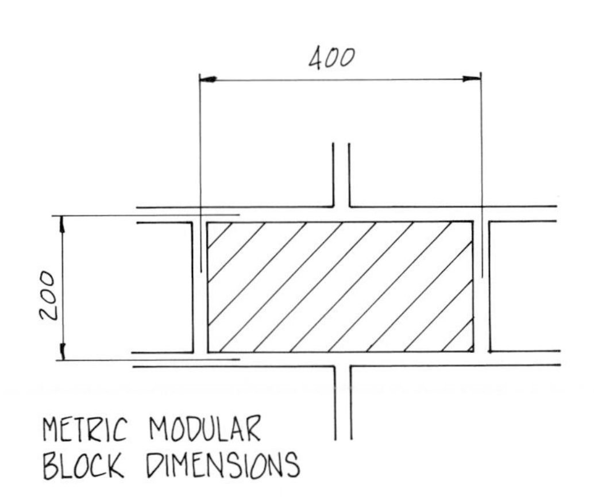

METRIC MODULAR BLOCK DIMENSIONS

Standard modular spacing for concrete blocks are shown here in both Customary and metric measure. Again, the standards are not totally compatible, making it difficult to mix the two products. Noncompatibility is the expected norm when comparing almost all Customary and metric standard sizes.

TRANSITION-PHASE PROBLEMS

Working within the conversion transition phase tends to present some interesting difficulties. Although the U.S. construction industry has made little headway in conversion, progress in other countries has been a little faster. The transition-phase experience in Canada points to some of the potential problems:

- Some material groups have been changed to a totally metric format, including concrete blocks, asphalt roof shingles, and some steel products. Some products such as windows, doors, stairs, and other fabricated items, are available in both Imperial and hard metric sizes. Other building products, such as lumber, plywoods, and plasterboards, are not available in metric sizes at all. Although all product-sector manufacturers have expressed their intentions of converting to metric sizing, the momentum to do so has slowed considerably in past years due to fluctuating governmental and economic pressures.
- In contrast, the majority of design work has been done in hard metric since the late 1970s. By government decree, all designs and working drawings for government projects must be in SI metric and most, but not all, private commercial designs have followed suit. Dimension and product sizes are generally all given in hard metric with very little soft conversion (see 2-7 for definitions of hard and soft metric measurements). In addition, most national and provincial building codes have been rewritten entirely in metric.
- Having working drawings and codes in metric but being unable to buy key metric building materials poses obvious problems. Depending on the type of construction, contractors must build in either metric or Imperial (after dimensional conversion) while making allowances for converted or unconverted products.

Since conversion in the United States is in its infancy and progress is likely to be as at least as slow as Canada's, it seems certain that similar problems can be expected. Although it is definitely not the aim of this book to make political statements, it is generally agreed that significant conversion headway will not be achieved without strong governmental insistence and/or suitable economic inducements. Neither appear imminent for a variety of reasons.

IN THIS BOOK

As can be seen, this book is written using U.S. Customary measure with equivalent metric dimensions in parentheses. This would seem to be the best compromise with regard to the foregoing circumstances, although it does create some difficulties. Technically, all metric dimensions should be given in hard SI metric, but this is not always possible since many standard metric sizes are yet to be agreed on. Consequently, in an attempt to minimize confusion, an asterisk is placed over a metric dimension, e.g., 25.4*, when that dimension is a soft conversion from Customary measure. Metric dimensions without asterisks are hard metric sizes generally agreed on by industry whether or not actual standards have been set. For readers in Canada and other countries using Imperial measure, specific reference will be made to any sizing different from U.S. Customary.

METRIC

BASE UNITS AND DERIVED UNITS

SI BASE UNITS

QUANTITY	NAME	SYMBOL
length	meter	m
mass	kilogram	kg
time	second	s
electric current	ampere	A
temperature	kelvin	K
amount of substance	mole	mol
luminous intensity	candela	cd
plane angle	radian	rad
solid angle	steradian	sr

SI BASE UNITS

There are nine base units used in the SI system as shown in the table at left. Technically, the units radian and steradian are classed as supplementary units since they have not yet been officially classified as base or derived units.

SI DERIVED UNITS with SPECIAL NAMES

All other units in the SI system are derived from combinations of base units. The table below shows the 15 most frequently used derived units which have been given names for the sake of convenience. In the "Description of Base Units" column, the dot placed midway between base unit symbols indicates that units are multiplied.

SI DERIVED UNITS with SPECIAL NAMES

QUANTITY	UNIT NAME	SYMBOL	DESCRIPTION of BASE UNITS
frequency	hertz	Hz	s^{-1}
force	newton	N	$m \cdot kg \cdot s^{-2}$
pressure, stress	pascal	Pa	$m^{-1} \cdot kg \cdot s^{-2}$
energy, work, quantity of heat	joule	J	$m^{2} \cdot kg \cdot s^{-2}$
power, radiant flux	watt	W	$m^{2} \cdot kg \cdot s^{-3}$
quantity of electricity, electric charge	coulomb	C	$s \cdot A$
electric potential	volt	V	$m^{2} \cdot kg \cdot s^{-3} \cdot A^{-1}$
electric capacitance	farad	F	$m^{-2} \cdot kg^{-1} \cdot s^{4} \cdot A^{2}$
electric resistance	ohm	Ω	$m^{2} \cdot kg \cdot s^{-3} \cdot A^{-2}$
electrical conductance	siemens	S	$m^{-2} \cdot kg^{-1} \cdot s^{3} \cdot A^{2}$
magnetic flux	weber	Wb	$m^{2} \cdot kg \cdot s^{-2} \cdot A^{-1}$
magnetic flux density	tesla	T	$kg \cdot s^{-2} \cdot A^{-1}$
inductance	henry	H	$m^{2} \cdot kg \cdot s^{-2} \cdot A^{-2}$
luminous flux	lumen	lm	$cd \cdot sr$
illuminance	lux	lx	$m^{-2} \cdot cd \cdot sr$

DERIVED UNITS AND NON-SI UNITS

SI DERIVED UNITS without SPECIAL NAMES

QUANTITY	DESCRIPTION	SYMBOL	DESCRIPTION of BASE UNITS
area	square meter	m^2	Indicated by symbol
volume	cubic meter	m^3	
speed - linear	meter per second	m/s	
- angular	radian per second	rad/s	
acceleration - linear	meter per second squared	m/s^2	
- angular	radian per second squared	rad/s^2	
density, mass density	kilogram per cubic meter	kg/m^3	
concentration (amount of substance)	mole per cubic meter	mol/m^3	
specific volume	cubic meter per kilogram	m^3/kg	
luminance	candela per square meter	cd/m^2	
moment of force	newton meter	N·m	$m^2 \cdot kg \cdot s^{-2}$
heat flux density, irradiance	watt per square meter	W/m^2	$kg \cdot s^{-3}$
heat capacity, entropy	joule per kelvin	J/K	$m^2 \cdot kg \cdot s^{-2} \cdot K^{-1}$
specific heat capacity, specific entropy	joule per kilogram kelvin	J/(kg·K)	$m^2 \cdot s^{-2} \cdot K^{-1}$
specific energy	joule per kilogram	J/kg	$m^2 \cdot s^{-2}$
thermal conductivity	watt per meter kelvin	W/(m·K)	$m \cdot kg \cdot s^{-3} \cdot K^{-1}$
energy density	joule per cubic meter	J/m^3	$m^{-1} \cdot kg \cdot s^{-3}$
electric field strength	volt per meter	V/m	$m \cdot kg \cdot s^{-3} \cdot A^{-1}$
electric charge density	coulomb per cubic meter	C/m^3	$m^{-3} \cdot s \cdot A$
surface density of charge, flux density	coulomb per square meter	C/m^2	$m^{-2} \cdot s \cdot A$
permittivity	farad per meter	F/m	$m^{-3} \cdot kg^{-1} \cdot s^4 \cdot A^2$
current density	ampere per square meter	$A \cdot m^{-2}$	indicated by symbol
magnetic field strength	ampere per meter	$A \cdot m^{-1}$	
permeability	henry per meter	H/m	$m \cdot kg \cdot s^{-2} \cdot A^{-2}$

NON - SI UNITS

CONDITIONS OF USE	UNIT	SYMBOL	VALUE IN SI UNITS
permissible universally with SI	minute	min	1 min = 60 s
	hour	h	1 h = 3600 s
	day	d	1 d = 86 400 s
	degree (of arc)	°	1° = (π/180) rad
	minute (of arc)	′	1′ = (π/10 800) rad
	second (of arc)	″	1″ = (π/648 000) rad
	liter	l or L*	1 L = 1 dm^3
	tonne	t	1 t = 10^3 kg
	degree Celcius	°C	
permissable for a limited time	ångström	Å	1 Å = 0.1 nm = 10^{-10} m
	are	a	1 a = 10^2 m^2
	hectare	ha	1 ha = 10^4 m^2
	bar	bar	1 bar = 100 kPa
	standard atmosphere	atm	1 atm = 101.325 kPa

* see 2-11

SI DERIVED UNITS without SPECIAL NAMES

The table above shows units derived either from base units alone or from a combination of derived, named, and base units. The oblique strokes (/) in the symbol column indicate that the first unit is divided by the second unit. Symbols in parentheses must be computed first.

NON – SI UNITS USED WITH SI

The table at the left shows non-SI units used within the SI system for various practical reasons. Some will be used indefinitely, while others will be used only until a suitable replacement can be found.

METRIC

MULTIPLES AND SUBMULTIPLES

SI MULTIPLIERS and PREFIXES

Whereas in the inch-pound system there are no consistent rules for forming larger or smaller quantities, in SI metric there exists a standard system for multiplying and dividing all units. The table at the right shows the multiplying factors plus the prefixes and symbols that allow a quantity to be written in two ways. For example, a meter (m) when multiplied by 1000 becomes a kilometer (km) and when divided by 1000 becomes a millimeter (mm).

SI PREFIXES

MULTIPLYING FACTOR	SI PREFIX	SI SYMBOL
1 000 000 000 000 = 10^{12}	tera	T
1 000 000 000 = 10^{9}	giga	G
1 000 000 = 10^{6}	mega	M
1 000 = 10^{3}	kilo	k
100 = 10^{2}	hecto	h
10 = 10^{1}	deca	da
1		
0.1 = 10^{-1}	deci	d
0.01 = 10^{-2}	centi	c
0.001 = 10^{-3}	milli	m
0.000 001 = 10^{-6}	micro	μ
0.000 000 001 = 10^{-9}	nano	n
0.000 000 000 001 = 10^{-12}	pico	p
0.000 000 000 000 001 = 10^{-15}	femto	f
0.000 000 000 000 000 001 = 10^{-18}	atto	a

CONSTRUCTION DRAWING DIMENSIONS

Although provision is made for multiplying and dividing the original unit (in this case a meter) by 10 or 100, the convention for construction drawings is to use *only* multiples and submultiples of 1000. On metric construction drawings almost all dimensions are given in millimeters (mm) since there are few items on a construction site that are smaller than this unit of measure. This allows the convenience of writing all dimensions in whole numbers and without the need to add the symbol (mm) since each dimension is presumed to be in millimeters unless otherwise noted. Meter dimensions are also used on drawings for some specific applications (see the next page). If so, they must be written with a decimal point and three decimal places, e.g., 48.500. Written in this way, the symbol (m) is not necessary. *Note*: In the U.S., meter is spelled met*er*, whereas in Canada and most other countries it is spelled met*re*. The same applies to liter (lit*re*)

RULES for WRITING SYMBOLS

- When the symbol is a letter or letters, leave a full space between the quantity and the symbol, e.g., 45 kg (not 45kg). When the symbol is not a letter, do not leave a space, e.g., 28°C (not 28 °C or 28° C).
- Symbols should not be pluralized, e.g., 50 m (not 50 ms).
- Do not use a period after a symbol unless the symbol is at the end of a sentence, e.g., 50 m or 50 m. The same rule applies to Customary dimensions, e.g., in and ft.
- To avoid confusion between units, write the symbol in upper or lower case as indicated in the unit tables. E.g., the symbol H is for henry, the symbol h is for hour.
- Use a period as a decimal marker and divide the digits on either side into groups of three by leaving a space between each group, e.g., 25 475.375 58. With four-digit numbers the space is optional, e.g., 2400 or 2 400. Do not use a comma to separate the groups since some countries use this as a decimal marker. A comma is still permitted in fiscal documents such as financial statements or on checks, e.g., $1,345,567.00.
- Where a number is a decimal without a whole number, a 0 (zero) must be placed before the decimal point, e.g., 0.457 (not .457).
- Angles should be expressed in decimalized degrees rather than in minutes and seconds, e.g., 18.43° (not 18° 26′).
- Other than degrees Celsius, unit names are not capitalized unless they occur at the start of a sentence.

ROUNDING-OFF CONVERSIONS

It is often desirable to round-off a converted quantity depending on the required accuracy of the conversion.

- If the digits to be rounded start with 5 or more, add one to the quantity. If the digits start with less than 5, leave the quantity as is. E.g., 34.3527 can be rounded to 34.353 or 34.35 or 34.4 or 34 depending on the required number of decimal places.
- Where maximum or minimum limits apply to a quantity, rounding must be in a direction that does not exceed that limit.
- For calculation purposes, mixed Customary units must be reduced to a single unit before conversion. E.g., 6′-4½″ should be reduced to 6.375 ft or 76.5 in before conversion.

SOFT AND HARD METRIC – CONSTRUCTION DRAWINGS

METRIC

2-2

SOFT METRIC CONVERSION

A soft metric conversion is when Customary dimensions and sizes (in feet and inches) are converted directly to metric dimensions using conversion factors. The illustrations on 2-2 and 2-3 show two such examples. Soft conversion is necessary during the transition phase of a total conversion of design dimensions and product sizes to hard metric measure.

HARD METRIC

Dimensions and product sizes that have not been soft converted from a Customary size, but have been designed directly in simplified and rounded-off SI metric units, are hard metric dimensions. Again, see the examples on 2-2 and 2-3. Because the 100-mm metric design module is slightly smaller than the 4-in Customary module (see 2-8), both drawing dimensions and product sizes based on the module will be slightly smaller than their Customary counterparts. This is not a problem provided that all building dimensions and product sizes are in hard metric sizes. It is a problem however, if metric materials are to be used on a Customary-sized building during, for example, a renovation project.

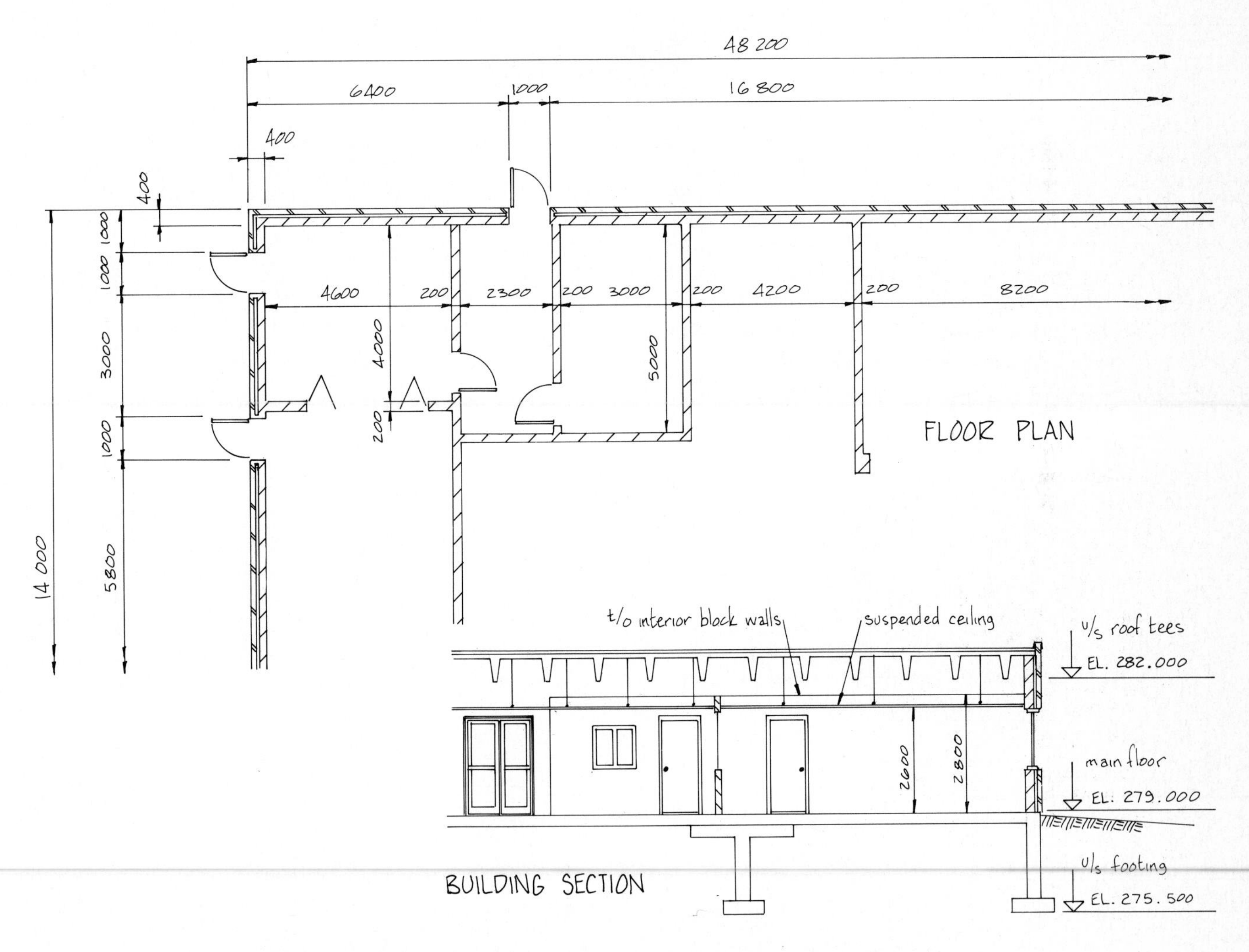

The construction drawing above is typical of a hard-metric designed project. All dimensions are given in millimeters, with the exception of the elevation levels on the section, which are in meters (there is a decimal marker in the figures). This is standard practice and no unit symbols are necessary. Although the elevation levels are in meters, they can be read as millimeters simply by removing the decimal marker.

METRIC

LINEAR MEASURE

LINEAR COMPARISONS

As stated previously, the two units of linear measure for construction purposes are the millimeter (mm) and the meter (m) with the 4-in Customary building module being replaced by the 100-mm metric module. At the right is a comparison of both modules plus a feet-to-meter comparison.

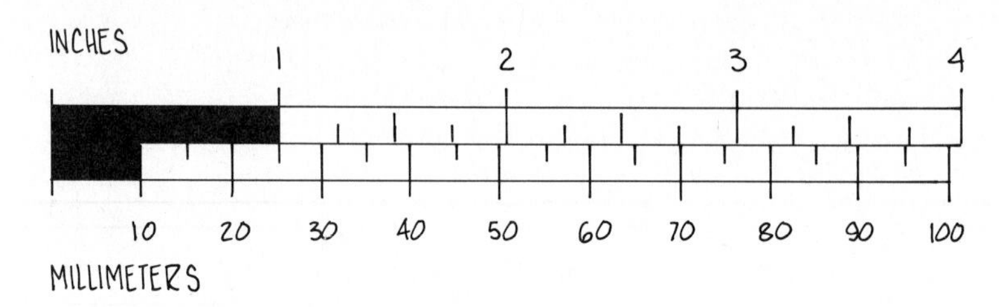

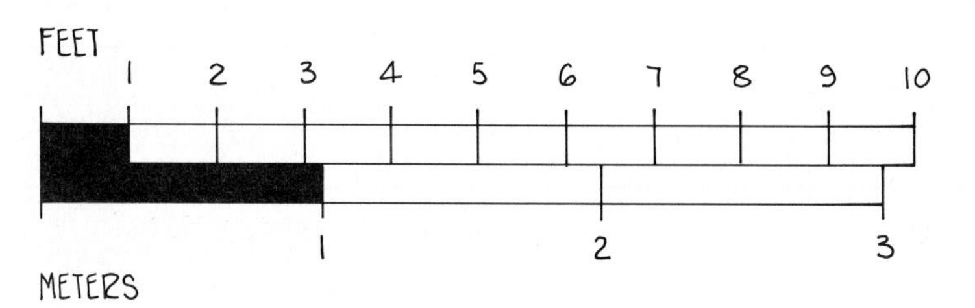

SCALES

Scales for construction drawings are expressed as a ratio rather than the representational scales of Customary measure. The table below shows the standard Customary scales with the nearest comparable metric scale. The relative size of some comparable scales are very similar. For example, the 1/8" = 1'-0" scale when converted to a ratio is 1:96 (there are 96 1/8"s in a foot). Therefore, although not a recommended practice, it is possible to read feet from a metric drawing or meters from a Customary drawing using the appropriate scales, provided that accuracy is not essential.

CUSTOMARY SCALE (EXPRESSED AS RATIO)	RECOMMENDED SI SCALE (EQUIVALENT REPRESENTIONAL)
full size (1 : 1)	1 : 1 (1000 mm = 1000 mm)
half size (1 : 2)	1 : 2 (500 mm = 1000 mm)
3" = 1'-0" (1 : 4)	1 : 5 (200 mm = 1000 mm)
1½" = 1'-0" (1 : 8)	see the note below
1" = 1'-0" (1 : 12)	1 : 10 (100 mm = 1000 mm)
¾" = 1'-0" (1 : 16)	see note below
½" = 1'-0" (1 : 24)	1 : 20 (50 mm = 1000 mm)
¼" = 1'-0" (1 : 48)	1 : 50 (20 mm = 1000 mm)
1/8" = 1'-0" (1 : 96)	1 : 100 (10 mm = 1000 mm)
1/16" = 1'-0" (1 : 192)	1 : 200 (5 mm = 1000 mm)
1/32" = 1'-0" (1 : 384)	1 : 500 (2 mm = 1000 mm)

Note: In the case of the 1½" = 1'-0" and the ¾" = 1'-0" scales a choice will have to be made between the next larger or smaller SI scales, depending on the type of drawing concerned.

CONVERSION FACTORS

CUSTOMARY TO METRIC			
1 inch	=	25.40 mm	
1 foot	=	304.8 mm	= 0.304 8 m
1 yard	=	914.4 mm	= 0.914 4 m
1 mile	=	1.609 344 km	

METRIC TO CUSTOMARY			
1 millimeter	=	0.039 370 in	= 0.003 280 839 ft
1 centimeter	=	0.393 700 in	= 0.032 808 398 ft
1 meter	=	39.370 07 in	= 3.280 839 8 ft
1 meter	=	1.093 613 2 yards	
1 kilometer	=	0.621 37 miles	

SQUARE MEASURE

COMPARISON of CUSTOMARY and METRIC SQUARE MEASURE

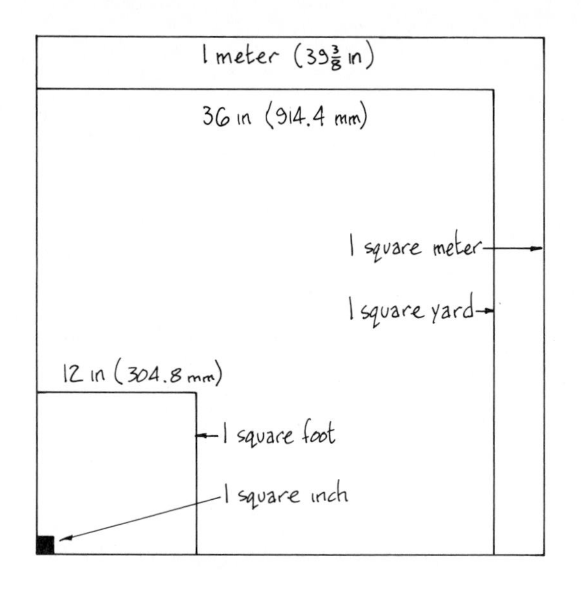

Note: Each drawing is a different scale and is not in proportion with the others.

SQUARE MEASURE (AREA)

Units of square measure are derived from linear base units. Symbols include:

- mm^2 for square millimeters
- m^2 for square meters
- km^2 for square kilometers

Square centimeters (cm^2) and square decimeters (dm^2) are not recommended units for construction purposes. The square hectare (ha^2) is not an SI unit but is acceptable as a supplemental unit for measuring surface areas of land and water only. In construction, mm^2 is used for objects having small areas, while m^2 replaces square feet and square yards.

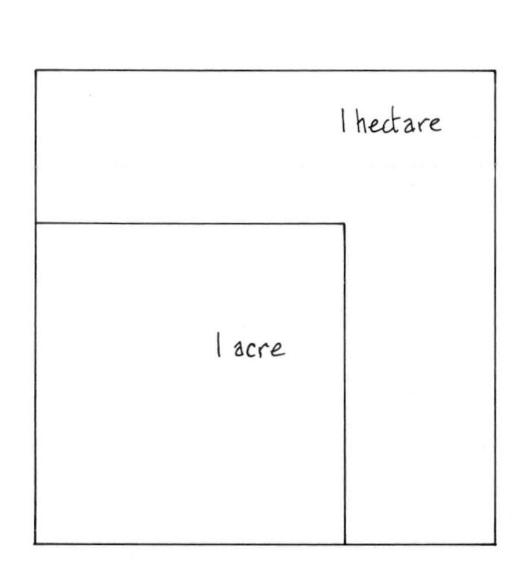

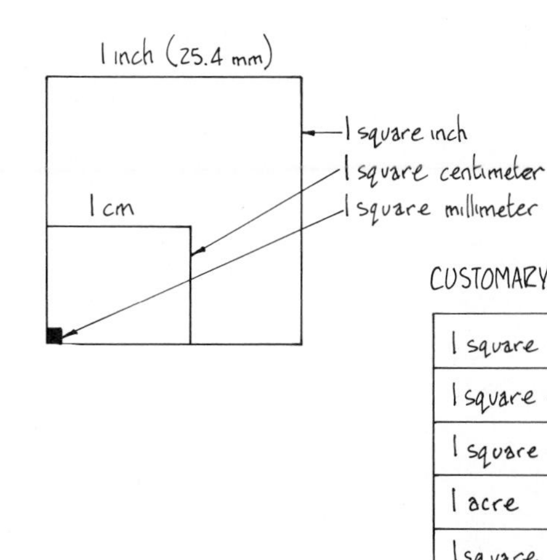

CONVERSION FACTORS

CUSTOMARY TO METRIC

1 square inch	= 645.16 mm^2	= 6.451 6 cm^2
1 square foot	= 0.092 903 04 m^2	= 929.03 cm^2
1 square yard	= 0.836 127 4 m^2	
1 acre	= 4 046.856 m^2	= 0.404 685 ha
1 square mile	= 2.589 988 km^2	= 259.0 ha

METRIC SQUARE MEASURE UNITS OF MEASUREMENTS

100 square millimeters	=	1 square centimeter
100 square centimeters	=	1 square decimeter*
100 square decimeters	=	1 square meter
10 000 square meters	=	1 hectare
100 hectares	=	1 square kilometer

* The decimeter is not a commonly used unit.

METRIC TO CUSTOMARY

1 square millimeter	= 0.001 550 sq in	
1 square centimeter	= 0.155 sq in	= 0.001 076 sq ft
1 square meter	= 10.763 91 sq ft	= 1.195 99 sq yd
1 hectare	= 11 959.9 sq yd	= 2.471 05 acres
1 square kilometer	= 0.386 102 sq mile	

METRIC 2-10

CUBIC MEASURE - VOLUME

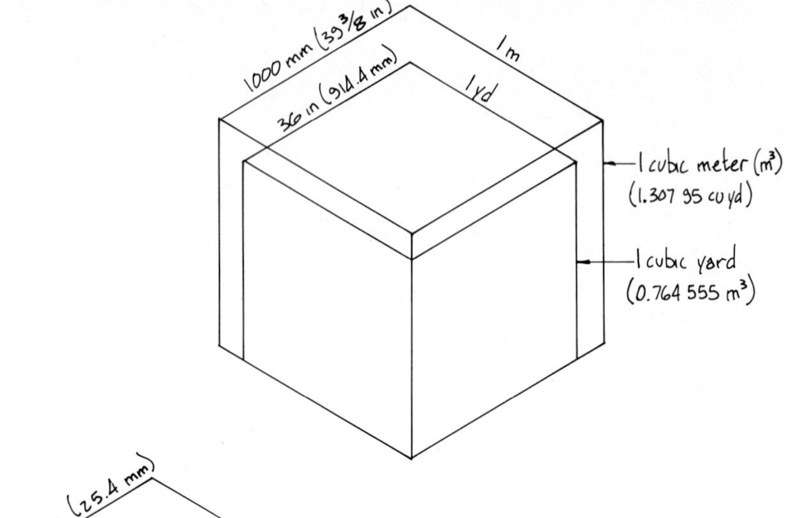

CUBIC MEASURE (VOLUME)

Units of cubic measure are derived from linear units, and for most construction purposes the cubic meter (m³) replaces the cubic foot and cubic yard. Cubic millimeters (mm³) and cubic centimeters (cm³) are used only for small quantities. In general, cubic volume refers to solids while cubic capacity (see the next page) refers to liquids or gases.

Lumber and related products are often Customary-measured in board feet (1 square foot of lumber 1 in thick) and cunits (100 cubic feet of lumber). Both units are replaced by the m³.

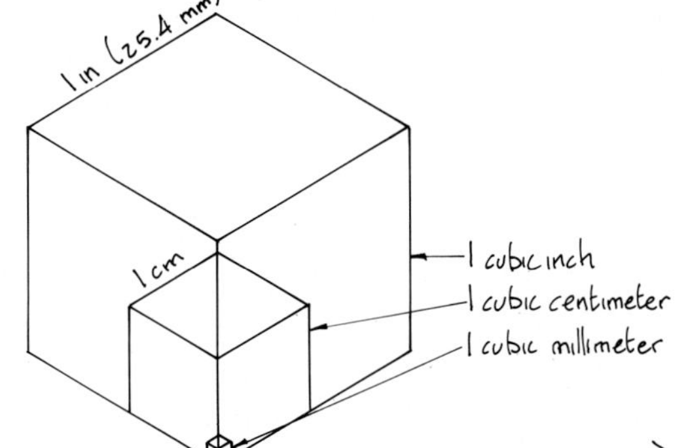

CONVERSION FACTORS

CUSTOMARY TO METRIC

1 cubic inch	= 16.387 064 cm³
1 cubic foot	= 28.316 85 dm³
1 cubic yard	= 0.764 555 m³

12 in (305 mm)

1 ft

10 cm (4 in)

1 dm

1 cubic foot (28.361 dm³)

1 cubic inch

1 cubic decimeter (0.035 3 14 cu ft)

METRIC TO CUSTOMARY

1 cubic millimeter (mm³)	= 0.000 061 024 cu in	
1 cubic centimeter (cm³)	= 0.061 023 74 cu in	
1 cubic decimeter (dm³)	= 61.023 74 cu in	= 0.035 314 76 cu ft
1 cubic meter (m³)	= 35.314 76 cu ft	= 1.307 95 cu yd

LUMBER CONVERSION FACTORS

1 board foot	= 0.002 359 737 5 m³
1 cunit	= 2.831 68 m³
1 cubic meter (m³)	= 423.777 12 board feet
	= 0.353 147 6 cunits

RELATIONSHIP of COMMON METRIC UNITS

1000 cubic millimeters (mm³) = 1 cubic centimeter (cm³)
1000 cubic centimeters (cm³) = 1 cubic decimeter (dm³)
1000 cubic decimeters (dm³) = 1 cubic meter (m³)

CUBIC MEASURE - CAPACITY

CONVERSION FACTORS

CUSTOMARY TO METRIC

1 US gallon	=	3.785 412 liters
1 Canadian gallon	=	4.546 090 liters
1 UK gallon	=	4.546 092 liters
1 US quart	=	0.946 353 liter
1 Canadian quart	=	1.136 522 liters
1 UK quart	=	1.136 523 liters
1 US pint	=	0.473 176 liter
1 Canadian pint	=	0.568 261 liter
1 UK pint	=	0.568 262 liter
1 US fluid ounce	=	29.573 53 ml
1 Canadian fluid ounce	=	28.413 062 ml
1 UK fluid ounce	=	28.413 08 ml

METRIC TO CUSTOMARY

1 liter	=	0.264 172 US gallon
1 liter	=	0.219 969 Canada, UK gallon
1 liter	=	1.056 688 US quarts
1 liter	=	0.879 877 Canada, UK quarts
1 liter	=	2.113 38 US pints
1 liter	=	1.759 75 Canada, UK pints
1 milliliter	=	0.033 814 US fluid ounces
1 milliliter	=	0.035 195 Canada, UK fluid ounces

LIQUID MEASURE (CAPACITY)

Comparison between Customary and Imperial units of liquid measure is extremely confusing and is a prime example of the benefits of metric conversion. Gallons, quarts, pints, and ounces are each a different size in U.S. Customary, Canadian and U.K. Imperial measure. The U.S. gallon is approximately 17% smaller than the Canadian and U.K. gallons, which, in turn, are not exactly the same but are considered to be so for most practical purposes. To add to the confusion, ounces may be "fluid", avoirdupois", "troy", or "apothecary" depending on what is being measured. Other odd items such as "barrels" also come in a variety of sizes.

The liter (L) is the main unit of metric capacity, with the kiloliter (kL) used for large capacities and the milliliter (ml) used for very small capacities. Technically, the symbol for liter is lowercase L (l), but this can easily be mistaken for the number 1. It is therefore recommended that uppercase L be used for all applications of the liter symbol, although some publications use L after a number (4.5 L) and l after a prefix (4500 ml). This text uses L throughout.

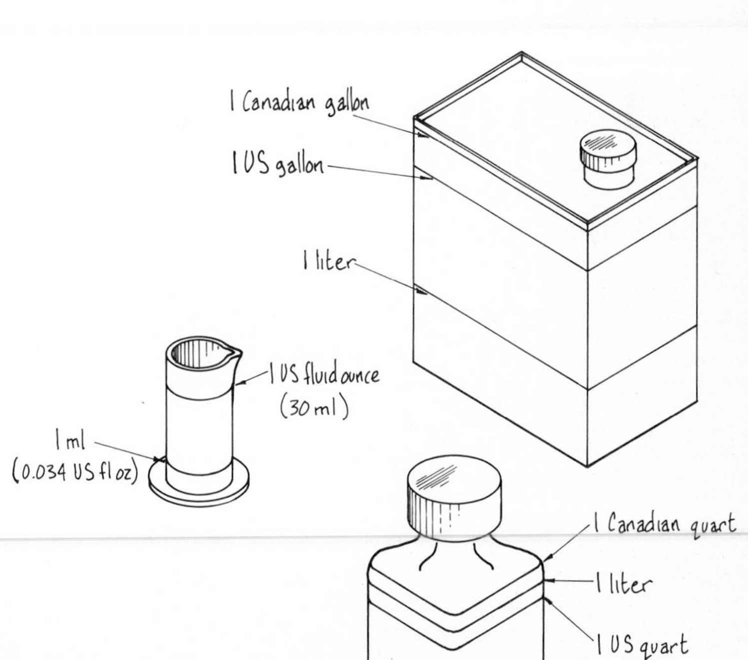

RELATION of COMMON METRIC UNITS

1 milliliter (ml)	=	1 cubic centimeter (cm^3)
1000 milliliters (ml)	=	1 liter (L)
1000 liters (L)	=	1 kiloliter (kL)

METRIC 2-12

MASS (WEIGHT)

MASS and WEIGHT

The confusion between Customary and Imperial units of weight is a close second to the confusion surrounding liquid measure. The pound is the only unit that is totally common to both systems. In SI there are three units: the gram (g), kilogram (kg), and the megagram (Mg)/tonne (t), all based on a standard of mass, the International Prototype Kilogram (I.P.K.). The mass unit cannot be derived without reference to this prototype metal cylinder, preserved in France by the International Bureau of Weights and Measures.

The SI system also differentiates between mass and weight:

- Mass is the quantity of matter in a body and is measured in the gram units noted above.
- Weight is a measure of the force acting on a body due to gravity or other forces. The base unit of force is the newton (N) and is defined on 2-15.
- When something is "weighed" it is actually the mass that has been ascertained as compared the known mass of the I.P.K. The mass of an object remains constant but its weight will vary due to the forces acting on it.

For most purposes the terms mass and weight are interchangeable and the use of the word "weight" to define the "weighed mass" of something will probably be retained. For other purposes, notably when calculating the forces acting on a structure or surface, the difference between mass and weight is significant.

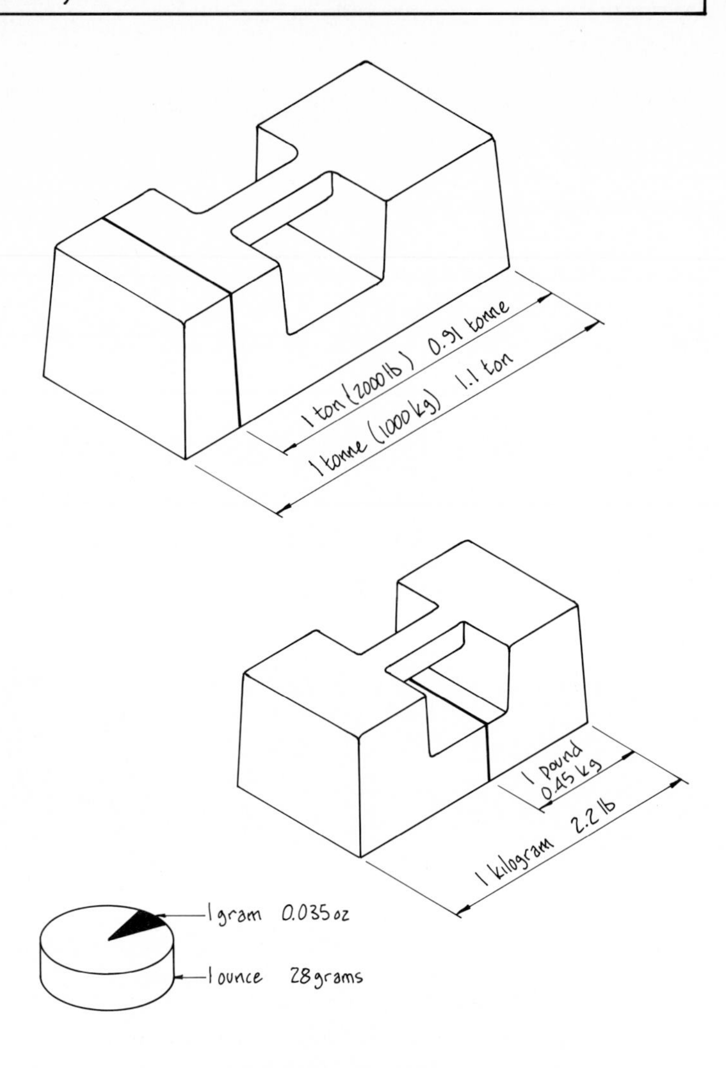

CONVERSION FACTORS

CUSTOMARY TO METRIC

1 grain	= 64.798 91 mg	= 0.064 798 18 g
1 ounce (avoirdupois)	= 28.349 523 g	= 0.028 349 523 kg
1 pound (avoirdupois) (16 oz)	= 0.453 592 37 kg	
1 ton (2000 lb)	= 0.907 184 74 t	= 907.184 74 kg

METRIC TO CUSTOMARY

1 milligram (mg)	= 0.000 035 274 oz (avoirdupois)	
1 gram (g)	= 0.035 274 oz (avoirdupois)	
1 kilogram (kg)	= 35.273 9 oz	= 2.204 623 lb
1 tonne (t)	= 1.102 311 ton (2000 lb)	

RELATIONSHIP OF METRIC MASS UNITS

1000 micrograms (µg)	= 1 milligram (mg)
1000 milligrams (mg)	= 1 gram (g)
1000 grams (g)	= 1 kilogram (kg)
1000 kilograms (kg)	= 1 tonne (t) or megagram (Mg)

VOLUME, CAPACITY, MASS, AND DENSITY

LIQUID MEASURE	CUBIC MEASURE	MASS (WEIGHT) of WATER
1 milliliter (ml)	1 cubic centimeter (cm³) — 1 cm, 3/8 in	1 gram (g)
1 liter (L)	1 cubic decimeter (dm³) — 10 cm, 4 in; cubic centimeter	1 kilogram (kg)
1 kiloliter (kL)	1 cubic meter (m³) — 1 m, 39 in; cubic decimeter	1 tonne (t)

One advantage of the metric system is the interrelation of the units for cubic, liquid, and mass measurement. Basically, 1 liter of water has a volume of 1 cubic decimeter and a mass of 1 kilogram. The relationship is constant through equal multiples of each measure. In actual practice the exact value of each quantity will vary slightly with water temperature but is nevertheless a useful relationship.

The same cannot be said for Customary measure, where 1 gallon of water weighs 8.33 pounds with a volume of 0.133 cubic foot, or Imperial with quantities of 1 gallon, 10 pounds, and 0.16 cubic foot respectively.

Density is a measure of the mass per unit volume of a substance. It is measured in kilograms per cubic meter (kg/m³) or grams per cubic centimeter (g/cm³) for low densities and megagrams per cubic meter (Mg/m³) for high densities. Mass per linear length or square area follow similar guidelines.

CONVERSION FACTORS

CUSTOMARY TO METRIC

UNIT/LENGTH	
1 ounce per lineal inch	= 1.116 12 kg/m
1 pound per lineal foot	= 1.488 16 kg/m
1 pound per lineal yard	= 0.496 055 kg/m
UNIT/AREA	
1 ounce per square inch	= 4.394 189 g/cm²
1 pound per square inch	= 703.0696 kg/m²
1 pound per square foot	= 4.882 43 kg/m²
UNIT/VOLUME (DENSITY)	
1 ounce per cubic inch	= 1729.994 kg/m³
1 pound per cubic inch	= 27.679 90 g/cm³
1 pound per cubic foot	= 16.018 46 kg/m³

METRIC TO CUSTOMARY

UNIT/LENGTH	
1 kilogram per meter (kg/m)	= 0.895 961 oz/lin in
	= 0.055 997 lb/lin in
	= 0.671 971 lb/lin ft
UNIT/AREA	
1 gram per square meter (g/m²)	= 0.003 277 oz/sq ft
1 kilogram per square meter (kg/m²)	= 0.001 422 334 lb/sq in
	= 0.204 816 lb/sq ft
UNIT/VOLUME	
1 kilogram per cubic meter (kg/m³)	= 0.000 578 oz/cu in
	= 0.000 036 lb/cu in
	= 0.062 427 974 lb/cu ft

METRIC

2-14

TEMPERATURE AND TIME

TEMPERATURE

The degree Kelvin (K) is the SI unit of temperature and has a base of absolute zero (-459.7°F or -273.16°C). However, for construction and most other applications, the Celsius (C) scale is in general use. Celsius degrees are equal to degrees Kelvin, but 0°C is set at the freezing point of water. The boiling point of water is 100°C. The comparison thermometer at the right gives generalized cross-reference information.

COMPARISON of FAHRENHEIT / CELSIUS

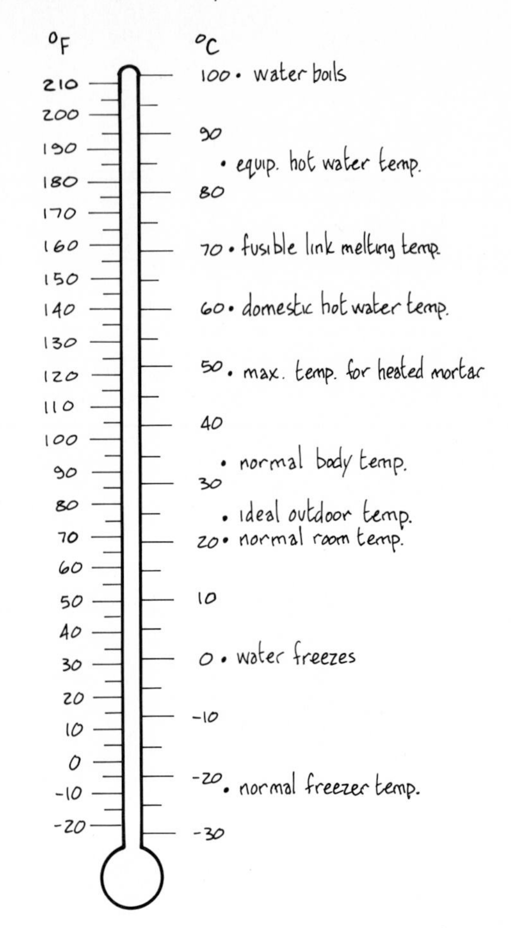

TEMPERATURE INTERVAL

1 degree F	=	0.5556 degree C
1 degree C	=	1.80 degree F

CONVERSION FACTORS

Fahrenheit to Celsius $(°F - 32) \times 5/9$
Celsius to Fahrenheit $(°C \times 9/5) + 32$

TIME

The measurement of time remains the same in SI metric, with the second being the base unit. Minutes are not recommended for construction purposes, and quantities such as flow rates should be given in cubic meters per second (m^3/s), liters per second (L/s), or cubic meters per hour (m^3/h).

International recommendations for writing time and dates are:

- Time – express by hour/minute/second for a 24-hour day, e.g., 14:25:45.
- Date – express by year/month/day, e.g., 1988-08-02. The hyphens may be omitted if desired but spaces should remain.

FORCE, PRESSURE, AND STRESS

FORCE

A force is anything that changes or tends to change the state of rest or motion in a body where Force = Mass x Acceleration. In Customary, it is generally measured in pound-force units and is the equivalent of the gravitational force acting on a mass of 1 pound. The ounce-force, kip-force, and ton-force, are also accepted units.

In SI metric, force is measured in newtons, one of the specially named derived units where $N = kg.m/s^2$. The newton (N) is approximately 23% of the pound-force and is used for light forces only. Kilonewtons (kN) are used for medium forces and meganewtons (MN) for high forces.

PRESSURE and STRESS

Pressure, or force per unit area, applies to gaseous or fluid matter and may be in positive or negative form. Stress is also force per unit area but applies to solid material and results when force is applied to a restrained object. In Customary, pressure and stress are measured in pound-force per square inch, kip-force per square foot etc. This is normally expressed as "psi", "psf", "ksf" etc.

In SI metric, pressure and stress are measured in pascals (Pa), another of the specially named derived units. A pascal is a force of 1 newton applied to 1 square meter ($Pa = N/m^2$). Since the pascal is a very small unit compared to the pound-force (1 Pa = 0.000 145 psi), pressure and stress are normally expressed in kilopascals (kPa) and megapascals (MPa).

FORCE UNITS

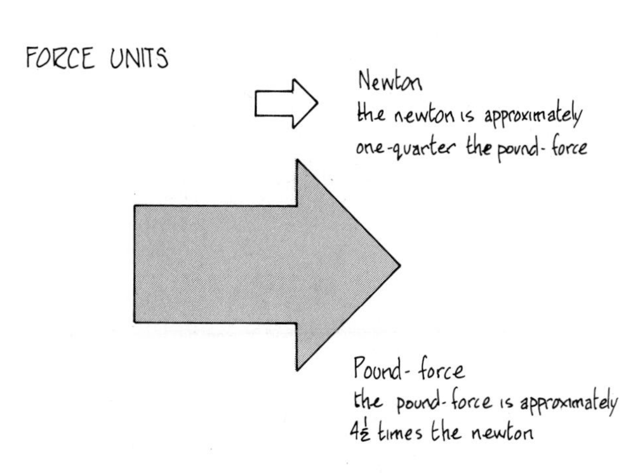

PRESSURE UNITS / STRESS UNITS

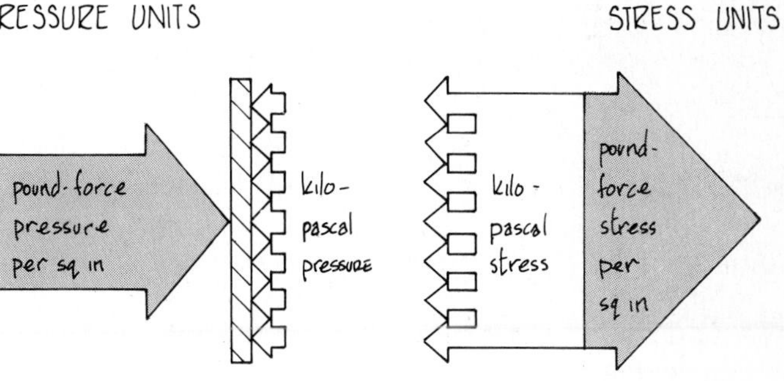

RELATIONSHIP of NEWTON MULTIPLES

kg·/m²	=	newton (N)
1000 newtons (N)	=	1 kilonewton (kN)
1000 kilonewtons (kN)	=	1 meganewton (MN)
1000 meganewton (MN)	=	1 giganewton (GN)

RELATIONSHIP of PASCAL MULTIPLES

N/m²	=	pascal (Pa)
1000 pascals	=	1 kilopascal (kPa)
1000 kilopascals	=	1 megapascal (MPa)
1000 megapascals	=	1 gigapascal (GPa)

CUSTOMARY TO METRIC — CONVERSION FACTORS — METRIC TO CUSTOMARY

Customary to Metric			Conversion Factors	Metric to Customary	
1 pound-force	= 4.448 222 N	= 0.004 448 kN	FORCE UNITS	1 newton (N)	= 0.224 809 pound-force
1 pound-force per foot	= 14.59 N/m	= 0.014 59 kN/m			= 0.000 224 808 kip-force
1 kip-force (1000 pf)	= 4 448.222 N	= 4.448 222 kN			= 0.000 112 404 ton-force
1 ton-force per foot	= 29 187.81 N/m	= 29.187 81 kN/m			= 7.223 015 poundal force
1 pound-force per square inch	= 6894.757 Pa	= 6.894 757 kPa	PRESSURE UNITS	1 kilopascal (kPa)	= 0.145 038 pf per sq in
1 pound-force per square foot	= 47.880 26 Pa	= 0.047 880 kPa			= 20.885 4 pf per sq ft
standard atmosphere (14.7 psi) (760 mm Hg)		= 101.325 kPa	ATMOSPHERE	1 kilopascal (kPa)	= 4.021 862 in water column
1 in of water (H_2O) column (20°C, 68°F)		= 0.248 641 kPa			= 0.296 374 in mercury column
1 in of mercury (Hg) column (20°C, 68°F)		= 3.374 110 kPa			= 0.01 bars

METRIC

2-16 ENERGY AND POWER

The energy content of fuels, the potential energy of a power source, or the heat units essential to physical process are important factors in construction. Equally important is the measurement of energy flow and energy consumption as the energy potential is converted to heat or work.

ENERGY UNITS

In construction, energy sources fall into three general categories:

- Heat energy, used for heating purposes and usually supplied in the form of gas and oil or related products. Also present in related units when calculating cooling/refrigeration capacity. Measured in British thermal units (Btu).
- Electrical energy, used for heating and cooling and as a power source for mechanical work (motors, etc.). Measured in kilowatt-hours (kW·h).
- Mechanical energy, applicable to the design of machines and equipment. Measured in horsepower or foot pound-force.

SI EQUIVALENTS

In the metric system the four units listed above are expressed in joules. A joule is a force of 1 newton ($kg \cdot m/s^2$) moving a distance of 1 meter (N·m). The kilojoule (kJ) is the preferred unit, with the megajoule (MJ) used for large quantities. Conversion tables are given on the next page.

POWER and HEAT FLOW UNITS

The quantity of power consumed or the heat energy output of heating or cooling equipment are also important considerations in construction. In Customary, the units of power measurements are:

- Heat power, expressed in British thermal units per hour (Btu/h), therms (100 000 Btu), or tons (12 000 Btu-refrigeration).
- Electrical power, expressed in "watts" in Customary measure, remains the same in SI metric.
- Mechanical power, expressed in "horsepower" for electric motors and combustion engines. One horsepower = 500 foot pounds-force per second.

SI EQUIVALENTS

In the metric system each of the above units will be expressed in watt (W) units and its multiples. A watt is a named derived unit and is equal to 1 joule per second (J/s). Power usage or heat flow is therefore measured on a per second basis (the SI time standard). Conversion tables are given on the next page.

RELATIONSHIP of JOULE UNITS

N·m	=	1 joule (J)
1000 joules (J)	=	1 kilojoule (kJ)
1000 kilojoules (kJ)	=	1 megajoule (MJ)

COMPARISON of CUSTOMARY / METRIC UNITS

1 Btu	=	1.055 kJ
1 kJ	=	0.948 Btu
1 watt (W)	=	1 joule/sec

HEAT ENERGY - HEAT FLOW

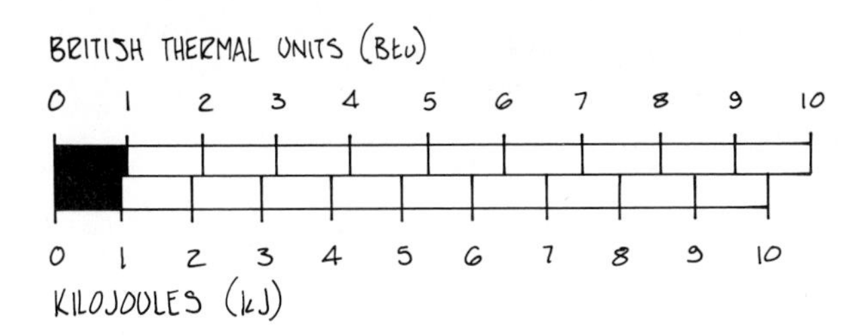

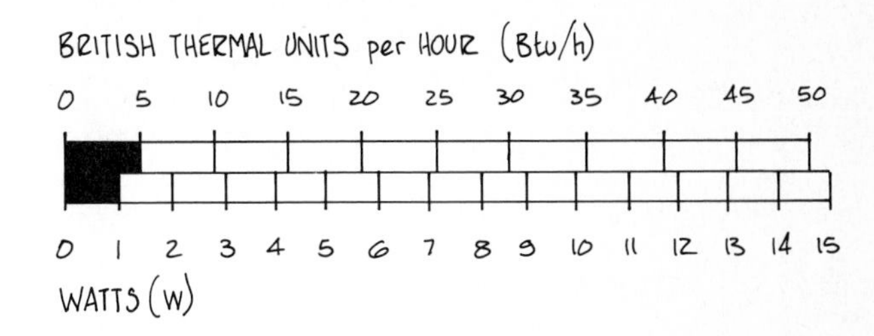

HEAT ENERGY and RELATED UNITS

PROPERTY	CUSTOMARY UNITS (and conversion factors to metric)		METRIC UNITS (and conversion factors to Customary)	
Quantity of heat	British thermal unit (Btu)	= 1055.06 J	Joule (J)	= 0.000 947 817 Btu
		= 1.055 06 kJ	kilojoule (kJ)	= 0.947 813 3 Btu
	therm	= 105.506 MJ	megajoule (MJ)	= 0.009 478 therm
Heat capacity and entropy	Btu per °F	= 1899 J/°C	Joule per °C (J/°C)	= 0.000 526 593 Btu/°F
Specific heat capacity and specific entropy	Btu per (pound °F)	= 4186.8 J/kg·°C	Joule per (kilogram °C) (J/kg·°C)	= 0.000 238 846 Btu/(lb·°F)
	Btu per (cubic foot °F)	= 67.066 1 kJ/(m³·°C)	kilojoule per (cubic meter °C)	= 0.014 910 6 Btu/(cuft·°F)
Heat density	Btu/square foot	= 11 356.53 J/m²	joule per square meter (J/m²)	= 0.000 088 055 Btu/ft²
Calorific value	Btu per pound (Btu/lb)	= 2.326 kJ/kg	Joule per kilogram (J/kg)	= 0.000 429 923 Btu/lb
	Btu per cubic foot (Btu/cuft)	= 37.259 1 kJ/m³	kilojoule per cubic meter (kJ/m³)	= 0.026 839 1 Btu/cuft

HEAT FLOW and RELATED UNITS

PROPERTY	CUSTOMARY UNITS (and conversion factors to metric)		METRIC UNITS (and conversion factors to Customary)	
Heat flow rate (Quantity of heat per unit time)	Btu / per hour	= 0.293 072 W	watt (W)	= 3.412 130 8 Btu/h
		= 0.000 293 kW	kilowatt (kW)	= 3412.130 8 Btu/h
	ton of refrigeration	= 3.517 kW	megawatt (MW)	= 3 412 130.8 Btu/hr
Heat flow intensity	Btu per (sq ft hour) [Btu/(sf·h)]	= 3.154 60 W/m²	watt per square meter (W/m²)	= 0.316 997 4 Btu/(sq.ft·h)

METRIC

2-18 HEAT TRANSFER - HEAT FLOW

HEAT TRANSFER IN BUILDINGS

In heat loss/heat gain calculations when sizing heating and cooling equipment or designing building assemblies, the values "U", "C", "K", and "R" are used to specify heat transfer or resistance to heat flow. These terms will be retained in SI metric but should be referred to as metric U (USI), metric R (RSI) etc. during the transition phase to avoid confusion. Since each transfer value is in watt units there is a direct and simple relationship between the calculated heat loss/heat gain of a building and the capacity of all types of heating/cooling equipment and fuels, which are also rated in watt units.

HEAT TRANSFER - HEAT FLOW and RELATED UNITS

PROPERTY	CUSTOMARY UNITS (and conversion factors to metric)		METRIC UNITS (and conversion factors to customary)	
Heat flow rate (quantity of heat per unit time)	Btu/hour	= 0.293 072 W	watt (W)	= 3.412 130 8 Btu/h
		= 0.000 293 kW	kilowatt (kW)	= 3412.130 8 Btu/h
				= 0.284 345 ton
	ton of refrigeration	= 3.517 kW	megawatt (MW)	= 3 412 308 Btu/h
Heat flow intensity	Btu per (square foot hour)	= 3.154 60 W/m²	watt per square meter (W/m²)	= 0.316 997 4 Btu/(sq ft·h)
U value (heat transfer) and C value (thermal conductance)	Btu per (square foot hour °F) (Btu/(sq ft·h·°F))	= 5.678 29 W/(m²·°C)	watt per (square meter °C) (W/(m²·°C))	= 0.176 109 3 Btu/(sf·h·°F)
K value (thermal conductivity per unit thickness)	Btu inch/(square foot hour °F) [Btu·in/(sq ft·h·°F)]	= 0.144 228 W/(m·°C)	watt per (meter °C) [W/(m·°C)]	= 6.933 466 4 Btu·in/(sf·h·°F)
R value (thermal resistance per given thickness)	(sq ft hour °F) per Btu (sq ft·h·°F)/Btu	= 0.176 109 3 (m²·°C)/W	(sq meter °C) per watt [(m²·°C)/W]	= 5.678 229 (sq ft·h·°F)/Btu
Thermal resistivity (resistance per unit thickness)	(sq ft hour °F) per Btu inch [(sq ft·h·°F)/Btu·in]	= 6.933 466 4 (m·°C)/W	(meter °C) per watt [(m·°C)/W]	= 0.144 228 (sq ft·h·°F)/Btu·in

ELECTRICAL UNITS

There are only four significant changes from Customary measure for electrical units or terminology:

- The name of the electrical conductance unit has changed from mho to siemens (S) but without a change in value.
- Luminance is now calculated on a square meter basis instead of square feet or square inches.
- Illuminance is now measured in lumen per square meter instead of square feet and is given the special name "lux" (lx).
- Alternating-current frequency, previously measured in "cycles per second" is now in hertz (Hz) but without a size conversion, e.g., 60 cps = 60 Hz.

CUSTOMARY and METRIC UNITS

PROPERTY	CUSTOMARY UNIT	METRIC UNIT	SI SYMBOL
electric current	ampere	ampere	A
electric charge or quantity	coulomb	coulomb	C
electric potential	volt	volt	V
electric resistance	ohm	ohm	Ω
electric capacitance	farad	farad	F
electric inductance	henry	henry	H
electric magnetic flux	weber	weber	Wb
electric conductance	mho	siemens	S
luminous flux	lumen	lumen	lm
luminous intensity	candela	candela	cd
luminance	candela per sq ft candela per sq in foot lambert	candela per sq meter	cd/m^2
illuminance	lumen per sq ft foot candle	lux	lx (lm/m^2)
a-c current frequency	cycles per second (cps)	hertz	Hz

Note: For watt units see 2-17

METRIC

2-20 VELOCITY, FLOW, AND SLOPE

VELOCITY and FLOW

In Customary, linear velocity is measured in miles per hour, fluids in gallons per minute, and gases in cubic feet per minute.
In SI metric, the preferred measurements are kilometers per hour (km/h) for wind and vehicle speeds, meters per second (m/s) for other linear speeds, cubic meters per second (m^3/s) for gases, and liters per second (L/s) for fluids.

CUSTOMARY TO METRIC	
1 inch per minute	= 2.54 cm/min
1 foot per second	= 0.304 8 m/s
1 foot per minute	= 0.005 08 m/s
1 mile per hour	= 0.447 04 m/s
	= 1.609 344 km/h
1 cubic foot per second	= 0.028 316 85 m^3/s
	= 28.316 85 dm^3/s (L/s)
1 cubic foot per minute	= 0.000 471 947 4 m^3/s
	= 0.471 947 4 dm^3/s (L/s)
1 cubic yard per minute	= 0.012 742 58 m^3/s
1 US gallon per minute	= 0.063 090 2 L/s
1 US gallon per hour	= 0.001 051 5 L/s
1 Imp. gallon per minute	= 0.075 768 L/s
1 Imp. gallon per hour	= 0.001 263 L/s

METRIC TO CUSTOMARY	
1 meter per second	= 3.280 84 ft/sec
	= 196.850 394 ft/min
	= 2.236 936 mph
1 kilometer per hour	= 0.621 371 mph
1 cubic meter per second	= 2 118.880 197 cu ft/min
	= 35.314 662 cu ft/sec
	= 78.477 043 cu yd/min
1 cubic decimeter (or liter) per second	= 2.118 880 cu ft/min
1 liter per second	= 15.850 322 US gal/min
	= 951.022 349 US gal/h
	= 13.198 184 Imp. gal/min
	= 791.765 637 Imp. gal/h

SLOPES

In Customary, slopes for roofs, ramps, etc. are expressed as the ratio of rise over run based on inch units. The run is always 12 units, while the rise varies with the slope, e.g., 4 in 12 or 7 in 12.
In SI metric, slope is expressed as a true ratio of rise over run but is based on unity (1) as follows:

- For slopes of less than 45° the first number (rise) is 1, e.g., 1:3
- For slopes greater than 45° the last number (run) is 1, e.g., 2:1. This allows easy verification of slope magnitude.
- Equal units must also be used. E.g., 1:5 can refer to a rise of 1 mm in a 5 mm horizontal run but not 1 mm in 5 cm.

Example:

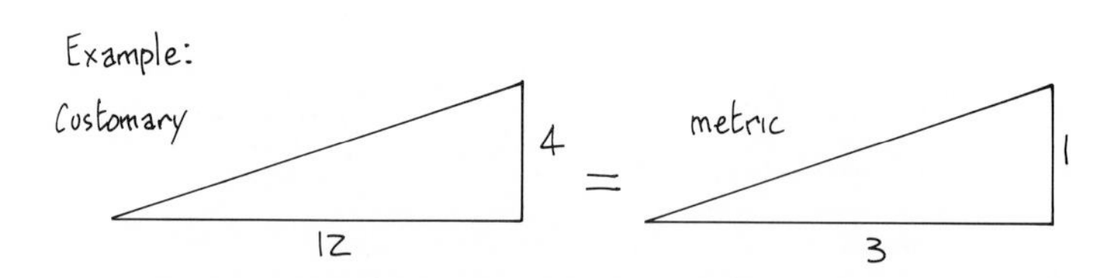

CUSTOMARY TO METRIC CONVERSION

CUSTOMARY	ANGLE		METRIC
1 in 12	4°-46'	4.77°	1 : 12
2 in 12	9°-28'	9.47°	1 : 6
3 in 12	14°-02'	14.03°	1 : 4
4 in 12	18°-26'	18.43°	1 : 3
5 in 12	22°-30'	22.50°	1 : 2.4
6 in 12	26°-34'	26.57°	1 : 2
7 in 12	30°-0'	30.00°	1 : 1.73
9 in 12	36°-52'	36.87°	1 : 1.33
12 in 12	45°-0'	45.00°	1 : 1
24 in 12	63°-26'	63.43°	2 : 1
36 in 12	71°-34'	71.57°	3 : 1
48 in 12	75°-58'	75.97°	4 : 1
72 in 12	80°-32'	80.53°	6 : 1

CHAPTER PAGE TITLES

DESIGN	
3-2	BASIC DESIGN - BUNGALOW

Most house designs tend to fall into one of four major dwelling types : bungalow, two-story, split level, and multiple. Some designs may follow one of these styles closely, while others may combine elements of two or more styles. In all cases the arrangement of floor levels and the positioning of stairs is of particular concern if good traffic flows and "livability" is to be achieved. In addition to the interior layout, there are many variations in style and finish that can be applied to the exterior elements which may radically change the visual aspects of the house.

Each of the four major dwelling types are illustrated on this and the next three pages. Refer to both the exterior pictorial and the floor plan of each example to fully appreciate the exact design configuration.

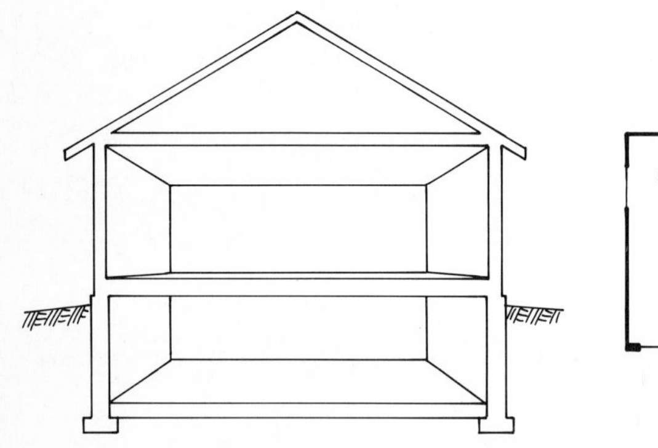

Typical Schematic Cross-Section

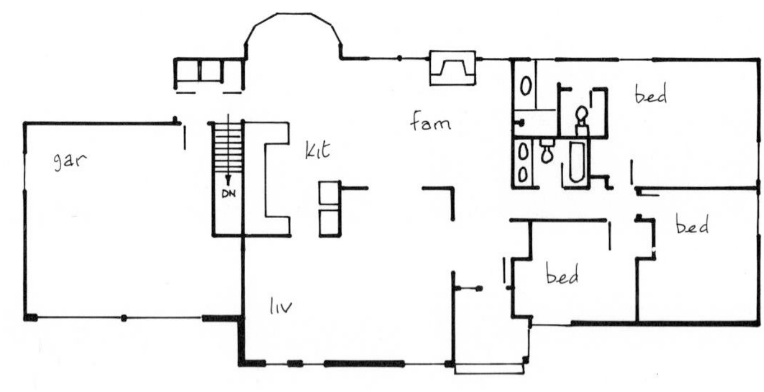

Typical Floor Plan

BUNGALOW (RANCH)

- Single main floor level containing all main high-usage rooms
- Secondary-use rooms such as a laundry room may also be located on the main floor level.
- A basement is optional.
- Most expensive in terms of construction and heating/cooling. Largest roof area in relation to living area (see 10-5).

DESIGN 3-3

TWO-STORY

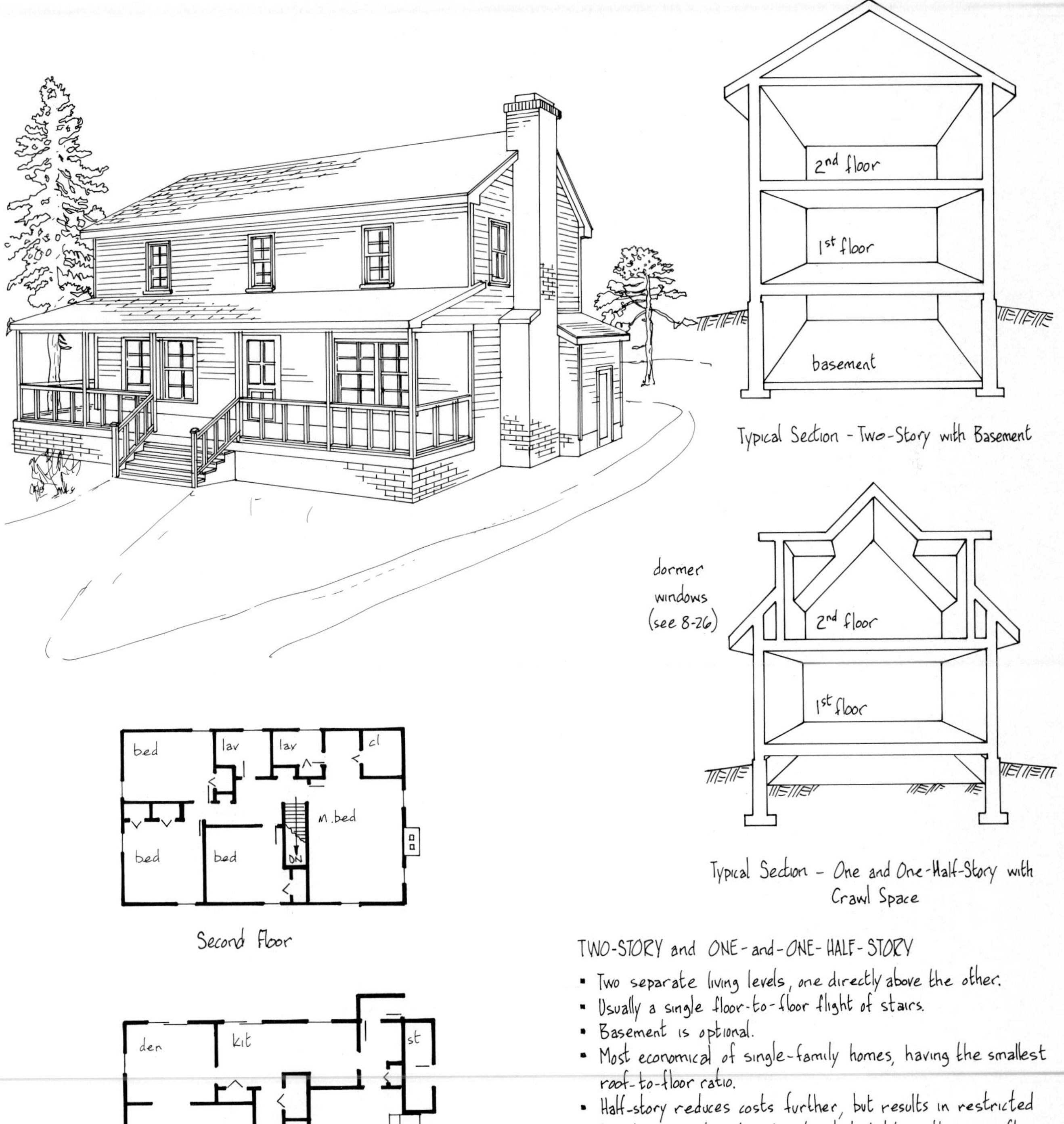

TWO-STORY and ONE-and-ONE-HALF-STORY

- Two separate living levels, one directly above the other.
- Usually a single floor-to-floor flight of stairs.
- Basement is optional.
- Most economical of single-family homes, having the smallest roof-to-floor ratio.
- Half-story reduces costs further, but results in restricted headroom and reduced natural daylight on the upper floor.

Typical First and Second Floor Plans for Two-Story House

DESIGN	
3-4	SPLIT-LEVELS

SPLIT-LEVEL

- Essentially a bungalow that has the main floor level split from side-to-side (called a back-split), or from front-to-back (a side-split).
- A full basement is standard under the higher level but is optional under the lower level (a crawl space or slab-on-grade may be substituted).
- One or both of the upper levels may be split again to create a "sunken room" effect.
- Sets of short stairs replace the single long stair of a bungalow.

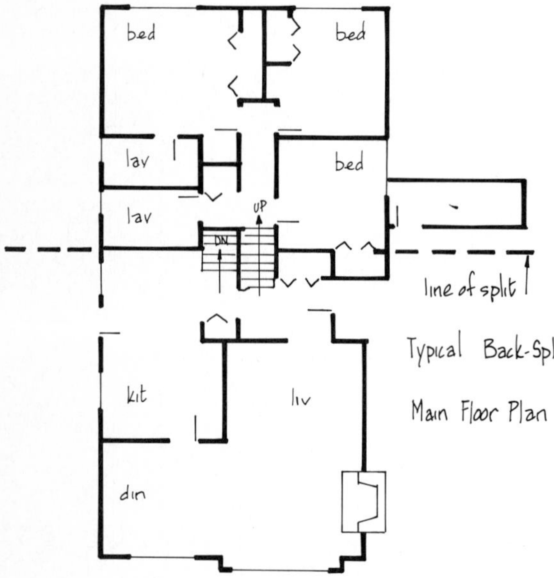

BACK-SPLIT DESIGN

bed
bed
lav
bed
lav
UP
DN
line of split
kit
liv
din

Typical Back-Split

Main Floor Plan

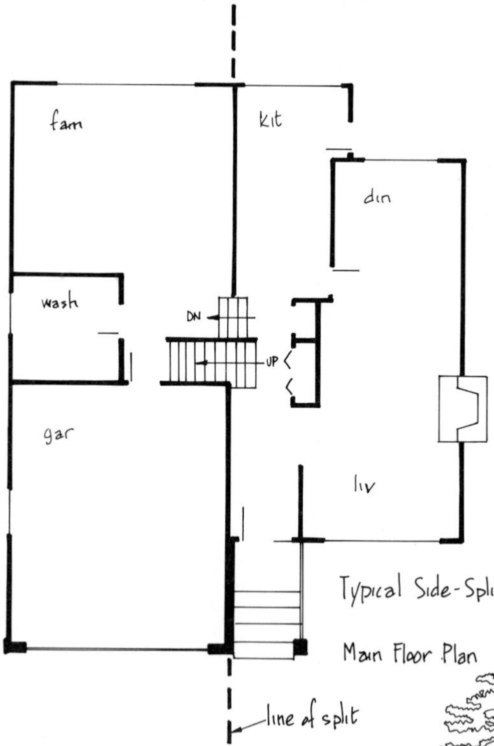

fam
kit
din
wash
DN
UP
gar
liv
line of split

Typical Side-Split

Main Floor Plan

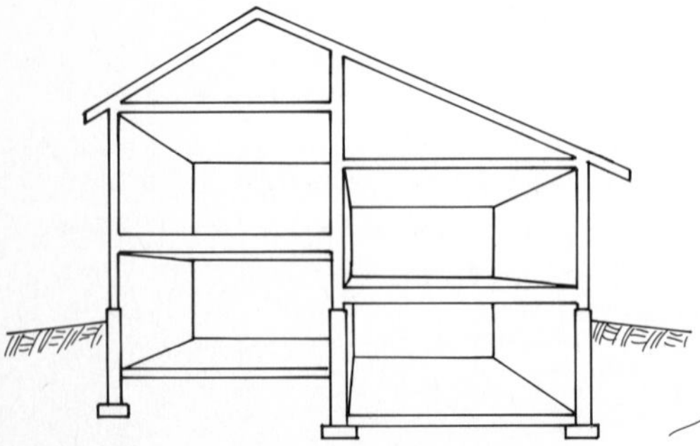

Typical Split Design Cross-Section

SIDE-SPLIT DESIGN

MULTIPLE DWELLINGS

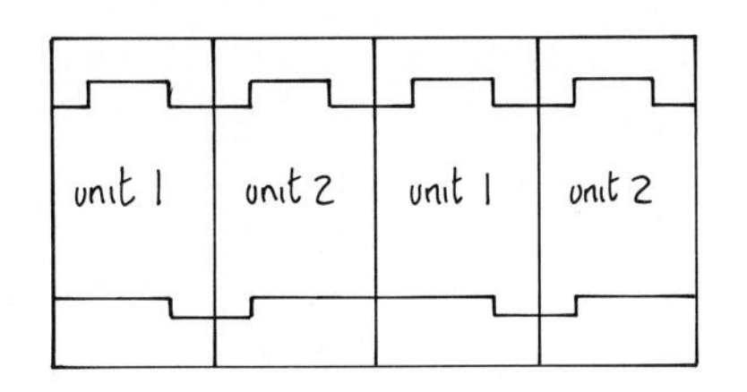

Site Plan - Units In-Line

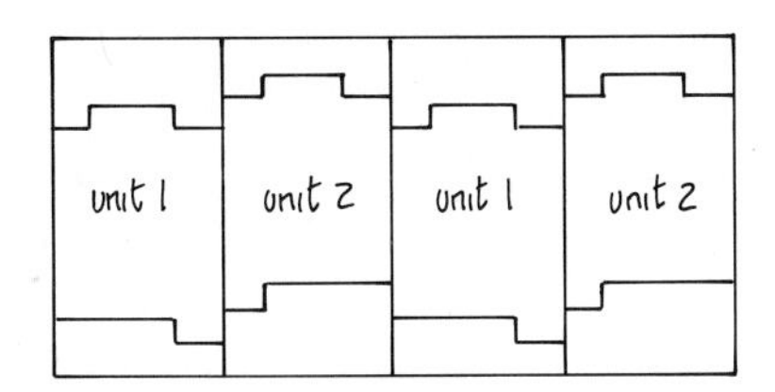

Site Plan - Units Staggered

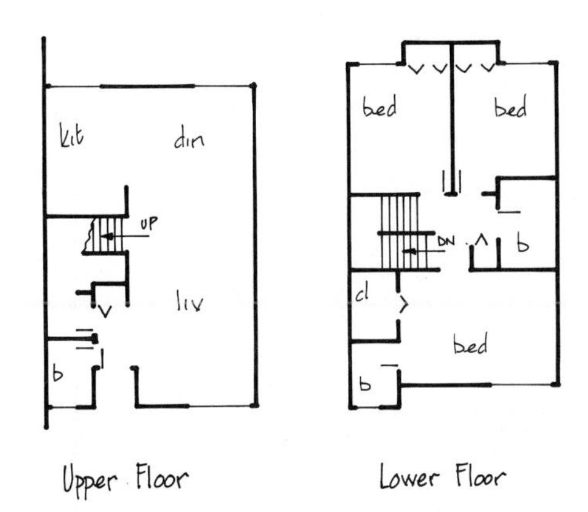

Upper Floor Lower Floor

MULTIPLE UNITS

SEMI-DETACHED (2-UNITS)

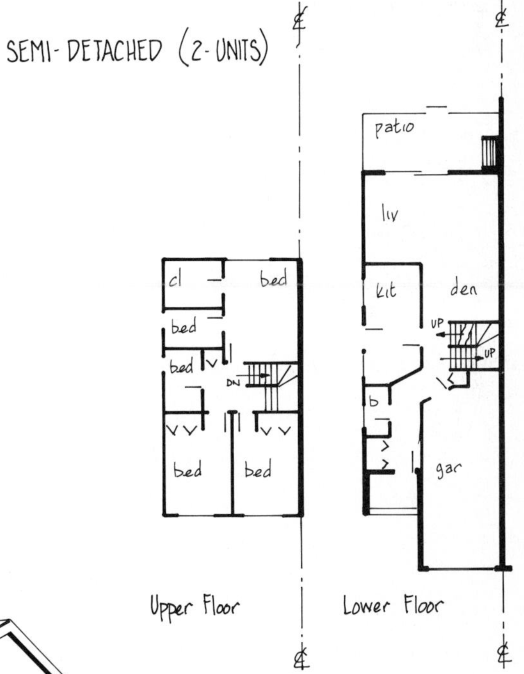

Upper Floor Lower Floor

MULTIPLE DWELLINGS

- Units are connected in pairs (called a semi-detached), or in multiples (row or town houses).
- Most units are some form of back-split or two-story design.
- Very cost-effective in terms of construction, heating/cooling, and land use.

3-6 SOLAR DESIGN - GREENHOUSE EFFECT

In architectural design, the deliberate use of the sun for heating purposes is not new. For centuries, cultures throughout the world have included passive solar design features in traditional home construction. A passive design is one that allows sunlight to enter the dwelling through extensive south-facing windows. Materials used in the dwelling's construction store and slowly release the captured solar heat. Rooms are carefully positioned to take best advantage of solar insolation and there is often no need to use mechanical devices for heat distribution. In essence, the entire structure functions as a complete and integrated system and must be designed as such.

In recent years there has been a move toward active solar systems that do employ extensive mechanical equipment to collect, store, and distribute solar heat. However, because of the extreme complexity and high cost of most active systems, it has proven very hard to justify their installation on a cost, maintainence, and payback basis. Consequently, this text will touch only briefly on active systems.

In addition, note that the design information contained in this section is intended as a basic primer only. Functional and economic solar design requires a deeper understanding of the subject, and further studies should be pursued through specialized solar texts.

A good passive solar design will take into consideration:

- The basic system – direct gain or indirect gain. (see 3-7 and 3-9).
- Building shape and orientation (see 3-12)
- Location of indoor spaces (see 3-13).
- Building materials and mass for heat storage (see 3-11).
- Thermal envelope and air/vapor barriers (see Chapter 10).

The "greenhouse effect" is the basis of all passive systems. Shortwave solar radiation can pass through glazing and into the enclosed space. This radiation is absorbed by, and raises the temperature of, structural materials and objects within the enclosure. The materials and objects, in turn, re-radiate the captured heat as longwave radiation. However, glass is opaque to longwave radiation (infrared), and the solar energy is trapped inside the enclosure.

A passive house functions in the same way, and, if sufficient mass is provided for heat storage, will function very efficiently.

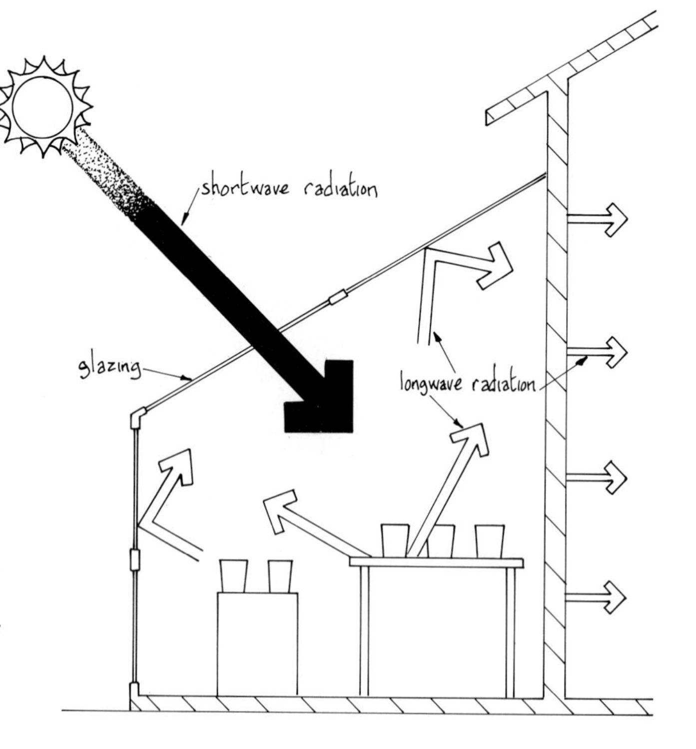

THE GREENHOUSE EFFECT

Human comfort. The re-radiation of solar energy in a living space has a significant effect on perceived comfort levels. Essentially, under conditions where the temperature of walls and floor of a room are higher than its air temperature, people tend to perceive that room as being warmer than its air temperature would indicate. As the mean radiant temperature (mrt) of the surrounding surfaces increases, lower air temperatures will be perceived as comfortable. The chart below gives combinations of mrt and air temperatures that will result in a perceived comfort level of 70°F (20°C).

MEAN RADIANT / AIR TEMPERATURES for perceived 70°(20°C) comfort level

MRT	65	66	67	68	69	70	71	72	73	74	75	76	77
AIR TEMP.	76.8	75.6	74.2	72.6	71.4	70	68.4	67.2	65.6	64.4	62.8	61.5	60

source: ASHRE

DIRECT GAIN - THE BASIC CONCEPT

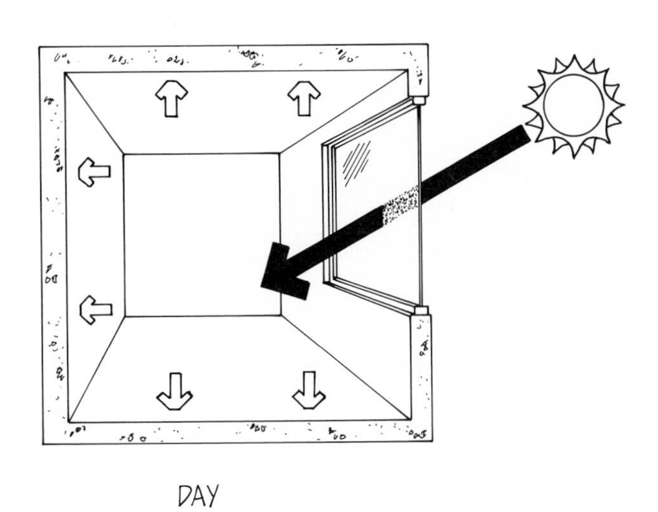

DAY

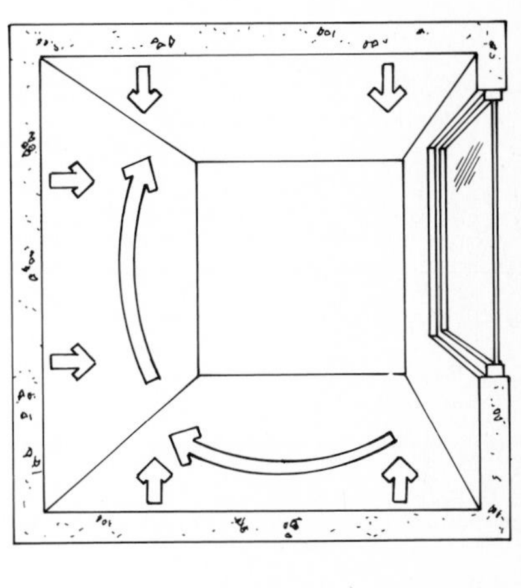

NIGHT

The most direct approach to passive solar heating, in its simplest terms, is that in which the actual living space is heated <u>directly</u> by sunlight. Mass located within the living space is used to store and slowly release solar heat. The system will collect and use both direct and diffuse light with high efficiency.

BASIC REQUIREMENTS

- Large south-facing glazing area to function as the collector with exposed living space directly behind.
- Sufficient mass located within the living space for heat storage. Mass must be placed so as to receive direct or diffuse solar energy. The quantity of mass is a function of window area and insulation values (see 3-20).
- High insulation values to conserve the retained heat. It is especially important to insulate the mass from exterior conditions to minimize heat loss (see Chapter 10).
- Heat distribution within the directly heated space is normally by radiation and convection, although a low-horsepower fan system may sometimes be necessary to move heat to other parts of the structure.

VARIATIONS

The most common variations found in direct gain systems center around the type and placement of mass in the heated space:

- Mass can be placed on the floor, walls, or ceiling, or can be freestanding within the space.
- Mass materials can be chosen to provide different heat storage capacities, conductive properties, and time-lag effects. Commonly used mass materials include brick, concrete, block, and tile.

source: Franklin Research Center

BASIC DIRECT GAIN DESIGNS

Direct gain house designs can take advantage of all standard traditional house shapes and configurations plus those designed specifically as solar houses. In each case the building must be elongated in the east-west direction to maximize solar gain. However, the same effect can be achieved by stacking, staggering, or adding clerestory windows, as shown below.

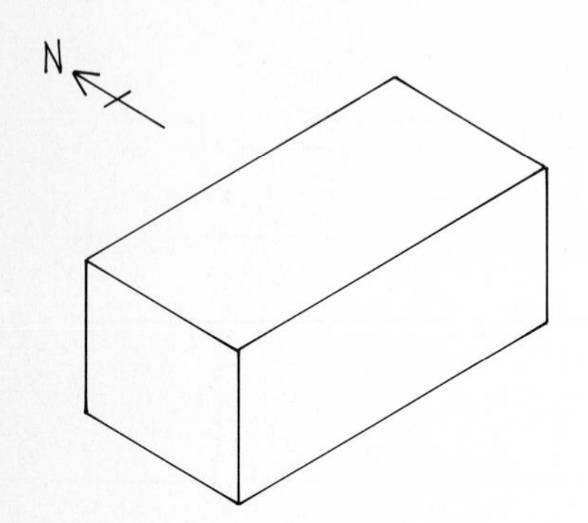

EAST-WEST ELONGATION
- Wide lots.
- Flat or gently sloping lots.
- When multilevels are not permissible.

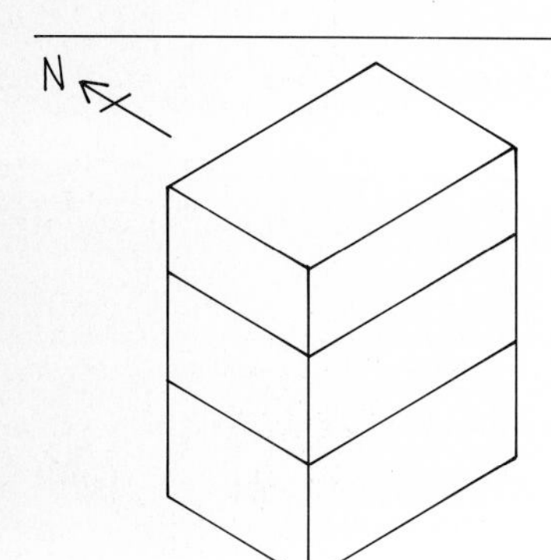

STACKED
- Narrow lots.
- Multiple housing.
- High thermal efficiency.

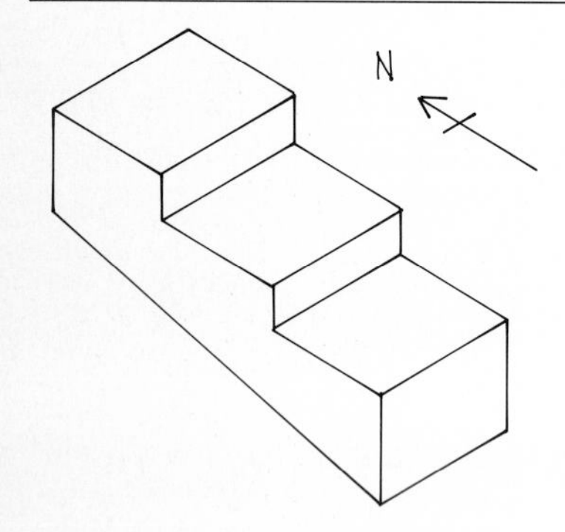

STAGGERED
- Medium-to-steep slopes.
- Multilevels
- Aesthetic effect.

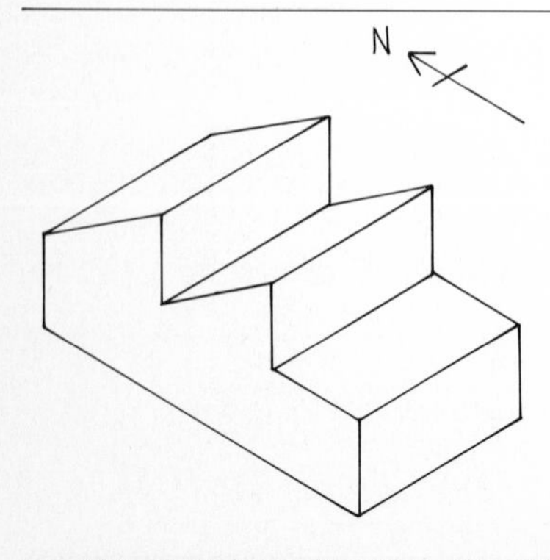

CLERESTORY
- Brings additional solar energy to north-side rooms.
- Narrow- or medium-width lots.
- Requires high ceiling spaces.

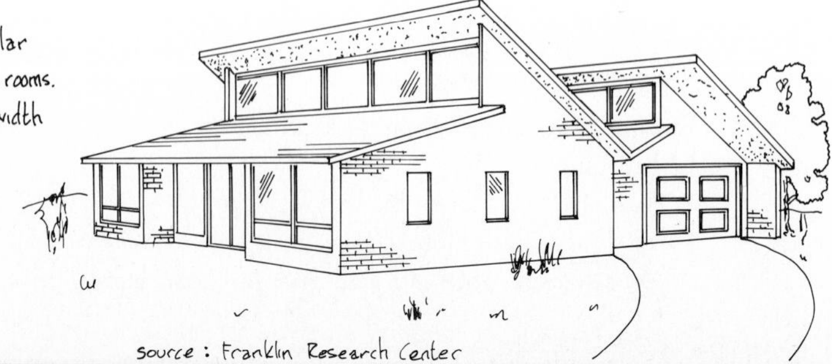

source: Franklin Research Center

INDIRECT GAIN

INDIRECT GAIN - THE BASIC CONCEPT

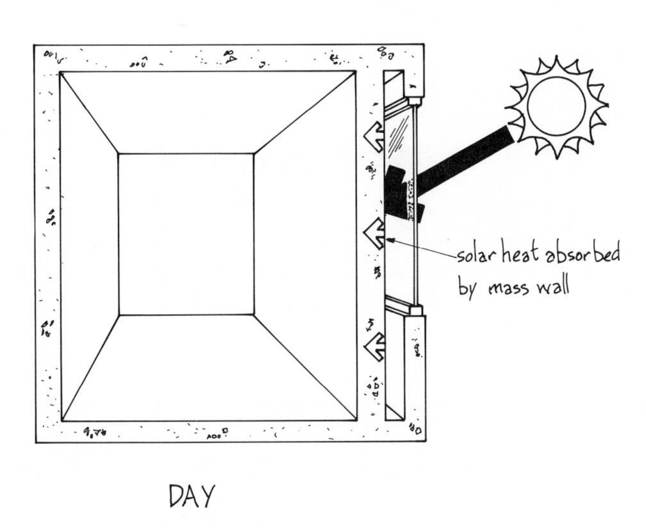

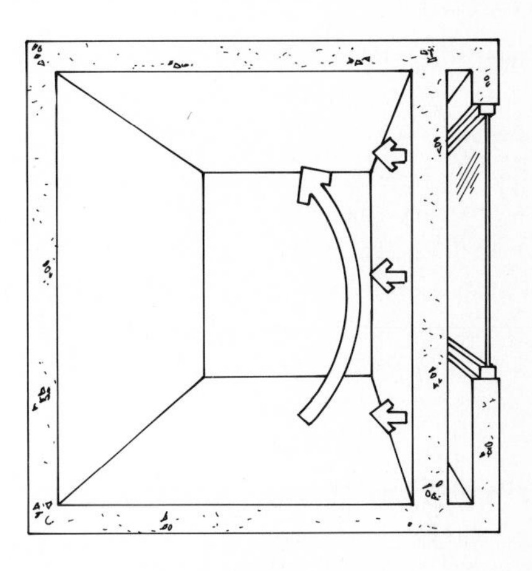

DAY

NIGHT

In an indirect gain system a thermal mass wall is placed between the space to be heated and the sun. In this way, heat is collected and stored by the interposed mass and slowly released to the space behind by radiation and convection. By varying the type and size of the mass wall, control can be exercised over the quantity and speed of heat transfer. Materials used for the wall construction will normally be masonry or concrete, or water in suitable containers.

Basic elements are:

- Radiation is the primary heat distribution mode, with convection as a secondary mode.
- For maximum radiation efficiency and human comfort, room dimensions must be carefully controlled. Depth of space behind a solar wall is limited to 15-20 ft (4500-6000).
- Daytime heat can be provided by vents in the mass wall that will allow hot air trapped between glazing and wall to escape to the space.
- Materials used in rooms behind a solar mass wall can be of any type and color.
- Since no natural light will be admitted to the room behind the wall, additional windows can be located in the wall itself.
- This system allows greater control of the large daily temperature swings normally experienced with direct gain systems.

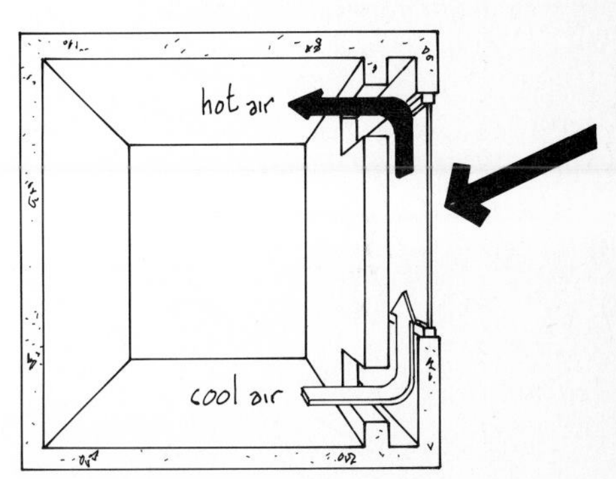

Wall With Operable Vents
Cool air from the space is convected between the glazing and the wall during the daytime.

The illustrations at the right show the use of operable vents located in the mass wall that provide immediate heating to the living space during the daytime.

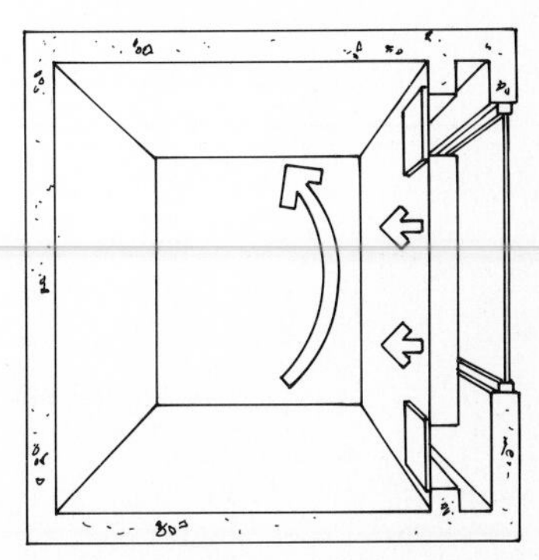

During the night time, vents must be closed to prevent reverse convection and cooling.

source: Franklin Research Center

DESIGN CONSTRAINTS - INDIRECT GAIN

Care must be exercised when integrating a mass wall indirect gain system into building design. Factors to be considered are:

- Low light levels behind the mass wall resulting in a closed-in feeling and lack of exterior views.
- To function properly, the wall must be placed 4"(100) or more behind the collector glazing and sealed at floor and ceiling. Provision must also be made for cleaning the interior face of the windows.
- The wall must be painted black or be of a similar dark color for maximum collector efficiency.
- The mass wall will need a separate foundation system if it is not part of the building's exterior wall system.
- The overall efficiency of the basic mass wall system is lower than that of the direct gain system. This is so because of the additional glazing necessary to bring light to the space behind the wall. Efficiency will increase if window insulation is applied at night.

To solve these design restraints, designers will compromise by:

- Partial use of mass walls.
- A combination of direct and indirect gain systems.

The residential design illustrated here is a good example of the compromise possible between direct and indirect passive systems.

SOUTH FACE OF INDIRECT GAIN SOLAR RESIDENCE

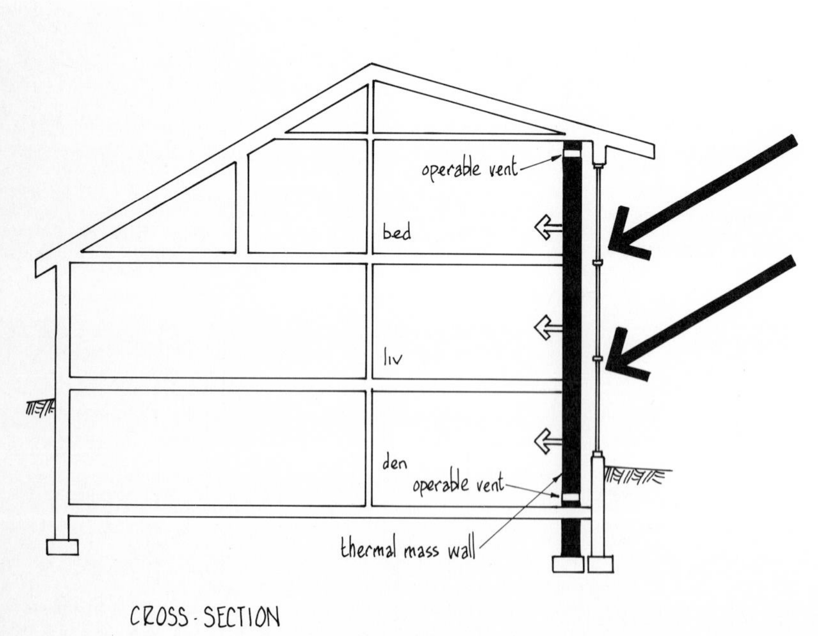

CROSS-SECTION

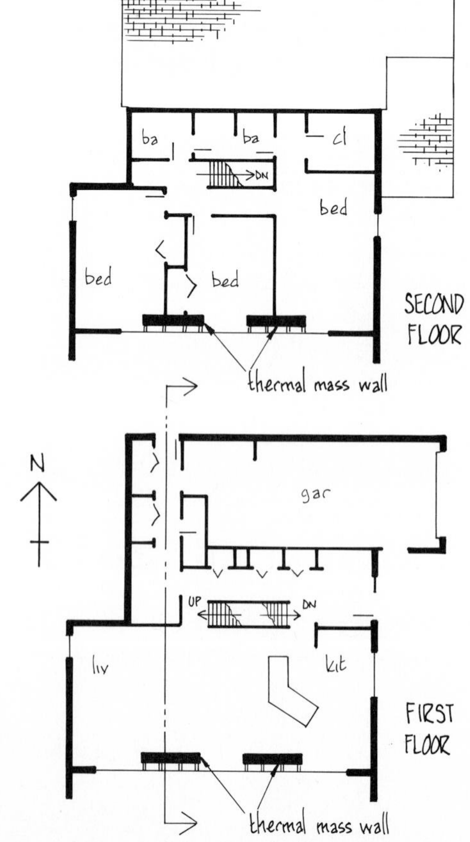

Source: Housing Section, Engineering Branch, Alberta Agriculture

SOLARIUMS

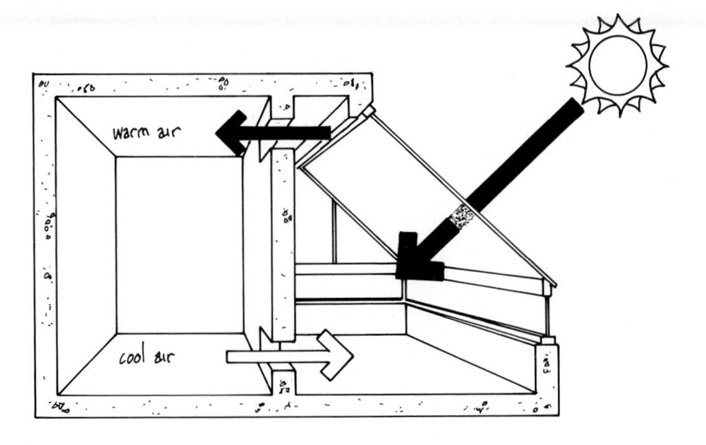

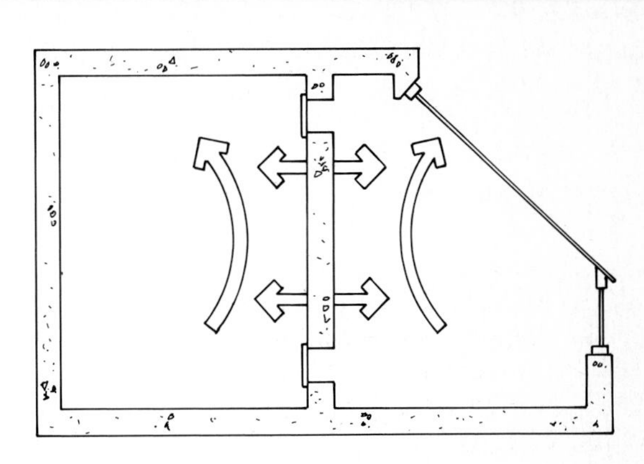

DAYTIME

A solarium can be defined as a greenhouse, sunroom, atrium, or sunporch. It functions as a hybrid of direct and indirect gain systems, utilizing features from each. During the day, the solarium space receives solar heat as direct gain and stores heat in the common mass wall as indirect gain. Operable louvers provide heated air to the living space behind.

NIGHT TIME

If the common wall, along with additional heat storage mass, is correctly sized, both the solarium and living space behind can be heated.

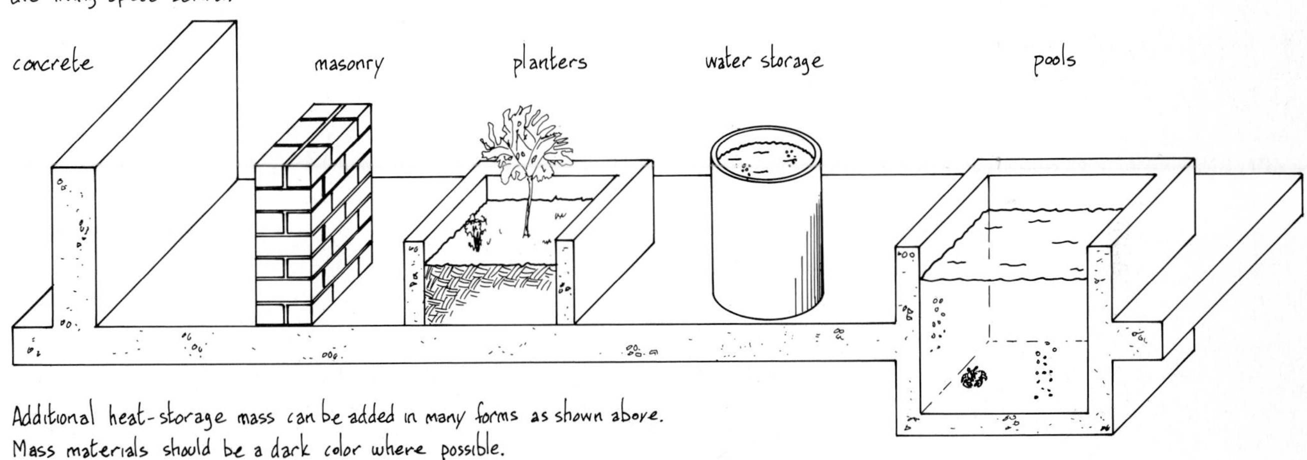

Additional heat-storage mass can be added in many forms as shown above. Mass materials should be a dark color where possible.

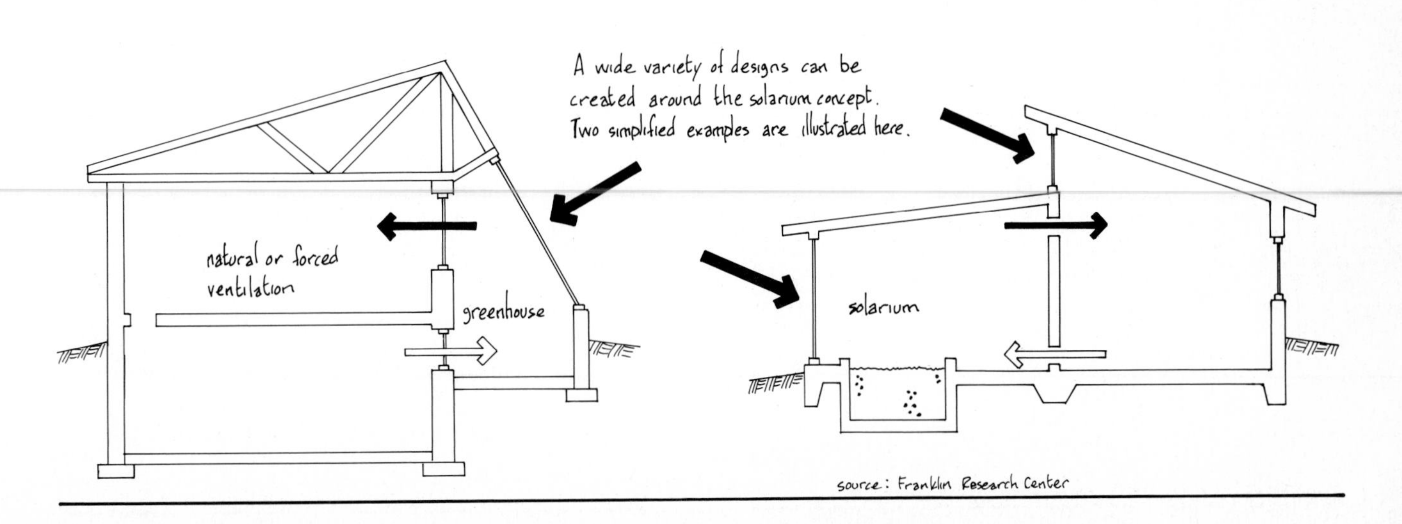

source: Franklin Research Center

BUILDING SHAPE AND ORIENTATION

To achieve the best solar efficiency for winter heating and summer cooling, the basic house shape should be elongated in the east-west direction. This shape is equally efficient in all climates. If it is not possible to elongate in the E-W direction because of building lot restrictions, i.e., narrow lot, then use clerestory, staggered, or stacked designs (see 3-8).

The two illustrations at the right compare the solar impact on each building face under winter and summer conditions. Note the following:

- In winter, the south face receives more solar energy than in the summer.
- In summer, east and west faces receive more solar energy than in the winter.
- In summer, roof areas receive considerably more solar energy than in the winter.
- Very little energy is received on the north side in any season.

Each of these effects is caused by the relative altitude-angle of the sun in the winter and summer seasons. See 3-18 for sun charts that show sun positions throughout the year.

Resulting design criteria:

- Maximum window area should be concentrated on the south side.
- Minimum window area on the east and west faces.
- Zero or minimum window area on the north face.

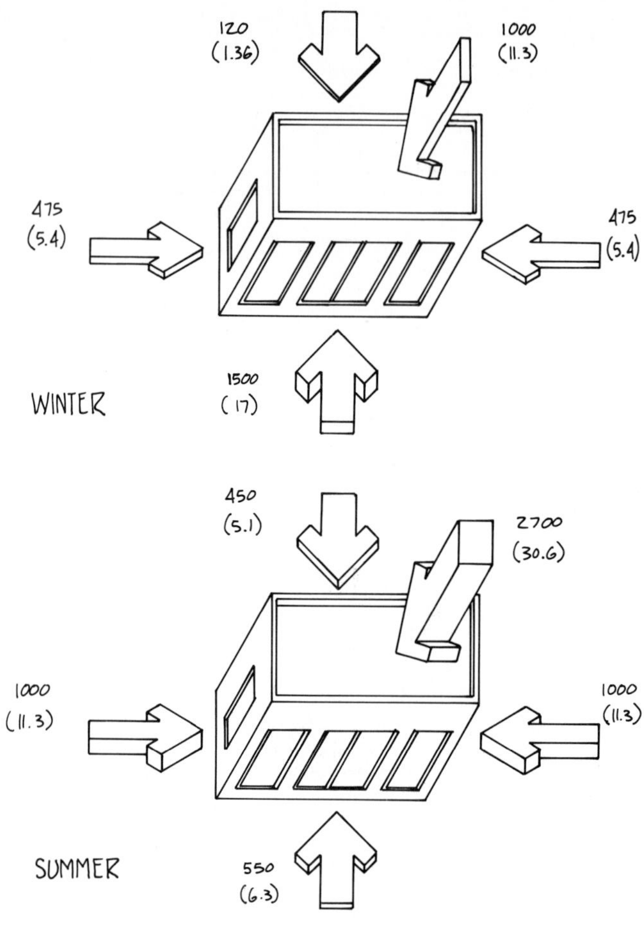

Approximate Solar Impact at 40°N Latitude
Impact figures are Btu/sf/day (MJ/m²)

Correct positioning of the house on the lot is very important. The south side of the lot will be the most valuable for human social purposes in all but hot desert areas. Additionally, to avoid serious solar blockage by buildings, trees, hills, etc., the house should be located in the northern portion of the building lot.

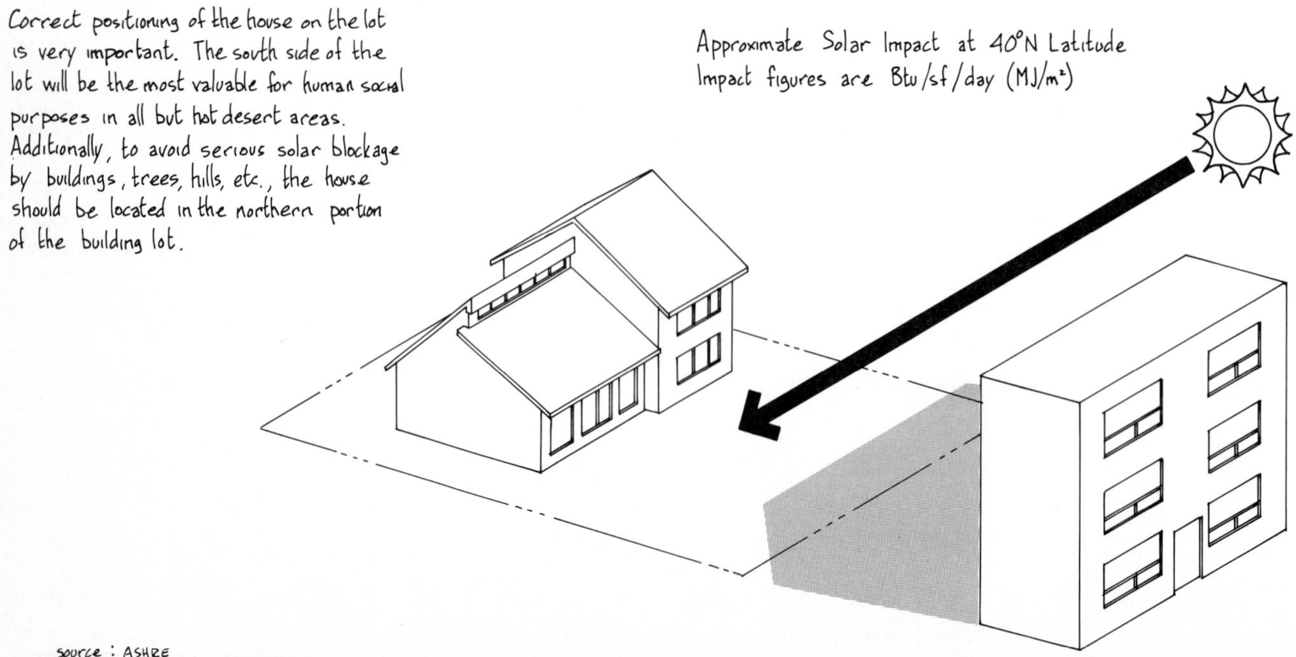

source: ASHRE

LOCATION OF INDOOR SPACES

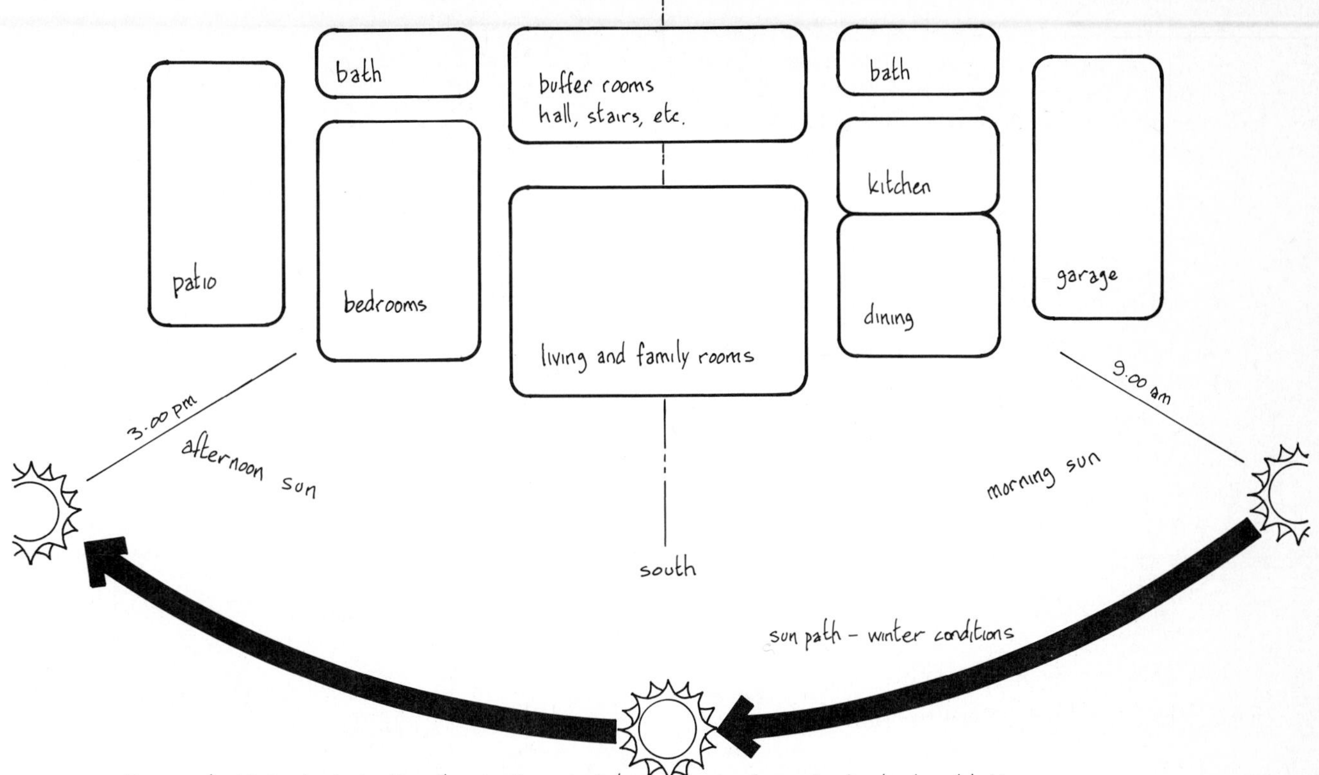

Rooms should be located within the structure to take best advantage of solar heat and light:

- High-usage rooms should be located to the south.
- Low-usage rooms to the north to act as buffer zones.

Refinements:

- On the east side, place bedrooms, kitchens, and dining rooms, to take advantage of early morning heat.
- On the south side, place living rooms, live-in kitchens, and family rooms, for all-day heat and light.
- On the west side, place afternoon and evening usage rooms, such as workshops, family rooms, and bathrooms.

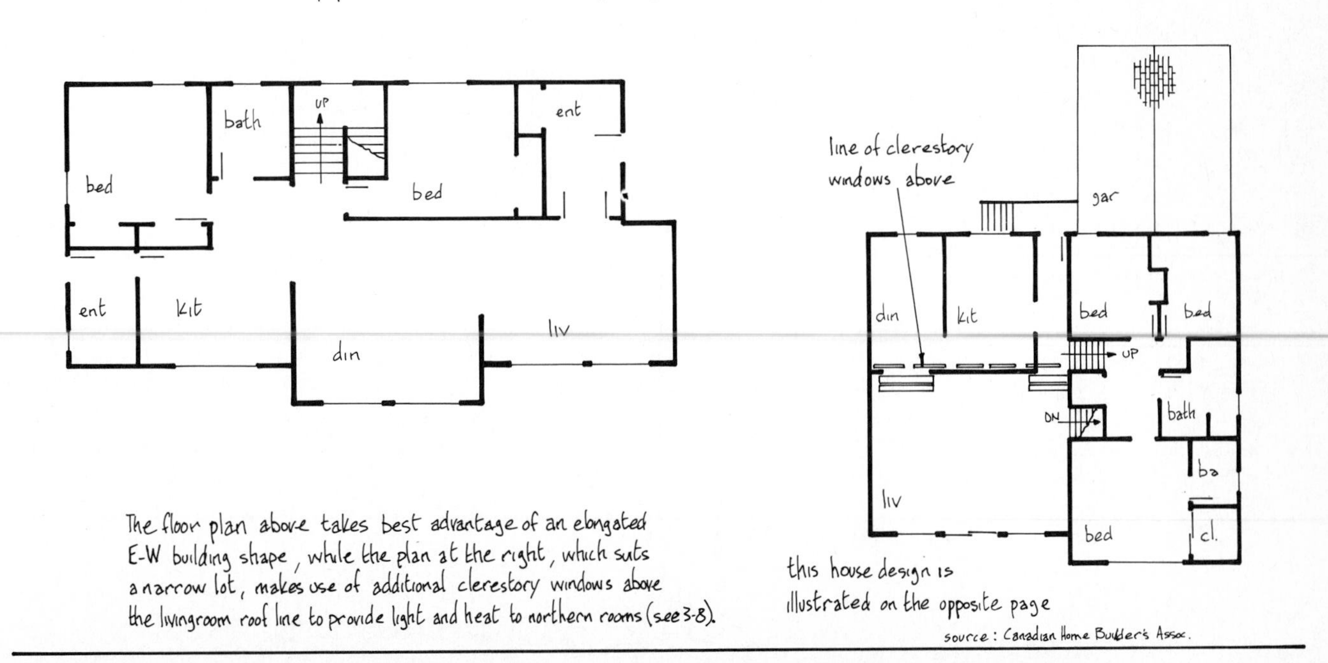

The floor plan above takes best advantage of an elongated E-W building shape, while the plan at the right, which suits a narrow lot, makes use of additional clerestory windows above the livingroom roof line to provide light and heat to northern rooms (see 3-8).

this house design is illustrated on the opposite page

source: Canadian Home Builder's Assoc.

PASSIVE SOLAR

3-14 WINDOW ORIENTATION

In the winter season, windows are an energy drain on a building regardless of the direction they face. For example, a single-pane window will lose over 20 times as much heat as will a standard stud wall with 3½ in (90) of fiberglass insulation. As the structure's energy efficiency increases, windows become major sources of heat loss. Therefore, windows in a solar structure are located so that their heat gain from sunlight during the day is greater than their heat loss during the night. In addition, adequate shading must be provided to avoid summer overheating.

Factors to consider when locating windows in a structure are:

- Transmission of light at varying incident angles.
- Window orientation and solar radiation values.
- Vertical or tilted window orientation.

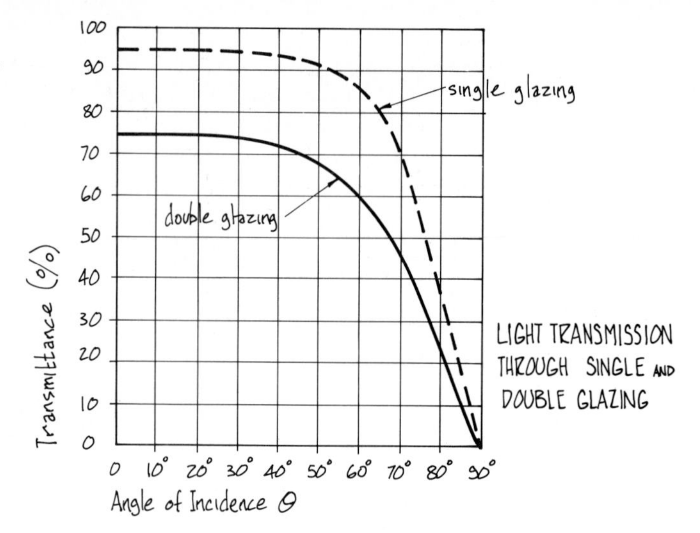

LIGHT TRANSMISSION THROUGH SINGLE AND DOUBLE GLAZING

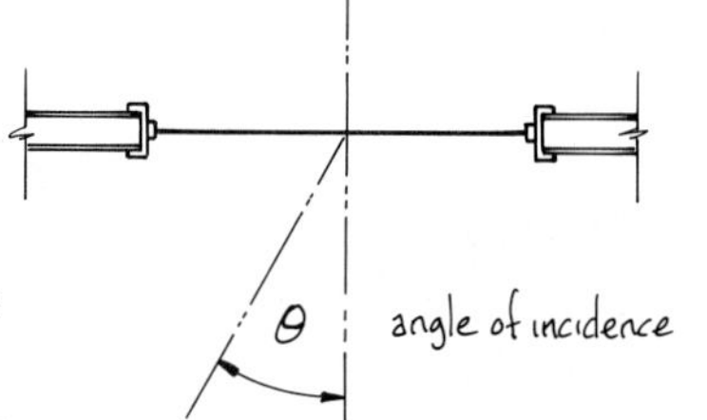

PLAN VIEW OF GLAZING

Light transmission through glazing declines as the angle of incidence moves away from the perpendicular (0°). Beyond 30°, transmittance starts to drop precipitously.

SOLAR RADIATION AND WINDOW ORIENTATION

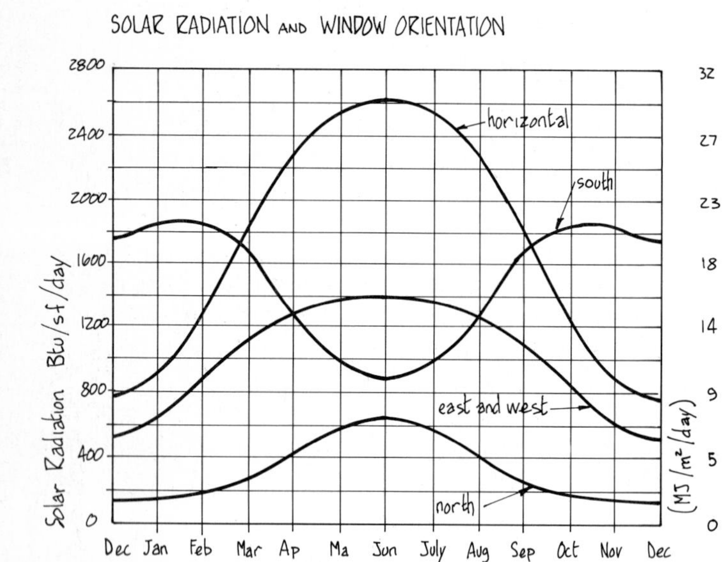

The direction a window faces has a significant effect on the amount and variation of solar radiation received. In the chart at left note that north, east, and west windows receive the lowest values in winter and maximum values in summer. In direct contrast, south windows receive maximum values in winter and minimum values in summer. This is an ideal situation for solar design (see 3-12).

THE IDEAL GLAZING SYSTEM

For optimum solar efficiency, glazing should face due south, although the house or the glazing may be positioned up to 30° east or west of south before performance is noticeably affected.

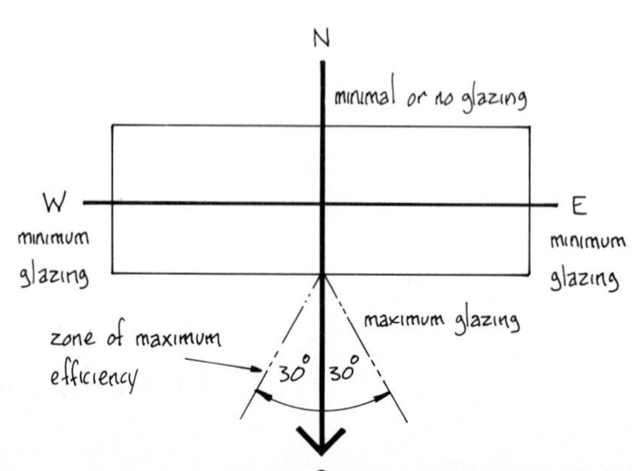

Glazing should be kept to a minimum on east and west faces. This is especially so on the west face, where evening summer sun plus high air temperatures can easily cause overheating. The north face should, if possible, contain no glazing at all.

source: Canadian Home Builders' Assoc., ASHRAE

STREET ORIENTATION - LANDSCAPING

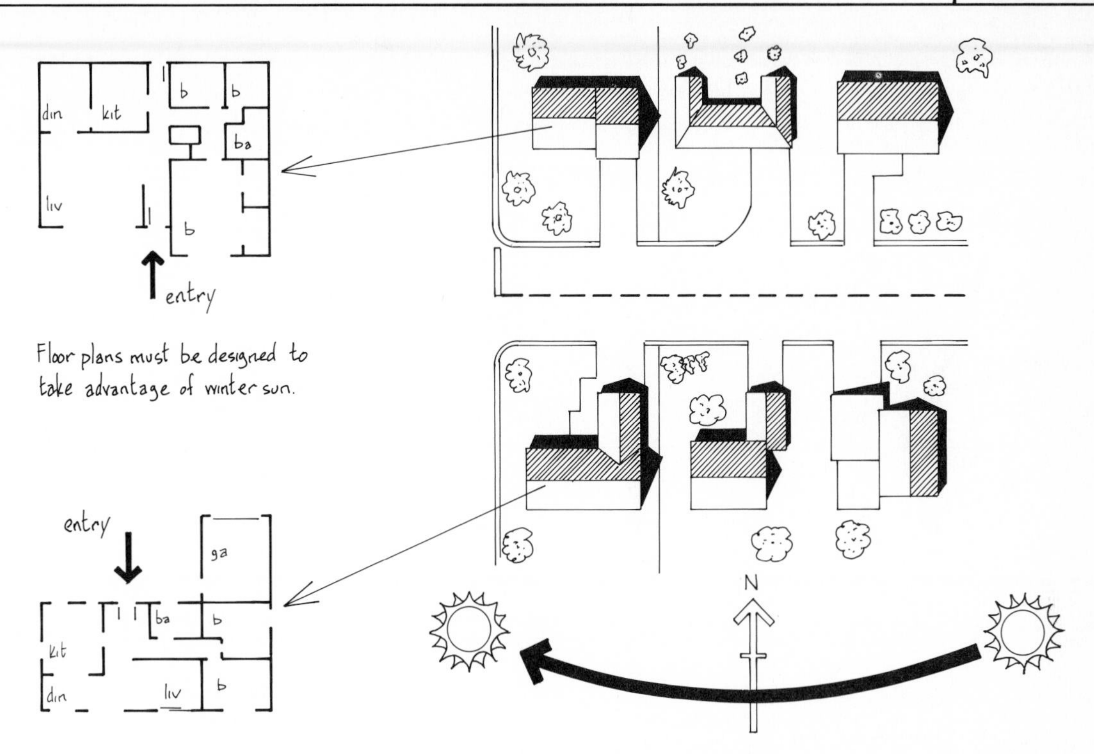

Floor plans must be designed to take advantage of winter sun.

The optimum window locations shown on the preceding page dictate the overall design of streets and house positions within a tract or survey. To achieve maximum solar potential, streets should be set out in the east-west direction. In addition, the solar effects and microclimate of each dwelling must be considered. Individual house designs must allow for solar impact and integration of driveways, garages, access doors, and room usage.

Exposure to prevailing winds, topography and landscaping all have a noticeable effect on the microclimate of a dwelling in addition to solar exposure.

- Prevailing winter winds are from the north and west. Use of coniferous trees on these sides will deflect winds up and over the dwelling, reducing air pressure on the windward side.
- Deciduous trees on the south side will provide shade and cooling.
- Sites with coniferous trees to the south are to be avoided unless solar obstruction is not a factor (see the suncharts on 3-18).
- The ideal topography is a south-facing slope.

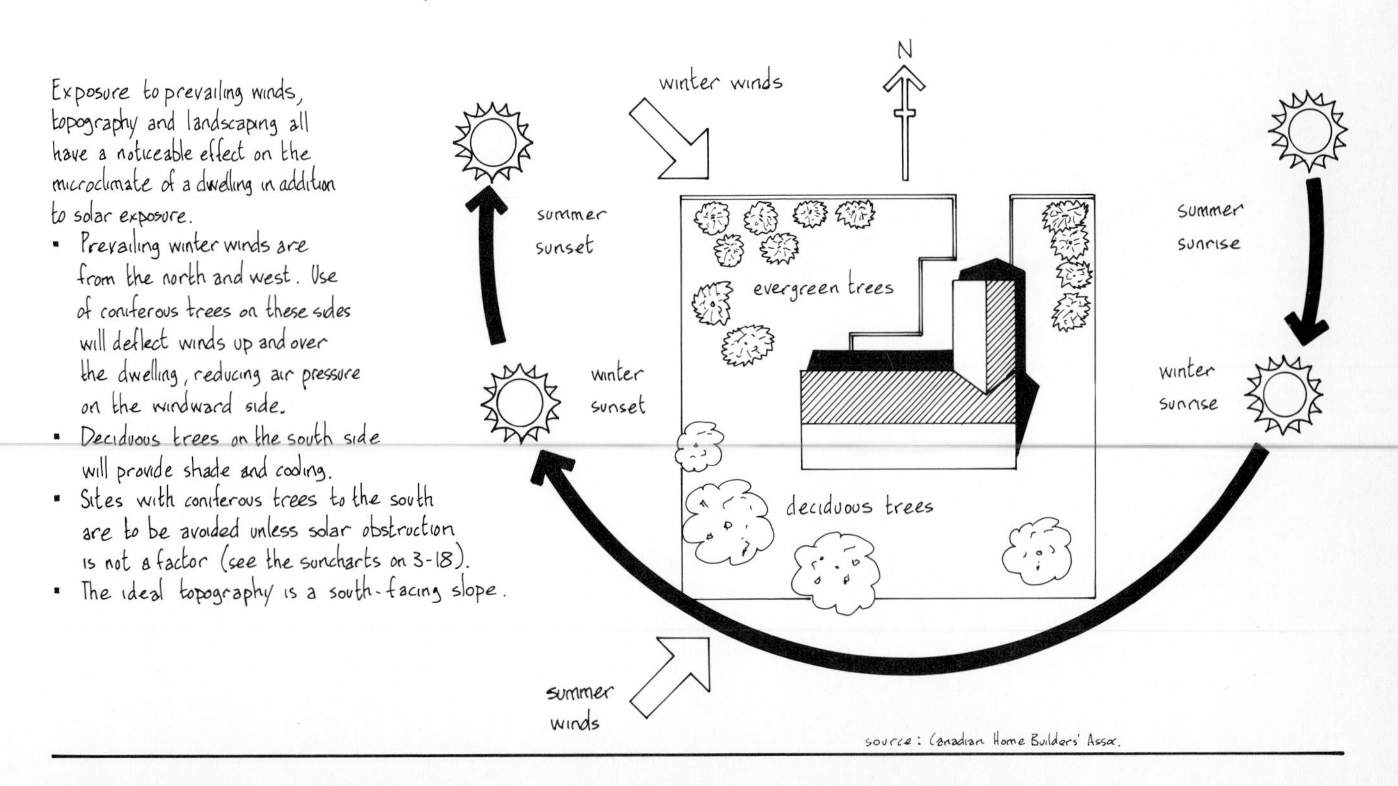

source: Canadian Home Builders' Assoc.

CLERESTORIES - SKYLIGHTS

The use of clerestory and skylight windows for collecting sunlight has several advantages. When properly designed, the toplighting effect eliminates glare from floors and walls by reducing the contrast between interior surfaces and windows. In addition, because they are in or above a roof line, they provide privacy and reduce the chance of solar shading by off-site obstructions.

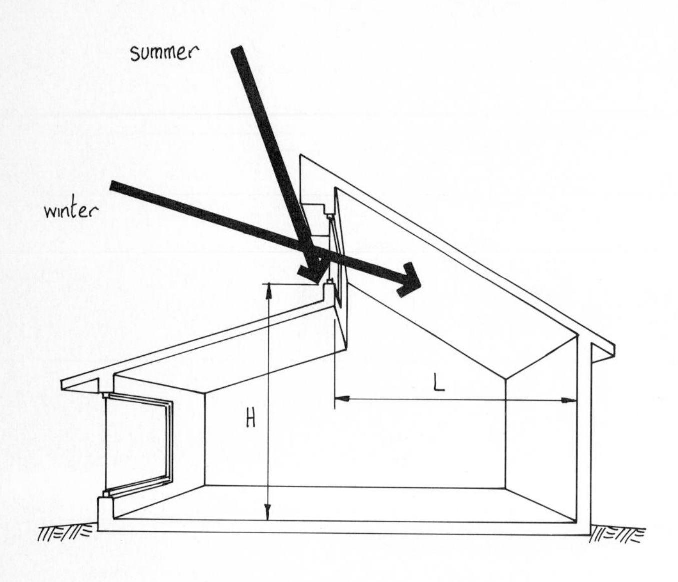

L = 1 - 1½ H

Typical clerestory configuration. Note the ratio of length to height for best distribution of sunlight on the north wall.

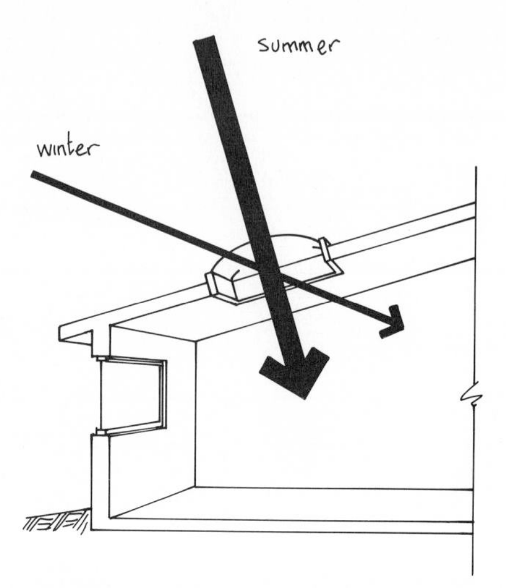

Skylights, although very efficient at bringing light to a space, also work against the solar ideal. The low winter sun has a small impact through the skylight when heat is needed, while the high summer sun can cause serious overheating. This problem increases as the roofline becomes flatter.

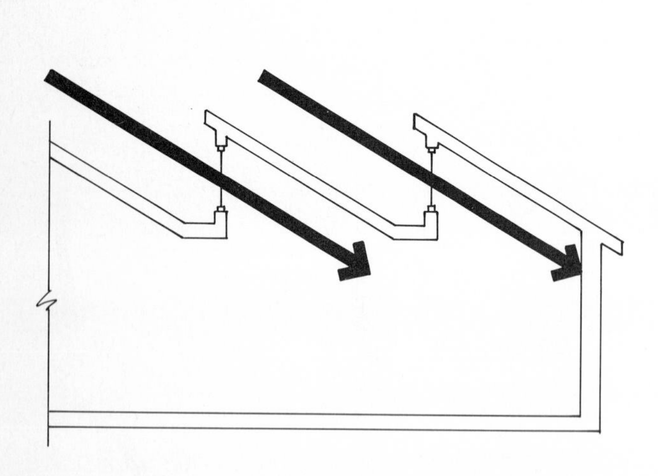

A sawtooth roof shape will distribute light and heat to the entire space. The angle of each roof should approximate the lowest sun angle on December 21st (see suncharts on 3-18).

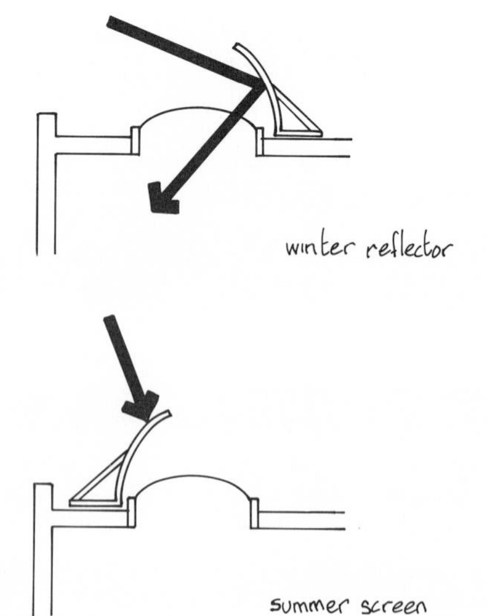

To alleviate winter underheating and summer overheating, the use of a reflector and screen, respectively, is recommended, especially for flat roofs. However, this approach requires the active participation of the house occupant, and may present problems in this regard.

source: Franklin Research Center, Canadian Home Builders' Assoc.

WINDOW SHADING

South-facing windows that are designed to accept full winter sun must be shaded in the summer months to avoid overheating problems. Effective shading of south windows can be achieved with overhangs or canopies, while east or west windows present a special problem.

In addition, the middle of the summer climatic season does not coincide with the longest day of the year on June 21st. Nor does the middle of the winter season coincide with the shortest day on December 21st. Because of this, a fixed shading device will provide the same shade on September 1, when it is still warm, as on March 1, when it is cool (see the suncharts on 3-18). If this factor is deemed important in a design, adjustable shading should be considered.

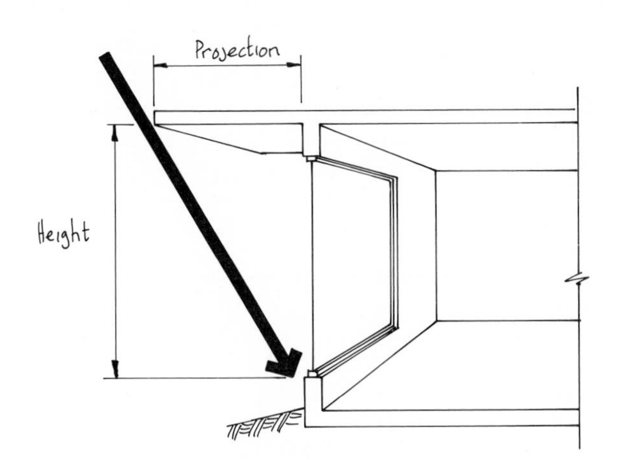

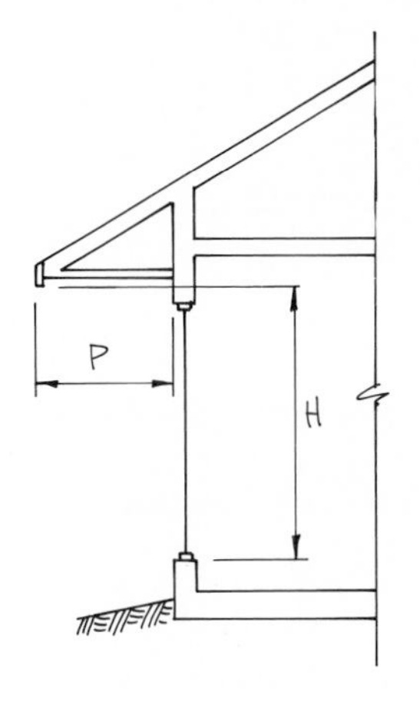

For pitched roofs the basic formula can be used provided that allowance is made for the depth and projection of the overhang.

The basic formula for overhang calculation:

$$\text{projection} = \frac{\text{height (window opening)}}{\text{factor}}$$

Factor table:

	100% shading at noon on:	
Latitude	June 21	Aug 1
30°	5.8	8.7
35°	3.0	4.5
40°	2.5	3.4
45°	1.9	2.6
50°	1.6	2.0
55°	1.3	1.5

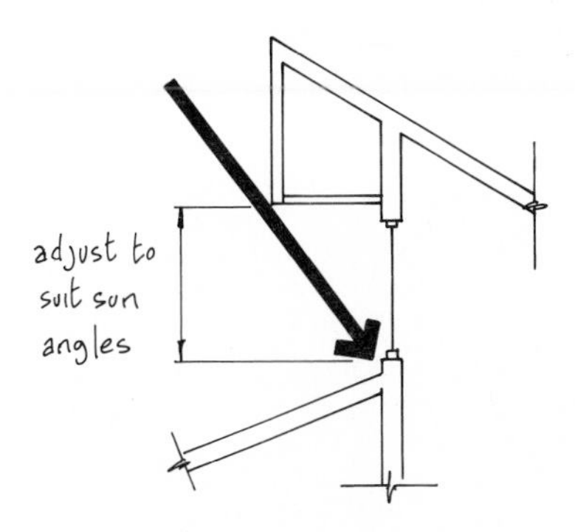

For more complex roof shapes, e.g., clerestories, it is advisable to ignore the formula and draw a scale diagram of the intended structure. Add the high and low sun angles (see 3-18) and adjust the overhang to suit.

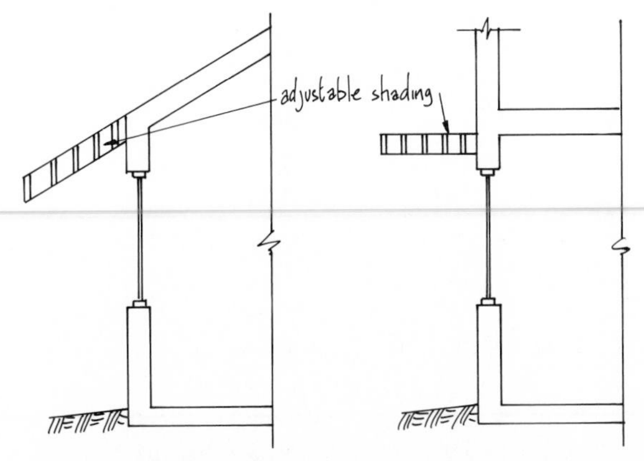

Use adjustable louvers to take maximum advantage of seasonal solar values.

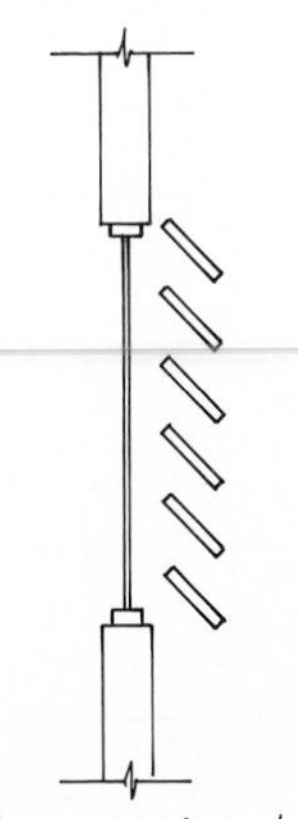

plan view of window

On east walls in the morning and west walls in the evening, sun angles will preclude the use of horizontal overhangs. Under these conditions, vertical adjustable louvers can be used to exclude direct sun but allow diffuse light to enter the room. Alternatively, trees or other buildings can be interposed between window and sun.

source: Franklin Research Center, Canadian Home Builders Assoc.

SUN CHARTS

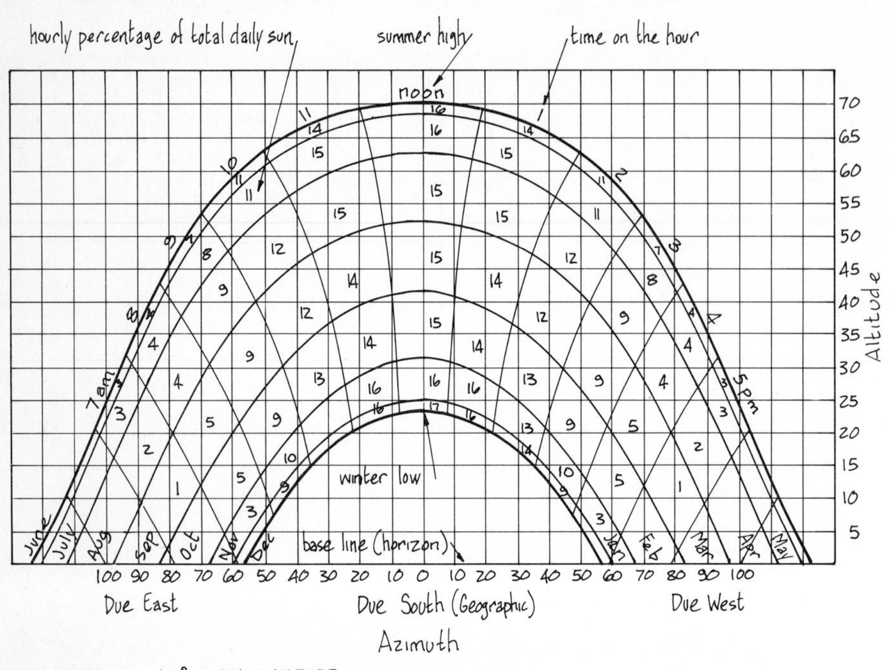

A sun chart is an extremely useful tool that helps evaluate a proposed solar site. It is a two-dimensional record of the annual movements of the sun across the sky, as observed from a specific latitude. All obstructions blocking the sun are mapped on the chart and the resulting loss of solar energy is clearly shown. The chart is orientated due south with the horizon as the baseline. Each arch traces the monthly sun path with month names centered on the 21st day of each month. The sun traces the same path twice each year with the high summer path on June 21st and the low winter sun path on December 21st. The radial lines indicate the time of day on the half hour and the numbers in each "square" give the hourly percentage of total daily sunshine.

SUN CHART FOR 43° NORTH LATITUDE

The sun chart for 43° N above shows hourly percentage of total daily sunshine. Example: from 10-30 am to 11-30 am on October 21st (or February 21st), 15% of the total daily sunshine is available in that hour, under perfect conditions.

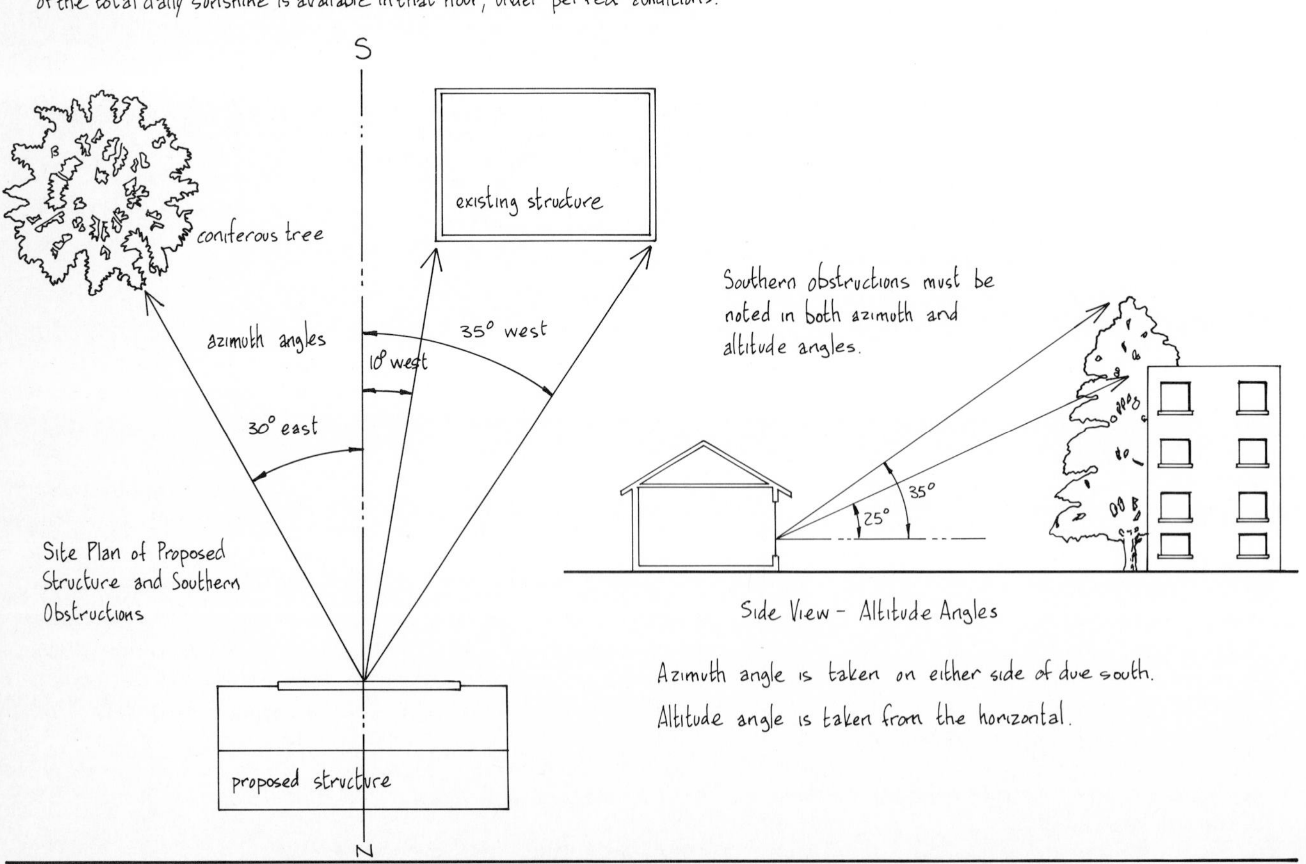

Azimuth angle is taken on either side of due south.
Altitude angle is taken from the horizontal.

SUN CHARTS

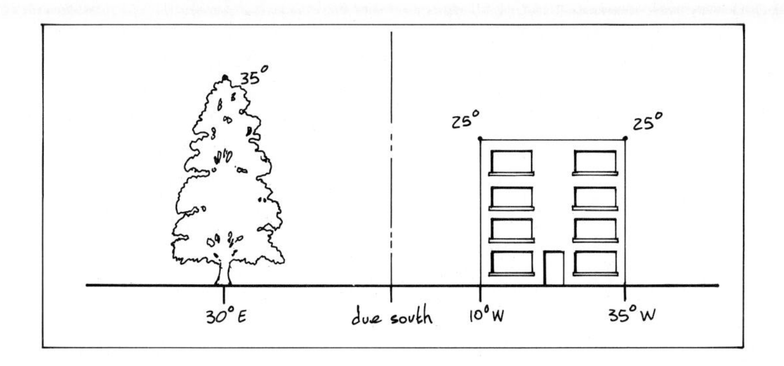

Sketch obstructions, indicating azimuth and altitude angles. Include any anticipated future obstructions.
Transfer the information to the sun chart using the grid information.

With the obstructions from the previous page drawn in place on the sun chart, the site's solar potential can be analyzed. For example, in November or January, the deciduous tree east of south will obstruct approximately 20% of available solar energy. However, this tree should drop its leaves in the Fall, leaving bare branches. Assume then that the tree's solar obstruction is reduced to 7%. The building west of south will obstruct perhaps 16%, giving a total of 23%. In other words, from November 1 to Jan 31 approximately 77% of the ideal daily solar energy is available to the building. Before November and after January there is very little obstruction.

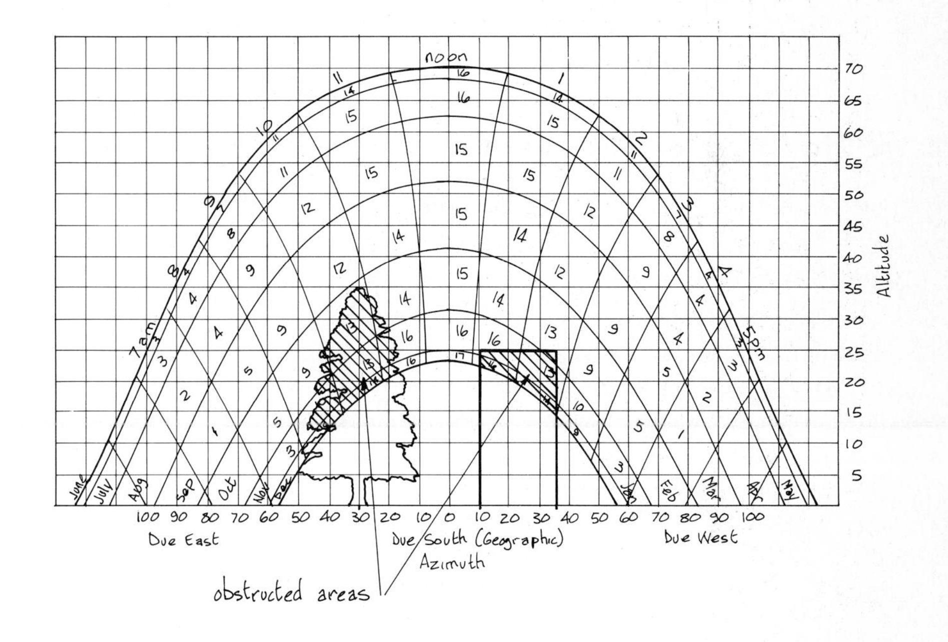

Additional notes:

- Compass south (magnetic) is not always the same as true south (geographic). Specific location determines the size of the variance, which can exceed 20°. For example, on a line from New York to Ottawa, Canada, there is a westerly deviation of approximatly 11°, meaning that true south is 11° west of magnetic south. Seattle and Vancouver, Canada, have a deviation of 22° to the east. Consult a surveyor to determine the exact variance in a specific region and adjust the azimuth readings accordingly.
- Care should be taken with coniferous trees since these do not drop their leaves in the Fall. Trees also grow, some quite fast, and allowance must be made for future obstructions.
- Check with local authorities for possible changes in zoning bylaws, or structures already approved but not yet built, that may become obstructions in the future.
- The higher above ground level the solar collection area is, the less chance of solar obstruction
- Sun charts can be used in reverse for the planning of landscaping, tree planting, and future building locations.
- Since the sun path will vary at different latitudes, use a chart drawn for each particular location. Most solar textbooks contain sun charts for a variety of latitudes.

SIZING GLAZING

Passive solar systems, when properly designed, can be extremely efficient in collecting and using solar energy for space-heating purposes. However, it must be clearly understood that to achieve this efficiency, very careful integration of all elements in the system is paramount. Depending on location, climate, and system efficiency, a good solar design should contribute 30-40% or more of total heating requirements. Major items of consideration are:

- Glazing area/floor area. This ratio is very critical and complex. Too much south glazing and overheating will result. Too little south glazing (or too much north, east, or west glazing) reduces cost efficiency.

- Storage mass. Basically, as much mass as possible should be incorporated into the structure. However, the mass should be of the correct type and thickness for the intended system and located in the correct position. There is also a size and position relationship between collector glazing and mass. Mass solar walls in an indirect gain system must be designed very carefully to provide optimum heat collection and distribution.

- Insulation values. As insulation and air/vapor barrier systems in a structure increase in value, the percentage of solar heat contribution to the structure will also increase, leading to possible overheating problems. Therefore, as thermal efficiency rises, collection glazing must decrease or mass storage capacity and distribution systems must increase. A structure that is "super insulated" may require very low levels of solar input to function economically (see Chapter 10 for insulation techniques).

- Heat distribution systems. The method by which solar heat is distributed through a building affects the size of the collector system. For example, if distribution is fan-forced throughout the building, a larger collection system can be used than if a natural convection cycle were employed. If the collection system is indirect gain, room dimensions behind the mass wall must be carefully controlled.

- Temperature swings. In direct gain systems, oversizing of collector glazing will result in room-temperature swings uncomfortable to the dwelling's occupants. Additionally, summertime ventilation must be incorporated into the design to avoid overheating.

- Cost efficiency. The general rule is that there must be a reasonable payback period wherein the specific costs of additional systems, over and above a standard structure, are returned through savings in space heating charges. Normally acceptable payback periods are in the range of 3-5 years.

RATIO SIZING OF SOLAR COLLECTOR GLAZING
DIRECT GAIN

Average Winter Degree Days °F (°C)	Sq Ft (m²) Of Window Area Per Sq Ft (m²) Of Floor Area		
	Lat. 35°N	Lat 45°N	
9000 (5022)	0.29	0.50	Adjust ratios to reflect latitude of site.
7300 (4070)	0.25	0.40	
6300 (3510)	0.21	0.34	
5000 (2780)	0.18	0.30	
4300 (2390)	0.15	0.25	
3000 (1662)	0.12	0.21	
2300 (1270)	0.10	0.17	

Notes to table:

- Table is based on south-facing glazing with a maximum deviation of 25° east or west.
- Table assumes an 8 Btu/day/°F/sq.ft floor area (8.4 kJ/day/°C/m²) heat loss. Adjust ratios downward for lower heat loss ratings.
- Ratios are based on individual south-facing room areas only. If there is mechanical distribution of solar heat to nonsouth-facing rooms, ratios can be increased.
- Example: A south-facing room of 300 sf (27.8 m²) in the 7300 (4070) degree day, 45°N region, could have [300(27.8) × 0.40] or 120 sf (11.2 m²) of collector glazing.

SYSTEM COMPARISON

SYSTEM COMPONENT	DIRECT GAIN	INDIRECT GAIN	GREENHOUSE
GLAZING	Main areas face south. Serves for light and view as well as collectors. Possible problems with glare.	Serves as solar collector only. Must be south-facing. Additional glazing may be needed for light and view.	Extensive south glazing required. Consequent high heat loss at night without movable insulation.
BUILDING SHAPE	Elongated in east-west direction unless clerestories, stacking, or staggering is employed. High-use rooms on south side.	Elongated east-west. Rooms behind mass wall limited to 15-20 ft (4500-6000) in depth. Stacking and staggering are possible options.	Requires large area of south wall. Must attach to adjoining room that are to be heated.
THERMAL CONTROL	Since room is collector, large temperature swings will occur. Size and location of mass will moderate swings. Use of mechanical heat distribution strongly recommended.	Mass wall controls temperature swings if properly sized. Daytime temperature controlled by wall vents.	Carefully sized and placed mass needed to control temperature swings. Fan-forced underfloor rock storage is effective.
CONSTRUCTION MATERIALS	Heated space must contain sufficient mass in floors and walls. Mass materials should be a dark color, light construction can be a light color.	Thermal mass wall of masonry or water placed directly behind glazing. Materials in space behind wall can be any type and color.	Common wall between greenhouse and living space must be a mass wall. Material in living space can be any type and color.
EFFICIENCY *	System efficiency can reach 75% if all glazing in that space is used as a collector. Glazing not used as a collector will reduce overall efficiency.	In the range of 30-45% since glazing is required in addition to collectors for light and view.	As a greenhouse, efficiency is 50-75%. When used as a heat source for a living space - 10-30%. Active rock heat-storage will increase efficiency slightly. Has the added advantage of food production.

* Efficiency is based on the total amount of solar energy available on the face of a collector under ideal conditions.

ACTIVE SOLAR SYSTEMS

An active system is one that uses mechanical equipment to collect, store, and distribute solar heat throughout a building. There are two basic types of systems in general use : air and water systems. Each has specific advantages and can be sized and designed to suit virtually any building shape. Their major disadvantage however, is the high capital cost and sophisticated engineering necessary to achieve system efficiency. There is quite likely to be an extremely long payback period (if any payback at all) and maintainance costs can be quite high, especially for the water system.
In general, there is very little justification for the use of active solar in residential construction, although a reasonable case can sometimes be made for an active domestic hot-water system in suitable climatic regions.

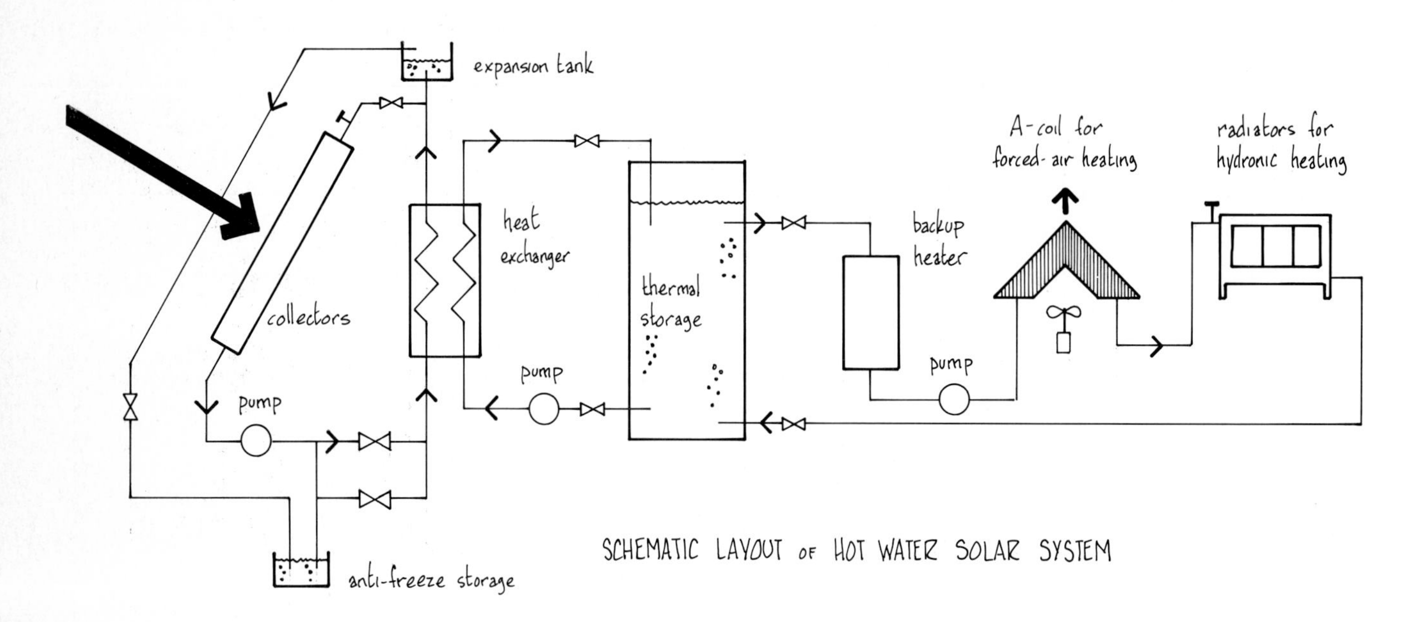

SCHEMATIC LAYOUT OF HOT WATER SOLAR SYSTEM

In each system, solar energy is collected by the exterior collector panels and stored in the appropriate water or rock thermal storage. In the water system, hot stored water can be pumped through an A-coil heat-exchanger for warm air heating, or directly through standard hot water individual room radiators. In the air system, hot air from the rock storage is passed directly through a standard furnace and into the house air circulation. Note that in each case a full-sized standard heating system is still required as backup heating for cloudy days or breakdown periods.

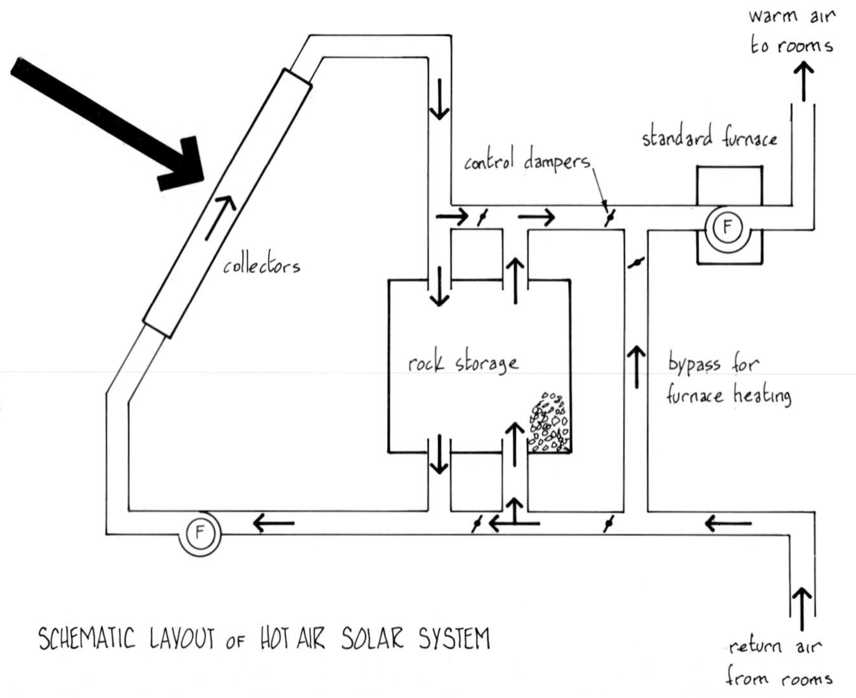

SCHEMATIC LAYOUT OF HOT AIR SOLAR SYSTEM

SITE CONDITIONS

CHAPTER PAGE TITLES

LEGALITIES AND DESIGN FACTORS

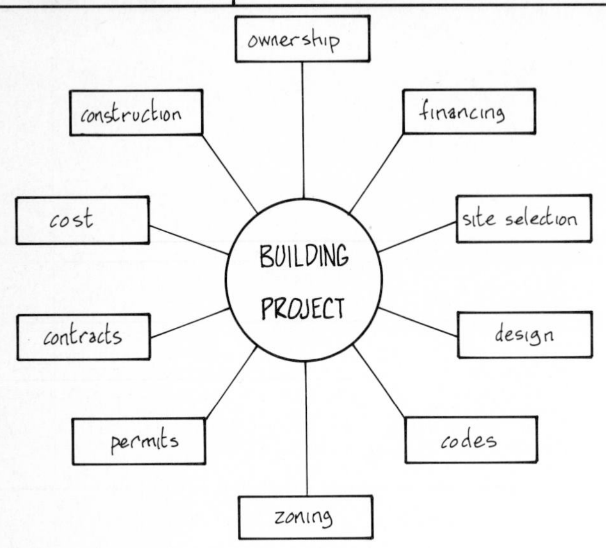

Before design details are completed or construction starts there are several important items that must be considered. The items fall into two broad categories : legal matters and building regulations . Each must be thoroughly investigated and taken into account if potentially difficult and often expensive problems are to be avoided during the building process. The chart at the left indicates the major items to be considered and each is briefly discussed below. Note that these are only the major items common to most localities and that special regional and local considerations must also be taken into account.

LEGALITIES

Ownership of the land must be clearly documented . This will involve a title search at the local land registration office , usually located at each county's court house or similar government building. The search will confirm the name of the registered owner and provide a legal description (precise location and size) plus a survey drawing or "Plat" of the land. Transfer of ownership is also registered at this office on presentation of the appropriate legal documentation. A search for liens that may be registered against the property should also be conducted. A lien is a claim for unpaid debts, and each must be cleared before ownership can be transferred.

A deed is a document that identifies the owner and describes the property. It may also contain a selection of restrictive clauses that control use of the property. Notable clauses are easements that allow utility companies access to portions of the land for inspection or repair of their services and encroachments that allow neighboring buildings or fences to extend beyond their own property lines. Other deed restrictions are similar to standard zoning regulations and may specify such things as minimum and maximum building sizes , setbacks , architectural style, and rights-of-way.

Financial considerations are equally important . Buying property will usually involve a mortage or other type of loan that must be carefully arranged and legally documented . The total cost of the entire project is also a major factor. Most owners or clients are working within a specific budget that requires coordination of design features and projected building costs.

DESIGN FACTORS

Site selection is made from a combination of factors , including land cost, type of neighborhood, personal preference , utility services available, proximity to schools and commercial areas , and any other input that is deemed significant by individual clients. The zoning regulations for a particular location are very important and are discussed on the next page. For larger developments where many housing units are planned, utilities, commercial , and social services tend to be part of the planning process , and site selection involves only choosing a lot within that housing tract or subdivision.

The actual design of the structure is also subject to many inputs. Client preference based on cost, size, quality, required amenities, and style generally dictate final design. It should be noted that clients tend towards higher expectations than is practical for their circumstances, forcing designers constantly to monitor budget restraints. House type and style were discussed in Chapter 3.

All buildings must be designed to meet the minimum standards specified in the building codes and building regulations for that location. It is important that the designer be totally familiar with local codes since they often have a major effect on the design process. If the design does not meet code requirements, it is unlikely that a building permit will be issued.

BUILDING CODES AND ZONING

BUILDING CODES

- As mentioned previously, the codes set the minimum property standards for each locality.
- Most regions use a combination of national, state (or provincial), and local codes, while some rely entirely on national codes. Enforcement, however, is the responsibility of the local authority and is monitored through the building permit system.
- Codes are constantly being updated. National codes change relatively slowly, while local codes can be changed quite quickly in response to specific situations.
- Separate codes are in force for the major divisions of general construction, plumbing, and electrical installations.
- The code extract shown at the right is typical of the style used to write most codes. As with any document of this kind, it is relatively difficult to read and understand and may be open to different interpretations if loosely written.

SECTION 10.10 MEANS OF EGRESS

Subsection 10.10.1 General

10.10.1.1.(1) An *exit* shall be provided from every *floor area*.

(2) An *access to exit* shall be provided from every roof intended for *occupancy* and from every podium, terrace, platform or contained open space.

(3) Where a roof is intended for an *occupancy* load of more than 60 persons, at least 2 separate *means of egress* shall be provided from the roof to stairs designed in conformance with the requirements for *exit* stairs and located remote from each other.

(4) Egress requirements from a podium, terrace, platform, or contained open space shall conform to the appropriate requirements for rooms, or *suites* in Article 10.10.8.5.

An Example of the Kind of Wording and Information In a Building Code

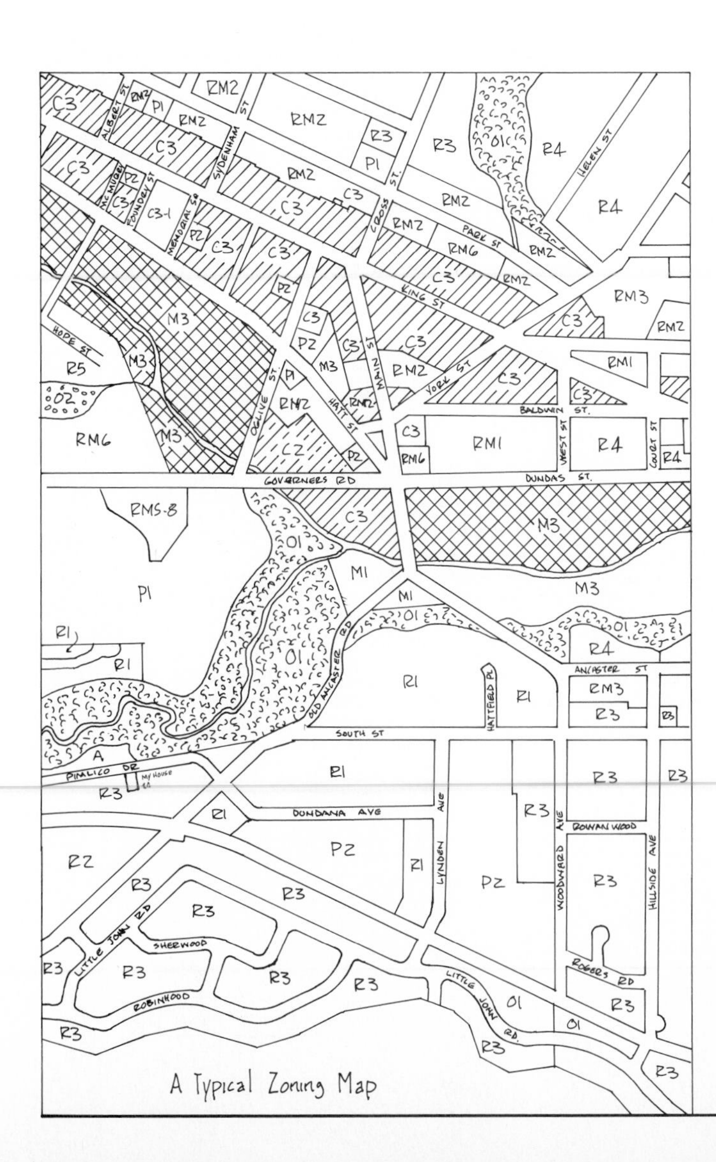

A Typical Zoning Map

ZONING REGULATIONS

- These regulations specify the type of structure that may be built in a given area. They are designed to control the growth and content of each area to maintain its desired character.
- Land usage will include single-family, multiple-family, high-rise, agriculture, and light, medium, or heavy industrial. A single or combination usage may be specified.
- Zones must be carefully researched when selecting a building site. Present and future trends should be considered since the zoning designation can be changed or special allowances made with relative ease in some regions.
- The zoning map and zoning classifications shown here are typical of many regions.

SECTION 11 – RESIDENTIAL ZONE R3

1. Uses Permitted
   (a) One-family detached dwellings.
   (b) Accessory buildings to the foregoing permitted use.
2. Schedule Of Standards
   (a) One-Family Detached Dwellings

(1)	Minimum lot frontage	50 ft
(2)	Minimum lot area	5000 sq ft
(3)	Maximum lot coverage	35%
(4)	Minimum front yard	20 ft
(5)	Minimum side yard	4 ft for 2 stories or less 6 ft for more than 2 stories

An Example of a Typical Residential Zoning Regulation

LEGALITIES

4-4 BUILDING PERMITS AND CONTRACTS

BUILDING PERMITS

Before construction can begin, application must be made for a building permit:

- The permit serves several purposes. It controls the type and quality of each structure, allows authorities to maintain neighborhood standards, and identifies new properties to be added to the tax rolls.
- The application will therefore require information about the type and size of structure to be built, the proposed starting date, intended use of the building, occupancy classification, and total cost of the structure.
- The name of the general contractor plus those of the sewer, plumbing, heating, and electrical contractors will probably be required in most regions since these are regulated trades and may require separate permits.
- Copies of the working drawings and specifications must be submitted with the application. These will be inspected by the planning department for zoning compliance and by the building department for code compliance.
- With all items in order a permit will be issued for each construction phase. In some regions this may add up to a suprising number of permits because of the complex nature of governmental and utility regulation.

THE INSPECTION PROCESS

The issuance of each building permit will trigger a variety of construction inspections by building authorities and utilities.

- Depending on locality, building inspectors will give approval for each phase of the work, including footings, footing drainage, floor structures, insulation installation, roofs, and the completed structure.
- Other inspectors may inspect installation of sewers, water supply connection, plumbing, heating equipment, and electrical wiring.
- It is normally the contractors' responsibility to ask for the appropriate inspections before proceeding with each subsequent construction phase.
- Upon final inspection and approval, an occupancy permit will be issued by the building inspector that allows the building to be used for its intended purpose.

CONTRACTURAL RELATIONSHIPS

There are certain sets of legal and contractural relationships between the parties involved in a building project. The nature of the project and the circumstances under which it is to be built determine the type and extent of the contracts. In most cases a contract will define the work to be done, the agreed price, and the projected completion date plus other pertinent items. The contracts are legal and binding between parties and include working drawings and specifications as part of the legal documentation.

RESIDENTIAL CONSTRUCTION

There are several possible relationships, depending on the parties involved, who include developers, individual clients, architects, general contractors, and subcontractors. The chart at the right shows definite relationships in solid lines and possible relationships in broken lines.

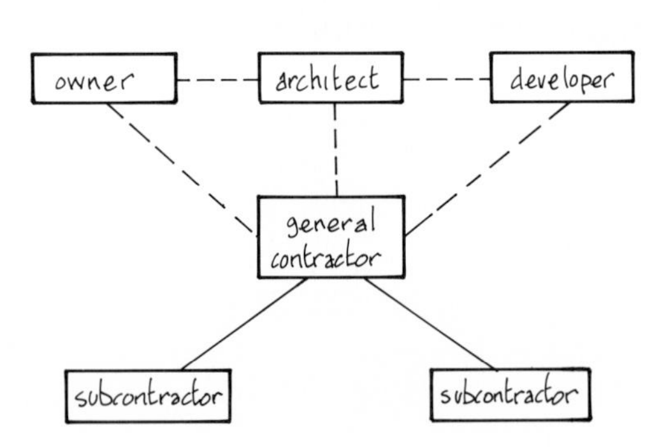

Residential Relationships

COMMERCIAL CONSTRUCTION

Although obviously more complex in terms of cost and size, most commercial projects are relatively simple in basic relationships. A typical project will have an owner-client, an architect acting as designer and construction agent, a general contractor responsible for all construction, and as many subcontractors as necessary for the type of project. A contractural chart is also shown at the right.

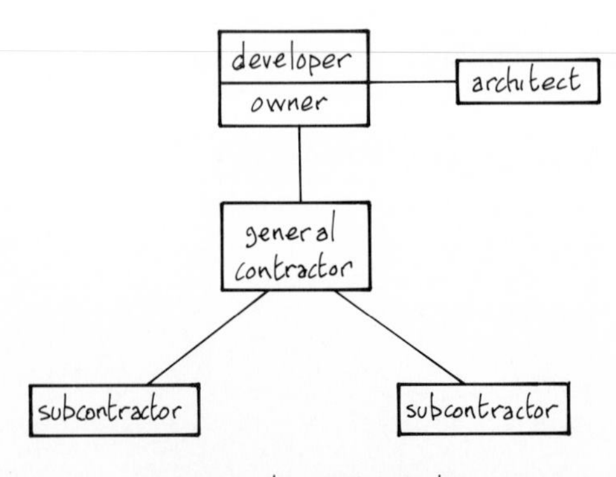

Commercial Relationships

SURFACE CONDITIONS

Each new building site will need some form of preparation before construction can begin. The extent of work to be done before completion of construction and final landscaping depends on several initial conditions:

- The lot may need to be graded by removing excess soil, rocks, trees, etc. Some house designs make deliberate use of existing slopes or surface conditions, which must then be preserved. Mature trees are a valuable asset to most sites provided they do not interfere with the building process, and many regions have bylaws prohibiting their removal without permission.
- Some lots require the addition of soil to raise the surface to the level of neighboring properties or to solve high water-table problems. The material used must be "clean fill" and not contain organic material that will rot or decompose to cause extensive settling. Excavated material from other building sites is ideal. Care must be taken with foundation systems on filled sites since load-bearing strength is reduced and excessive settling is likely. See the next page for more information.
- Removal of existing buildings will require a demolition or moving permit from the local authority. Existing gas, water, sewer, or electrical services may also need to be removed and replaced, depending on their condition. Special attention must be paid to the basement cavities of demolished structures since it is unlikely that the existing foundation system can be reused. The cavity can be filled and compacted or enlarged and used for the new foundations and basement or crawl space.
- Of particular importance is surface-water runoff. Provision must be made in the initial and final grading to direct rain or snow-melt water away from the structure to a suitable drainage area. Most regions require a site survey showing drainage patterns and grading to be submitted with the building permit application.

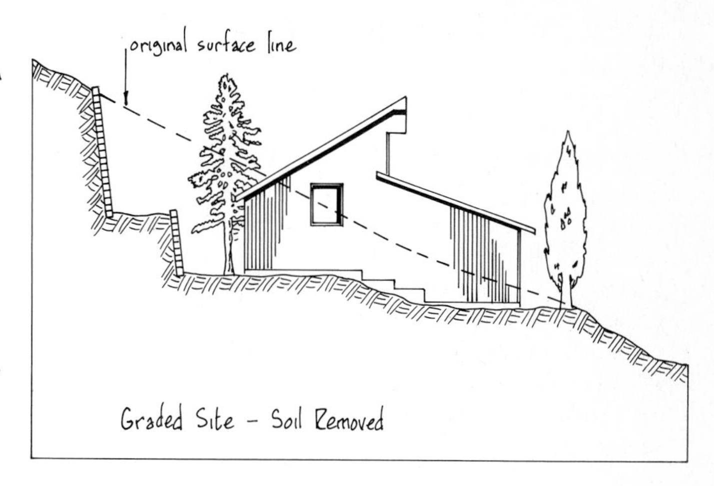

Graded Site – Soil Removed

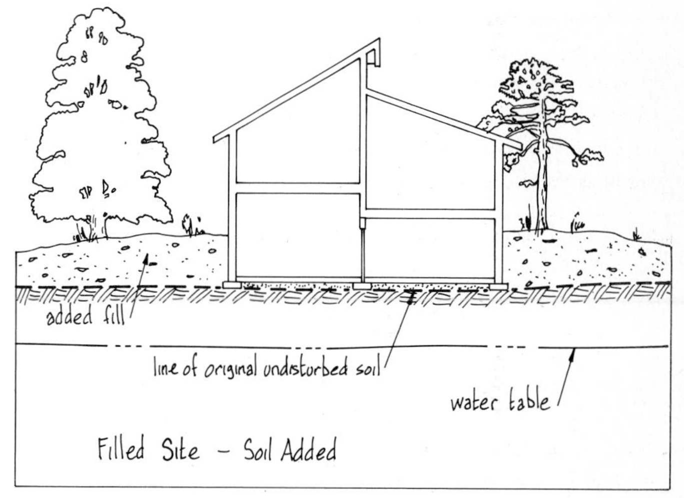

Filled Site – Soil Added

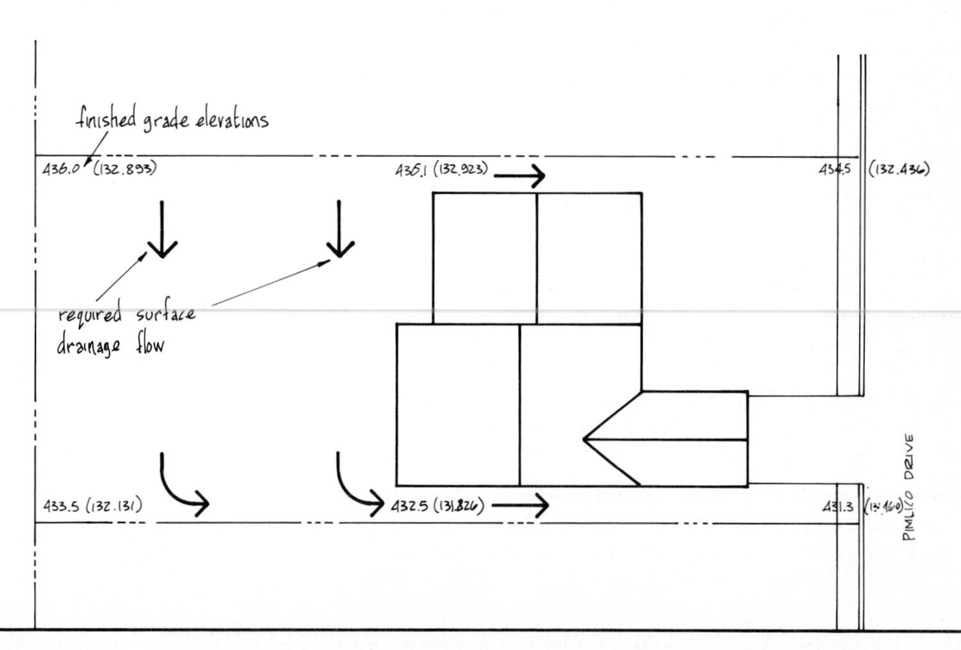

Typical surface drainage plan showing required water runoff. In this case the water must be directed to the street at the front of the house.

SITE CONDITIONS - SUBSURFACE

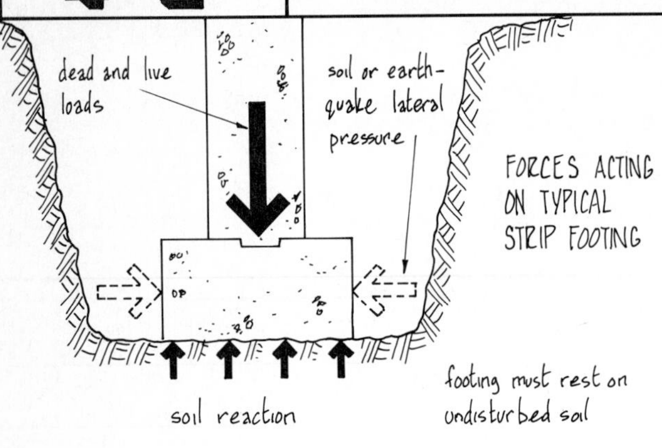

SUBSURFACE CONDITIONS

Subsurface conditions are often more important than surface conditions since the foundations of the building must be supported by soil that can withstand all loads imposed upon it without crumbling, settling, or extruding.

- Structural and live loads are transmitted to the soil as shown at the left. Although most loads are axial (vertical), the footing may also be subject to lateral (sideways) loading from wind, earth pressure, liquid pressure, and earthquakes.
- Frost heave is a significant problem in cold regions unless footings are located below the local frost line.
- The level of the local water table is also important. Some soil types are changed to a plastic state with a greatly reduced bearing strength in the presence of excessive water. Special footings or extensive dewatering techniques must be used under these conditions.
- Any of the problems noted above can result in uneven settlement of the foundation system across the footing or along the footing length, as shown at the left. This will cause cracking of the foundation and upper walls, sticking doors, and damage to interior and exterior finishes.

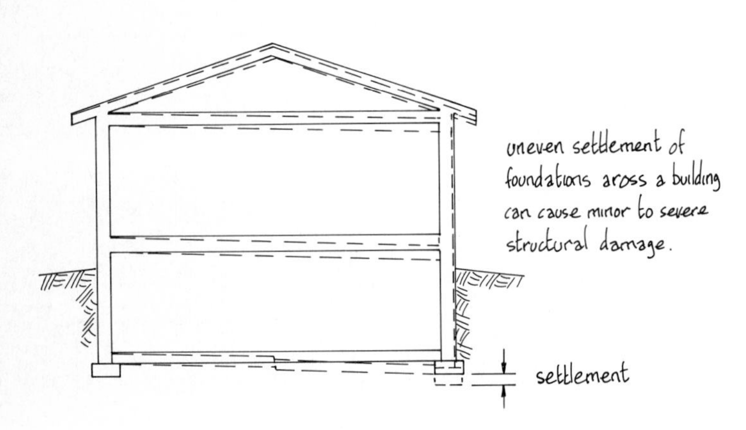

SOIL TESTING

Subsoil information can often be obtained from contractors who have already built in the immediate area or from the local authority's building inspectors who should also be familiar with local problems. Failing this, subsoil conditions will have to be tested in areas where problems are thought to exist or when heavy foundation loads are anticipated. There are two test methods:

- A hand or mechanically dug test pit that extends below the expected foundation depth. Soil layer type and water content are easily seen with this method.
- For deeper foundations a test bore made with a powered auger or rock drill is standard. Core samples show soil layers and give some indication of water content.

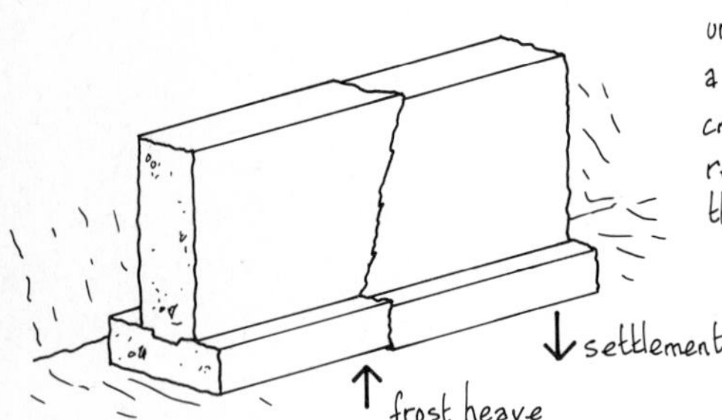

FILLED SITES

Sites with added fill pose a special problem since it takes many years before the fill is compact enough to support standard foundations systems. There are two basic choices, as illustrated at the right:

- Take the footing down to undisturbed soil beneath the fill. This may prove impossible with standard footings if the fill depth is too great.
- Design footings to suit the load-bearing strength of the fill. Where this also proves impractical, a pile and beam foundation (see 5-20) may be necessary.

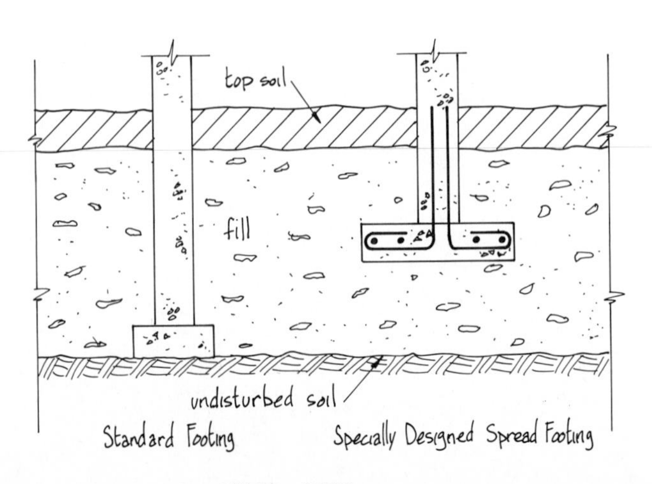

FILLED SITES

SITE SERVICES

The availability of sewage disposal, water supply, and utility services on any particular lot depends on its proximity to population centers:

- City or town lots in the major centers usually have sewers, water, electricity, gas, telephone, and television services available at or near the lot line, as illustrated at the right.
- Rural lots may have only electricity and telephone services in close proximity, and residents must rely on wells (see 16-3) and on-site disposal systems.
- Remote lots are unlikely to have any services, and the cost of bringing even a power supply to the site may be prohibitive.

The exact location of each service on a lot is of significance. Sewer lines are the most important since their depth below grade will determine the lowest level of plumbing and drain lines in the building (see 4-9 and 16-8). The water supply main is located below the local frost line, while power, gas, telephone, and television services are generally buried within 4 ft (1200) of grade or carried above ground on power poles. Utility companies generally provide free location services.

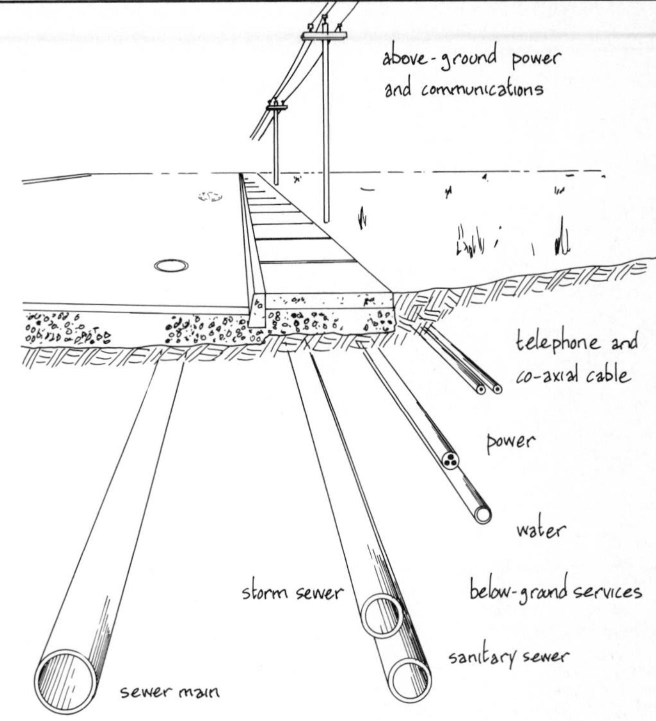

Check plot plans, survey drawings, or local building engineer's office for exact location of all services, especially sewers.

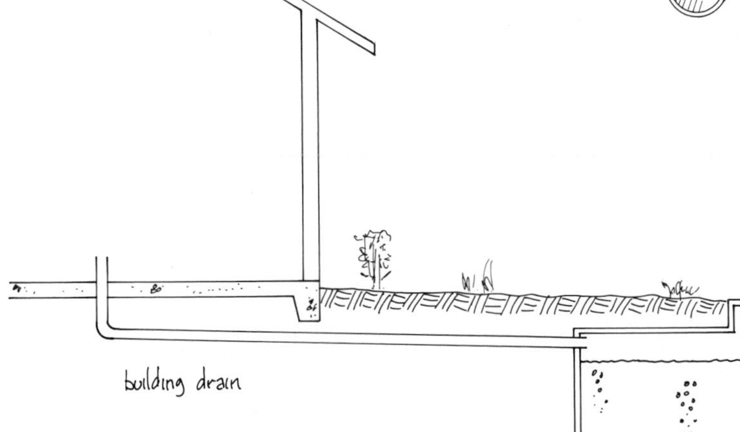

GRAVITY-FED SEPTIC WASTE DISPOSAL SYSTEM

ON-SITE WASTE DISPOSAL

For lots without public sewers, subsoil conditions will determine the best method of on-site waste disposal. Where water percolation tests are satisfactory, a septic tank/leach field system is commonly used. With this method waste is biologically treated and allowed to drain into the surrounding soil. In regions where water will not drain through a soil, a simple holding tank is often used. Waste is held for later disposal at a safe site (see 16-14 to 16-17 for more information).

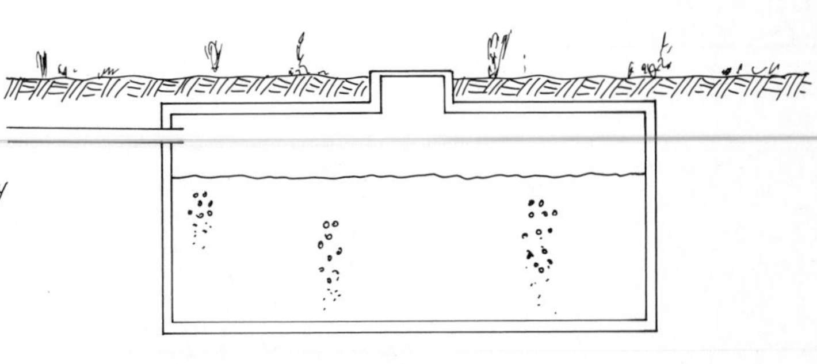

HOLDING TANK

BUILDING SITE

4-8

SITE PLAN

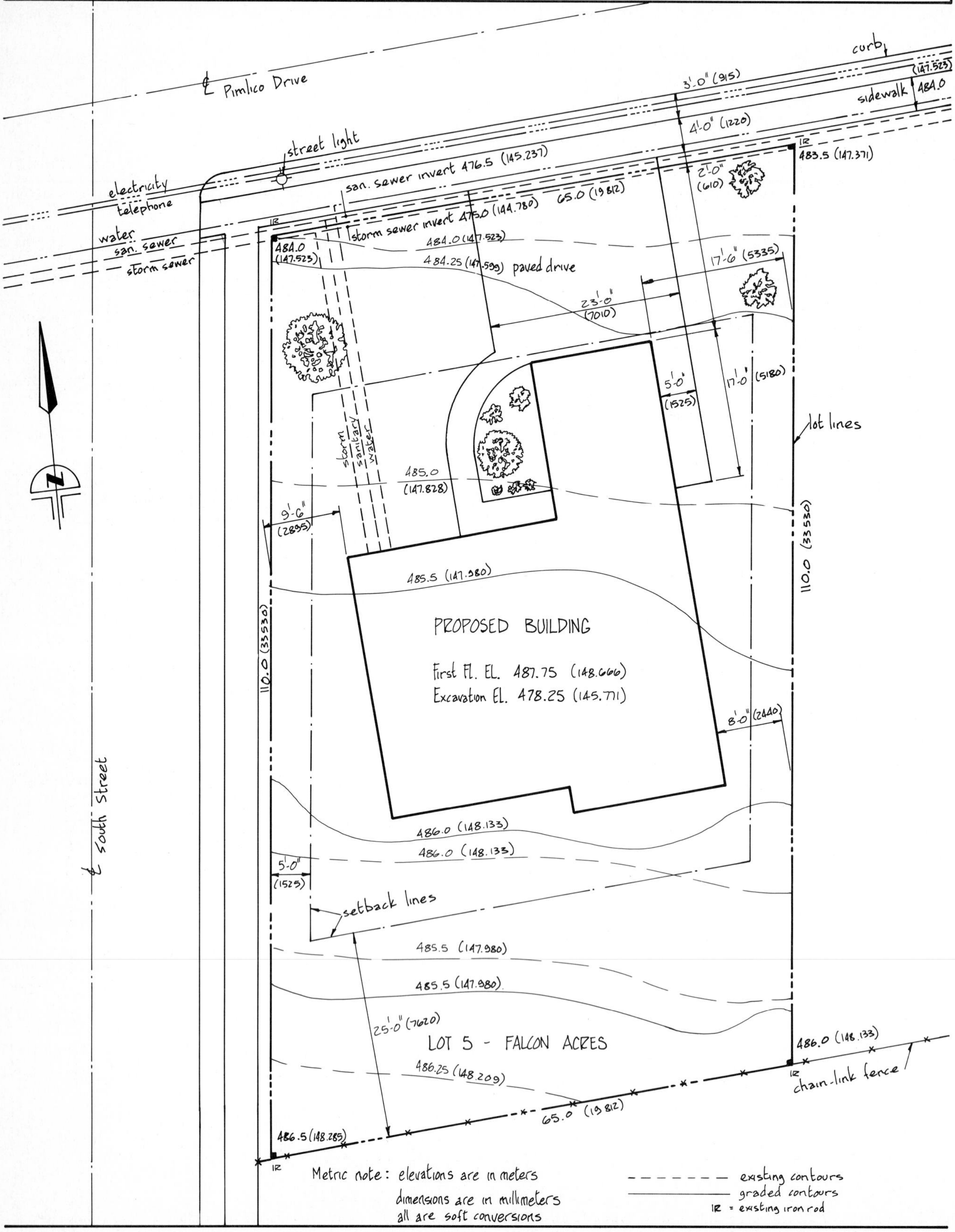

℄ Pimlico Drive

curb

(147.523)

484.0

sidewalk

3'-0" (915)

4'-0" (1220)

street light

electricity

telephone

water

san. sewer

storm sewer

san. sewer invert 476.5 (145.237)

storm sewer invert 475.0 (144.780)

65.0 (19 812)

IR

483.5 (147.371)

2'-0" (610)

IR

484.0 (147.523)

484.0 (147.523)

484.25 (147.599) paved drive

17'-6" (5335)

23'-0" (7010)

5'-0" (1525)

17'-0" (5180)

lot lines

storm

sanitary

water

485.0 (147.828)

9'-6" (2895)

110.0 (33530)

485.5 (147.980)

PROPOSED BUILDING

First Fl. EL. 487.75 (148.666)

Excavation EL. 478.25 (145.711)

110.0 (33530)

8'-0" (2440)

℄ South Street

486.0 (148.133)

486.0 (148.133)

5'-0" (1525)

setback lines

485.5 (147.980)

485.5 (147.980)

25'-0" (7620)

LOT 5 - FALCON ACRES

486.0 (148.133)

486.25 (148.209)

IR

chain-link fence

65.0 (19 812)

486.5 (148.285)

IR

Metric note: elevations are in meters
dimensions are in millimeters
all are soft conversions

- - - - - - existing contours

——— graded contours

IR = existing iron rod

SITE PLAN, SEWERS AND EXCAVATION

The typical site plan opposite shows the location of a house on a midsized individual lot. The information on the plan relating to existing features has been obtained by a surveyor, while the finished elevations and features are the responsibility of the designer. If this house were part of a housing tract or subdivision, a much larger plan would also exist showing the entire area. Notable information includes:

- Setbacks, which specify the minimum dimensions the house must be set back from the lot lines. Setback dimensions are part of the zoning bylaws and will vary with zone classification and location. Careful dimensional planning by the designer and accurate building layout by the contractor is very important. If the structure is located closer to a lot line than the minimum setback dimension allows, the structure may, under extreme conditions, be ordered demolished unless a special allowance is granted. It is good practice always to exceed the setback dimensions to allow a margin for error.

- The contour lines allow the designer to take advantage of the natural slope of the lot if desired.

- Existing and finished grade elevations allow the contractor to determine whether grading or filling is required to achieve the desired surface drainage flow.

- This is a serviced lot and the roads, sidewalks, and services are already in place before construction starts. The location and depth of the sewers plus the sidewalk elevation are shown. This information is used by the designer to relate the lowest interior drain level to the public sewer and by the contractor for surveying information to set the excavation depth for the foundations. The illustrations at the right show the general approach for each procedure.

- This lot has a rear southern exposure that is ideal for a passive-solar-designed structure (see 3-6 onward).

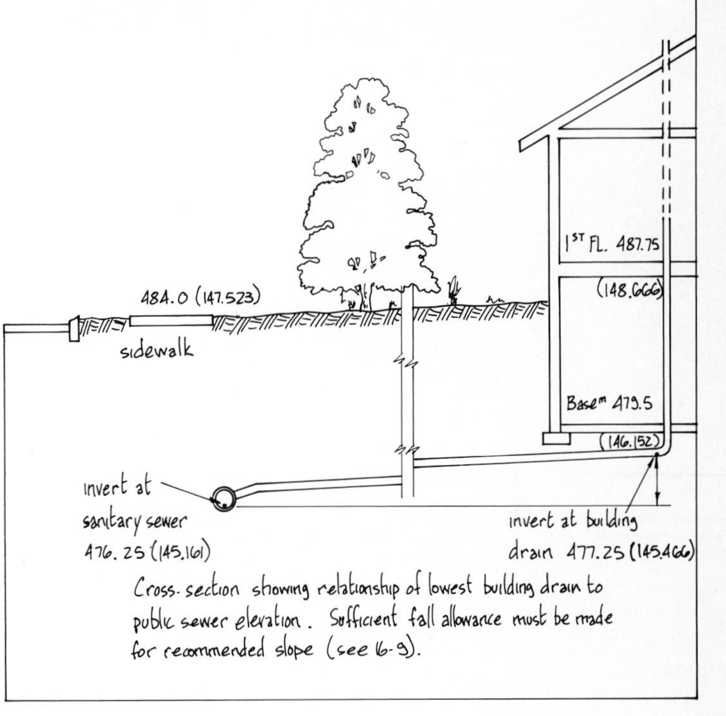

Cross-section showing relationship of lowest building drain to public sewer elevation. Sufficient fall allowance must be made for recommended slope (see 6-9).

The information in the cross-sections relates to the site plan opposite. All dimensions and elevations are in feet. Conversion factor: ft × 0.3048 = m.

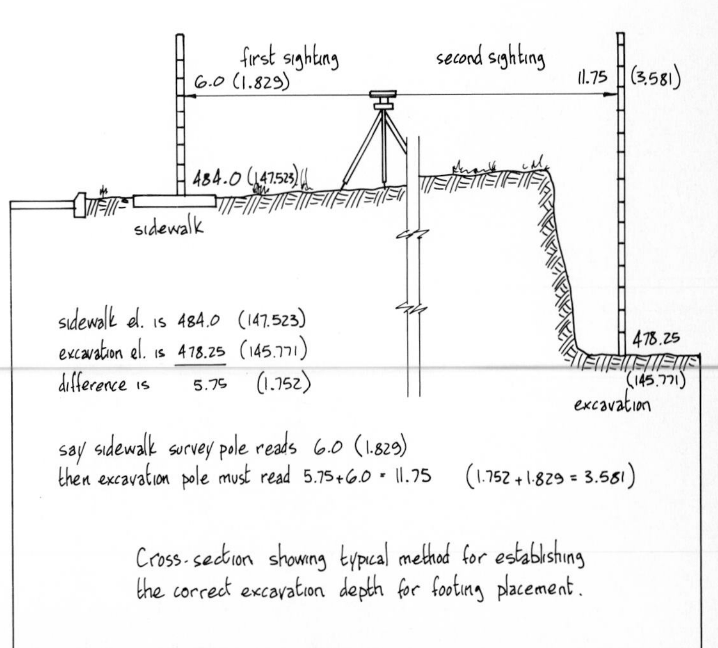

Cross-section showing typical method for establishing the correct excavation depth for footing placement.

SETTING OUT

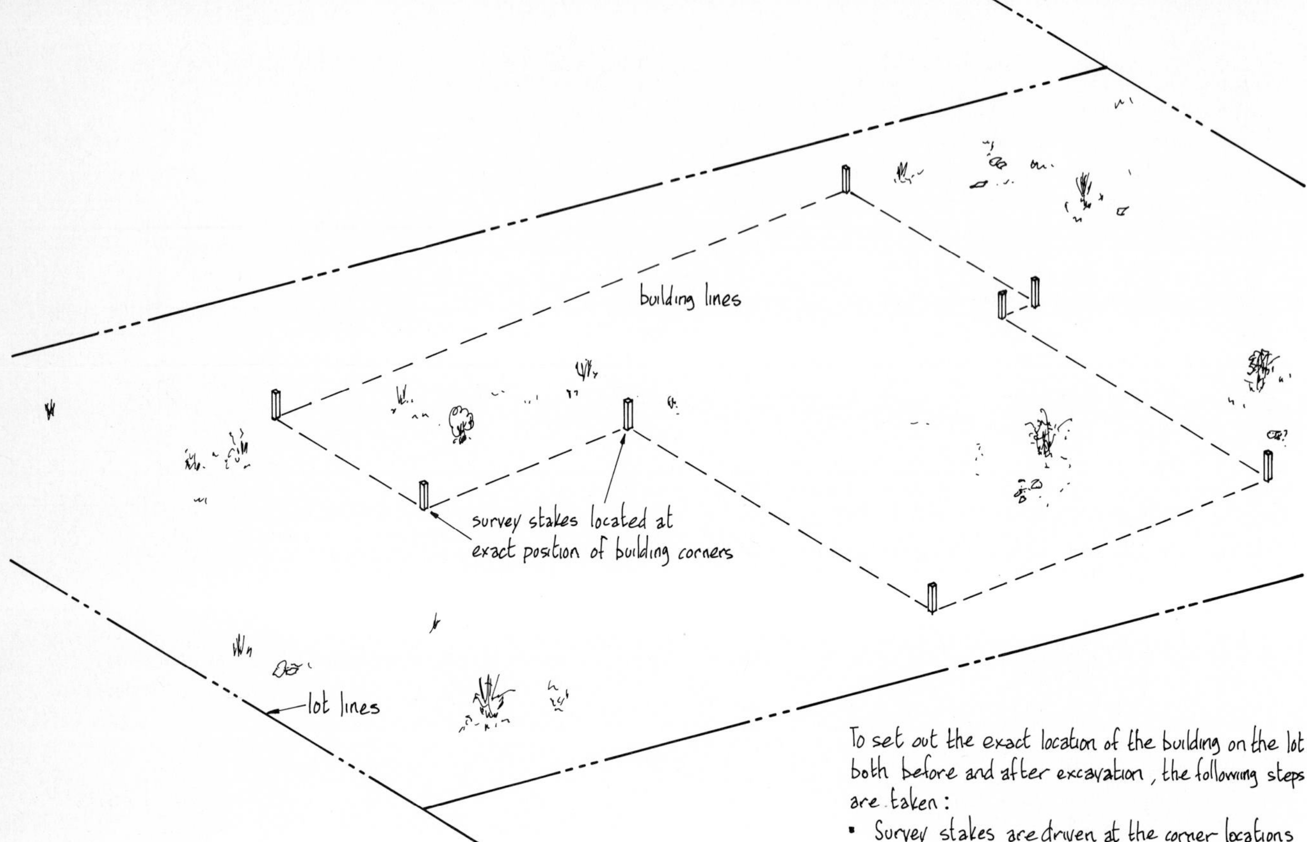

To set out the exact location of the building on the lot both before and after excavation, the following steps are taken:

- Survey stakes are driven at the corner locations of the building's foundation walls. A nail on the top of the stake indicates the exact position. The elevation of each stake is not important at this time. It is usual to have a licensed surveyor to do this work so that the layout is accurate.
- Since the stakes will have to be moved before excavation starts, the locational information must be recorded for use after excavation. To do this, batter boards are erected on the outside of each stake as shown at the left, and string is stretched between the boards in line with the stakes. Saw kerfs are made in the batter boards so that the string can be replaced after excavation.
- The tops of all the batter boards should be at approximately the same level so that the string is stretched horizontally.

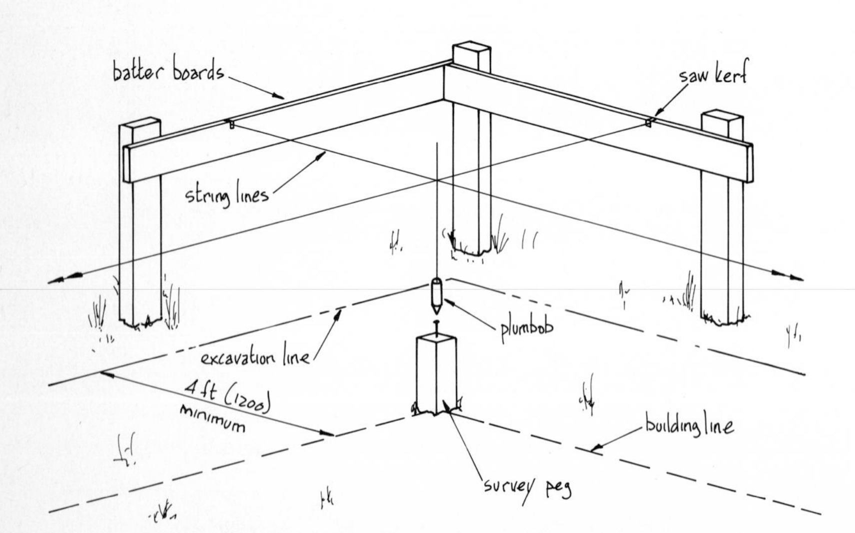

Batter Boards and Building-Corner Survey Stake

EXCAVATION

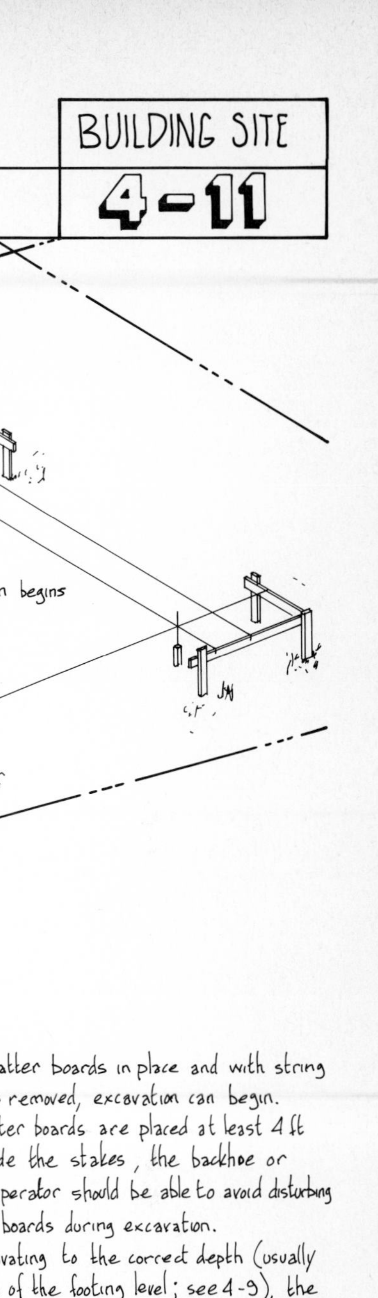

Batter Boards In Position

- With all batter boards in place and with string and stakes removed, excavation can begin.
- If the batter boards are placed at least 4 ft (1200) outside the stakes, the backhoe or bulldozer operator should be able to avoid disturbing the batter boards during excavation.
- After excavating to the correct depth (usually the bottom of the footing level; see 4-9), the string is replaced in the batter board saw kerfs and the corner stakes relocated on the excavation floor, as shown at the left.
- The stakes are now leveled with surveying instruments to the top of the footing elevation, and layout of the footing formwork can be started (see 5-6).

Survey Peg Relocated After Excavation

FOUNDATIONS

CHAPTER PAGE TITLES

FOUNDATION SYSTEMS - GENERAL

The foundation walls of a structure serve two basic functions: to support and carry the loads from walls, floors, and roof and to form an enclosure for basements or crawl spaces. There are four foundation systems in common use today:

1. Basement wall systems using poured concrete, unit-masonry concrete blocks, or preserved wood.
2. Perimeter wall systems using materials as in option 1.
3. Ground slab systems using poured concrete.
4. Pile and beam systems using concrete and/or preserved wood.

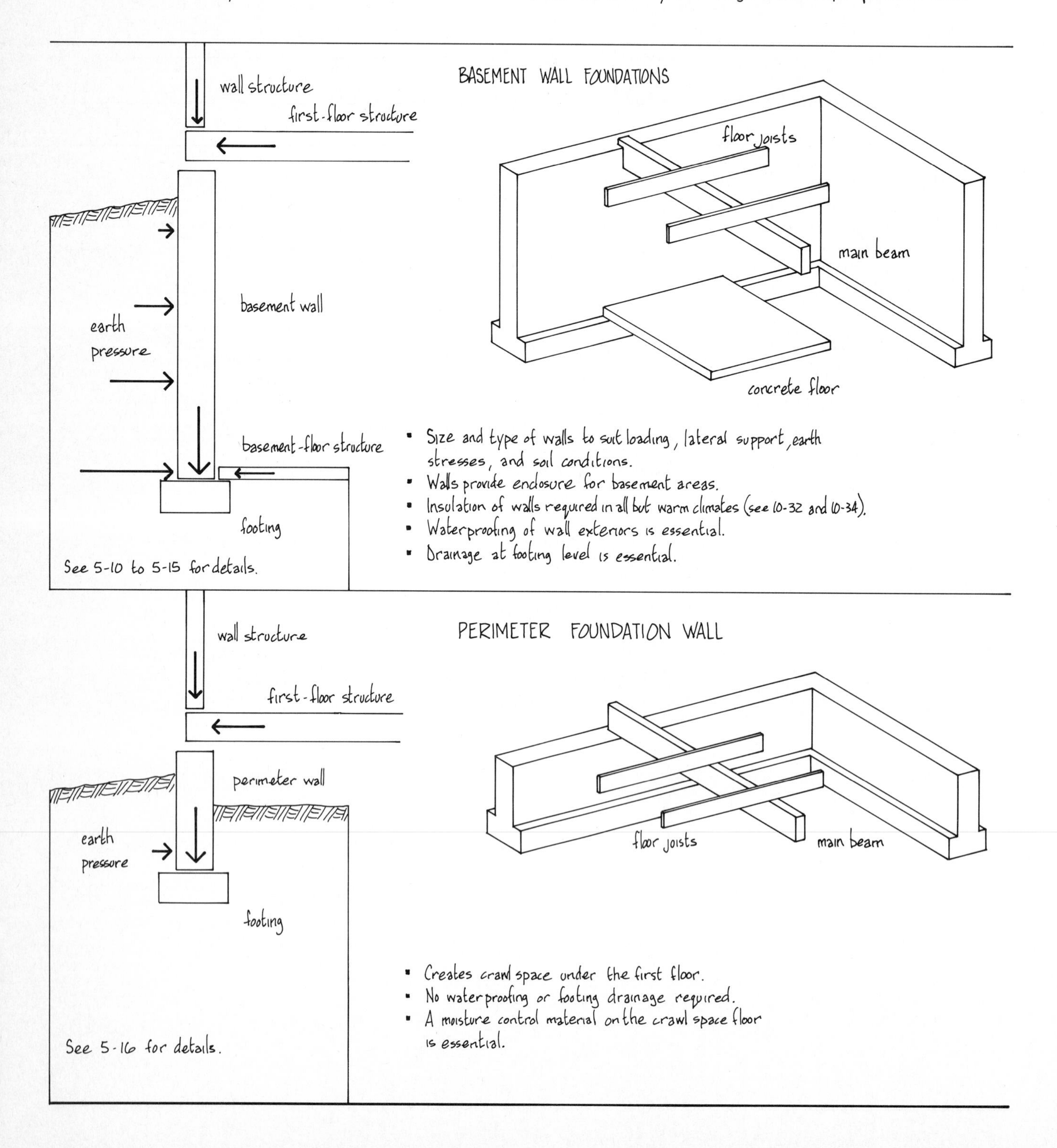

See 5-10 to 5-15 for details.

- Size and type of walls to suit loading, lateral support, earth stresses, and soil conditions.
- Walls provide enclosure for basement areas.
- Insulation of walls required in all but warm climates (see 10-32 and 10-34).
- Waterproofing of wall exteriors is essential.
- Drainage at footing level is essential.

See 5-16 for details.

- Creates crawl space under the first floor.
- No waterproofing or footing drainage required.
- A moisture control material on the crawl space floor is essential.

FOUNDATION SYSTEMS - GENERAL

The type of foundation chosen for a particular structure depends on several factors:

- Most important is the load-bearing capacity of the soil which can vary considerably, depending on soil material and water content. See 5-6 for more information.
- Soil plasticity and soil movement in earthquake and other regions may exert significant lateral pressure on foundations.
- In high wind regions, wind uplift forces may require additional anchorage of the upper structure to the foundation, and increased resistance to racking forces.
- Deep frost conditions are not normally a problem provided footings are located below the frost line, but exceptional conditions such as permafrost will require special foundations.

GROUND SLAB FOUNDATION

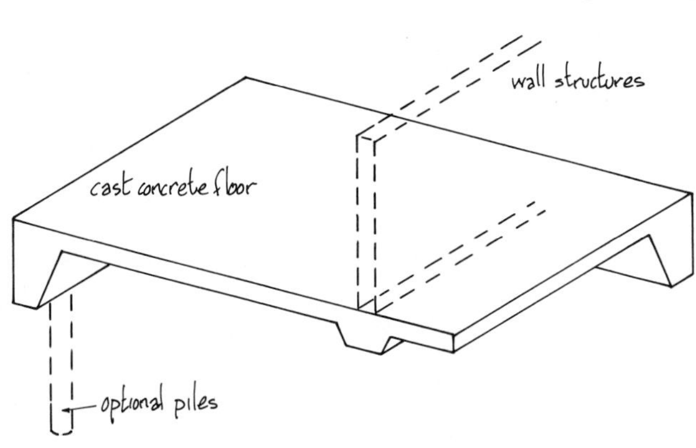

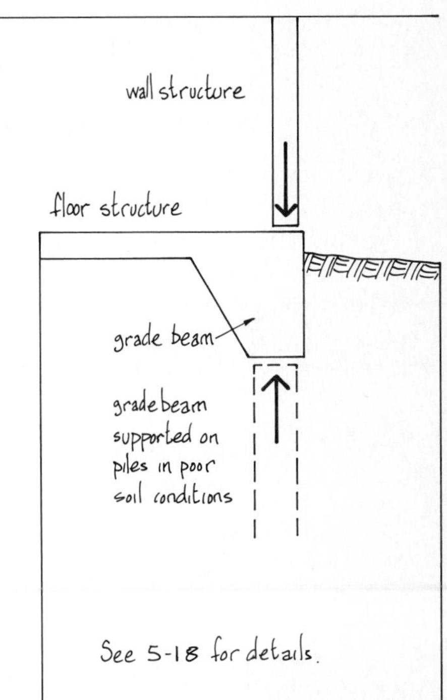

- Reinforced thickened edge applicable only to warm climates.
- Cool and cold climates require perimeter wall or grade beam and piles for slab support under frost conditions.
- Stable, compacted soil preferred as base for slab.
- A "floating", reinforced slab is necessary on poor soil.

PILE AND BEAM FOUNDATION

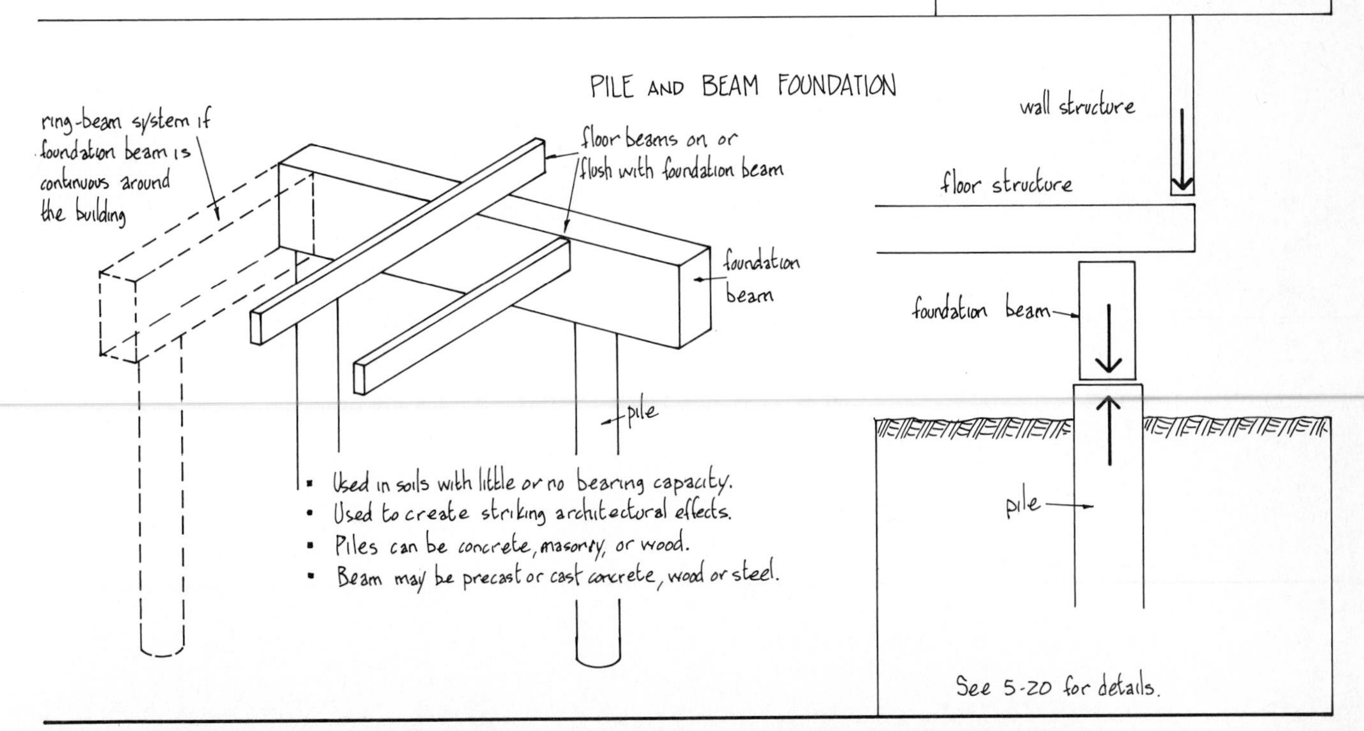

- Used in soils with little or no bearing capacity.
- Used to create striking architectural effects.
- Piles can be concrete, masonry, or wood.
- Beam may be precast or cast concrete, wood or steel.

CONCRETE AND FORMWORK

CONCRETE

Concrete is used for foundations in footings, foundation walls, and floor slabs. It is made from a combination of portland cement, sand, and coarse aggregate in varying proportions. The final cured strength of the concrete may be adjusted to suit design considerations up to certain limits. Thereafter, as expected stresses increase, steel reinforcing must be added to the concrete structure.

In most urban areas ready-mixed concrete is available and can be delivered to the job in almost any quantity and quality. If this service is not available, on-site mixing is necessary. Care must be taken to select the correct grade and size of sand and aggregate and to ensure a supply of good-quality water. The proportions of the mix should yield a concrete liquid enough to allow it to fill all voids created by corners, angles, or reinforcing without the mix separating or allowing free water to collect on the surface. The exact ingredients of the mix depend on the strength and type of concrete required and the manufacturer's instructions must be followed carefully.

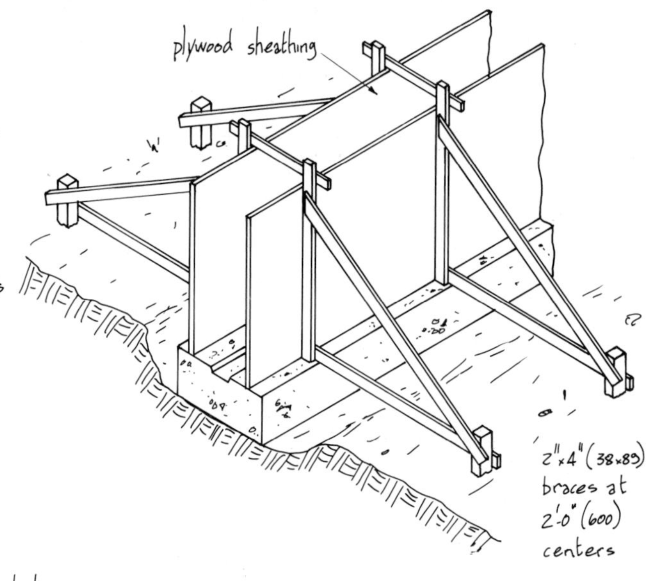

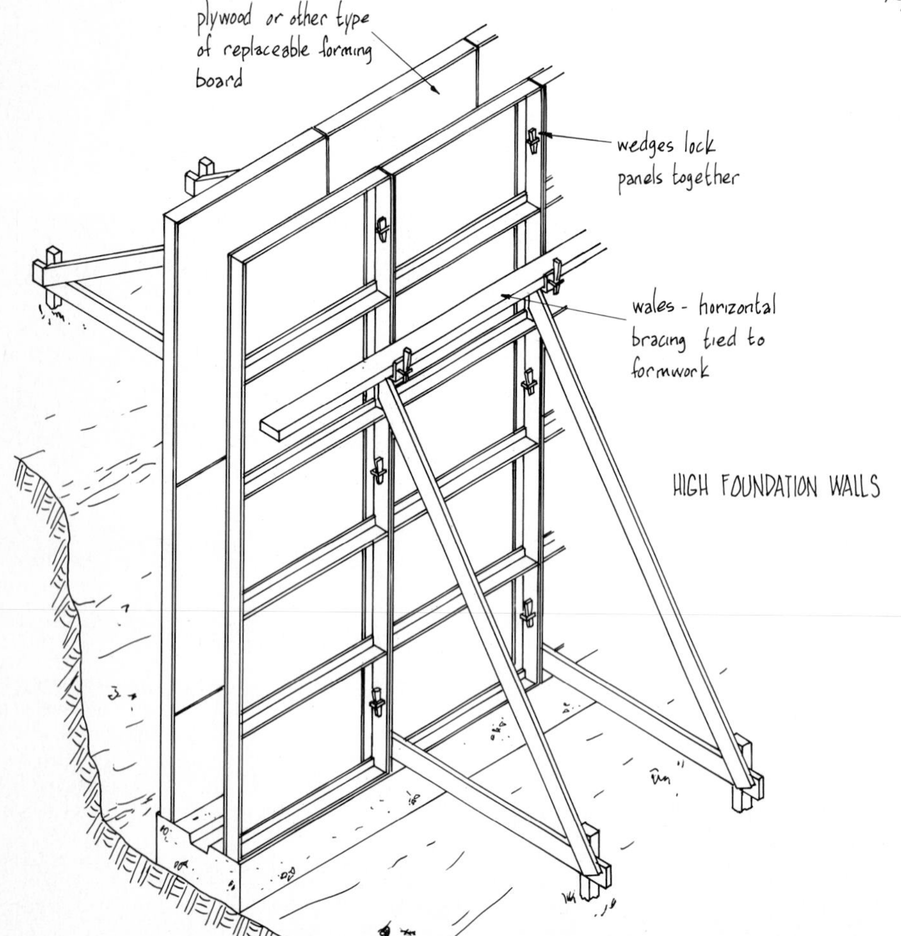

FORMWORK

Formwork can be made on site from a variety of materials or assembled from prefabricated commercial formwork units. In either case, to achieve good-quality work, the forms must be defect free, smooth, tightly jointed, and properly aligned. Careful attention must also be paid to proper bracing of the forms. Wet concrete exerts considerable side pressure as well as having an unstable, top-heavy effect on the forms.

Shown above is a simple form for short foundation walls while at the left is a metal framed forming system assembled from a series of interlocking panels.

CONCRETE BLOCKS

CONCRETE PLACING

Ideally, concrete should be poured continuously and placed evenly and symmetrically in all forms. Uneven placement can cause failure by imposing unbalanced stresses on the formwork. Use vibrators, rods, or shovels to work the concrete around obstructions, remove air pockets, and force the concrete into all areas of the forms. Avoid impact loads on filled forms by buckets, trolleys, or other equipment. If pouring operations must be interrupted for more than a few hours, clean, score, and wet the top surface of previously poured concrete before continuing the pour. Do not add water before or during a pour to increase plasticity — this will greatly reduce design strength.

CURING

Under ideal conditions, concrete will reach full design strength approximately 28 days after pouring, although forms can be stripped much earlier. Concrete cures not by a "drying-out" process but through a chemical reaction called "hydration". In fact, allowing concrete to be air-dried hinders the hydration process. Exposed concrete surfaces should be kept damp during the initial curing process, especially in hot weather. In air temperatures below 40°F (4°C) the concrete must be both mixed and maintained at a minimum of 50°F (10°C) for the first 72 hours.

CONCRETE BLOCK

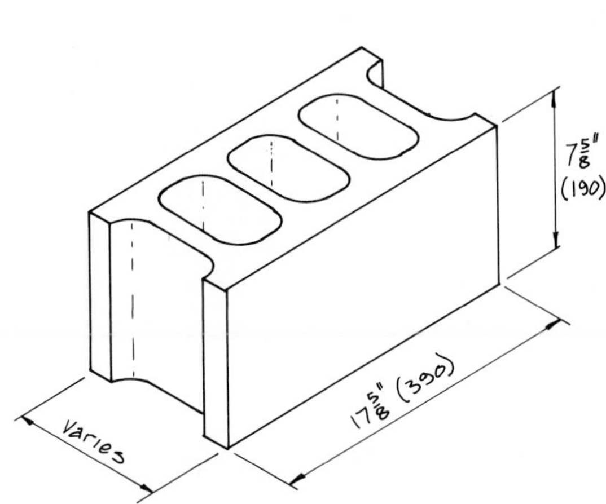

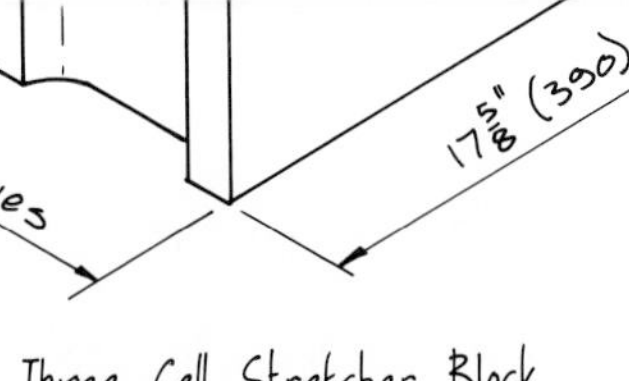

Three-Cell Stretcher Block

Two-cell blocks are also standard.

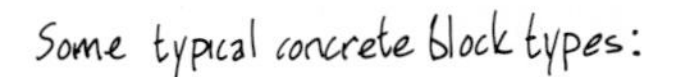
Some typical concrete block types:

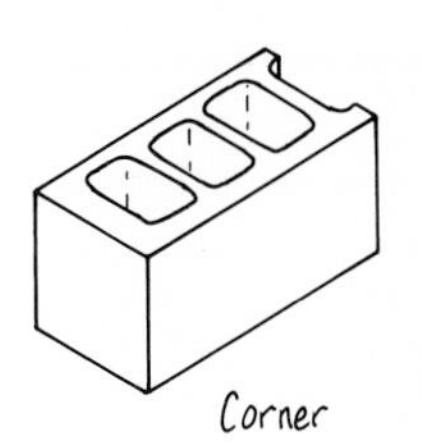
Solid

Corner

Lintel

Jamb

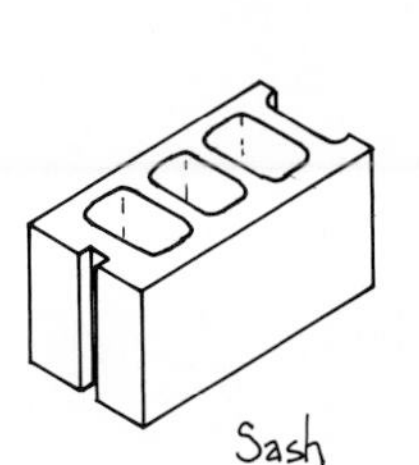
Sash

Half-Sash

Concrete blocks are used extensively for foundation walls and are manufactured in a wide variety of shapes. Standard weight blocks are made from portland cement, sand, and fine gravel or crushed stone. Lightweight blocks substitute cinders, pumice, or slag as aggregate.

Sizes of units generally used for foundations:

	Customary	Metric
nominal face	8" × 16"	200 × 400
actual face	7 5/8" × 15 5/8"	190 × 390
nominal width	4", 6", 8", 10" and 12"	100, 150, 200, 250, 300
actual width	nominal less 3/8"	nominal less 10

- Check with local manufacturers for full range of block products.
- For information on nominal and modular sizing, see 7-22.
- For insulating type blocks see 10-38.

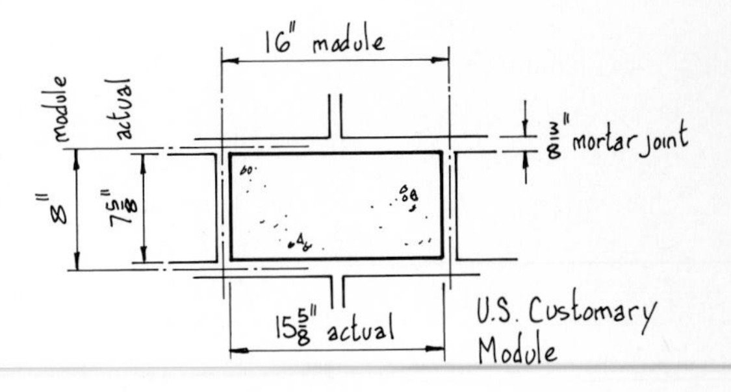

U.S. Customary Module

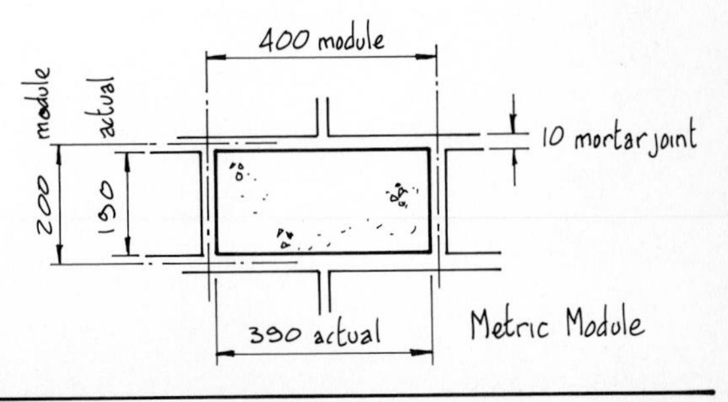

Metric Module

FOOTINGS

SETTLEMENT

Footings transmit the combined dead and live loads of a structure to the soil beneath. The footing must be designed to support this weight without excessive settlement or lateral motion. Particularly important is the avoidance of <u>differential</u> settlement. Whereas a uniform settlement of 1 in (25) or more can be tolerated, a differential settlement of over 1/4 in (6) can cause serious cracking in the supported structure. Factors to consider are: soil bearing capacity, footing size, frost conditions, drainage, and sloping sites.

Typical Bearing Capacities Under Normal Conditions

MATERIAL	TONS/SQ FT	MPa
medium soft clay	1.0 to 1.5	0.1 to 0.15
medium stiff clay	2.5	0.25
hard clay	4.0	0.40
fine sand	2.0	0.20
course dense sand	3.0	0.30
compacted fine sand	3.0	0.30
sand-gravel compacted	4.0 to 6.0	0.4 to 0.6
hard rock	40 to 60	4.0 to 6.0

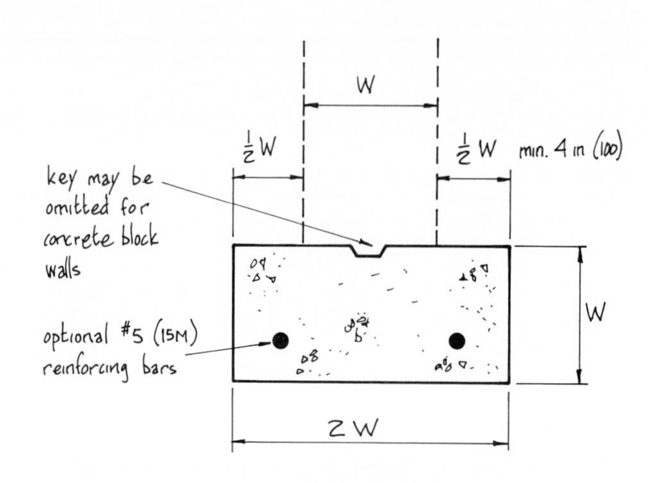

SOIL BEARING CAPACITY

Because the total loading for a residential or small commercial structure is not large, a standardized footing size can be used provided that the soil is of at least average bearing capacity. Bearing capacity varies considerably and is dependent upon soil material and water content. Where bearing capacity is suspect or soil type is not consistent in all footing areas, a soil analysis should be undertaken and footings designed accordingly.

STANDARD FOOTING SIZE FOR AVERAGE SOILS

- Footing size is determined by wall thickness (W).
- Footing should project a minimum of 4 in (100) beyond each wall face.
- Never backfill a too-deep footing excavation. Make up the extra depth with concrete.
- Tops of footings must be level.
- Where footing passes over a backfilled area (e.g., trenches for sewers or other services), or where soil is considered to be slightly unstable, reinforcing with two 5/8 in (15M)-diameter steel rebars is recommended.

FOOTING FORMWORK

- Stakes should already be established on the excavation floor at foundation wall corners (see 4-11).
- Drive additional grade stakes along the footing line with top of stakes level with top of footings.
- Install the outside boards first, spaced out from the building lines a distance equal to the footing extension (minimum 4 in (100) - see above right).
- Use a carpenter's level to transfer grade stake level to top of formwork.
- When all outside forms are in place, install inner forms and remove grade and corner pegs prior to pouring.

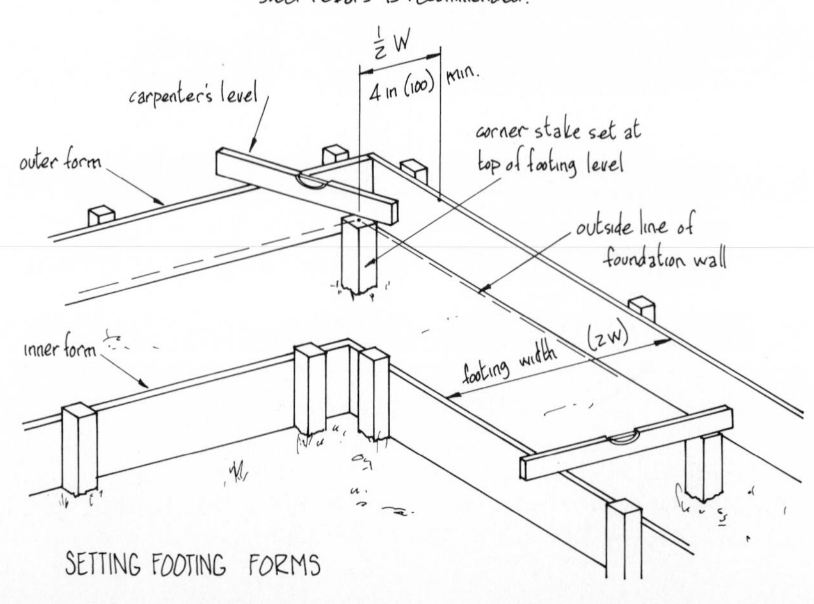

SETTING FOOTING FORMS

FOOTINGS

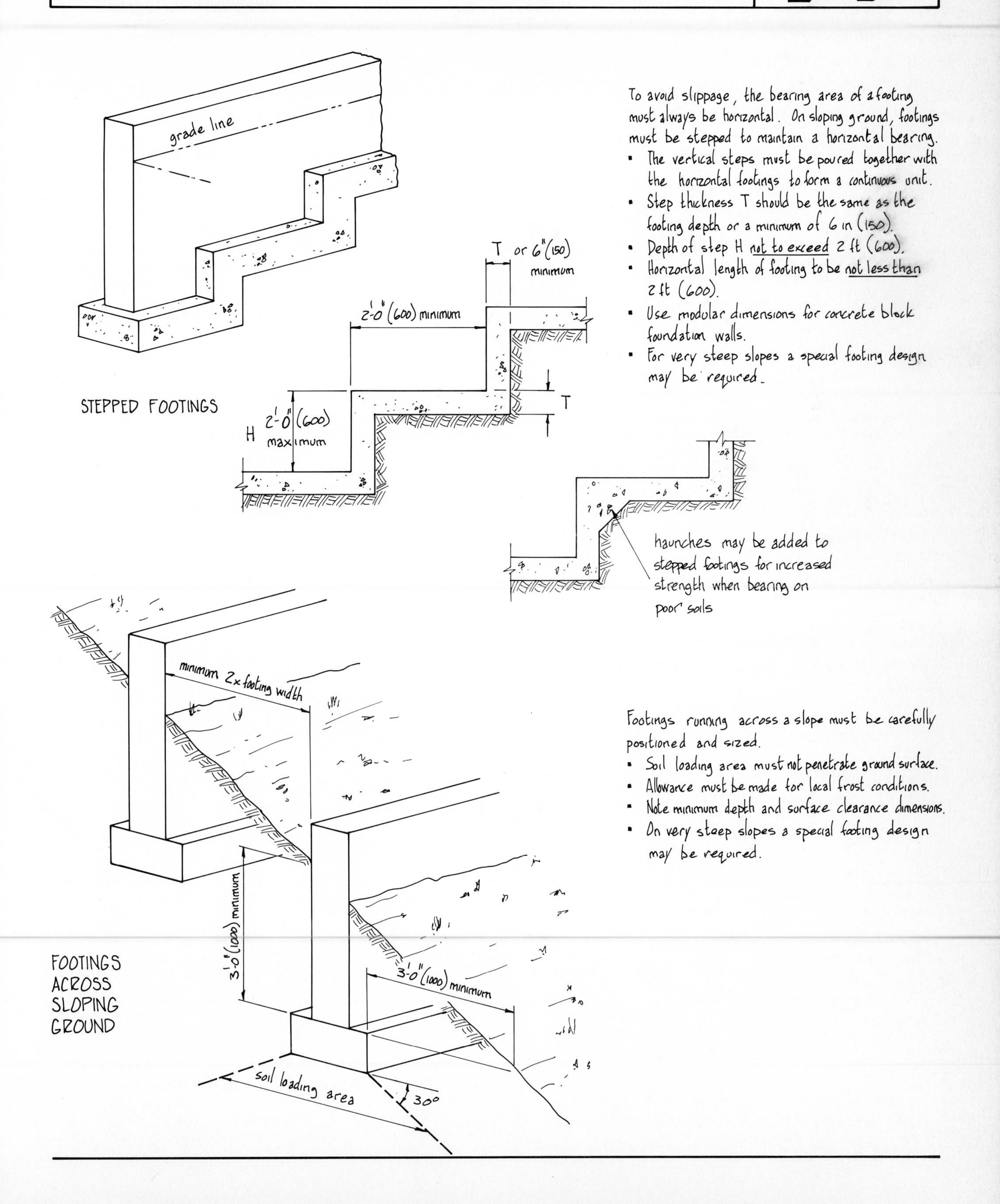

To avoid slippage, the bearing area of a footing must always be horizontal. On sloping ground, footings must be stepped to maintain a horizontal bearing.

- The vertical steps must be poured together with the horizontal footings to form a continuous unit.
- Step thickness T should be the same as the footing depth or a minimum of 6 in (150).
- Depth of step H not to exceed 2 ft (600).
- Horizontal length of footing to be not less than 2 ft (600).
- Use modular dimensions for concrete block foundation walls.
- For very steep slopes a special footing design may be required.

Footings running across a slope must be carefully positioned and sized.

- Soil loading area must not penetrate ground surface.
- Allowance must be made for local frost conditions.
- Note minimum depth and surface clearance dimensions.
- On very steep slopes a special footing design may be required.

FOOTINGS, COLUMNS, PILES, AND PIERS

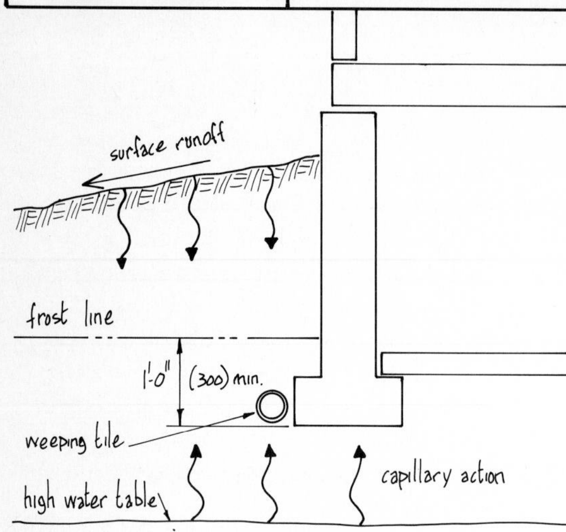

In all cool or cold areas where there is frost penetration into the ground, footings must be extended below the frost line. Footings located above the frost line will be subject to frost "heave" when water in the soil freezes and expands. Even though a soil may be naturally or artificially drained, water can rise by capillary action from the water table to form "ice lenses", (discontinuous layers of ice) above the frost line. The heaving action is extremely powerful and causes structural damage by first lifting and then dropping the foundation during the freeze/thaw cycle.

Additional notes:

- Check codes for minimum footing depths in each region.
- Never pour footings on ground that is in a frozen or heaved condition. Uneven settlement will occur on thawing.
- All soils are subject to frost heave to a greater or lesser degree.
- Good surface and subsurface drainage is essential.
- In very poor soils, backfill excavation around foundation walls with gravel or crushed rock to within 1 ft (300) of grade.
- See 10-32 onward for foundation insulation.

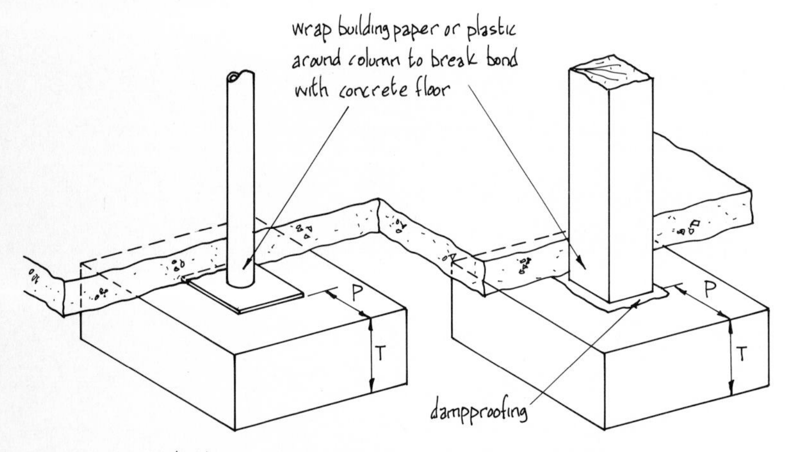

Steel Column and Base Plate

Wood Column

COLUMN FOOTINGS

Minimum size of column footing pads for average stable soil:

Supporting —	One Story	Two Story
	2'-0" x 2'-0" (600 x 600)	2'-6" x 2'-6" (800 x 800)

Thickness T is 6" (150) minimum but not less than P for unreinforced pads.

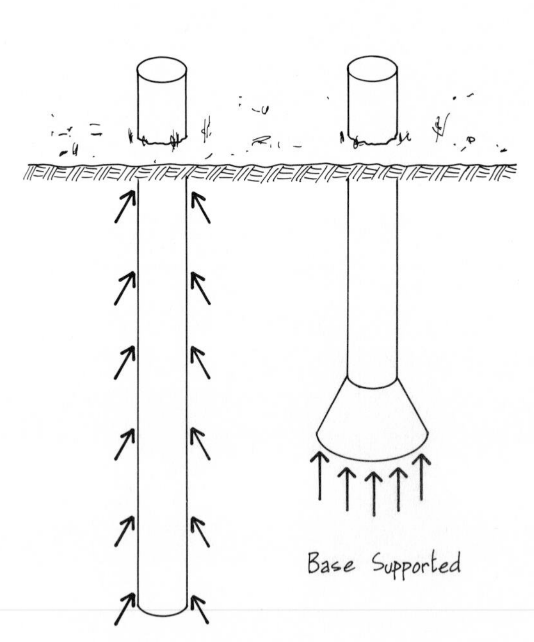

Base Supported

Friction Supported

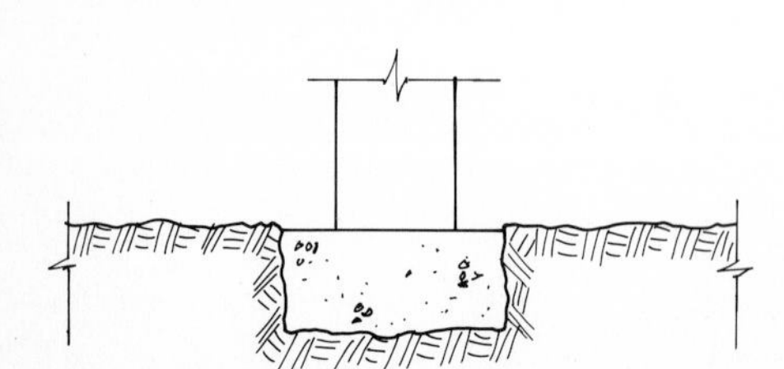

If the soil is firm enough, a footing trench can be dug by hand or machine, and the footing poured without need of formwork.

Columns, piles, or piers support point loads imposed by some foundation systems. They may be friction supported or base supported and be of concrete, steel or treated wood. In poor soil areas, they must be designed to suit the specific conditions.

FOOTINGS - DRAINAGE

In basement areas, groundwater must be prevented from seeping into or through the completed structure. Groundwater can also exert large hydrostatic pressures on foundation walls and basement slabs, sufficient in some cases to damage the structure. Water must therefore be channeled away from foundations and floors with a suitable drainage system.

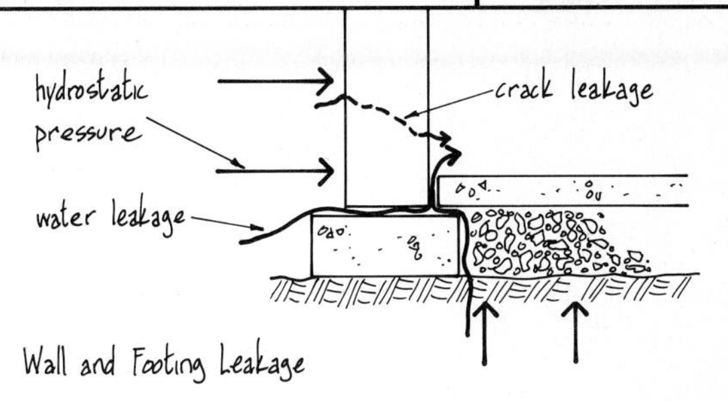

Wall and Footing Leakage

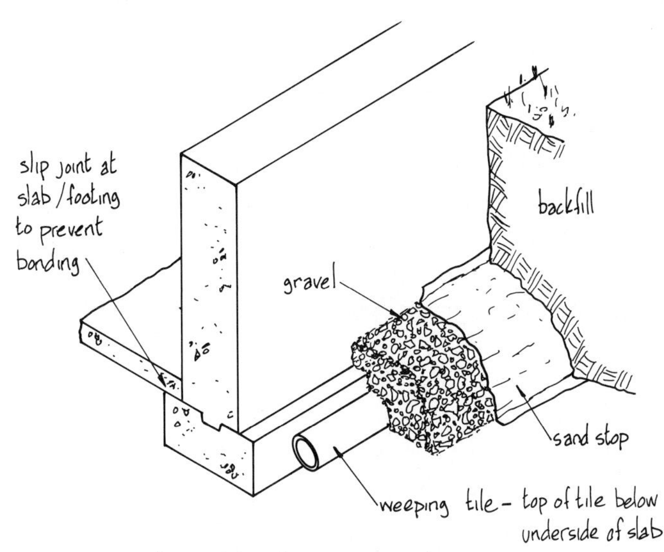

Typical Foundation Drainage Detail

THE BASIC FOUNDATION DRAINAGE SYSTEM

System consists of:

- 4 in (100) diameter drain tile (weeping tile) laid around the perimeter of footings. Bottom of tile to be level with underside of footing and top of tile to be no higher than level of basement floor or crawl space.
- Tile is usually either perforated continuous plastic pipe, or, for unusual soil conditions, clay/concrete pipe sections installed with open joints.
- Tile should be slightly sloped to an outlet.
- Preferred water disposal is by gravity flow to a storm drain system or a ditch, drywell, or pool located below foundation level (see 16-18).
- Where gravity drainage is not possible, water must be directed to an internal sump pit and pumped to a disposal location.
- Where very wet soil conditions are expected, a second tile line can be located on the inside of the footings.
- Course gravel acts as a filter to prevent sand filling the tiles.
- Sand stop prevents mass sand movement into tile before backfill consolidation has occurred. Suitable materials are building paper, plastic sheet, burlap, or straw.

For preserved wood foundation drainage, see 5-15.
For foundation wall waterproofing, see 5-11.
For window wells and areaways, see 5-17.

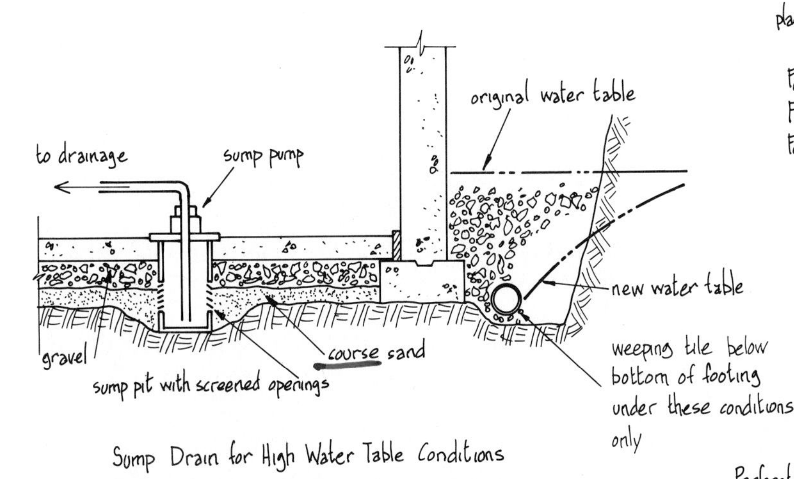

Sump Drain for High Water Table Conditions

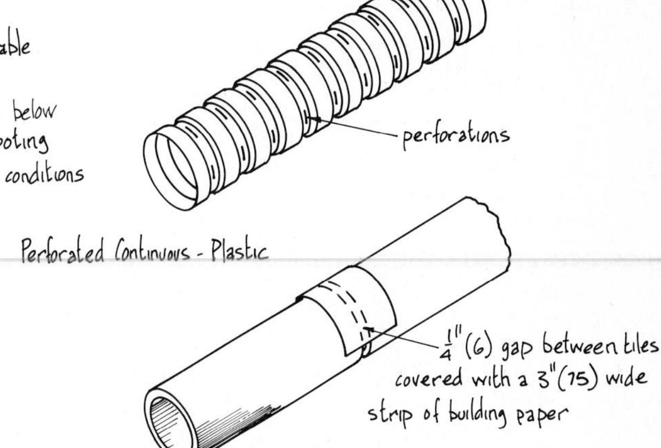

Perforated Continuous - Plastic

Clay or Concrete Tile

Drain Tiles

Although it is not normally considered economical or practical to position a basement or crawl space floor below the water table, it can be done provided that sufficient drainage is employed to minimize hydrostatic pressures. In the example above, one or more sump pits in the floor provide an escape path for groundwater, thus reducing uplift pressures and seepage through the floor.

FOUNDATION WALLS - POURED CONCRETE - BASEMENT

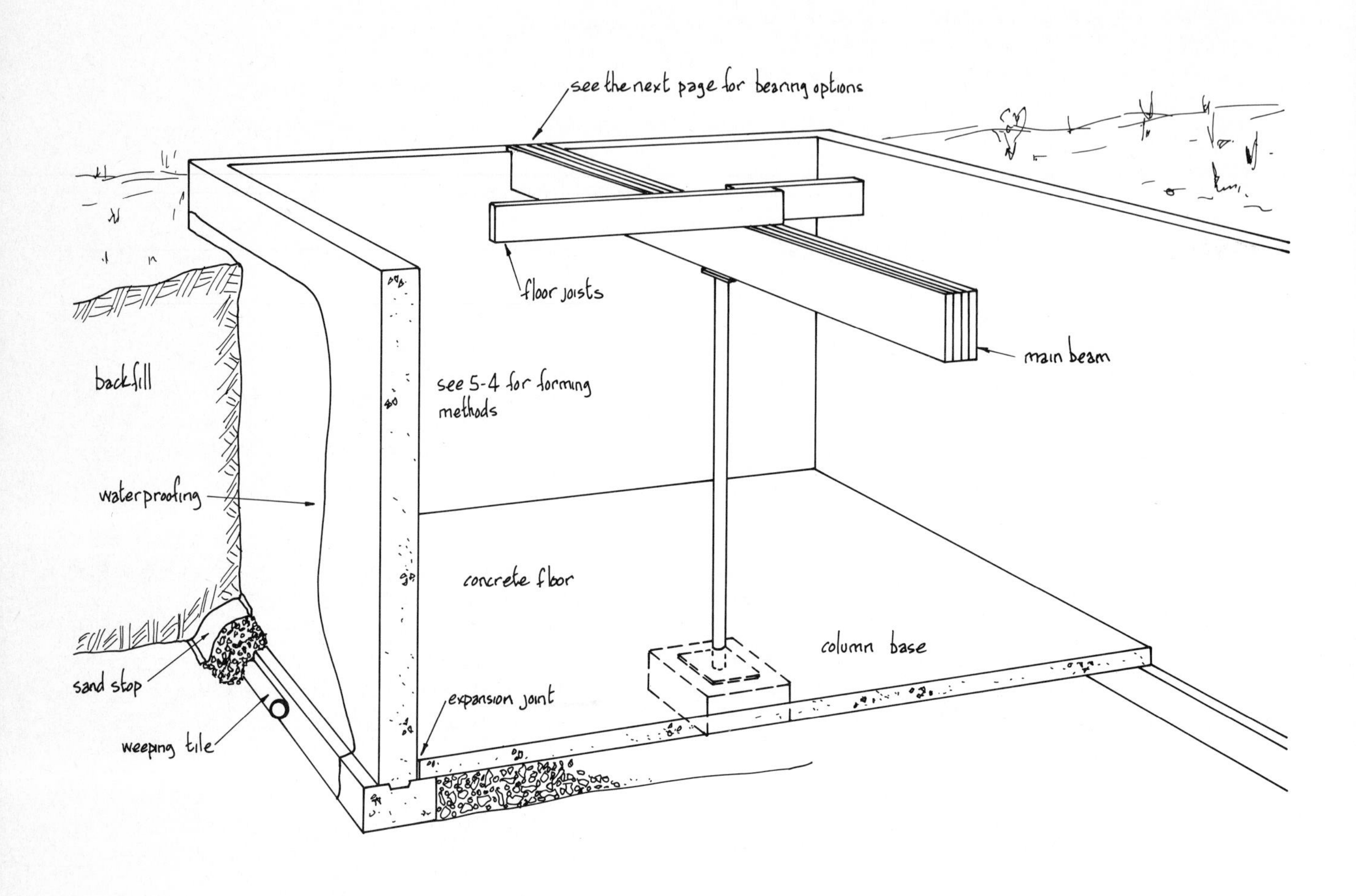

THICKNESS OF CONCRETE FOUNDATION WALLS
Typical code requirements

Material	Wall Thickness	Maximum Height of Finished Grade Above Basement Floor: Wall Laterally Unsupported	Maximum Height of Finished Grade Above Basement Floor: Wall Laterally Supported
concrete 2000 psi (14 MPa) minimum	6" (150)	2'-6" (760)	5'-0" (1520)
	8" (200)	4'-0" (1220)	7'-0" (2130)
	10" (250)	4'-6" (1370)	7'-6" (2290)
	12" (300)	5'-0" (1520)	7'-6" (2290)

It is extremely important that all deep foundation walls be laterally supported at the top. Foundation walls are considered laterally supported at the top under the following conditions:

- If the floor joists are embedded in the top of the foundation wall.
- If the floor joists are properly tied to a sill plate that is clamped to the top of the wall with anchor bolts.
- When solid masonry upper walls are supported by the foundation, the foundation is considered laterally supported by the first floor.
- Check local codes for maximum permissible unreinforced opening sizes.
- See 5-22 for more information.

POURED CONCRETE FOUNDATION WALLS

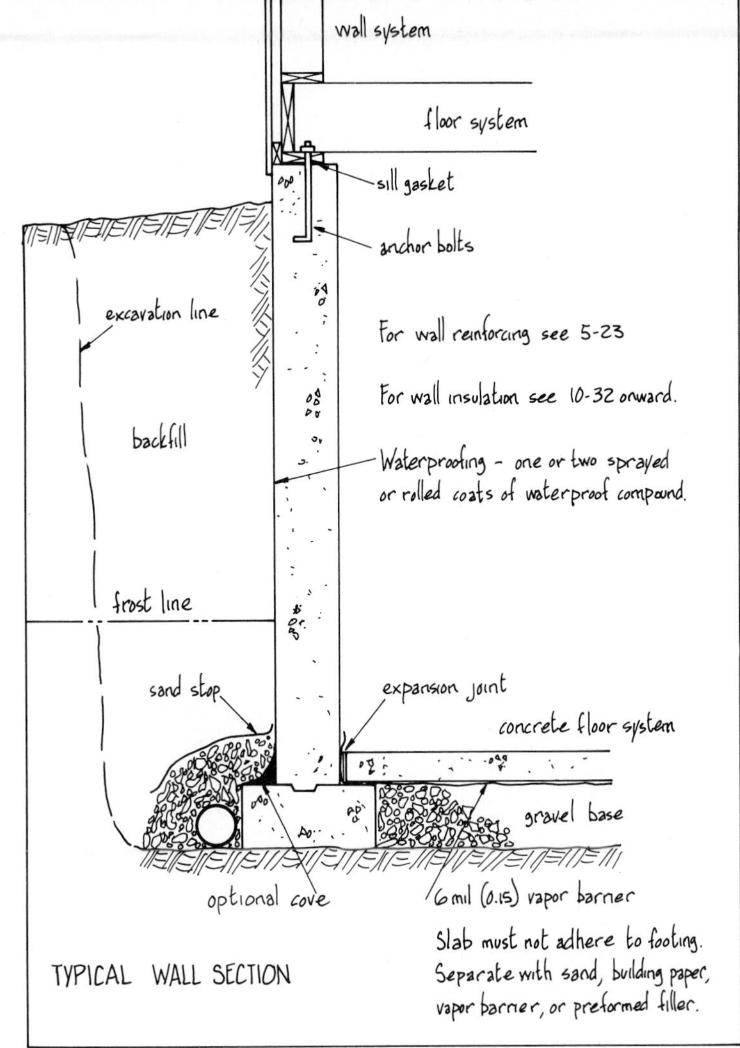

TYPICAL WALL SECTION

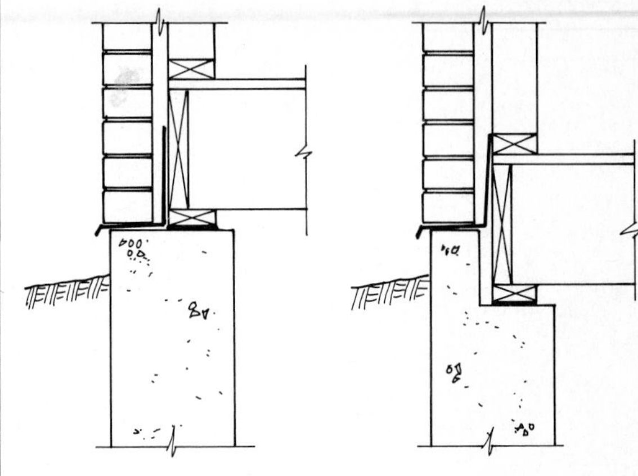

Alternate Sill Details

Considerable variation in construction details are to be expected. Floor and wall systems, exterior finishes, and sizing are key factors. See 10-28 onward for air/vapor barrier and insulation details.

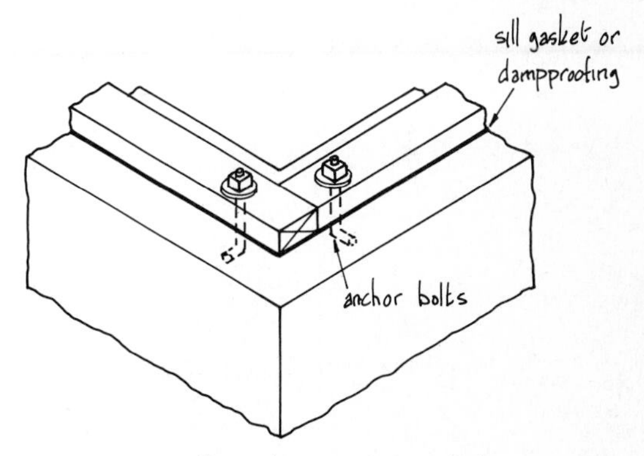

Sill Plate Corner Detail

Sill plates are anchored to the wall with preset anchor bolts. Set bolts in place immediately after pouring and levelling of concrete in the forms. Space bolts at 8ft (2400) centers with a minimum of 2 per sill. See 6-10 for sill treatment.

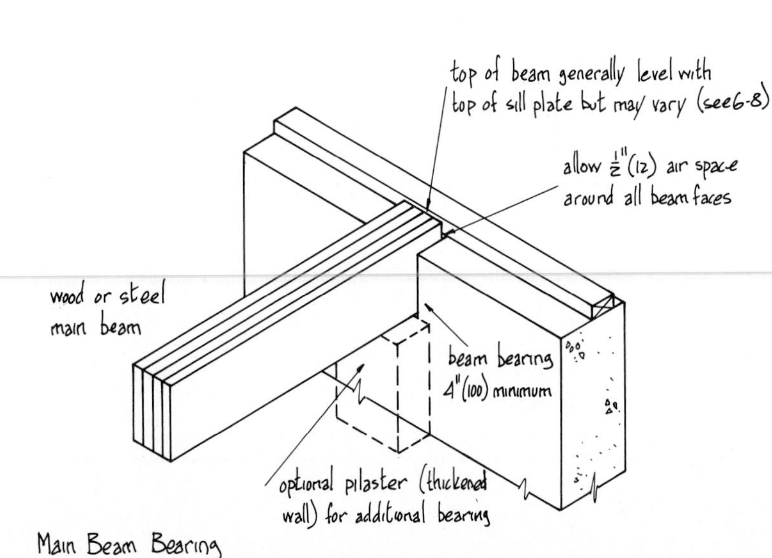

Main Beam Bearing

See 5-17 for windows and openings in foundation walls.
See 6-7 for main beam details.

MASONRY (BLOCK) FOUNDATION WALLS - BASEMENT

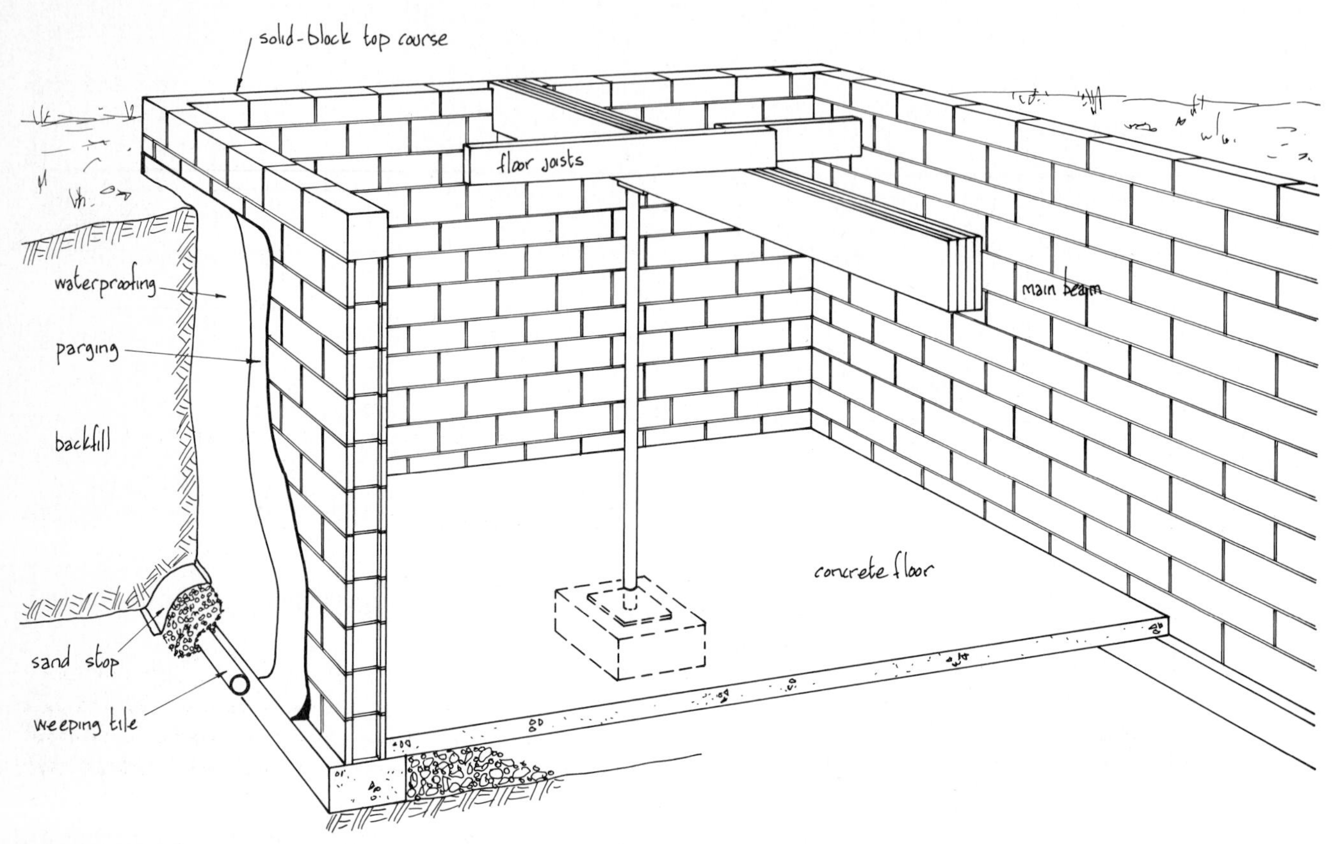

CONCRETE BLOCK FOUNDATION WALLS

Typical code requirements.

Material	Wall Thickness	Maximum Height of Finished Grade Above Basement Floor: Wall Laterally Unsupported	Maximum Height of Finished Grade Above Basement Floor: Wall Laterally Supported
concrete unit masonry	6" (150)	2'-0" (610)	2'-0" (610)
	8" (200)	3'-0" (910)	4'-0" (1220)
	10" (250)	4'-0" (1220)	6'-0" (1830)
	12" (300)	4'-6" 1370)	7'-0" (2130)

The ultimate strength of a masonry wall is derived from the combined masonry unit and mortar strengths. For most residential and light commercial applications, both standard and lightweight units are acceptable. Most manufacturers produce units having a minimum compressive strength of 1000 psi (7.5 MPa) for hollow units and 1800 psi (12.5 MPa) for solid units. Higher-strength units are available for multistory and heavy commercial structures.

Mortars can be mixed to specific requirements as shown at the right.

MORTAR TYPES

Mortar Type	Average Compressive Strength At 28 Days PSI (MPa)	Construction Suitability
M	2500 (17.25)	Masonry subjected to high compressive loads, severe frost action, or high lateral loads from earth pressures, hurricane winds, or earthquakes. Structures below grade, manholes, and catch basins.
S	1800 (12.5)	Structures requiring high flexural bond strength, but subject only to normal compressive loads.
N	750 (5.0)	General use in above-grade masonry, residential basement construction, interior walls and partitions. Concrete masonry veneers applied to frame construction.

A.S.T.M. mortar type designation system.

For materials and proportions of mortar mix, refer to cement manufacturer's specifications.

BLOCK FOUNDATION WALLS

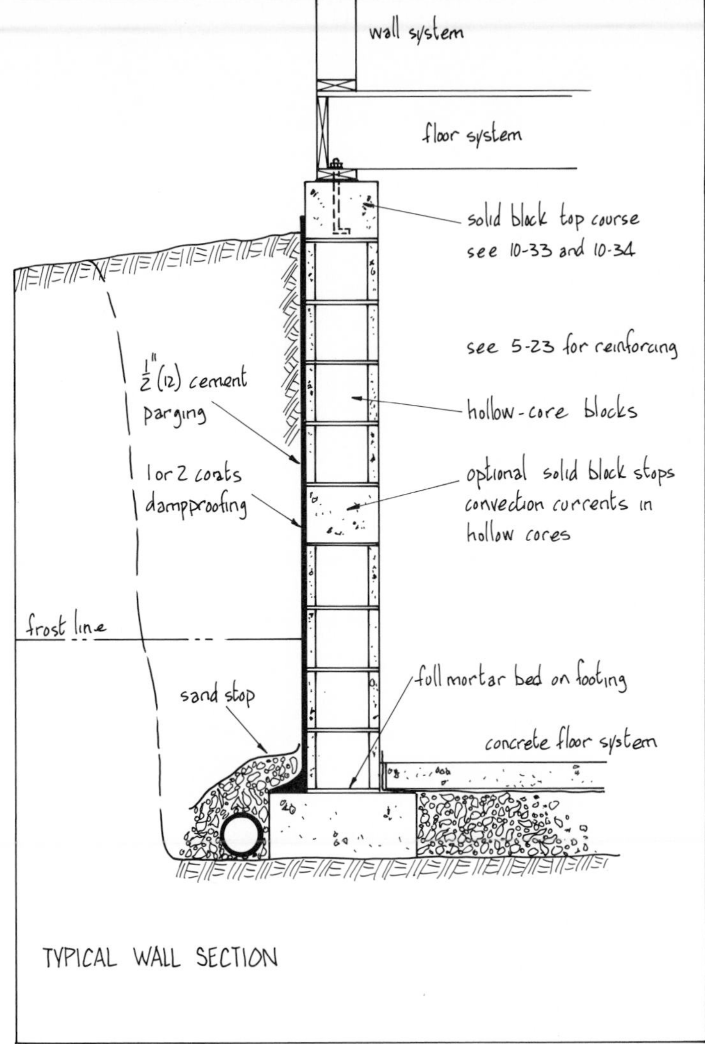

TYPICAL WALL SECTION

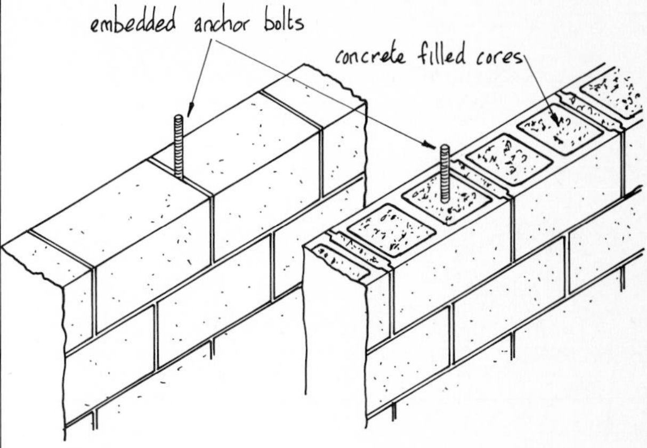

Solid-Block Top Course Hollow-Block Top Course

Walls should be capped with either solid units or by filling the cores of the top course with mortar. Sill anchor bolts spaced as described on 5-11 and placed in joints of solid caps or in filled core of hollow caps.

No formwork is required for concrete block walls since each unit is laid individually in horizontal courses. The first course (or row) is laid on the footing in a full mortar bed. Subsequent courses can be laid with mortar only on contact surfaces. For modular spacing and joint size, see 5-5. All joints must be tooled smooth to resist water penetration.
The exterior face of the wall must be parged with a minimum 1/4" (6) of cement plaster to provide a base for dampproofing. Apply at least one heavy coat of bituminous liquid dampproofing material over the parging, including the cove at the wall and footing joint. In extremely wet soils, two or three layers of a bituminous saturated membrane are applied to the wall and then coated overall with a bituminous liquid material.

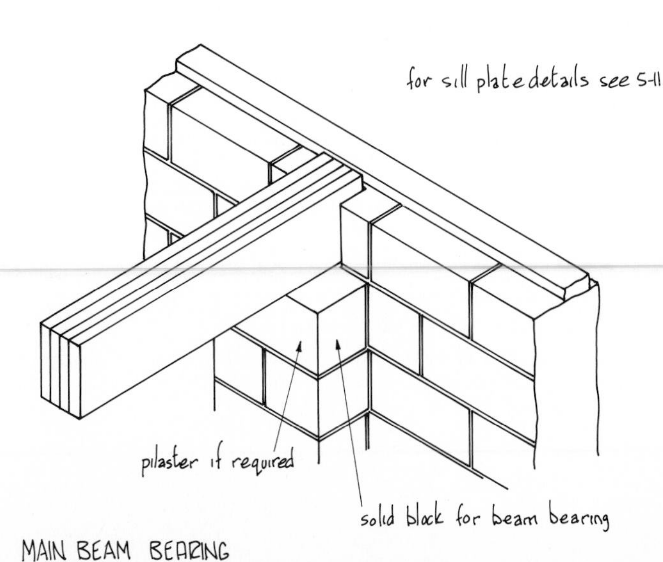

MAIN BEAM BEARING

See 5-17 for windows and openings in the wall.
See 10-32 to 10-34 for wall insulation.

PRESERVED WOOD FOUNDATIONS – BASEMENT

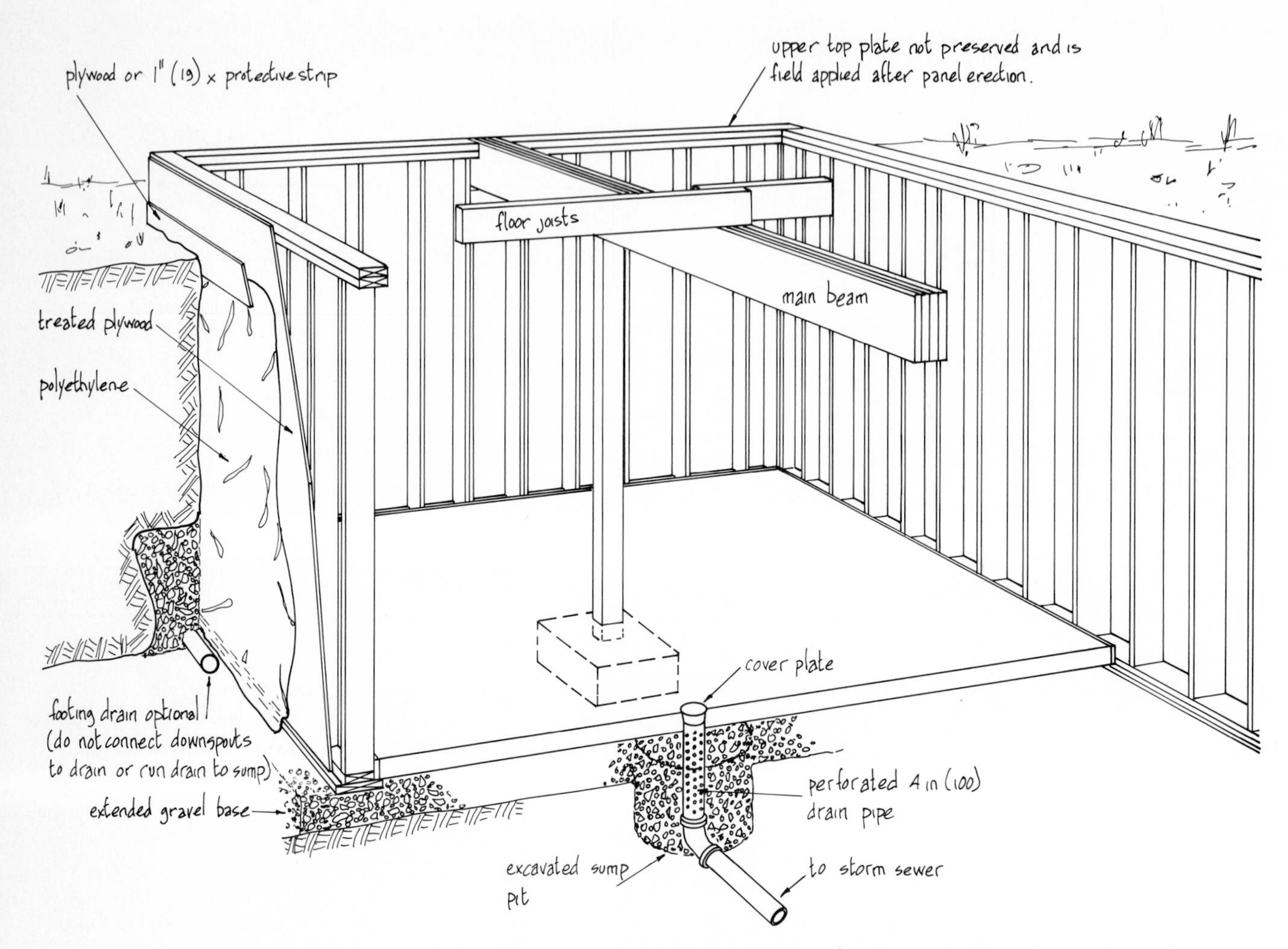

PWF systems must be designed to carry all upper structure and live loads plus lateral stresses imposed by earth pressures. In addition, the species, grade, and size of PWF members, stud spacing and backfill height all have bearing on the final design. Consequently, it is important to consult design information provided by specific trade associations, such as:

- In the United States — National Forest Products Association, 1619 Massachusetts Ave. NW, Washington, D.C.
- In Canada — Canadian Wood Council, 85 Albert Street, Ottawa, Ontario.

A preserved wood foundation (PWF) is essentially the below-grade continuation of upper stud wall framing, acting as a foundation wall. Because PWF wall construction details are similar to those for standard stud walls, this type of foundation system can be assembled and erected very quickly under virtually all weather conditions.

- All wood products used in PWF (except the upper top plate) must be pressure-treated with an approved wood preservative.
- Fasteners in preserved wood must be corrosion resistant and meet applicable codes.
- Walls are prefabricated either on or off site and erected in whole panels.
- Poured concrete footings are not required.
- Good foundation drainage is very important.
- For insulation of PWF systems follow the same procedures as for upper wall studs (see 10-46).
- Basement floors can be of either concrete or wood.

source: National Forest Products Assoc.

PRESERVED WOOD FOUNDATIONS

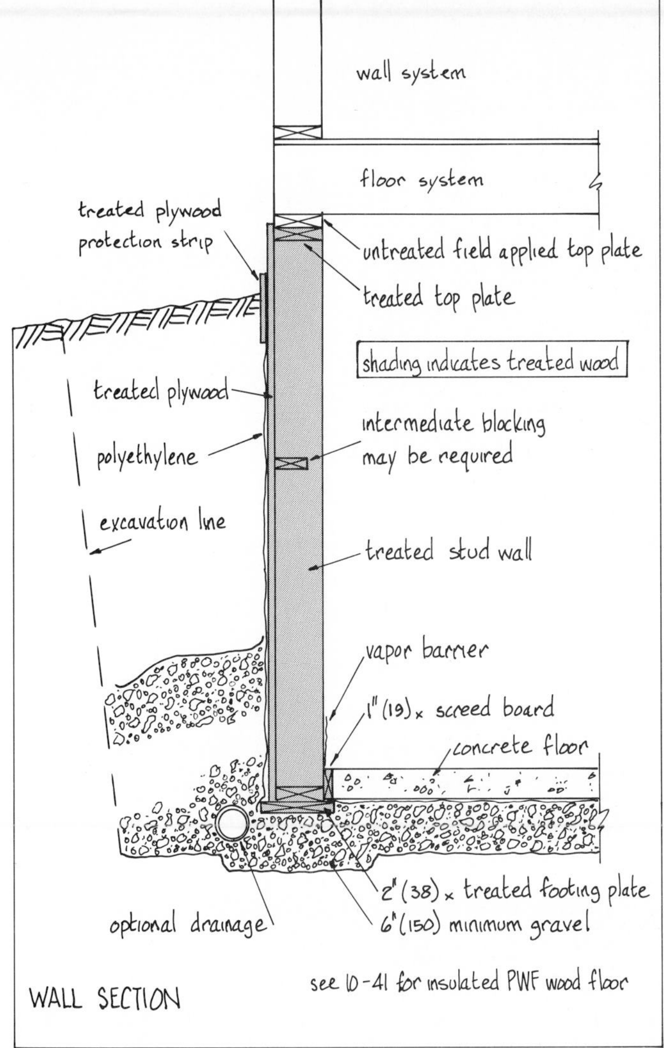

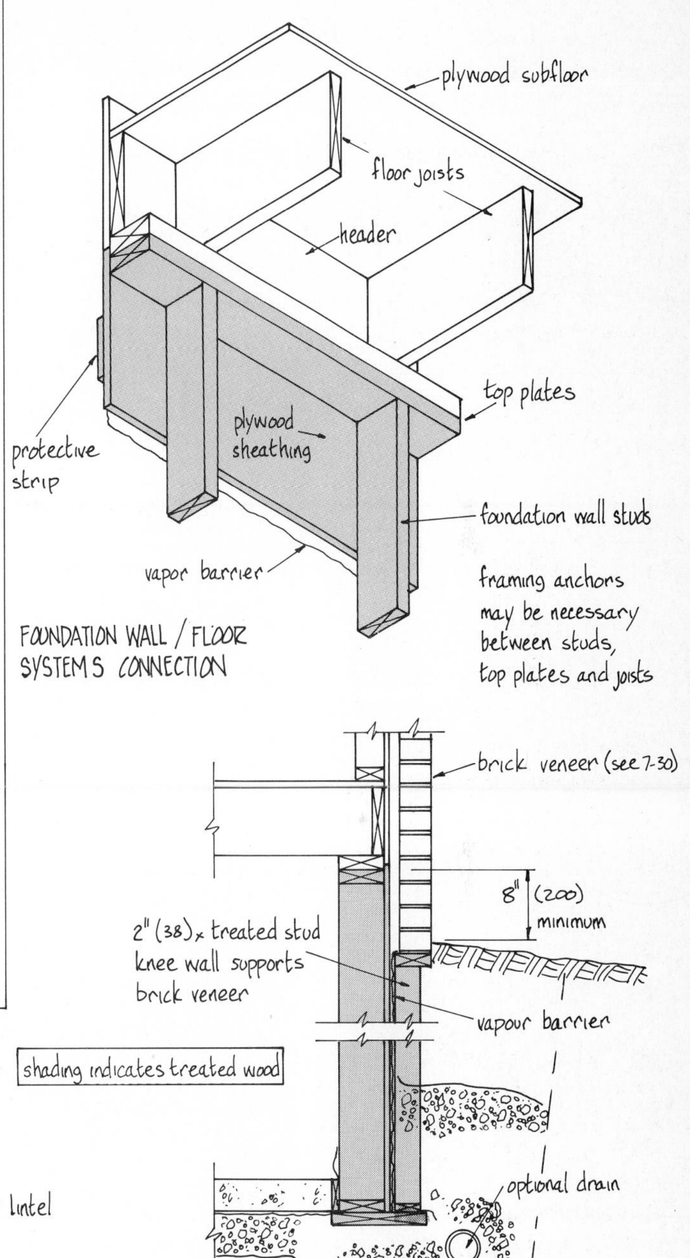

BASEMENT WALL WITH BRICK VENEER SUPPORT

cut top plate to suit

two or three ply lintel

main beam

wall stud (spacing not to be broken)

support stud continued to bottom plate

MAIN BEAM BEARING

Notes:

- all exterior joints in plywood and studs to be caulked
- joints in polyethylene lapped minimum 6" (150) and caulked
- refer to design books for nailing and fastening specifications

see 10-36 for insulation practice
see 5-17 for window and openings

source: National Forest Products Assoc.

PERIMETER WALLS - CRAWL SPACES

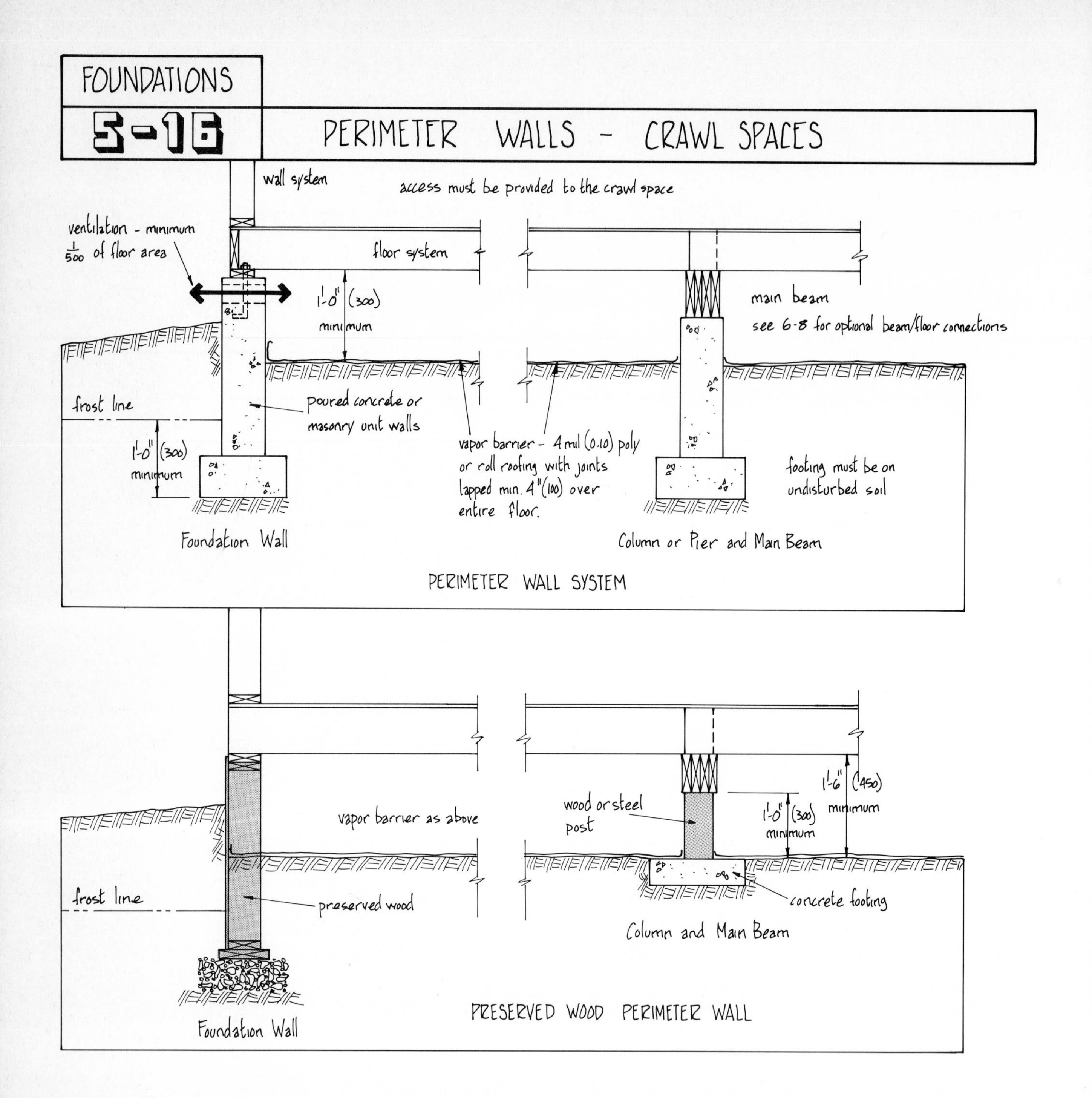

A perimeter wall system is similar in design to a full basement system. It is used when building designers do not call for bulk storage space, in some split-level structures, and when extensive mechanical equipment is not required. Additionally, poor soil conditions or a high water table may dictate the use of this system.

- For wall thicknesses see tables on 5-10 and 12. For PWF walls, refer to design manuals supplied by the associations mentioned on 5-14.
- The crawl space will be designed as either a heated or an unheated space, depending on climate and utility design. For appropriate insulation options see 10-44 and 10-45.
- In heated crawl spaces, vents must be operable and weather-stripped. Vents are opened in summer and closed in winter. Unheated spaces are permanently vented.
- If ground clearance is less than 24 in (600), access trenches must be provided to service heating equipment and plumbing.
- In high-water-table or poor-drainage soils and where the crawl space floor is significantly below outside grade, water-proofing of walls and drainage of footings may be necessary (see 5-9).

BASEMENT WINDOWS

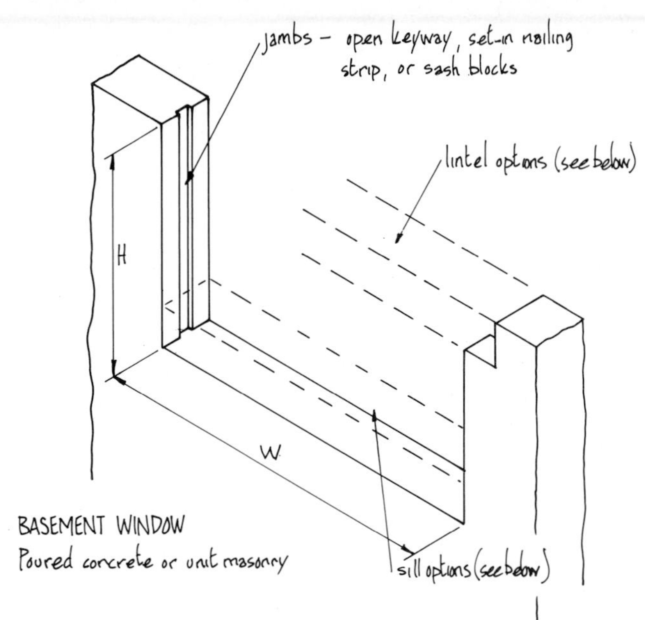

BASEMENT WINDOW
Poured concrete or unit masonry

Dimensions – H and W can be any size in poured wall.
H and W should be modular size in block for modular-sized windows.
For nonmodular windows, set frame in place during block laying or give rough opening size.

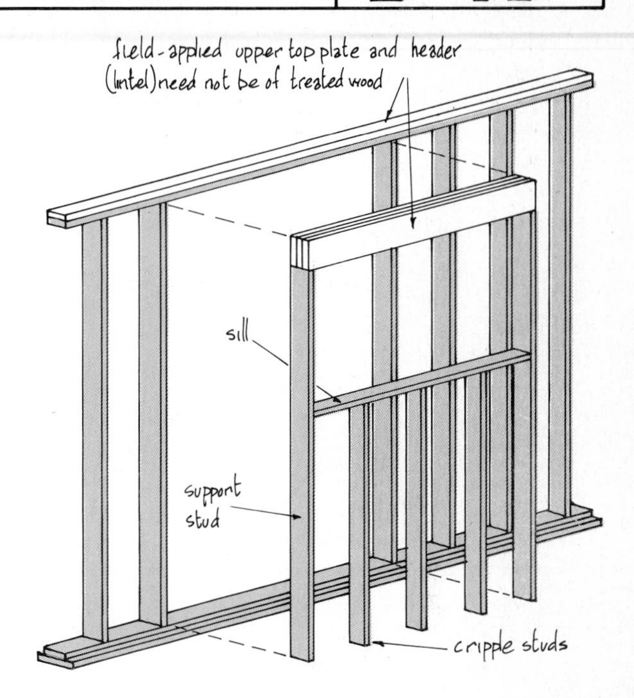

PWF BASEMENT WINDOW
Window framing details similar to standard stud walls

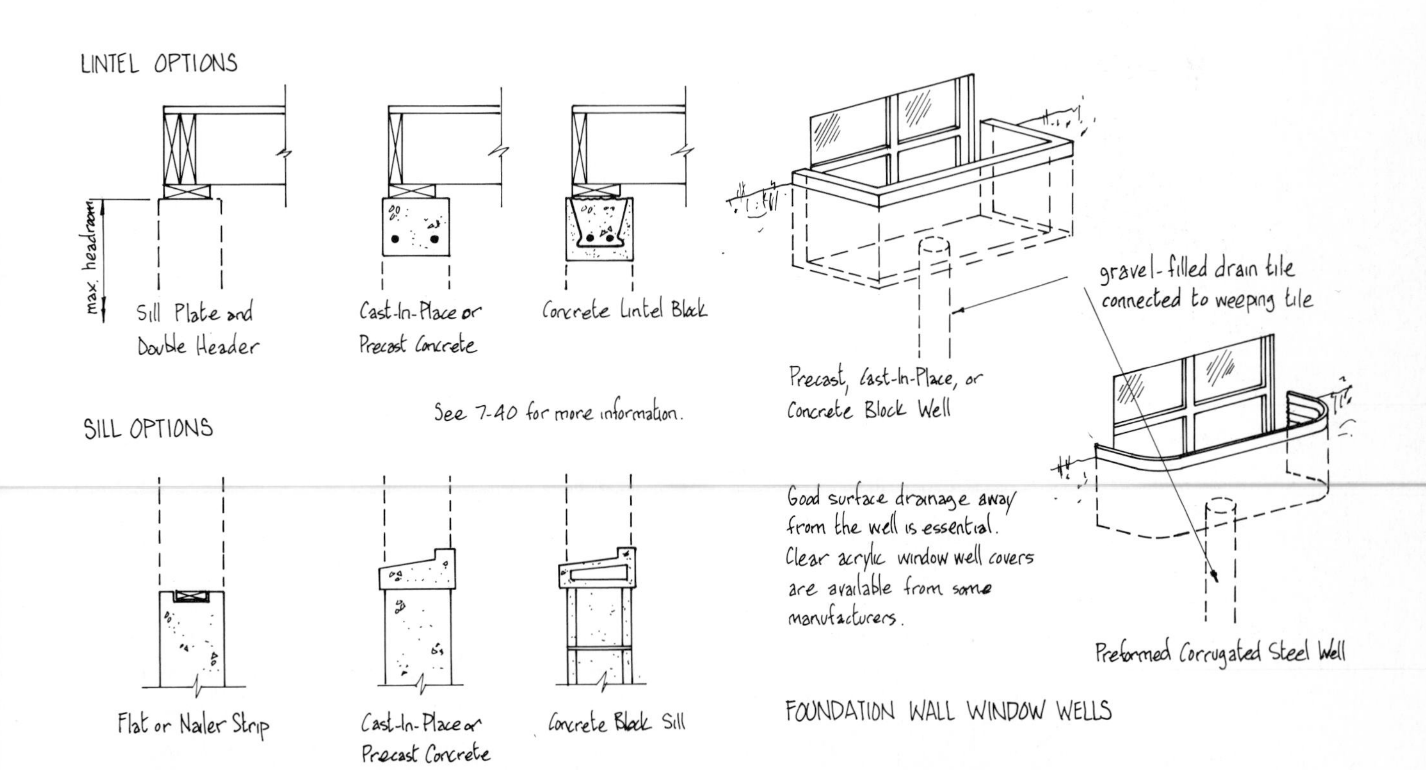

SLAB-ON-GRADE FOUNDATIONS

Floor slabs at grade level (ground slabs) are very common in residential and commercial construction, especially in warm and temperate regions. They are not generally recommended for cool and cold climates because of potential frost problems. In addition, problems can occur where wet, low-bearing-strength, or unpredictable soil conditions exist.

- A smooth soil surface must be provided, free of debris, stumps, and organic matter. Loose soil must be well tamped.
- All utility and mechanical services must be installed before the slab is poured. Backfill in trenches must be compacted.
- Slab must not carry any upper superstructure loads.
- Slab thickness is minimum 4 in (100). Gravel thickness is minimum 5 in (125).
- Reinforcing is necessary for independent slabs. Use 6×6 in (150×150) wire mesh or No. 3 (10M) bars at 24-in (600) centers both ways.
- A vapor barrier is essential to stop soil moisture migrating through the slab. Use 6-mil (0.15) plastic with joints lapped minimum 6 in (150).
- Insulate perimeter either externally or internally using rigid insulation suitable for below grade applications (see 10-42 for details).

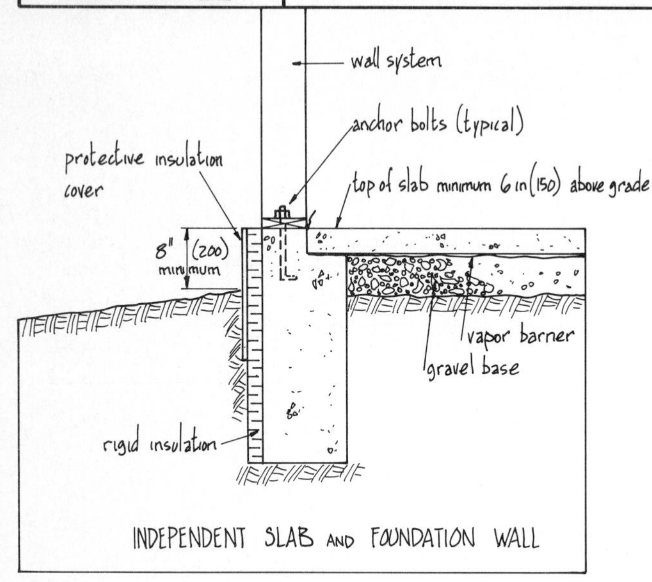

INDEPENDENT SLAB AND FOUNDATION WALL

refer to general notes above

heating duct
vapor barrier

COMBINED SLAB AND FOUNDATION

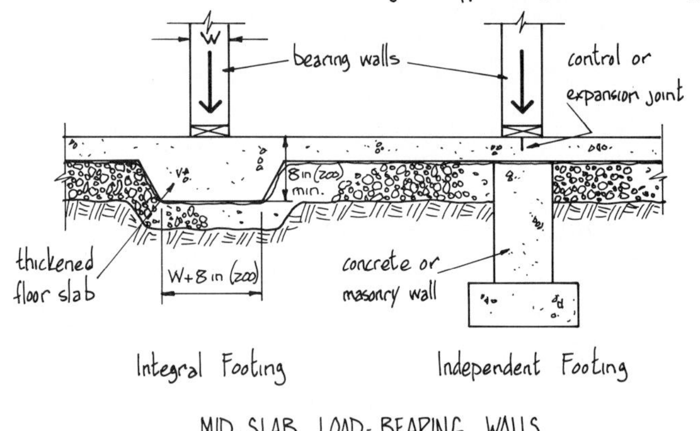

MID-SLAB LOAD-BEARING WALLS

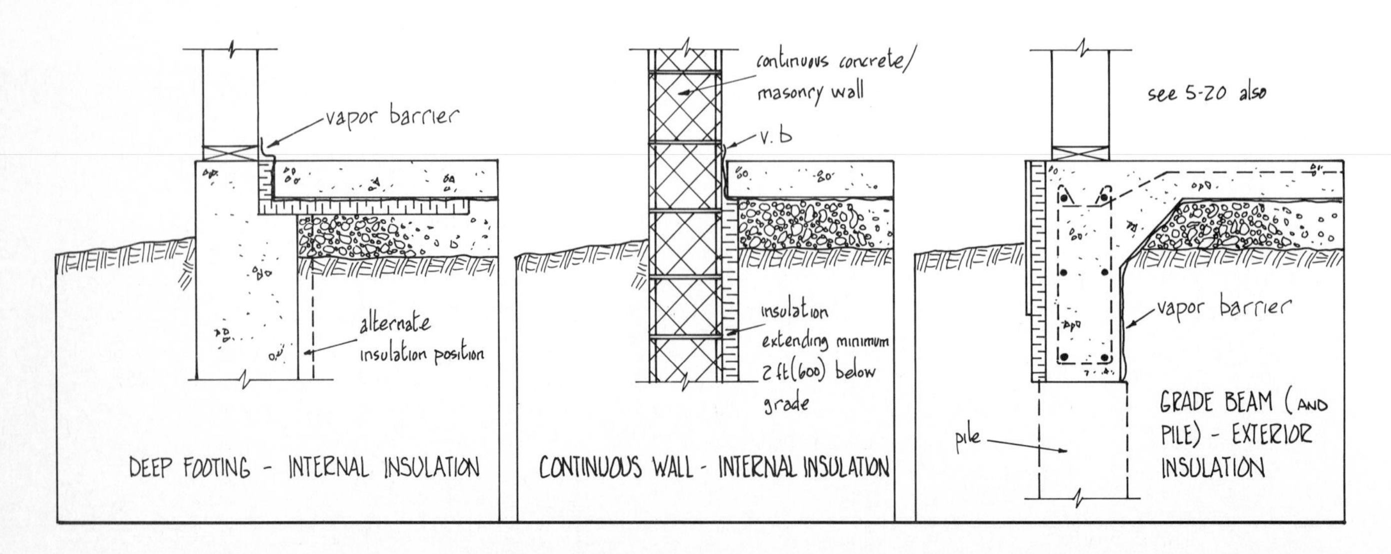

DEEP FOOTING - INTERNAL INSULATION

CONTINUOUS WALL - INTERNAL INSULATION

GRADE BEAM (AND PILE) - EXTERIOR INSULATION

HEATING AND CRACK CONTROL

Since there is no basement or crawl space below slab-on-grade construction, heating services are normally incorporated into the floor slab itself. A standard forced air system can be used with ducts embeded in a combined slab and foundation. Or the slab itself can be used as a heat distribution medium by embeding in the slab small copper or plastic pipes for hot-water radiant heating or resistance wires for electric radiant heating.

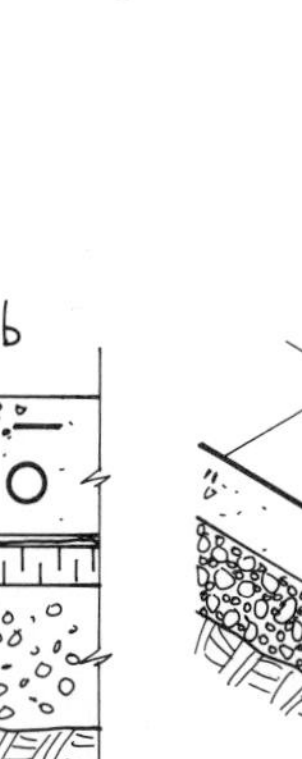

SLAB HEATING

hot-air registers
finish flooring
hot-air supply duct
2 in (50) minimum
concrete all around
additional supply or return ducts
(see 14-13 for more information)

AIR DUCTS IN SLAB

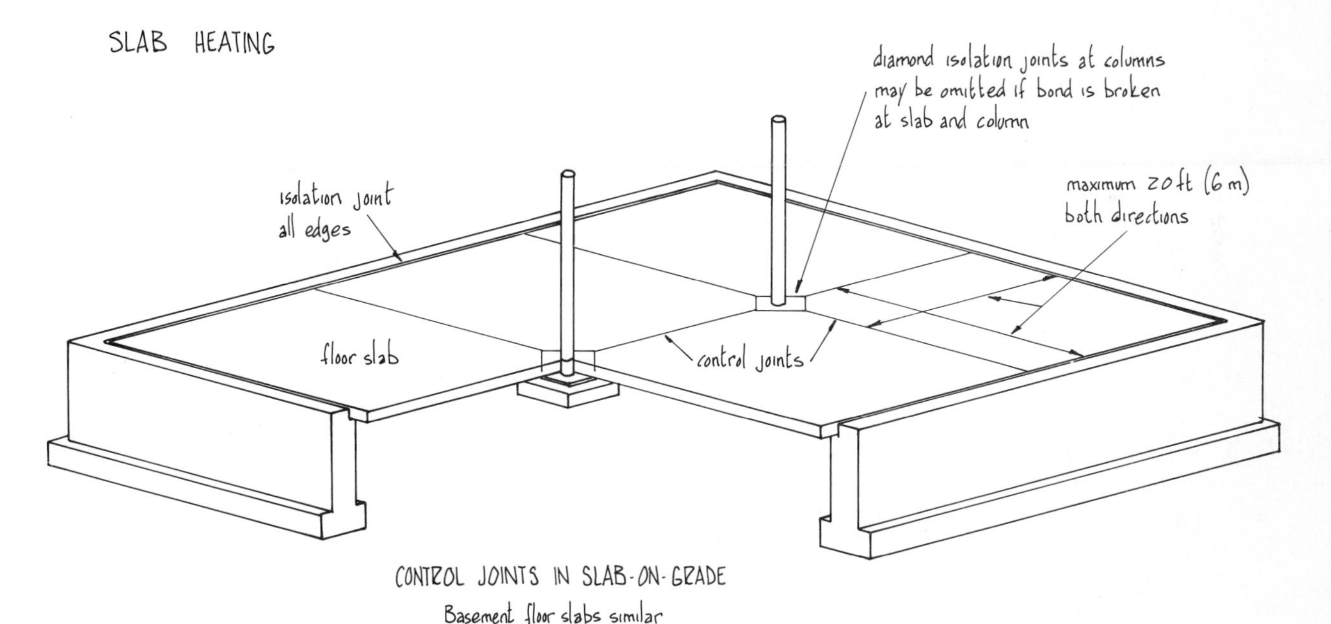

CONTROL JOINTS IN SLAB-ON-GRADE
Basement floor slabs similar

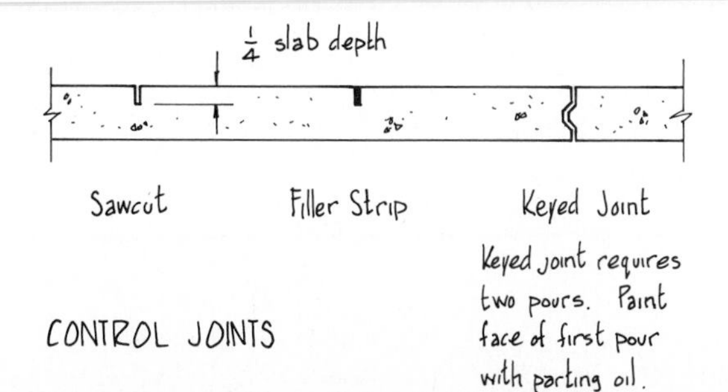

Keyed joint requires two pours. Paint face of first pour with parting oil.

CONTROL JOINTS

A floor slab will develop random cracks, first during the curing process and later through normal expansion and contraction. To avoid unsightly cracks in exposed floors, control joints are necessary. The joints are simply grooves cut into the slab surface before the concrete has set. A scribing tool or a power saw may be used and the grooves are cut to one-quarter of the slab thickness. Or, if specified, metal or premolded strips can be set in the concrete during pouring.

PILE FOUNDATIONS

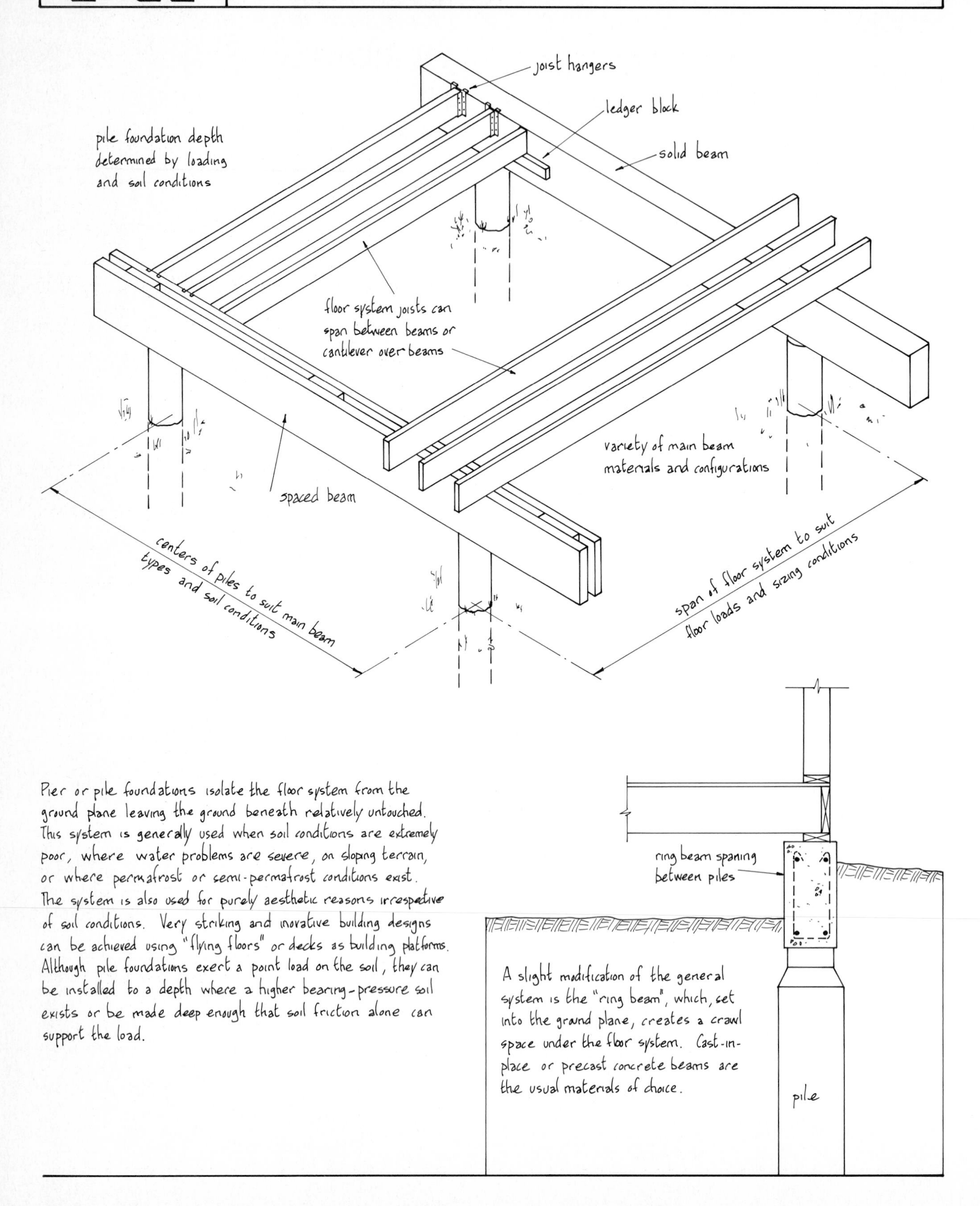

Pier or pile foundations isolate the floor system from the ground plane leaving the ground beneath relatively untouched. This system is generally used when soil conditions are extremely poor, where water problems are severe, on sloping terrain, or where permafrost or semi-permafrost conditions exist. The system is also used for purely aesthetic reasons irrespective of soil conditions. Very striking and inovative building designs can be achieved using "flying floors" or decks as building platforms. Although pile foundations exert a point load on the soil, they can be installed to a depth where a higher bearing-pressure soil exists or be made deep enough that soil friction alone can support the load.

A slight modification of the general system is the "ring beam", which, set into the grand plane, creates a crawl space under the floor system. Cast-in-place or precast concrete beams are the usual materials of choice.

PILES, COLUMNS AND PIERS

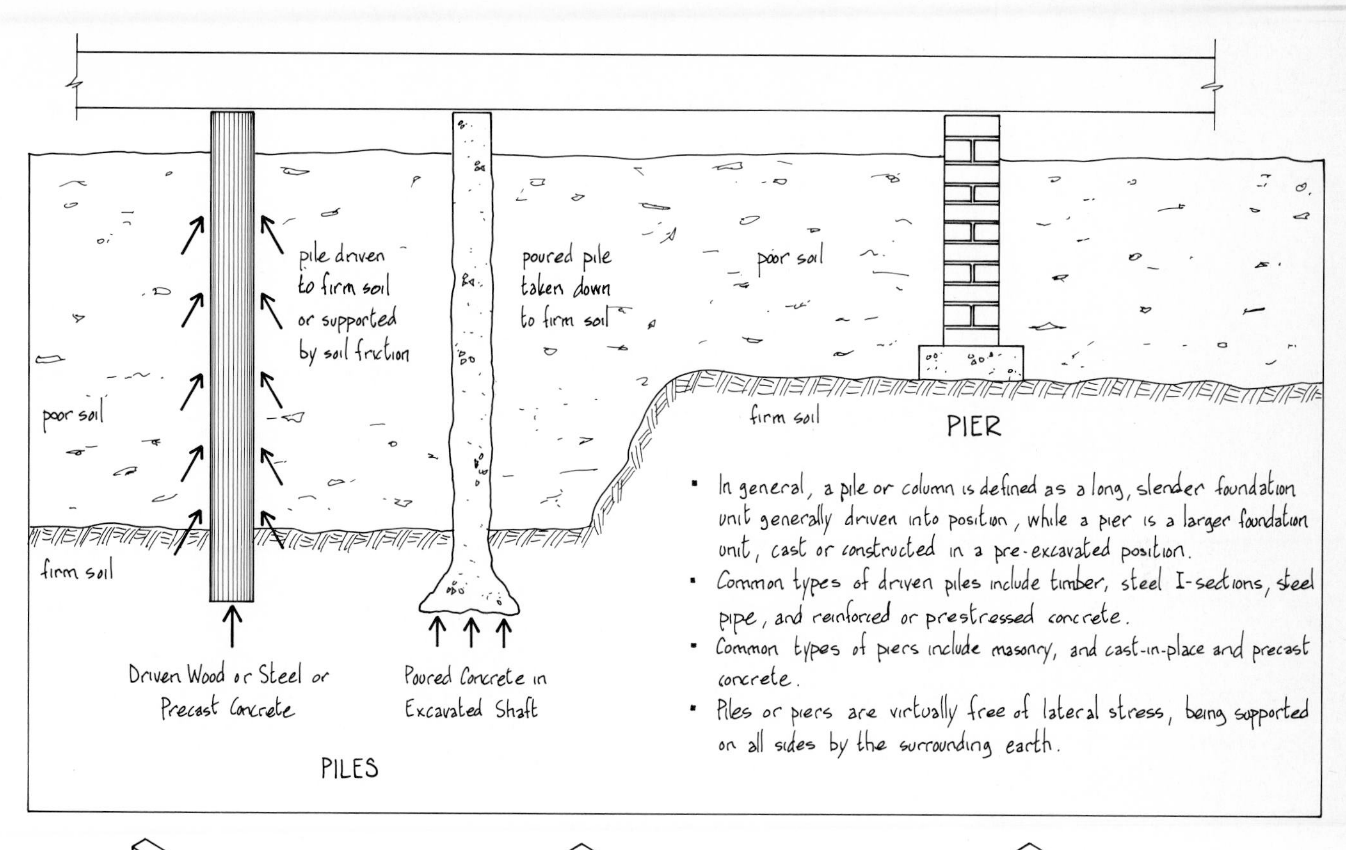

- In general, a pile or column is defined as a long, slender foundation unit generally driven into position, while a pier is a larger foundation unit, cast or constructed in a pre-excavated position.
- Common types of driven piles include timber, steel I-sections, steel pipe, and reinforced or prestressed concrete.
- Common types of piers include masonry, and cast-in-place and precast concrete.
- Piles or piers are virtually free of lateral stress, being supported on all sides by the surrounding earth.

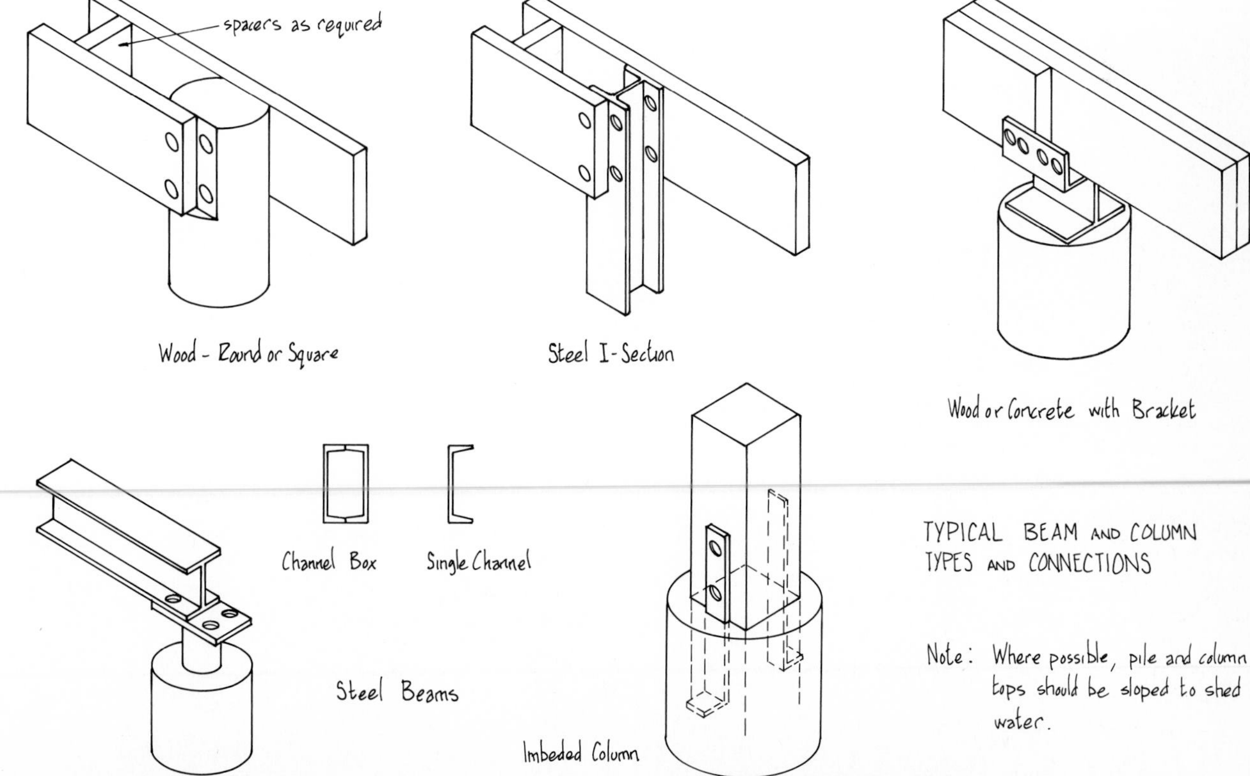

TYPICAL BEAM AND COLUMN TYPES AND CONNECTIONS

Note: Where possible, pile and column tops should be sloped to shed water.

WALL LOADING AND LATERAL RESTRAINT

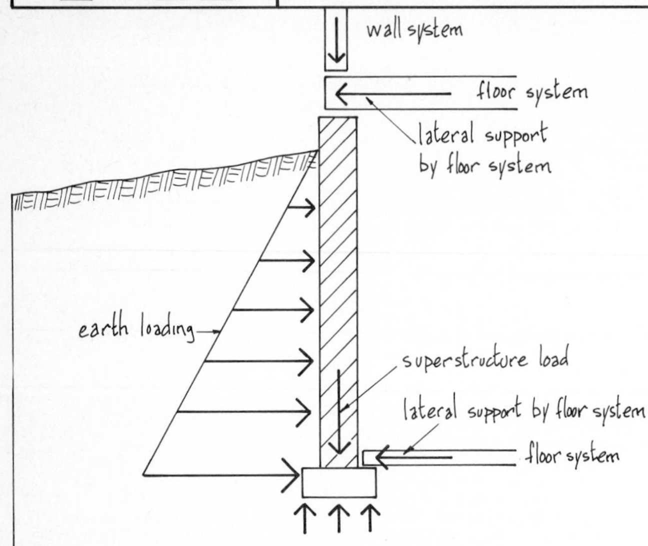

LATERAL EARTH PRESSURES ON FOUNDATION WALL

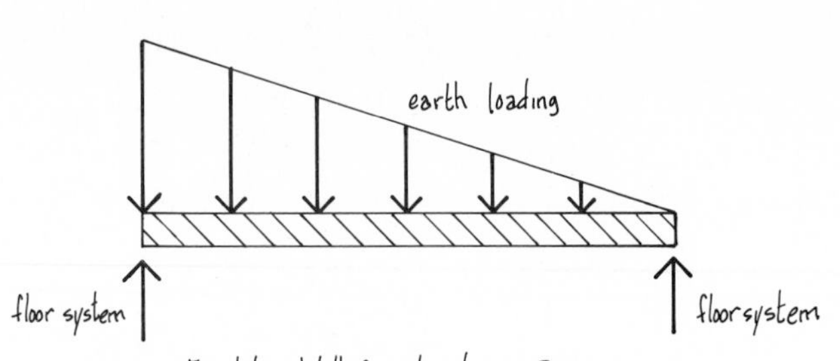

Foundation Wall Considered as a Beam

A foundation wall is subject to a variety of dynamic forces upon it by the elements it supports.
Earth pressure is a major load factor. The foundation must support lateral earth pressure, which increases with depth. Under these conditions the wall acts as a beam supported at each end and with an imposed triangular load. Actual loads will vary with type and condition of soil; wet soils will impose a greater load than dry soil, for example. The wall will be laterally supported at its base by the floor slab, but care must be taken to ensure lateral support at the top by correct anchorage of the first floor system to the wall (see 5-10). Vertical loads are imposed on the wall by superstructure weight and by compaction or expansion of soil under the footings.
Foundation walls are therefore sized to adequately support the forces acting on them, and under normal conditions the wall thickness charts on 5-10 and 5-12 are applicable. However, in the presence of a poor or wet soil condition, a back-filled site, a heavy superstructure, or other adverse load conditions, reinforcing of the wall may also be necessary. The information presented on this page and the next indicate typical foundation wall (and superstructure wall) reinforcement techniques to be used as dictated by design considerations.

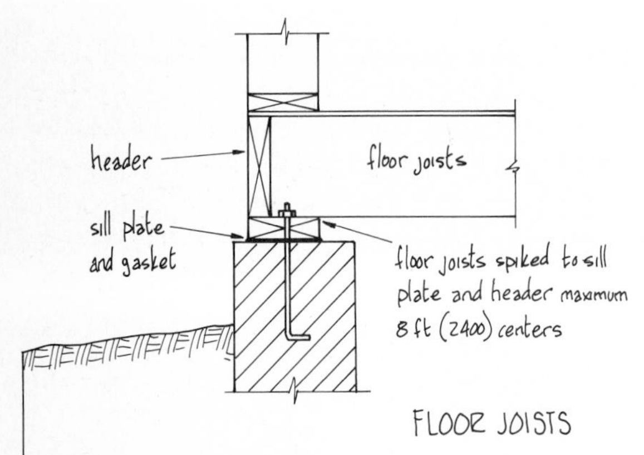

FLOOR JOISTS

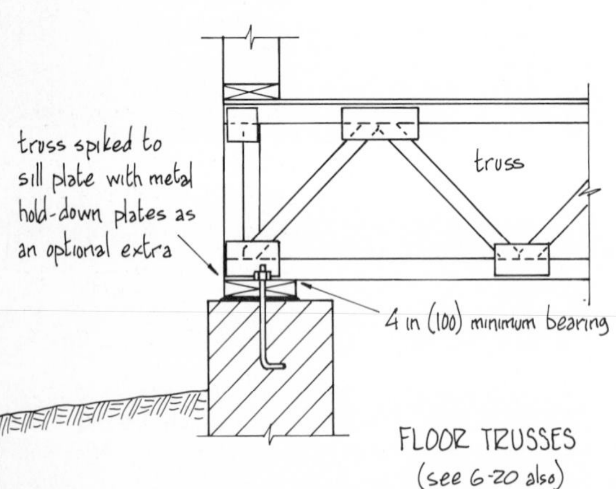

FLOOR TRUSSES
(see 6-20 also)

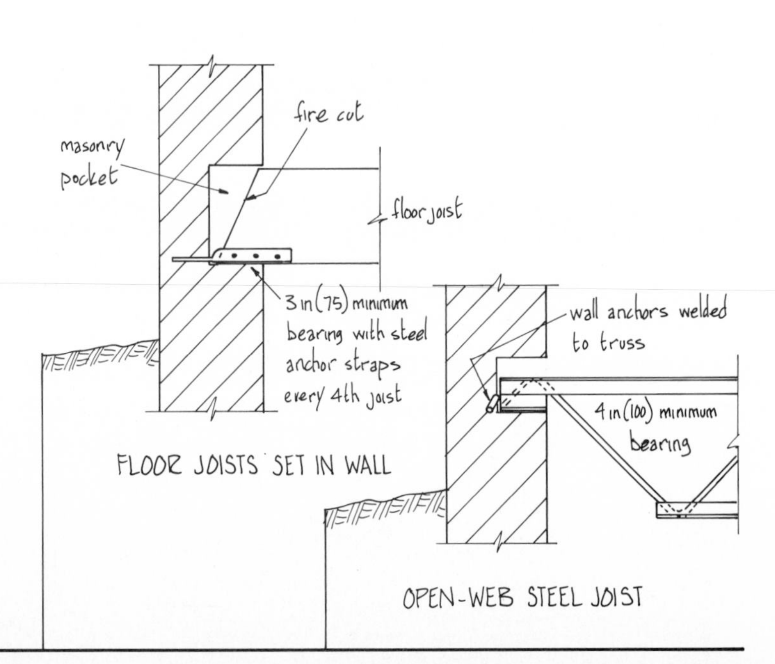

FLOOR JOISTS SET IN WALL

OPEN-WEB STEEL JOIST

The details above and right show several methods of anchoring floor framing systems to a foundation wall in order to provide sufficient lateral restraint.

FOUNDATION WALL REINFORCEMENT

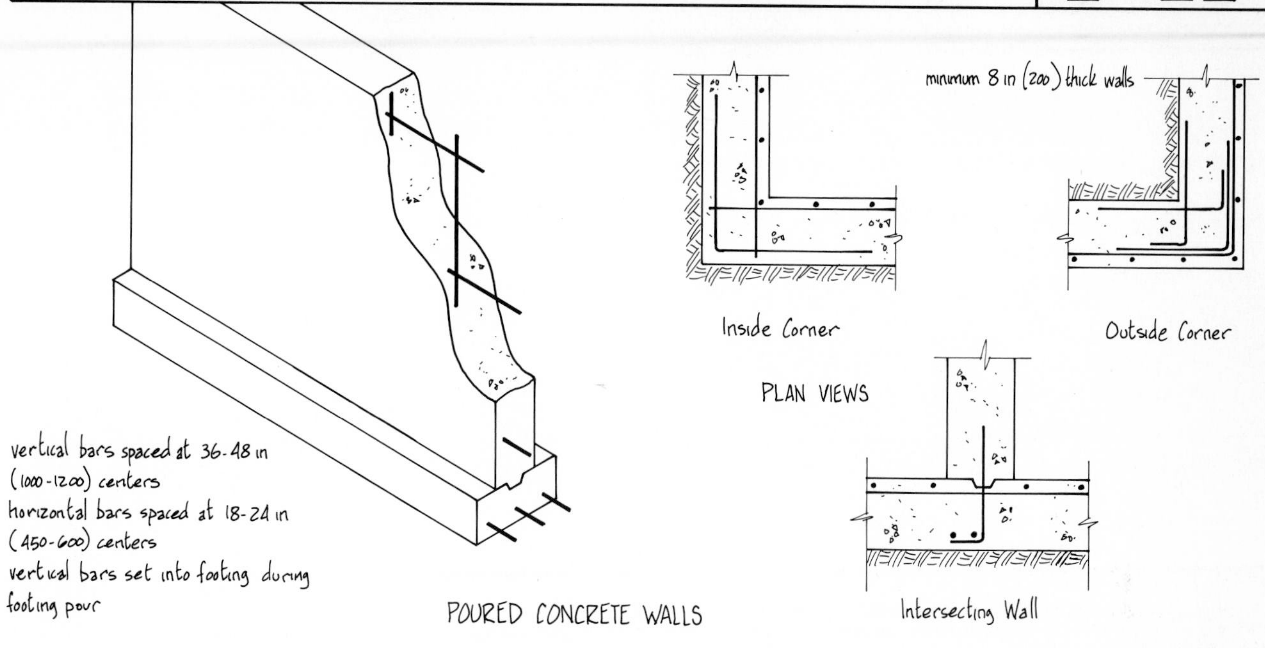

POURED CONCRETE WALLS

CONCRETE BLOCK WALLS

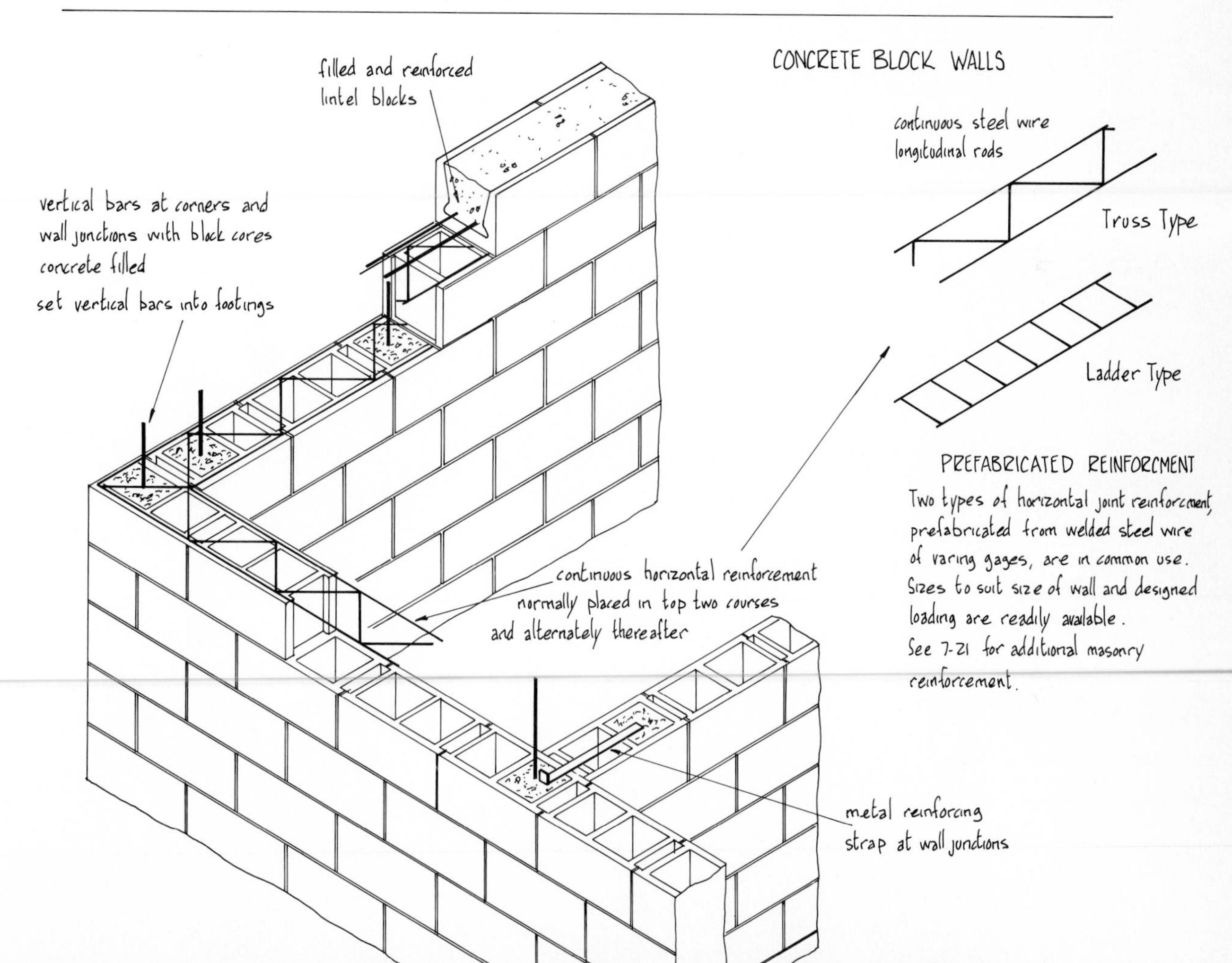

PREFABRICATED REINFORCMENT

Two types of horizontal joint reinforcement, prefabricated from welded steel wire of varing gages, are in common use. Sizes to suit size of wall and designed loading are readily available.

See 7-21 for additional masonry reinforcement.

RETAINING WALLS - GRAVITY TYPES

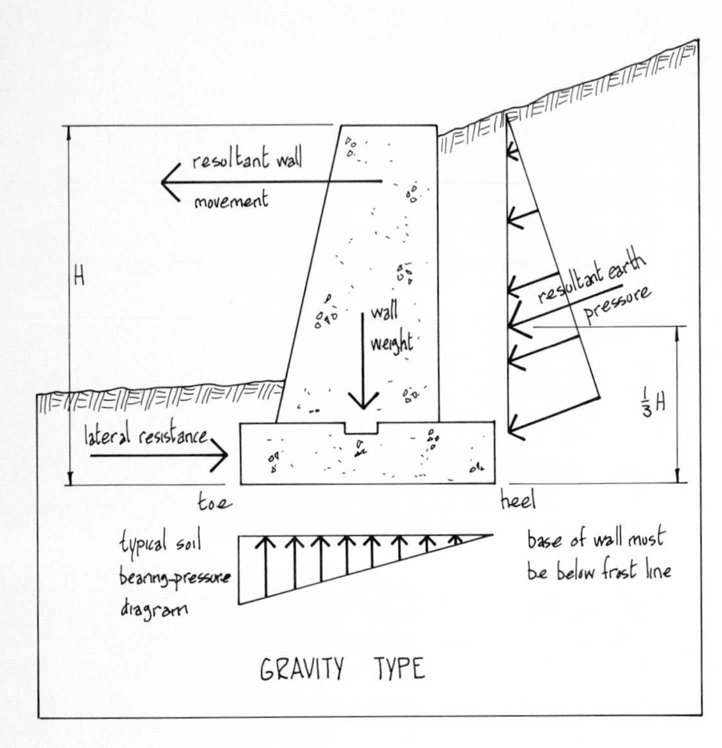

GRAVITY TYPE

The design of a retaining wall must consider the following:

- The wall must be safe against overturning and lateral sliding.
- The wall structure must resist bending moments and shear pressures imposed by earth loads.
- The bearing capacity of the foundation soil must be sufficient to resist undue settlement and distortion of the wall.

In essence, a retaining wall is similar to a basement wall without lateral support at the top. For design purposes, the wall is considered to be acting as a cantilever beam.

Retaining walls can be divided into two general categories:

- Gravity walls, the simplest type, rely on the wall's weight to provide the required stability. Height restrictions limit the practicality of this type since cross-sectional area increases significantly as height increases.
- Cantilever walls (see the next page), which use the retained earth to contribute stability to the structure. In doing so, the wall requires less material than a gravity wall but also requires careful design of its reinforcing members.

On this page and the next are shown typical examples of both retaining wall types.

RETAINING WALLS - CANTILEVER TYPES

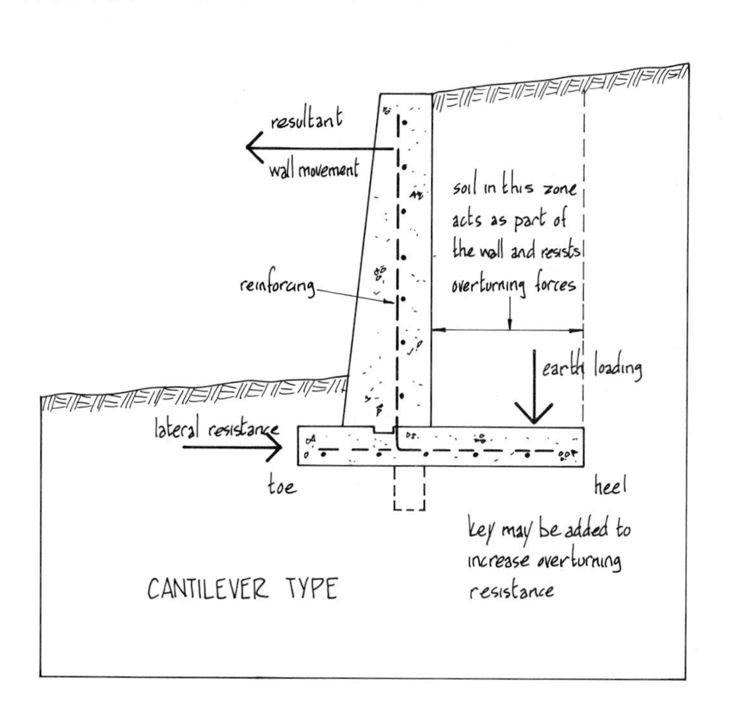

CANTILEVER TYPE

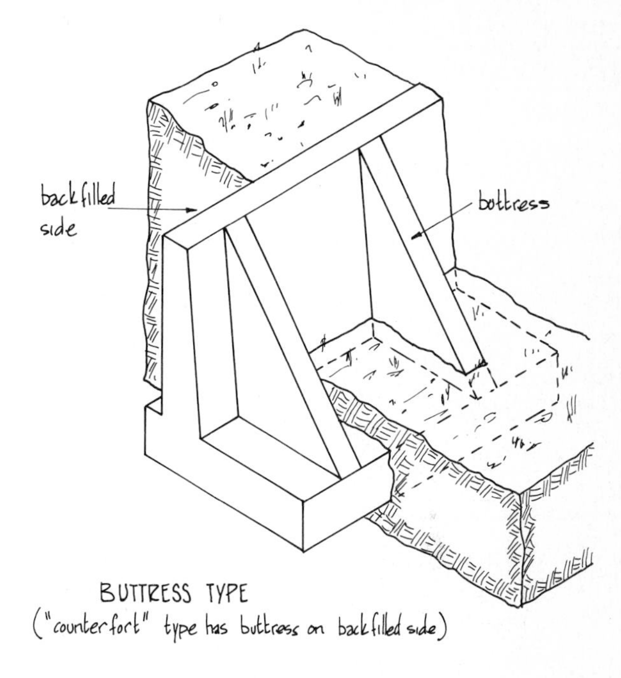

BUTTRESS TYPE
("counterfort" type has buttress on backfilled side)

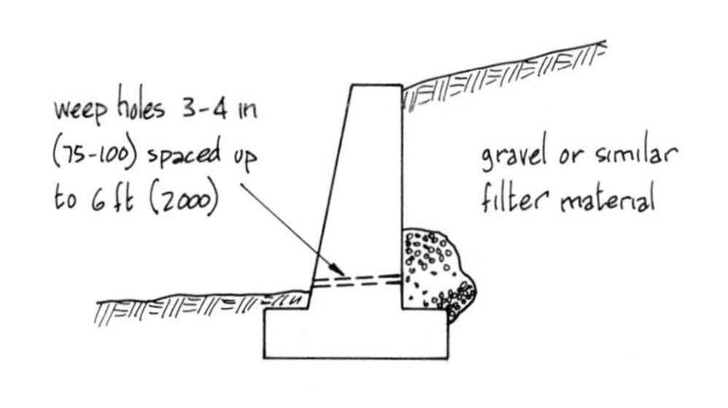

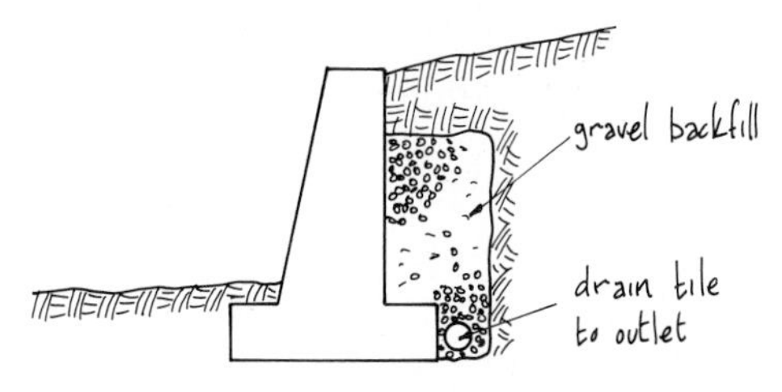

RETAINING WALL DRAINAGE

It is important to reduce hydrostatic pressures on the wall by providing a suitable drainage system. The two examples shown above are typical.

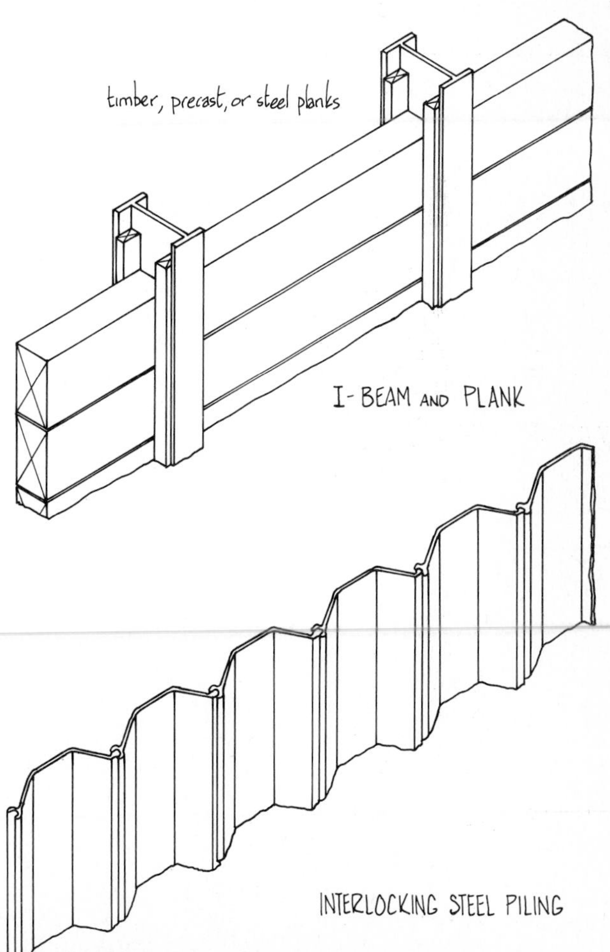

I-BEAM AND PLANK

INTERLOCKING STEEL PILING

FLOORS

CHAPTER PAGE TITLES

FLOORS

MATERIALS - WOOD PRODUCTS

Lumber products for construction purposes are classified into various groupings, reflecting size and usage. The chart below gives nominal and dressed sizes for commonly used lumber. Standard lengths for each product are normally in 2'-0" (0.26 m*) increments with a 6'-0" (1.83 m*) minimum. Some products are also available in 1'-0" (0.13 m*) increments.

PRODUCT	DESCRIPTION	NOMINAL THICKNESS INCHES	NOMINAL WIDTH INCHES	DRESSED THICKNESSES AND WIDTHS DRY in (mm)	DRESSED THICKNESSES AND WIDTHS UNSEASONED in (mm)
Framing	S4S	2	2	1 1/2 (38)	1 9/16 (40)
		3	3	2 1/2 (64)	2 9/16 (66)
		4	4	3 1/2 (89)	3 9/16 (91)
			6	5 1/2 (140)	5 5/8 (143)
			8	7 1/4 (184)	7 1/2 (191)
			10	9 1/4 (235)	9 1/2 (242)
			12	11 1/4 (286)	11 1/2 (293)
			over 12	off 3/4 (19)	off 1/2 (12)
Boards	S4S	1	as for	11/16 (17)	3/4 (19)
		1	framing	3/4 (19)	25/32 (20)
		1 1/4	nominal and	1 (25)	1 1/32 (26)
		1 1/2	dressed	1 1/4 (32)	1 9/32 (33)
				DRESSED - DRY THICKNESS	DRESSED - DRY WIDTH
Decking	2 in single T&G	2	6	1 1/2 (38)	5 (114)
			8		6 3/4 (171)
			10		8 3/4 (222)
			12		10 3/4 (273)
	3" and 4" double T&G	3	6	2 1/2 (64)	5 1/4 (133)
		4		3 1/2 (89)	
Flooring	(D&M) (S2S & CM)	3/8	2	5/16 (8)	1 1/8 (29)
		1/2	3	7/16 (11)	2 1/8 (54)
		5/16	4	9/16 (14)	3 1/8 (79)
		1	5	3/4 (19)	4 1/8 (105)
		1 1/4	6	1 (25)	5 1/8 (130)
		1 1/2		1 1/4 (32)	

Notes: Chart information is based on a moisture content of 19%.
Lumber with a 20% mc or above is termed "green" and if used, shrinkage will be excessive.
All metric dimensions are soft conversions.

The moisture content (mc) of wood both during and after construction is an important consideration. Seasoned, kiln-dried lumber sold for general construction has an mc of 15-19%. After construction the lumber will continue drying until reaching equilibrium with the surrounding air. The equilibrium mc for most of North America averages 8% with regional variations. As wood dries it also shrinks, a factor that must be allowed for in construction practice.

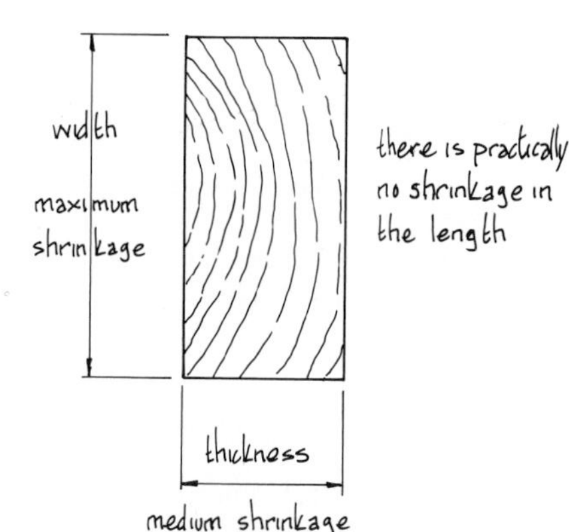

Typical Shrinkage Pattern

METRIC SIZING

Although the effect of SI metric on products and working dimensions was discussed in Chapter 2, a review as it relates to wood products is appropriate here:

CROSS-SECTION SIZE

- All dimensions are expressed in millimeters.
- The concept of "nominal size", (lumber size before being dressed or surfaced), is not used. Sizes given refer to actual thickness and width after surfacing.
- To date, no "hard metric" sizes have been standardized. All lumber is manufactured to Customary sizes but expressed in mm. (Note that some products, eg., plywood sheets, are offered in hard metric sizes.)
- Example: a 2 x 4 in (nominal) stud is actually 1 1/2 x 3 1/2 in when surfaced. The converted metric size is 38.1 x 88.9 but is referred to as 38 x 89.

LENGTH

- Lumber is cut to Customary lengths but expressed in meters to two decimal places.
- Example: a standard 8'-0" stud length is referred to as 2.44 m.

source: Western Wood Products Assoc.

JOIST SPANS

The sizing of wood beams and joists used in floors (and other areas of a building) is dependent on span, loading, spacing, species, and grade of lumber, and allowable deflection. When designing and choosing lumber sizes, applicable codes must be consulted together with a knowledge of locally available materials. The typical span table at the right shows allowable spans for spruce lumber. Other species and grades will have different properties and the appropriate span tables should be consulted.

If a joist size is chosen that is close to its limit in the span table, the floor will tend to have a bouncy or springy feel even though it is structurally adequate. To counteract this effect:

- Install bridging or blocking between joists (see 6-11).
- Shorten span, increase joist size, or space joists at closer centers.

TYPICAL FLOOR JOIST SPAN TABLE

SPECIES	GRADE	JOIST SIZE in (mm)	JOIST SPACING AND SPAN					
			12 in (300)		16 in (400)		24 in (600)	
			ft-in	(m)	ft-in	(m)	ft-in	(m)
Spruce	Select Structural	2×4 (38×89)	6-5	(1.98)	5-10	(1.79)	5-1	(1.57)
		2×6 (38×140)	10-1	(3.11)	9-2	(2.82)	8-0	(2.46)
		2×8 (38×184)	13-4	(4.10)	12-1	(3.72)	10-7	(3.25)
		2×10 (38×235)	17-0	(5.23)	15-5	(4.75)	13-6	(4.15)
		2×12 (38×286)	20-8	(6.36)	18-9	(5.78)	16-5	(5.05)
	No. 1	2×4 (38×89)	6-5	(1.98)	5-10	(1.79)	5-1	(1.57)
		2×6 (38×140)	10-1	(3.11)	9-2	(2.82)	7-9	(2.41)
		2×8 (38×184)	13-4	(4.10)	12-1	(3.72)	10-3	(3.18)
		2×10 (38×235)	17-0	(5.23)	15-5	(4.75)	13-1	(4.06)
		2×12 (38×286)	20-8	(6.36)	18-9	(5.78)	15-11	(4.93)
	No. 2	2×4 (38×89)	6-2	(1.91)	5-7	(1.73)	4-10	(1.49)
		2×6 (38×140)	9-9	(3.00)	8-7	(2.65)	7-0	(2.16)
		2×8 (38×184)	12-10	(3.96)	11-4	(3.49)	9-3	(2.85)
		2×10 (38×235)	16-5	(5.05)	14-6	(4.46)	11-10	(3.64)
		2×12 (38×286)	20-0	(6.15)	17-8	(5.42)	14-5	(4.43)
	No. 3	2×4 (38×89)	5-2	(1.58)	4-5	(1.37)	3-8	(1.12)
		2×6 (38×140)	7-5	(2.33)	6-5	(2.02)	5-3	(1.65)
		2×8 (38×184)	9-9	(3.07)	8-6	(2.66)	6-11	(2.17)
		2×10 (38×235)	12-6	(3.92)	10-10	(3.40)	8-10	(2.77)
		2×12 (38×286)	15-2	(4.77)	13-2	(4.13)	10-9	(3.37)

Notes: Floor joists - living quarters and supporting ceiling
Live load 40 psf (1.9 kN/m²)
Maximum deflection 1/360 of span

Joist spacing is based on a 4'-0" module with joists at 12", 16", or 24" centers.

4 spaces at 12"
12"
12" o/c
3 spaces at 16"
16"
16" o/c
2 spaces at 24"
24"
24" o/c

Customary

Joist spacing is based on a 1200 mm module with joists at 300, 400, and 600 mm centers.

4 spaces at 300
300
300 o/c
3 spaces at 400
400
400 o/c
2 spaces at 600
600
600 o/c

Metric

FLOOR JOIST SPACING

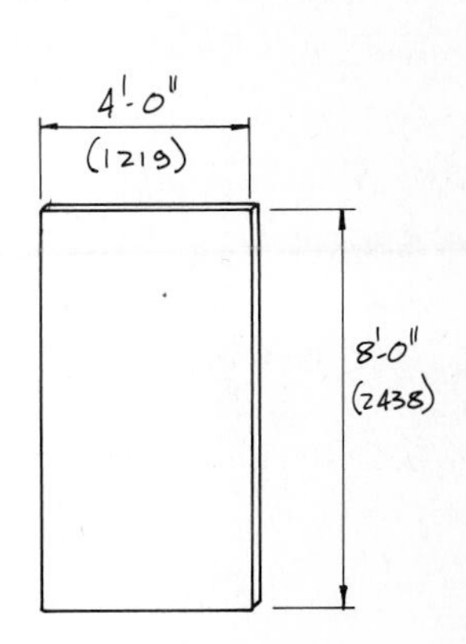

Customary Sheet

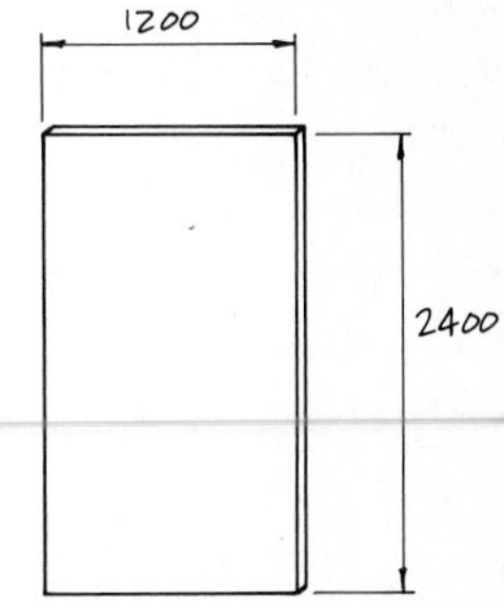

Metric Sheet

Comparison between Customary and metric building sheets (plywood, drywall, etc.). Note that a metric sheet is smaller than a Customary sheet. It cannot be used on Customary spaced studs, and vice versa.

source: CMHC

FLOORS

6-4 PLATFORM FRAMING

Before floor framing can be discussed in detail, it is first necessary to distinguish between the two types of frame construction: platform (or western) framing and balloon framing. Each method is described on this page and the following.

PLATFORM FRAMING

- The first floor is built on top of the foundation walls as a complete platform.
- First-floor stud walls are framed horizontally in units either on the platform or off-site. They are then raised into position by hand.
- The second floor is built as a separate platform on top of the first-floor stud walls.
- Second-floor stud walls are framed and erected as for the first floor.
- The roof structure is then erected on top of the second-floor walls (or additional-floor walls).

ADDITIONAL DETAILS

- High safety factor – most work is performed on a solid platform.
- Firestopping is provided by top and bottom plates of the stud walls.
- Shrinkage occurs evenly throughout the structure.

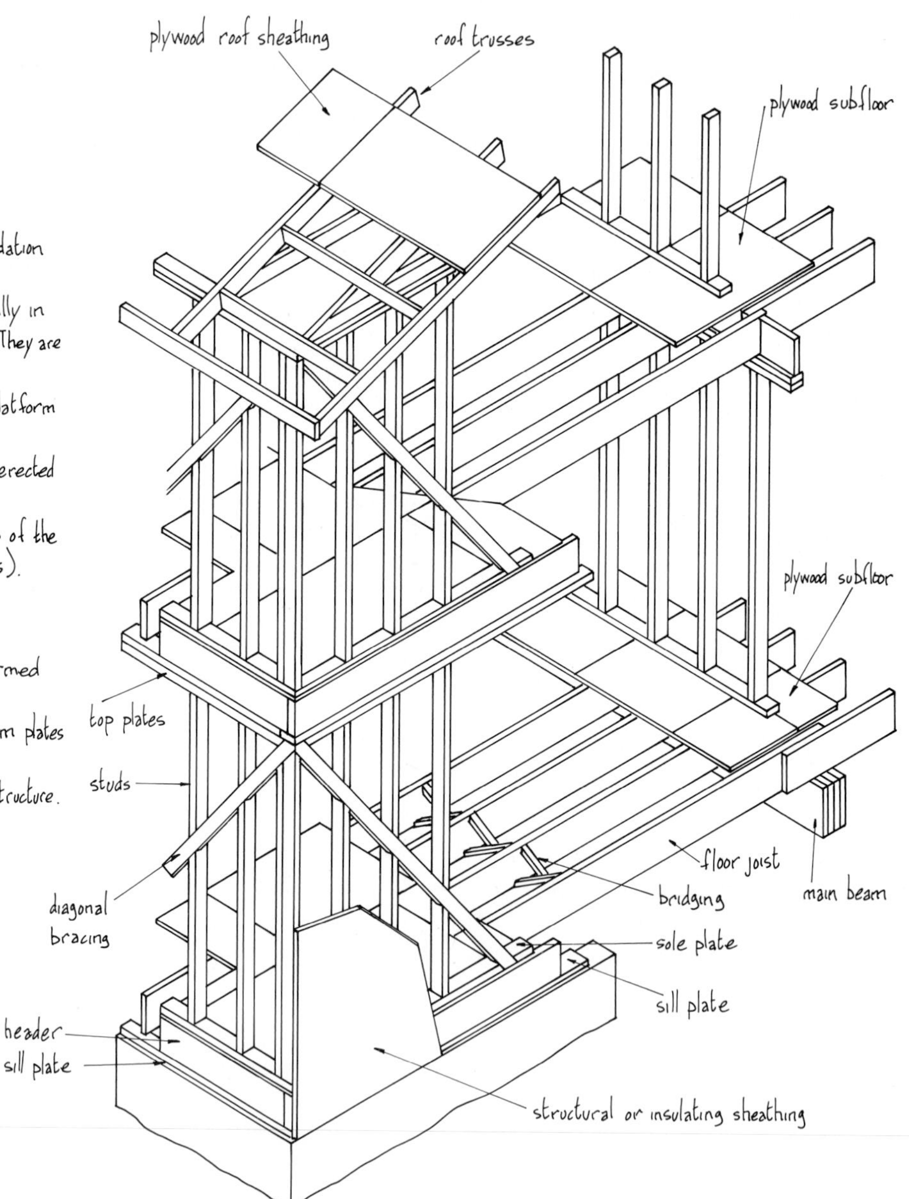

IMPORTANT NOTE

Because of its many advantages, platform framing is used almost exclusively in modern construction. Even so, the basics of balloon framing should still be studied since renovation work on older buildings will often reveal this type of framing.

Source: National Forest Products Assoc.

BALLOON FRAMING

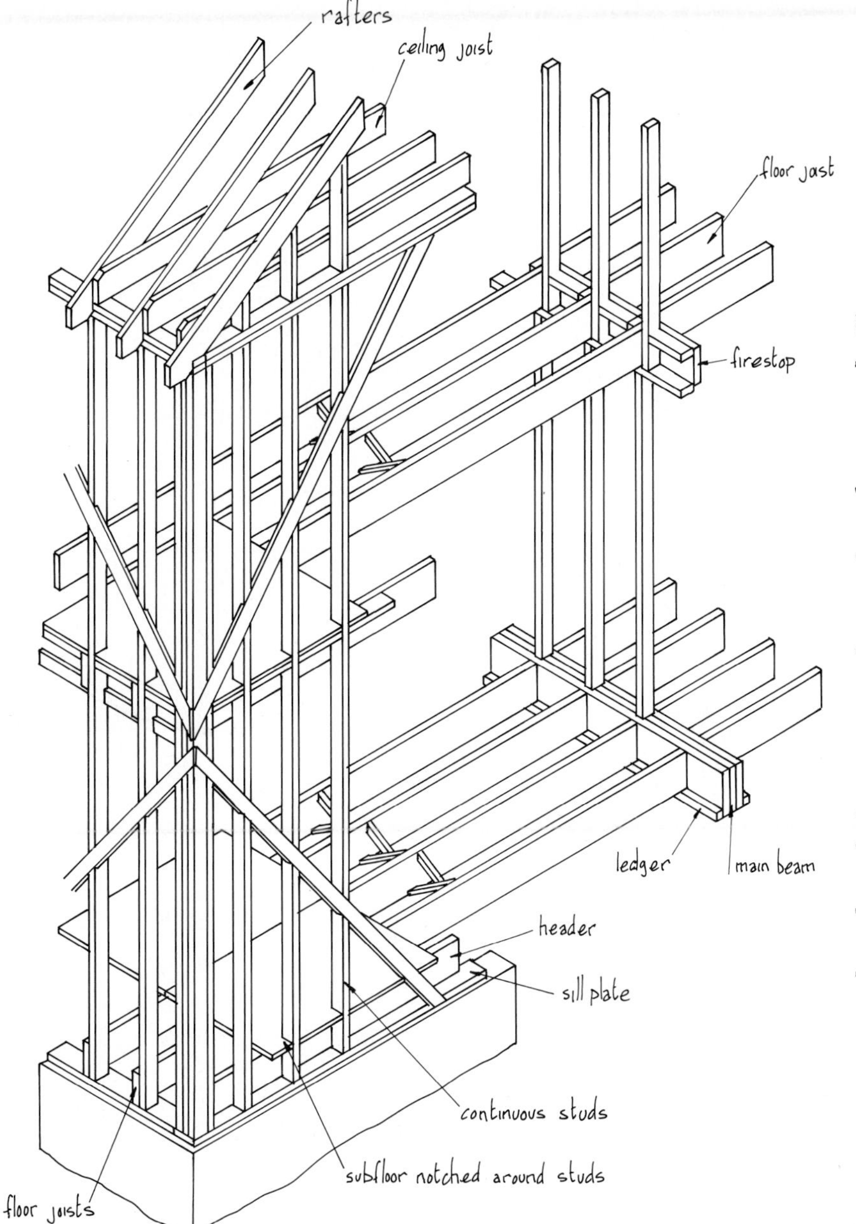

BALLOON FRAMING

- The studs used for exterior walls (and some interior walls) are continuous between the sill plate and the top plate that supports the roof structure.
- First-floor joists bear on the sill plate and are nailed to the vertical studs.
- Second-floor joists bear on a 1×4 in (19×89) ribbon let into the interior stud face. The joists are also nailed to the stud.
- Since stud spaces are continuous from foundation to roof, firestops must be installed in spaces at each floor level. Lumber of 2-in (38) thickness should be used.

ADDITIONAL DETAILS

- Shrinkage is minimal since lumber shrinks least along its length.
- Construction time is lengthened since pre-fabrication and rapid assembly is difficult.
- A modified balloon method can be used for sunken floors, dropped ceilings, or offset floors in split-level structures.

It should be noted that within each of the two types of framing there can be considerable variation in specific construction detail. Individual carpenters will employ a variety of methods to frame a structure, and because of this, the framing details that follow (floors, walls, and roofs), should be regarded as standard good practice but not the final word in construction technique. Other methods that are structurally sound, cost-effective, and meet code requirements are equally acceptable.
It should also be noted that some of the details illustrated on the following pages may not meet code requirements in every region. Consequently, local codes will always govern the final construction details.

source: National Forest Products Assoc.

PLATFORM FRAMING - TYPICAL 1ST FLOOR

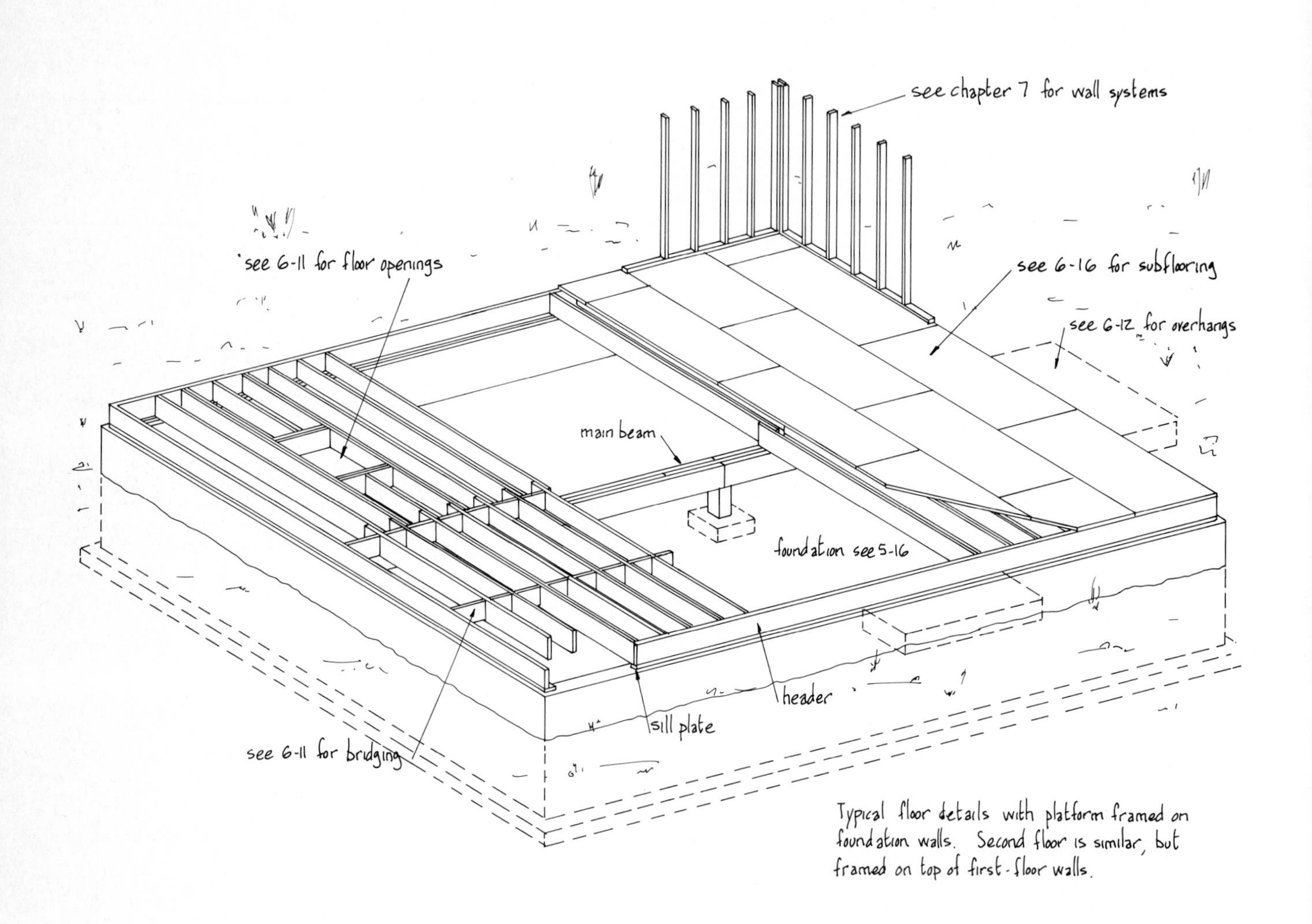

Typical floor details with platform framed on foundation walls. Second floor is similar, but framed on top of first-floor walls.

The floor is designed to act as a structural unit that must support superstructure dead loads and anticipated live loads. In addition, the floor must meet certain stiffness requirements so that deflection under moving live loads (i.e., movement of people) is kept to a minimum. The maximum deflection allowed by codes is normally 1/360 of the span. On a 15 ft (4570) span this amounts to a 1/2-in (13) deflection at the center of the span.

Lumber used for floor joists is generally of 2-in (38) thickness. Joist depth varies with span and loading but is usually 6, 8, 10, or 12 in (140, 184, 235, 286) (see 6-3 for a typical span table). Standard joist spacing is 16 in (400) o/c but in restricted-headroom areas, shallower joists at 12 in (300) centers may be used. Conversely, deeper joists at 24-in (600) centers may be used if headroom allows.

FLOOR SUPPORT BEAMS

Because floor joists must meet strength and stiffness requirements, the allowable joist span is often shorter than the width of the building. In such cases the joists must be supported between spans by beams running at right angles to the joists. The beams are in turn supported by the foundation walls and intermediate columns or posts (see 5-8 for column bases). Beams are usually either built-up members made from 2-in (38)-thick lumber or steel I-sections. Solid wood beams are sometimes used but large timbers are often hard to find. Check local codes for minimum beam sizes, allowable deflections, and recommended lateral restraint methods.

SPECIES AND GRADE	SUPPORTED JOIST LENGTH ft (m)	SIZE OF BUILT-UP BEAM in (mm)					
		3- 2×8 (38×184)	4- 2×8 (38×184)	3- 2×10 (38×235)	4- 2×10 (38×235)	3- 2×12 (38×286)	4- 2×12 (38×286)
		ft-in (m)	ft-in (m)	ft-in (m)	ft-in (m)	ft-in (m)	ft-in (m)
spruce (all species) no. 1 grade	8 (2.4)	6-4 (2.16)	8-0 (2.71)	8-0 (2.75)	10-3 (3.46)	9-9 (3.35)	12-5 (4.21)
	10 (3.0)	5-3 (1.80)	6-8 (2.28)	6-9 (2.30)	8-6 (2.91)	8-2 (2.79)	10-4 (3.53)
	12 (3.6)	4-7 (1.56)	5-9 (1.96)	5-10 (1.99)	7-4 (2.50)	7-2 (2.42)	8-11 (3.04)
	14 (4.2)	4-1 (1.39)	5-1 (1.73)	5-3 (1.77)	6-6 (2.21)	6-5 (2.16)	7-11 (2.69)
	16 (4.8)	3-9 (1.26)	4-9 (1.56)	4-9 (1.61)	5-10 (1.99)	5-10 (1.96)	7-2 (2.42)

Typical Span Table – Wood Composite Beams – Two Story Houses

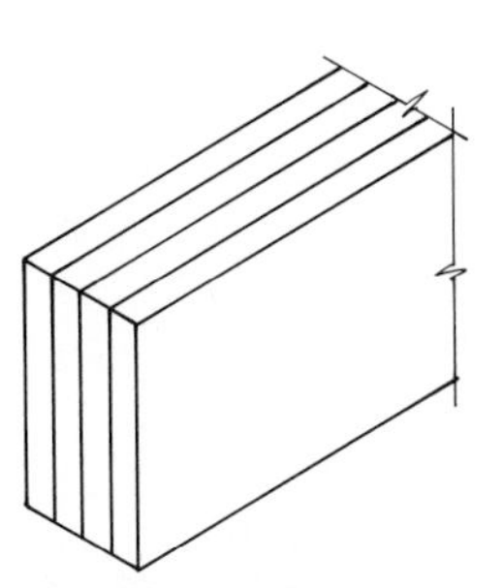

BUILT-UP BEAM
- 3 or 4 2-in (38) thick lumber.
- depth 6, 8, 10, or 12 in (140, 184, 235, 286).

Joints in built-up beams must be staggered and made at column positions.

NUMBER OF STORIES	MINIMUM DEPTH in (mm)	MIN. WEIGHT lb/ft (kg/m)	WIDTH OF FLOOR TO BE SUPPORTED				
			8 ft (2.4)	10 ft (3.0)	12 ft (3.6)	14 ft (4.2)	16 ft (4.8)
1½ or 2	4 (100)	7.7 (11.46)	8 (3.08)	7.5 (2.74)	7 (2.52)	6.5 (2.34)	6 (2.18)
	5 (125)	10.0 (14.88)	10.5 (3.89)	9.5 (3.48)	8.5 (3.18)	8 (2.94)	7.5 (2.74)
	6 (150)	12.5 (18.60)	12.5 (4.77)	11.5 (4.27)	10.5 (3.91)	9.5 (3.61)	9 (3.38)
	8 (200)	18.4 (27.38)	17.5 (6.63)	16 (5.96)	15 (5.44)	14 (5.03)	13 (4.72)

Typical Span Table – Steel I-Beams

Typical Beam Span Tables

The tables above are based on the "supported joist length" which is half the sum of the joist spans on both sides of the beam (see the illustration below). All metric sizes are soft conversions.

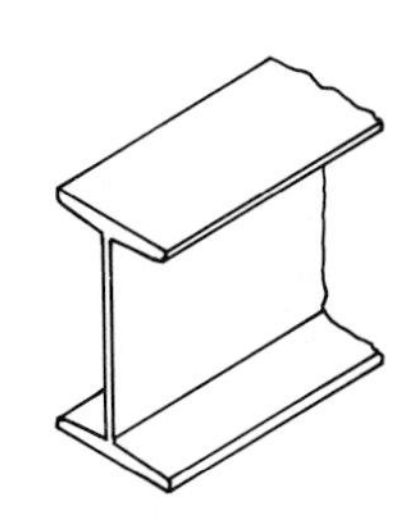

STEEL I-BEAM

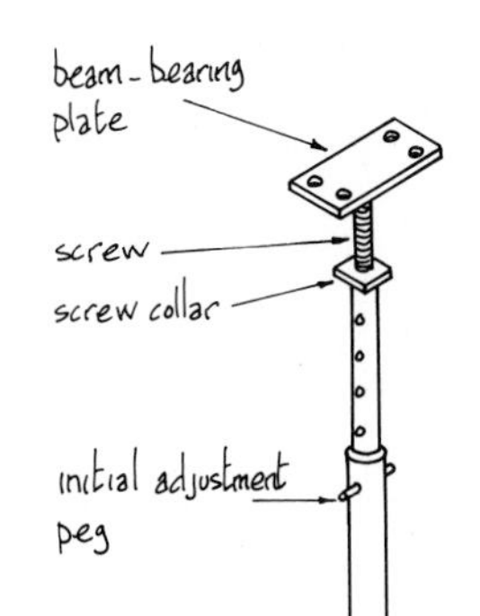

TYPICAL ADJUSTABLE STEEL COLUMN

Although solid wood or built-up wood columns can be used, steel columns are preferred so that height adjustments can be made when wood superstructure members shrink.

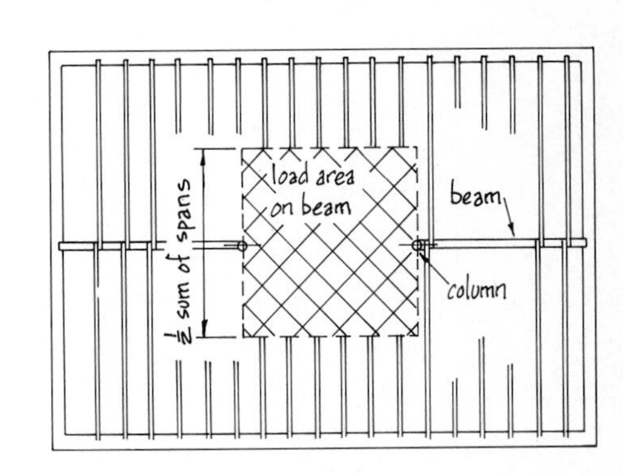

Floor Plan Showing Beam Load Area

See 5-11 and 5-13 for beam/foundation wall details.

source: CMHC

BEAM - JOIST CONNECTIONS

A selection of beam/joist connections is illustrated on this page. In all cases the joists must be securely nailed to each other, the beam, and/or whatever support materials are used.

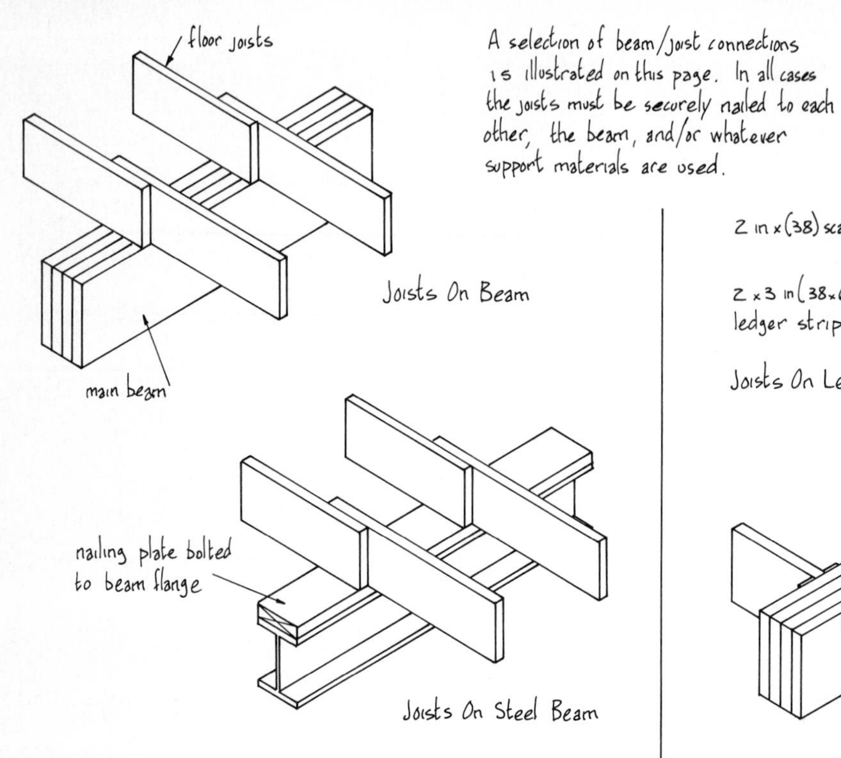

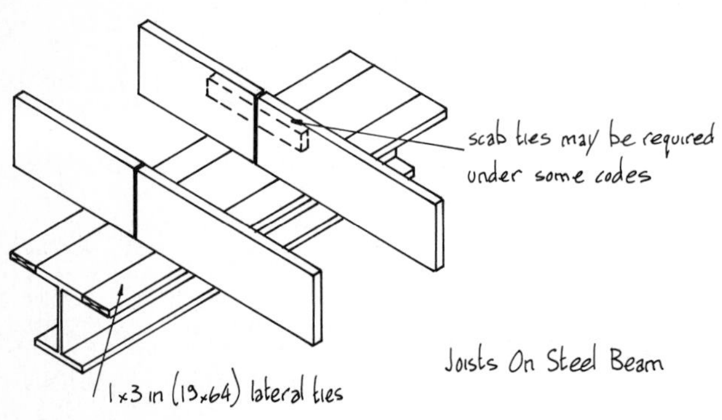

JOISTS OVER BEAMS

The simplest beam/joist connection, used where the basement height will provide adequate headroom below the beam.

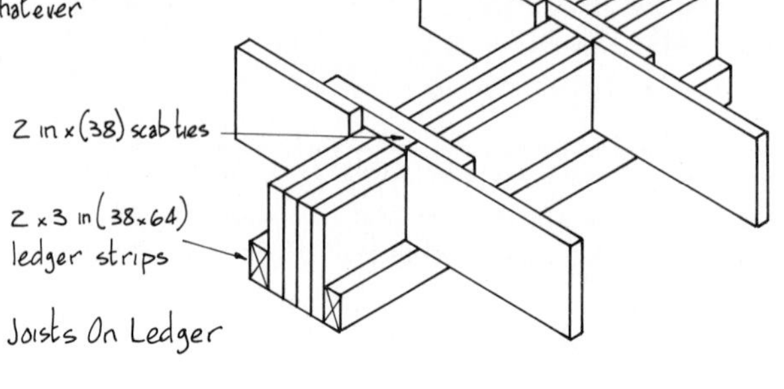

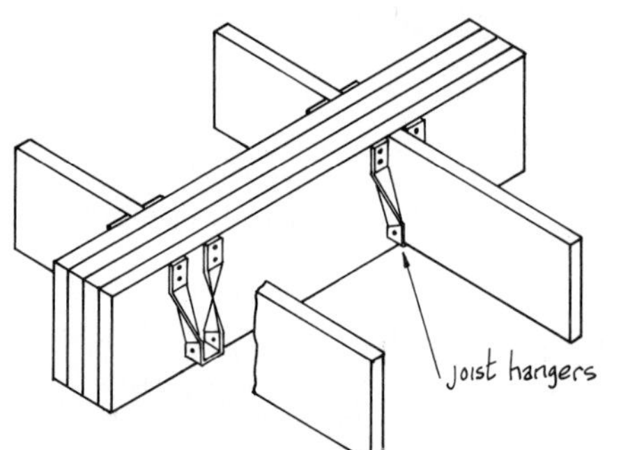

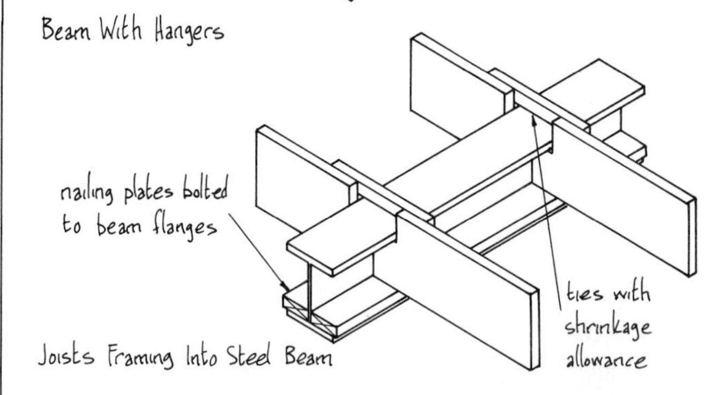

JOISTS INTO BEAMS

More labor-intensive and generally used when greater headroom is required under the beam.

In addition to beams, load-bearing stud walls can be used to support floor joists. For normal loading, standard 2×4-in (38×89) at 16-in (400) o/c studs are quite adequate provided that a separate footing or deepened floor slab is specified (see 4-18, center bearing walls).

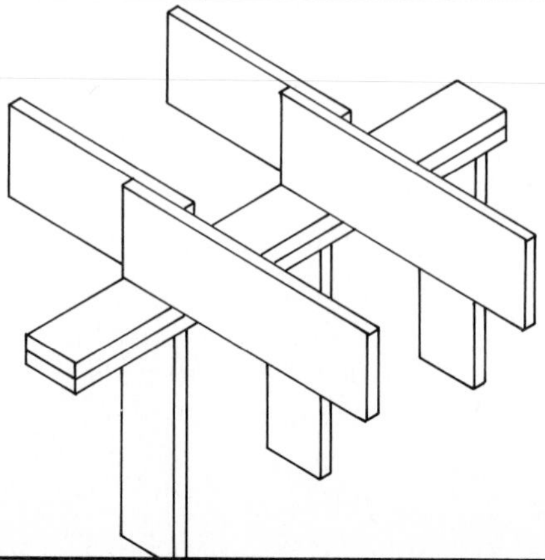

FOUNDATION WALL - JOIST CONNECTIONS

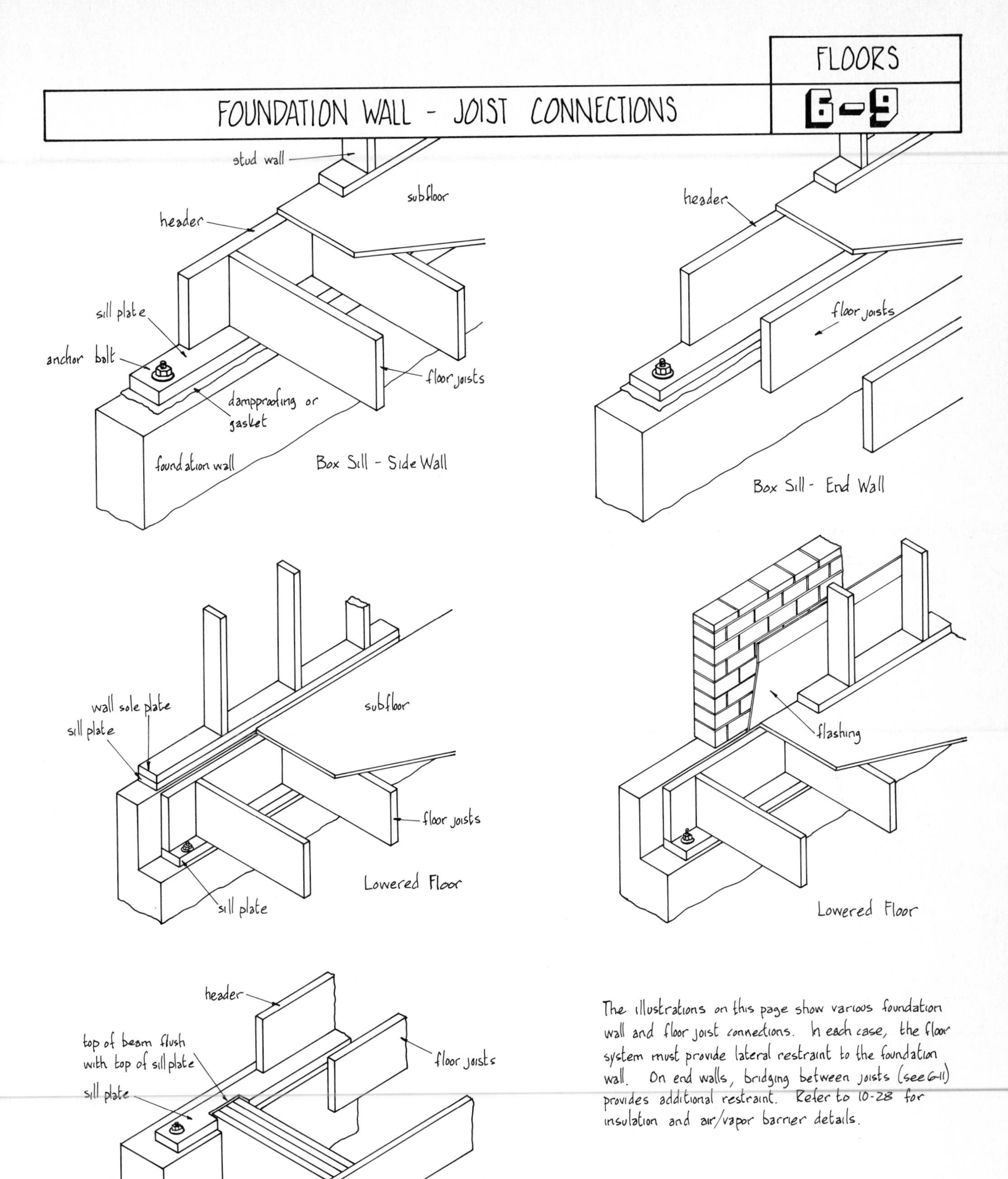

The illustrations on this page show various foundation wall and floor joist connections. In each case, the floor system must provide lateral restraint to the foundation wall. On end walls, bridging between joists (see 6-11) provides additional restraint. Refer to 10-28 for insulation and air/vapor barrier details.

BALLOON FRAME - PWF - SILL PLATES

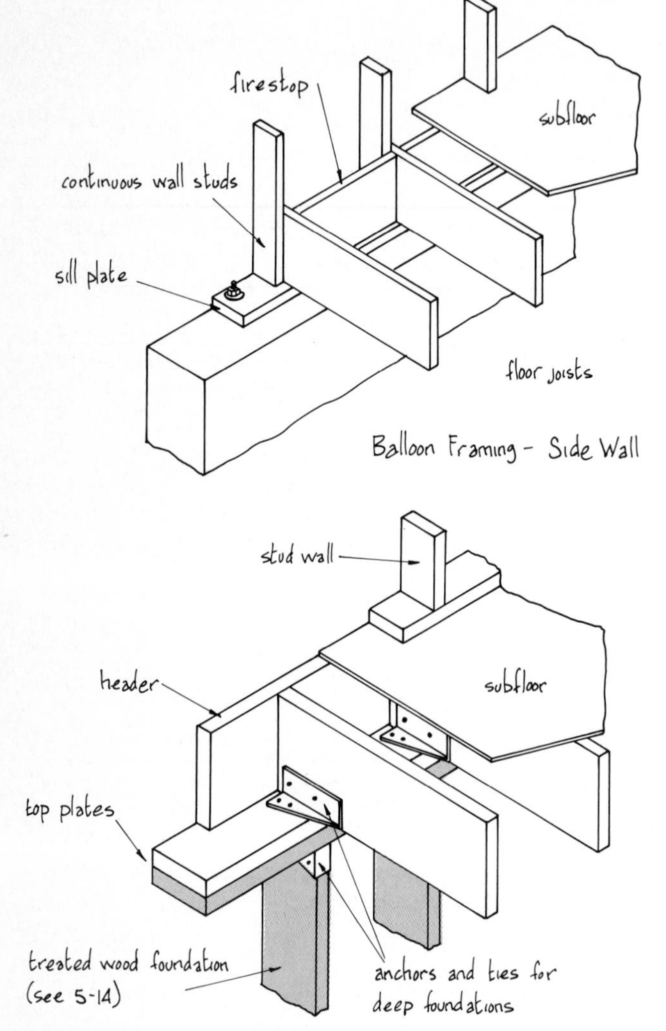

Balloon Framing - Side Wall

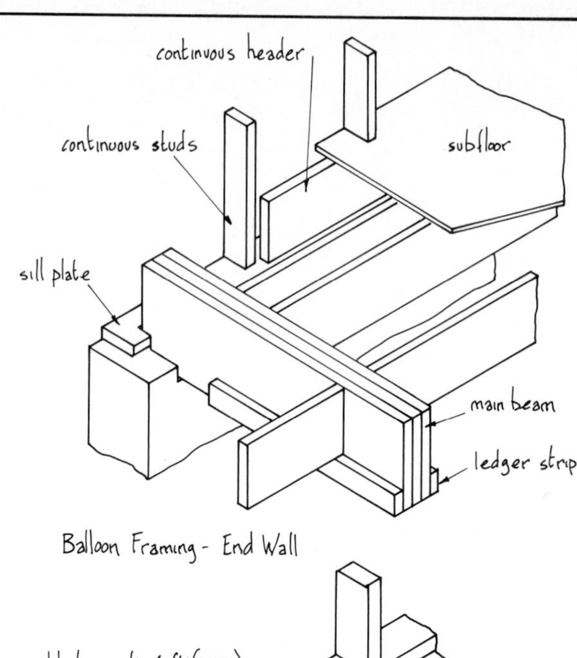

Balloon Framing - End Wall

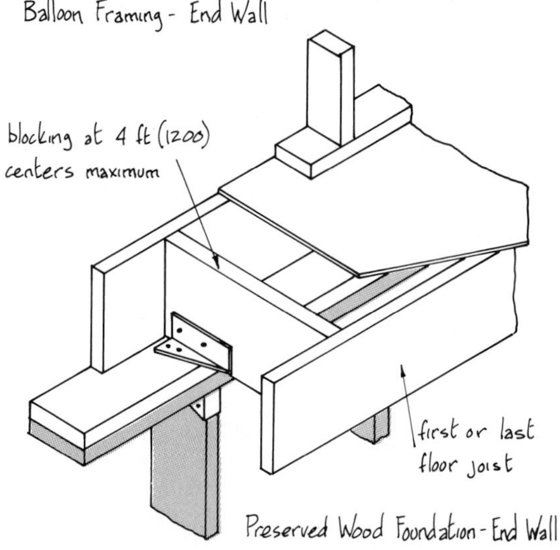

Preserved Wood Foundation - End Wall

Preserved Wood Foundation - Side Wall

On the right are four options for sealing and leveling sill plates, in addition to the standard method of positioning a strip of polyethelyne or roofing felt between the wall and the sill.

- Use caulking, mineral wool, or foamed plastic for sealing.
- Use a mortar bed for leveling the sill on uneven walls.
- A termite shield is required in some regions, with the addition of a treated-wood sill and soil poisoning in high termite-damage areas.

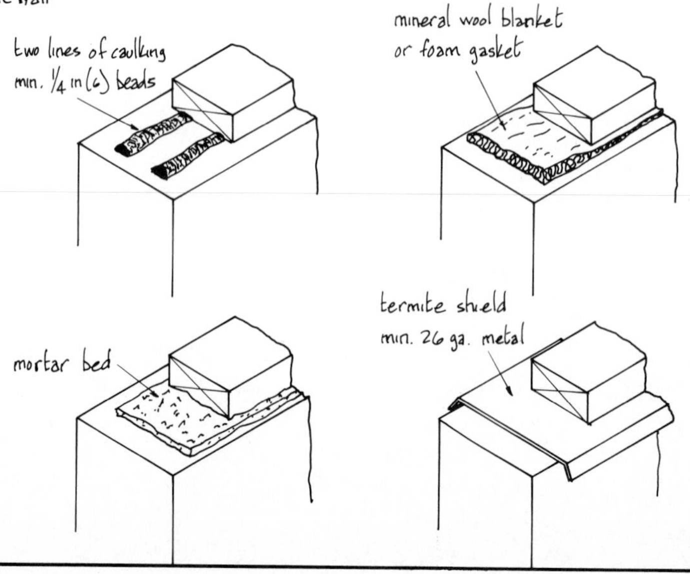

FLOOR OPENINGS - JOIST RESTRAINT

Floor Openings

- Trimmer joists run parallel to joist span and support loads imposed by headers. Trimmer joists must be doubled when supporting headers that exceed 32 in. (800) span. Trimmers that support headers over 6ft. 8in. (2000) span must be engineered.
- Header joists run at right angles to joist span and support loads imposed by tail joists. Headers longer than 4ft. (1200) span must be doubled. Headers over 10ft. (3200) span must be engineered.
  - Tail joists are floor joists shortened by the opening. When tail joist span exceeds 6ft. (1800) use metal joist hangers at headers.

The use of metal joist hangers for all header and tail joist connections is recommended whatever the size of the span.

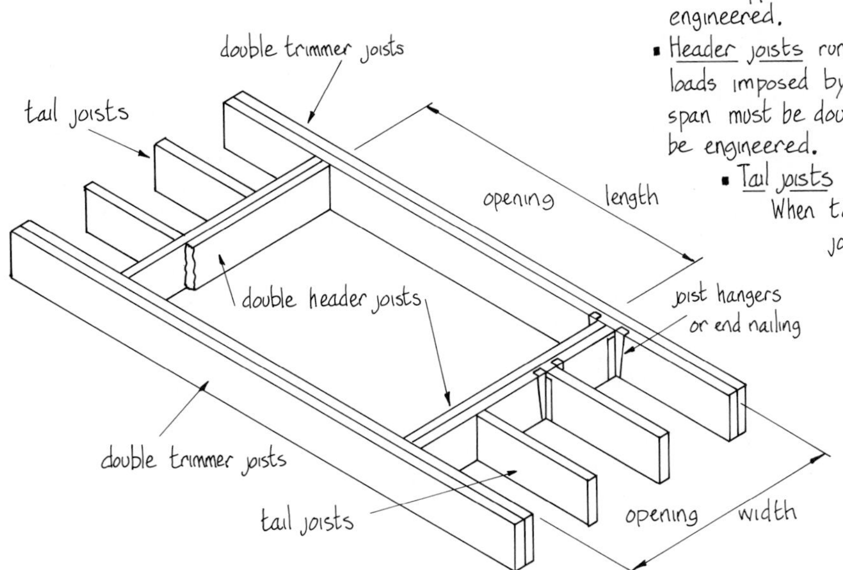

Joist Restraint — Bridging

Floor joists must be tied to each other so that uneven loads can be distributed laterally across the floor. The bridging also keeps joists in alignment as drying proceeds and loads are applied. Restraint should be provided at joist end supports and at maximum 8ft (2400) between supports. Bridging also acts with the subfloor (see 6-16) to stiffen the floor.

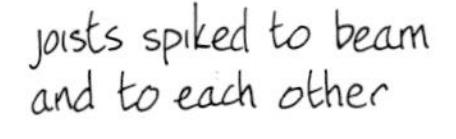

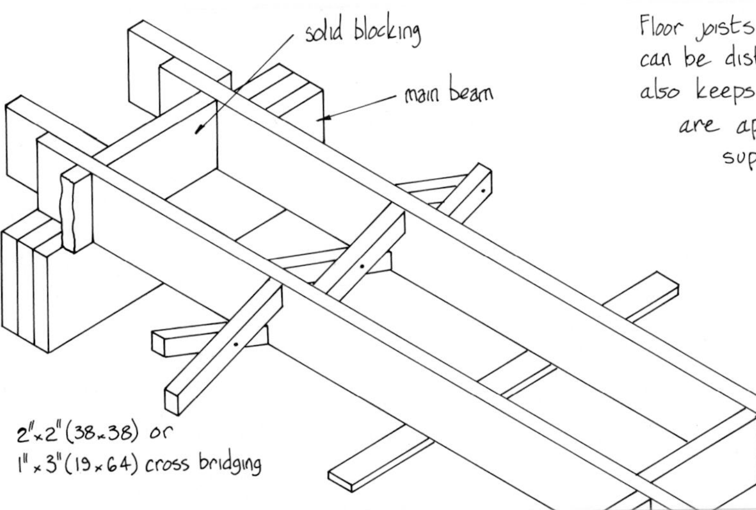

JOIST RESTRAINT

FLOORS

6-12

OVERHANGS AND CUTOUTS

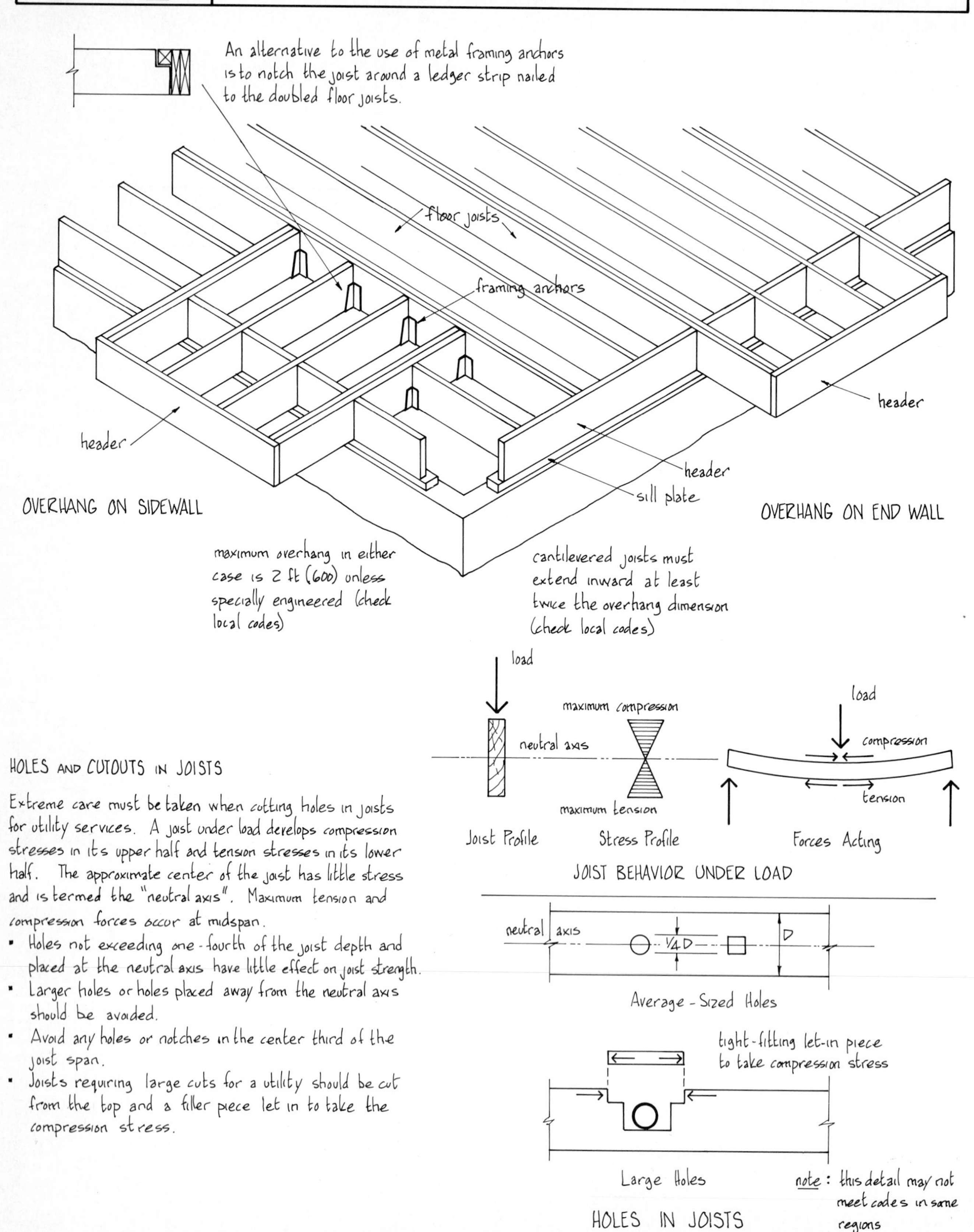

HOLES AND CUTOUTS IN JOISTS

Extreme care must be taken when cutting holes in joists for utility services. A joist under load develops compression stresses in its upper half and tension stresses in its lower half. The approximate center of the joist has little stress and is termed the "neutral axis". Maximum tension and compression forces occur at midspan.

- Holes not exceeding one-fourth of the joist depth and placed at the neutral axis have little effect on joist strength.
- Larger holes or holes placed away from the neutral axis should be avoided.
- Avoid any holes or notches in the center third of the joist span.
- Joists requiring large cuts for a utility should be cut from the top and a filler piece let in to take the compression stress.

SECOND-FLOOR DETAILS – FIRESTOPPING

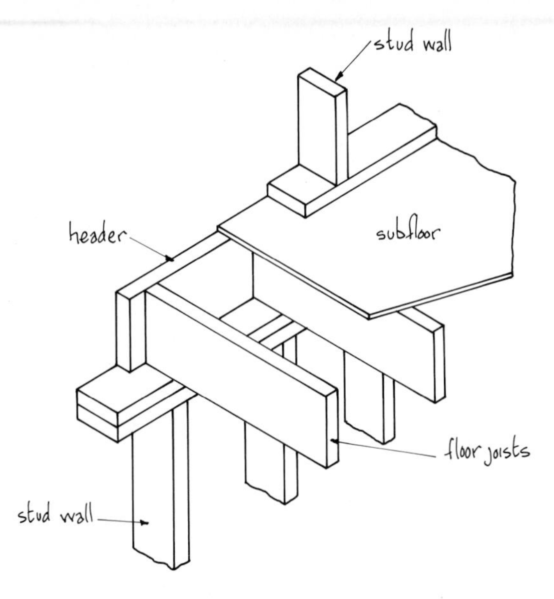

Second Floor Framing at Exterior Wall

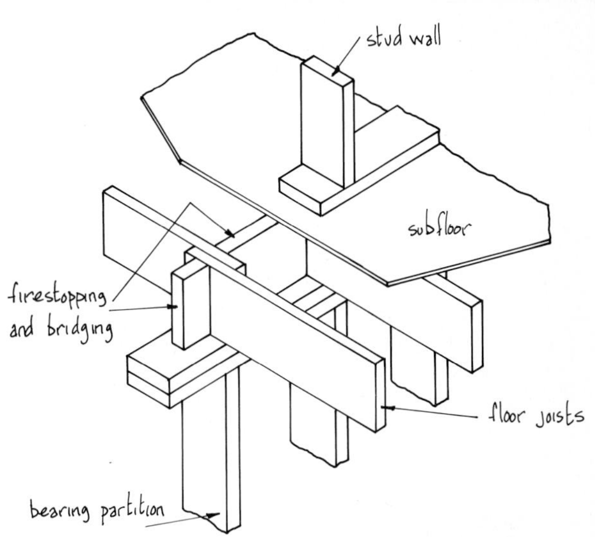

Second Floor Framing at Bearing Interior Partition

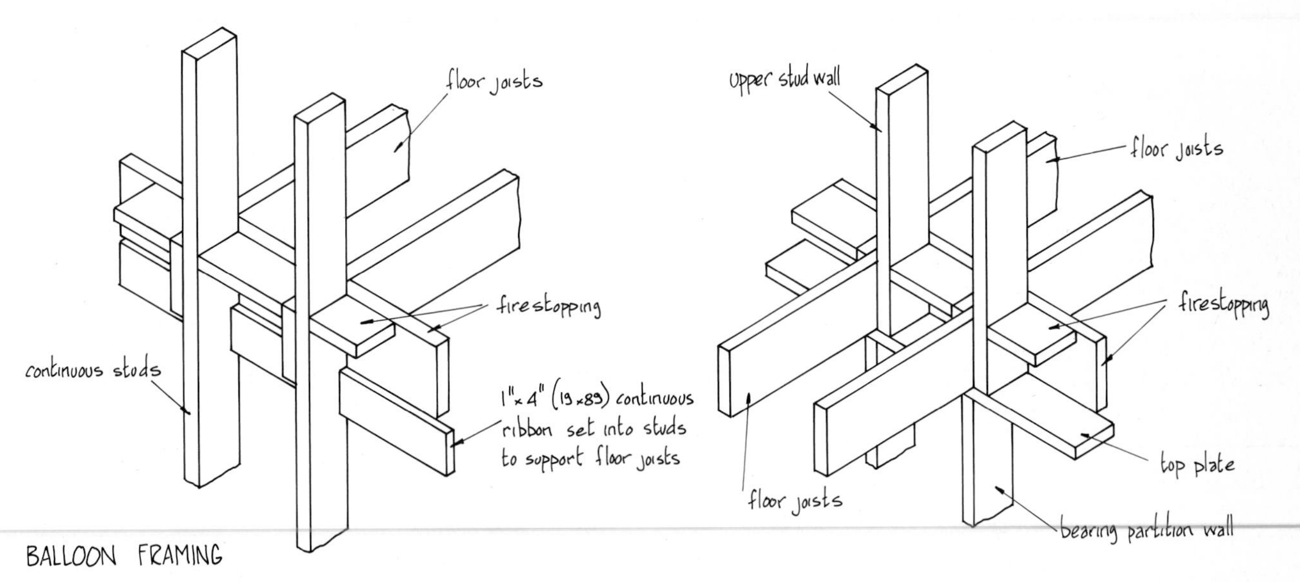

BALLOON FRAMING

Second Floor – Exterior Wall

BALLOON FRAMING

Second Floor Framing at Bearing Partition

The illustrations on this page show typical details of second-floor floors at exterior walls and interior partitions. It is important to note that firestopping with 2-in (38) lumber is essential.

FLOORS WITH BEARING AND NON-BEARING PARTITIONS

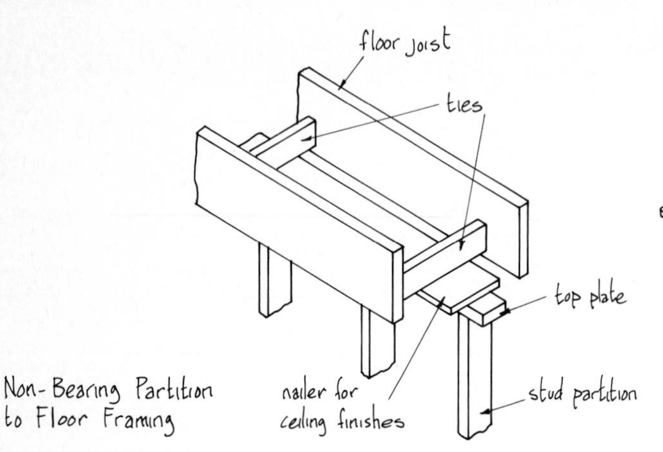

Non-Bearing Partition to Floor Framing

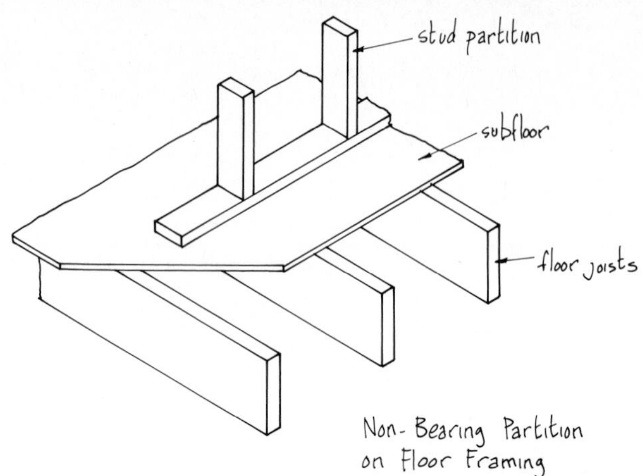

Non-Bearing Partition on Floor Framing

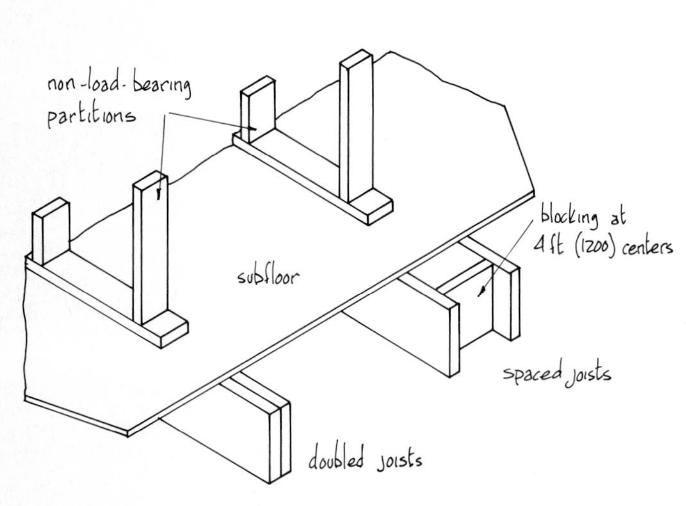

Non-Bearing Partitions Parallel to Floor Joists

Bearing Partitions Parallel to Floor Joists

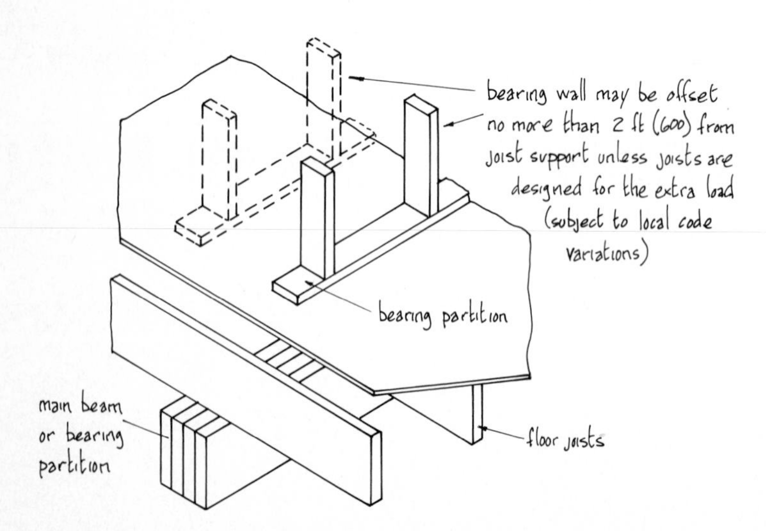

Bearing Partition at Right-Angles to Floor Joists

The drawings on this page illustrate structural support for bearing and non-bearing second-floor partitions. In each case, two major requirements must be met:

- All loads must be transmitted to the foundation system through structural members that are adequately designed and positioned.
- Allowance must be made for attachment of finished ceilings at partition heads.

STAIR FRAMING

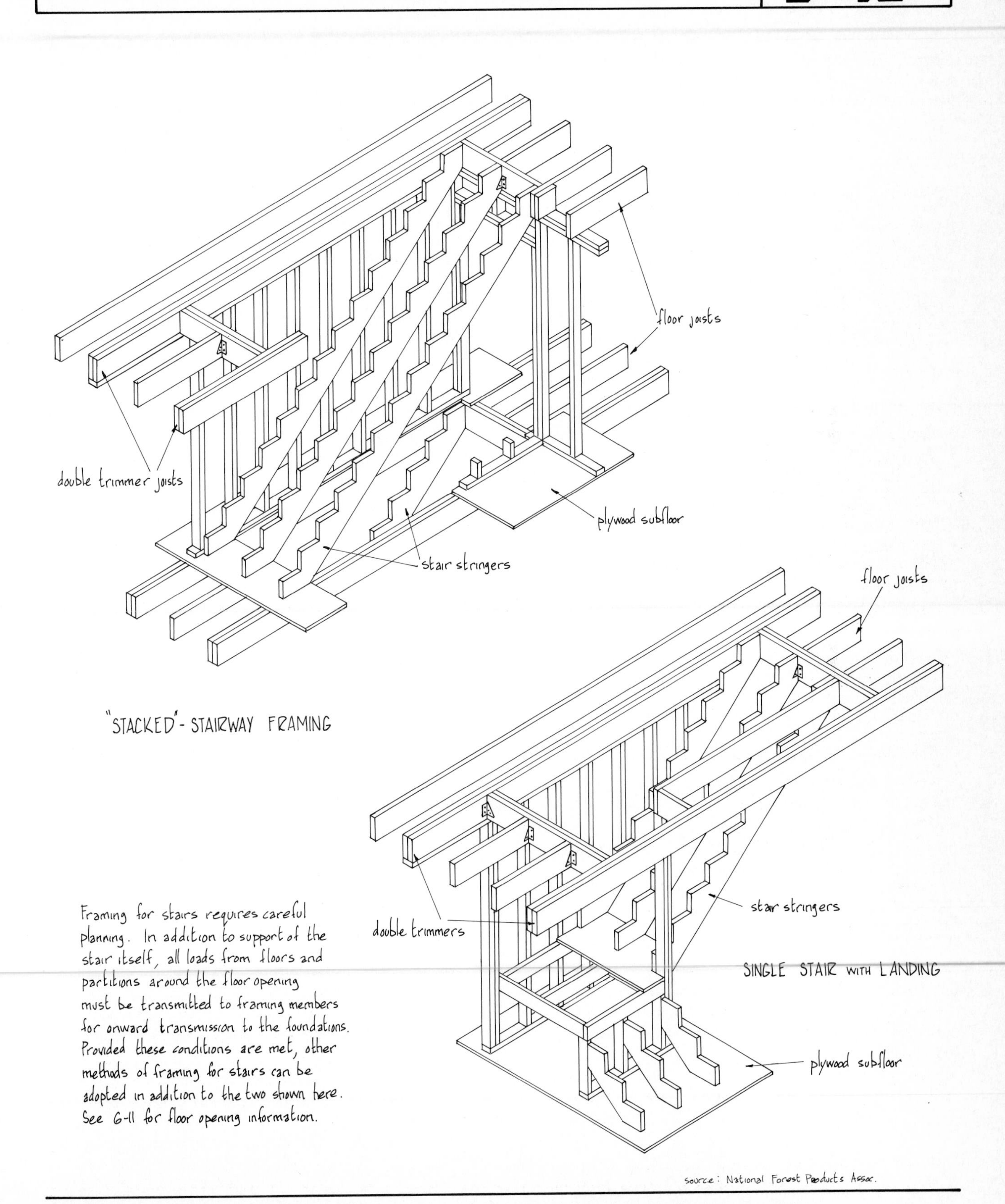

Framing for stairs requires careful planning. In addition to support of the stair itself, all loads from floors and partitions around the floor opening must be transmitted to framing members for onward transmission to the foundations. Provided these conditions are met, other methods of framing for stairs can be adopted in addition to the two shown here. See 6-11 for floor opening information.

source: National Forest Products Assoc.

FLOORS

SUBFLOORING

Subflooring normally consists of sheet material (plywood, particleboard) or solid material (boards). Plywood is the preferred material for most applications since it offers important advantages in strength, application, and performance. Subflooring must be considered as a structural member of the floor system and designed accordingly, especially with regard to the subfloor span between floor joists.

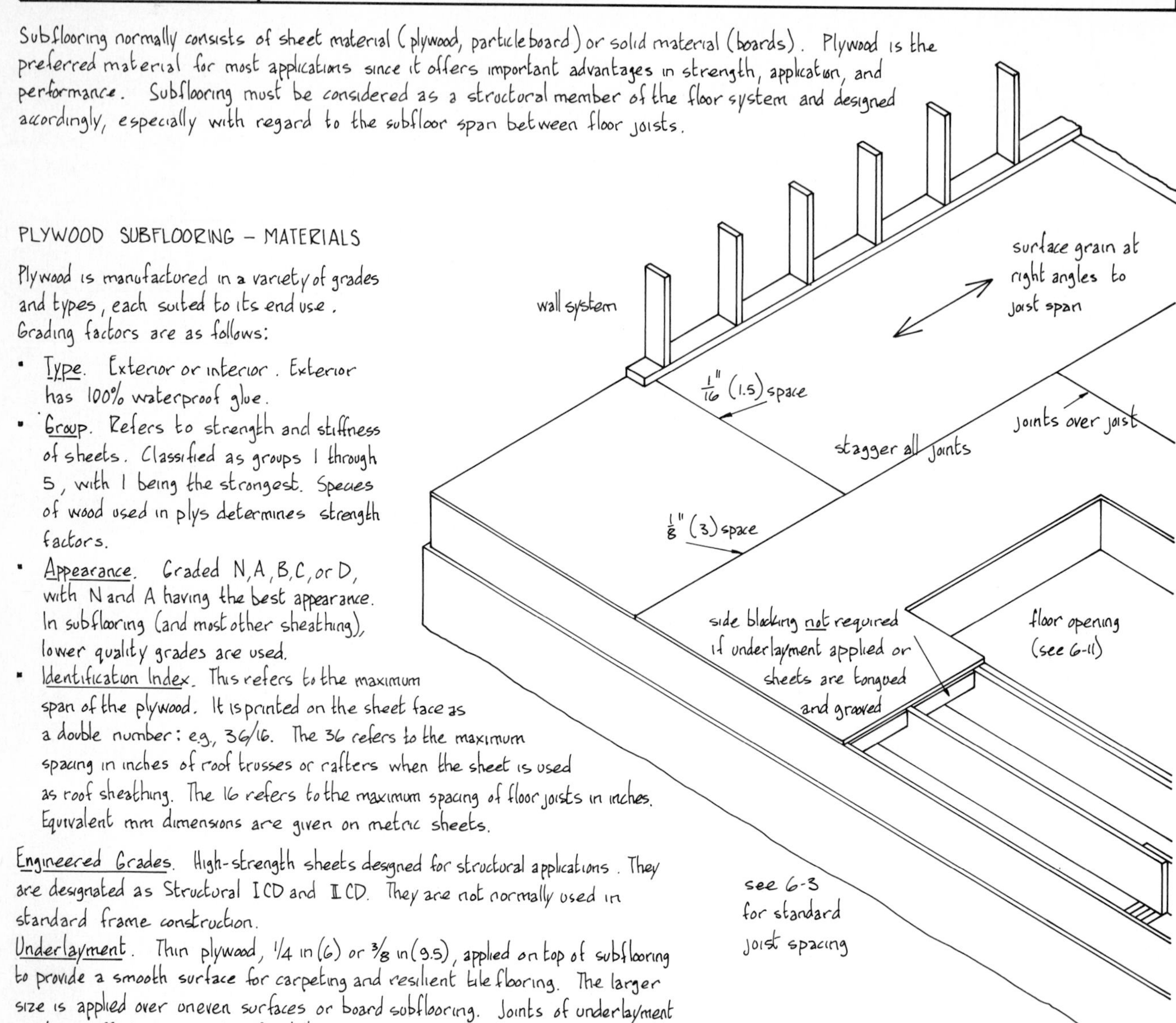

PLYWOOD SUBFLOORING – MATERIALS

Plywood is manufactured in a variety of grades and types, each suited to its end use. Grading factors are as follows:

- Type. Exterior or interior. Exterior has 100% waterproof glue.
- Group. Refers to strength and stiffness of sheets. Classified as groups 1 through 5, with 1 being the strongest. Species of wood used in plys determines strength factors.
- Appearance. Graded N, A, B, C, or D, with N and A having the best appearance. In subflooring (and most other sheathing), lower quality grades are used.
- Identification Index. This refers to the maximum span of the plywood. It is printed on the sheet face as a double number: e.g., 36/16. The 36 refers to the maximum spacing in inches of roof trusses or rafters when the sheet is used as roof sheathing. The 16 refers to the maximum spacing of floor joists in inches. Equivalent mm dimensions are given on metric sheets.

Engineered Grades. High-strength sheets designed for structural applications. They are designated as Structural I CD and II CD. They are not normally used in standard frame construction.

Underlayment. Thin plywood, 1/4 in (6) or 3/8 in (9.5), applied on top of subflooring to provide a smooth surface for carpeting and resilient tile flooring. The larger size is applied over uneven surfaces or board subflooring. Joints of underlayment must be offset from joints of subfloor.

Combined Subfloor–Underlayment. Provides for both systems in a single sheet and has tongue and groove edges.

PLYWOOD SUBFLOOR SYSTEMS

1. Standard plywood subflooring used as a base for wood strip flooring. Requires support blocking under side joints.
2. Standard plywood subflooring plus underlayment as base for carpet, resilient flooring, etc. No blocking required at sides.
3. Combination subfloor-underlayment sheet. T & G edges requiring no blocking at sides.

GLUED SUBFLOORS

To greatly increase floor stiffness and reduce squeaks, subfloors can be attached directly to floor joists using elastomeric glue.

- The joists and plywood act as a composite T-beam structural unit, minimizing differential deflection between joists.
- Stiffness may be increased beyond 50%, depending on floor composition.
- Labor costs are reduced since only minimal nailing is required.
- Additional stiffness is acquired by gluing tongue-and-groove joints.
- Under some codes, bridging may be omitted if the subfloor is glued.

SUBFLOORING

PANEL IDENTIFICATION INDEX	PLYWOOD THICKNESS in	(mm)	MAX. SPAN in	(mm)
30/12 (760/300)	$\frac{5}{8}$	(15.5)	12	(300)
32/16 (800/400)	$\frac{1}{2}, \frac{5}{8}$	(12.5, 15.5)	16	(400)
36/16 (900/400)	$\frac{3}{4}$	(18.5)	16	(400)
42/20 (1060/500)	$\frac{5}{8}, \frac{3}{4}, \frac{7}{8}$	(15.5, 18.5, 22)	20	(500)
48/24 (1200/600)	$\frac{3}{4}, \frac{7}{8}$	(18.5, 22)	24	(600)
$1\frac{1}{8}$ (28.5) groups 1 & 2	$1\frac{1}{8}$	(28.5)	48	(1200)
$1\frac{1}{4}$ (32) groups 3 & 4	$1\frac{1}{4}$	(32)	48	(1200)

PLYWOOD SUBFLOORING

PLYWOOD GRADES AND SPECIES GROUP	APPLICATION	MINIMUM PLYWOOD THICKNESS
groups 1,2,3,4,5 underlayment INT-APA (with interior, intermediate, or exterior glue), or underlayment EXT-APA (C-C plugged)	over plywood subfloor	$\frac{1}{4}$ in (6)
	over lumber subfloor or other uneven surface	$\frac{3}{8}$ in (9.5)
same grades as above, but group 1 only	over lumber floor up to 4 in (89) wide. Face grain must be perpendicular to boards	$\frac{1}{4}$ in (6)

PLYWOOD UNDERLAYMENT

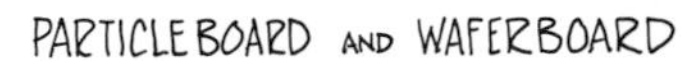

PARTICLEBOARD AND WAFERBOARD

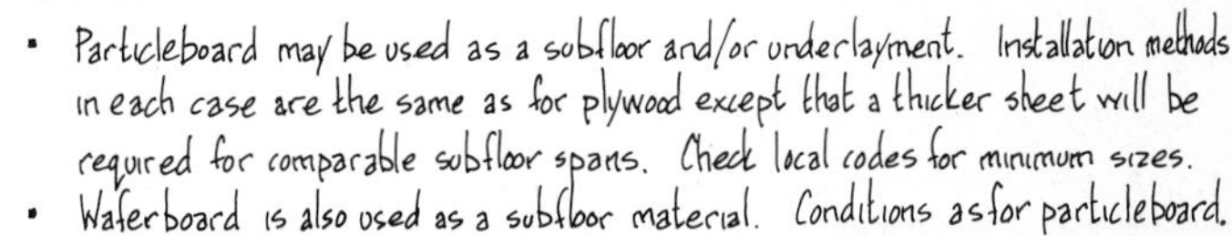

- Particleboard may be used as a subfloor and/or underlayment. Installation methods in each case are the same as for plywood except that a thicker sheet will be required for comparable subfloor spans. Check local codes for minimum sizes.
- Waferboard is also used as a subfloor material. Conditions as for particleboard.
- Waferboard when used as underlayment may require a vapor barrier between it and the subfloor. Check codes and manufacturer's instructions.

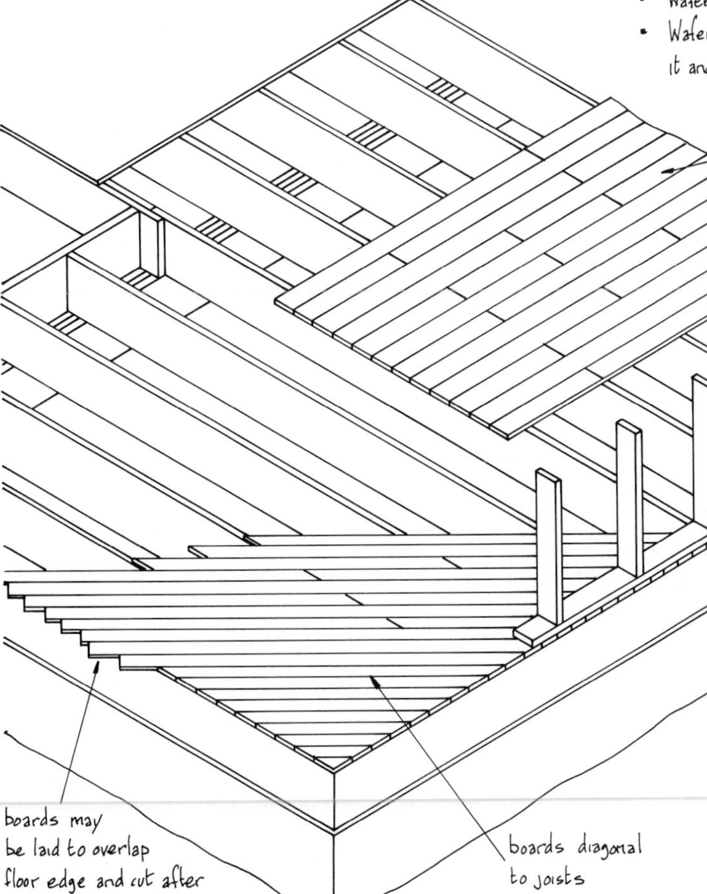

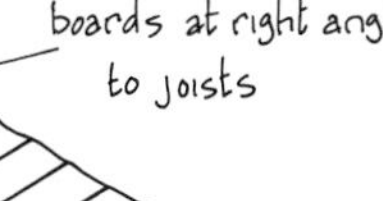

SOLID LUMBER (BOARD) SUBFLOORS

- Square edge, shiplap, or tongued-and-grooved lumber no wider than 8 in (184) is generally used.
- Board thickness is normally $\frac{3}{4}$ in (19) although $\frac{11}{16}$ in (15) boards can be used when joist spacing is 16 in (400) or less.
- End joints of square edge and shiplap boards must occur over a joist. End joints of t & g boards need not occur over a joist.
- Boards may be laid diagonally or at right angles to the joists. If laid diagonally, finish strip flooring can be laid either parallel or across the joist span. If laid at right angles, strip flooring must be laid at right angles to subfloor boards unless an underlay is used. Note that some codes allow only diagonal installation.

Note: While in modern construction board subflooring is rarely used, it will be very evident in older structures undergoing renovations or repair.

FABRICATED JOIST SYSTEMS – WOOD I-BEAMS

Prefabricated wood-component I-shaped beams offer several advantages over traditional solid lumber joists.

- Longer clear spans with greater loading capacities.
- Lighter and straighter than solid lumber.
- Wide nailing and bearing surfaces for subfloor.
- Dimensionally stable with little or no shrinkage after erection.
- Provides a uniform ceiling beneath the floor without the need for strapping.
- Allows larger ducts to pass through the web.
- Spacing of beams is normally at 19.2 in (480) or 24 in (600) o/c (19.2 in is used when working to an 8-ft (2400) module).

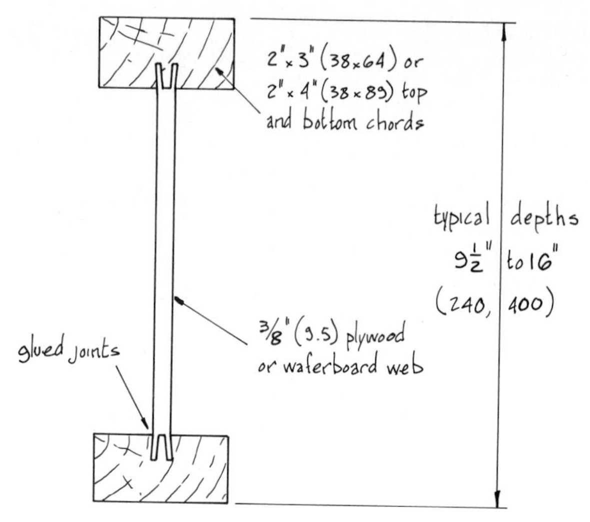

Section Through Typical Wood I-Beam

Design and erection of wood I-beams is similar in nature to those for solid lumber joists, although attention must be paid to bearing and stress details. Stiffening blocks must be attached to each side of the web at all load points as illustrated below. Check manufacturer's specifications and local codes for permissible loading and erection details.

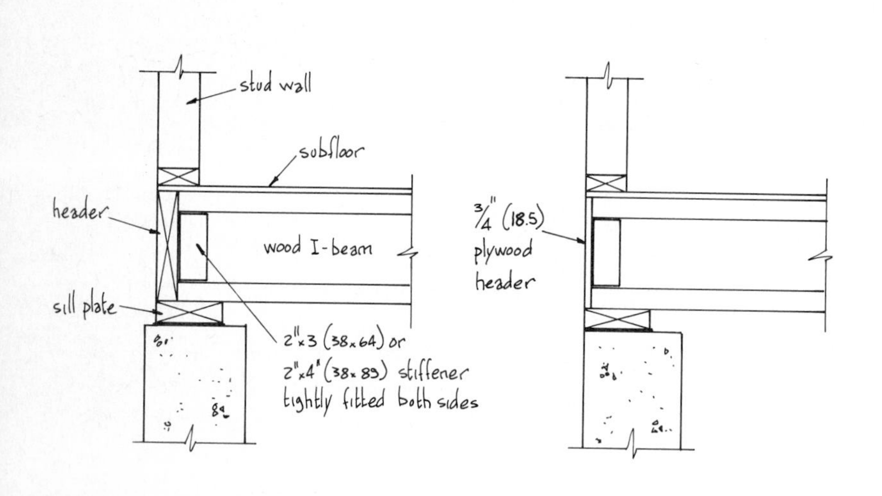

TYPICAL WOOD I-BEAM END DETAILS

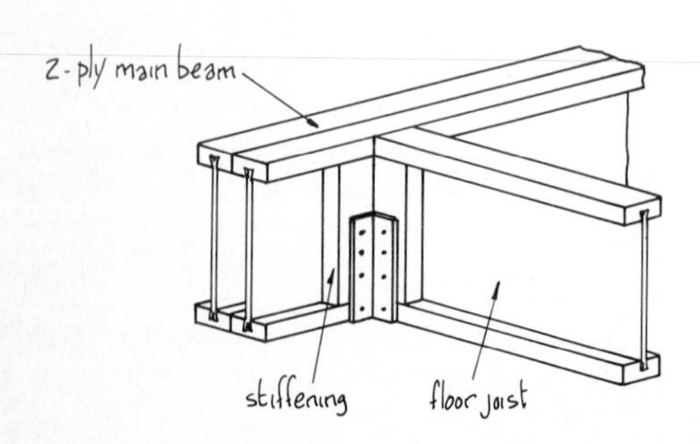

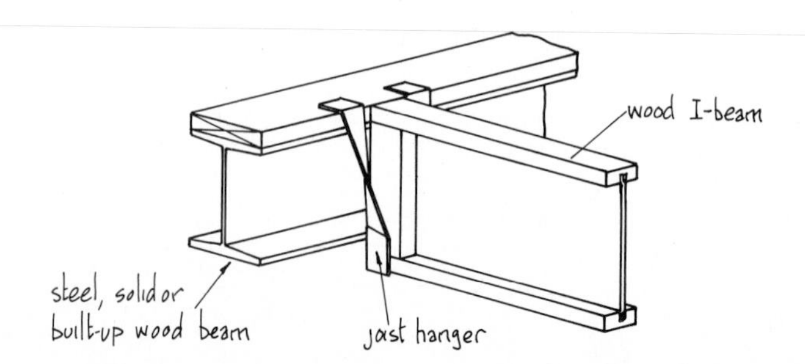

source: Jager Industries Inc.

WOOD I-BEAMS

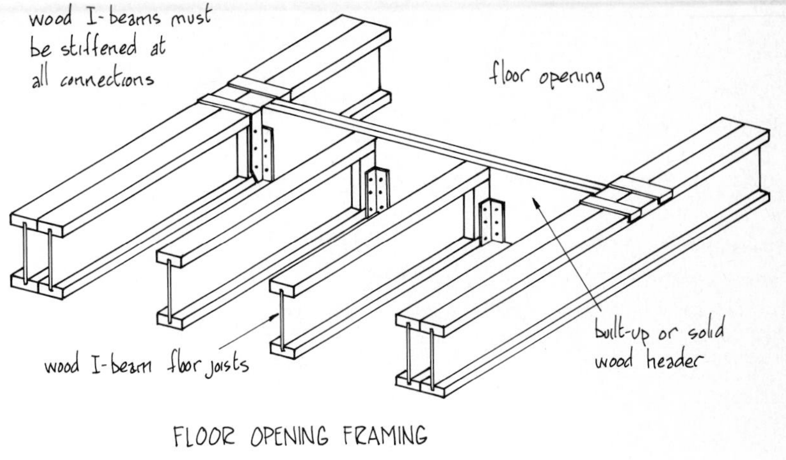

FLOOR OPENING FRAMING

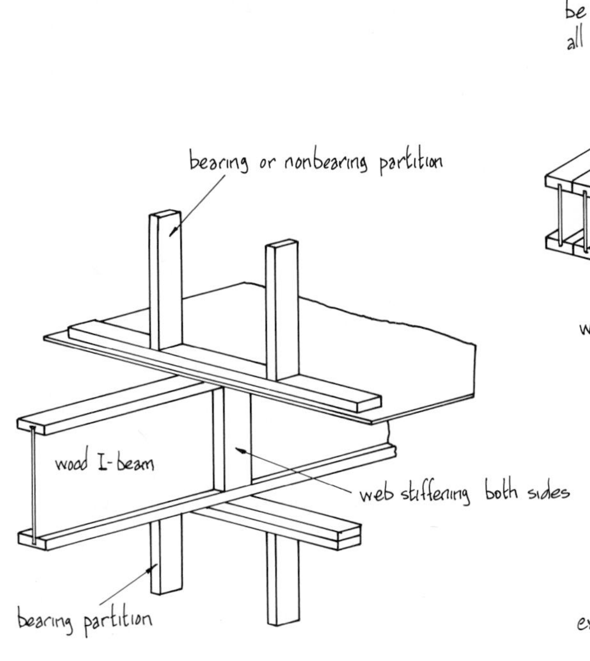

BEAM STIFFENING AT PARTITION WALLS

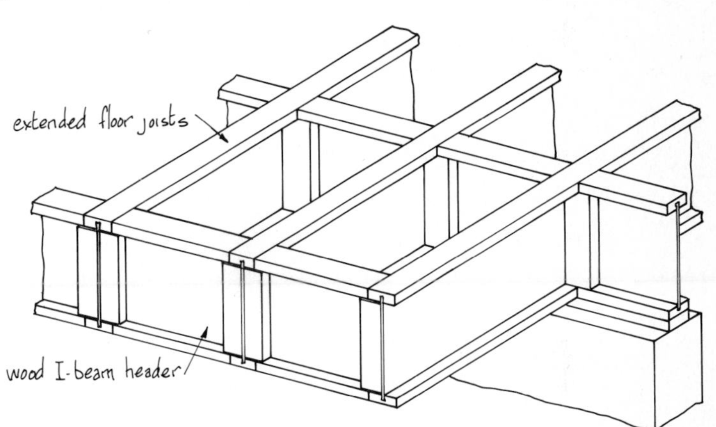

SIDE WALL CANTILEVER OVERHANG

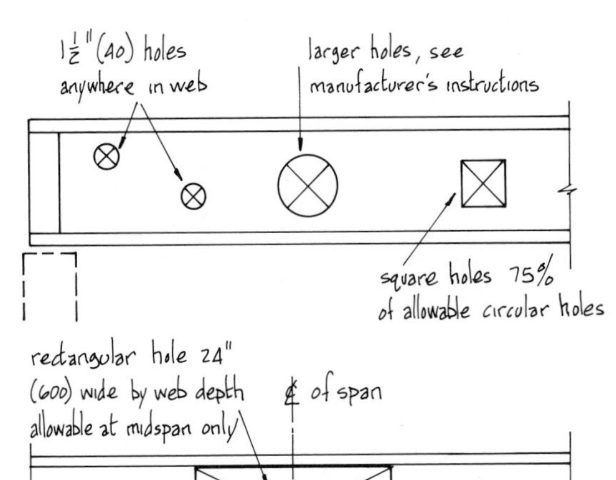

TYPICAL ALLOWABLE HOLES

refer to manufacturer's specifications for exact details

SPECIES	JOIST DEPTH	JOIST SPACING		
spruce pine or fir no. 1 grade	in (mm)	16 in (400)	19.2 in (480)	24 in (600)
	11½ (292)	19'-6" (5.94)	17'-9" (5.41)	15'-11" (4.85)
	12½ (317)	20'-6" (6.25)	18'-9" (5.71)	16'-9" (5.10)
	16 (406)	23'-11" (7.29)	21'-10" (6.65)	19'-6" (5.94)

TYPICAL SPAN TABLE

based on: live load 40 psf (1.92 kN/m²)
deflection 1/360 of span
soft converted metric dimensions

source: Jager Industries Inc.

PREFABRICATED WOOD TRUSSES

Wood truss systems are extremely flexible and allow wide latitude in the design of interior spaces. In most cases, trusses will eliminate the need for interior supporting beams, columns and footings and interior walls. They provide a flat uniform ceiling and floor plane and allow all mechanical services to be concealed in the truss depth. Trusses are commonly fabricated with lumber chords and webs and use a metal plate connection system. Other systems use flat or tubular steel webs.

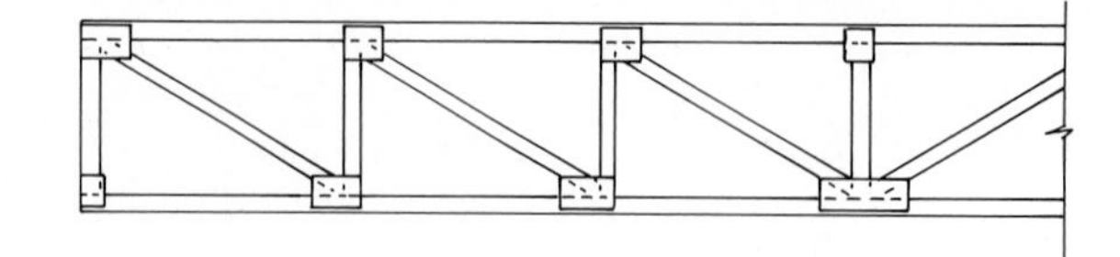

Pratt Type
- most rigid type
- easy connection of bridging to verticals
- larger openings for mechanical services

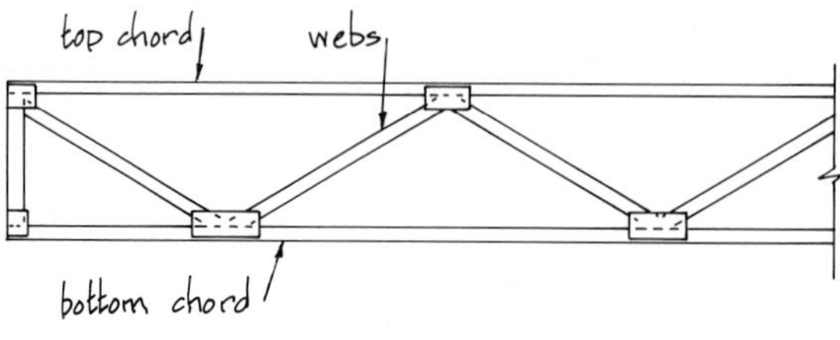

Warren Type
- least expensive in production

BASIC FLAT TRUSS TYPES
each is used extensively for floor and roof systems

Trusses can bear on top or bottom chords. Cantilevers must be engineered

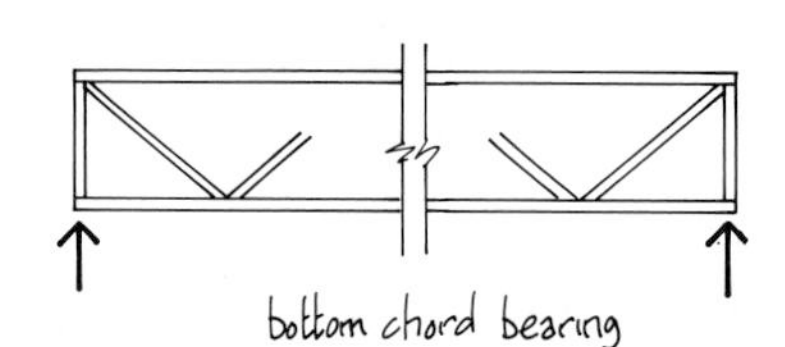
bottom chord bearing

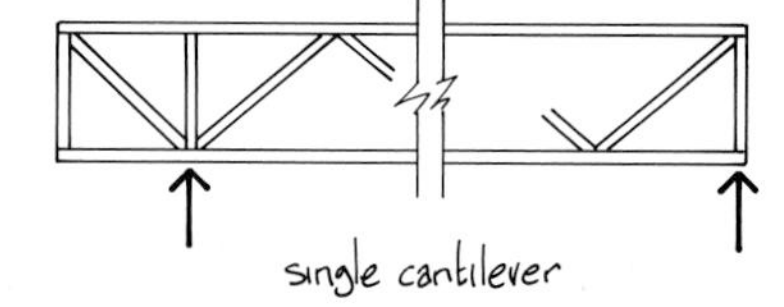
single cantilever

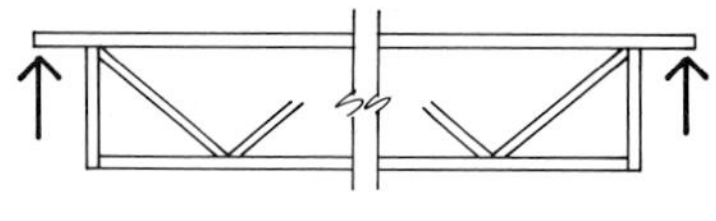
top chord bearing

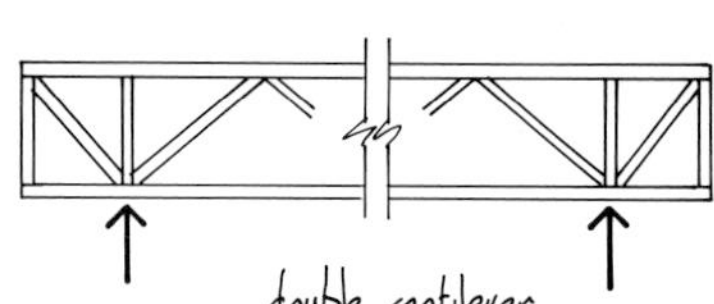
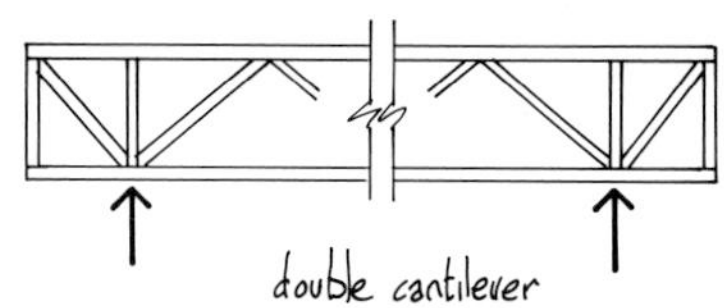
double cantilever

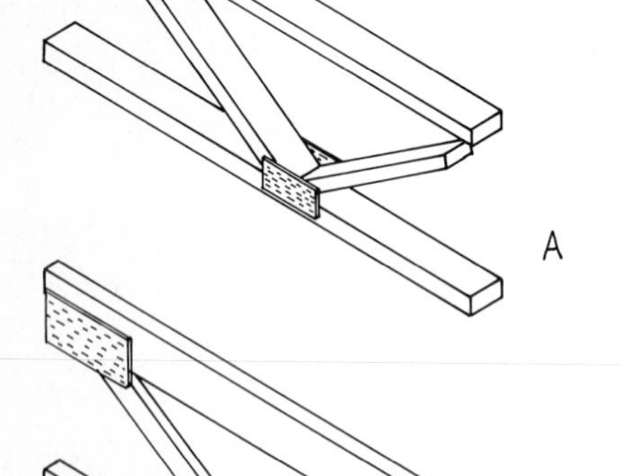

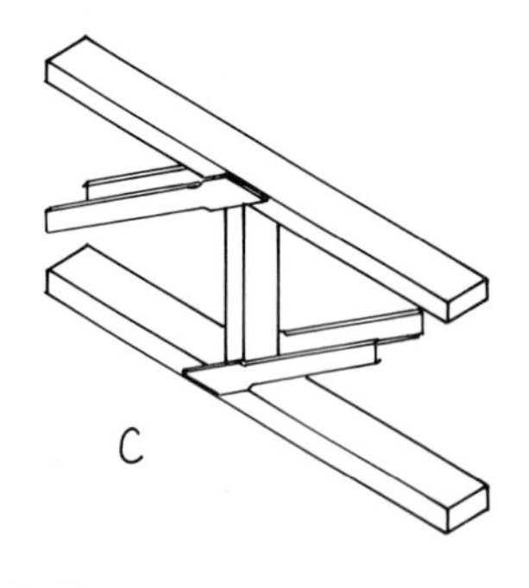

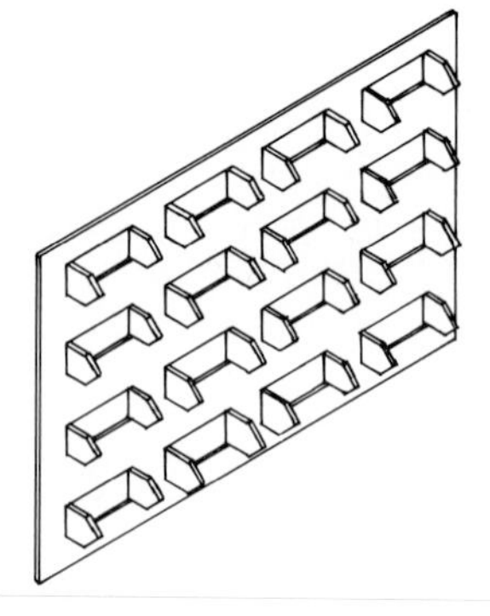

Illustrated here are three common types of floor trusses:

A. Flat chords and webs
B. Vertical chords and webs.
C. Flat chords and vertical webs with flat metal diagonals.

Connector plates for wood trusses are produced in a variety of styles and sizes. When hydraulically pressed into both sides of a truss joint, they form a strong antirotational connection. Trusses fabricated in this way are extremely economical and can be custom designed to suit virtually any loading or shape condition.

WOOD TRUSSES

CHORD SIZE	SPACING in (mm)	DEPTH in (mm)	SPRUCE, PINE, FIR No. 1 GRADE ft (m)	DOUGLAS FIR No. 1 GRADE ft (m)
2" × 3" (38 × 64)	19.2 (480)	13.8 (350)	16.2 (4.93)	17.3 (5.27)
	24 (600)	13.8 (350)	14.1 (4.30)	15.2 (4.64)
	19.2 (480)	19.7 (500)	19.7 (6.00)	21.0 (6.41)
	24 (600)	19.7 (500)	17.2 (5.23)	18.5 (5.64)
2" × 4" (38 × 89)	19.2 (480)	11.8 (300)	17.5 (5.35)	18.5 (5.64)
	24 (600)	11.8 (300)	15.5 (4.71)	16.4 (5.00)
	19.2 (480)	15.7 (400)	20.4 (6.23)	21.7 (6.60)
	24 (600)	15.7 (400)	17.8 (5.43)	19.1 (5.84)

TYPICAL FLOOR TRUSS SPAN TABLES
Note: tables based on 55 psf D.L+LL. (2.64 kN/m²)

The table at the left indicates typical spans for flat-chord floor trusses. Spans are based on 1/360 maximum deflection. As a rough guide to truss depth, a ratio of 1:12-15 for depth of truss to span can be used. Above 1:15 the floor will have a "bouncy" feeling unless trusses are engineered for extra stiffness. Since trusses are pre-fabricated units, each manufacturer's specifications must be consulted for exact truss performance.

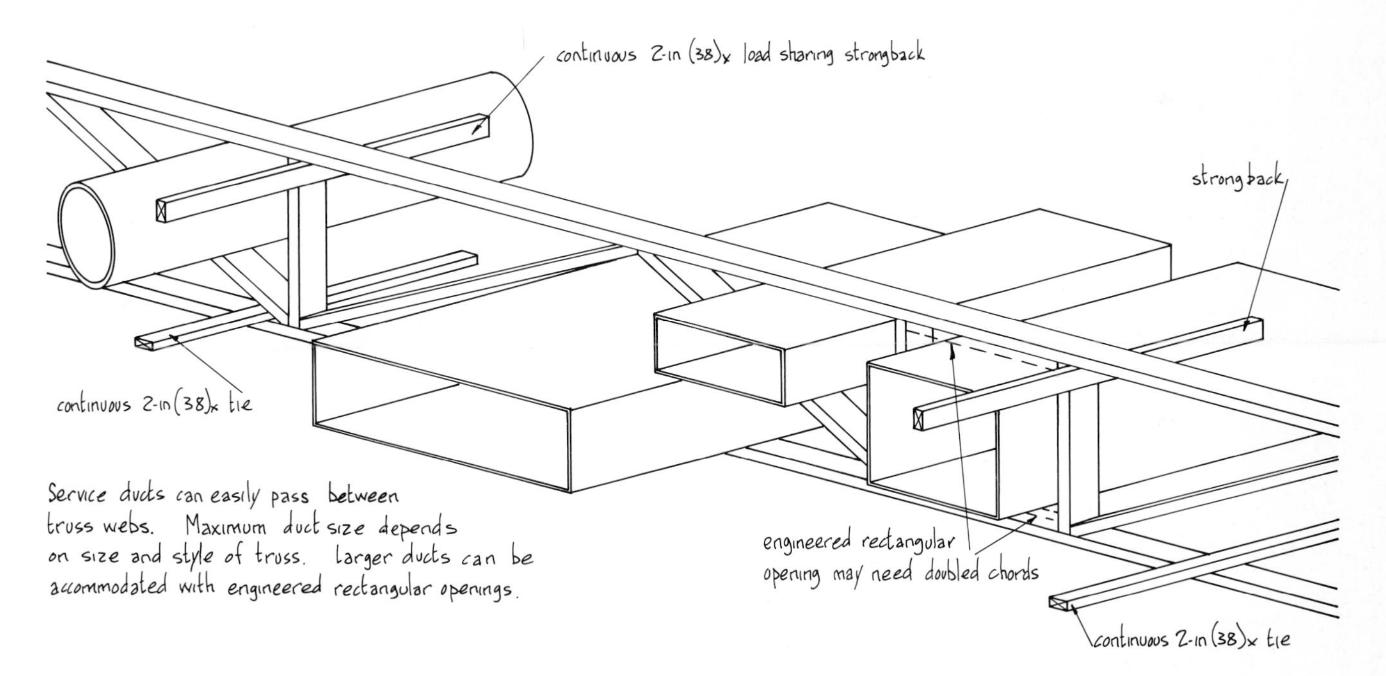

Service ducts can easily pass between truss webs. Maximum duct size depends on size and style of truss. Larger ducts can be accommodated with engineered rectangular openings.

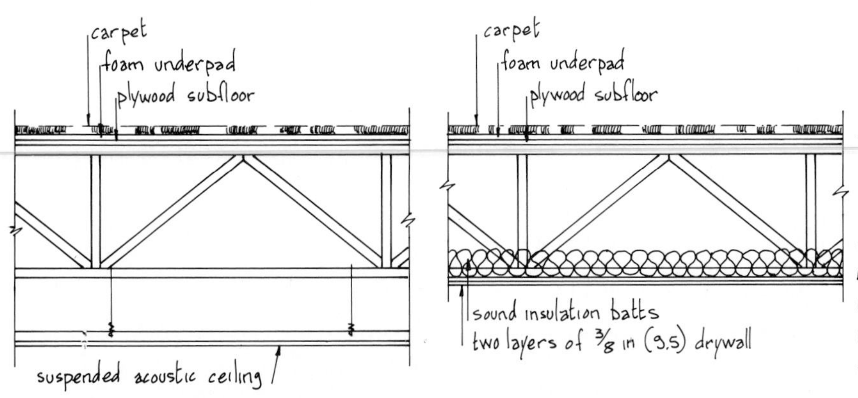

Two Examples of Sound-Rated Floor Assemblies

Trusses offer good sound resistance but where acoustical performance is critical, special floor assemblies can be designed to increase sound ratings. In general, blown-in loose-fill acoustic insulation is better than batts and a thick carpet underpad is better than a thick carpet. The truss manufacturer's specs will indicate rated floor assembly details.

source: Jager Industries Inc.

WOOD TRUSS FLOOR DETAILS

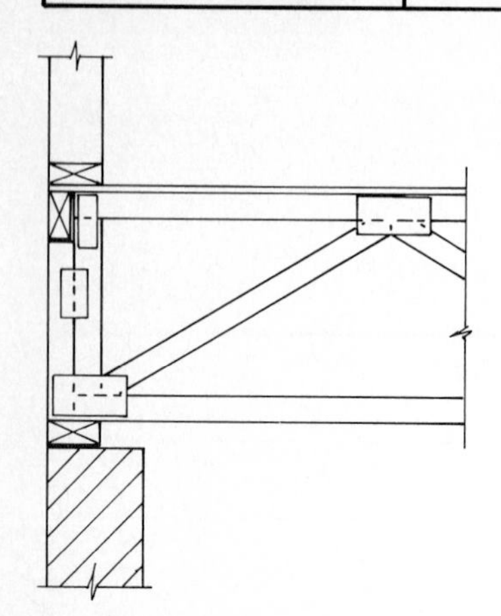

Bottom Chord Bearing

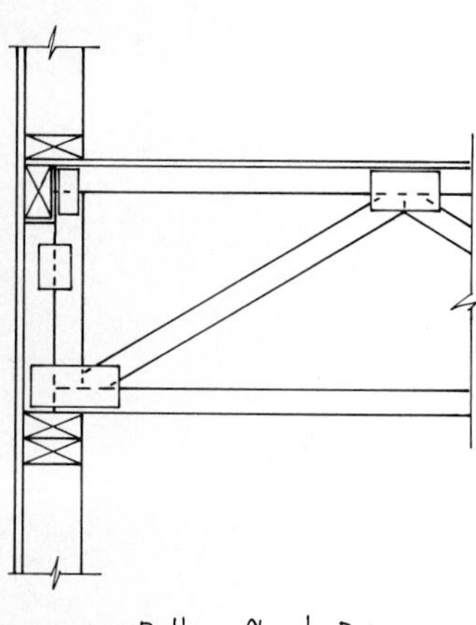

Bottom Chord Bearing

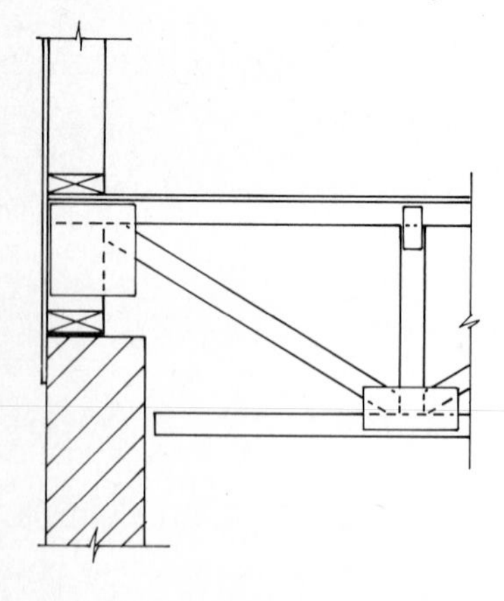

Top Chord Bearing

TYPICAL END BEARING

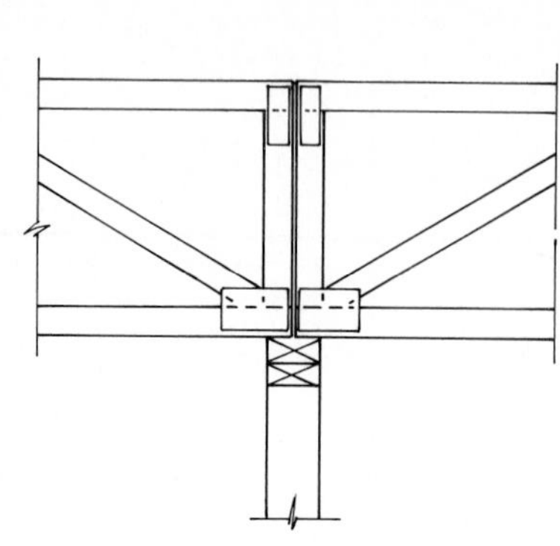

Bottom Chord Bearing

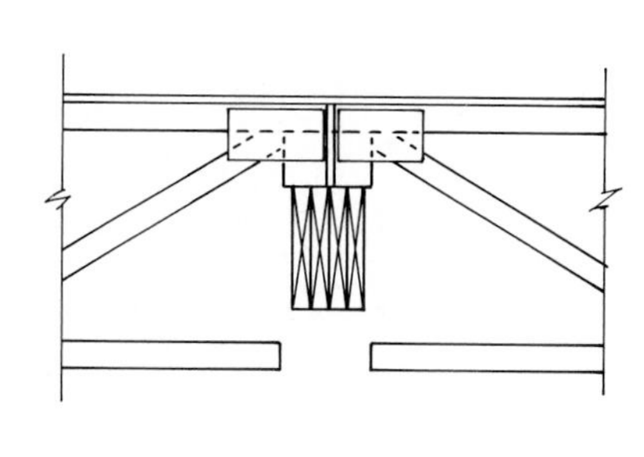

Top Chord Bearing

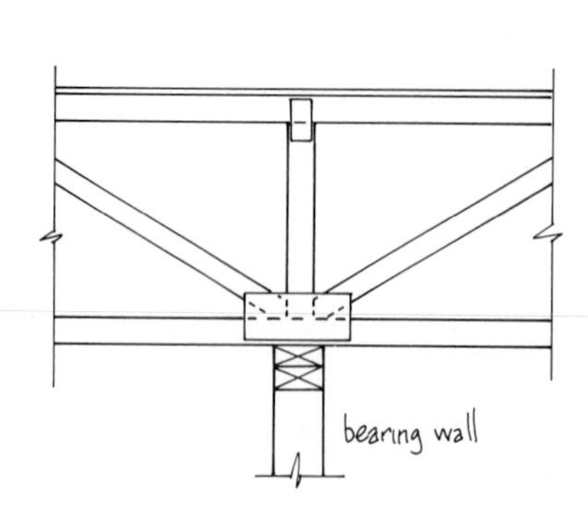

Multi-Span Truss

TYPICAL INTERIOR BEARING

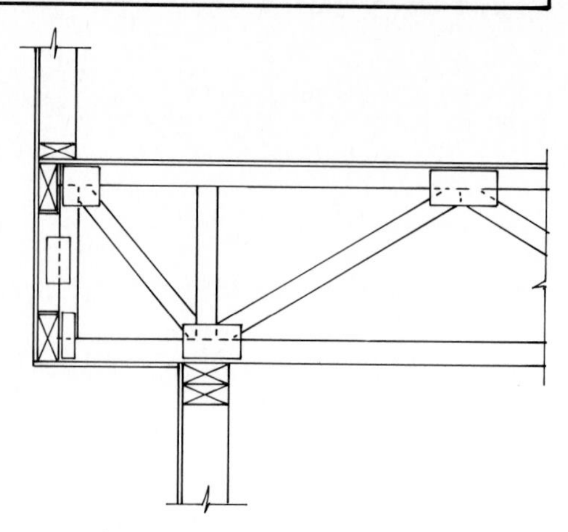

Cantilever Carrying Wall Above

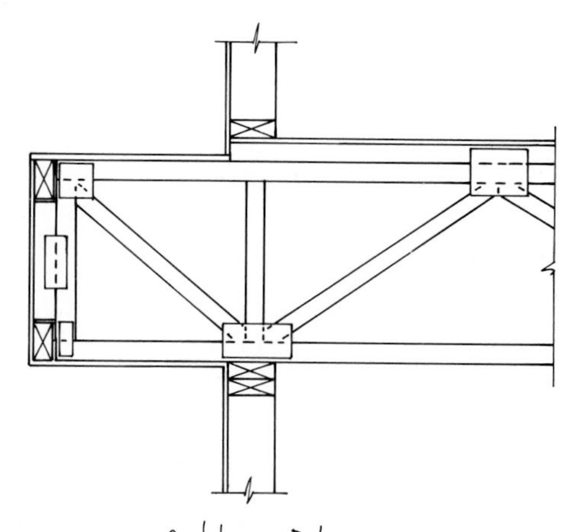

Cantilever Balcony

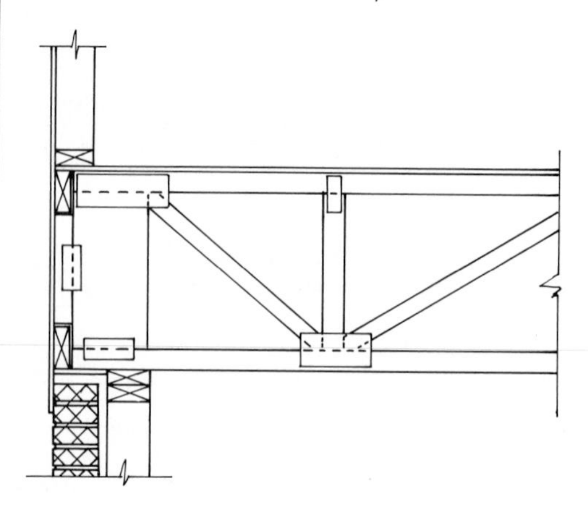

Short Cantilever for Brick Veneer

TYPICAL CANTILEVERS

source: Jager Industries Inc.

WOOD TRUSS FLOOR DETAILS

Wood trusses can be fabricated to meet the requirements of virtually any type or style of floor design. Details on this and the preceding page indicate some of the possible configurations. Most truss manufacturers provide a design service and will engineer all truss configurations on a project and provide design data for submission to local authorities.

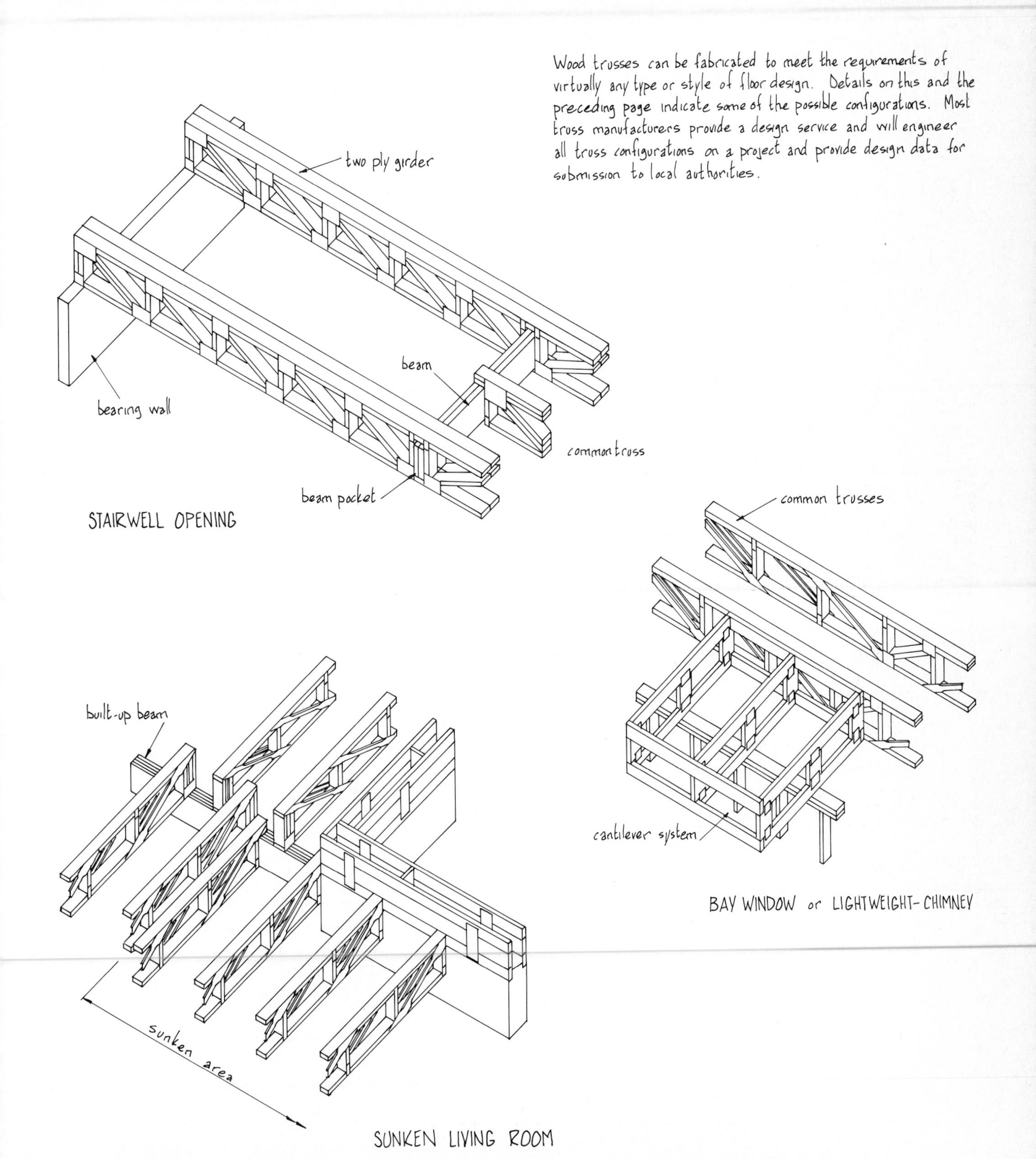

source: Jager Industries Inc.

LIGHTWEIGHT OPEN-WEB STEEL JOISTS

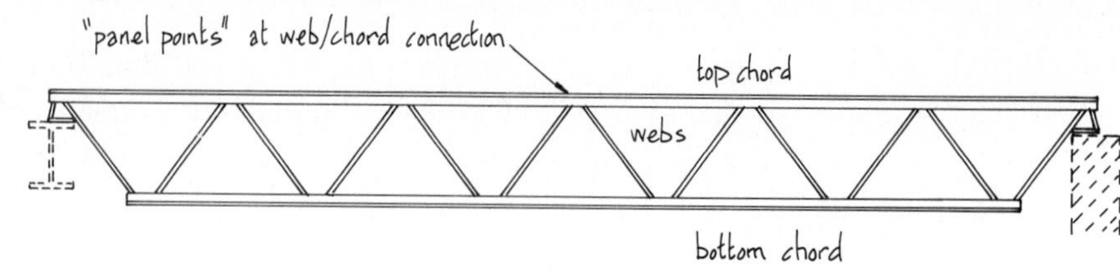

Lightweight steel truss systems (open web steel joists) are used extensively in all types of construction and are not limited to small or light-occupancy structures. Designers can specify OWSJs in two forms:

- Custom fabricated, where trusses are individually designed to suit specific structural conditions. Wide variations in shape, size, span, and load-bearing capacity can be accommodated.
- Standardized OWSJs, where depth, span, and strength of each unit conforms to standard specifications set by the Steel Joist Institute. In this form, designers must tailor the structure to best utilize the standard units, while fabricators, who have developed many styles of manufacture, must produce units meeting those specifications.

In actual practice most OWSJs are custom fabricated since computerized design programs allow each truss to be designed on an individual basis. In addition, rapid fabrication methods can produce customized trusses at little, if any, extra cost over standardized trusses.

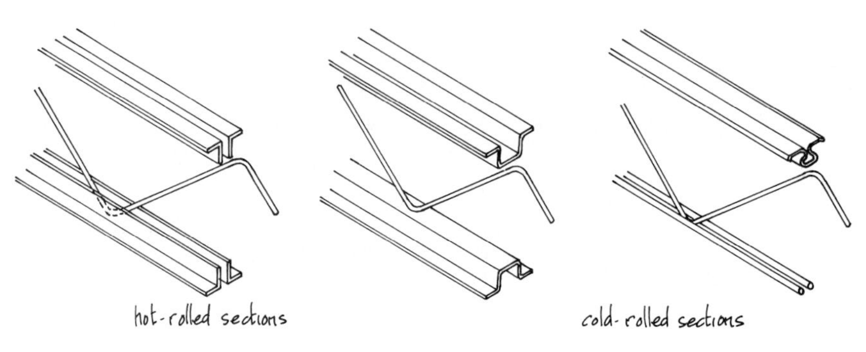

There are a variety of truss fabrication styles, three of which are shown above. In general, hot-rolled structural steel (angles, channels, bars, etc.) or cold-rolled steel sections are used. See individual fabricators' specs for specific joist composition.

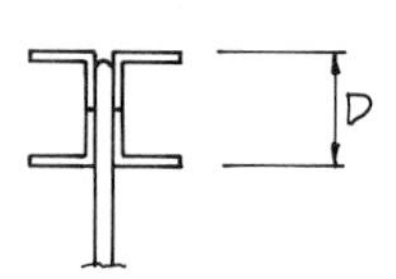

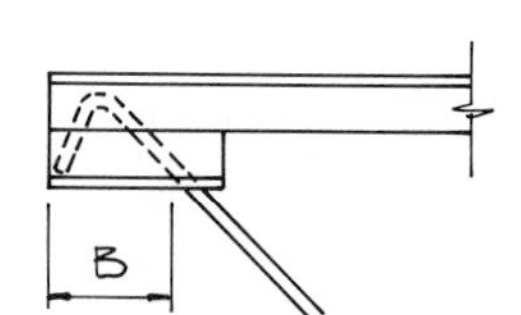

Typical End Detail
(varies with manufacturer)

STANDARDIZED JOIST SERIES
(Steel Joist Institute specifications)

SERIES	JOIST DEPTH (3)	SPAN	SHOE DEPTH D	BEARING B (4) steel	conc./mas.
H standard	8 in to 30 in in 2 in increments	up to 60 ft	$2\frac{1}{2}$ in	$2\frac{1}{2}$ in	4 in
LH long span	18 in then 20 in to 48 in in 4 in increments	up to 96 ft	5 in	4 in	6 in
DLH (1) deep longspan	52 in to 72 in in 4 in increments	up to 144 ft	5 in or $7\frac{1}{2}$ in	4 in	6 in
Joist Girder (2)	20 in to 72 in in 4 in increments	up to 60 ft	6 in	4 in	6 in

The above examples are typical only; refer to S.J.I. specifications for exact details.
Metric joist series are not available at this time.

Notes to chart:

1. Deep long-span joists are generally used only as roof trusses.
2. Joist girders are designed as main beams supporting floor trusses at panel points.
3. Each joist depth size is available in several weight-per-lineal-foot options. E.g., 26-in-deep H-series — options of 12.8, 14.8, 16.2, or 17.9 lb/ft (approx.). Strength of joist increases with weight.
4. Bearing load should not exceed 250 psi for masonry and 750 psi for concrete. Adjust the bearing area to suit.

LIGHTWEIGHT STEEL TRUSSES

CEILINGS

- For ceilings attached directly to the underside of joists, a ceiling extension should be specified. Ceiling composition in this case is usually plaster on metal lath or plasterboard on furring strips.
- Ceilings can be suspended from truss bottom chords. A metal grid system supporting fiber or acoustic tiles is standard.
- For general acoustical treatment of trussed floors, see 6-21.

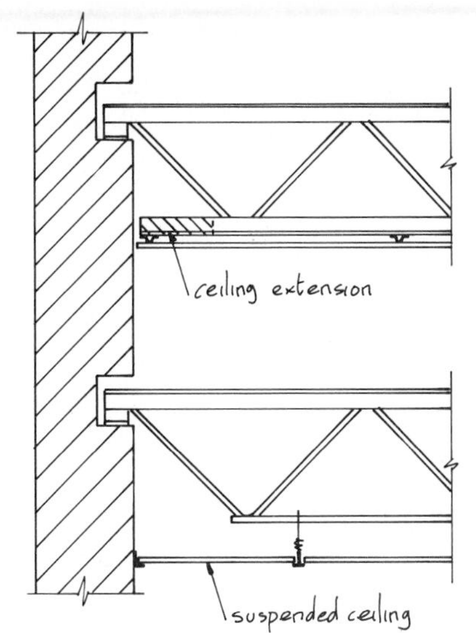

FLOOR JOIST DESIGN

OWSJs are most economical when designed for uniformly distributed floor loads and when spaced at uniform centers. Concentrated loads can be applied (normally at panel points) provided that joists are engineered for such loads.
The load of a lightweight partition at right angles to the joist is normally considered to be evenly distributed by the floor slab. Parallel partitions can be supported by doubling or tripling joists under the partition.
As a rule of thumb for standardized joists, the joist depth-to-span ratio should not exceed 1:24, except for H series joists, which should not exceed 1:20.

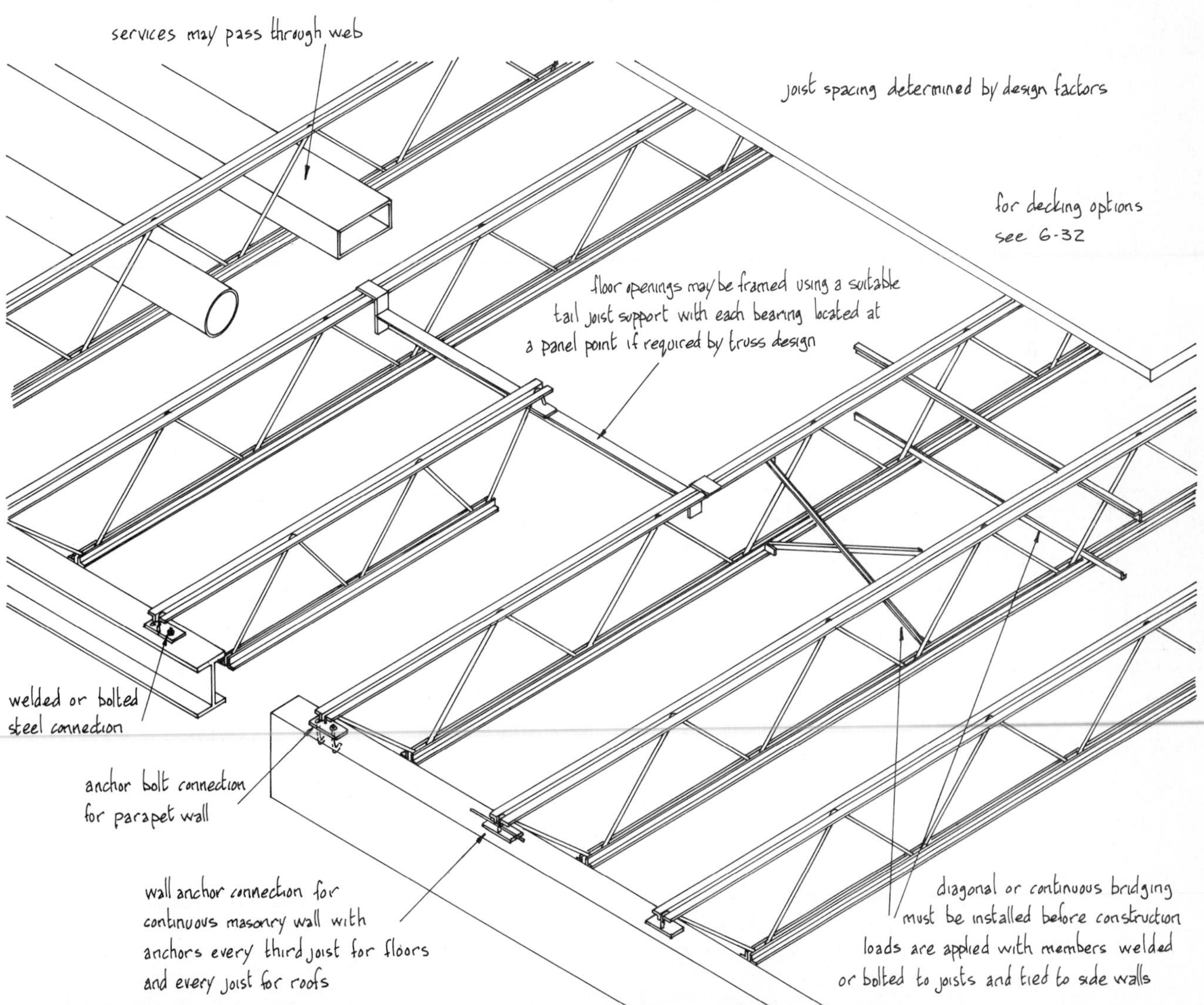

STRUCTURAL STEEL

Structural steel floor systems can be supported by steel columns or on concrete or masonry walls. When part of a steel-framed building, all elements of the structure must be designed as an integral unit.

- Structural steel floors are often used when heavy loads are to be applied.
- Steel structures allow rapid on-site erection. However, members must be accurately shop fabricated since field alterations are difficult to make.
- Steel beams are used to support a variety of floor types. Steel decking and cast-in-place or precast concrete floors are common (see 6-32 for steel decking and 6-28 for precast beams).
- Shape and size of floor members is determined by design stresses, span, economy, and in some cases, asthetics.

Connections between steel members can be welded, bolted, or riveted. The use of welds or bolts, either singly or in combination, is standard practice. All forces acting on a connection are normally shown on the structural drawings so that the members and their connections can be correctly designed. Shop fabrication drawings show all information necessary to fabricate and erect the structural members.

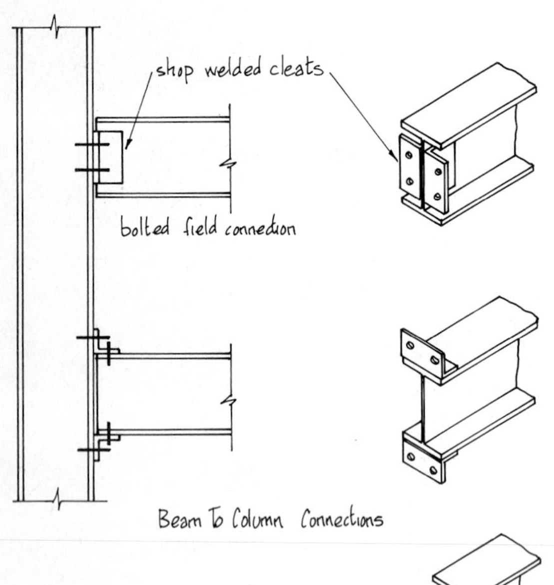

Beam To Column Connections

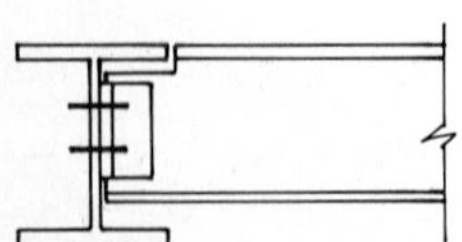

Beam To Beam Connection

Typical examples of simple beam connections. Exact connection design is subject to the type and size of shear, moment, axial, and other forces acting on it.

COMMONLY USED STRUCTURAL SHAPES

Each shape is available in a variety of sizes and weights. Sections can be used singly or combined to form composite sections.

SHAPE	GROUP	IDENTIFICATION
	beam	W, S, M, or HP
	channel	C or MC
	angle	L equal or unequal legs
	tee	ST, WT or MT
	tubing	TS round, rectangular or square
	bar	BAR flat, square or round

Abbreviations: W = wide flange; M = miscellaneous; HP = bearing piles; S = American standard; TS = tubing structural

SECTION SIZING

By depth and weight:

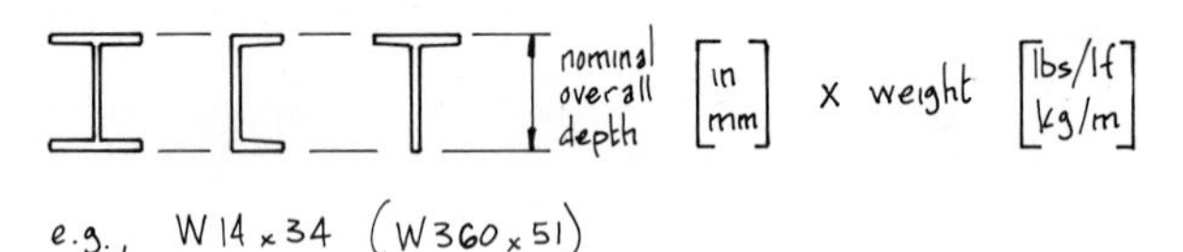

e.g., W 14 x 34 (W 360 x 51)

By dimension:

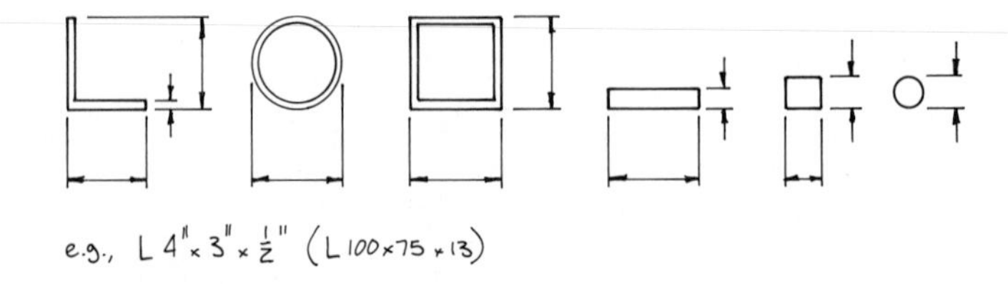

e.g., L 4" x 3" x ½" (L 100 x 75 x 13)

For the full range of shape sizes and their properties see American Iron and Steel Institute (AISI) product classification manuals. In Canada, see the Canadian Institute of Steel Construction (CISC).

STRUCTURAL STEEL

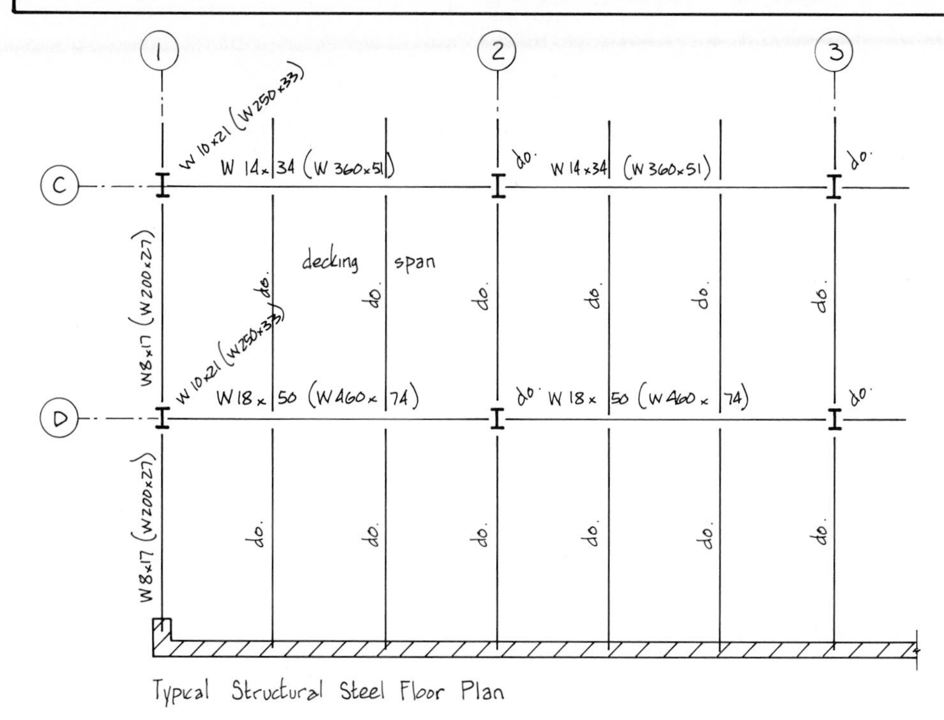

Typical Structural Steel Floor Plan

- This simplified structural steel floor plan shows the shape and size of steel members in the plane of the floor.
- Dimensions are normally given to column centerlines. Fabricators will detail and manufacture each member to suit design requirements.
- Loading information will be given either as notations on the structural plans or in table form elsewhere.

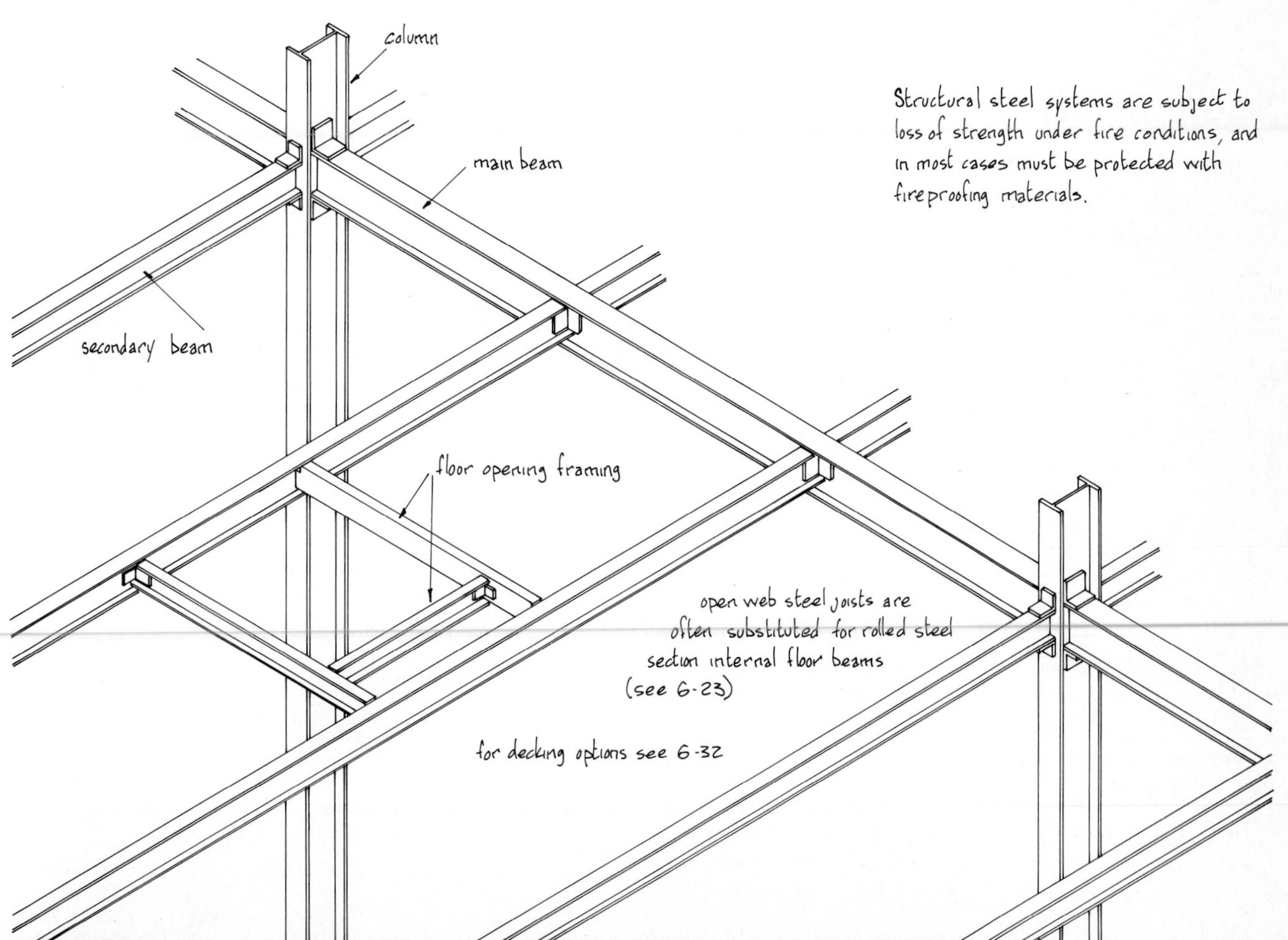

Structural steel systems are subject to loss of strength under fire conditions, and in most cases must be protected with fireproofing materials.

PRECAST CONCRETE FLOOR UNITS

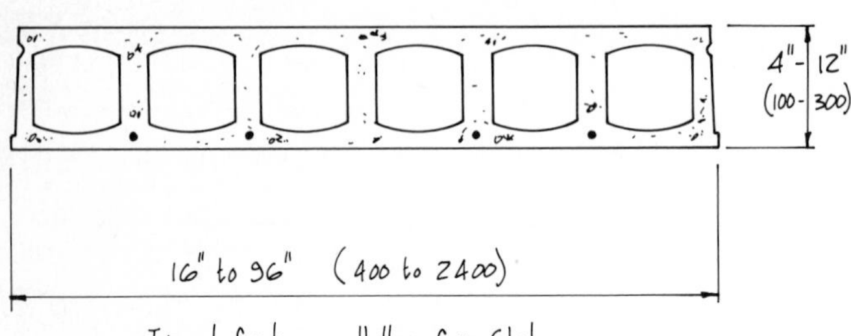

Typical Section — Hollow-Core Slab

There are a variety of precast concrete floor systems available for the commercial, industrial, and some residential markets. Two slab types are in common use: the hollow-core slab and the tee or ribbed slab. All such systems have similar attributes:

- Because slabs are prefabricated under factory conditions, quality is carefully controlled. High-strength concrete is normally used and units can be unreinforced, reinforced, or prestressed, as required. Mass production with continuous casting equipment and minimal formwork allows competitive pricing.
- Fire rating of units is high – 2-3 hours for most units.
- Sound rating of most units is also high.
- The units are normally erected by crane. On some units this erection method may determine design conditions since erection stresses can exceed in-service loads.
- Clear spans exceeding 50 ft (15 m) are possible with the largest units.

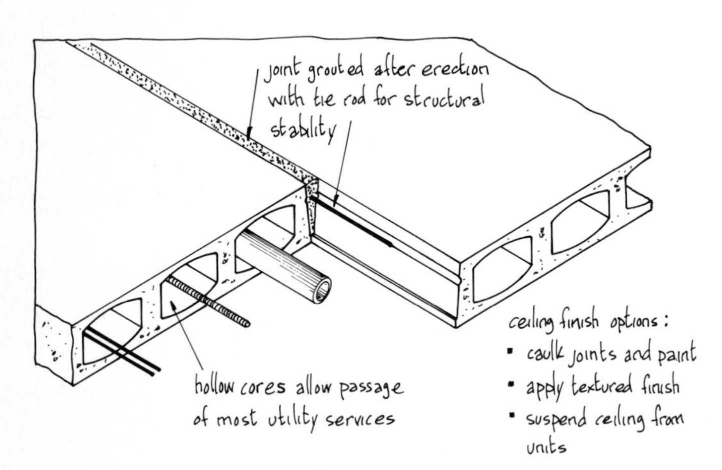

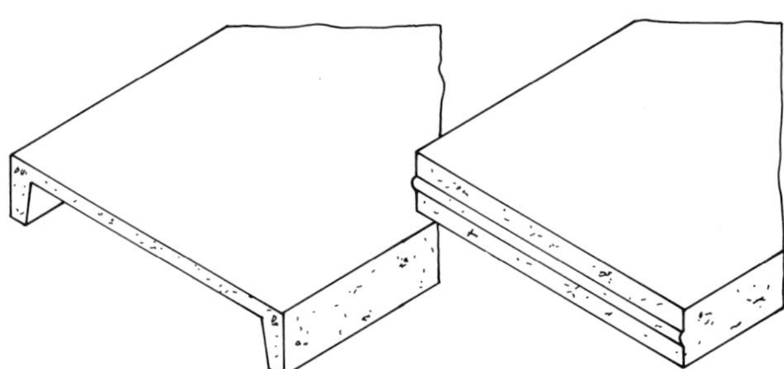

Alternative slab to the single- or double-tee system shown opposite.

These units are normally 2 in (50) or 2¾ in (70) thick, have galvanized reinforcing mesh on both sides and are reversible, allowing either side to face up. They are also nailable to the supporting members. Maximum span of units 4 ft (1200) or less.

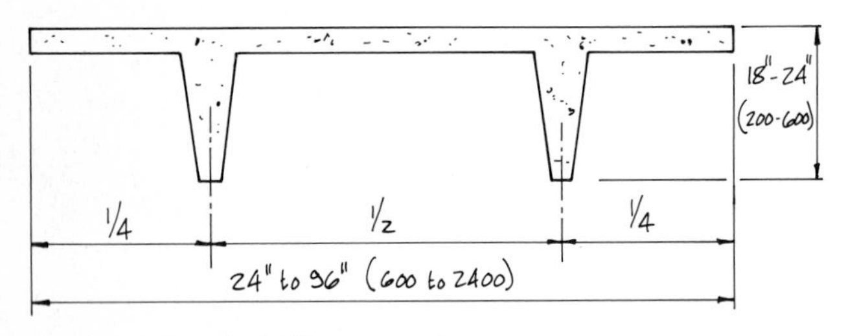

Typical Double-Tee Slab
(single tees also available)

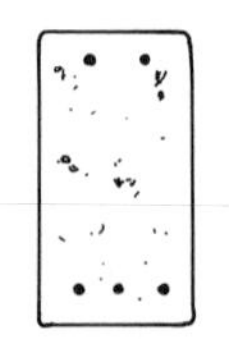

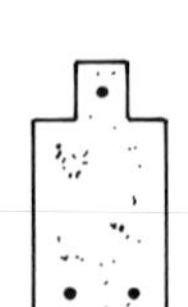

Precast Concrete Beams / Joists

Precast concrete beams and joists are manufactured in many styles and sizes. They can function as main structural beams directly supporting a floor load or as joists spanning between main beams. All beams of this type are prestressed during manufacture, allowing a one-half to two-thirds reduction in section depth when compared to cast-in-place beams.
See manufacturer's specs for available sections and spanning capabilities.

ECONOMICS

Prefabricated units are most economical when used in low or high-rise structures with repetitious design features. Buildings using these units should be engineered to take advantage of the prefabricated components.

UNIT BEARINGS

Under most conditions units require a minimum of 3 in (75) end bearing. Bearing is normally on steel, masonry, or concrete supports. Wood studs are not used as a bearing because of the high total loading of these floor systems.

HOLLOW CORE SLABS

Manufactorer's specifications must be consulted for engineering details of slab systems.

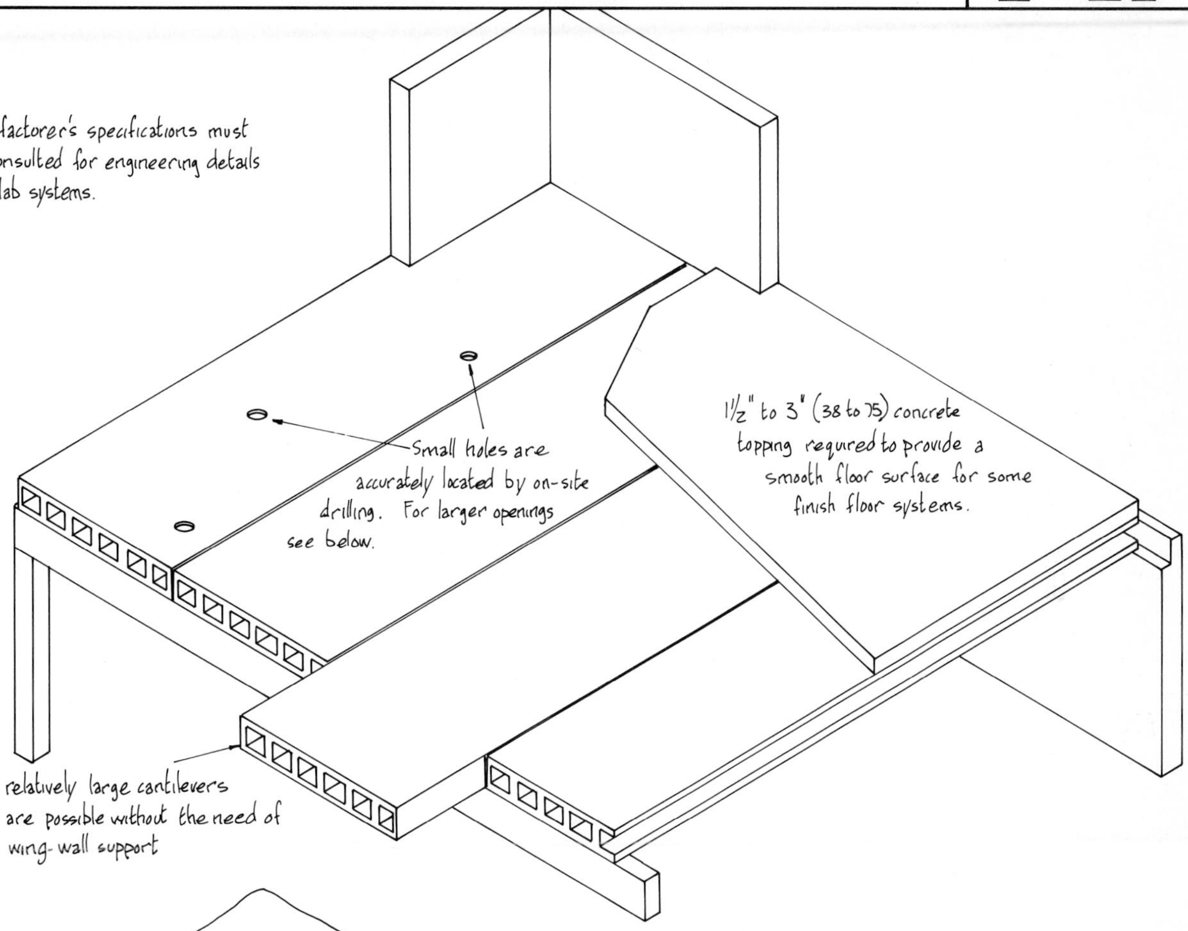

Small holes are accurately located by on-site drilling. For larger openings see below.

1½" to 3" (38 to 75) concrete topping required to provide a smooth floor surface for some finish floor systems.

relatively large cantilevers are possible without the need of wing-wall support

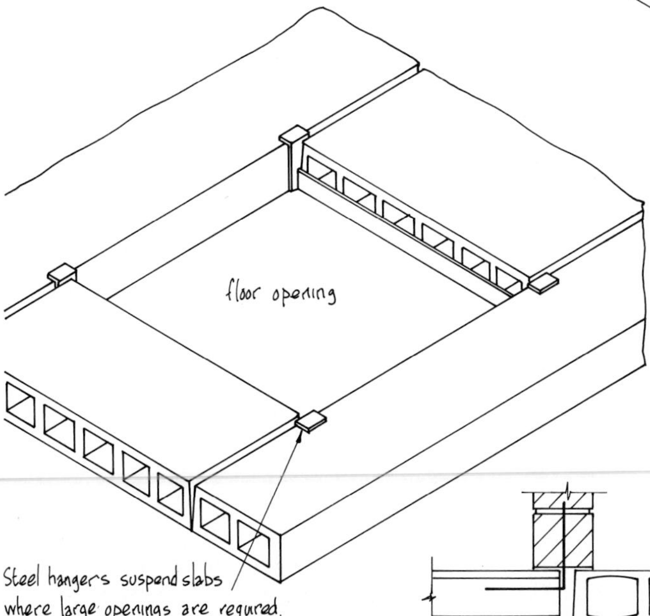

floor opening

Steel hangers suspend slabs where large openings are required. Openings more than one unit wide must be engineered.

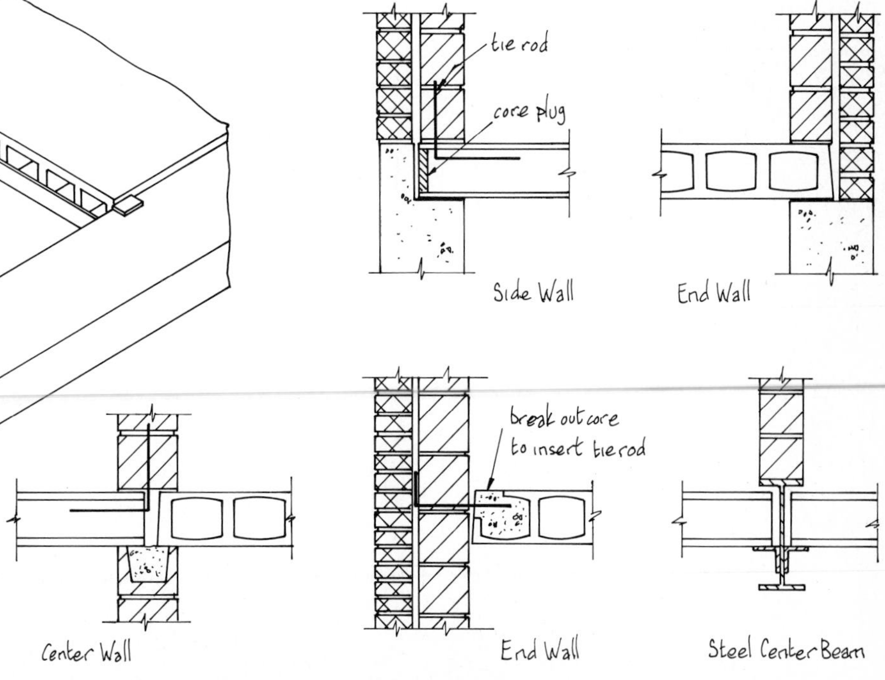

tie rod

core plug

Side Wall

End Wall

break out core to insert tie rod

Center Wall

End Wall

Steel Center Beam

source: Coreslab Ltd.

CAST-IN-PLACE CONCRETE FLOORS

There are a number of cast-in-place concrete floor systems in general use. All are suspended systems, i.e., floors that span from one support to another, and all can be divided into two groups: one-way slabs and two-way slabs. The one-way and two-way designation refers to the slab reinforcement running in either one direction only or in two directions at right angles.

The floor systems are constructed by pouring concrete into suitable formwork after reinforcement is placed. Formwork is removed after concrete is cured.

The type of floor system used depends on loads, spans, available ceiling heights, and visual aspects. Since the floor is designed as an integrated structure in conjunction with whatever girders, beams, and columns are required, careful structural analysis of all elements is essential.

Holes in slabs for utility services must be carefully engineered. Exact positions and sizes of holes must be shown on the drawings so that the necessary formwork can be constructed.

For complete technical information, refer to the American Concrete Institute design manuals.

ONE-WAY SOLID SLAB AND BEAM

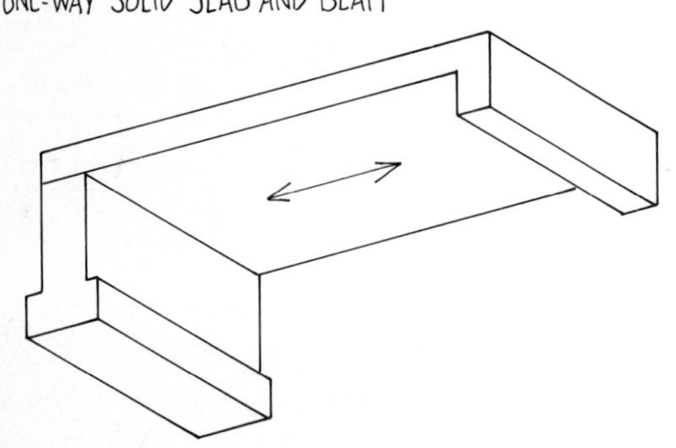

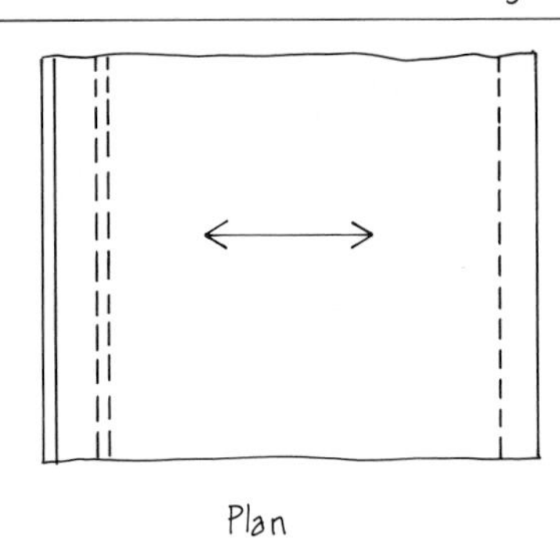

Plan

One-Way Solid Slab and Beam

- Most common form of concrete floor.
- Slab spans between two parallel supports, walls or beams.
- Slab is uniform in thickness, normally 4 to 6 in (100-150), and reinforcing runs between beams.
- Effective span is limited to 12 ft (3650)

ONE-WAY RIBBED SLAB

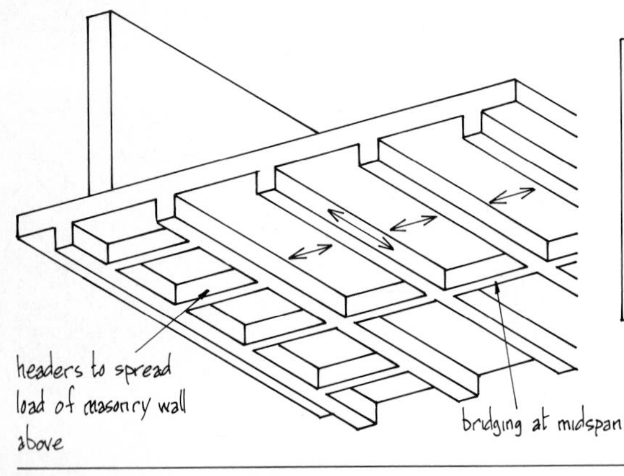

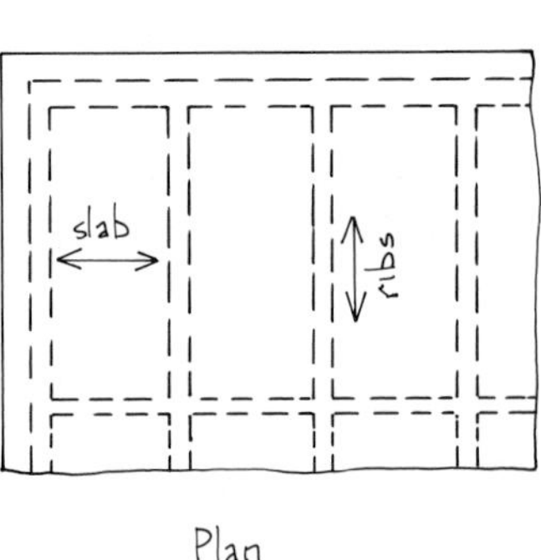

Plan

One-Way Ribbed Slab

- Ribs are spaced close together, normally 24- to 35-in (600-1000) centers, allowing a thin, economical slab to be used.
- Main reinforcement is placed at the bottom of the rib with minimal reinforcement in slab.
- Spans up to 30 ft (9000) possible in rib direction. Bridging required at midspan.
- Where masonry walls above run parallel to ribs, headers are required to carry the concentrated load.

ONE-WAY SOLID SLAB with BEAMS and GIRDERS

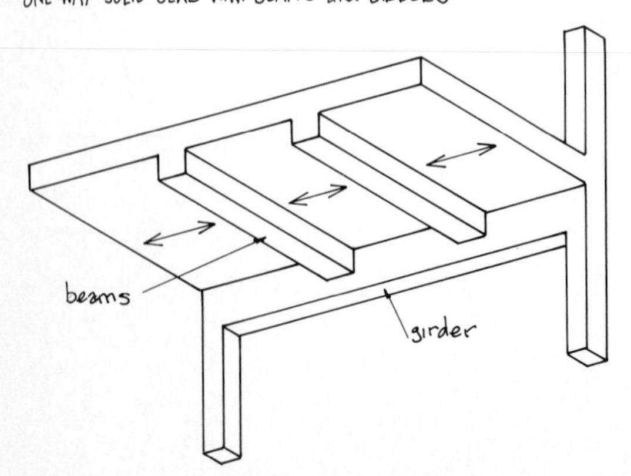

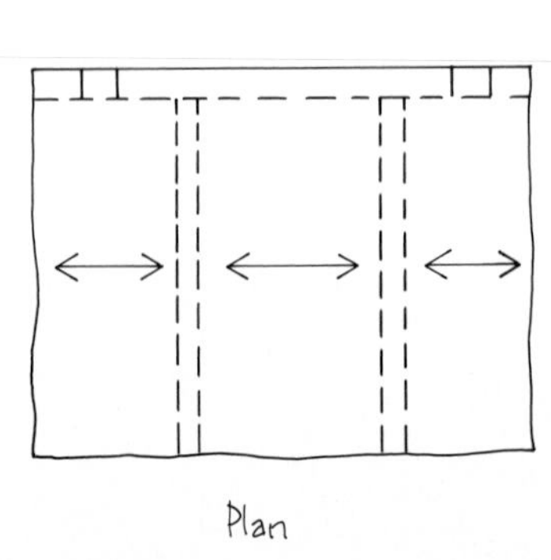

Plan

One-Way Solid Slab with Beams and Girders

- An extension of the solid slab and beam shown above and used for concrete-framed buildings.
- The slab spans between beams, which in turn are supported by girders.
- Maximum slab span is still limited but can be designed as a continuous element.

CAST-IN-PLACE CONCRETE FLOORS

TWO-WAY SOLID SLAB AND BEAMS

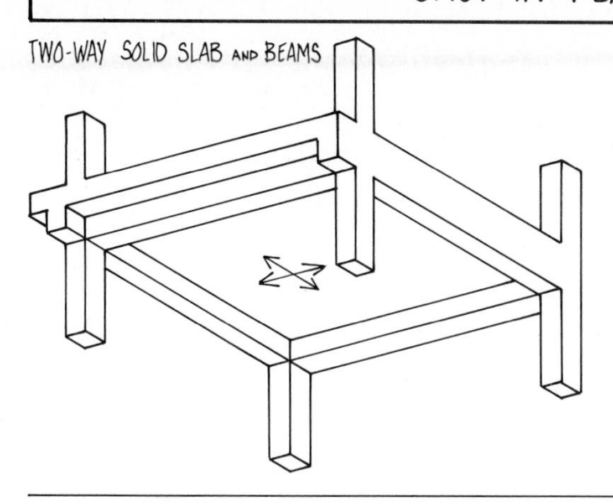

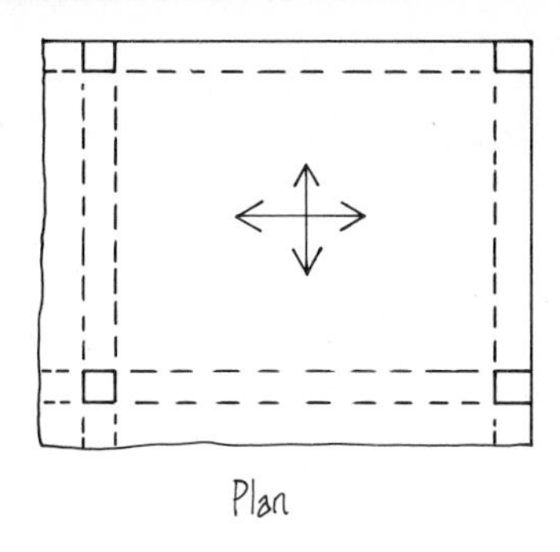

Plan

Two-Way Solid Slab and Beams

- The floor panel is square or nearly square and is supported on both sides.
- Reinforcing runs in both directions, allowing a stronger slab with a greater span.
- Support beams run between columns.
- Girders are not required in most designs.

TWO-WAY WAFFLE SLAB

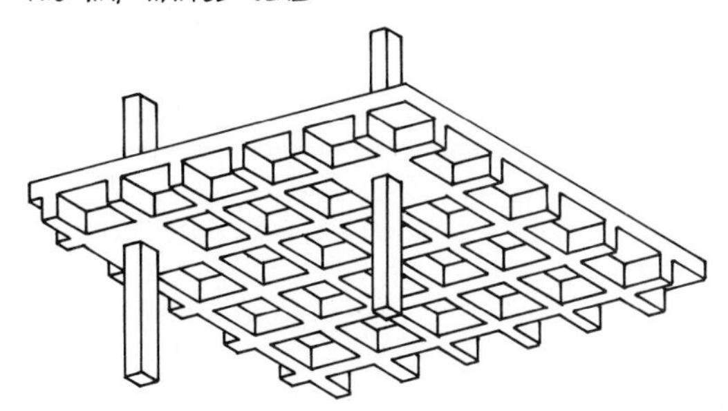

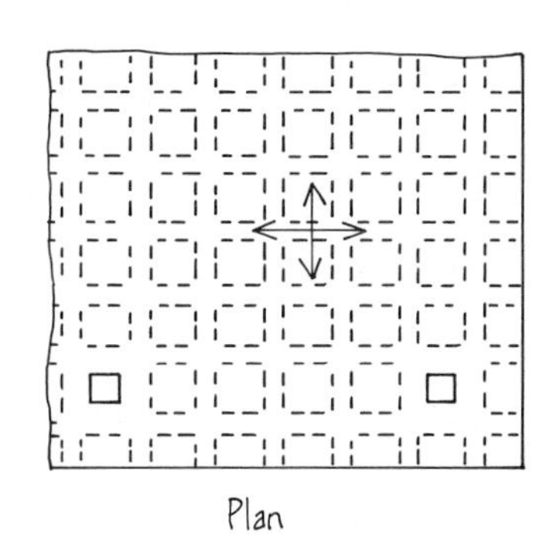

Plan

Two-Way Waffle Slab

- Used where an extra-thick slab would be demanded by span or loading conditions.
- Two-way ribs are formed by removable steel forms. Areas around columns are left solid to carry loads.
- The waffle system reduces total load and provides an interesting ceiling effect.

TWO-WAY FLAT PLATE

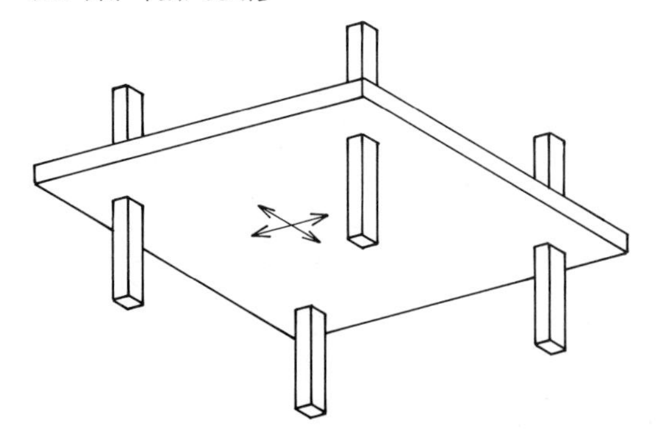

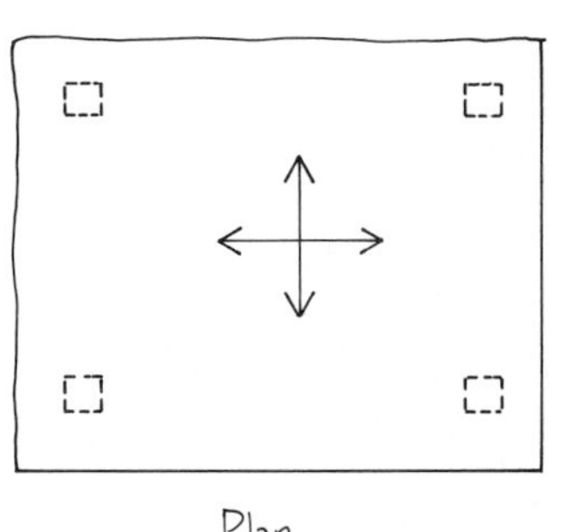

Plan

Two-Way Flat Plate

- A uniform thickness slab supported by the columns alone.
- Two-way reinforcement is arranged to transmit all loads to the columns.
- Used for medium spans and loads and is common in high-rise residential and commercial buildings.

TWO-WAY FLAT SLAB

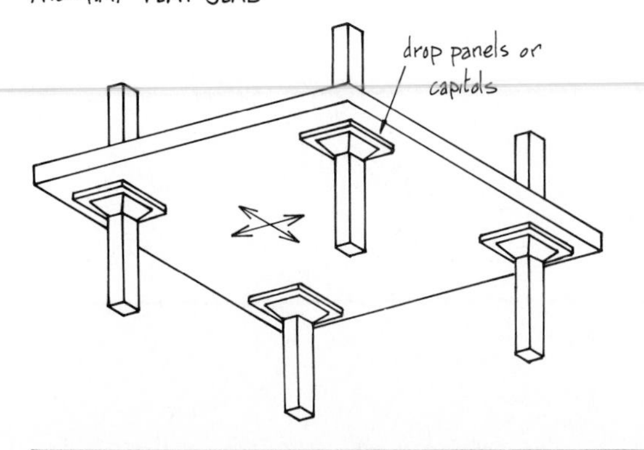

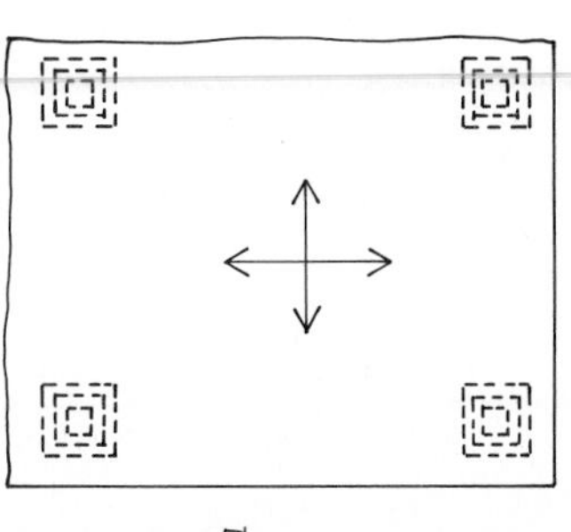

Plan

Two-Way Flat Slab

- Used mainly in industrial buildings with heavy loading.
- The square drop panels or column capitals carry the heavy loads and prevent the column from punching through the slab.
- Easy to form since no beams or girders are required.

CELLULAR-STEEL FLOOR DECKS

Celluar-steel floors made with light-gage steel decking are used extensively in structures framed with structural steel or open-web steel joists. The examples shown on this page are typical of the many types and styles produced by manufacturers. Floor systems of this type have many advantages:

- Rapid erection. Floor units are spot welded to the supporting steel frame and thereafter provide a working platform for following trades.
- Services. When a closed cellular section is used, power light and communication services can be run inside the cells. These services can be accessable to any position on the floor above or ceiling below.
- Permanent forms. The steel deck is topped with concrete to fill the cells and provide a flat, stable surface for the finish flooring. Since the units are not removed, costly formwork labor is eliminated.
- Fire resistance. The concrete topping provides an excellent fire rating for the floor structure. A fire-rated ceiling can also be added to increase the total fire rating.
- Dead load. The total weight of cellular steel floor systems is low in comparison to most other systems and allows spans of up to 13 ft. (4 m) under some conditions.

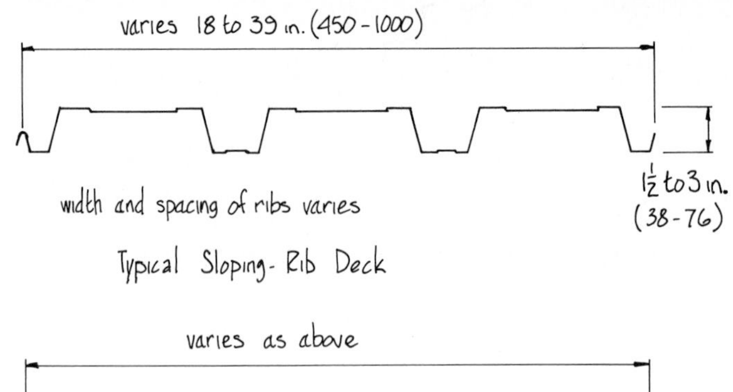

Typical Sloping-Rib Deck

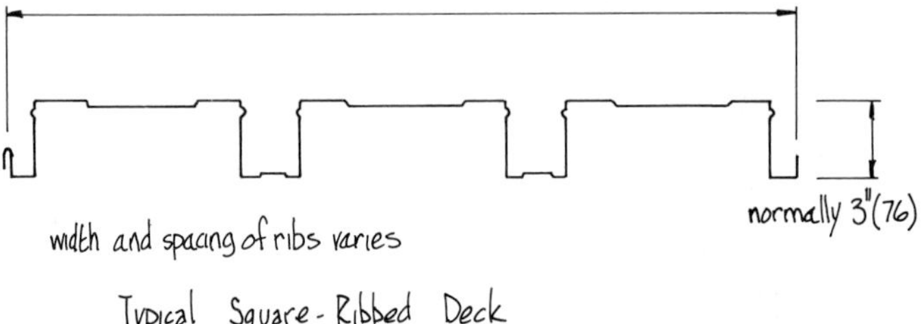

Typical Square-Ribbed Deck

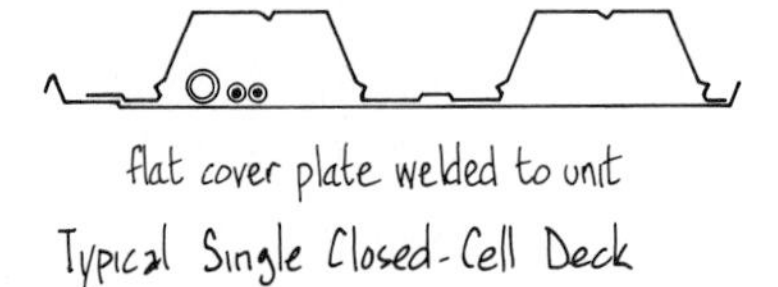

Typical Single Closed-Cell Deck

Typical Double Closed-Cell Deck

Closed-cell units can be used to run electrical and communication services to any location on the floor when used with raceway headers (see next page). At left are two examples of closed-cell decks.

Composite Cellular Steel and Concrete Floor.

Most cellular steel floor systems are designed so that the concrete topping is mechanically bonded to the steel units, thus creating a composite floor structure. Bonding is made by indentations and embossments formed in the steel unit during fabrication that lock the concrete to the steel. With vertical separation prevented and horizontal stress transferred the floor becomes a highly efficient composite structure. The concrete acts as an effective compression member while the units act as steel beams and resist tensile forces. In most designs the steel units are chosen to support the dead load of the concrete with the total live load being supported when the concrete has reached full strength.

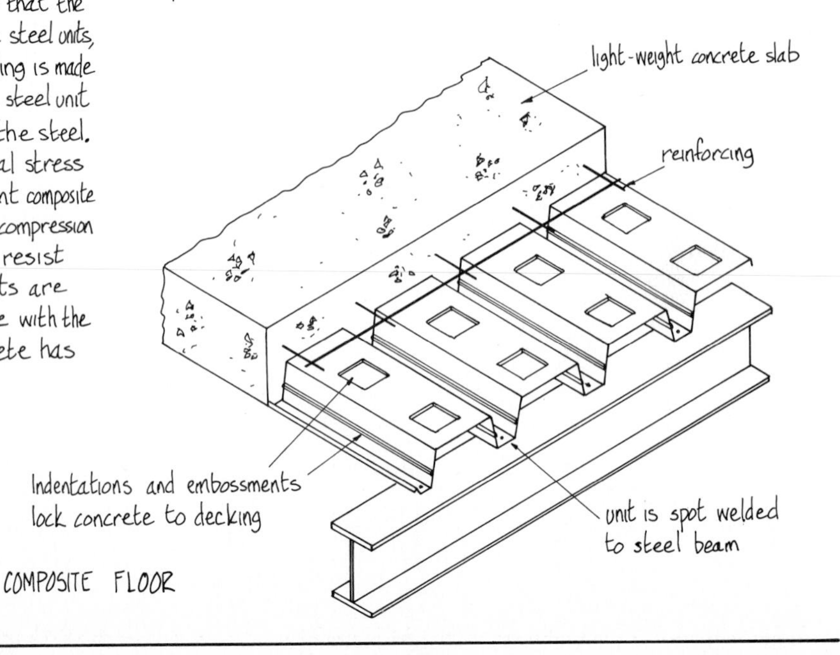

COMPOSITE FLOOR

CELLULAR DUCT SYSTEMS

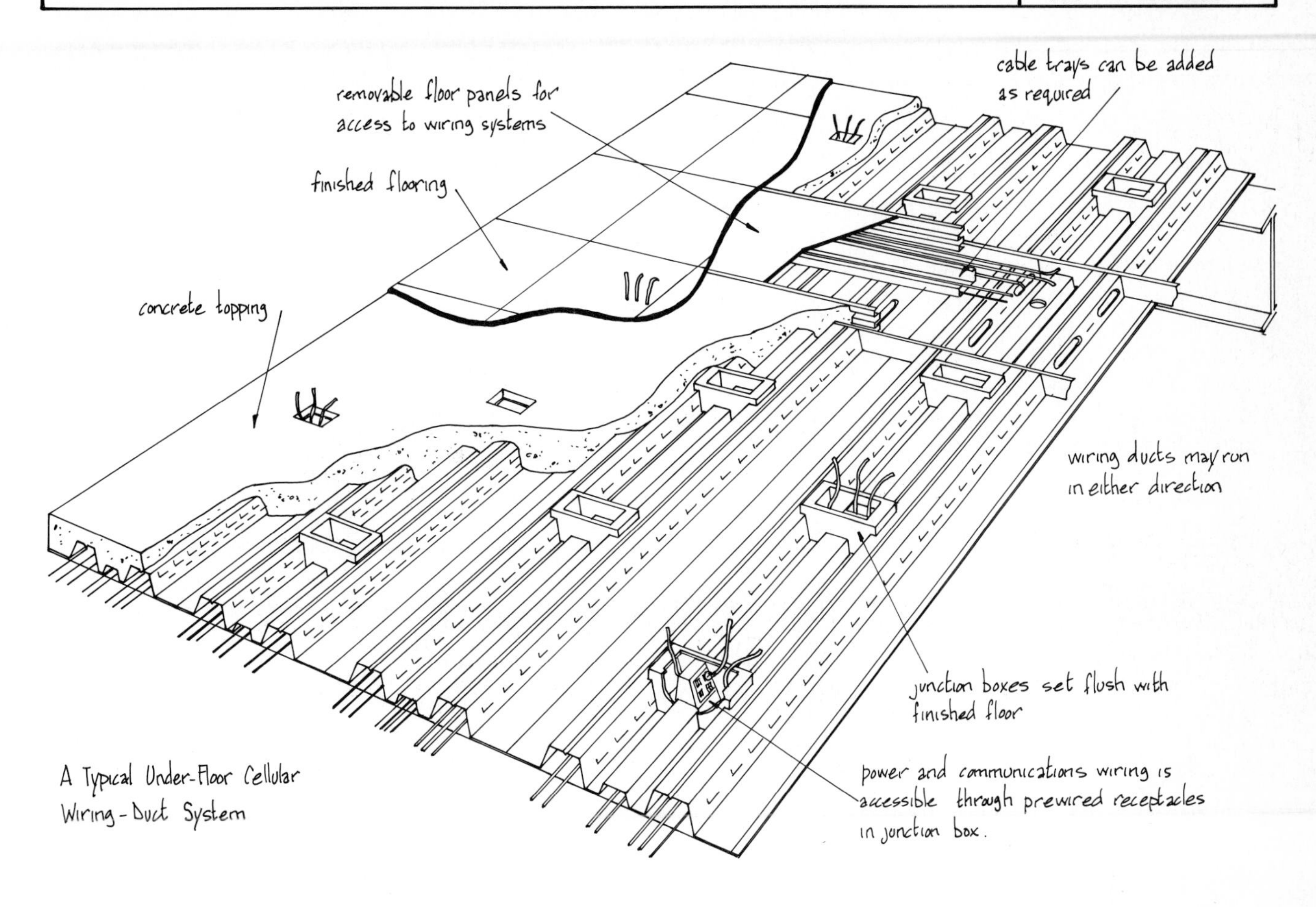

A Typical Under-Floor Cellular Wiring-Duct System

All large office complexes require some type of underfloor duct system to serve the wide range of electronic and communications equipment in use today. The system shown above is typical of the many commercially available cellular steel floor designs. It allows all service connections to be hidden beneath the floor finishes. Other systems provide "tombstone"-type connection devices above the finished floor. In most systems, the outlet locations must be preplanned in the initial design stage. However, provided that outlets are set in a suitable grid pattern, the initial workstation locations and inevitable future changes are easily accommodated.

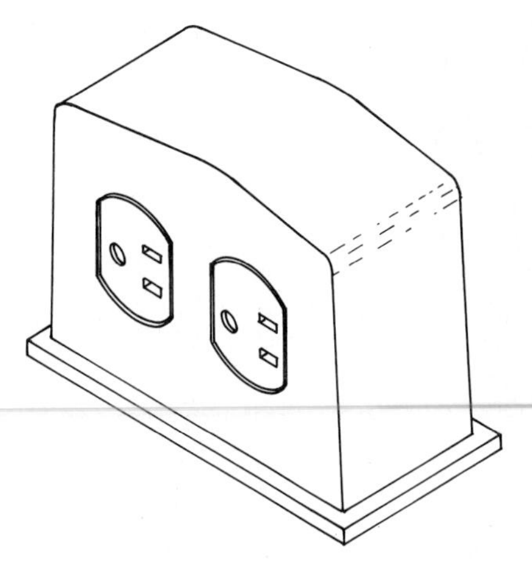

Example of the "tombstone" type of floor service connection.

source: H.H. Robertson Inc.

ACCESS FLOORS (COMPUTER FLOORS)

ACCESS FLOORS (COMPUTER FLOORS)

This type of flooring is composed of individual floor panels supported above the subfloor by an adjustable pedestal system. Each panel is easily removable, allowing total access to any part of the floor cavity. All communication, electrical, control, and mechanical services can be directed to any part of the room with extreme flexibility both in the initial setup and during later modifications. Floor finishes duplicate the full range of standard materials and can vary in general appearance from individual squares to totally monolithic.

The floor systems, which can be designed to withstand heavy equipment loadings, employ three basic panel attachment methods:

- A system of rigid stringer bars that snap-lock to the adjustable pedestals, supporting and positioning the floor panels, and providing good lateral stability. This type may also have a vinyl strip on the stringer top which seals the panel joints and allows the floor cavity to be used as an air plenum. The illustrations on this page are of this type.
- Panel attachment directly to freestanding pedestals by means of corner bolts or other mechanical fasteners. Stringers are not required in this case.
- Simple placement of panels on the pedestals without an attachment fastener other than corner guides. Stringers may or may not be used.

Manufacturers of these systems also supply a wide range of floor outlets for electrical and mechanical services. Outlets are usually built into special floor panels that can be relocated easily and quickly. Floor heights range from 4 in (100) to 36 in (1000) with a minimum 3-in (150) adjustment. Panel size is a 24-in (600) modular square.

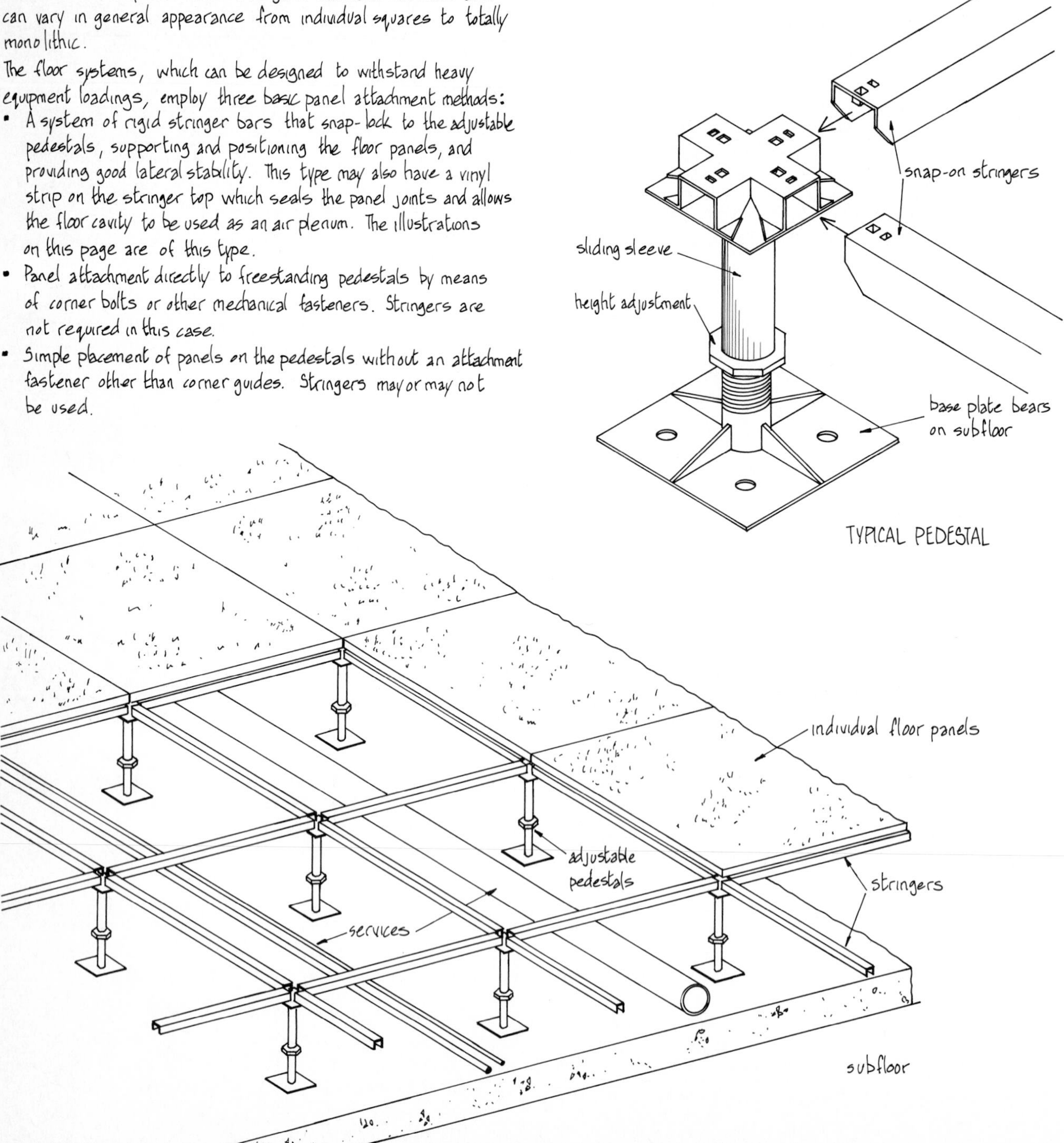

WALLS

STUD FRAMING

Wall systems perform two functions; they support the floors and roof and provide a nailing surface at regular intervals for the interior and exterior finishes. The exact construction details used by individual carpenters will vary with individual training and availability of local materials, but the basic concepts of stud wall framing remain the same. There are local variations in the names and terms used for some framing members, but again, the basic framing concepts hold true.

Because platform framing is the preferred method in modern construction, all details in this section deal with that method, with the one exception of some balloon framing details on 7-11. Before starting this section, re-read 6-4 and 6-5 for a review of both framing systems and 6-2 and 6-3 for a review of the materials used in the framing.

Together with the traditional framing techniques, (based on a 16 in (400) module), you will find information on a cost-saving framing system that is now standard construction in many areas of the country. This system uses a 24 in (600) o/c module for all elements of the frame and requires that the framing be preplanned before construction starts (see 7-12 for details).

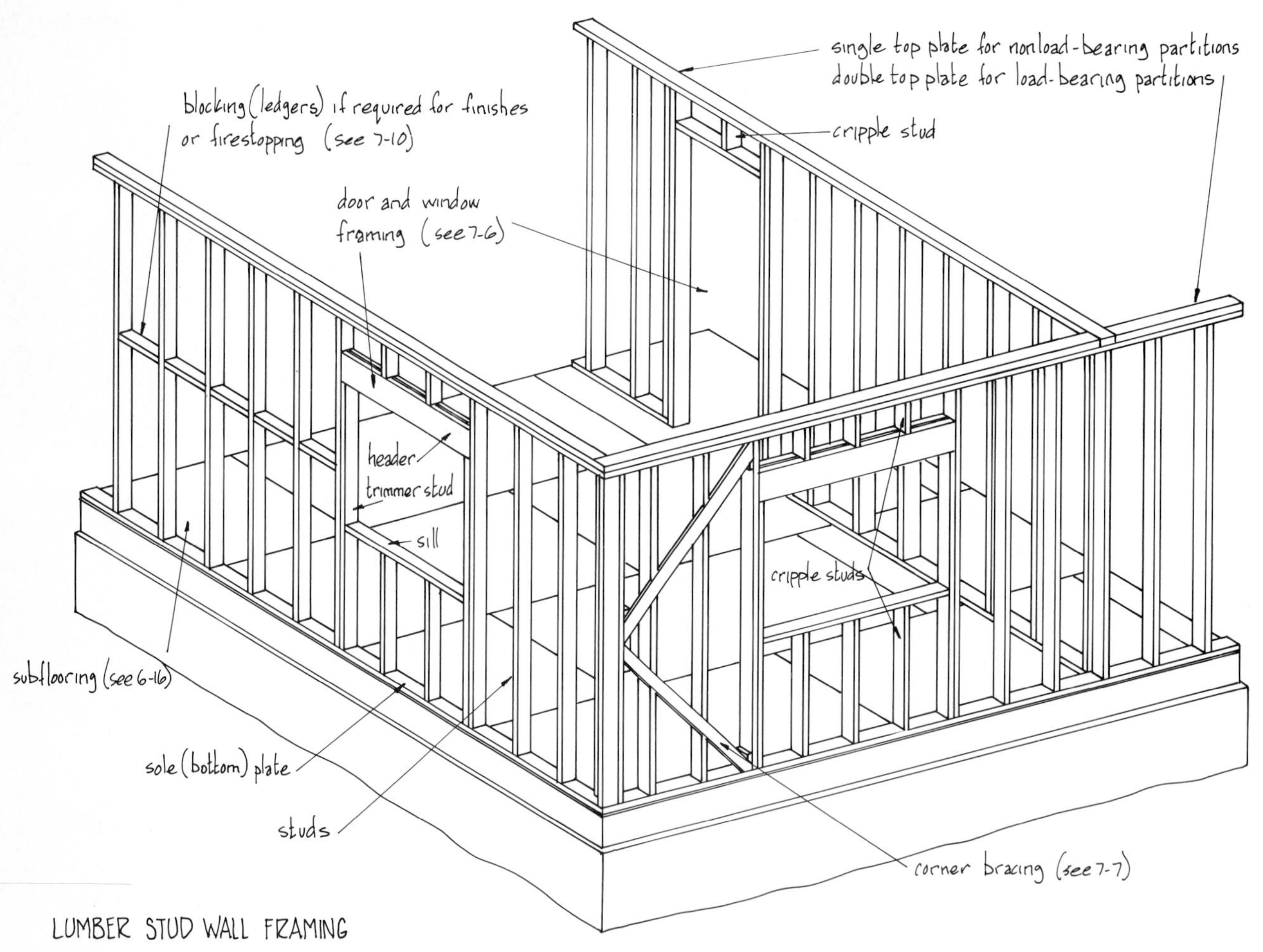

LUMBER STUD WALL FRAMING

The wall shown above are typical of standard traditional stud wall construction for residential and light commercial structures.
The walls are assembled with horizontal and vertical members comprising:

- Horizontal sole (bottom) plate.
- Vertical studs.
- Horizontal top plates.
- Other members – cripples, trimmers, blocking, headers, and bracing.

Each part of the wall structure is detailed on the following pages.

STANDARD STUDS

STANDARD WALL HEIGHTS

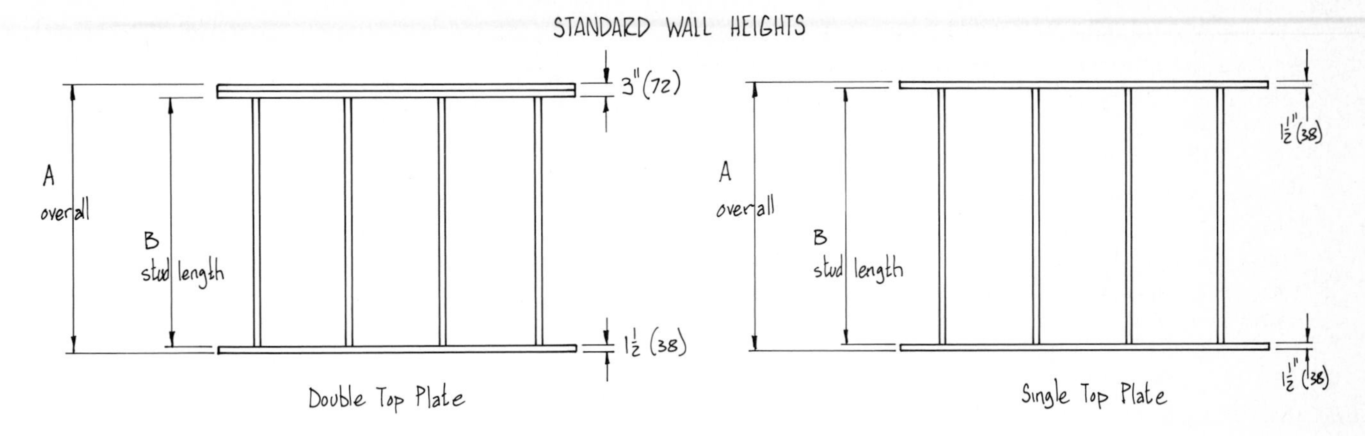

	FINISHED CEILING HEIGHT 8'-0" (2400)		FINISHED CEILING HEIGHT 7'-6" (2300)	
	A	B	A	B
double top plate	8'-1" (2425)	7'-8½" (2310)	7'-7" (2325)	7'-2½" (2210)
single top plate	8'-1" (2425)	7'-10" (2350)	7'-7" (2325)	7'-4" (2250)

Notes to chart

- Wall heights are based on the use of standard 8'-0" (2400) drywall or paneling sheets. Structure height is normally 8'-1" (2425) to allow maneuvering of sheets into place.
- The metric sizes given are "hard".
- Use of double or single top plates is determined by the loading on the wall. Load-bearing external walls and partitions have double plates; non-load-bearing partitions can be single plate.
- For these standard wall heights, precut (precision end trimmed) studs can be ordered.
- Walls over 10 ft (3000) high must have firestops in the stud spaces.

General Fabrication and Erection Procedures:

The general methods of stud framing have developed through the years into a fast and economical system that allows designers and contractors to create lightweight structures in a vast variety of styles and forms. Under most conditions, unless precise positioning of items such as windows or doors is required, the designer does not need to specify exact construction details. The carpenters will interpret the blueprints and frame the structure in their own way, with due allowance for economics and codes.

Other points relating specifically to platform construction are:

- All walls are normally set-out and fabricated horizontally on the floor deck before being raised into vertical position. A single top plate is used at this time. Exterior sheathing can also be applied to the wall before erection, eliminating the need to scaffold for this operation.
- After the walls are raised, temporary bracing is needed until the walls have been plumbed and nailed together at the corners or intersections.
- With the walls in their final positions, the second top plate is added so that it overlaps all joints and corners in the first plate.

STUD WALLS - CORNER DETAILS

Before prefabrication on the subfloor platform, stud walls are designated as either butt-walls or by-walls, since this determines corner framing details and erection sequence. Butt-walls butt against by-walls as shown at the right.

Other items to consider are:

- The corner configuration must provide support for the interior finishes. This can be achieved in a number of ways using studs, plates or clips.
- Energy conservation should be taken into account. The greater the number and volume of studs in a corner, the greater the heat loss. Ideally, the stud arrangement should provide the smallest possible path for heat conduction.

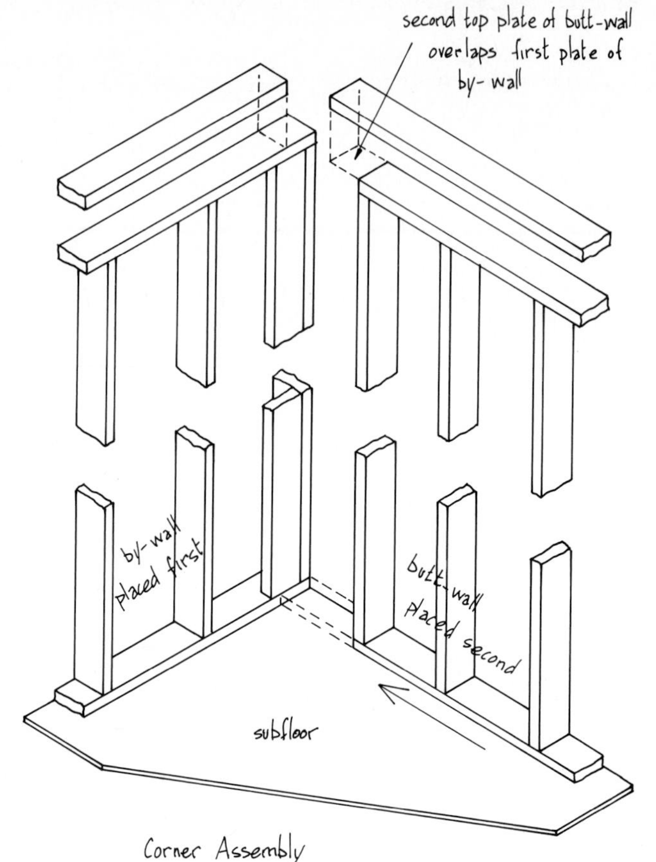

Corner Assembly

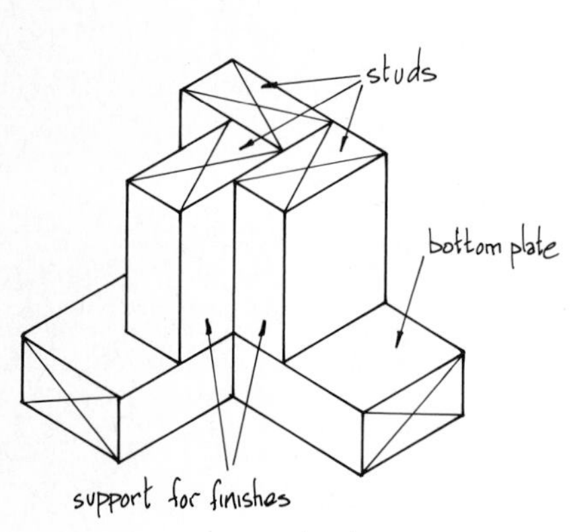

Corner Pictorial

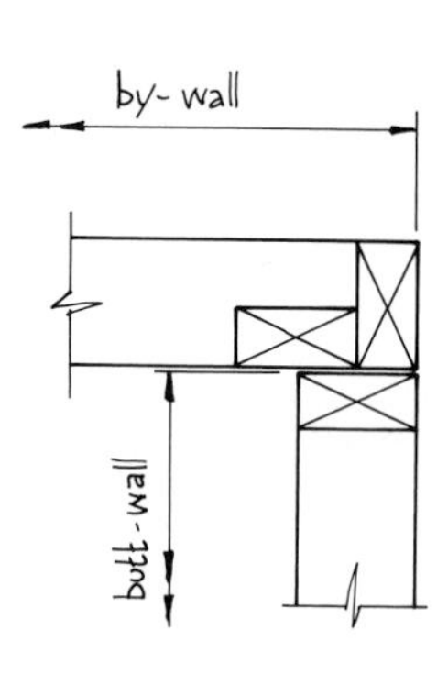

Plan at Corner

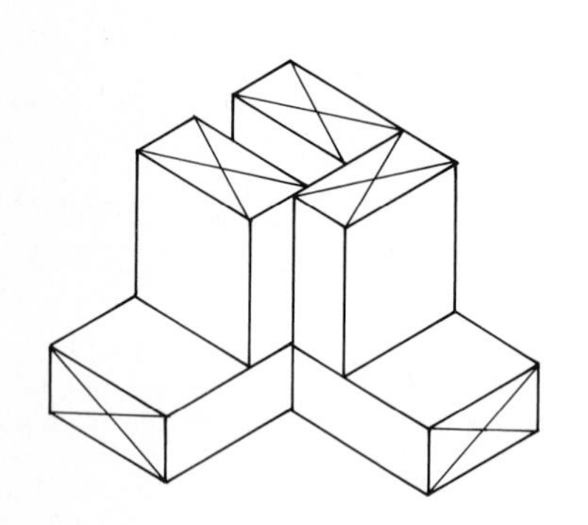

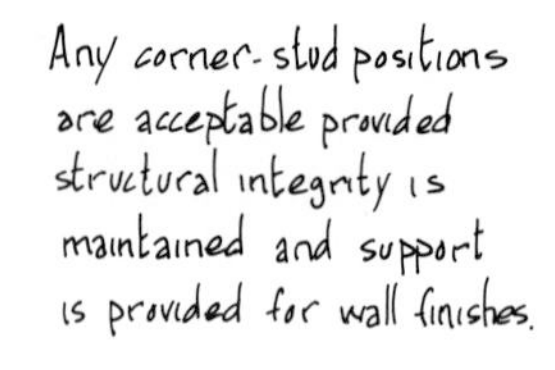

Any corner-stud positions are acceptable provided structural integrity is maintained and support is provided for wall finishes.

Alternate Corner-Stud Positions

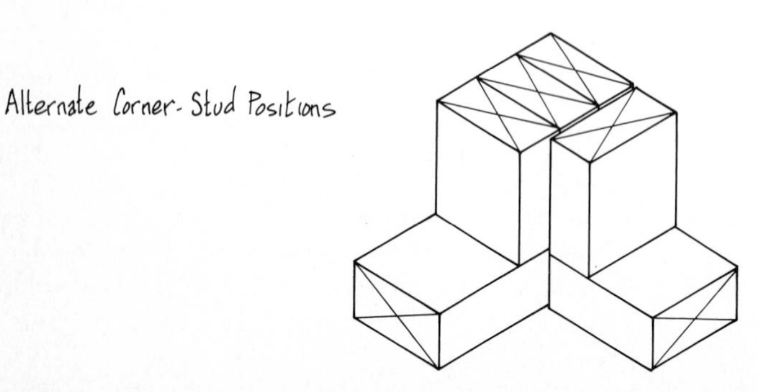

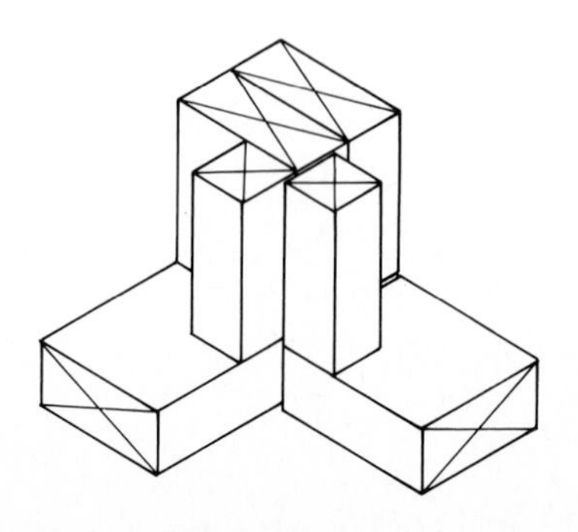

INTERSECTIONS – STUD SPACING

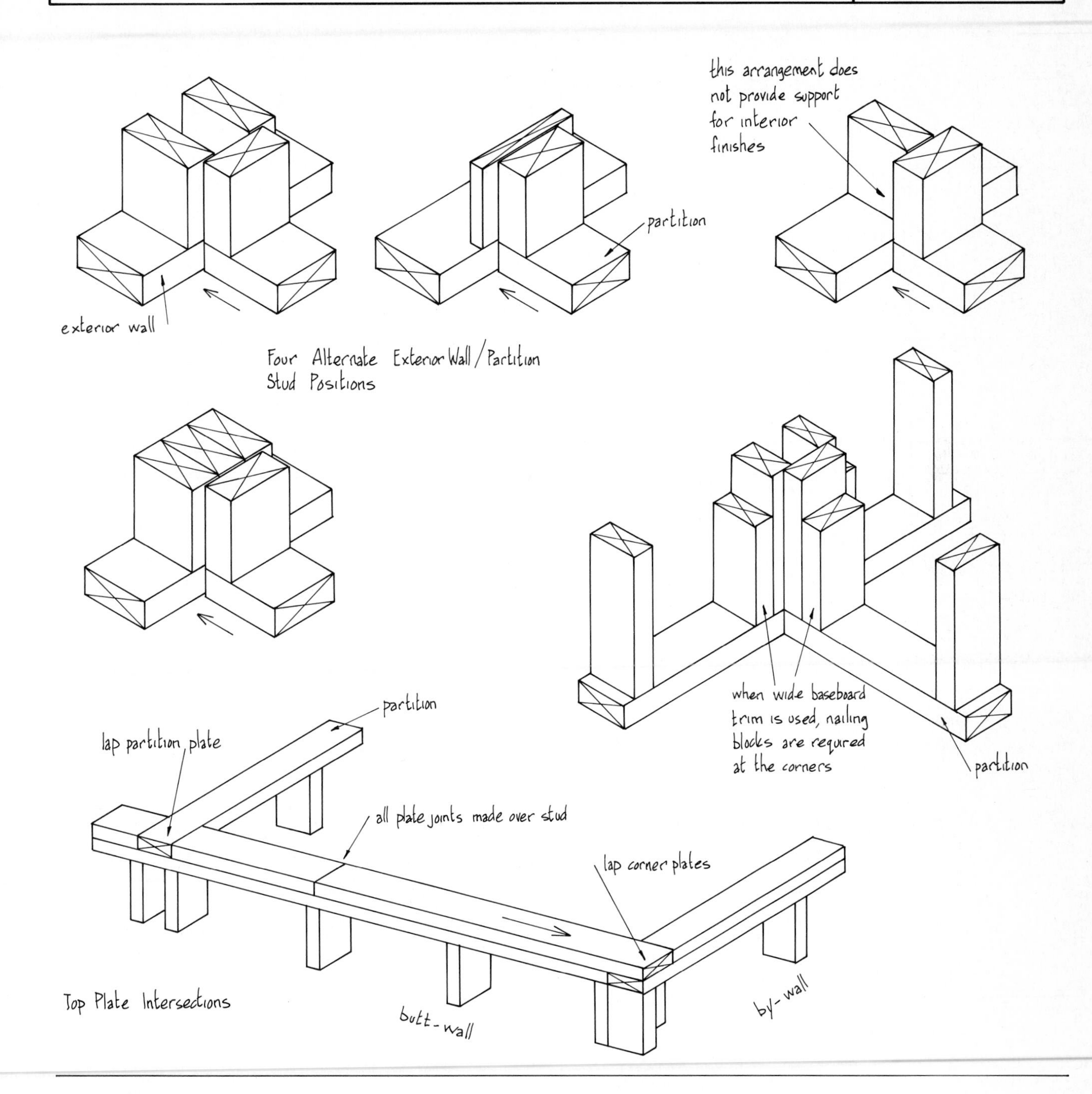

STUD SPACING

Stud sizes for exterior walls are usually 2×4 in (38×89) or 2×6 in (38×150) with spacing at 16- or 24-in (400, 600) centers to suit standard-width panel type finishes. A 12-in (300) spacing may be required for high walls or heavy load conditions. The choice of stud size, spacing, and maximum unsupported length generally depends on the number of supported floors, and is specified in local codes.

Thermally upgraded walls may require some deviation from traditional methods (see 10-18 onward for details).
Loadbearing interior stud walls are sized and framed in the same way as exterior walls. Non-loadbearing interior walls are generally framed with 2×4 in (38×89) studs at 16- or 24-in (400, 600) centers, although 2×3 in (38×64) studs may be used under some conditions.

DOOR AND WINDOW OPENINGS

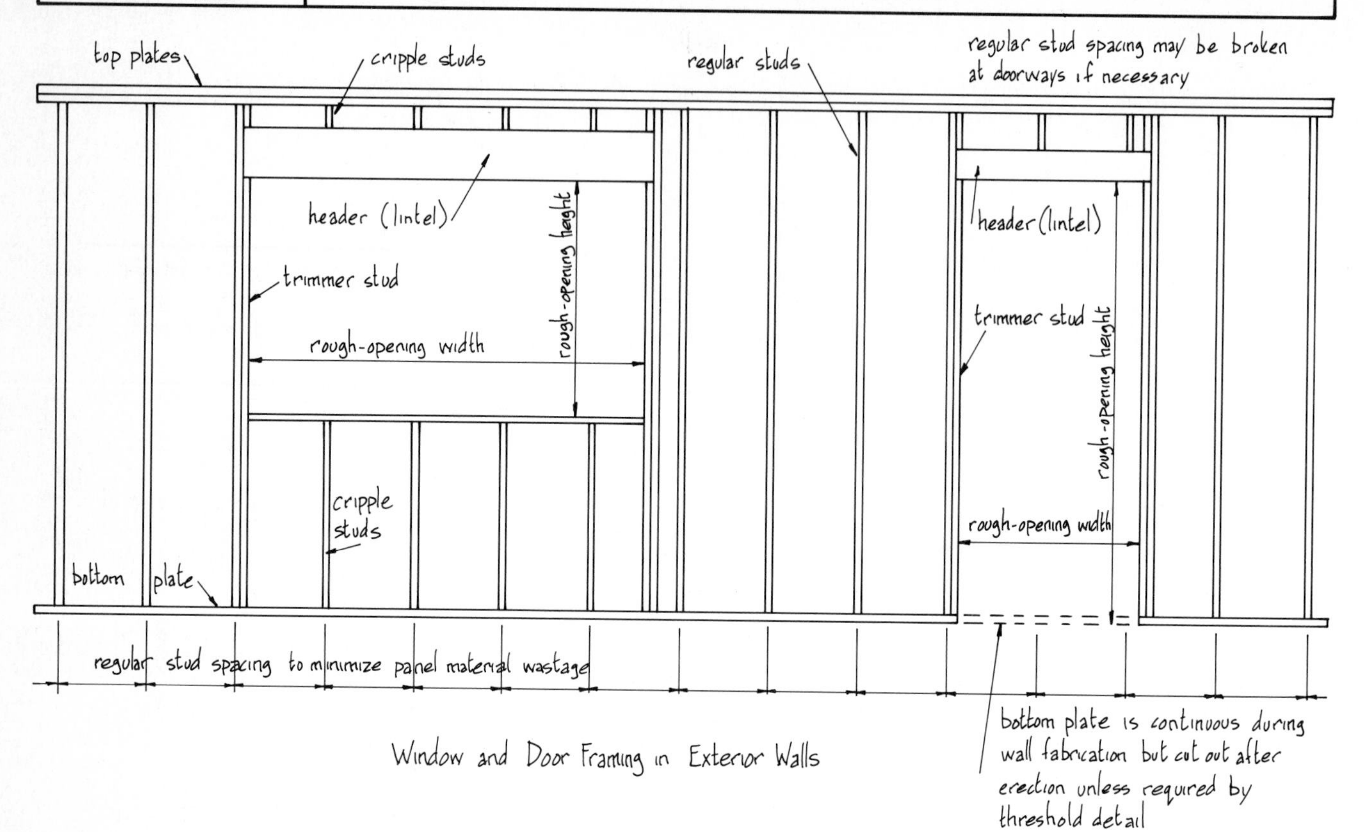

Window and Door Framing in Exterior Walls

Window and door units must be isolated from all structural loads. As shown above, a combination of headers (or lintels), trimmer studs, and cripple studs are used to transfer loads to the foundations. The framing members are also used as a nailing base for the window or door unit. See Chapter 12 for complete window and door information and 10-60 for installation methods.

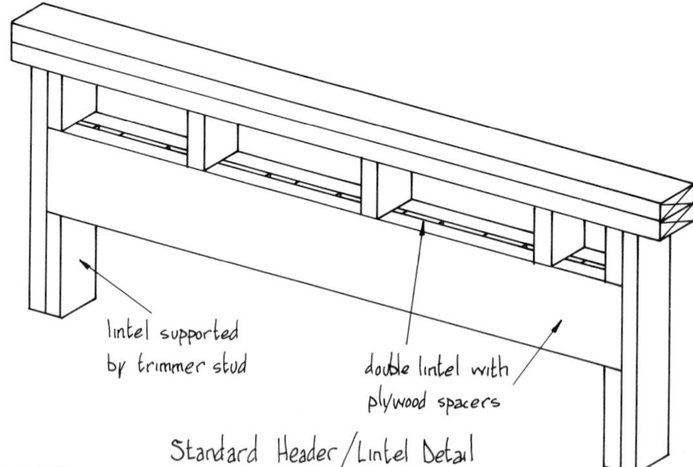

Standard Header/Lintel Detail

Typical Header/Lintel Spans - Exterior Walls

Nominal Depth of Lintel	Supporting One Floor, Ceiling, and Roof	Supporting Ceiling and Roof Only
4 in (89)	2'-0" (610)	4'-0" (1220)
6 in (140)	5'-0" (1525)	6'-0" (1830)
8 in (184)	7'-0" (2130)	8'-0" (2440)
10 in (235)	8'-0" (2440)	10'-0" (3050)
12 in (286)	9'-0" (2750)	12'-0" (3660)

Notes to table:

- Spans shown are for a standard double member of 2-in (38) nominal thickness lumber each.
- For all types of headers/lintels or loading conditions, refer to local building codes.
- Metric dimensions are "soft".

source: CMHC

HEADER/LINTEL TYPES

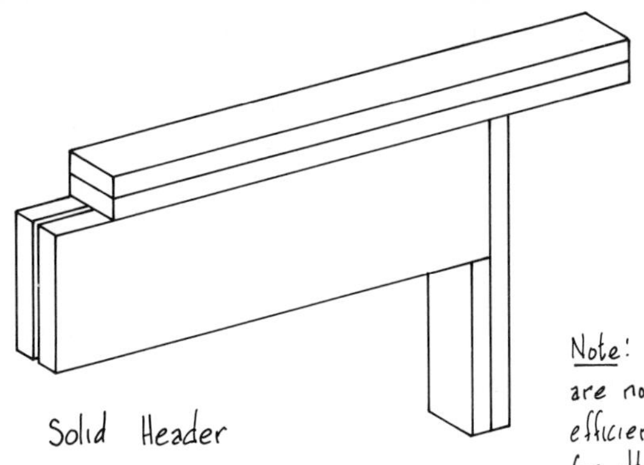

Solid Header

When the rough opening is close to the top plate, the header can be sized to completely fill the space between the opening and the plate, eliminating the short cripple studs. It is also common practice to use the same sized header for each series of openings, whatever the opening sizes. In some regions, a solid 4x12 in (89x286) header is used.

Note: solid or continuous headers are not recommended for energy efficient structures (see 10-47 for alternatives).

In walls where there are a large number of window and door openings, a continuous header is sometimes economical. Acting as a combined header and top plate, the continuous header allows a simplified rough-opening configuration.

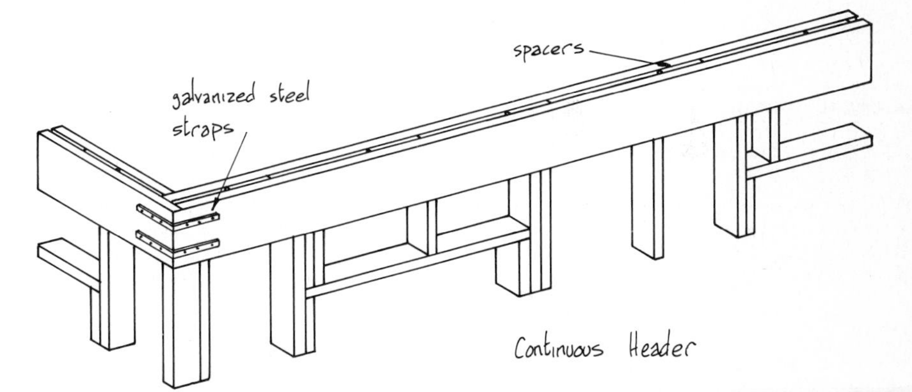

Continuous Header

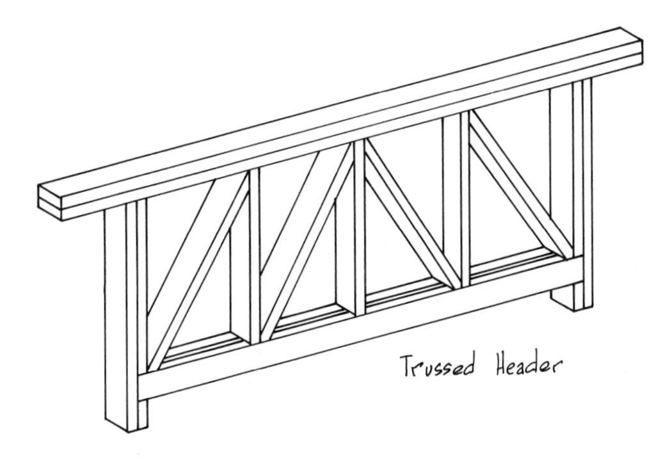

Trussed Header

In bearing walls where the loading is unusually heavy or where the span of an opening is unusually large, a trussed header may be necessary. This type of header should be engineered and is suitable only when enough headroom is available in the structure.

When long span, heavy loading, and minimum headroom applies, use of a steel header is necessary. As shown, the beam can be supported by jack posts, built-up studs, or solid lumber. Check codes for permissible supports.

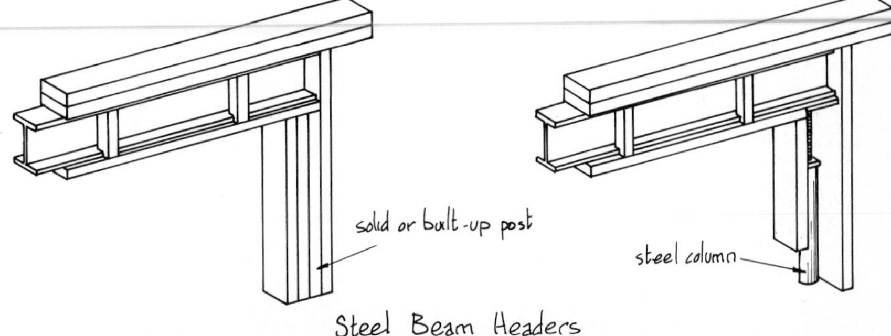

Steel Beam Headers

GABLE (RAKE) WALLS

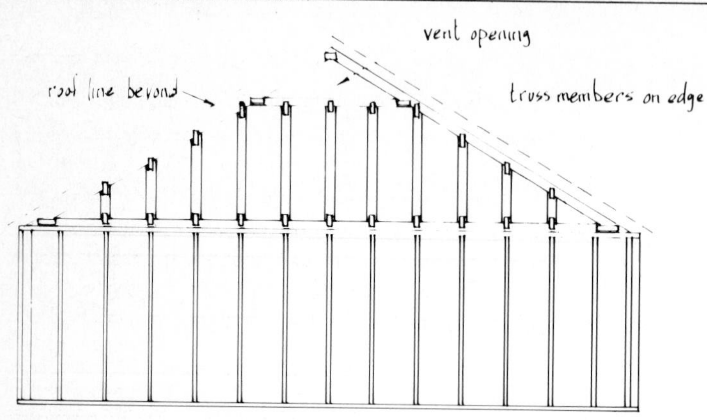

STANDARD-HEIGHT END WALL WITH GABLE TRUSS

Gable end (rake) walls can be framed in several ways:

- The standard-height side walls can be continued across the ends and a separate pitched frame added to form the gable. This is the commonly used method since standard stud lengths can be used for the most part.
- If the roof is framed with prefabricated trusses, special gable end trusses can be ordered to suit design requirements (see 8-10).
- The wall studs can be continued up to the sloping top plate. If the studs exceed 10 ft (3 m) in length, fitted horizontal firestops are required.

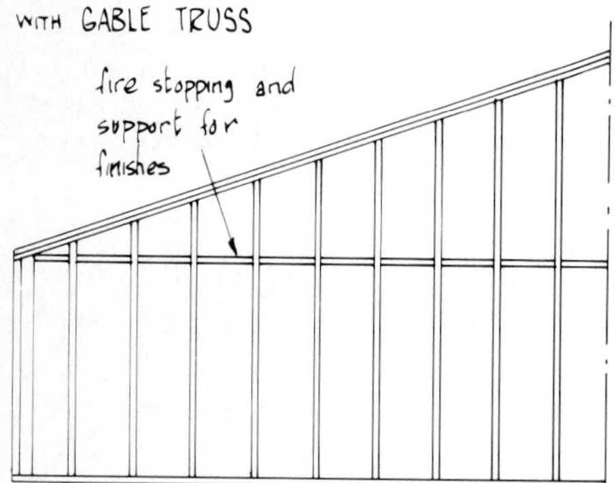

GABLE WALL WITH CONTINUOUS STUDS

The size and position of the top plate must also be considered. Rake overhangs are generally supported on the top plate, usually by a system of short "lookout" members. Three examples of typical top plate configurations are shown below.

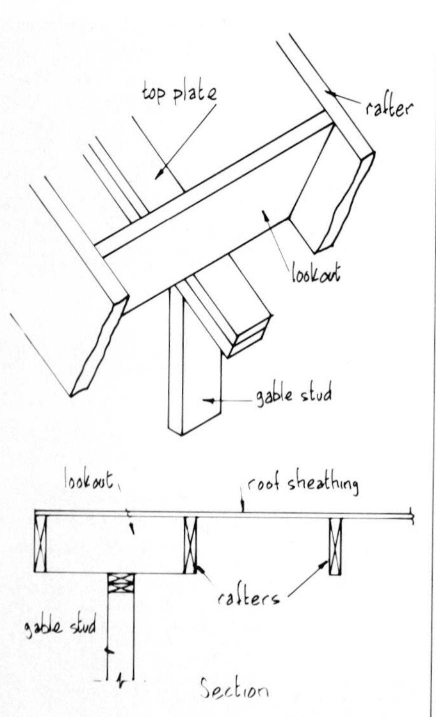

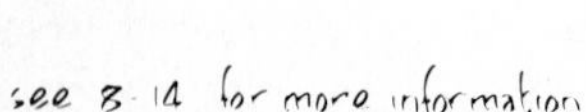

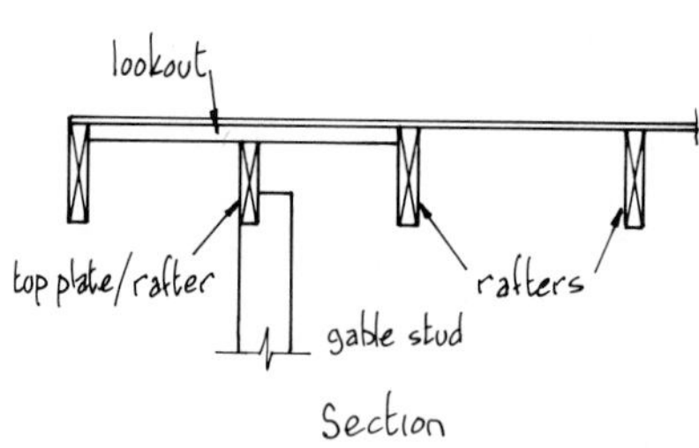

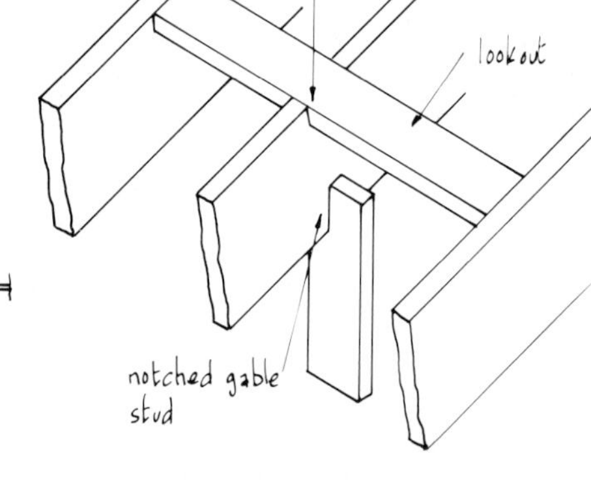

See 8-14 for more information.

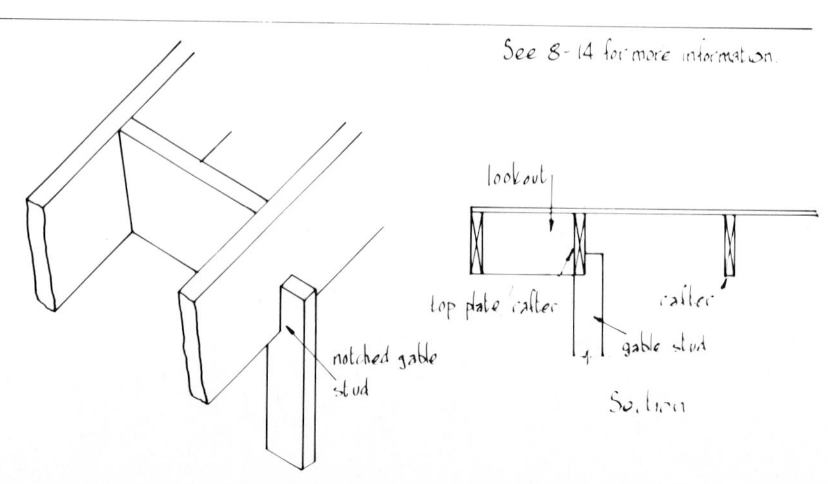

DIAGONAL BRACING

All stud structures must have some form of sway or wind bracing to provide lateral restraint to the walls. Restaint can be provided by exterior plywood or composition-board wall sheathing but these materials are now often replaced by nonstructural materials such as fiberboard, gypsum board, and semirigid insulation boards (see 11-16). To maintain lateral restraint, bracing members must be added to the walls when non-structural sheathing is used.

Bracing, installed on both sides of each corner, is normally added in two forms:

- Let-in bracing where the outside faces of the studs are notched to accept a 1×4 in (19×89) board set flush with each stud face. This method is fairly fast and allows almost full insulation in the stud space. Metal T-shaped bracing, requiring minimal saw cuts, is also used extensively.
- Cut-in bracing where 2-in x lumber is fitted between each stud at a 45° angle. This method is slower and displaces insulation in the stud space.

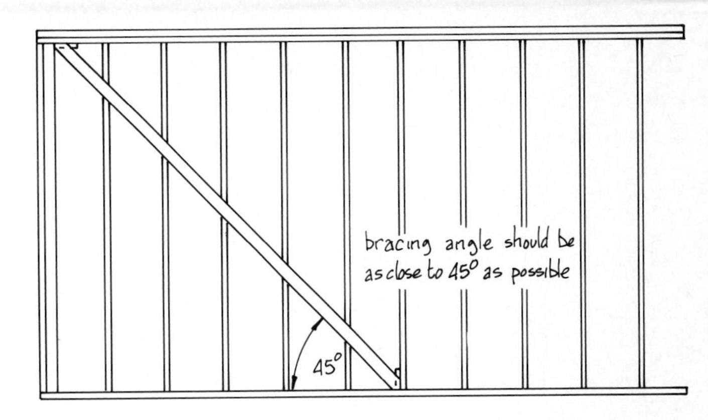

CUT-IN DIAGONAL BRACING

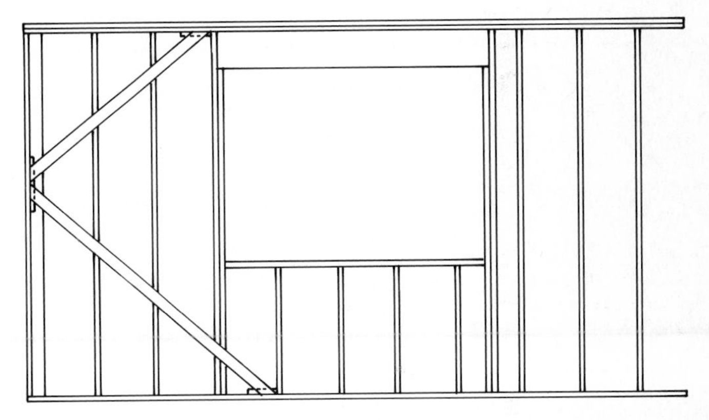

CUT-IN BRACING WHERE OBSTRUCTIONS OCCUR

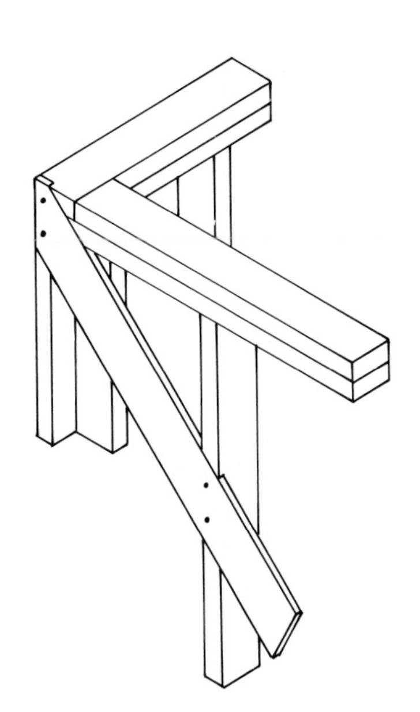

LET-IN 1"×4" (19×89) BRACING

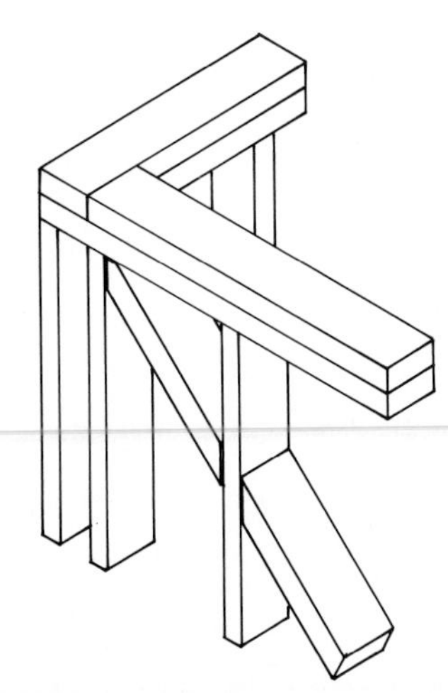

CUT-IN 2"× (38×) BRACING

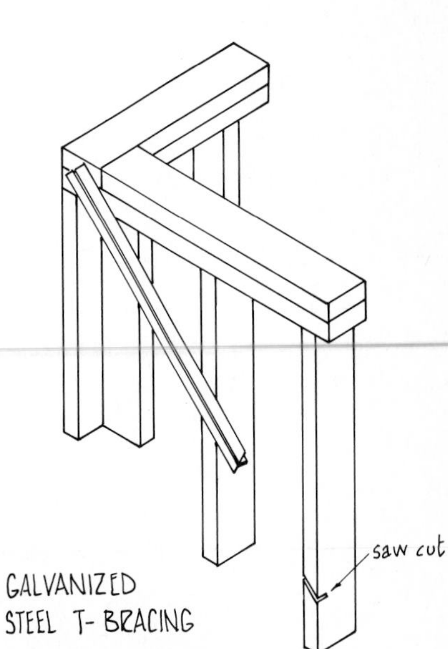

GALVANIZED STEEL T-BRACING

SPECIAL CONSTRUCTION AND LEDGERS

Special construction in this case refers to details such as bath and bathroom fixture supports, holes for heating equipment, etc., all of which must be properly framed to provide adequate strength and stiffness. Examples on this page show typical framing methods.

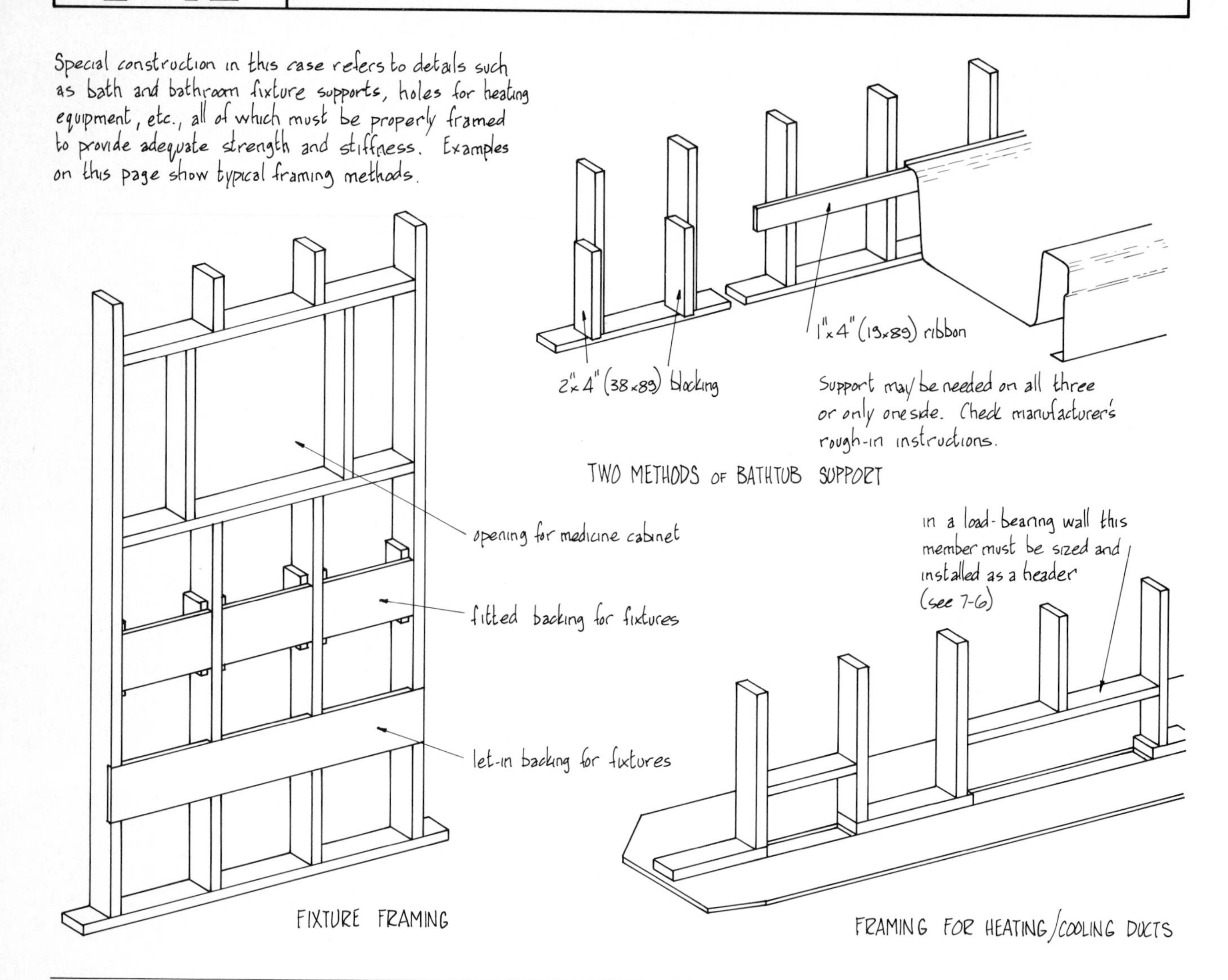

Ledgers are framing members inserted between the vertical wall studs. They are required on all walls over 8 ft (2400) high but on standard-height walls only when called for by the finishing product. Standard ledger types are:

Type 1. Provides nailing surface on both sides of stud but no space for insulation. Staggered for easy nailing.

Type 2. Provides nailing surface on one side of stud and space for insulation.

Type 3. Provides a continuous, level, one-sided nailing surface with room for insulation. Requires letting-in to stud face.

Ledgers must not be staggered where panel-type finishes or sheathing are to be used.

LEDGERS

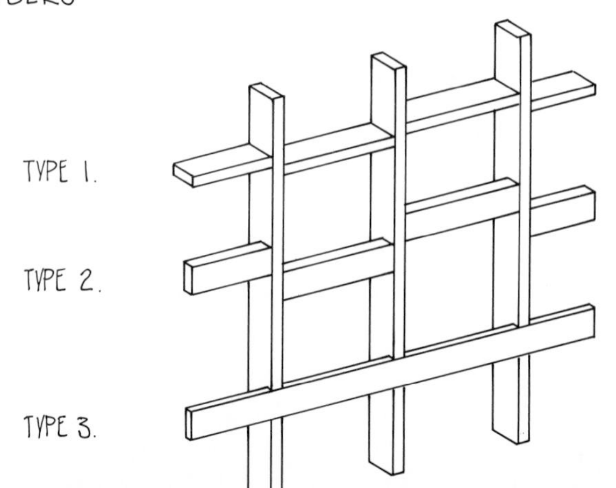

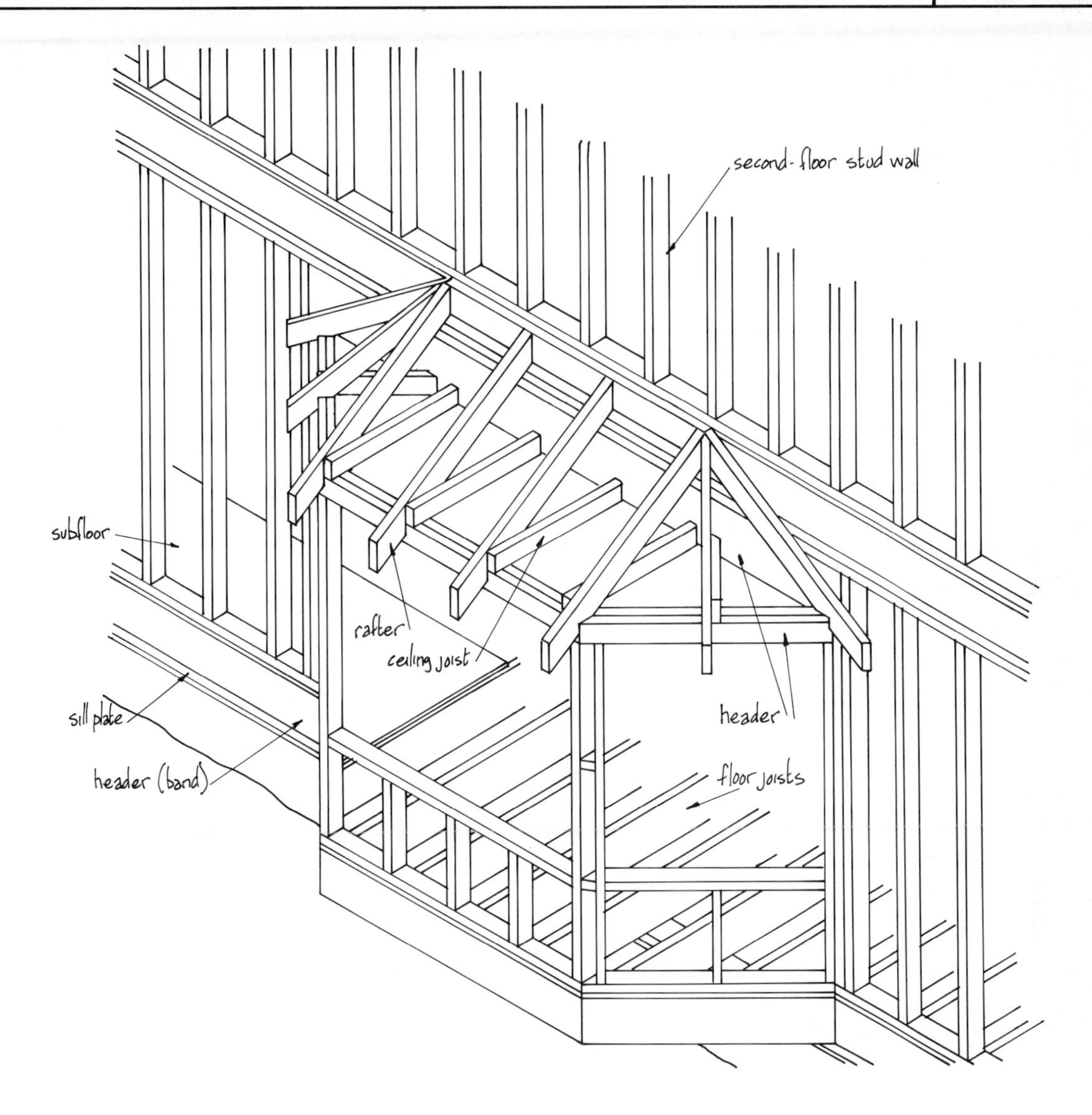

Stud wall systems are extremely flexible and a variety of special walls can be constructed. A bay window is shown here as an example. In this case, provided that the designer gives rough opening sizes for the windows together with a minimum of layout information, the exact construction details can be left to a competent carpenter. Other, more complex structures may require considerably more information from the designer.

Balloon Frame Construction

Balloon framing wall details can be derived from the floor framing illustrations on 6-5, 6-10 and 6-13.

source: National Forest Products Assoc.

PREPLANNED 24 IN (600) MODULE

The traditional framing techniques described so far have been based on a 16-in (400) module. By using a 24-in (600) spacing module and carefully preplanning the placement of all elements of a structure, significant cost savings can be realized. In general, savings accrue from:

- Less lumber used in each main element: floors, walls, and roofs.
- Less labor required for all phases of construction: inventory, handling, and installation.

The design and construction of this system is approved by all major regulatory authorities and is a well-established technique in many regions. Although it is particularly effective in mutiple housing projects where economy of scale is an advantage, low-multiple, or custom housing will also see cost benefits.

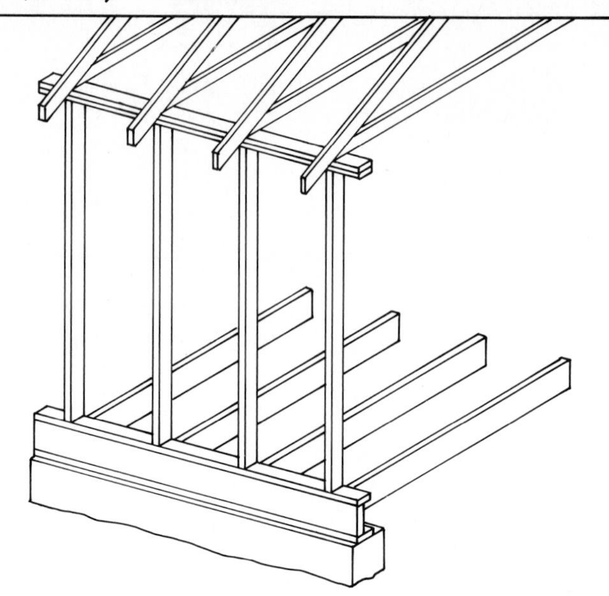

Framing members — joists, studs, and trusses — are aligned with each other, thus forming a series of in-line frames to best utilize the strengths of the structural system.

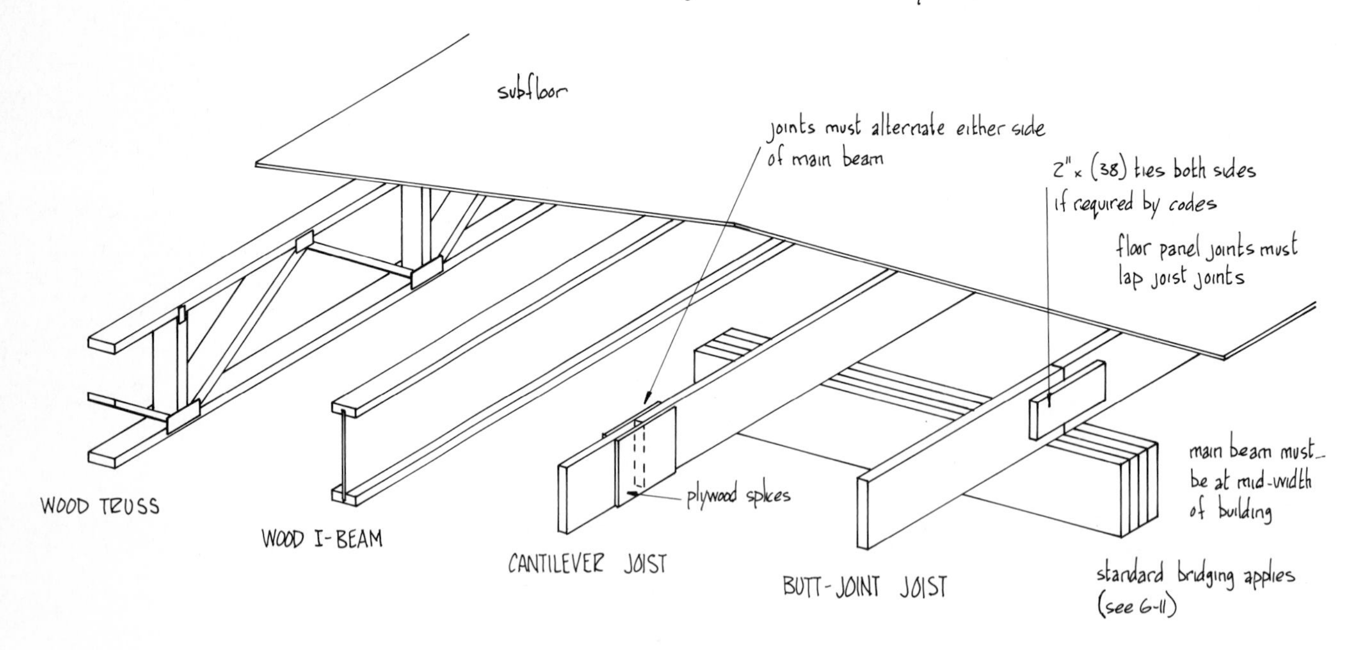

FLOORS

- Floor members should be in-line between exterior wall bearings. Where codes allow, lumber joists can be either butt jointed at center bearing or cantilevered over center bearing.
- Wood I-beams and wood trusses are ideally suited to the 24-in (600) module since they are designed for long continuous spans (see 6-18 and 6-20).
- Standard subfloor systems can be used provided that the material chosen is adequate for the 24-in (600) span. Glued and nailed plywood panel systems are generally recommended by lumber industry associations (see 6-16).
- Span tables should also be carefully consulted when sizing lumber joists; the wider spacing imposes additional loading on the floor.

PREPLANNED 24 IN (600) MODULE

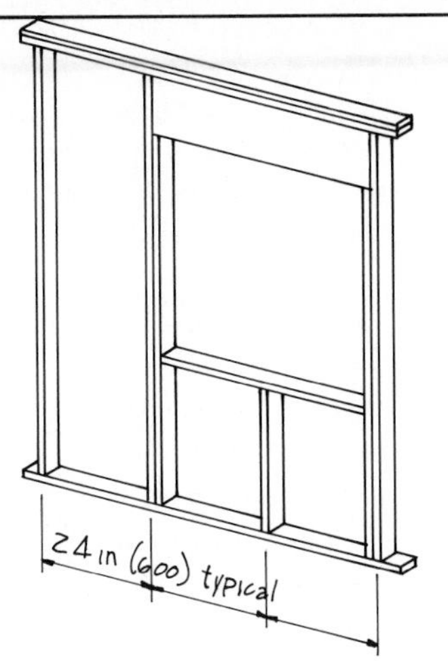

Where possible, all openings are located to coincide with the 24 in (600) module.

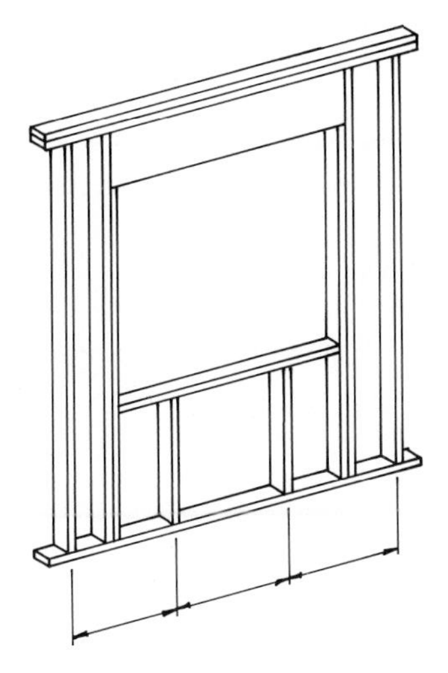

Openings can be located off the module but will require greater quantities of framing material.

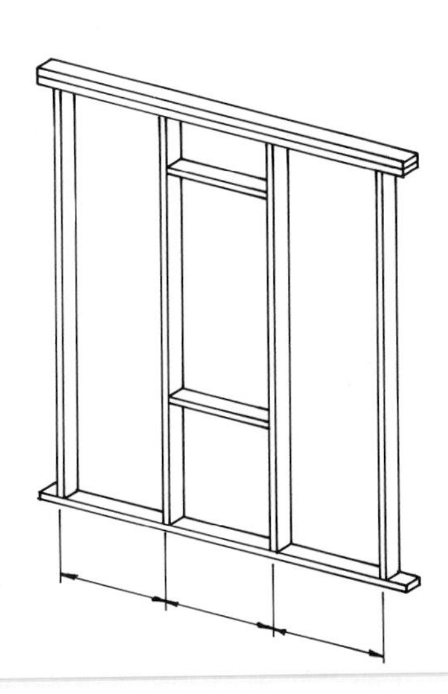

Windows or other items that fit within the 24 in (600) module can be framed without headers, trimmers, or cripples.

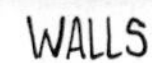

WALLS

In general, walls are framed using details similar to the traditional 16-in (400) module with the following additional steps:

- Studs are located directly in-line with floor joists and roof trusses.
- Under most codes, 2×4 in (38×89)s at 24 in (600) o/c are limited to 10-ft (3000)-high walls in single-story structures.
- Where codes allow this stud size and spacing in the lower walls of two-story structures, wall height should be limited to 8 ft (2400). In some areas, 2×6 in (38×140) studs or reduced spacing may be required for the lower story.
- Where possible, all openings for windows and doors are located so as to minimize the number of studs required. The drawings at the left show two typical examples.
- Code requirements for wall stiffening using structural sheathing or lumber wind bracing apply (see 11-16 for sheathing, and 7-9 for wind bracing).

ROOFS

In most modern construction, prefabricated roof trusses at 24 in (600) o/c are standard and provided that trusses are located directly over wall studs, little change in this existing system is required.

- When a traditional rafter system is used, check span tables to accommodate increased spacing.
- See 8-8 for trusses, and 8-16 for rafters.

ADDITIONAL INFORMATION

- The 24-in (600) module lends itself to increased energy efficiency. Wood framing members conduct heat faster than the cavity insulation, but since less framing is required in this system, there is also less heat loss through the structure. For more information on the importance of 24-in (600) stud spacing in energy efficient structures see 10-46 onward.
- Cost savings of the 24-in (600) module system over the traditional 16-in (400) system will vary depending on a contractor's current methods but should average 5-15%.
- Additional savings can be gained if the entire structure is designed on a rigid modular basis. Using a grid system based on the standard 4×8 ft (1200×2400) building sheet, all components of the building are aligned with dimensions of 4, 16, 24, or 48 in (100, 400, 600, or 1200). In this way, each component can be used to maximum advantage with a minimum of waste.

STEEL STUD SYSTEMS

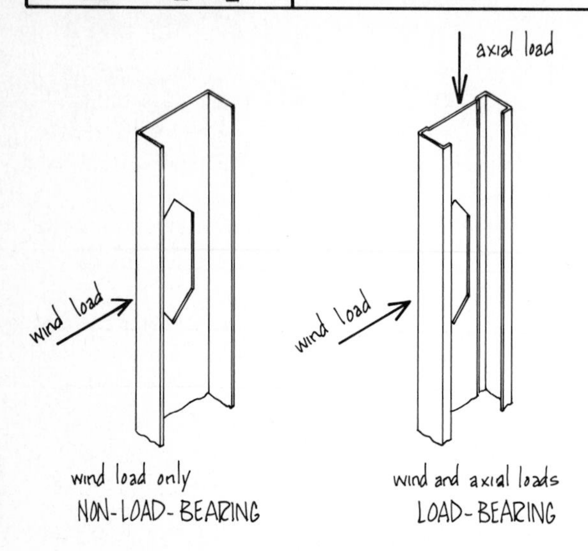

Studs are generally manufactured in two ways:

- Non-load-bearing, with the stud designed to carry a lateral wind load only. This type is fabricated from light-gage sheet and may or may not have a return lip on the flange. It is used for interior non-load-bearing partitions and for exterior curtain wall support (see 7-17 and 7-30) under wind-load-only conditions.
- Load-bearing, with the stud designed to carry wind and vertical (axial) loads. Fabricated from heavier gage sheet, this type has a larger return lip, web ribs, and can be used under most loading conditions.

Each type has factory-punched holes in the web to accommodate stiffening members and services.

Because of their many advantages, lightweight galvanized steel framing systems tend to be the preferred framing method in commercial and industrial construction. In residential markets the systems tend to be limited to apartment and multiresidence construction, although inroads are being made into the single-family and custom housing markets in some regions.

Although the materials and installation methods are different, the basic lightweight steel framing concept is very similar to the traditional wood stud system already described; in each case the wall is designed to support specific loads and to provide attachment for finishes. Advantages of steel systems are as follows:

- Because they are a manufactured product the components can be fabricated in a wide variety of types and sizes. Studs can be precut to any length and provided with uniformly spaced punched holes in stud webs for installation of bridging and services.
- Steel is not subject to insect infestation nor will it shrink, rot, or warp.
- Components are always of uniform quality, shape, and size, and are dimensionally stable.
- The sections are easily cut and quickly erected with screwed or welded connections.
- Components have a high strength-to-weight ratio, resulting in lightweight sections that reduce total loading on foundations.
- Cost-competitive with traditional lumber materials.

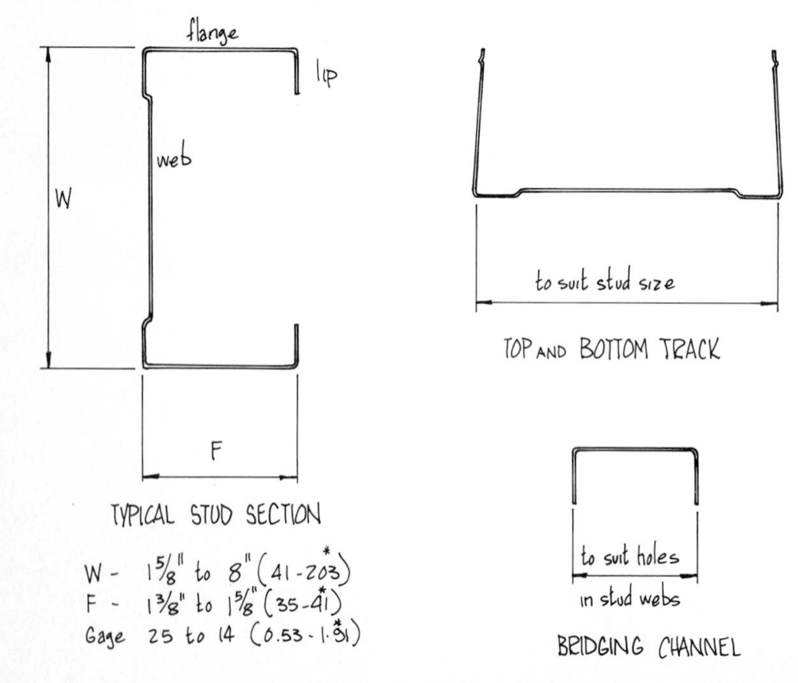

TYPICAL STUD SECTION

W - 1⅝" to 8" (41-203*)
F - 1⅜" to 1⅝" (35-41*)
Gage 25 to 14 (0.53-1.91*)

TOP AND BOTTOM TRACK

BRIDGING CHANNEL

TYPICAL SYSTEM COMPONENTS

The sections shown here are typical of most manufacturer's products. Refer to product specification tables for exact dimensions and loading capacities. The top or bottom channel track is sized to accept the studs in a friction fit and the bridging channel is sized to fit the punched holes in the stud's web.

STEEL STUD SYSTEMS

Studs may be connected to top and bottom channel tracks by one of four methods depending on design requirements:

- Friction fit — no mechanical fastening. Finishes screwed or glued to the studs provide the necessary stiffening.
- Screwed through channel and stud flanges on both sides.
- Cut and crimped channel flanges after stud installation.
- Welded connection.

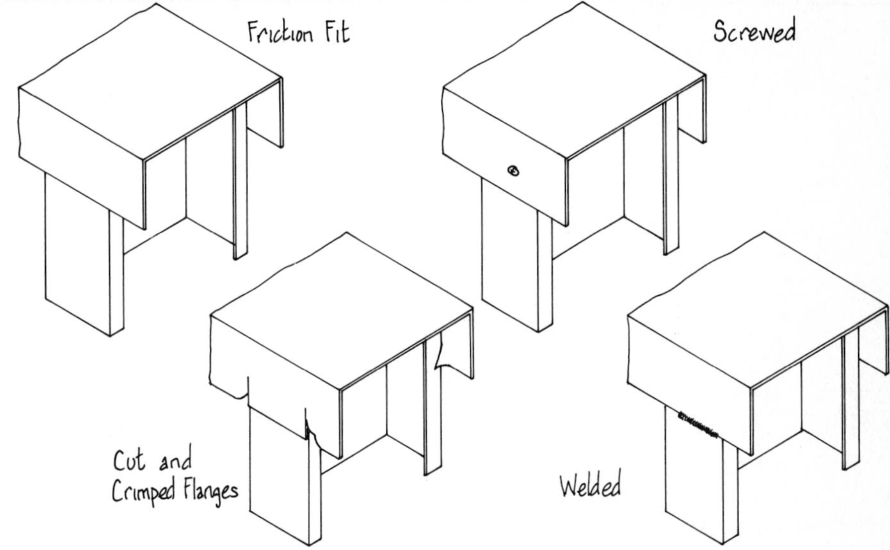

Bridging channels, added to increase wall stiffness, are attached to the stud in several ways. The three methods shown at the right are:

- Fitted through web holes and welded to hole sides.
- Fitted through holes and welded to clip angle.
- Bridging flanges cut and notched to fit stud flanges.

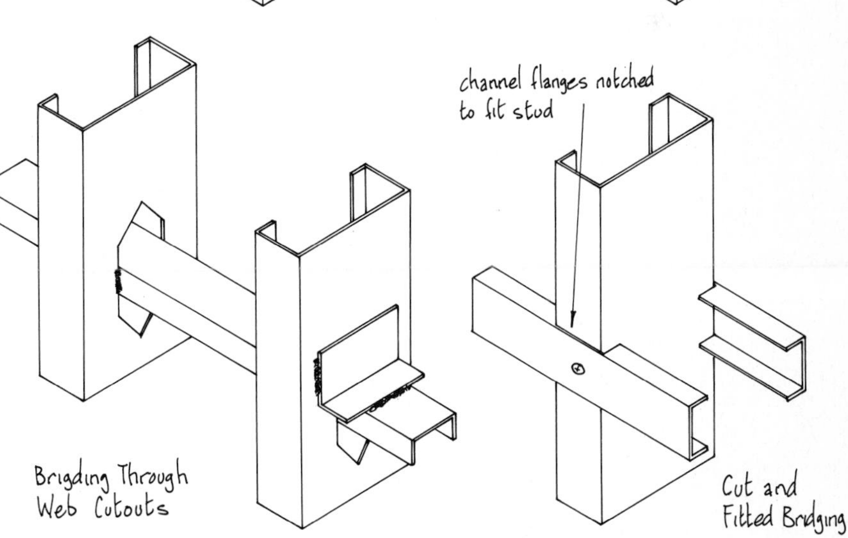

Diagonal wind bracing is usually added in the form of tension straps welded to stud and channel flanges. This bracing must therefore be applied in both directions.

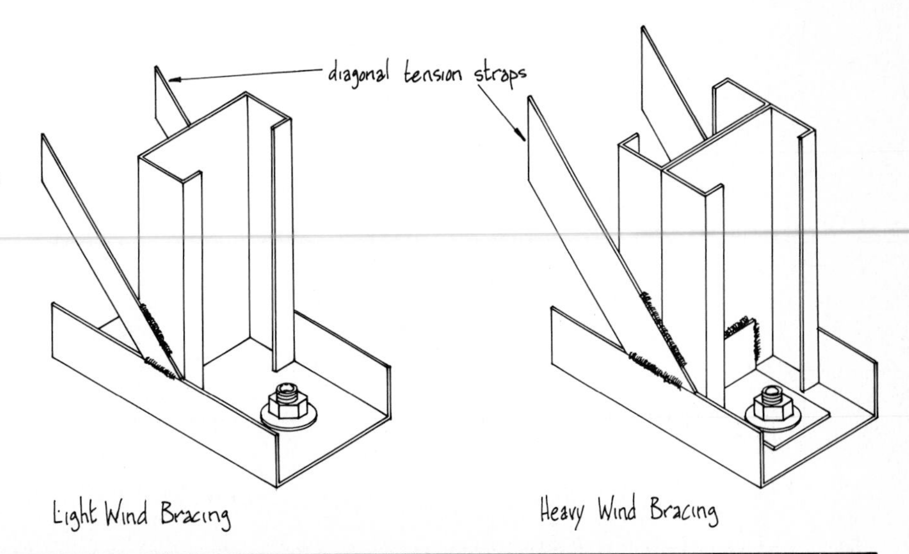

STEEL STUD SYSTEMS

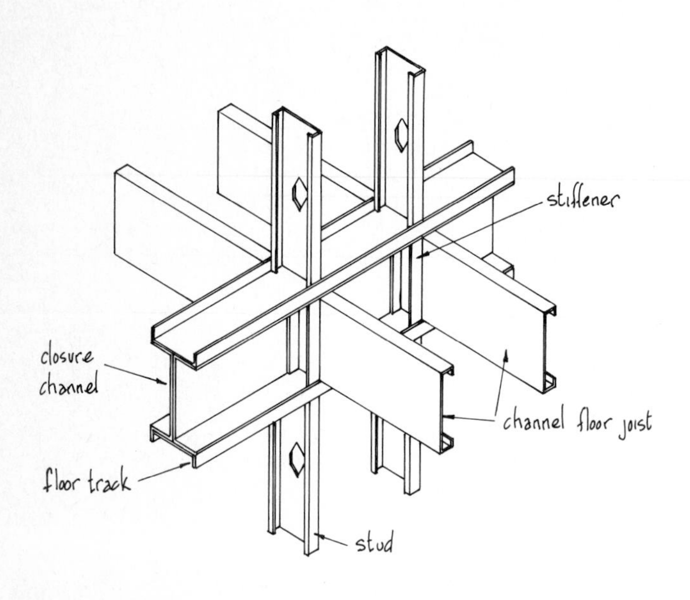

LOAD BEARING WALL WITH CHANNEL FLOOR JOISTS

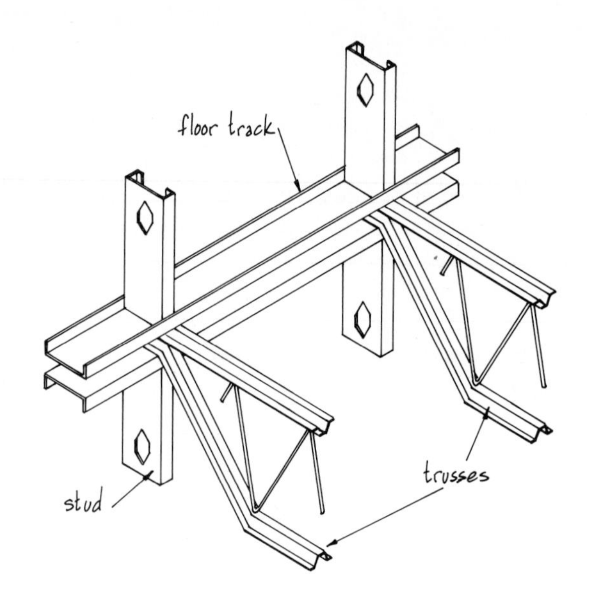

LOAD BEARING WALL WITH OPEN WEB STEEL JOISTS

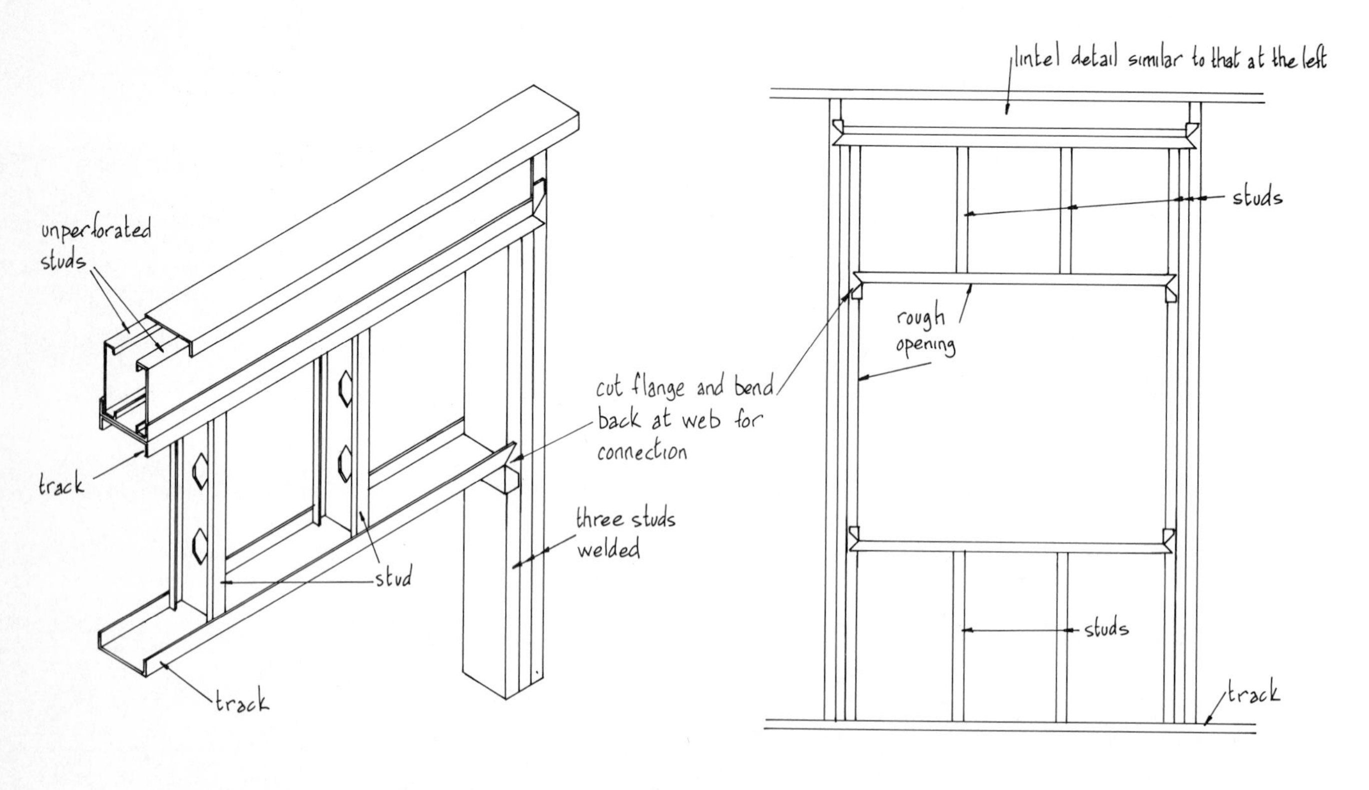

DOUBLE CHANNEL LINTEL

WINDOW OR DOOR OPENING

source: Bailey Metal Products

STEEL STUDS – CURTAIN WALLS

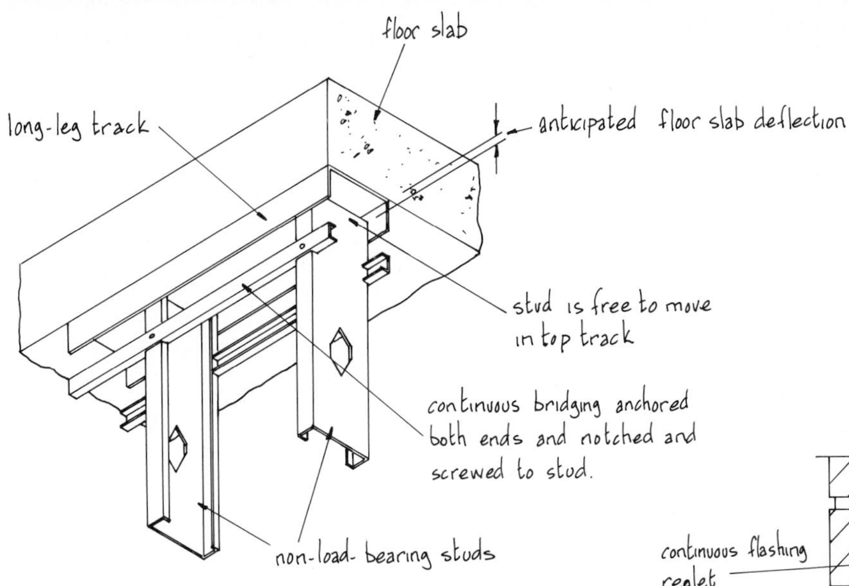

WIND-BEARING STUD CURTAIN WALL

floor slab

continuous flashing reglet

flashing

shelf angle and anchor

weep holes and elastic joint seal

compressible material

track

stud

brick tie

BRICK VENEER CURTAIN WALL WITH SHELF ANGLE

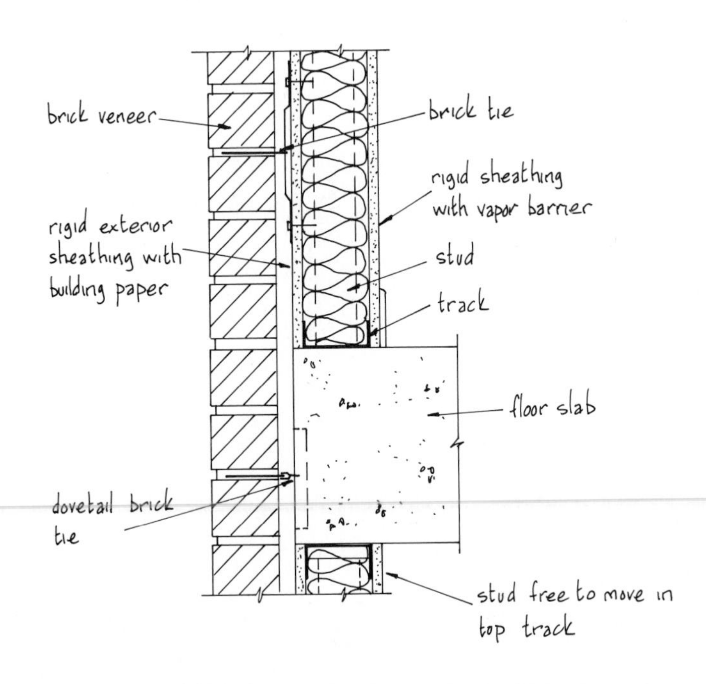

BRICK VENEER AND STUD BACKUP CURTAIN WALL

source: Bailey Metal Products

MASONRY WALLS - MATERIALS

Masonry walls are commonly used on all types of buildings. They provide a maintenance-free finish and allow a wide variety of visual effects, colors, and styles. The materials and design details shown here can be considered typical, but local variations in products and style must be investigated.

STRUCTURAL CLAY PRODUCTS - BRICK

- Brick is manufactured from clay or shale and is a fired (ceramic) product.
- Most brick is classified as "solid", i.e., being solid over 75% or more of its area in any parallel plane. Units are either totally solid or have cores that reduce weight and strengthen mortar joints.
- Most brick are produced in modular sizes based on the standard 4-in (100) module (see 7-22 for modular system details). Some examples of standard bricks are shown at the right.
- An extremely wide variety of brick finish, color, and texture is available through the use of local materials, chemical additives, and production processes.

BRICK CLASSIFICATION

Brick is generally classified into two groups, each with three grades:

Group 1 - Building brick or common brick. Used when durability and structural strength are more important than appearance.

- Grade SW. Extremely resistant to frost action, used in wet and cold conditions above and below grade.
- Grade MW. Resistant to freezing but used in dry conditions above grade.
- Grade NW. Used as backup and filler in above grade solid or composite walls.

Group 2. - Facing brick. This brick is strong and durable and is graded by face appearance. Subject to rigid manufacturing standards to ensure uniformity.

- Grade FBX. Uniform in color and size and used where a high degree of perfection is required.
- Grade FBS. Greater variations allowed in color and size.
- Grade FBA. Nonuniform in color, size, or texture, and used to create architectural design effects.

TYPICAL MODULAR BRICKS

dimensions are modular sizes - actual brick sizes are smaller (see 7-22)

Standard Modular: 2 2/3" (67), 8" (200), 4" (100)

Engineer: 3 1/5" (80), 8" (200), 4" (100)

Economy 8: 4" (100), 8" (200), 4" (100)

Norman: 2 2/3" (67), 12" (300), 4" (100)

Norwegian: 3 1/5" (80), 12" (300), 4" (100)

Roman: 2" (50), 12" (300), 4" (100)

"SCR" Brick: 2 2/3" (67), 12" (300), 6" (150)

6" (150) Jumbo: 4" (100), 12" (300), 6" (150)

8" (200) Jumbo: 4" (100), 12" (300), 8" (200)

NON-MODULAR BRICKS

dimensions are actual brick sizes

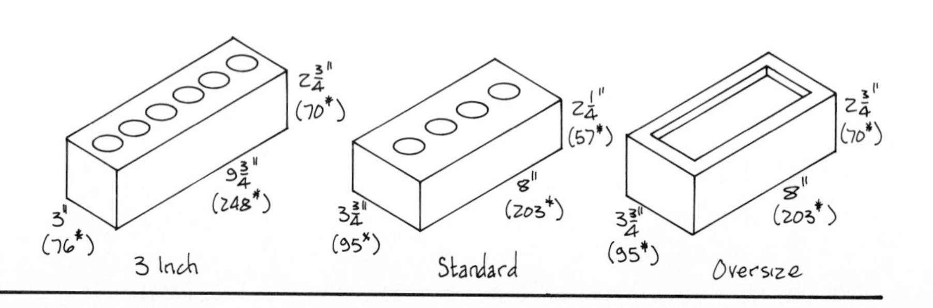

* soft conversion

MASONRY WALLS - MATERIALS

VERTICAL-CELL STRUCTURAL CLAY TILE

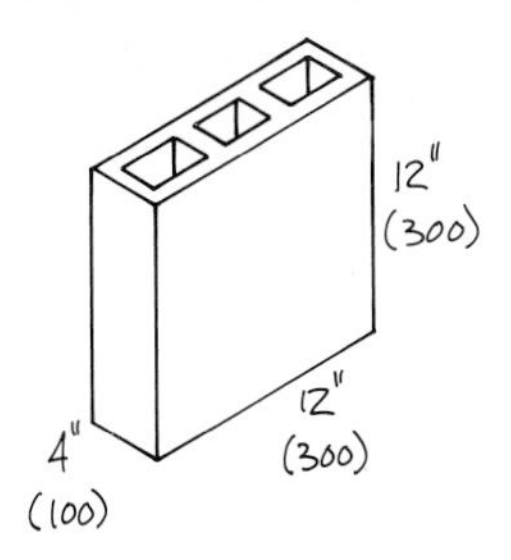

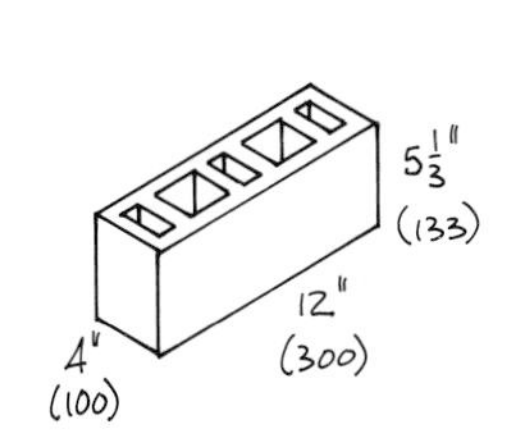

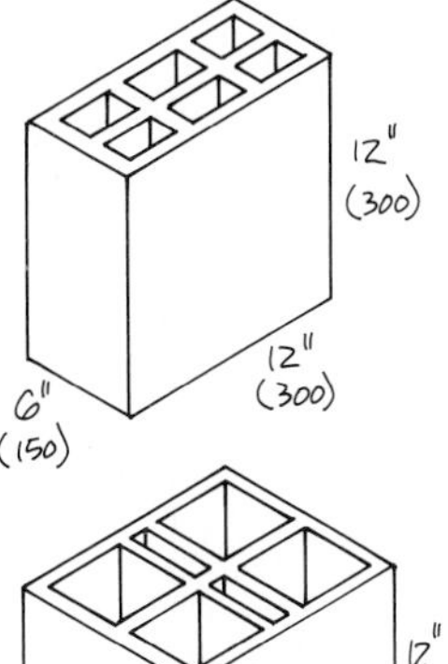

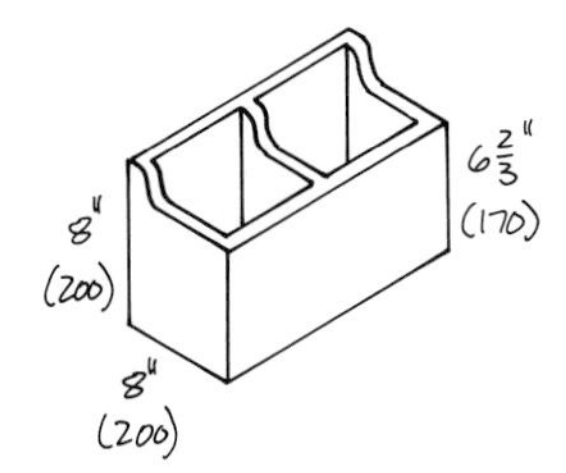

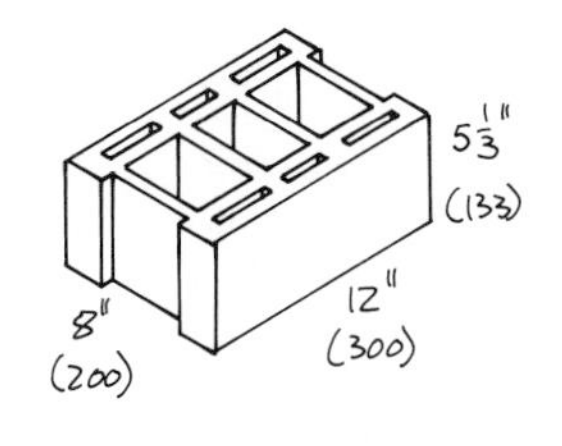

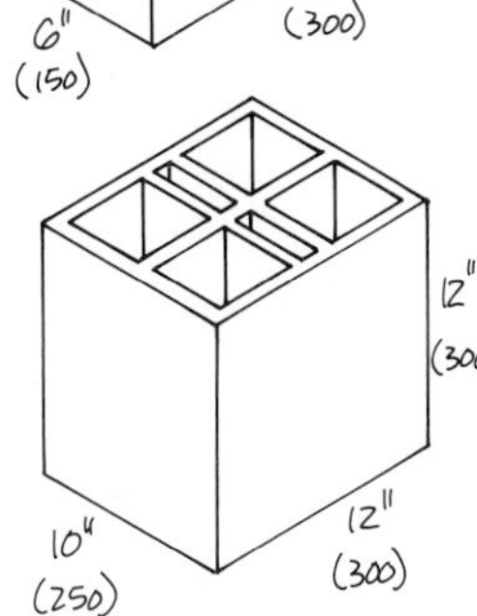

HORIZONTAL-CELL STRUCTURAL CLAY TILE

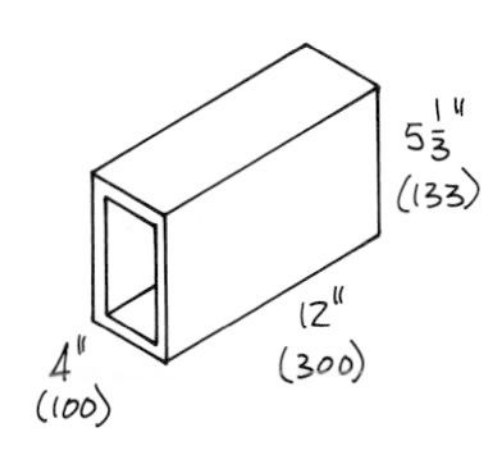

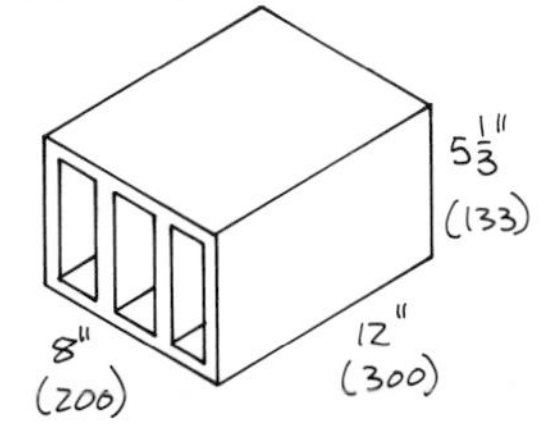

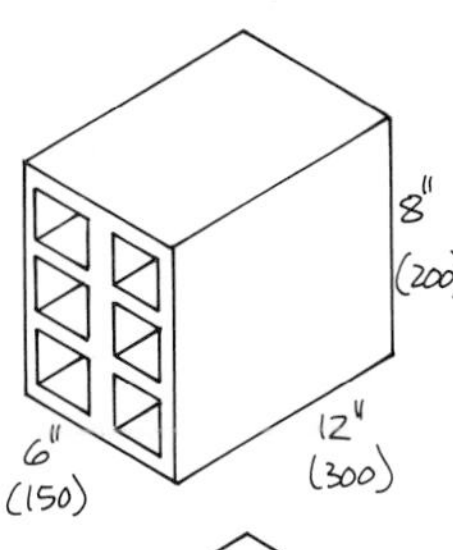

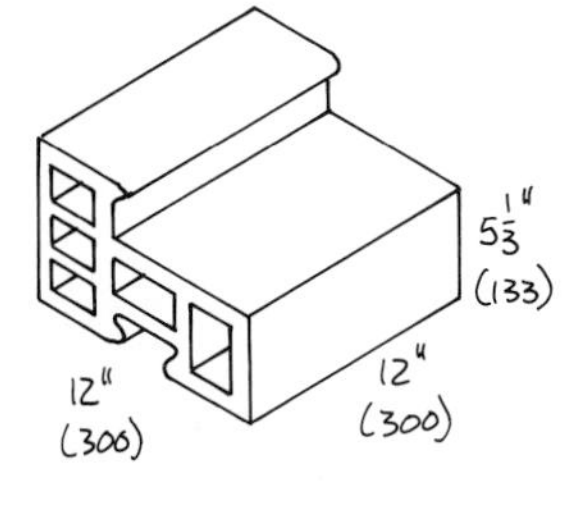

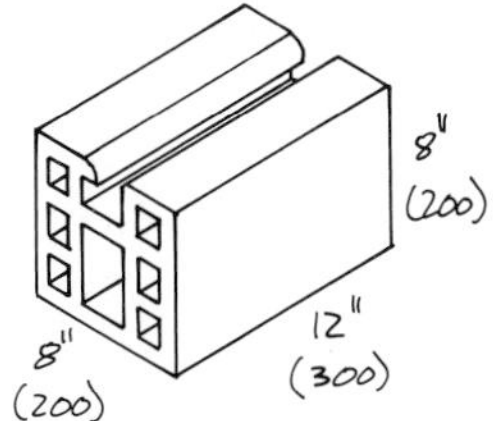

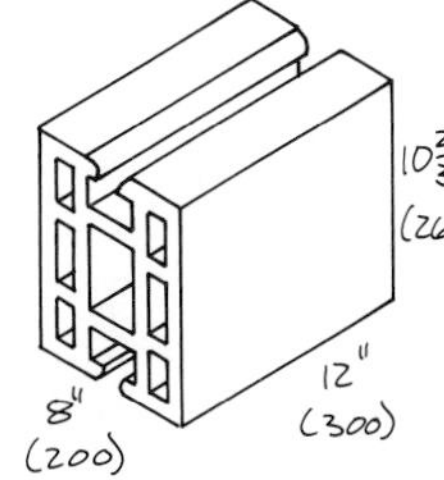

STRUCTURAL CLAY FACING TILE

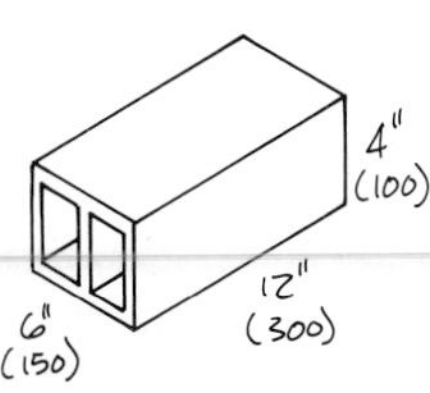

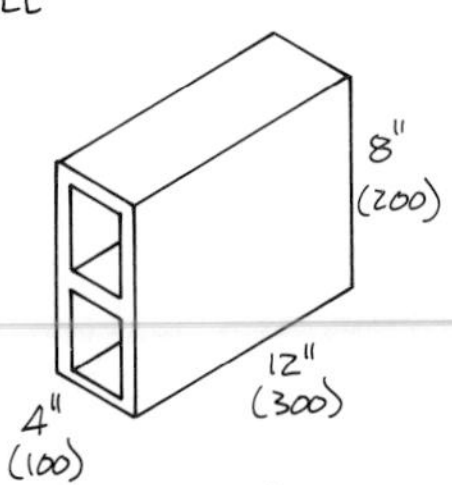

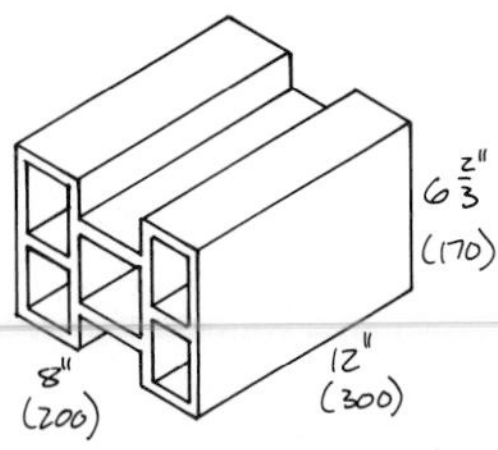

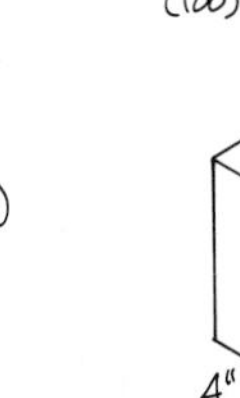

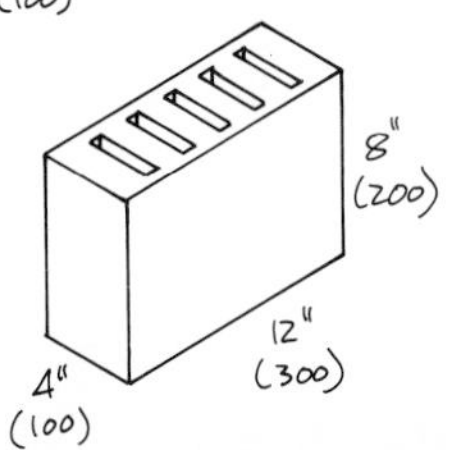

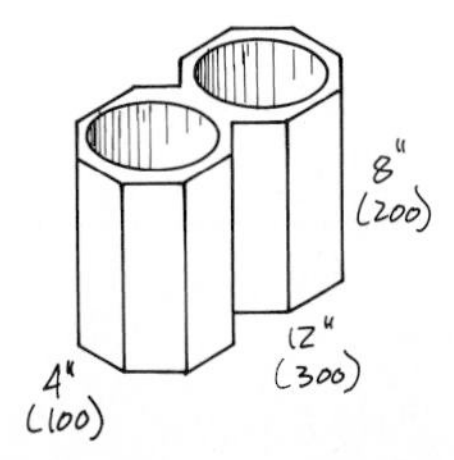

STRUCTURAL CLAY PRODUCTS — TILE

- The manufacturing process of tile products is similar to brick.
- Tile is classified as "hollow", i.e., being solid over less than 75% of it's area in any parallel plane.
- Tile is manufactured in an extensive variety of shapes, sizes, and finishes.
- Most tile is produced to modular sizing.

TILE CLASSIFICATION

Tile is classified into two groups and several grades. Group 1 – Structural clay tile. Tile can be obtained with vertical cells (designated as end construction) or horizontal cells (designated as side construction). Tile cells can be used as conduits for utility services. Tile are produced as:

- Load-bearing, for use in composite walls or bearing partitions. Two grades are available: LBX for use in exposed frost conditions and LB for nonfrost conditions.
- Non-load-bearing, for use in partitions, fire walls, and furring spaces. Available in NB grade only.

Group 2 – Structural facing tile. Used as the finished face for walls and produced in many different varieties.

- May be glazed or unglazed, single faced or double faced, and with or without exact dimensions.
- Most facing tile is produced to modular sizing.
- Produced in several grades to suit different service conditions.

MASONRY WALLS - MATERIALS

CONCRETE MASONRY UNITS

- Units fall into three groups: concrete block, special units, and concrete brick.
- All are concrete products and either air cured (subject to possible shrinkage) or steam cured (dimensionally stable).
- The type of aggregate used determines the weight of each unit. Standard weight units (40 lb/18 kg) contain sand, gravel, or limestone. Lightweight units (average 27 lb/12 kg) contain slag, cinders, or pumice.
- Standard-weight units tend to absorb less water than do lightweight units
- Blocks are produced in modular sizes. Face size is 8 × 16 in (200 × 400) with thicknesses available from 2 in (50) to 12 in (300) in 2 in (50) increments. Actual block size is nominal less 3/8 in mortar joint. Metric joint size is 10 mm.

CONCRETE BLOCK

Produced in three types:

- Solid load-bearing.
- Hollow load-bearing.
- Hollow non-load-bearing.

Definition of hollow: the unit is hollow over 25% or more of its total cross-sectional area.

In addition, units can be obtained in two grades:

- Grade N, used for wet and freezing conditions above or below ground.
- Grade S, for use above ground only and must have a protective coating if exposed to exterior conditions

The majority of units produced are hollow load-bearing grade N. Solid units are generally used where very heavy loads apply, to provide bearings for floor or roof beams, or to close off the hollow cores of a wall. Blocks generally have two or three cores depending on manufacturer.

TYPICAL CONCRETE BLOCK PRODUCTS

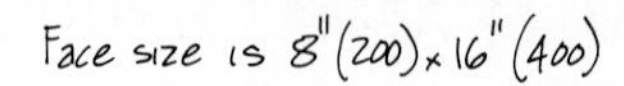

Face size is 8" (200) × 16" (400)

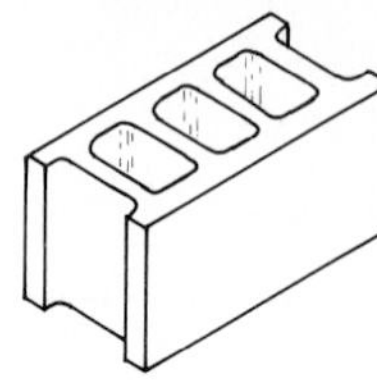

Stretcher

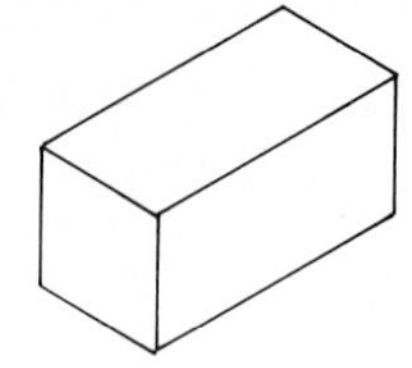

Solid

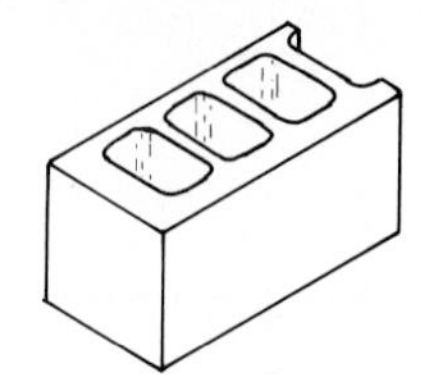

Corner

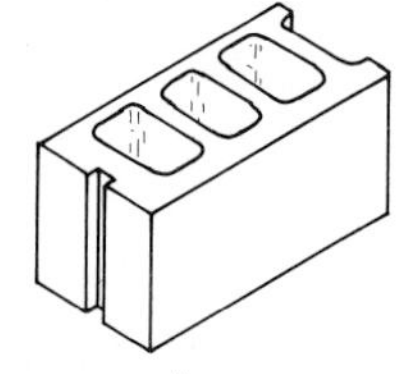

Sash

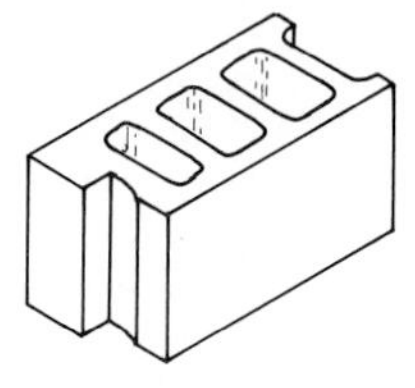

Jamb

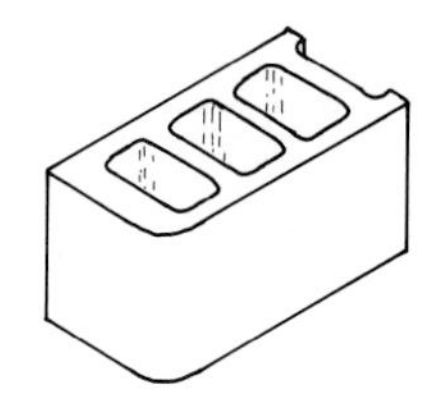

Bull Nose

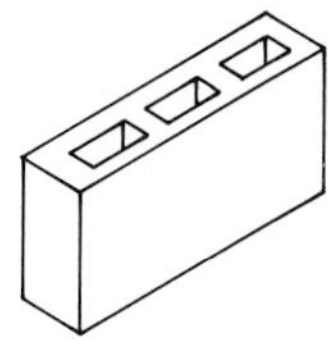

Partition

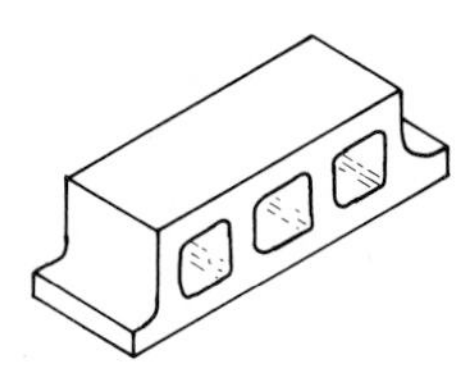

Soffit Floor

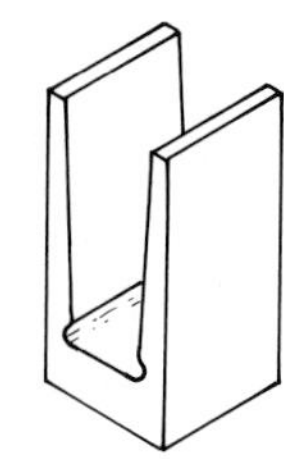

Lintel

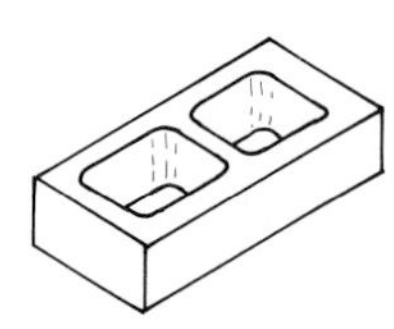

Half High

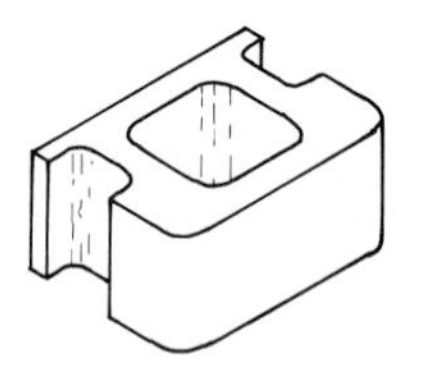

Pilaster

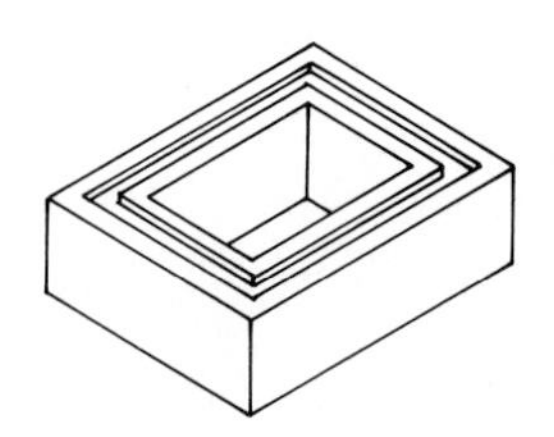

Chimney

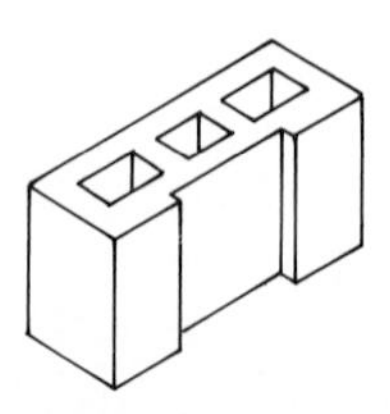

Offset Face

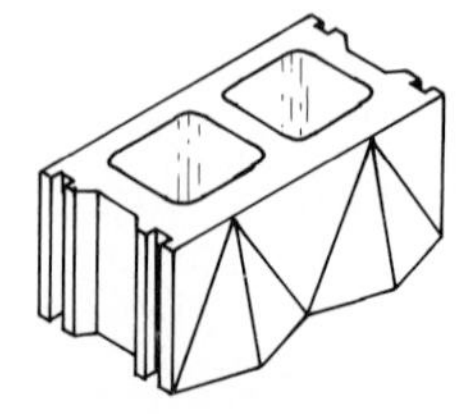

Sculptured

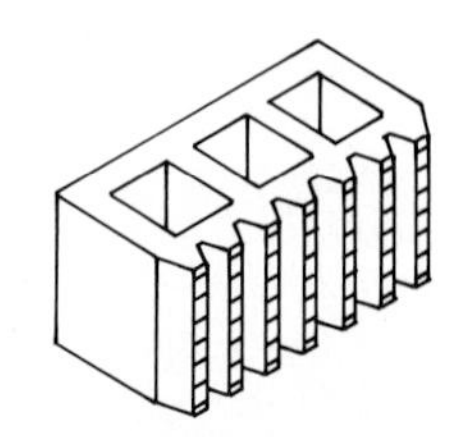

Fluted

MASONRY WALLS - MATERIALS

TYPICAL DECORATIVE SCREEN BLOCK

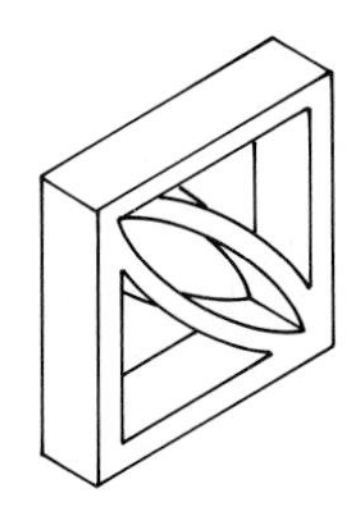

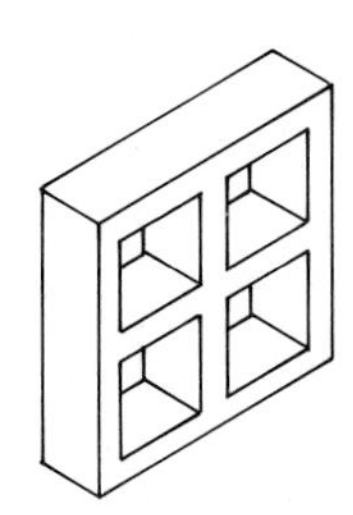

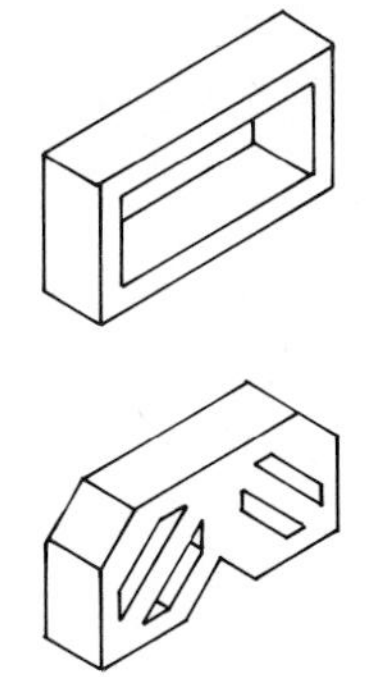

SPECIAL UNITS

Under this category can be found the following:

- Blocks with sculptured, offset, or depressed faces that, when laid, producing interesting visual effects.
- Blocks with flutes or ribs that produce the effect of vertical lines and patterns.
- Other face treatments for concrete units include hammered, glazed, or quarried stone effects.

CONCRETE BRICK

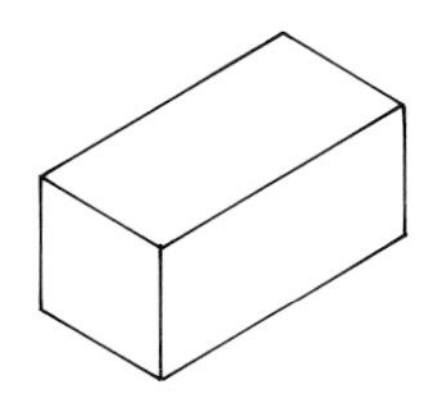

Solid Brick

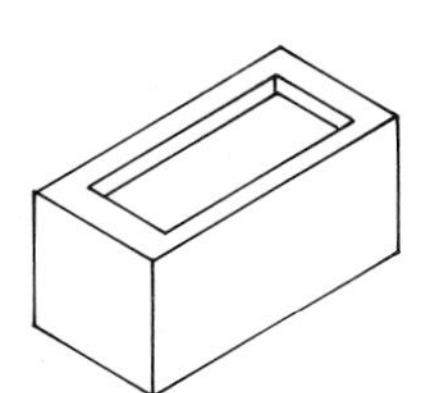

Frogged Brick

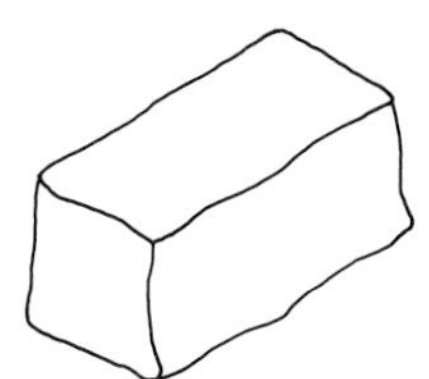

Slump Brick

CONCRETE BRICK

These units are similar in size and function to clay brick. Two types are commonly produced:

- Solid brick.
- "Frogged" brick. The "frog" is a shallow depression in the top and bottom faces that reduce weight and increase mortar bond.
- Slump brick. After molding, units are allowed to slump before curing, producing a variety of shapes and sizes.

See 5-5 and 5-12 for foundation wall block usage.

GLASS BLOCK

Plain Hollow

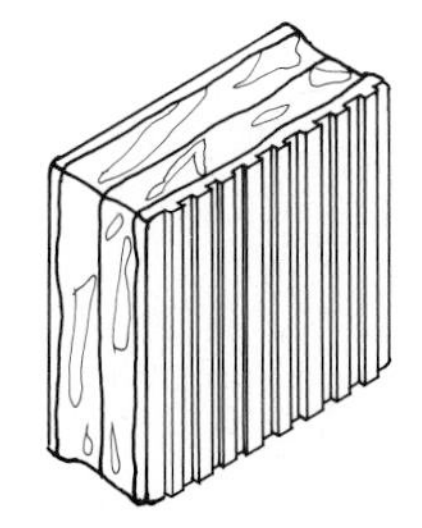

Ribbed Hollow

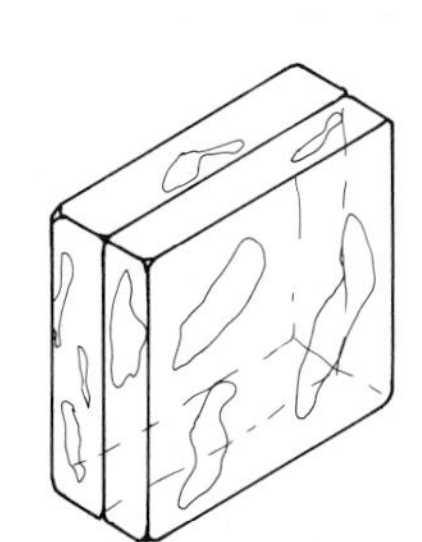

Solid Glass

GLASS BLOCK

These are pressed hollow glass units with a partially evacuated cavity.

- They transmit diffused light to internal spaces while acting as non-load-bearing masonry units.
- A variety of glass block types are available including clear, pebbled, striated, and colored.
- Glass blocks are laid in the same way as concrete units.
- There are limits to the total areas, heights, and widths that glass blocks are allowed to cover. Refer to manufacturers' specifications for exact limitations.
- In addition to light transmission, they also have some insulation value, low maintenance costs, and can be decorative as well as functional.

BLOCK REINFORCING

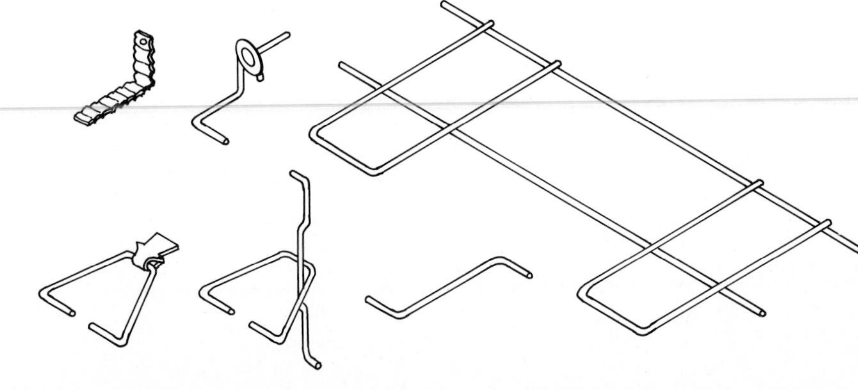

TIES AND REINFORCEMENT

Some masonry walls, notably composite, cavity, and veneer walls, require that the wythes be tied together to maintain structural integrity. Single-wythe walls often require some form of reinforcement for lateral stability. The ties and reinforcement shown here are typical of those produced in great variety by manufacturers. Installation details are shown on the following pages where applicable.

MASONRY - MODULAR LAYOUTS

Modular dimensions are used extensively in masonry construction. The use of a module for setting out building dimensions greatly assists the design process. For masonry, the base modular dimension is 4 in (100) for both horizontal and vertical directions. Using the module, building dimensions can be set without the need for unit fitting and cutting on site, and the vertical coursing of wythes can be aligned for ties and headers.

nominal thickness
joint thickness
actual thickness
1/2 joint

PLAN

nominal length
joint
joint
nominal height
actual height
actual length

ELEVATION (FRONT)

ACTUAL SIZE

The actual manufactured size of a masonry unit is a function of its nominal size and the size of the mortar joint specified by the manufacturer.

- A unit's nominal dimensions are multiples of, or fractions of, the standard 4 in (100) module.
- When laid with the correct joint size, vertical courses of nominal dimension will "course out" at multiples of the standard 4 in (100) module. The number of courses required to do this will depend on the nominal size of the unit. Look at the chart on 7-23 for modular brick sizes and coursing plus typical examples of coursing between different units.
- Joint sizes are normally 3/8 or 1/2 in for brick, and 3/8 in for block. Metric joint size for all units is 10 mm.
- Actual size of units is nominal dimensions less one joint size.

NOMINAL MODULAR SIZE OF BRICKS

UNIT DESIGNATION	THICKNESS in (mm)	FACE DIMENSIONS HEIGHT in (mm)	LENGTH in (mm)	NUMBER OF COURSES IN 16 IN (400)
standard	4 (100)	2 2/3 (67[1])	8 (200)	6
engineer	4 "	3 1/5 (80)	8 "	5
economy 8 or jumbo closure	4 (100)	4 (100)	8 (200)	4
double	4 "	5 1/3 (133[2])	8 "	3
roman	4 "	2 (50)	12 (300)	8
norman	4 "	2 2/3 (67[1])	12 "	6
norwegian	4 "	3 1/5 (80)	12 "	5
economy 12 or jumbo utility	4 (100)	4 (100)	12 (300)	4
triple	4 "	5 1/3 (133[2])	12 "	3
SCR brick	6 (150)	2 2/3 (67[1])	12 "	6
6-in (150) norwegian	6 "	3 1/5 (80)	12 "	5
6-in (150) jumbo	6 "	4 (100)	12 "	4
8-in (200) jumbo	8 (200)	4 (100)	12 "	4

① Rounded dimension - actually 66.666
② Rounded dimension - actually 133.333
Actual brick dimensions - modular less one joint size.

Source: Brick Institute of America

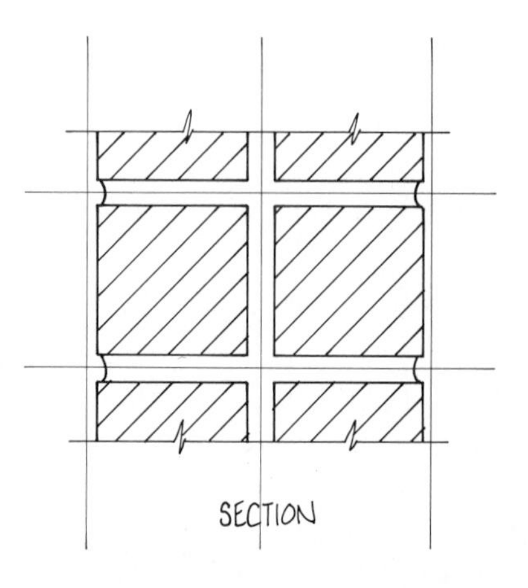

SECTION

NOMINAL MODULAR LAYOUT
based on a 4 in (100) module

see the next page for other brick sizes

MASONRY - MODULAR LAYOUTS

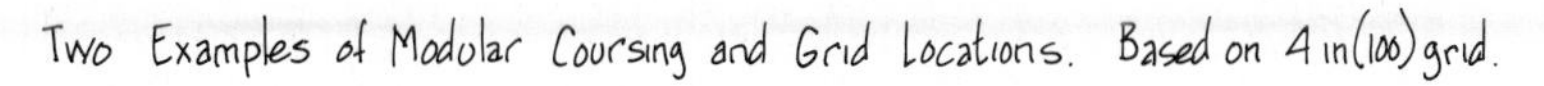

Two Examples of Modular Coursing and Grid Locations. Based on 4 in (100) grid.

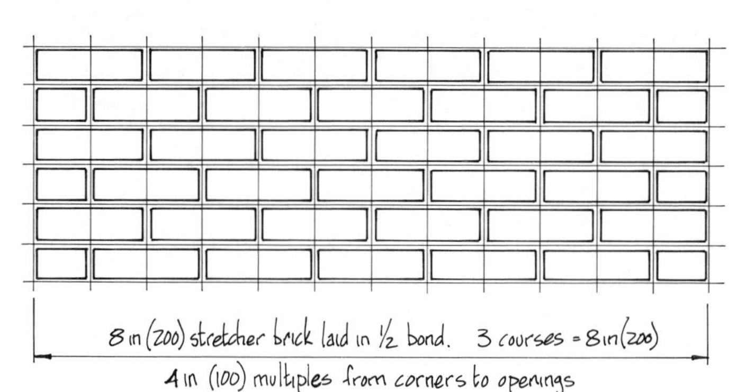

8 in (200) stretcher brick laid in ½ bond. 3 courses = 8 in (200)
4 in (100) multiples from corners to openings

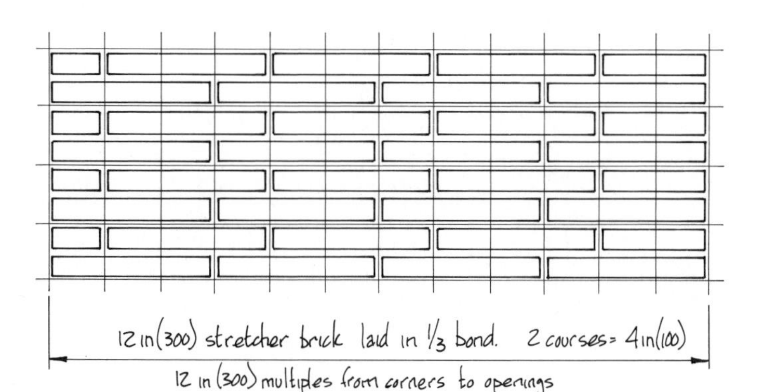

12 in (300) stretcher brick laid in ⅓ bond. 2 courses = 4 in (100)
12 in (300) multiples from corners to openings

Brick Cavity — Brick and Concrete Block Cavity — Solid Brick and Tile

Three typical examples of course alignment. In each case, the vertical joints in each wythe coincide at constant intervals to allow structural bonding. Bonding is achieved by metal ties or masonry headers (see 7-26).

Definition: Wythe - a continuous vertical section of a wall one masonry unit in thickness.

2⅔ in (67) Courses

Course	Height
22	4'-10⅔" (1467)
21	4'-8" (1400)
20	4'-5⅓" (1333)
19	4'-2⅔" (1267)
18	4'-0" (1200)
17	3'-9⅓" (1133)
16	3'-6⅔" (1067)
15	3'-4" (1000)
14	3'-1⅓" (933)
13	2'-10⅔" (867)
12	2'-8" (800)
11	2'-5⅓" (733)
10	2'-2⅔" (667)
9	2'-0" (600)
8	1'-9⅓" (533)
7	1'-6⅔" (467)
6	1'-4" (400)
5	1'-1⅓" (333)
4	10⅔" (267)
3	8" (200)
2	5⅓" (133)
1	2⅔" (67)

4" (100)

3⅕ in (80) Courses

Course	Height
18	4'-9⅗" (1440)
17	4'-6⅖" (1360)
16	4'-3⅕" (1280)
15	4'-0" (1200)
14	3'-8⅘" (1120)
13	3'-5⅗" (1040)
12	3'-2⅖" (960)
11	2'-11⅕" (880)
10	2'-8" (800)
9	2'-4⅘" (720)
8	2'-1⅗" (640)
7	1'-10⅖" (560)
6	1'-7⅕" (480)
5	1'-4" (400)
4	1'-0⅘" (320)
3	9⅗" (240)
2	6⅖" (160)
1	3⅕" (80)

Two Examples of Vertical Coursing Charts
Modular Brick

source: Brick Institute of America

MASONRY THICKNESS - WALL HEIGHTS

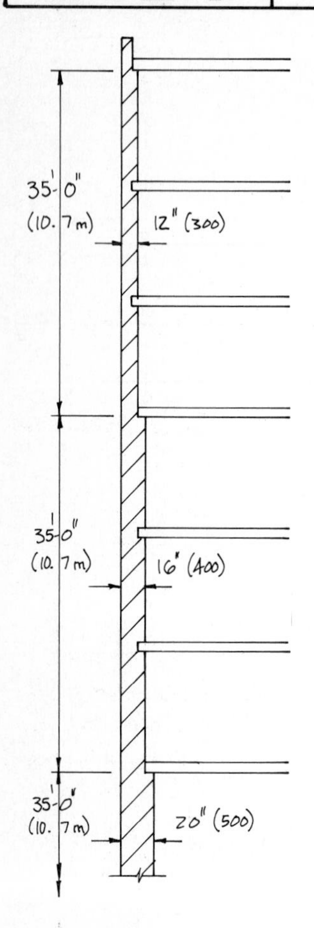

SOLID MASONRY LOAD-BEARING

- 12 in (300) thick for uppermost 35 ft (10.7 m).
- Increase thickness 4 in (100) for each successive 35 ft (10.7 m) or fraction thereof, measured downwards from the top.

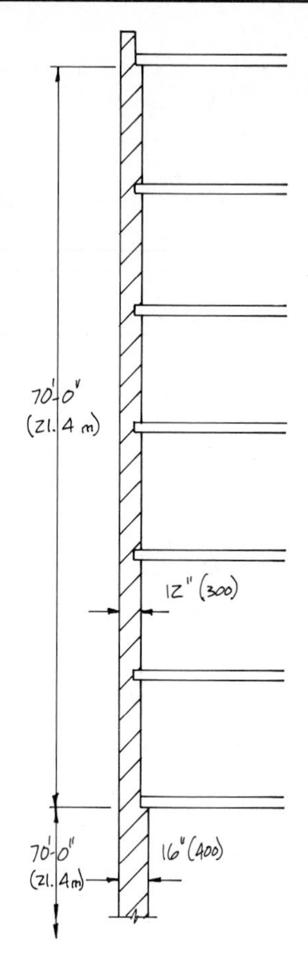

STIFFENED

- 12 in (300) thick for upper 70 ft (21.4 m), increased by 4 in (100) for each successive 70 ft (21.4 m).
- Wall must be stiffened by solid masonry cross walls spaced at maximum 12 ft (3.7 m) centers, or by reinforced concrete floors.

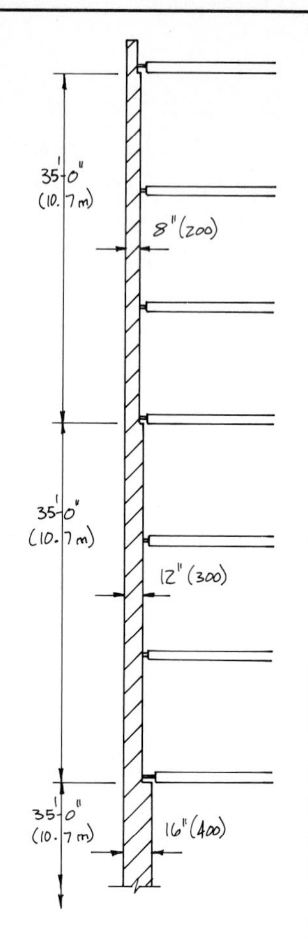

SOLID MASONRY NON-LOAD-BEARING

- In non-load-bearing exterior walls, thickness may be 4 in (100) less than that required for load-bearing walls.
- Minimum thickness not less than 8 in (200) except where 6 in (150) walls are permitted.

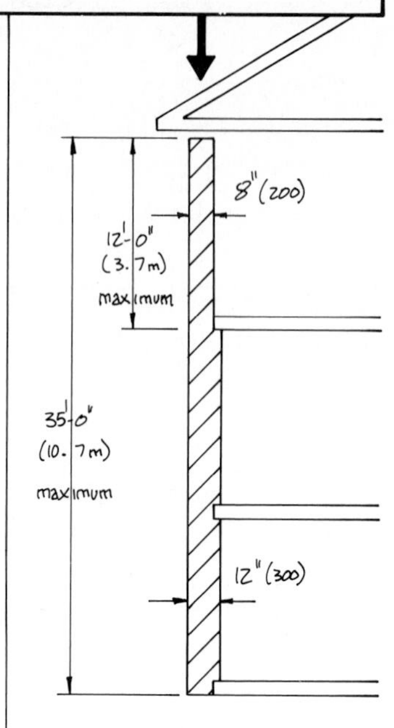

TOP-STORY WALLS

- The top story bearing wall of a maximum 35 ft (10.7 m) high building can be 8 in (200) provided:
- The top story does not exceed 12 ft (3.7 m) in height, and:
- There is no lateral thrust from the roof structure.

The thickness and height details shown on these two pages meet most national codes. Check with local codes for variations. In addition, the details shown are based on empirical or nonengineered design principles. Thickness and height for engineered walls will be specific to a particular project.

MASONRY THICKNESS - WALL HEIGHTS

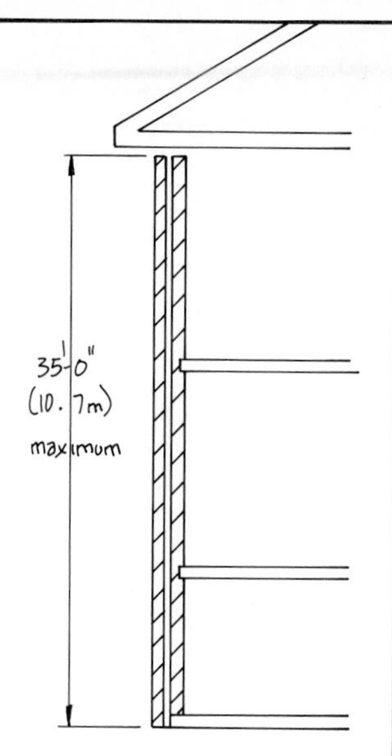

HOLLOW WALLS

- Cavity (metal tied) or masonry bonded.
- Not over 35 ft (10.7 m) except for 10 in (250) walls which should not exceed 25 ft (7.6 m).
- Facing and backing wythes should be minimum 4 in (100) thick.
- Cavity width to be 2 in (50) minimum, 3 in (75) maximum.

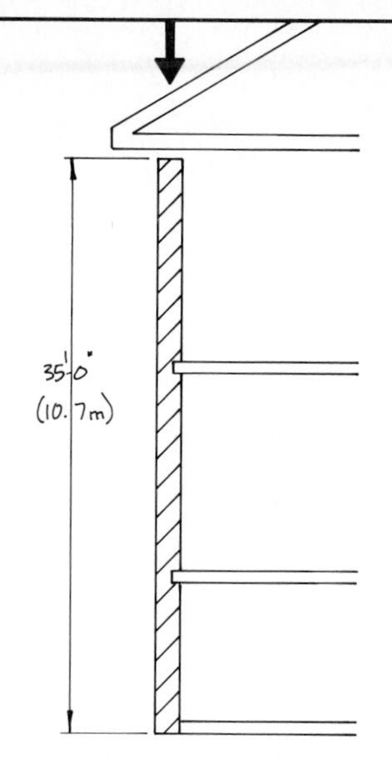

RESIDENTIAL WALLS - SOLID

- In dwellings up to three stories and no more than 35 ft (10.7 m) high, walls may be 8 in (200) thick.
- Roof structures must not subject walls to lateral thrust.

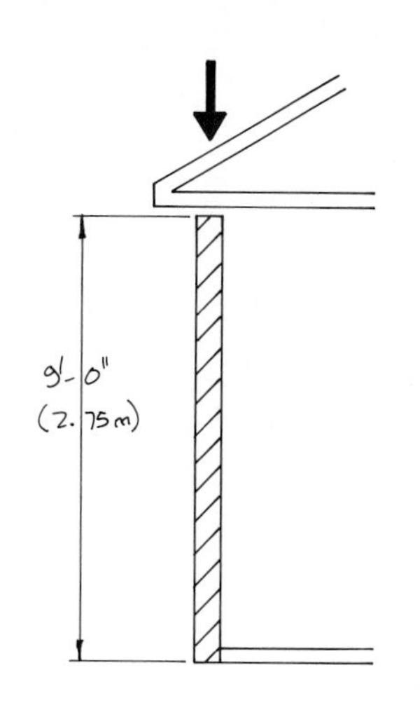

RESIDENTIAL WALLS - SOLID

- Dwellings with one story may have 6 in (150) walls provided:
- Maximum wall height is 9 ft (2.75 m) except for gable peaks which may be 15 ft (4.6 m) maximum.

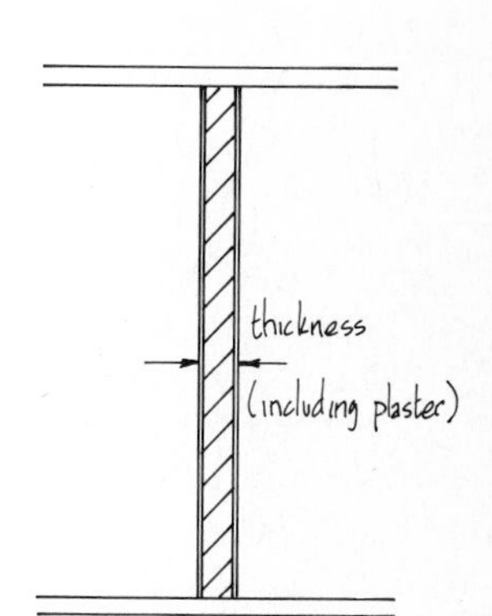

MASONRY PARTITIONS NON-LOAD BEARING

- Lateral support for partition must not exceed 36 times actual thickness of the wall, including plaster finishes.
- Partition anchorage must be sufficient to transmit all forces.

For non-load-bearing veneer construction see details on 7-30.

LATERAL SUPPORT

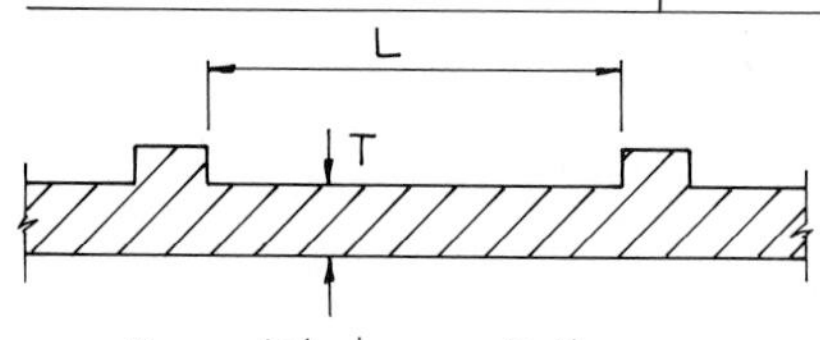

Piers, Pilasters, or Buttresses

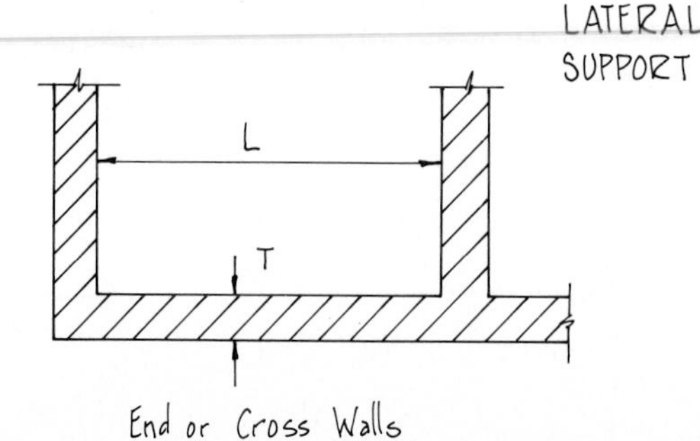

End or Cross Walls

Vertical Supports

Maximum Ratio of Unsupported Height or Length to Nominal Thickness

Type of Masonry	Ratio $\frac{L}{T}$
solid masonry bearing walls	20
hollow unit masonry bearing walls	18
cavity walls	18 ①
non-bearing walls	36 ②

① Thickness equal to the sum of the nominal thickness of the inner and outer wythes.
② Actual thickness of partition plus plaster.

MASONRY BONDING

The term "masonry bonding" has three meanings:

- Structural bond – where the units in the wall are interlocked or tied together so that the wall acts as a single structural unit.
- Pattern bond – where the units are laid to create a specific visual effect. In most walls, the pattern created is also a function of the structural bond.
- Mortar bond – where the strength and adhesion properties of the mortar bonds units and reinforcing together. The type of mortar joint used on the wall's surface also adds to the visual effects.

Each type of bond is discussed on this and the next two pages.

STRUCTURAL BONDING

Structural bonding can be achieved in three ways:

1. Overlapping or interlocking units. In solid walls one method is the use of "headers". A header is a unit that is laid across the width of the wall to tie the wythes together. Headers may be placed in many positions in a wall and therefore become part of the visual pattern of the wall face. See the next page for bonding patterns. Most building codes require that headers comprise at least 4% of the wall surface and the distance between headers must not exceed 24 in (600) either vertically or horizontally.

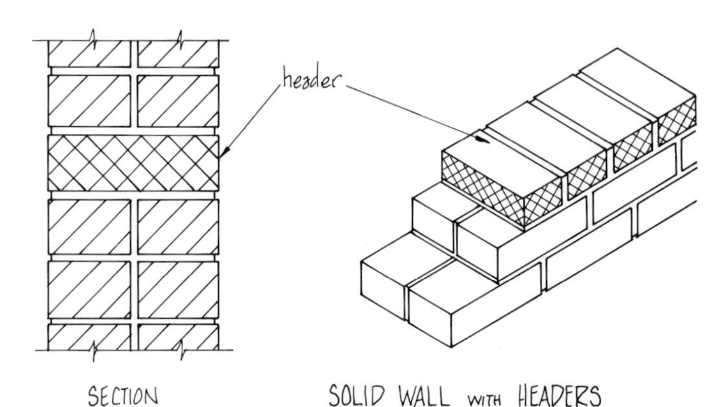

SECTION   SOLID WALL WITH HEADERS

2. Metal ties (see 7-21 for examples). These can be used for both solid and cavity walls and are manufactured to suit virtually every wall arrangement. Metal ties are preferred for all exterior wall types since they accommodate slight differential movement in the wall units and provide better water resistance than do solid headers. Most codes require one tie per 4½ sq ft (0.4 m²) with ties spaced apart no more than 24 in (600) vertically and 36 in (1000) horizontally. Ties must be at least 3/16 in (4.76*) in diameter or the equivalent thereof and be corrosion resistant. For ties in veneer walls, see 7-30.

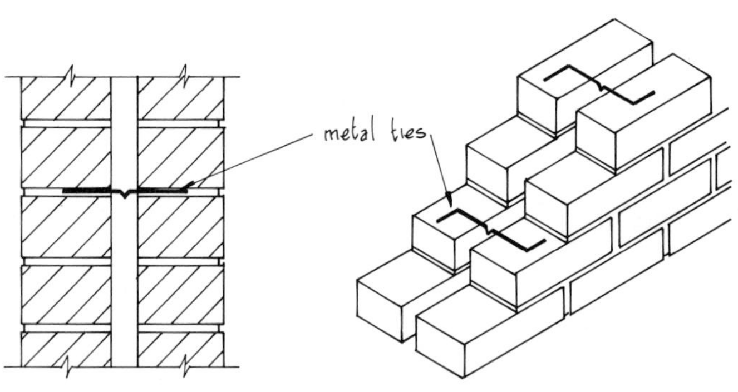

SECTION   CAVITY WALL WITH METAL TIES

3. Mortar bonding. Grout (liquid mortar) is poured into the joint between two wythes to provide a structural bond. The joint should be at least ¾ in (20) wide and the wythes need not be tied. This type of construction is often used for vertically reinforced walls (see 7-38).

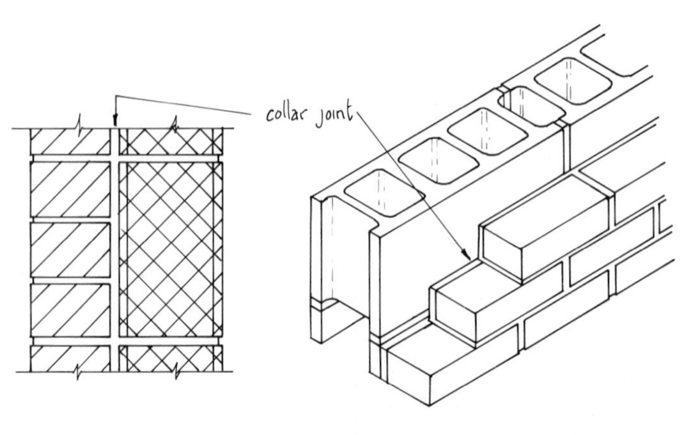

SECTION   BLOCK AND BRICK COLLAR JOINT

PATTERN BONDING

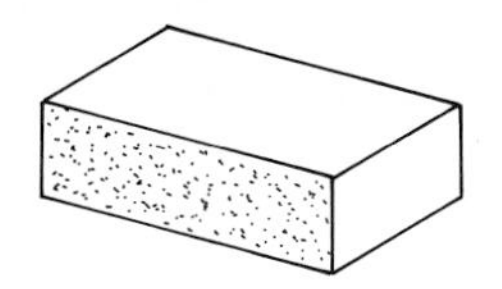
Stretcher

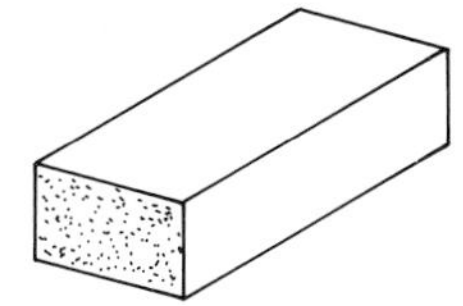
Header

Rowlock - Stretcher

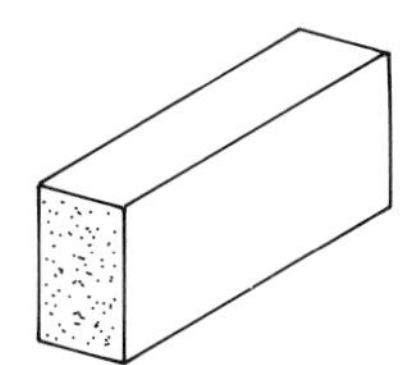
Rowlock

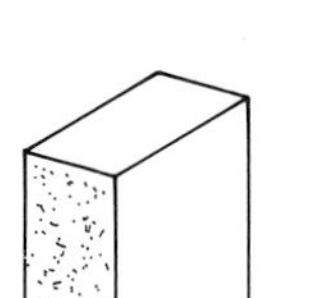
Soldier

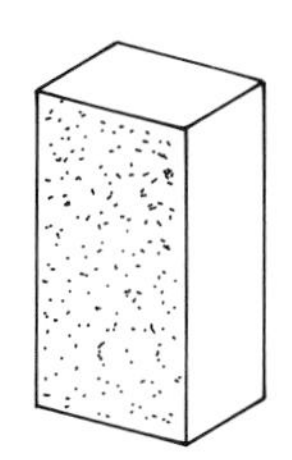
Sailer

BRICK POSITIONS

Brick can be laid in a variety of positions to achieve a visual effect or a structural bond. In the positions shown at the left, the header and rowlock can be used for bonding across the wall, while the stretcher and rowlock-stretcher are used for bonding along the length of the wall.

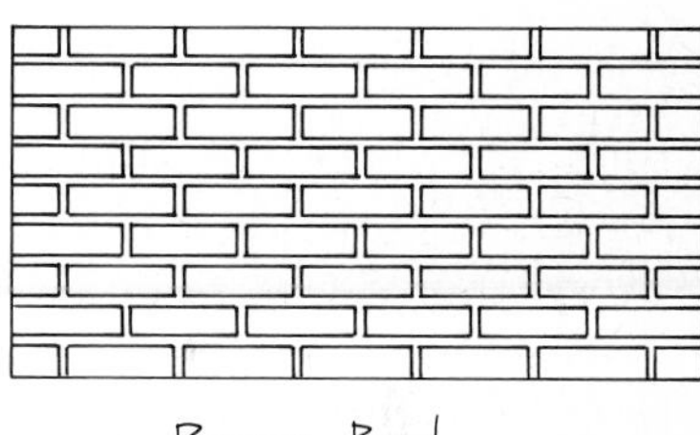
Running Bond

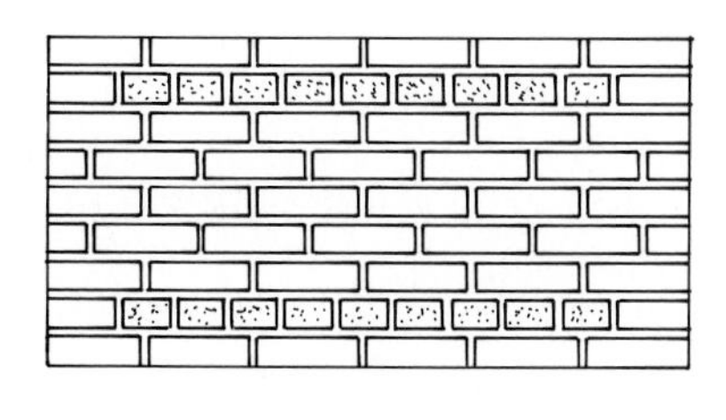
Common Bond
6th course headers

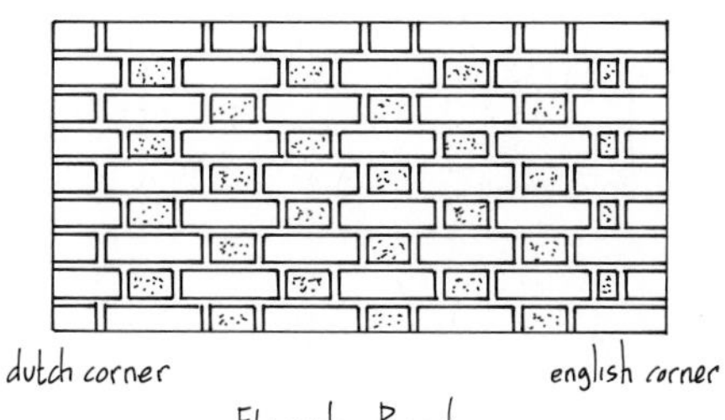

Flemish Bond
2nd course headers

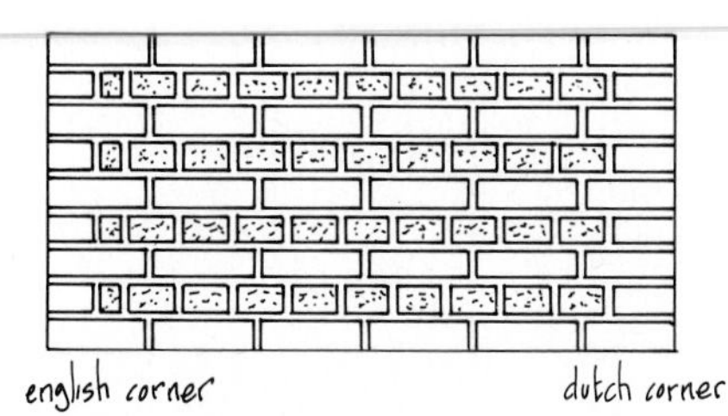

English Bond

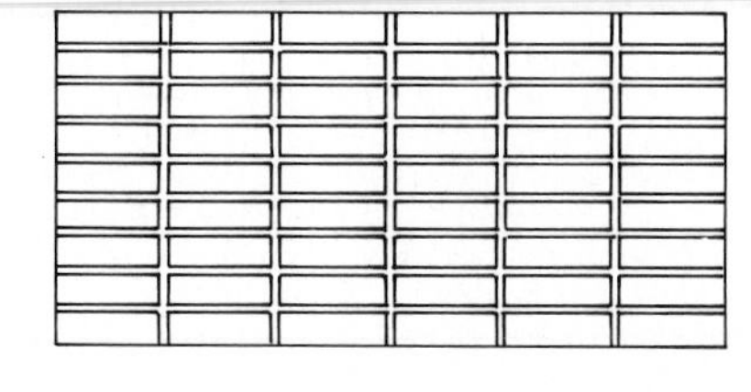
Stack Bond

BONDING PATTERNS

Shown here are the five most common pattern bonds in use today, each giving a specific visual effect. With the exception of the running and stack bonds, each pattern also provides headers for across-wall structural bonding (headers are shown shaded). As well as many other bonding patterns, colored, textured, recessed, or projecting units are often used to create specific patterns or interesting contrasts.

Note. In modern construction, veneer masonry and metal tied walls prevail in most regions. Since masonry headers are not required for such construction, a running bond is generally used for simplicity and economy.

MORTAR JOINTS

The mortar used to "lay up" a masonry wall serves several functions:

- The mortar bonds together the masonry units and reinforcing steel or ties. The strength of the bond is determined by the type of mortar used (see 5-12 for mortar types).
- The mortar compensates for slight variations in the actual sizes of masonry units as produced by manufacturers.
- The mortar seals the spaces between units and, depending on the type of exterior joint, provides weather resistance to a greater or lesser degree. The common joint types are shown at the right.
- The mortar color and joint type allow a wide range of visual effects to be created on the wall face.

JOINT TYPES

WATER RESISTANCE			
HIGH	Concave	V-Shaped	Weathered
MEDIUM	Struck	Flush or Rough Cut	
LOW	Raked	Extruded	

WATER RESISTANCE

One of the major concerns in masonry wall design, except perhaps for desert regions, is the wall's resistance to water. Although both masonry and mortar will absorb water by capillary action to some degree, water is most likely to penetrate a wall through cracks in the mortar joints. Cavity and veneer type walls allow water to drain down the back of the exterior wythe to weepholes in the wall's base. Solid walls have no means of drainage and are more susceptable to leakage. Water penetration can be prevented in several ways:

- The recommended practice for filling joints should be followed. Head and bed joints must have adequate mortar to completely fill the joint space when the unit is firmly pushed into position. Adding mortar to these joints after the unit is in position is not good practice.
- Choose a tooled joint that sheds water. The concave or V-joint is best.
- Avoid masonry contact with the finished grade, especially in regions with a freeze/thaw cycle.
- In cavity walls, if water penetration into or through the cavity must be stopped, parge, or back-plaster the cavity face of one wythe, depending on which wythe is built ahead of the other.

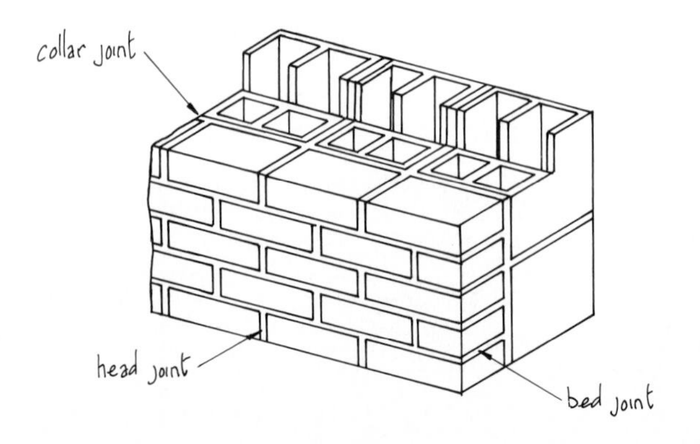

JOINT NAMES

EXPANSION JOINTS

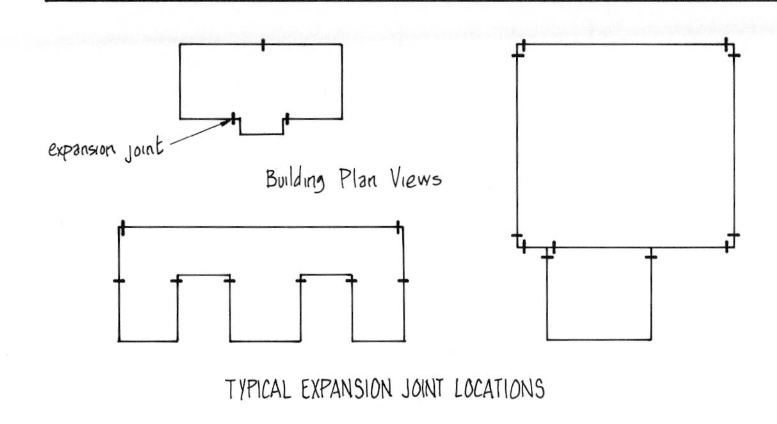

TYPICAL EXPANSION JOINT LOCATIONS

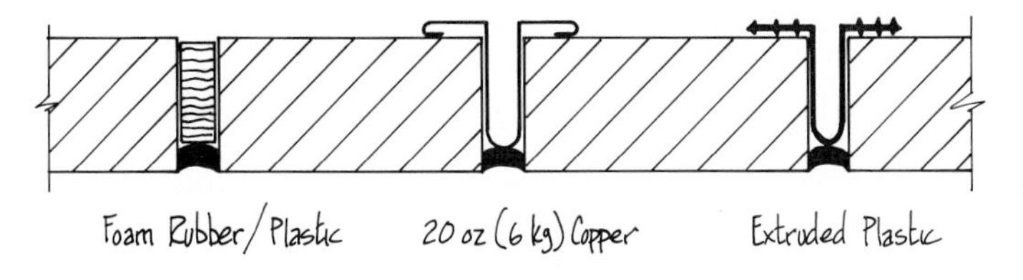

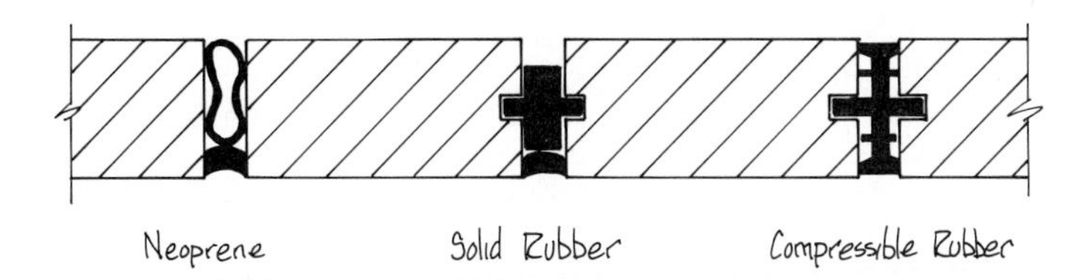

TYPICAL EXPANSION JOINT MATERIALS

It is important to understand that a building is a dynamic structure, the elements of which are in constant <u>differential</u> motion. Forces acting on the structure's elements and materials include temperature change, moisture absorption, loading stress, and plastic flow, factors which are themselves constantly changing.

To accommodate this movement in rigid masonry walls, expansion joints are located at strategic points to break the structural and mortar bond in, or between, wythes.

- The plan views at the left show typical expansion joint placement in three building shapes, and the details on this page give a selection of joint designs.
- Actual joints and their locations must be designed specifically for each structure to best accommodate the forces acting on it.
- Even though the masonry bond is broken across an expansion joint, lateral stability must still be maintained.
- Most residential structures using veneer construction do not need expansion joints since wall areas and heights are relatively small (see 7-30).

SOME TYPICAL EXPANSION JOINT CONDITIONS

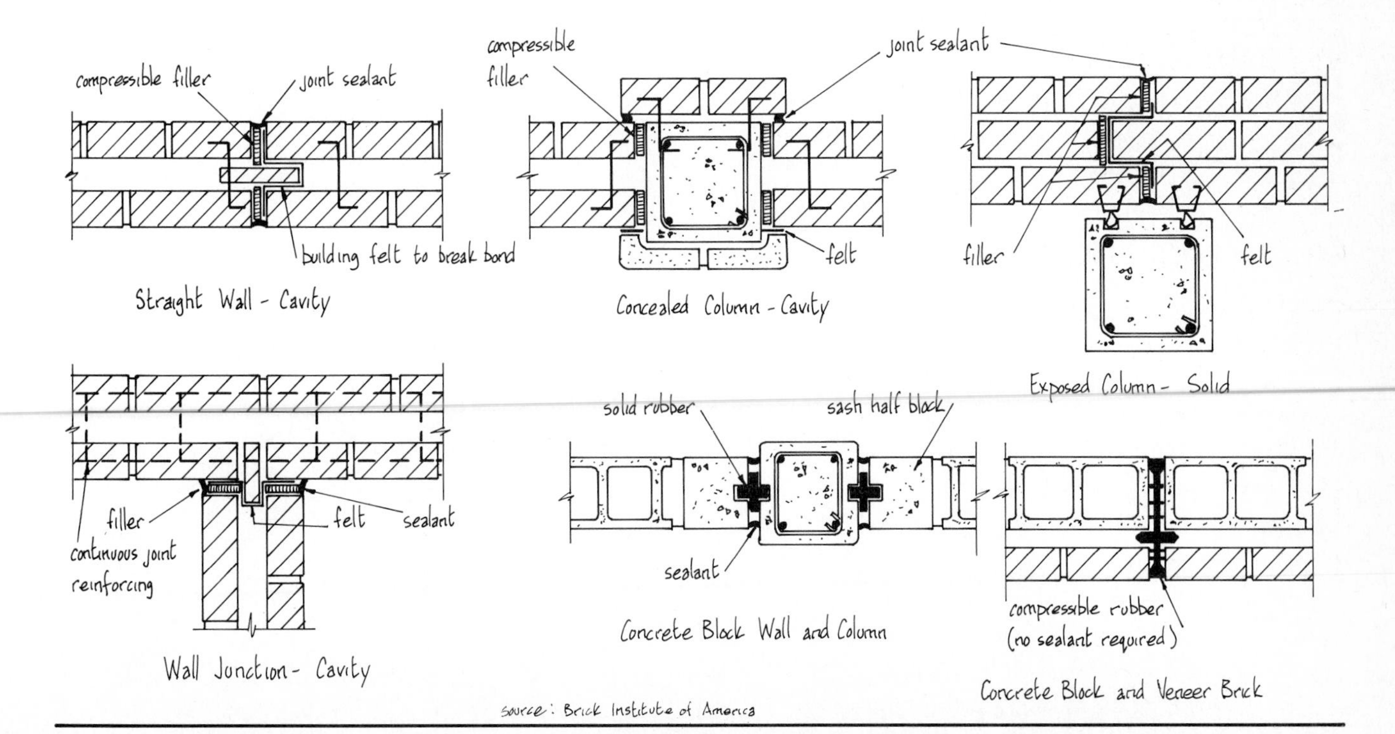

source: Brick Institute of America

BRICK VENEER

A veneered wall is one that has a facing of masonry units securely tied, but not bonded, to a backing. The facing carries no vertical loads other than its own weight, but lateral loads (wind forces) are shared between face and backing.

- Veneer walls are extremely common in both residential and commercial construction. The system allows wide flexibility in wall design and is commonly used with wood stud, metal stud, or masonry backing.
- Empirical (nonengineered) design methods can be applied to walls up to three stories in height. Above this, a more detailed analysis must be made of the structure, especially in steel- and concrete-framed buildings.
- 3-in (75) or 4-in (100) nominal facing masonry is generally used.
- A minimum 1-in (25) clear cavity must be provided between face and backup to facilitate drainage. Careful detailing of flashings at grade level and at all openings is also necessary.
- In most cases, the facing bears directly on the foundation walls.
- See 11-20 for stone veneer.

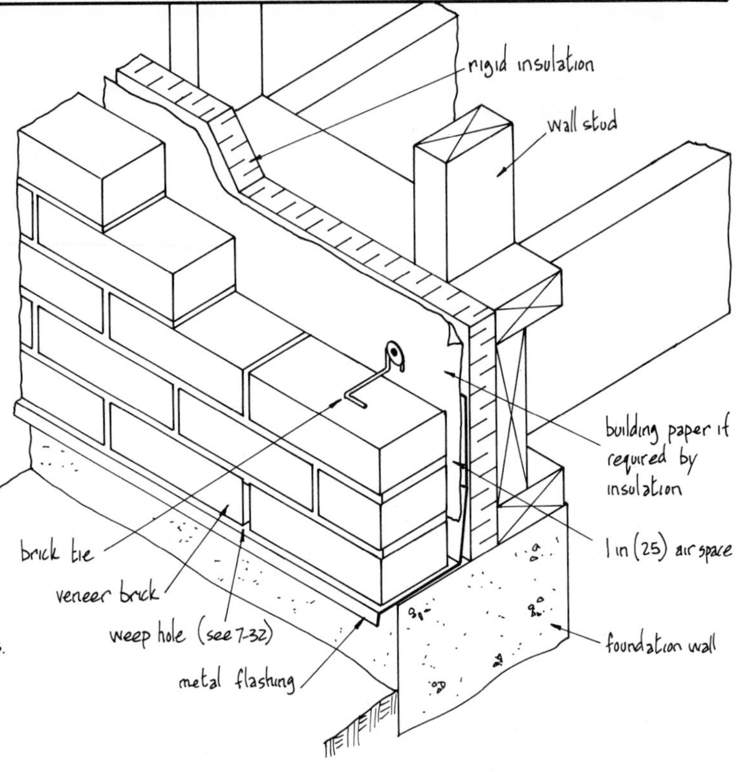

TYPICAL BRICK VENEER DETAIL

BRICK VENEER - EMPIRICAL HEIGHT LIMITATIONS

HEIGHT	NOMINAL BRICK THICKNESS 3 in (75)	4 in (100)
stories	2	3
height at plate	20 ft (6.1 m*)	30 ft (9.2 m*)
height at gable	28 ft (8.5 m*)	38 ft (11.6 m*)

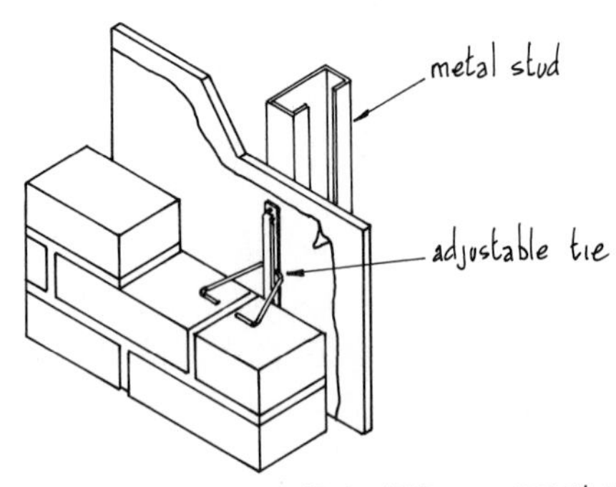

METAL STUD AND BRICK VENEER

BRICK VENEER TIE SPACING

APPLICATION	WALL AREA COVERED PER TIE	TIE MAXIMUM SPACING
one- and two-family dwellings not exceeding one story in height	3¼ sf (0.30 m²)	24 in (600)
all other structures	2⅔ sf (0.25 m²)	24 in (600)

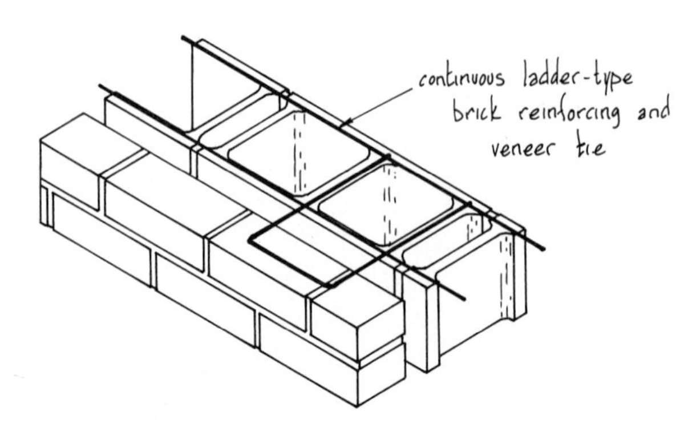

CONCRETE BLOCK AND BRICK VENEER

source: Brick Institute of America

BRICK VENEER

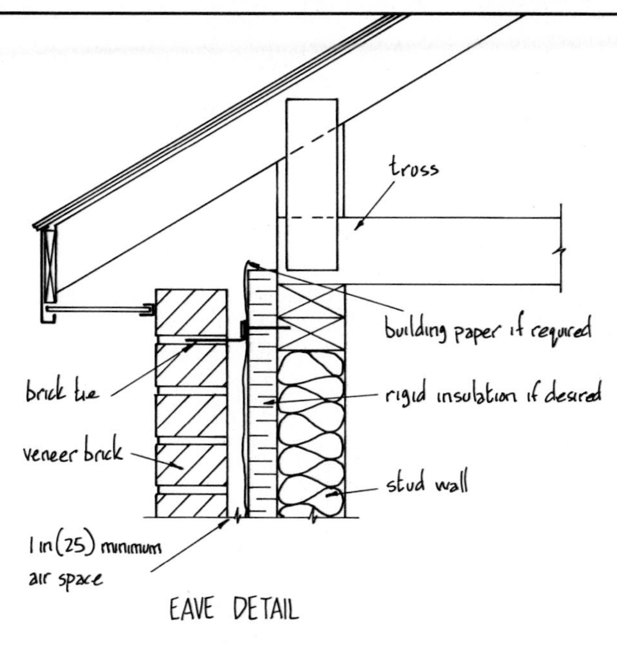

EAVE DETAIL

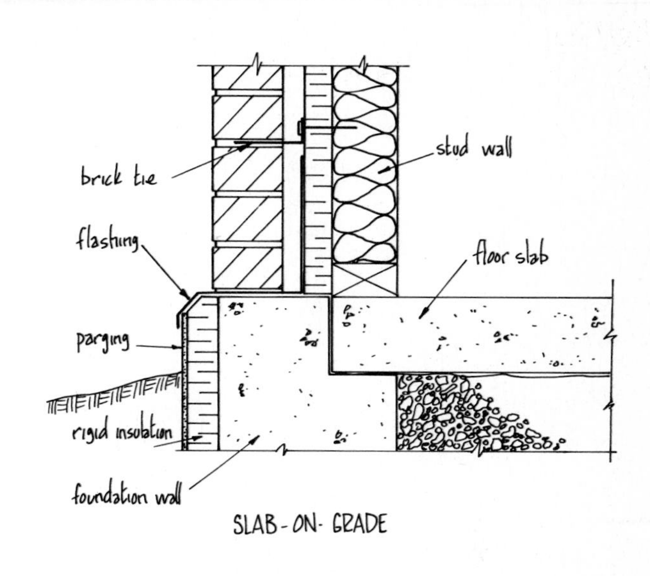

SLAB-ON-GRADE

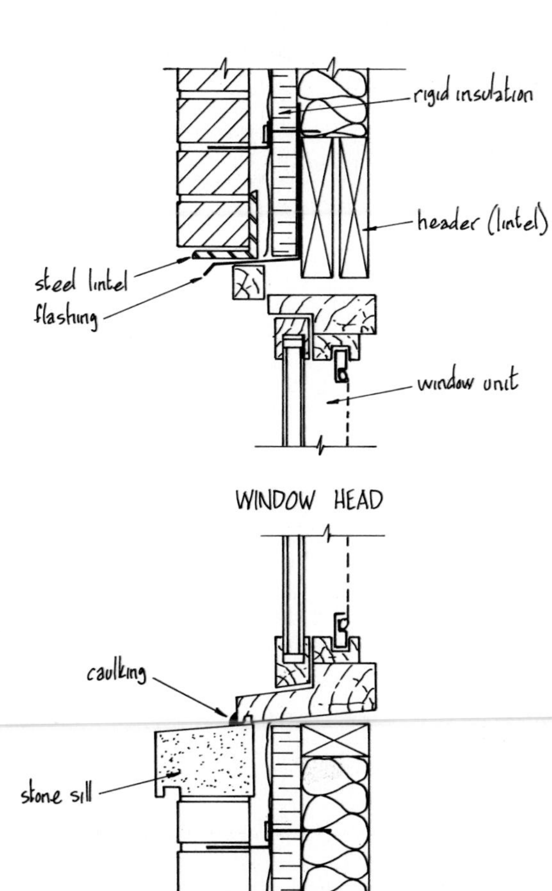

WINDOW HEAD

WINDOW SILL

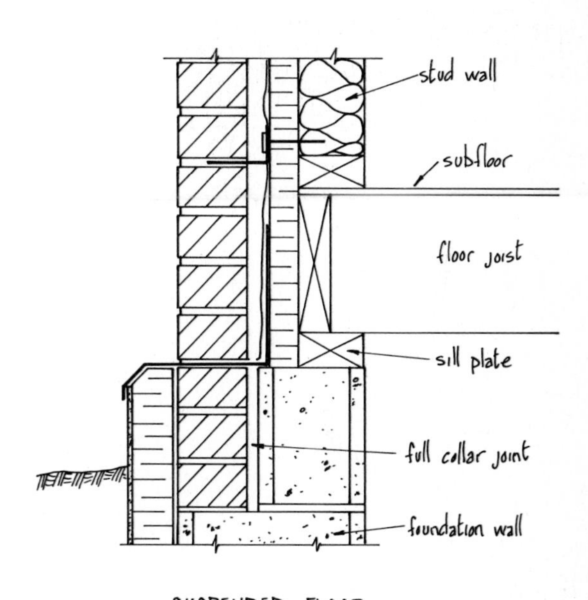

SUSPENDED FLOOR

The illustrations on this and the next page show typical construction details for veneer brick. In each case the tie used should be suitable for the construction materials employed. Each illustration is simplified with interior finishes and air/vapor barriers omitted. Refer to 10-60 for window installation and air sealing techniques.

BRICK VENEER

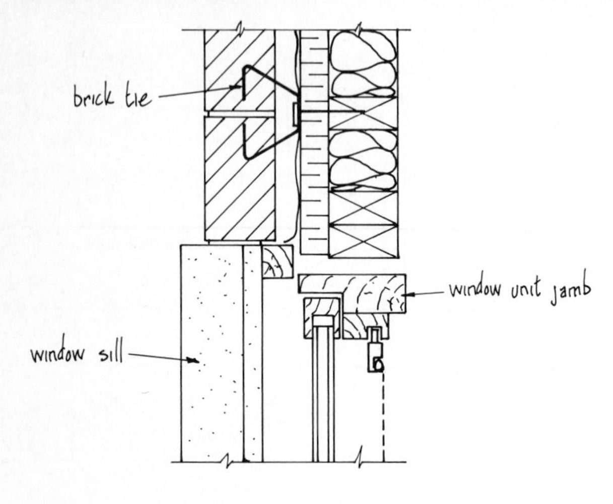

WINDOW JAMB (PLAN)

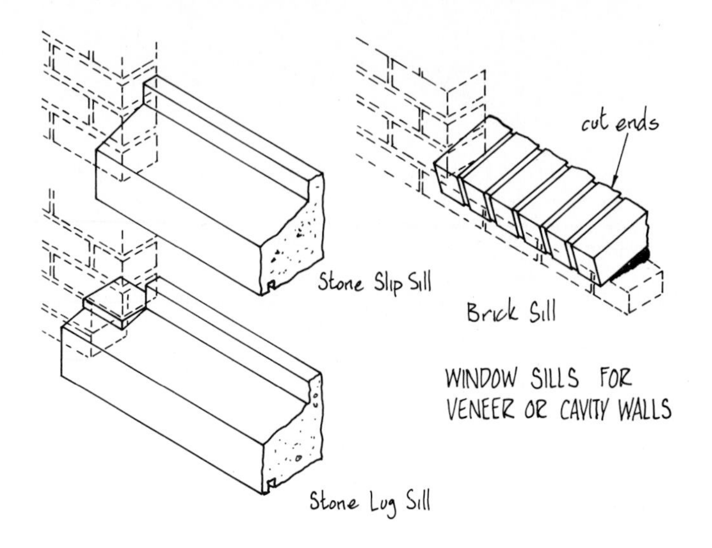

WINDOW SILLS FOR VENEER OR CAVITY WALLS

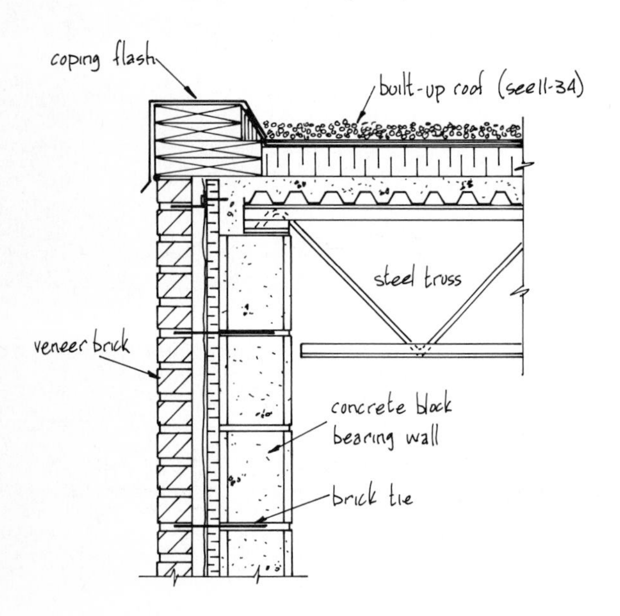

FLAT ROOF

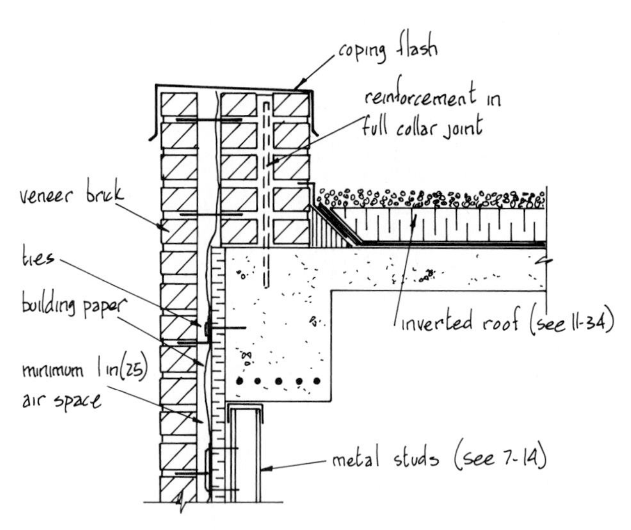

REINFORCED PARAPET WALL

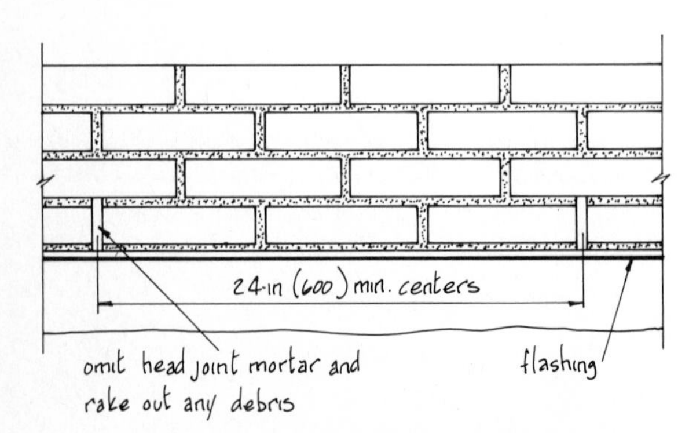

WEEPHOLES AT FLASHINGS

WEEPHOLES AND FLASHINGS

- Weepholes are required at minimum 24-in (600) spacing immediately above all flashings to drain water that may have penetrated the veneer cavity. Care must be taken to avoid surplus mortar being dropped into the cavity and accumulating on the flashing. The simplest type of weephole is made by omitting the mortar in the head joints as shown at the left.
- Flashings should be located above grade level at the wall base and at the head of all openings. They must be secured to the backing structure and should extend through the face of the veneer. Only top-quality long-life material should be used.

source: Brick Institute of America

6-IN (150) "SCR BRICK" AND 4-IN (100) RBM

6-IN (150) "SCR BRICK" LOAD-BEARING WALLS

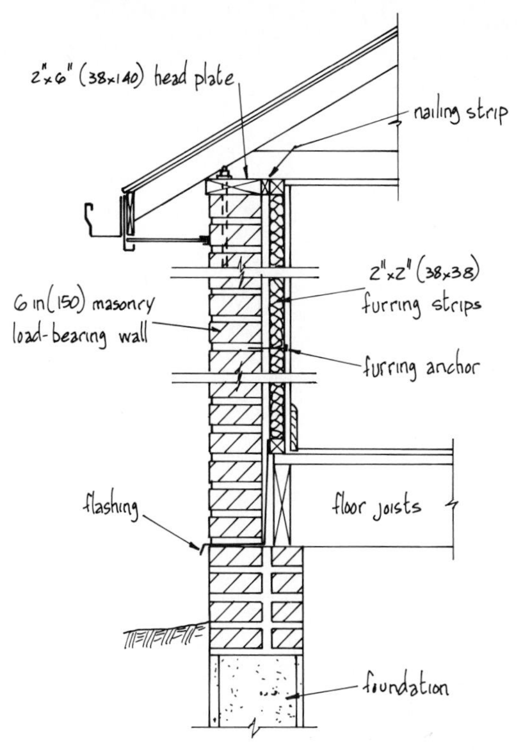

Typical Wall Section

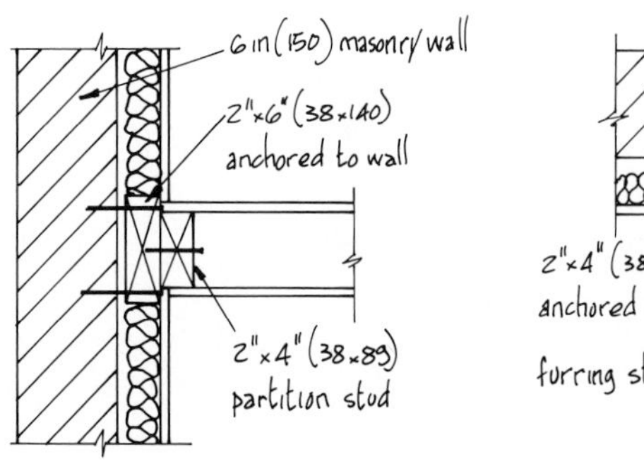

Partition Intersection (Plan)

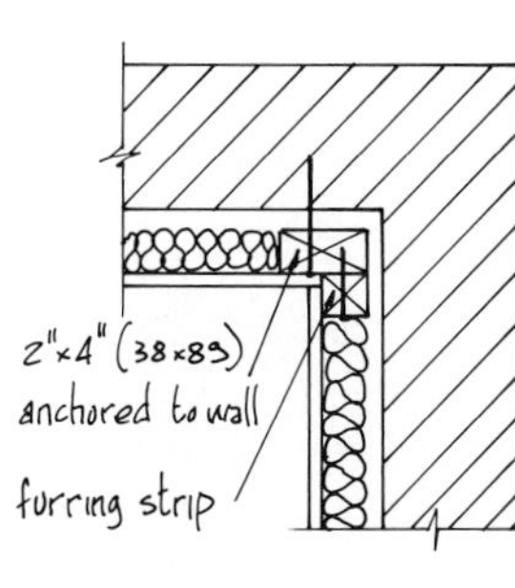

Corner Plan

4-IN (100) REINFORCED-BRICK MASONRY PANELS

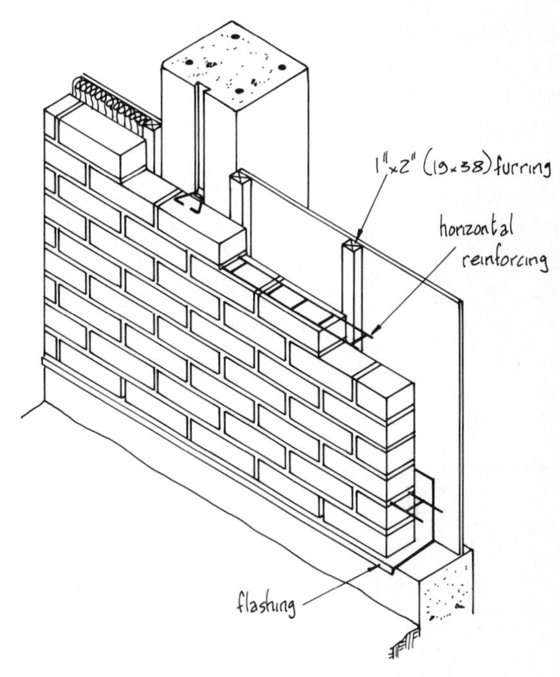

Typical Drainage Type Wall

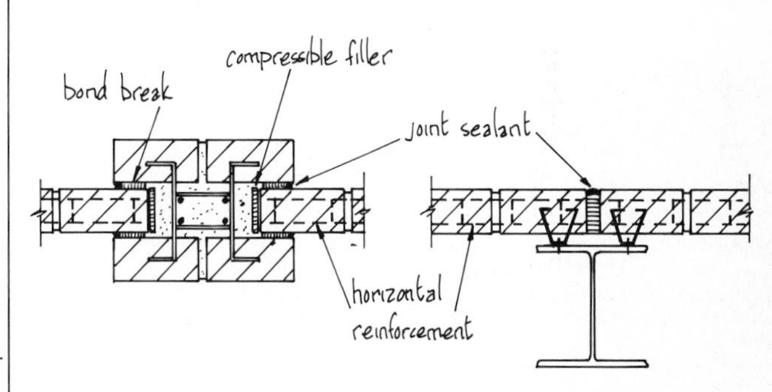

Reinforced Masonry Column

Steel Column

SCR BRICK

- Load bearing 6-in (150) nominal masonry walls used for single-story residential, commercial, and industrial applications.
- Either solid "SCR brick" or hollow tile generally used.
- Minimum 2x2-in (38x38) vertical furring strips clipped to masonry to support interior finishes. Firring strips should stand off the wall at least 1/4 in (6) to allow for drainage.
- Low insulation values and high air infiltration – use only in warm regions or where insulation factor is not important. For higher thermal values, increase size of vertical furring, add thicker insulation, and install an effective air/vapor barrier.

4-IN (100) RBM

- Non-load-bearing 4-in (100) nominal masonry used for commercial and industrial applications.
- Panels are engineered to withstand lateral forces and span horizontally between supports. Use of horizontal reinforcement at maximum 24-in (600) vertical spacing allows large panel sizes without intermediate lateral support.
- Can be insulated as for SCR brick and is subject to the same thermal limitations.

source: Brick Institute of America

CAVITY WALLS

Cavity walls are composed of two separate wythes of masonry tied together with metal ties. The cavity or space between wythes, 2 to 4½ in (50 to 115) wide, allows excellent drainage of the wall and provides space for insulation.

- The exterior wythe is normally brick, while the interior wythe can be brick, structural clay tile, or concrete masonry units.
- The wall can be load bearing on one or both wythes depending on design requirements.
- Empirical design methods are acceptable for one- and two-story structures with light to medium loading or multistory structures without excessive unsupported wall heights.

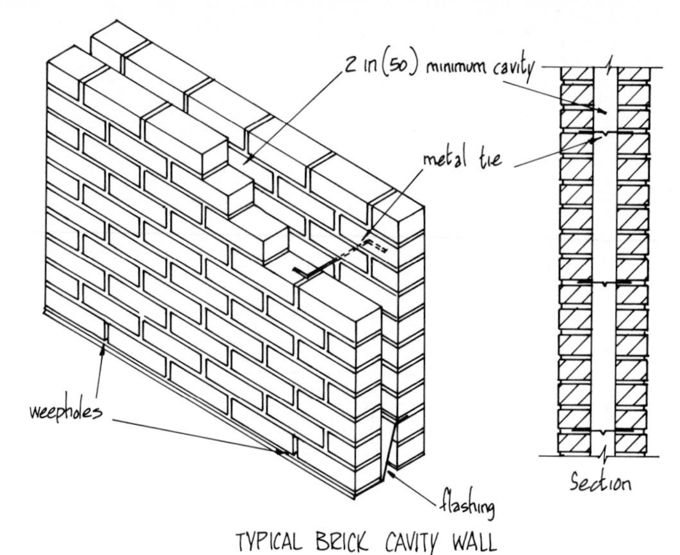

TYPICAL BRICK CAVITY WALL

TIES AND ANCHORAGE

The type of ties used depends on wall design, masonry materials, and expected differential movement. All are corrosion resistant, flexible, and capable of resisting tension and compression loads. Tie spacing is shown below, together with several examples of anchor types.

Ladder Reinforcing with Welded Rectangular Ties

Flexible Column Anchor

Inner Wythe Anchor and Z-Cross Ties

24" (600) maximum vertical spacing

36" (900) maximum horizontal spacing

4½ sq.ft (0.4 m²) maximum surface area per tie

SPACING AND STAGGERING OF WALL TIES

Source: Brick Institute of America

CAVITY WALL DETAILS

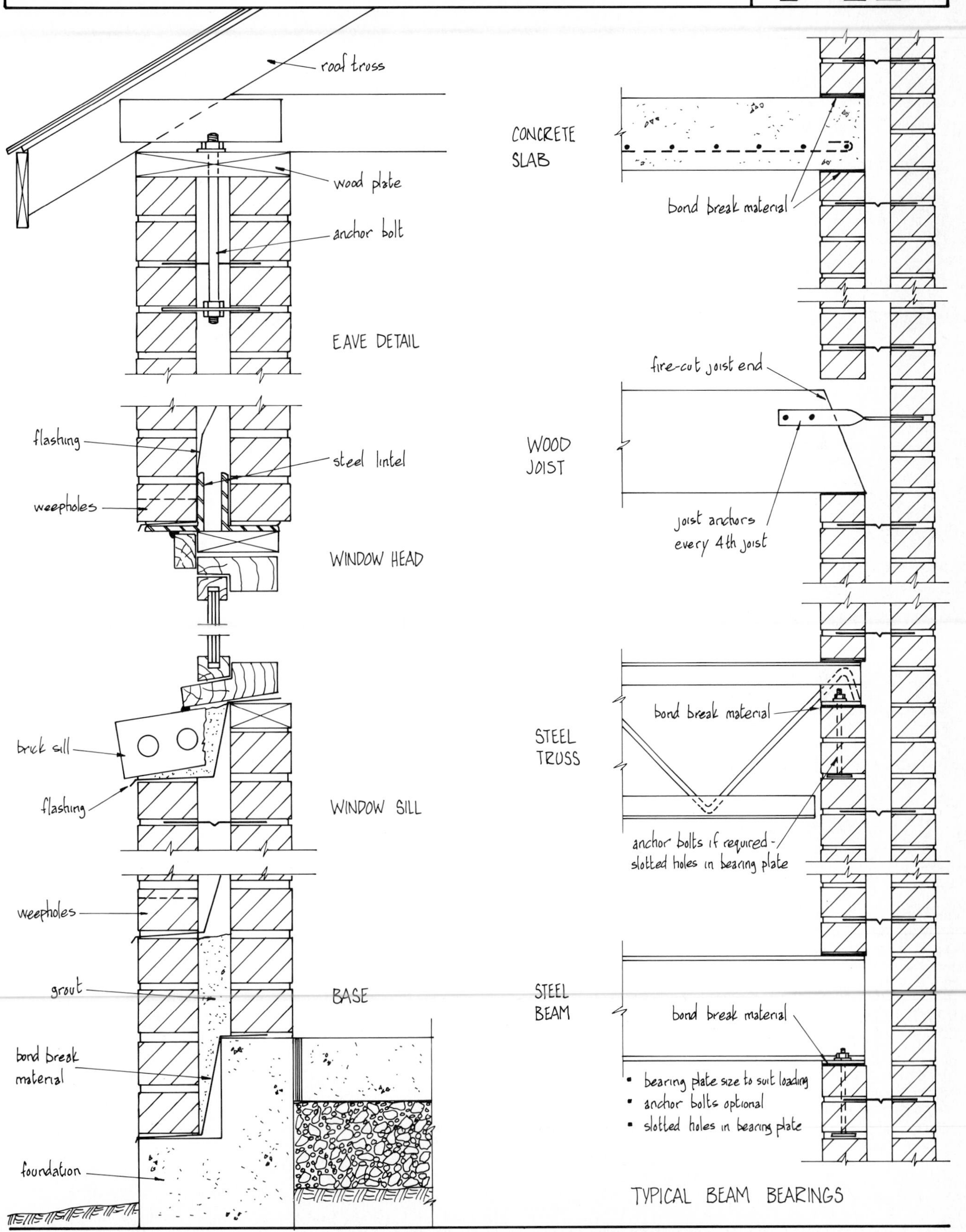

CAVITY WALL DETAILS

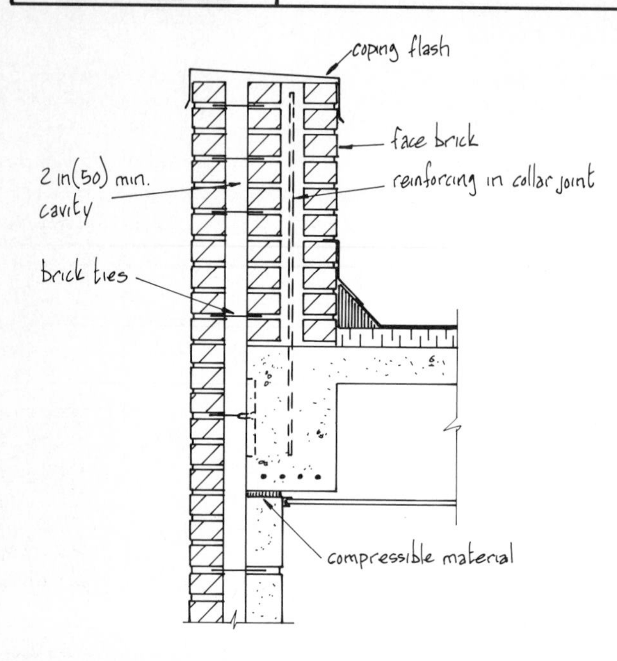

REINFORCED PARAPET WALL

PARAPETS

Parapets in any type of masonry wall present problems. They are subject to excessive water penetration and are exposed to weather extremes on both sides. The detail above is a typical solution.

- The wall cavity extends into the parapet to main flexibility.
- The inner cavity is a reinforced grouted collar joint (see 7-28).
- The coping is watertight and has drips on both sides of the wall.
- Vertical expansion joints should continue up to coping level.

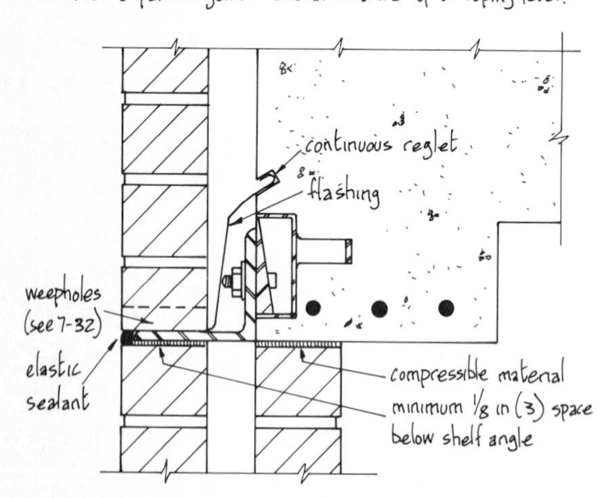

SHELF ANGLE DETAIL

When shelf angles are used to support the outer wythe of individual stories (as in brick veneer curtain walls), a horizontal pressure-relieving joint is required below the angle. In addition, slot the holes for the anchor bolts horizontally and leave a small gap between angle lengths.

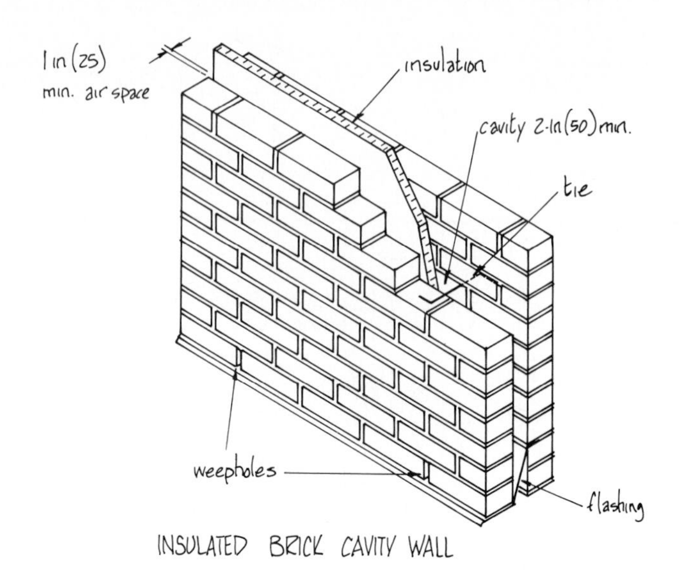

INSULATED BRICK CAVITY WALL

CAVITY WALL INSULATION

Cavity walls can be insulated internally in two basic ways:

- With rigid boards such as extruded polystyrene either glued to the inner wythe or held in place by the metal ties (see below).
- With granular material such as vermiculite or perlite poured into the cavity as wall construction proceeds.

In each case, however, certain conditions must be met. The insulation must:

- Allow the cavity to continue its drainage function.
- Not absorb water itself, or loss of insulation value and interior leakage will occur.
- Be resistant to rot, vermin, and fire.
- Be capable of supporting its own weight without settlement.

For additional information, see 10-15 onward.

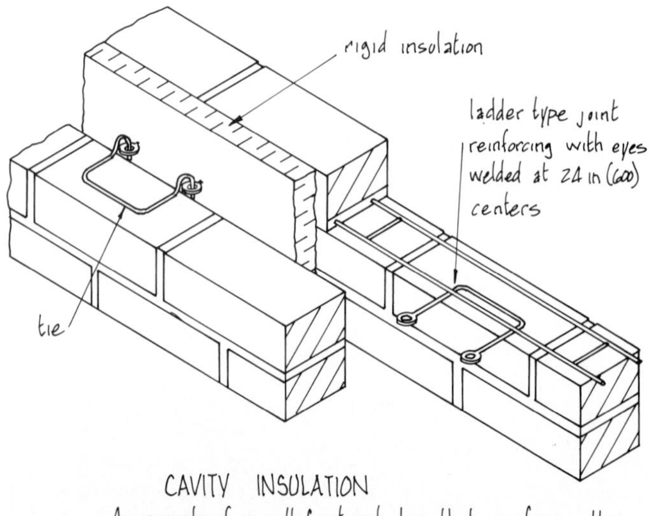

CAVITY INSULATION

An example of a multifunctional tie that reinforces the inner wall and locks the insulation to the wall face.

source: Brick Institute of America

SOLID MASONRY

In a solid wall, the masonry units are laid close together and all the joints are filled with mortar. There are several types of solid walls:

- Solid unit walls composed of solid units of brick or concrete masonry.
- Hollow unit walls composed of hollow units of clay or concrete masonry.
- Composite walls (faced walls) having wythes of different material but bonded so that both wythes act together under load.
- Reinforced brick masonry (RBM) having steel reinforcement (other than ties) bonded with grout in the collar joint (see the next page).

In addition, walls may be load-bearing or non-load-bearing and be bonded with masonry headers (see 7-26) or metal ties. For most walls, metal ties are the recommended bonding system because of their superior resistance to water penetration.

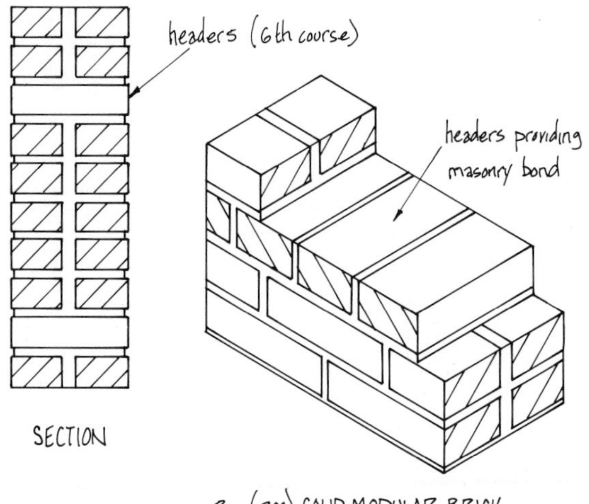

8 in (200) SOLID MODULAR BRICK MASONRY BOND

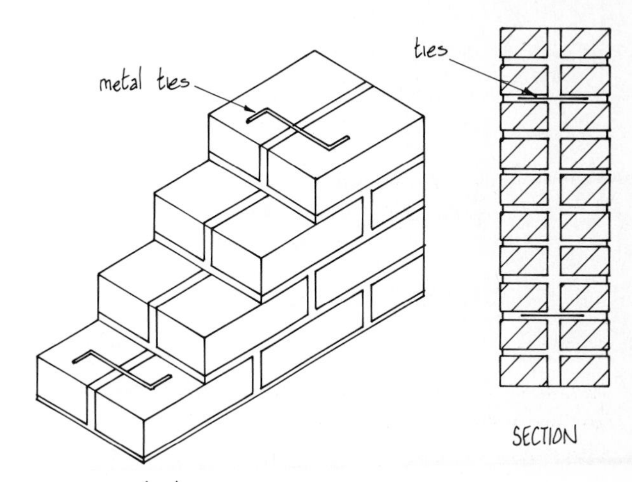

8 in (200) SOLID MODULAR BRICK METAL TIED

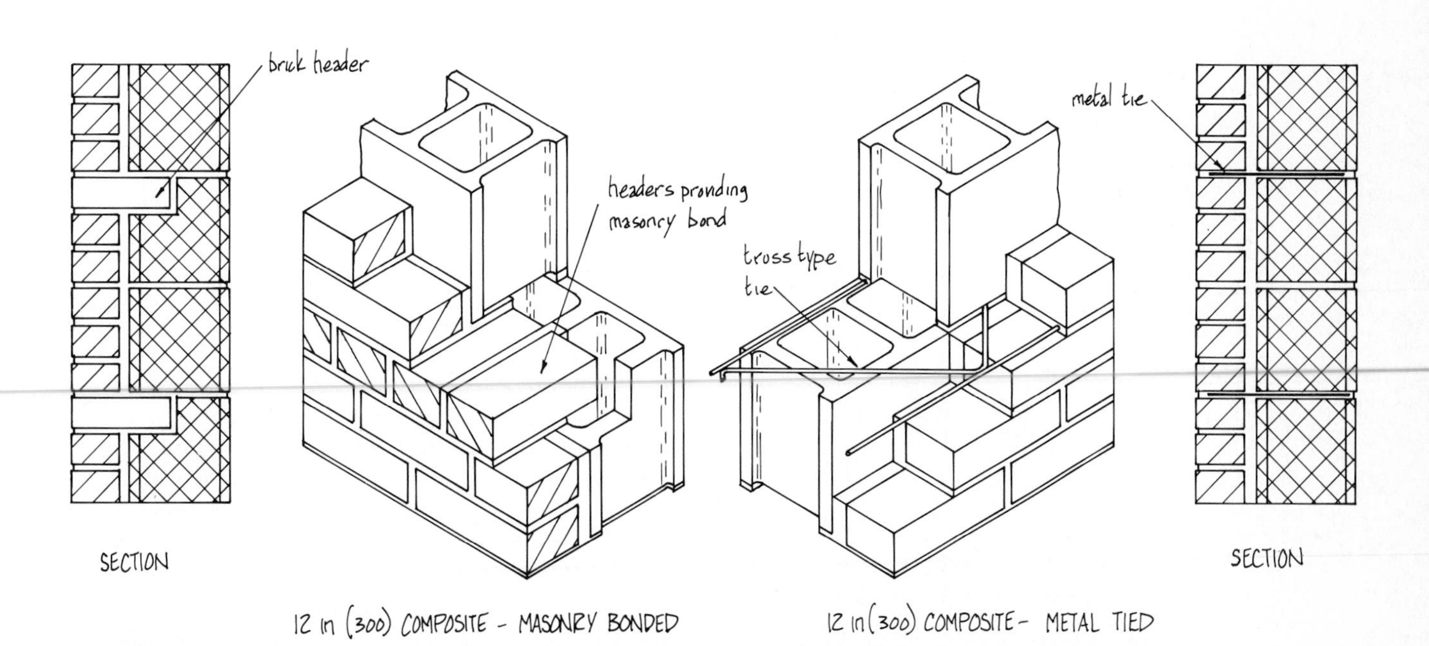

12 in (300) COMPOSITE - MASONRY BONDED

12 in (300) COMPOSITE - METAL TIED

SOLID MASONRY

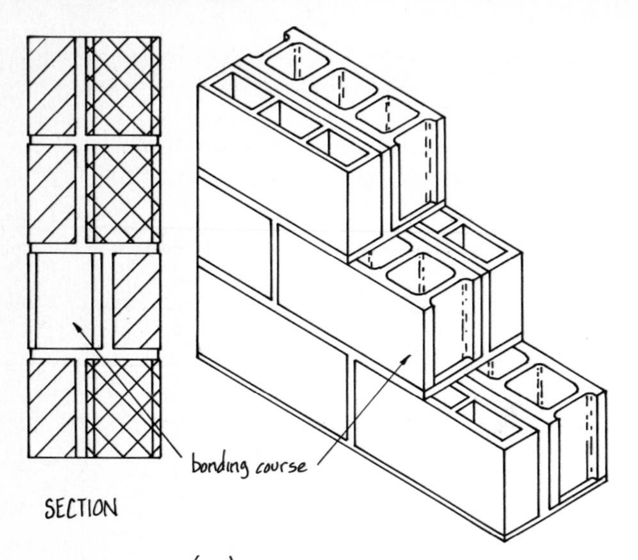

10 in (250) COMPOSITE HOLLOW CONCRETE BLOCK MASONRY BOND

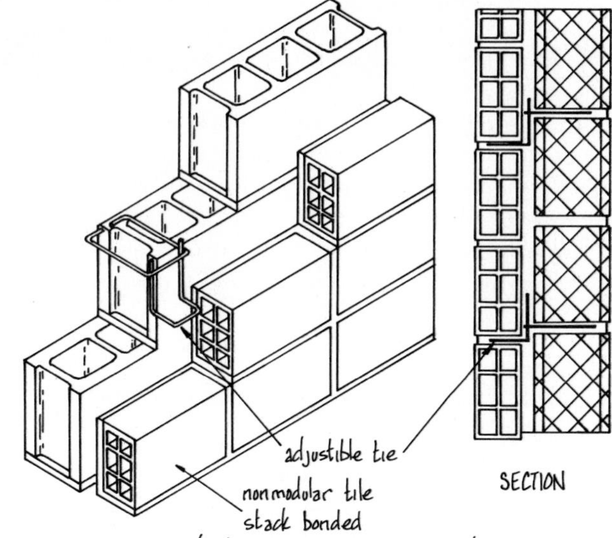

10 in (250) COMPOSITE CONCRETE BLOCK / HOLLOW TILE METAL BOND

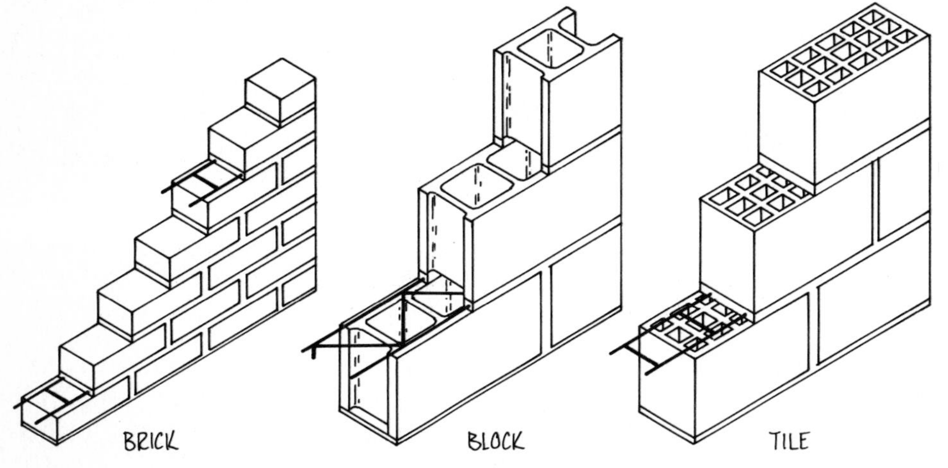

SINGLE-WYTHE SOLID MASONRY WALLS

Single-wythe walls can also be classed as "solid" in most cases, even though they may be composed of hollow units. Generally, only when there is a hollow or cavity in the wall (e.g., cavity or veneer walls) can the wall be classed as "hollow".

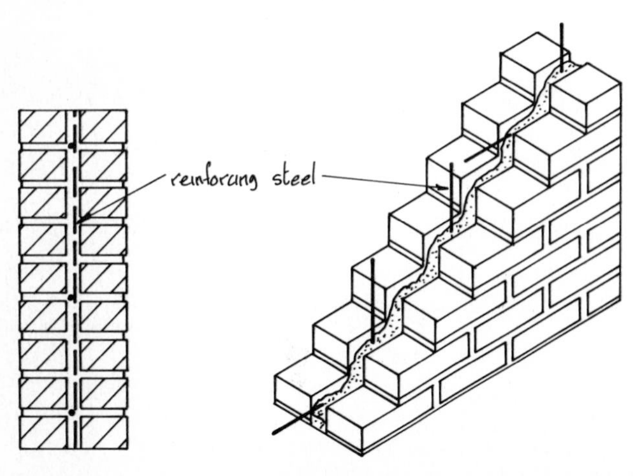

REINFORCED BRICK MASONRY

REINFORCED BRICK MASONRY SOLID WALLS

- Generally, an engineered wall similar in principle to reinforced concrete design, RBM walls reduce wall size and resist high lateral forces.
- Used extensively in earthquake-prone regions.
- RBM design is applied to columns, pilasters, beams, and lintels as well as walls.
- Reinforcing steel is grouted in the wall cavity, and metal ties or headers are not required provided that a strong mortar bond is developed.

MASONRY WALL LINTELS

LINTELS

Lintels are structural members that support wall loads above an opening. For masonry walls, lintels can be of steel, precast concrete, or reinforced clay tile, clay brick, and concrete block. Masonry arches also act as lintels, although in a more specialized form. Loads must be calculated and lintels carefully sized to avoid excessive deflection and subsequent wall cracking above openings. Empirical or "rule of thumb" design is normally acceptable only in single-wythe, non-load-bearing walls such as veneer and panel construction.

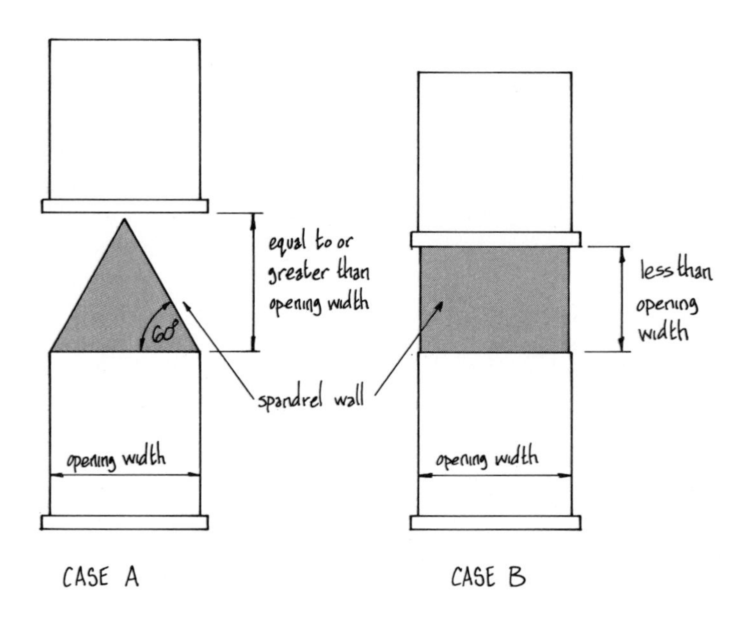

LINTEL LOAD CALCULATION

- Masonry units above an opening tend to support each other (arching) reducing the total load on the lintel provided that certain conditions are met.
- In case A, the vertical height of masonry between openings is equal to or greater than the opening width. Lintel load is the weight of material contained within the triangle only.
- In case B, vertical masonry height is less than opening width, and lintel load is the weight of material in the rectangle.
- For walls with uniform floor loads or point beam loads, consult design handbooks.
- Temporary shoring of lintel during curing or use of horizontal masonry reinforcement may be necessary to ensure arching action.

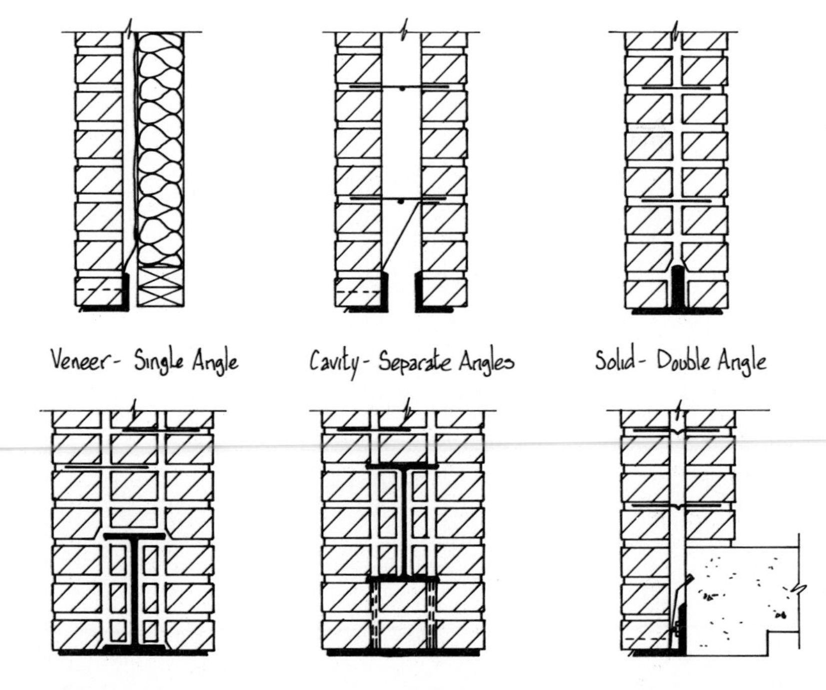

STRUCTURAL STEEL LINTELS

- Steel lintels can take the form of single angles in veneer walls to beam and plate combination lintels in heavy solid walls.
- The lintel is designed to best support the wall and is sized to carry all loads without excessive deflection.
- Angle horizontal leg size is 3½ in (90) for 4-in (100) wythes and 2½ in (65) for 3-in (75) wythes, each with a minimum ¼-in (6) leg thickness.
- Vertical leg size is determined by load factors.
- Length of the bearing on either side of the opening is normally 3-6 in (75-150) depending on load.
- Steel lintels require periodic maintenance if exposed to weather.

MASONRY LINTELS

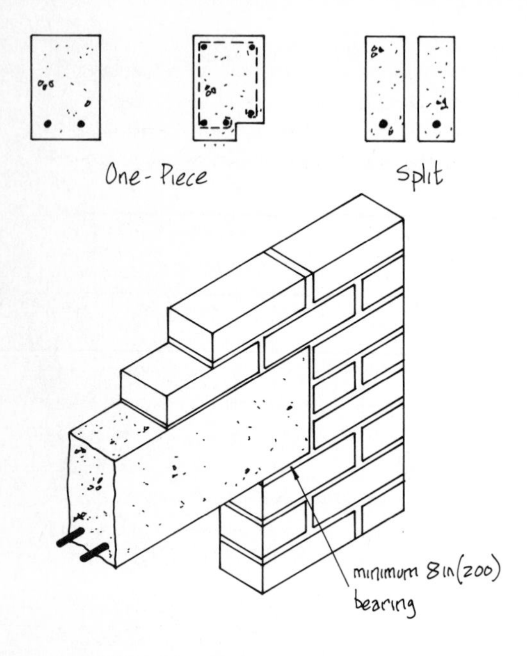

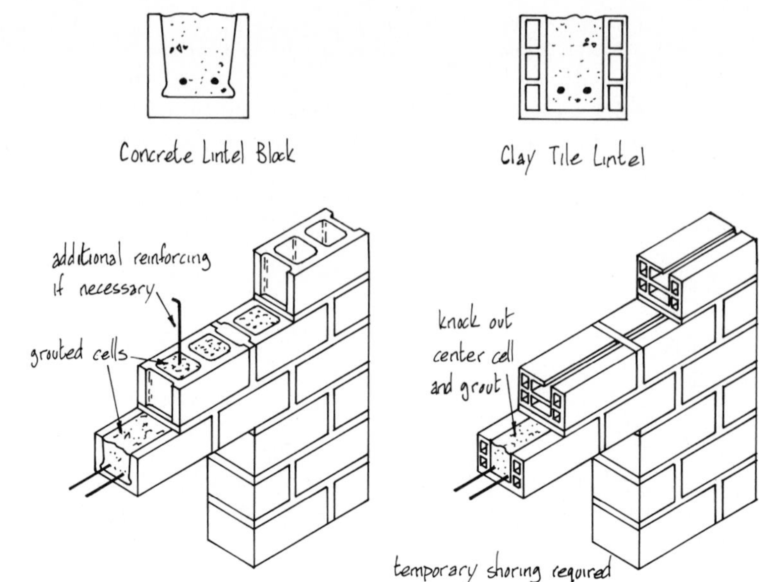

PRECAST CONCRETE LINTELS

Advantage – temporary shoring not required.

CONCRETE BLOCK LINTEL

CLAY TILE LINTEL

Advantage – lintel becomes part of the wall with no break in line or form.

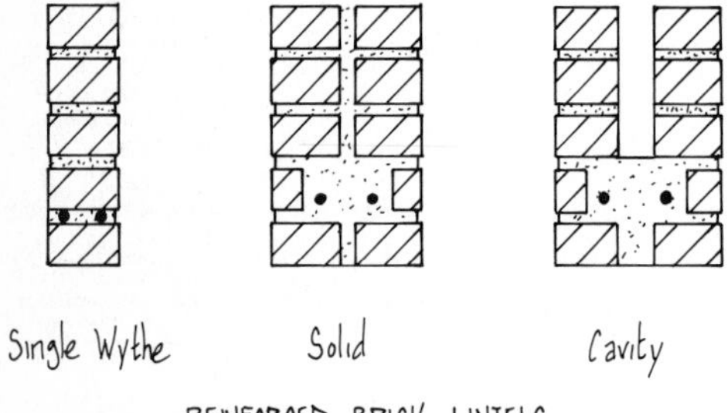

REINFORCED BRICK LINTELS

REINFORCED BRICK MASONRY LINTELS

- RBM lintels have the advantages of strength, economy, and aesthetics.
- Steel reinforcing is not exposed to weather and is sized to suit loading stresses.
- Beams and girders can also be constructed with RBM, utilizing design principles similar to those used for reinforced concrete. A typical example is illustrated at the right.

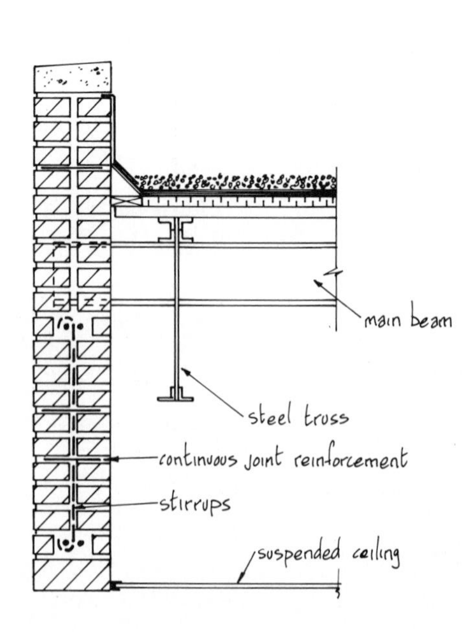

SECTION OF TYPICAL REINFORCED MASONRY GIRDER

Source: Brick Institute of America

MASONRY ARCHES

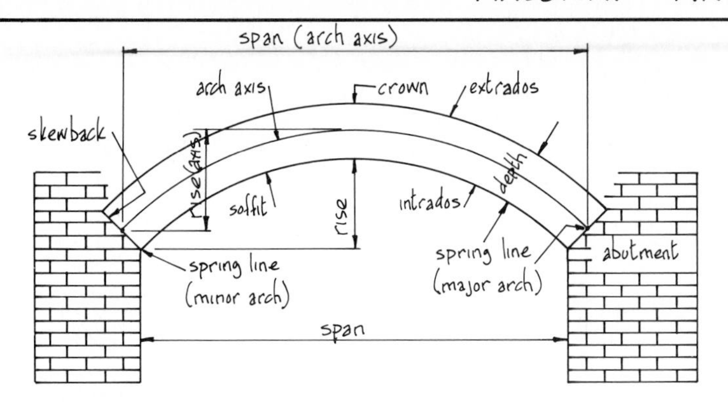

ARCH TERMINOLOGY

ARCHES

- An arch is essentially a curved lintel beam that is subject to almost entirely compressive forces. It therefore utilizes masonry's compressive strength to best advantage, as well as having a strong aesthetic appeal.
- An arch may be constructed using standard masonry units with a variable mortar joint thickness or with special tapered units and a uniform joint thickness.
- There are two classes of arches: minor and major. Minor arches have a maximum 6 ft (1800) span and a span-to-rise ratio of less than 0.15. Major arches exceed 6 ft (1800) span and have a ratio above 0.15.
- Arch terminology is given on the left, and a selection of arch types are shown below.

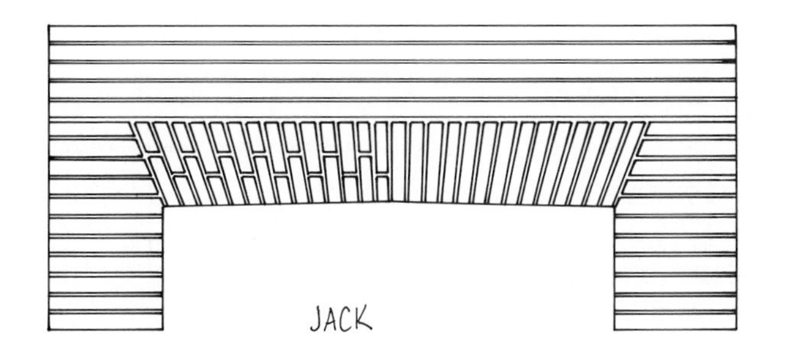

JACK

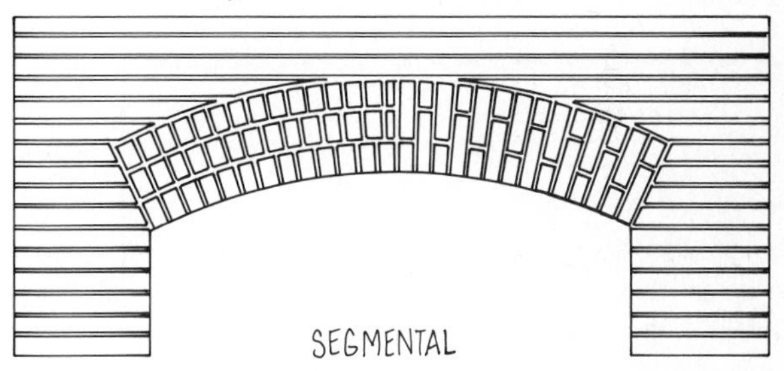

SEGMENTAL

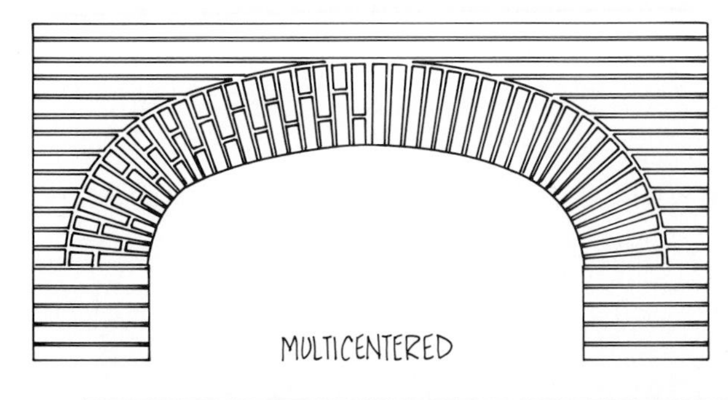

MULTICENTERED

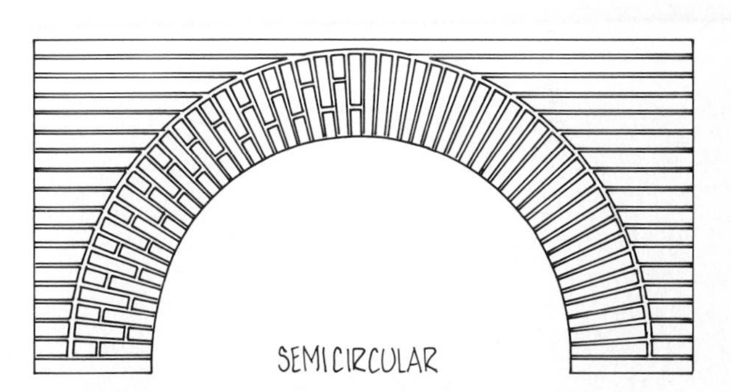

SEMICIRCULAR

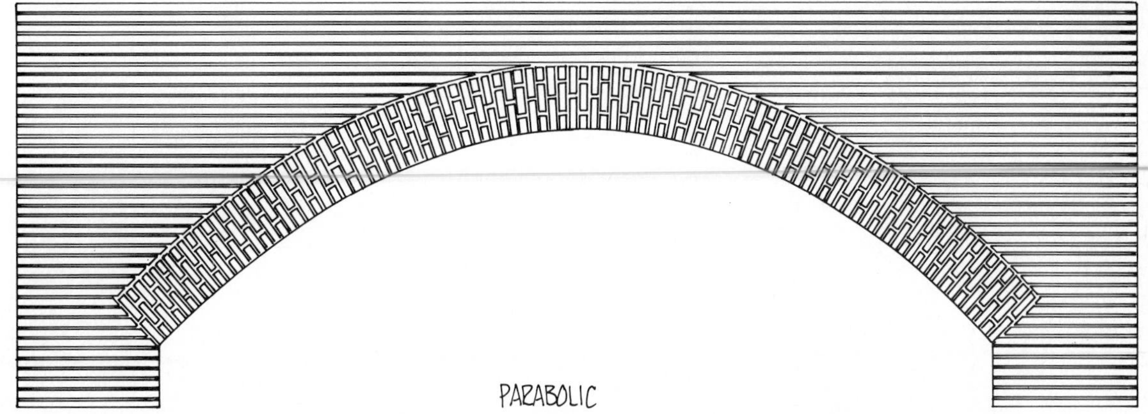

PARABOLIC

source: Brick Institute of America

CURTAIN WALLS

A curtain wall is broadly defined as a thin, exterior, non-load-bearing wall whose weight is carried directly by the building's structural frame. Virtually all types of materials are used in the extremely wide variety of curtain wall systems offered by manufacturers. However made, all systems have a number of points in common:

- Weatherproof. Any system must be either watertight or have provision for water drainage.
- Durable. The system must be durable enough to last at least as long as the building itself and with a minimum of maintenance.
- Flexibility. The system will be subject to movement in the building's structure in addition to its own expansion and contraction factors.
- Light control. Window units, which are a major part of most systems, control light intensity and heat gain within a building.
- Insulation. Modern systems incorporate suitable insulation values for both cold or hot regions. For thin sheet panels, rigid insulation is normally bonded to the sheet back to add strength.
- Aesthetic value. A major consideration in most systems; with modern materials, designers can create a wide range of visual effects. Six typical basic effects are shown at the right.

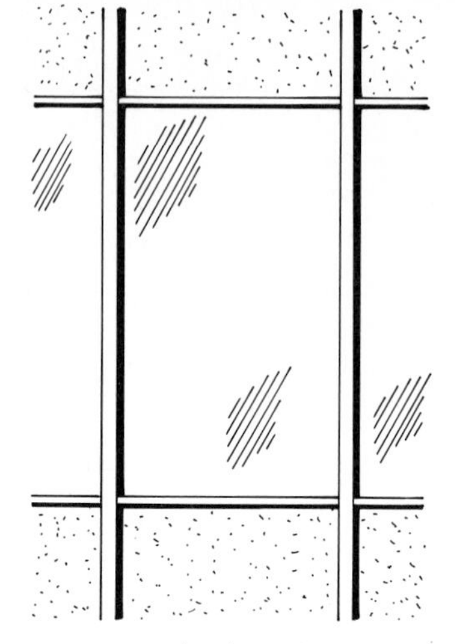

Vertical Emphasis

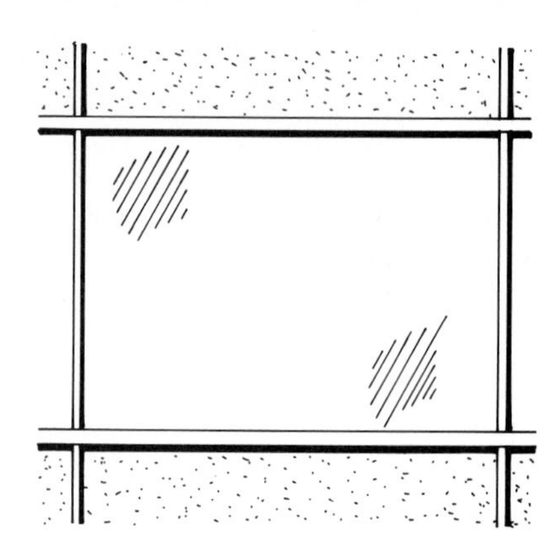

Horizontal Emphasis

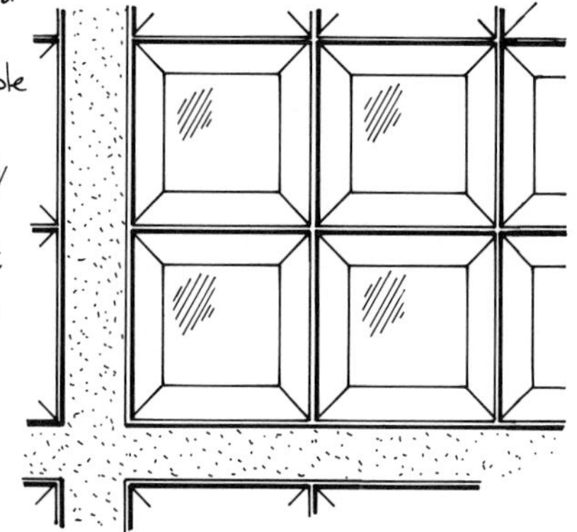

Square Emphasis

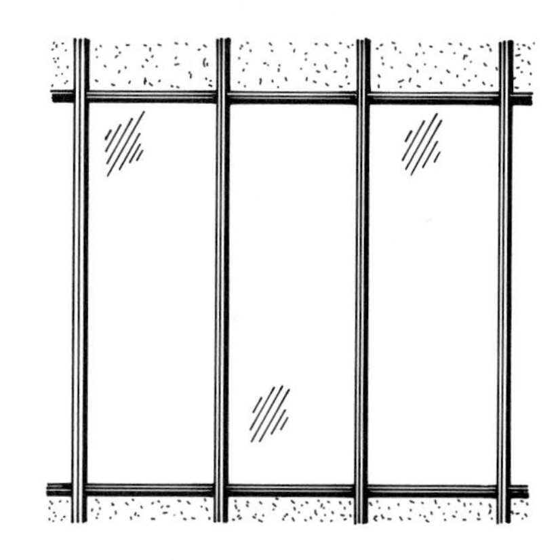

Rectangular Emphasis

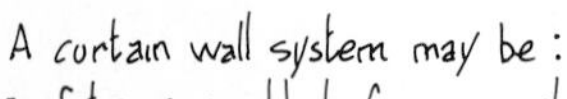

A curtain wall system may be:

- Site-assembled from a number of individual pieces (stick system).
- Part factory and part site-assembled (mullion and panel system).
- Almost totally factory assembled with a minimum of site work (panel system).
- A totally "custom" design for a single large project or an "off-the-shelf" standard system for smaller projects.

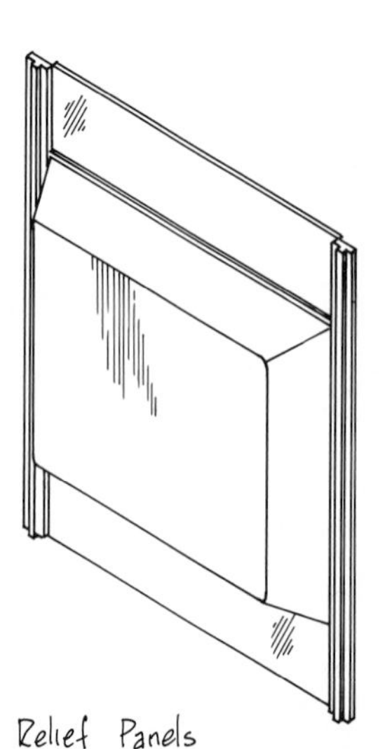

Relief Panels

Flat Panels

TYPICAL EXAMPLES OF CURTAIN WALL VISUAL EMPHASIS

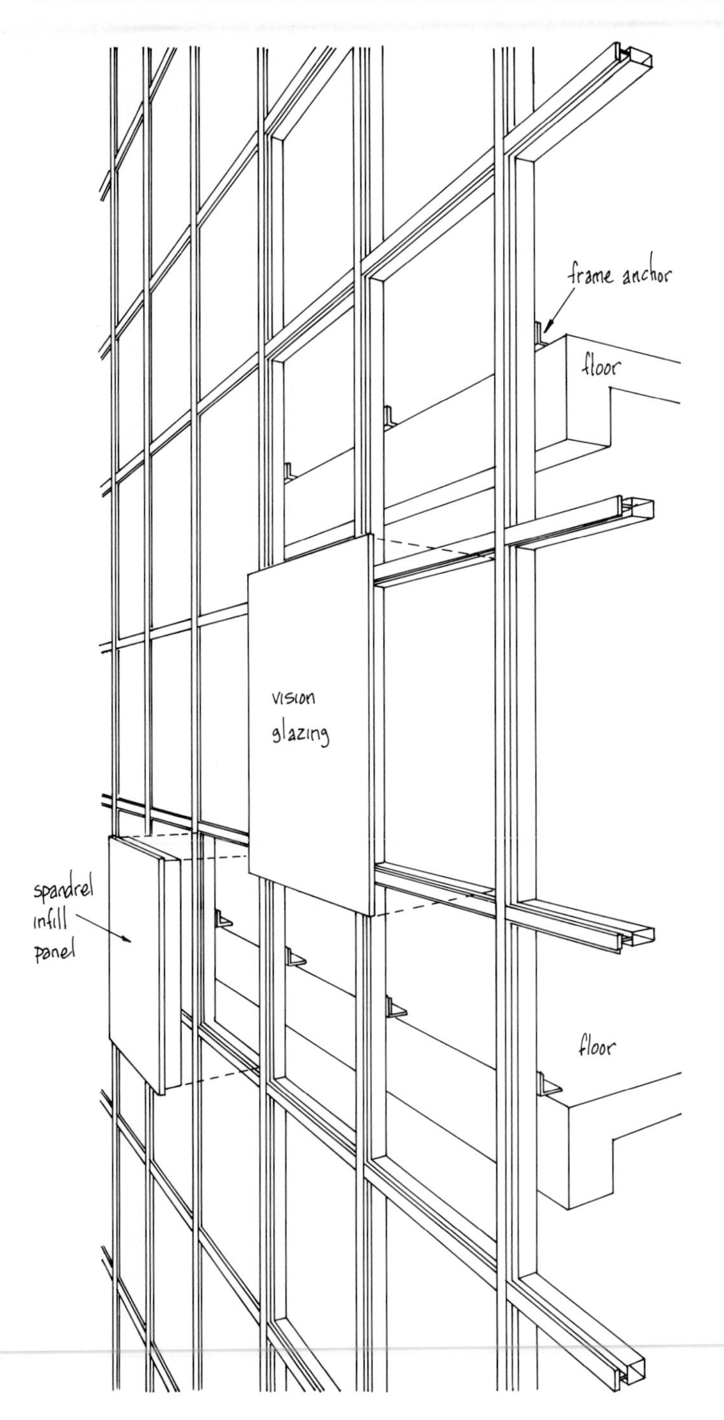

The above illustration shows a typical curtain wall frame with a flat, rectangular emphasis. Vision glazing runs in a horizontal band between floors, while infill spandrel panels hide floor construction and frame anchorage from exterior view.

CURTAIN WALL SYSTEMS - MATERIAL TYPES

Metal   Metal products are used extensively in curtain walls for both panels and fittings. Typical examples are:

- Stainless steel for all or part of the wall system.
- Porcelain-enamel on a steel or aluminum base.
- Aluminum with a plain, painted, or anodized finish.
- Copper and copper alloys with various finishes.
- Steel panels with a baked enamel or other finish. See 7-44 for an alternative system.

Concrete

- Precast or hollow-core panels similar to precast concrete floor slabs (see 6-29) but designed for wall applications.
- Solid precast panels in a variety of shapes and finishes.

Glass

- Used as clear glazing or with a "smoked" coating for light transmission control.
- Used as opaque panel with a finish such as a ceramic color fire-fused to the panel face.
- Glass blocks (see 7-21) or similar products laid as masonry or as an integral part of a wall system.

Cement-Asbestos

- Usually, a laminated panel is used comprising a core of rigid insulation material with cement-asbestos sheets bonded on either side. The exterior asbestos sheet may have a finish such as polyester glass-fiber cloth to give color and extra strength to the panel.

Masonry

- Clay masonry products, including brick, tile, and special glazed units, normally supported at each floor level on structural steel members (see 7-36, lower left).

source: H.H. Robertson Company

METAL SIDING CURTAIN WALLS

COMPONENT WALL SYSTEM
Wide range of liner and face sheet combinations in style, color, material, and fastening systems.

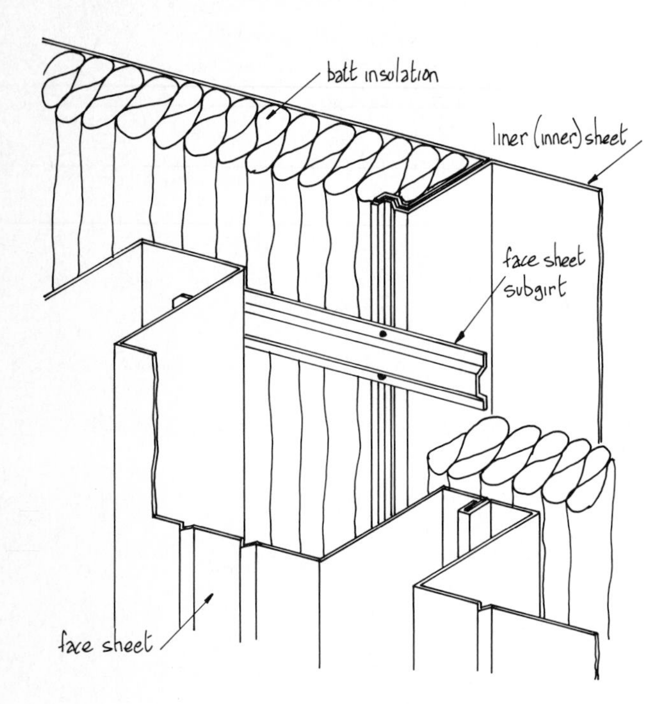

COMPOSITE WALL SYSTEMS
A self-contained, rigid, and lightweight panel system with liner and face sheets bonded to a foamed rigid insulation core. Less variety in color and style but features lower cost and speedy erection.

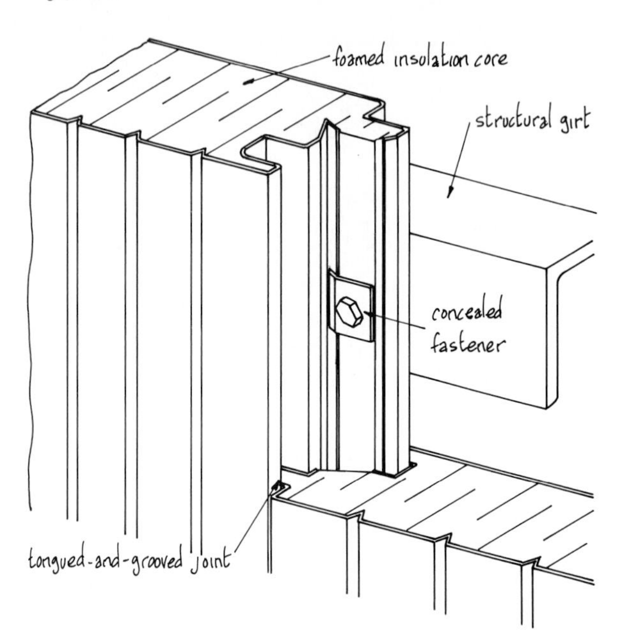

Metal siding systems are commonly used for many commercial and industrial applications. There are two basic system types as illustrated above. Each generally consists of an inner liner panel, an insulated core and an exterior face sheet. As a curtain wall system it has many advantages, especially when glazing is not the primary architectural function.

- Insulation values can be as high as R25 (RSI 3.5) in some systems.
- Strong or subtle architectural effects can be created with the many types of facing sheets available.
- A wide selection of standard or custom colors with a plain or textured finish is available to designers.
- Systems are lightweight and reduce dead load on wall structure and foundations.
- Long spans between structural supports of up to 20 ft (6 m) are possible.
- Since face sheets and liners are roll formed, long lengths of siding can be used to reduce end joints and weather penetration.
- Alterations or expansions to existing buildings are relatively easy, or, if necessary, the entire system can be demounted and re-erected elsewhere.
- Both component systems and composite panels can be designed for vertical or horizontal application.

MATERIALS
The three primary materials used in siding systems are:

- Galvanized steel, prepainted before roll forming.
- Aluminum, either plain, coated, or prepainted.
- Stainless steel, normally left unfinished.
- Any of the above but with a special finish for specific industrial applications.

METAL SIDING - DETAILS

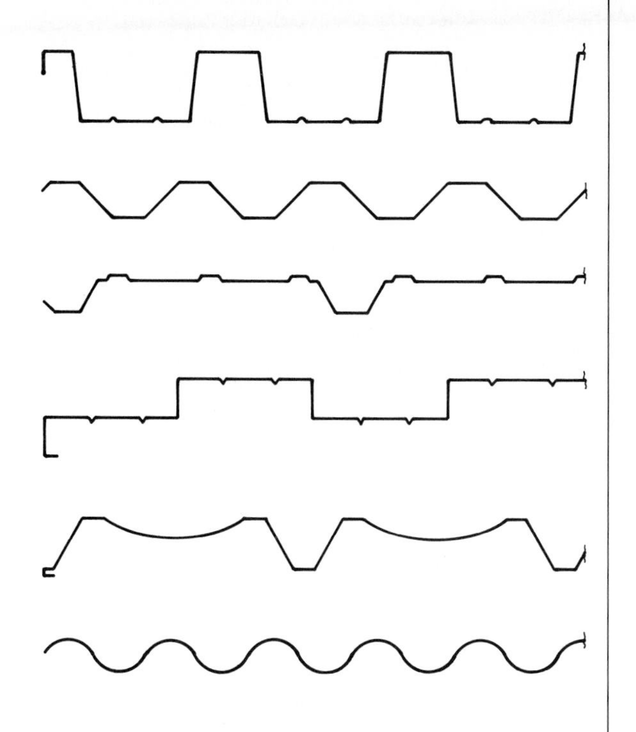

TYPICAL FACE SHEET PROFILES FOR COMPONENT SYSTEMS

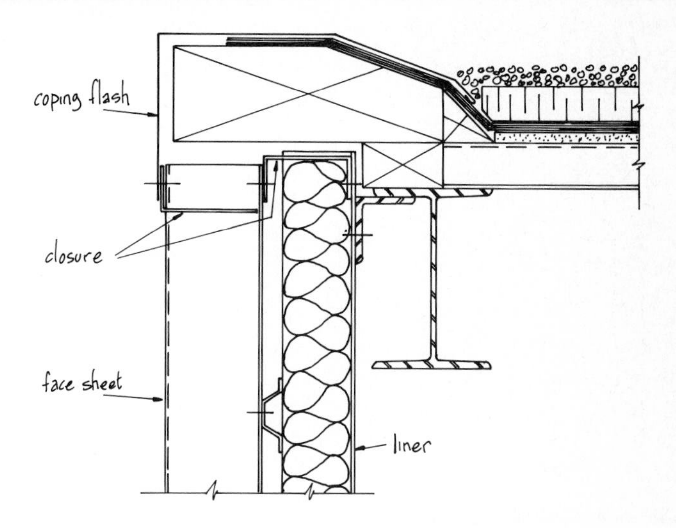

COPING DETAIL - COMPONENT SYSTEM

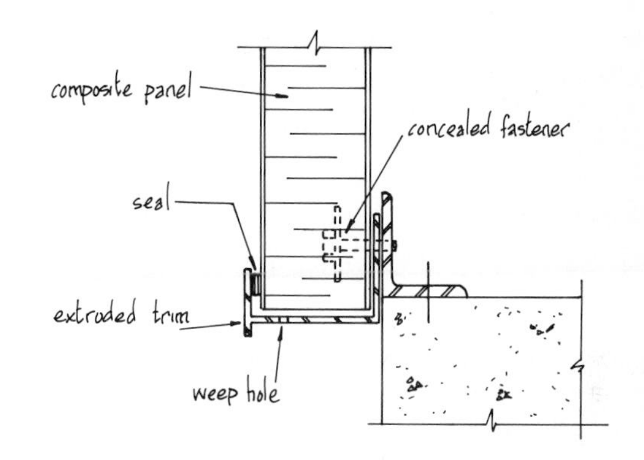

BASE DETAIL - COMPOSITE PANEL

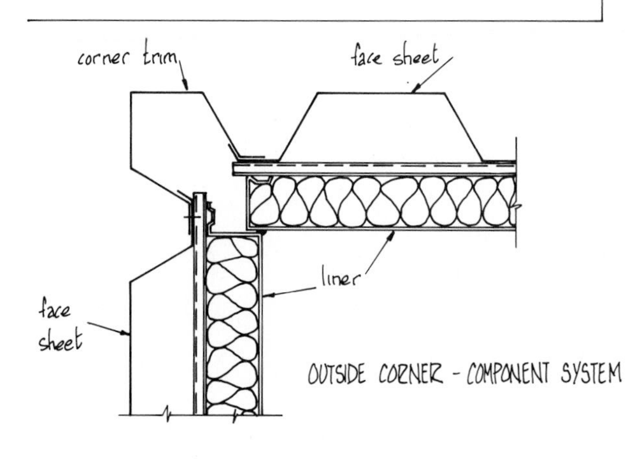

OUTSIDE CORNER - COMPONENT SYSTEM

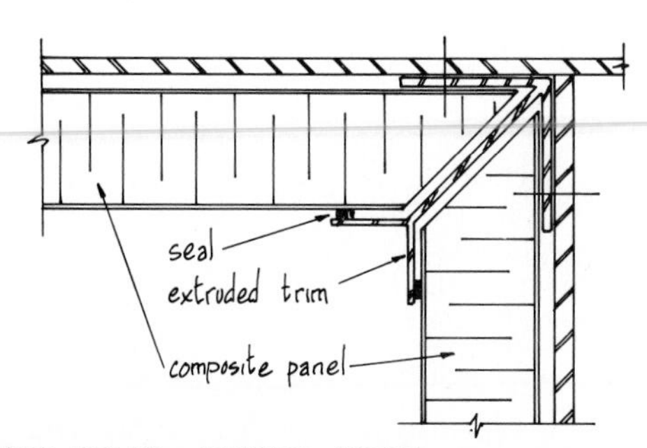

INSIDE CORNER - COMPOSITE SYSTEM

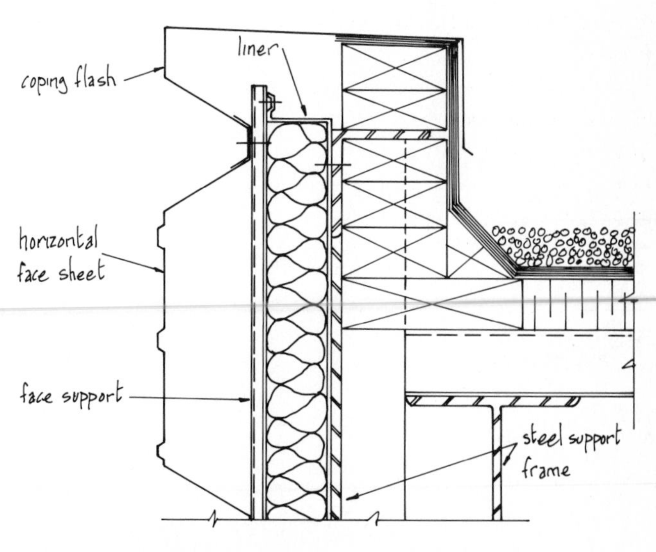

HORIZONTAL SIDING - COMPONENT SYSTEM

source: H.H. Robertson Company

CHAPTER PAGE TITLES

ROOF SHAPES

Roofs, both residential and commercial, can be designed with a variety of shapes and structural systems. All, however, have a number of points in common.

- The roof structure must be designed to support dead, live, and wind loads.
- The slope of the roof determines to a large degree the choice of exterior finish (see 11-25 onward).
- Roof shape has a strong effect on the overall character of a building and requires careful consideration.
- The roof structure may also be required to provide support for ceiling finishes.
- In some roof designs, glazing that provides light to interior roof spaces may be included.

BASIC ROOF SHAPES

The many variations of roof shapes described on 8-4 through 8-7 can all be derived from the three basic roof forms shown here.

PITCHED

- Pitch or slope can vary greatly (see the next page).
- Provides easy water runoff for rain or snow.
- Allows for ceiling plane in horizontal or pitched positions
- Steep pitches allow interior roof living or storage spaces.
- Pitched roofs require a tie system to counteract horizontal reaction forces at bearings. Tie can be at or above bearings.

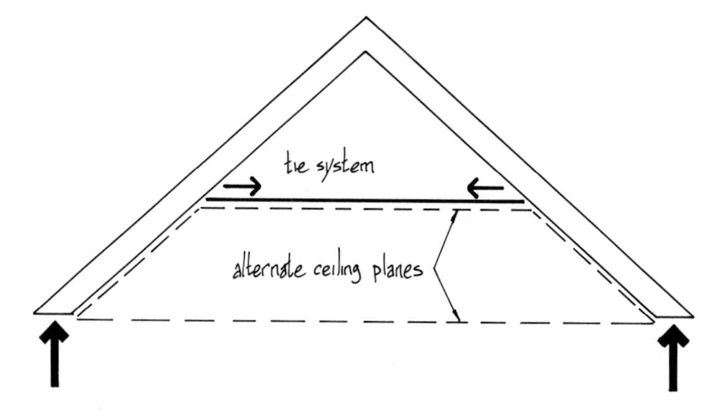

PITCHED

SHED

- Single slope roof with many pitched roof advantages.
- Horizontal tie system not normally required.
- Allows both flat and sloping ceiling applications.
- Used extensively in passive solar structures where light penetration through a high level wall is required (see 3-8 and 3-16).

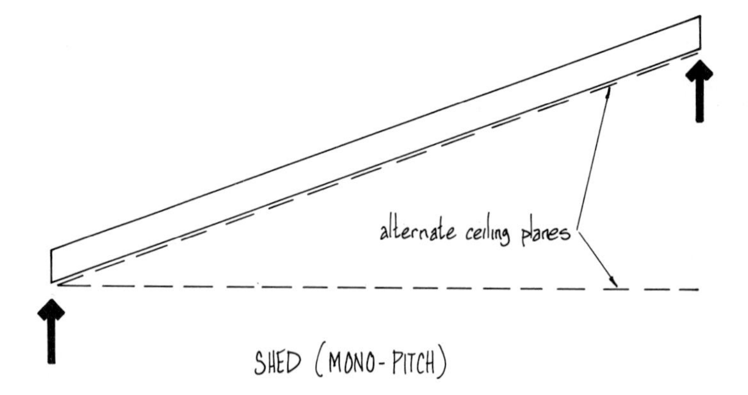

SHED (MONO-PITCH)

FLAT

- "Flat" roof will have a slight slope to facilitate drainage.
- Solid rafters are normally set level (to support ceiling) but have a tapered top edge or are fitted with a tapered strip to provide a drainage slope.
- Trusses for flat roofs have a sloping top chord (see 8-9 and 8-32).
- Roof finish is limited to a continuous membrane system (see 11-34).
- In highly insulated structures, special care must be taken with flat roofs to avoid condensation problems (see 10-10).

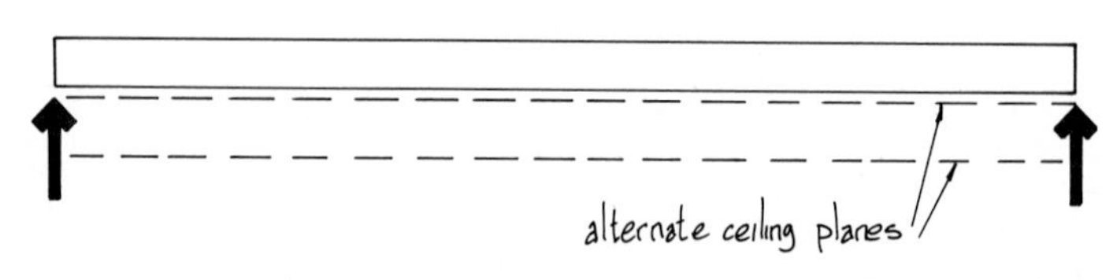

FLAT

BASIC ROOF SHAPES

ROOF PITCH AND SLOPE

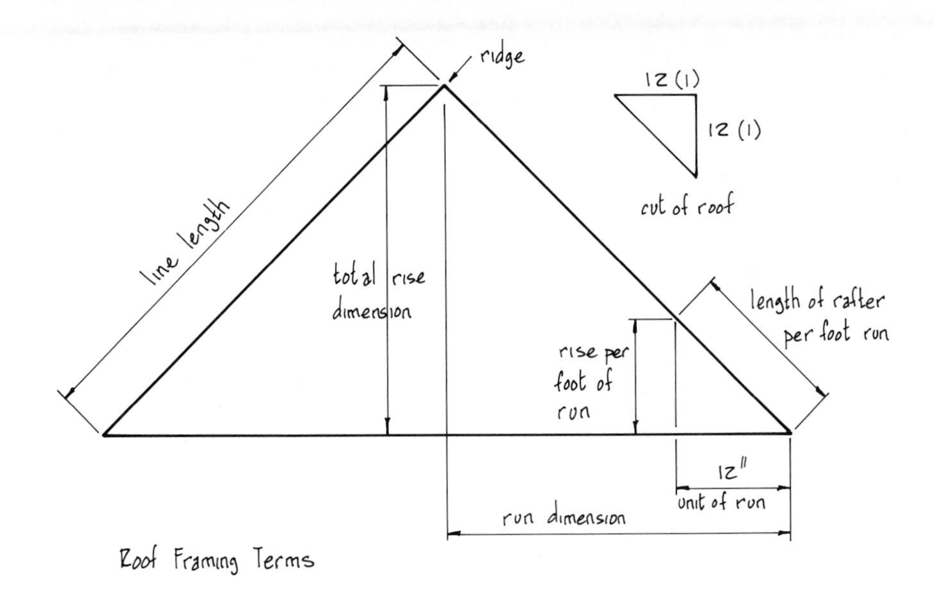

Roof Framing Terms

The framing terms shown here are those typically used in most regions of North America. The dimensional points do not take into account the thickness or width of lumber stock.

Slope

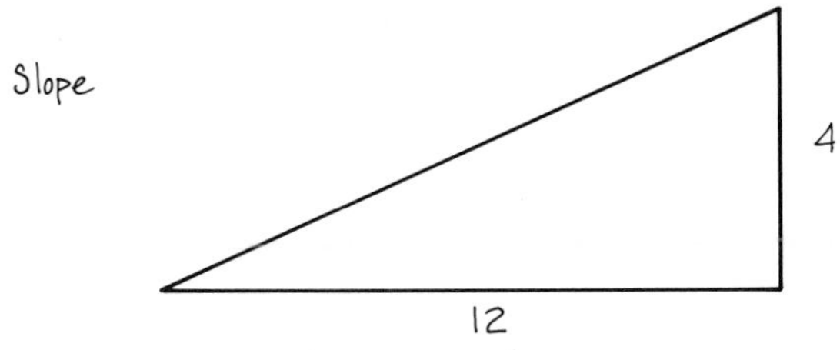

slope expressed as 4 in 12

Pitch

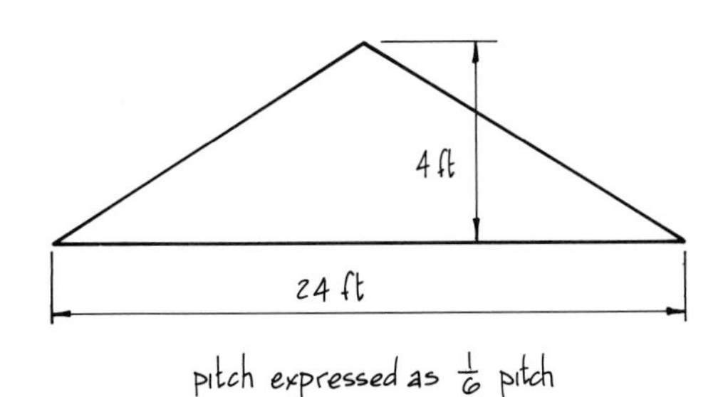

pitch expressed as $\frac{1}{6}$ pitch

Metric

slope is 1:5

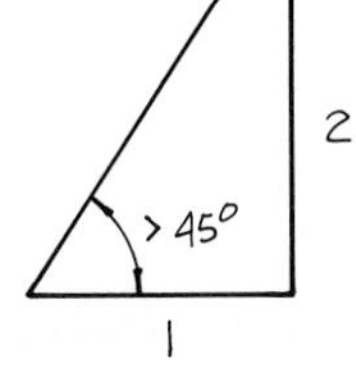

slope is 2:1

EXPRESSING ANGLE of ROOF

The angle or inclination of a roof from the horizontal is expressed in two different ways: slope and pitch.

SLOPE

- Slope is expressed as a ratio between the vertical rise and the horizontal run.
- In the example at the left the slope is expressed as 4 in 12, meaning that the slope has a vertical dimension of 4 in for each 12 in of horizontal run. (Be careful not to say 12 in 4 – this indicates a totally different slope).
- Slopes can range from 1/4 in 12 for flat roofs to 12 in 12 and greater for highly sloped roofs.

PITCH

- Pitch is expressed as a fraction using the dimensions of total rise over span.
- In the example at the left the total rise is 4 ft and the span is 24 ft. The fraction is therefore 4/24 or 1/6. This is expressed as 1/6 pitch.
- Pitch will vary in size as determined by roof dimensions.

SLOPE – METRIC MEASURE

- Only slope ratios are used for metric applications – not pitch.
- For roof slopes of less than 45° the first number (rise) is always 1 and the second number (run) varies. See example at the left.
- For slopes over 45° the first number (rise) varies and the second number (run) is 1. See the example.
- Slope is written as a proper ratio, i.e., 1:5.
- See 2-20 for more metric slope information.

GABLE ROOF

Gable Roof - The Basic Shape

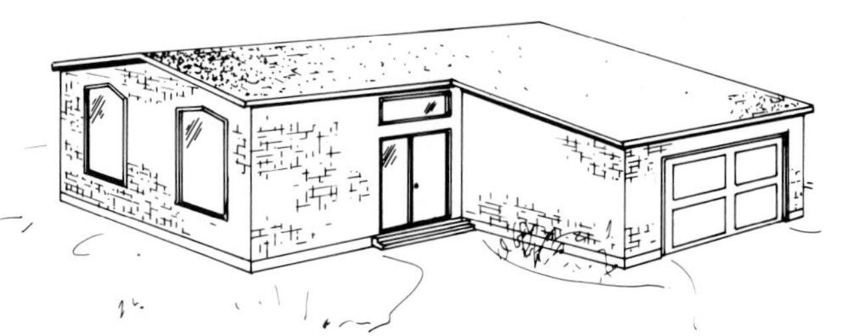

Continuous Low-Slope Gable

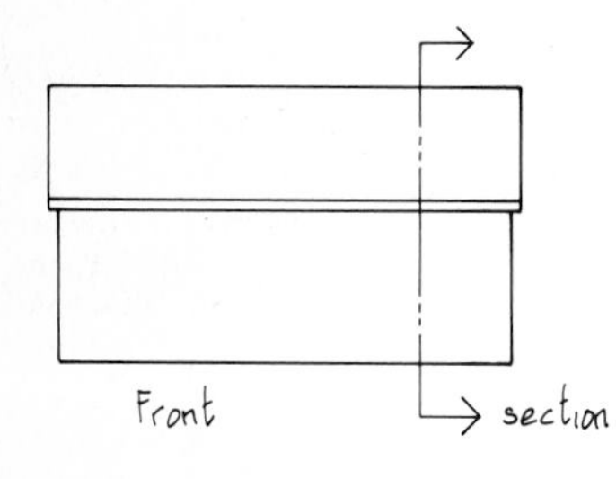

Front

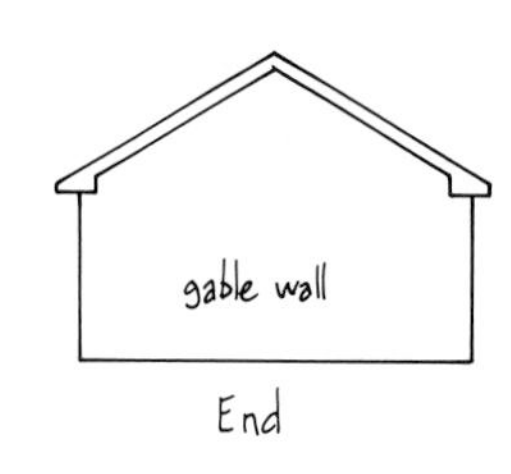

End

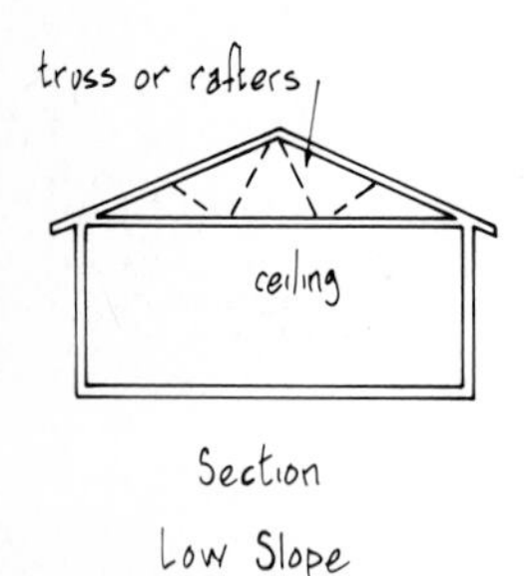

Section
Low Slope

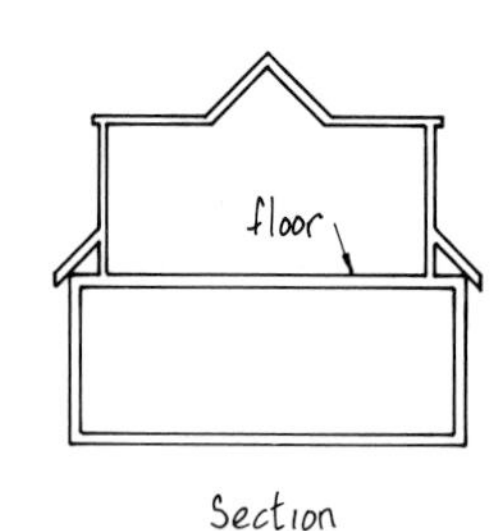

Section
High Slope with Dormers

L-Shaped Gable

GABLE ROOF

- Simplest and most economical of the double-pitched roofs.
- Used extensively in residential construction.
- With the ridge line centered on the structure, all trusses or rafters are the same and erection is straightforward.

Note

The roof shapes shown on this and the next three pages are basic shapes only. Each may be used either singly or in combination with each other to create the many interesting roof designs seen in modern construction.

HIP ROOF

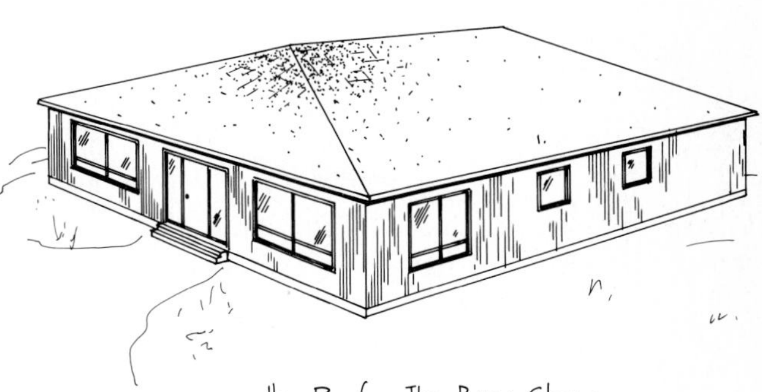

Hip Roof - The Basic Shape

L-Shaped Hip Roof

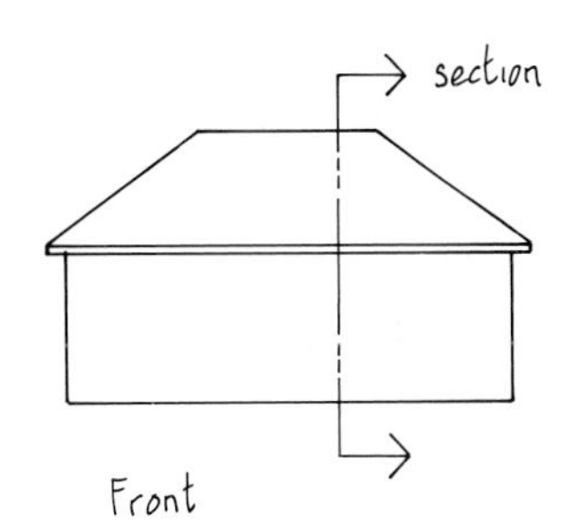

Front

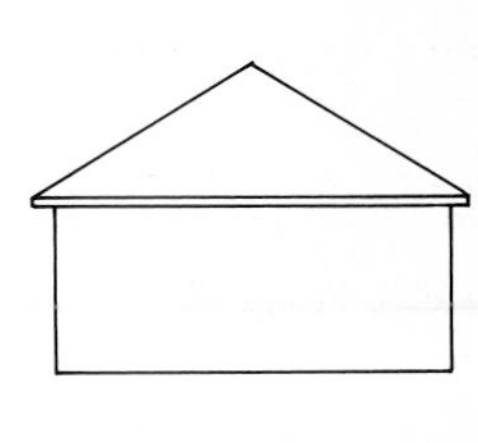

End

truss or rafters

Section

Dormers are not normally used with hip roofs because of the generally low slope.

Dutch Hip Roof

HIP ROOF

- In its simplest form, has four sloping sides meeting at the ridge line.
- Adjacent sides intersect at the hip.
- This type of roof provides an overhang on all four sides of the structure.

ROOFS

8-6

SHED AND FLAT ROOFS

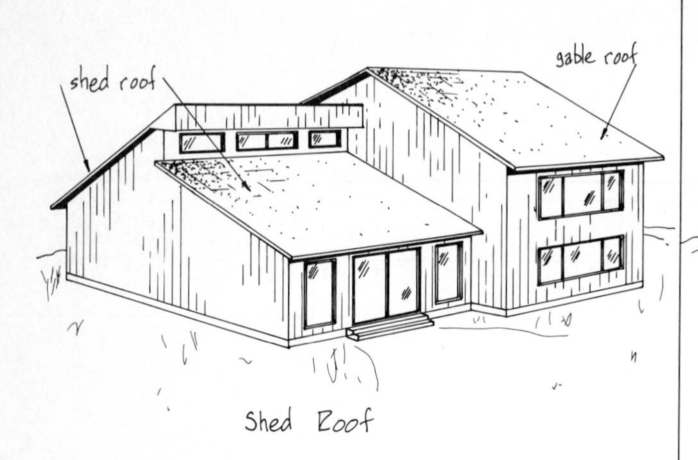

Shed Roof

Flat Roof

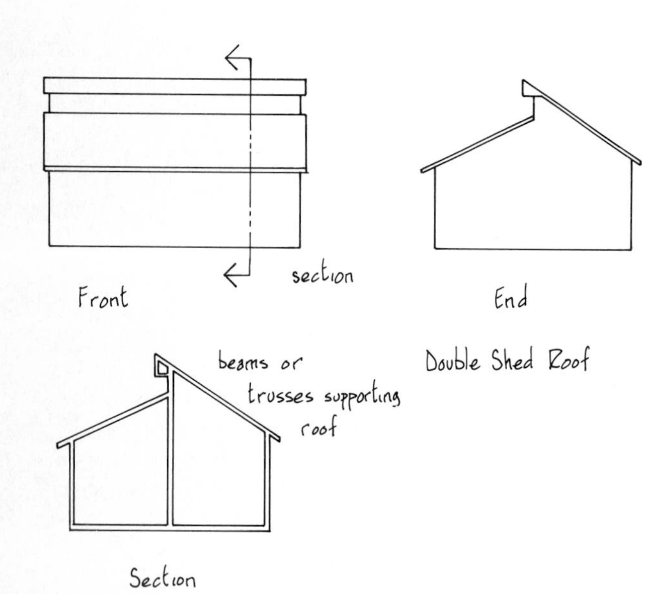

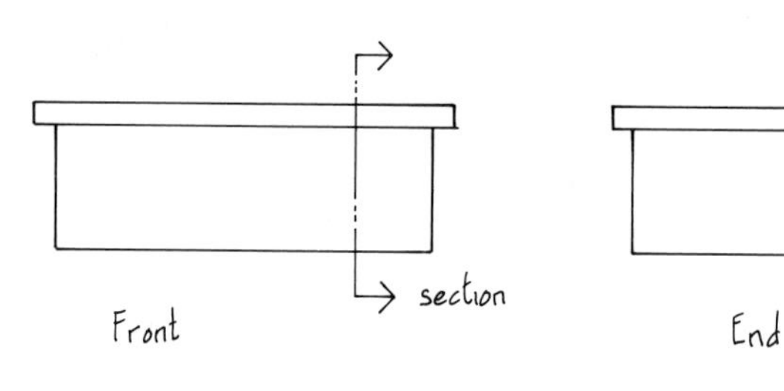

slight slope required for drainage

Section

SHED ROOF

- Also called a "lean-to roof".
- Is often used as an attachment to a larger structure or as a main roof when high ceilings or high-wall light entry is required.
- In most designs, the ceiling finish is attached directly to the roof structure.
- Trusses or rafters can be used.

FLAT ROOF

- Trusses are ideal for this type of roof for both drainage and insulation purposes (see 10-21 for insulation).
- Rafters are referred to as roof joists and carry both roof and ceiling loads.
- Overhangs can be provided over each wall of the structure.

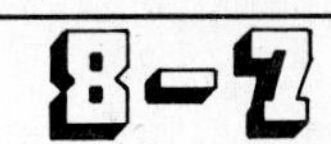

GAMBREL AND MANSARD ROOFS

Gambrel Roof

Mansard Roof

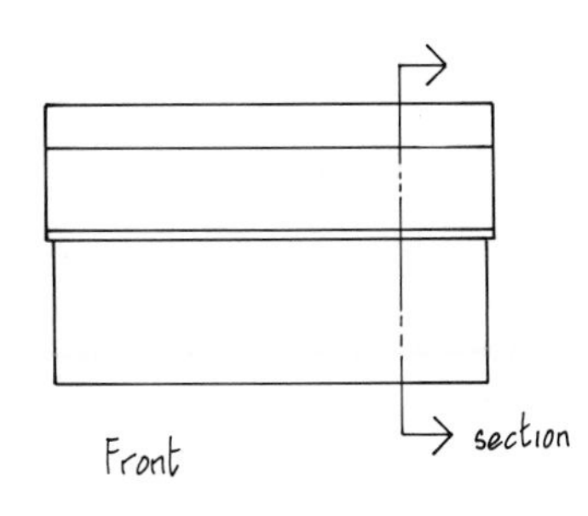

Front

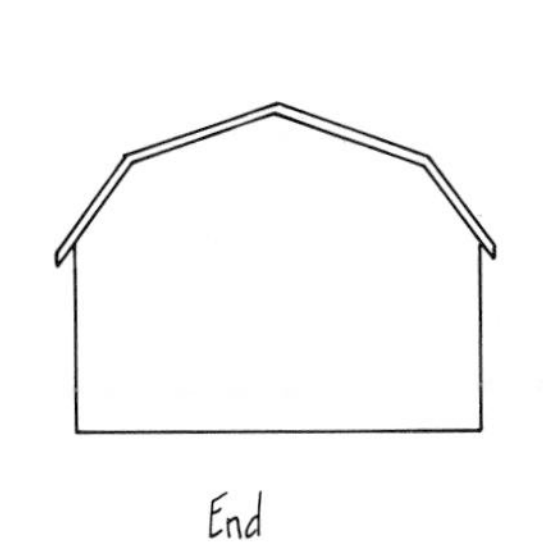

End

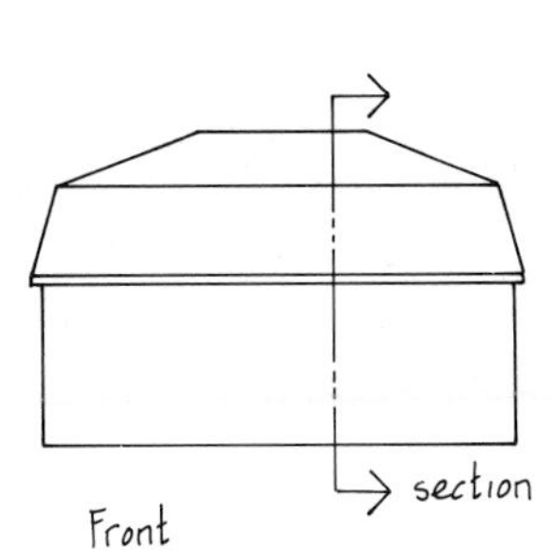

Front

End

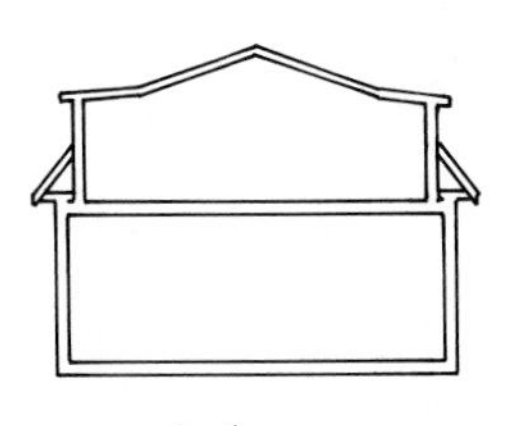

Section with dormers

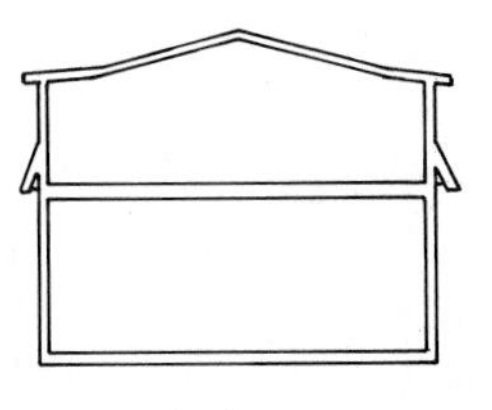

Section with dormers

GAMBREL ROOF

- Similar to a gable roof except that each slope is broken into two parts, usually at the center.
- Provides efficient use of the second story by increasing roof slope and headroom.
- Dormers are commonly used on this type of roof.

MANSARD ROOF

- Similar to the hip roof but with all four sides broken into two different slopes.
- On most designs, the lower slope is almost vertical and the upper slope almost horizontal.
- Dormers are also commonly used in this type of roof.

PREFABRICATED WOOD TRUSSES

In many regions the use of prefabricated roof trusses has supplanted the traditional rafter system in most types of residential and commercial construction. Wood trusses have wide design flexibility, with truss size and style limited only by manufacturing capabilities and shipping and handling considerations. Some of the principal advantages are:

- Most trusses are custom designed and can accommodate almost any roof shape or loading condition.
- Truss members are precut and factory assembled on a jig using galvanized connector plates (see 6-20). This allows fast production and accurate repetition.
- Site erection is very fast and does not require the skilled labor that a rafter system demands.
- The long span capability of trusses eliminates the need for interior load-bearing walls, allowing greater flexibility in floor layouts.
- General economies are received through the reduction of site pilferage, material shortages, and weather delays.

DESIGN AND ERECTION

- Using span and loading information contained on the working drawings, the truss manufacturer will design and fabricate each truss to suit the service conditions. Computer-generated truss design and shop drawings are standard in the industry.
- Copies of the truss design drawings will be supplied to the contractor for submission to local building inspection authorities.
- Careful handling is required in the erection and bracing of trusses. Excessive lateral bending during erection must be avoided or joint damage will result. Temporary and permanent bracing must be installed correctly to avoid instability. Refer to fabricators' instructions for exact erection details.
- In very cold regions a condition called "seasonal truss uplift" may occur. Refer to 10-54 for information.

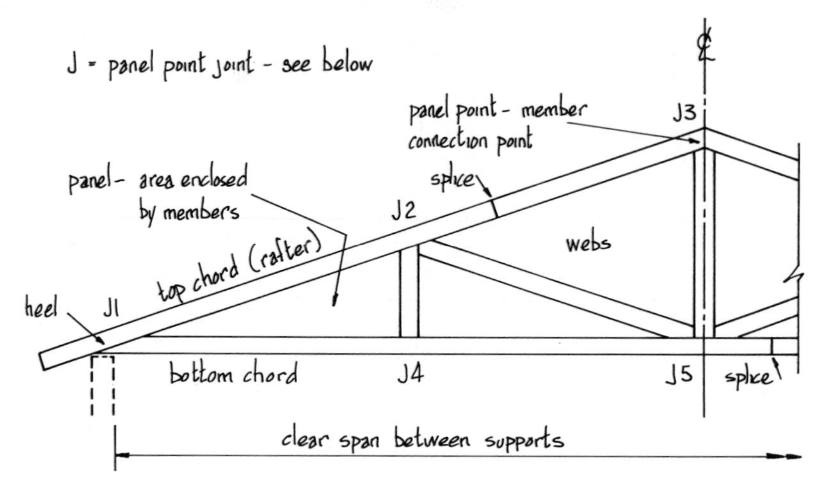

TYPICAL PREFABRICATED TRUSS symmetrical about centerline
connector plates not shown (see 6-20)

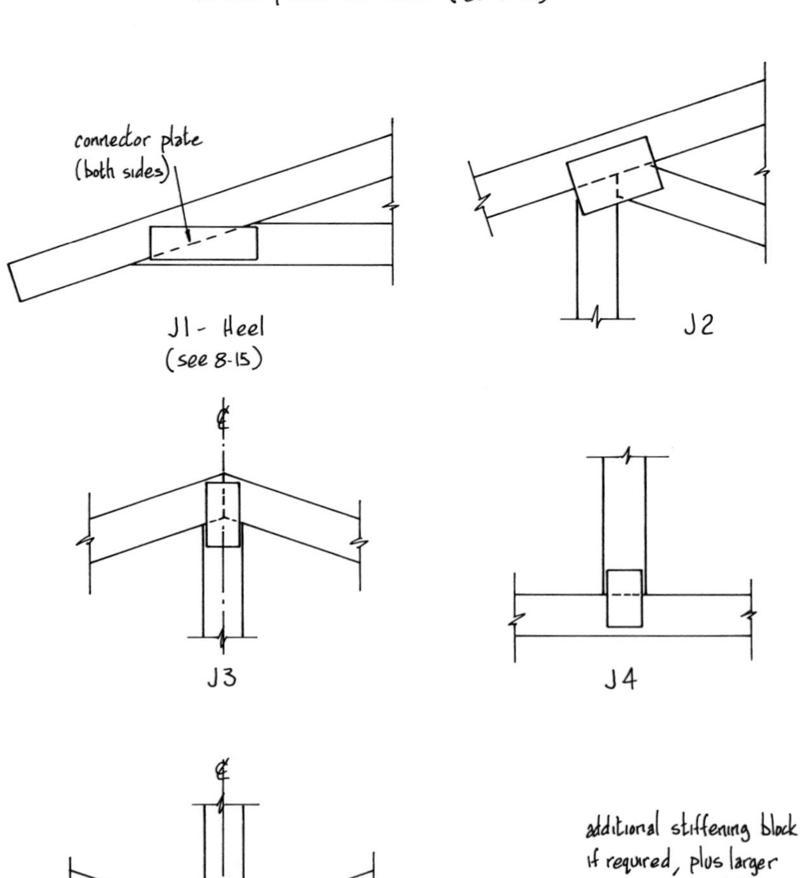

TRUSS TYPES

STANDARD TRUSSES

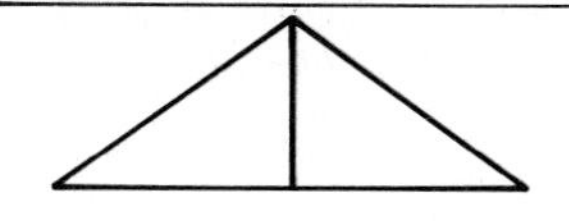

KING POST
short span – all construction types

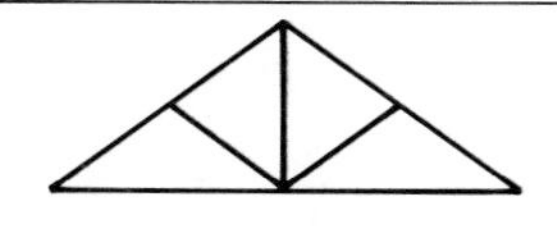

QUEEN POST
short span - heavy top chord loading

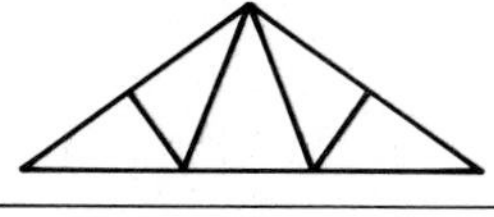

FINK
all construction types up to 40 ft (12 m)

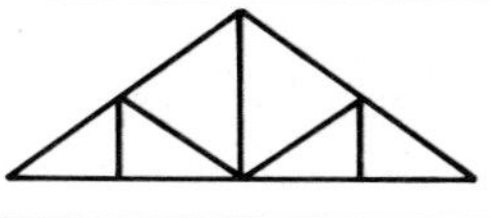

HOWE
designed for lower chord support

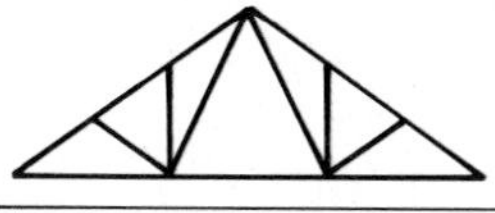

FAN
heavy top chord loading

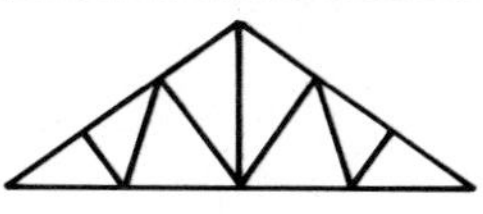

SPECIAL KING POST
up to 60 ft (18 m) and light bottom loads

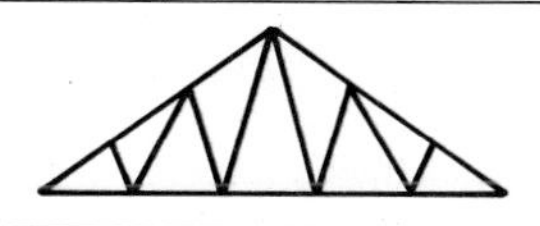

BELGIUM (or DOUBLE FINK)
long span applications to 60 ft (18 m)

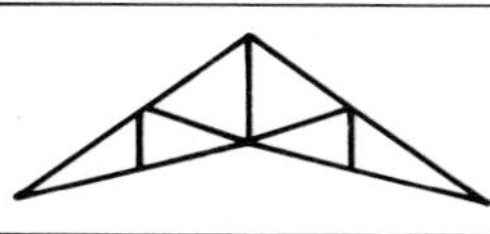

SCISSOR
vaulted ceilings

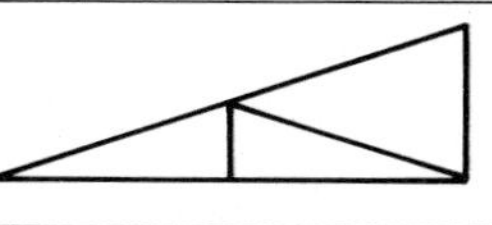

2-2 MONO
short span
single slope

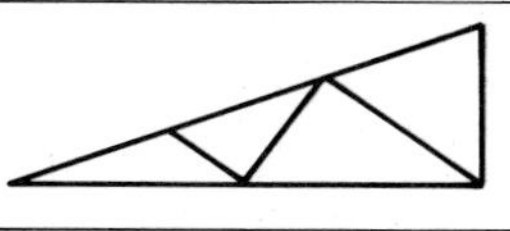

3-2 MONO
heavy top chord loads
single slope

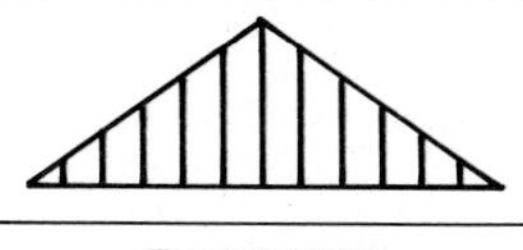

GABLE END
gable roof end walls

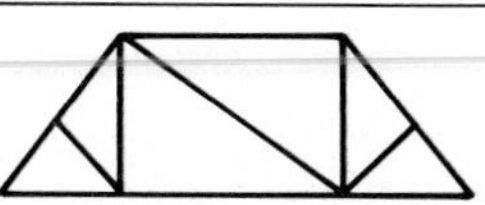

HIP or MANSARD
see 8-10 for hip detail

PRATT

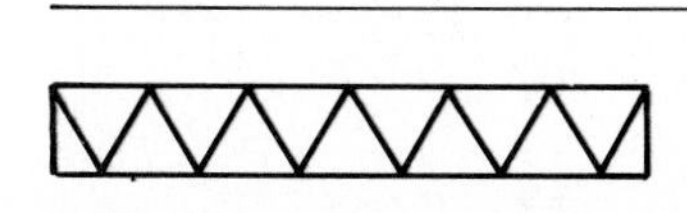

WARREN

Flat Trusses

top chord may have a slight slope in either mono or dual pitch

SPECIAL TRUSSES

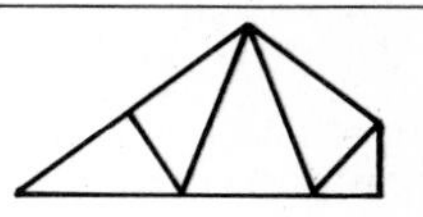

BOBTAIL
used at chimney openings or in split-level roofs

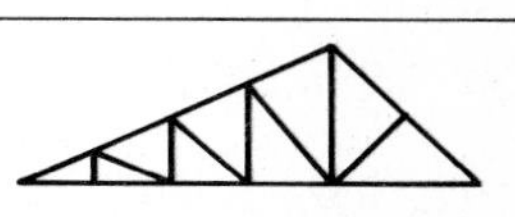

DUAL SLOPE
provides dual slope roof profile

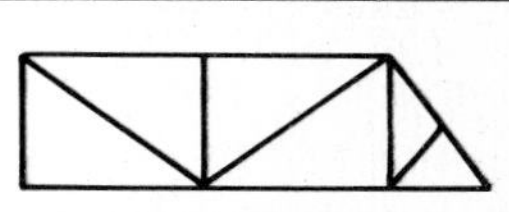

SINGLE SLOPE HIP/MANSARD
hip or mansard roofs

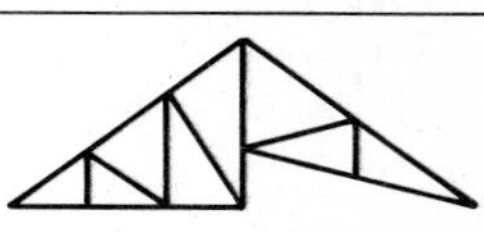

DUAL SLOPE ROOF/CEILING
provides a combination of roof and ceiling slopes

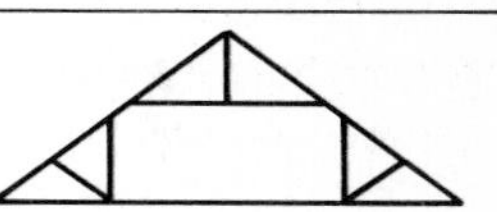

ATTIC FRAME
provides living or storage space in attic

TRUSS BEARING POSITIONS

There are five basic truss bearing positions as illustrated here.
Each truss must be engineered to suit design conditions.

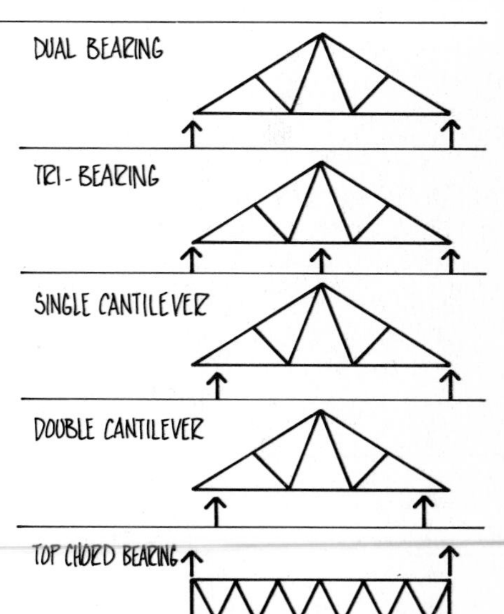

TRUSS SYSTEMS

GABLE ROOF

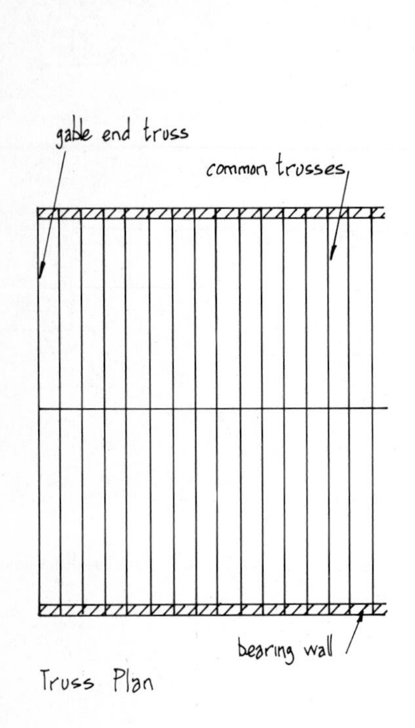

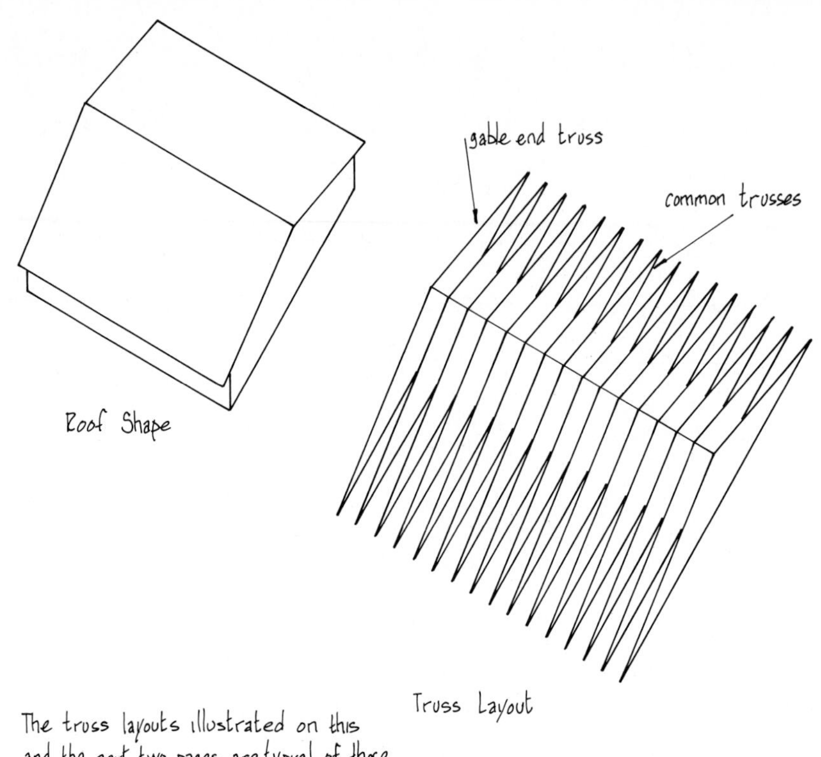

The truss layouts illustrated on this and the next two pages are typical of those used in most regions.

HIP ROOF

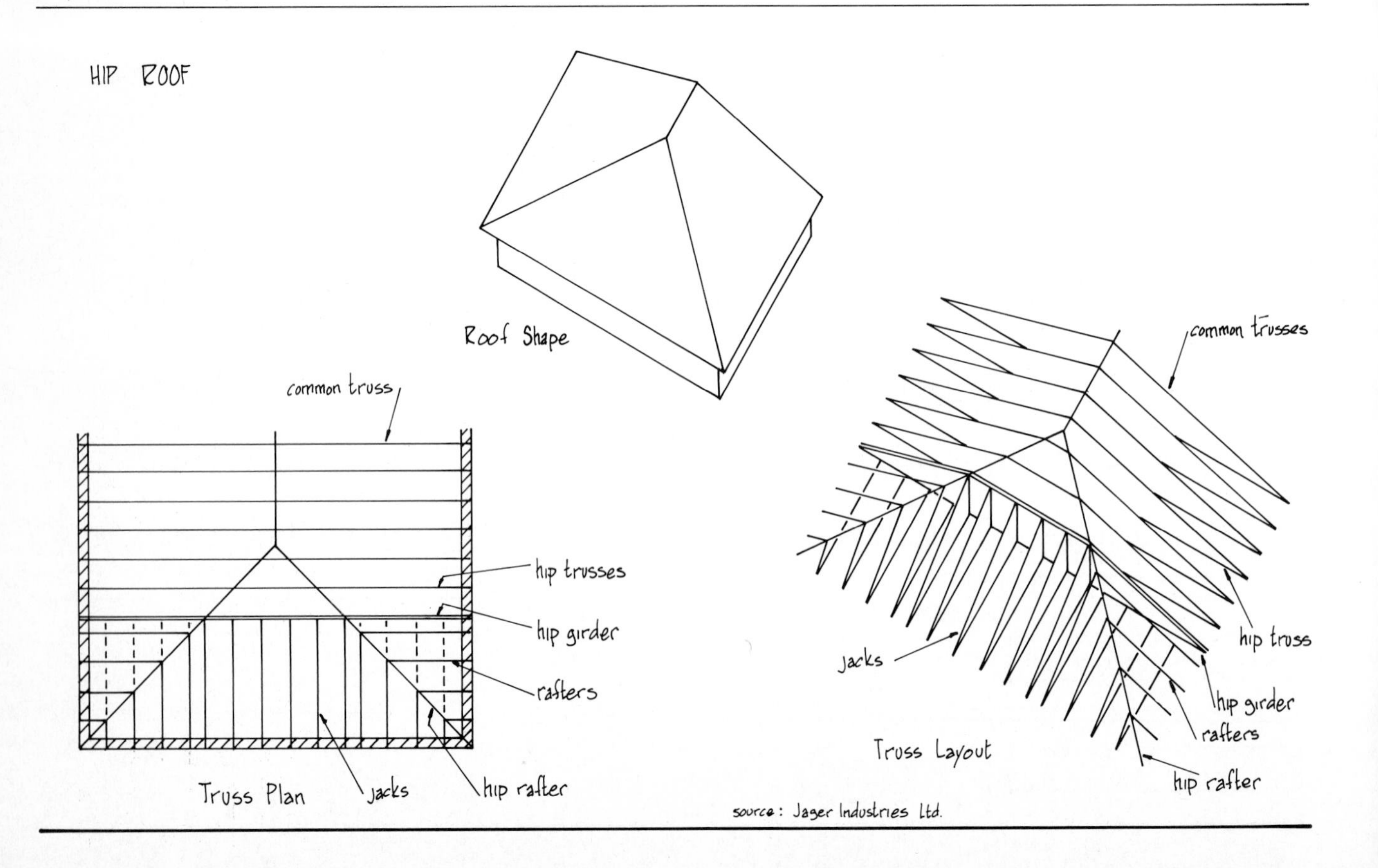

source: Jager Industries Ltd.

TRUSS SYSTEMS

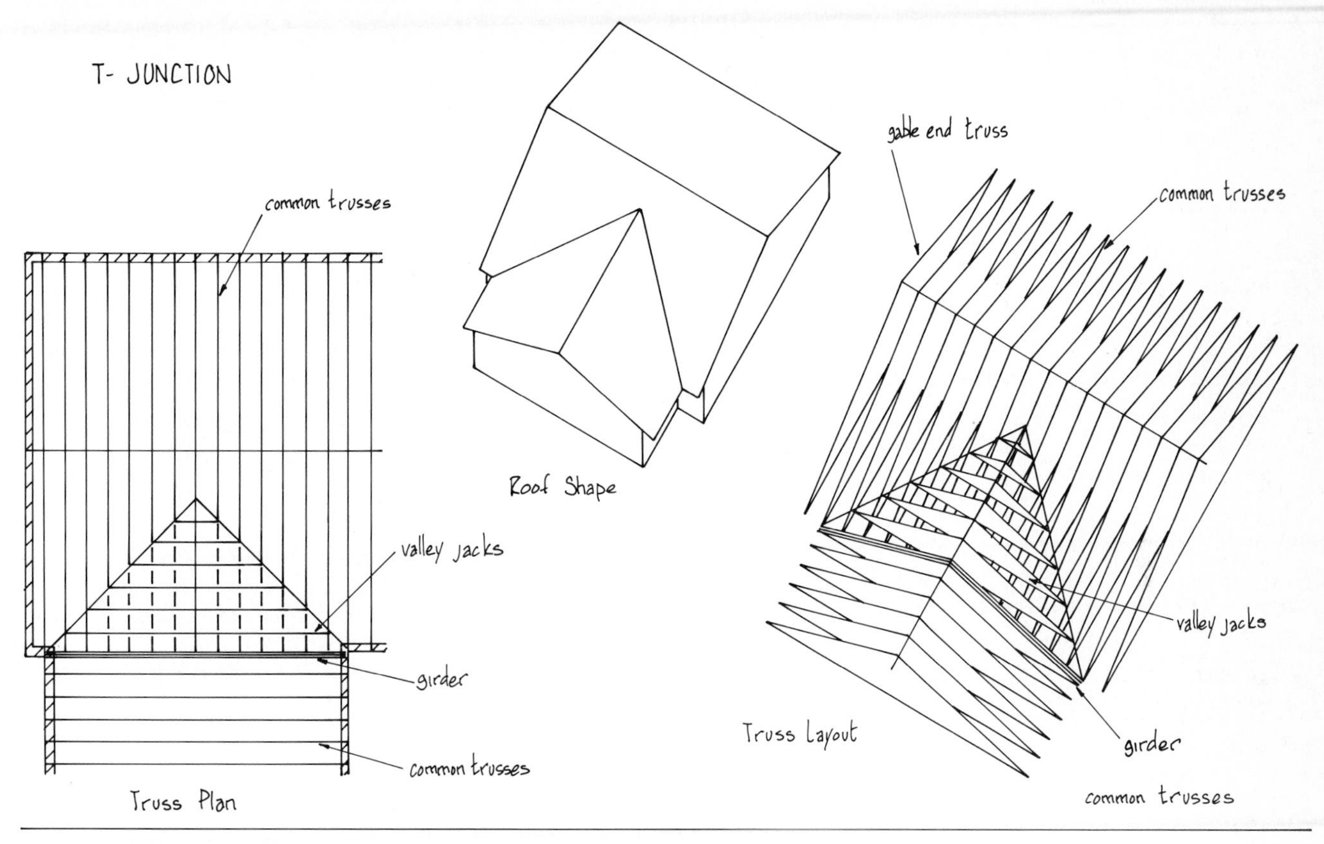

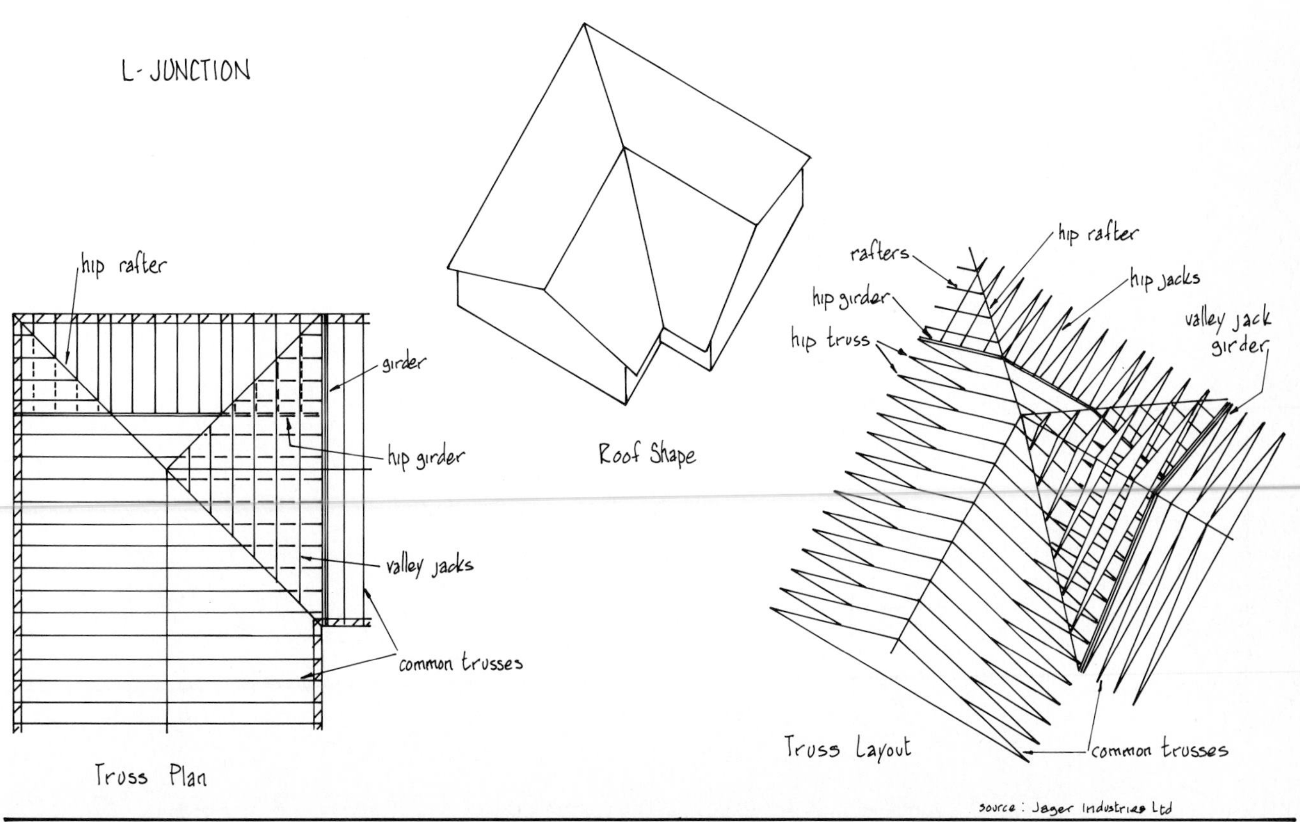

source: Jager Industries Ltd

TRUSS SYSTEMS

L- JUNCTION
Unequal Roof Heights

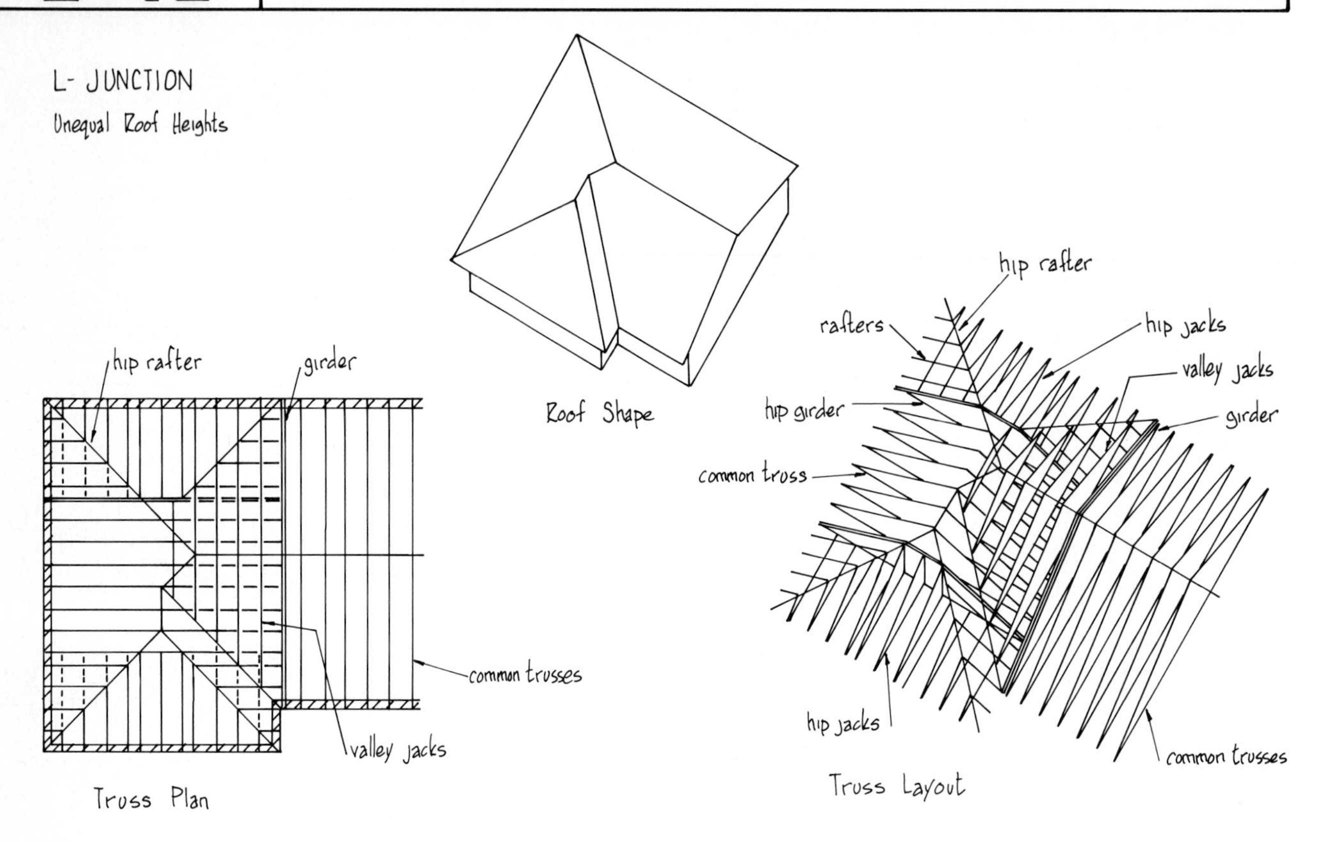

FLAT ROOF

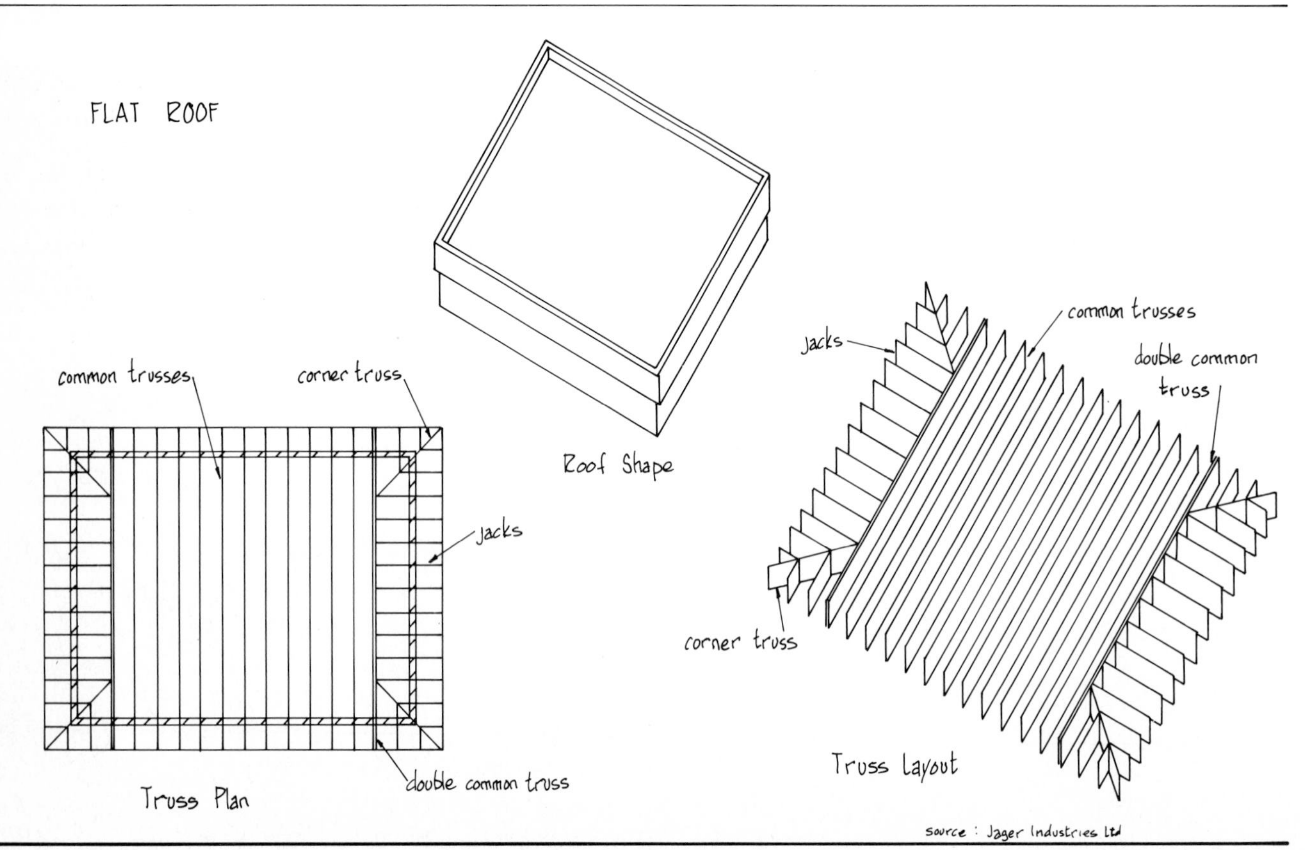

Source : Jager Industries Ltd

TRUSSES AT CHIMNEYS / SOLID BEAMS

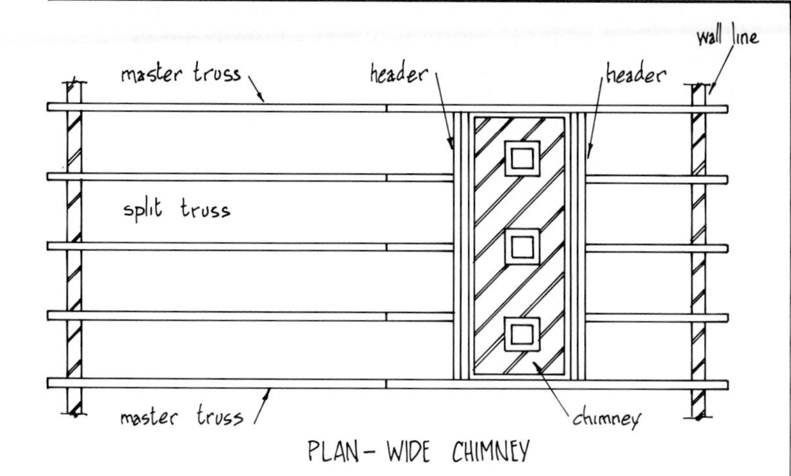

PLAN – WIDE CHIMNEY

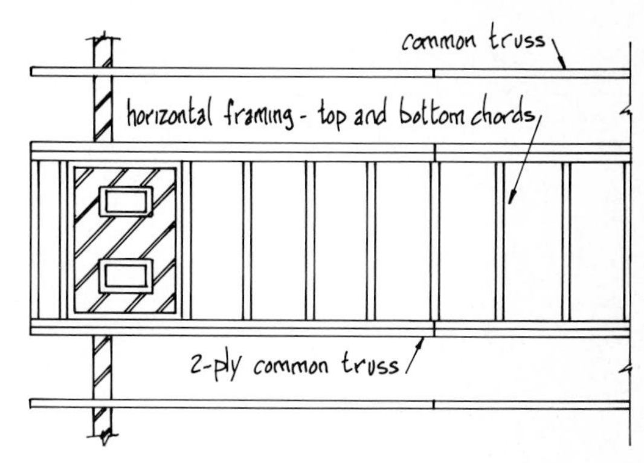

PLAN – NARROW CHIMNEY

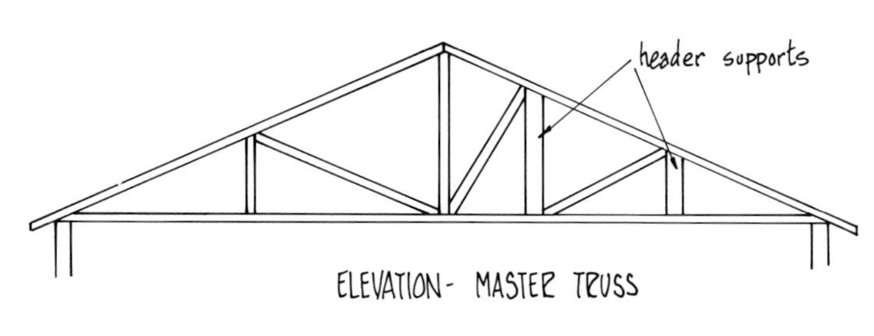

ELEVATION – MASTER TRUSS

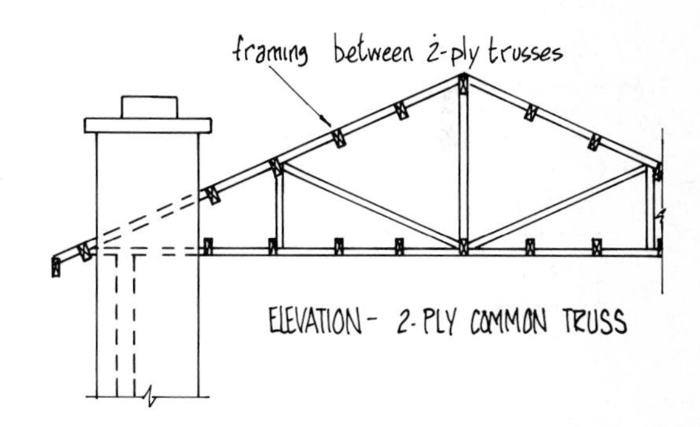

ELEVATION – 2-PLY COMMON TRUSS

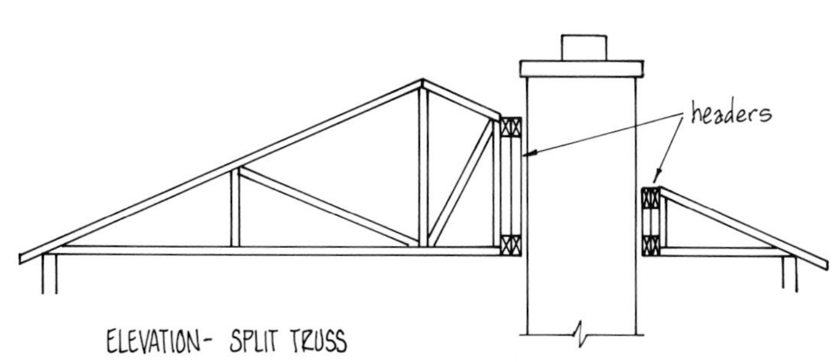

ELEVATION – SPLIT TRUSS

Note
A 2-in (50) air space is required between the framing and the chimney to meet most fire codes.

CHIMNEYS

Where chimneys project through a trussed roof, either inside or outside the building lines, the framing used is determined by the opening size.

- For large chimneys a system of special trusses and headers is available that continues the roof line without interuption. Care must be taken in dimensional accuracy when ordering the special trusses.
- For narrow chimneys, separate framing bridging between doubled-up standard trusses is used. This eliminates the need for special trusses and simplifies construction.

BUILT-UP BEAMS

In addition to trusses, built-up beams are also fabricated using pressure-applied metal connector plates. As illustrated at the right, both single and mutiple beams are standard. Beams of this type are generally stronger than single-member beams of equal depth and are particularly economic when a beam depth of over 12 in (286) is required.

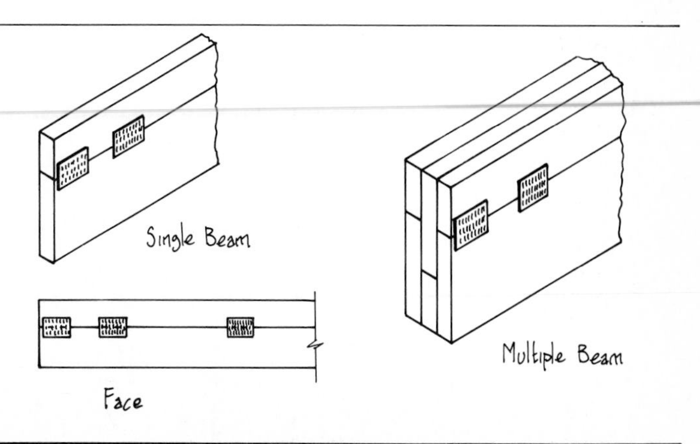

ROOF OVERHANGS

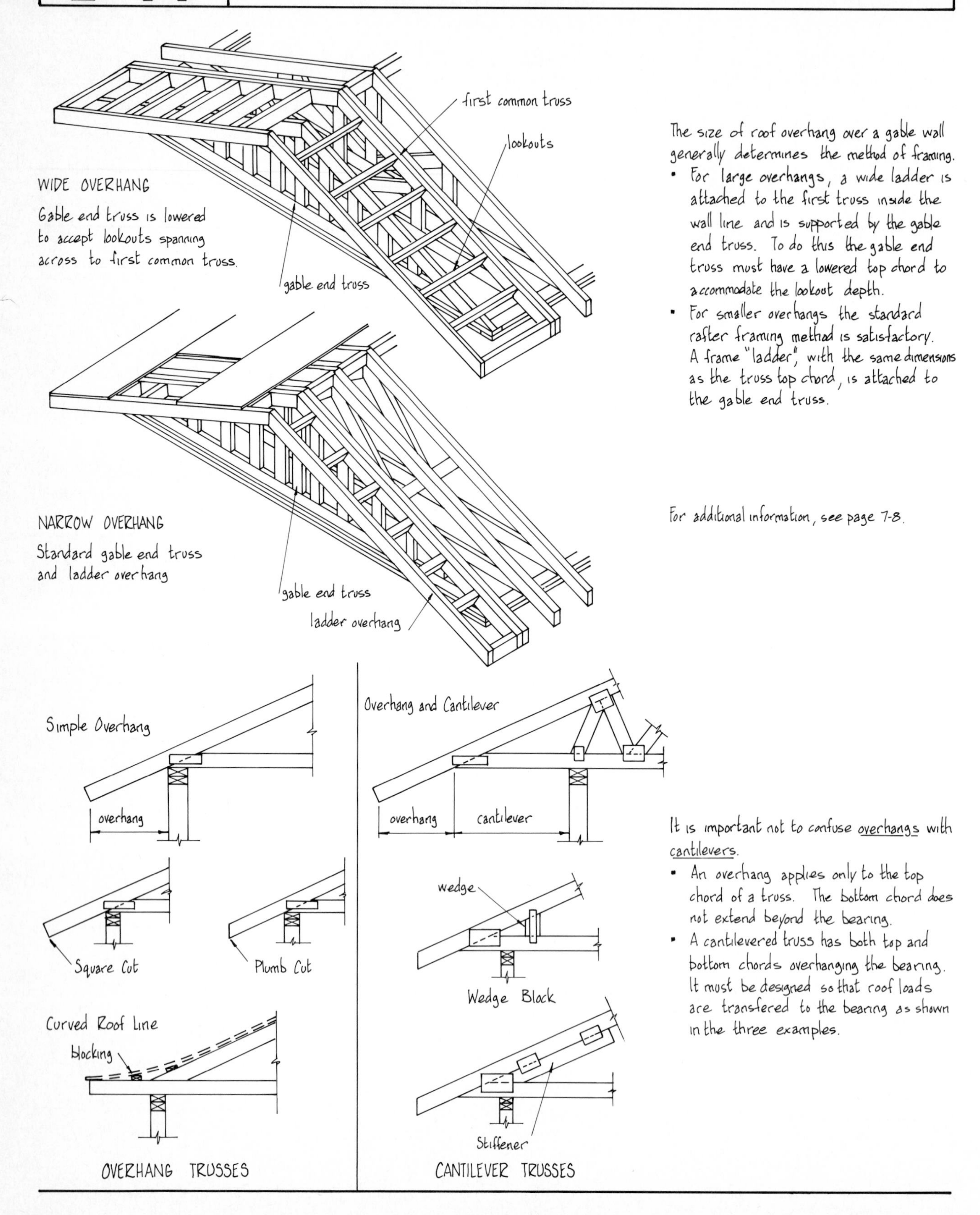

The size of roof overhang over a gable wall generally determines the method of framing.

- For large overhangs, a wide ladder is attached to the first truss inside the wall line and is supported by the gable end truss. To do this the gable end truss must have a lowered top chord to accommodate the lookout depth.
- For smaller overhangs the standard rafter framing method is satisfactory. A frame "ladder", with the same dimensions as the truss top chord, is attached to the gable end truss.

For additional information, see page 7-8.

It is important not to confuse overhangs with cantilevers.

- An overhang applies only to the top chord of a truss. The bottom chord does not extend beyond the bearing.
- A cantilevered truss has both top and bottom chords overhanging the bearing. It must be designed so that roof loads are transfered to the bearing as shown in the three examples.

SOFFIT RETURNS / HEEL CUTS

SOFFIT RETURNS

A soffit return added to the overhang of a truss provides framing for a horizontal soffit. The size and design of the return is determined by the wall conditions, as shown in these examples.

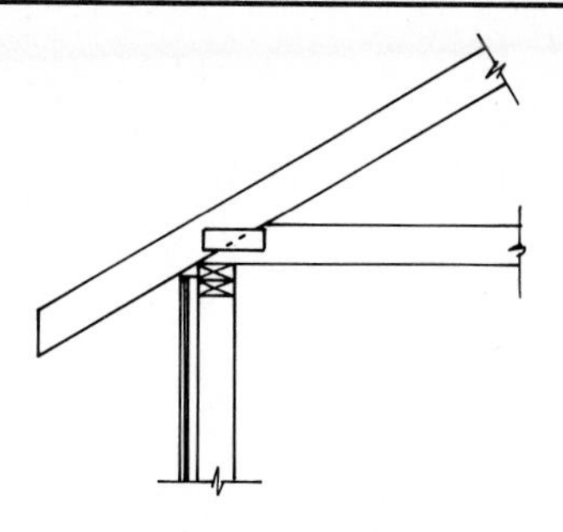

Exposed Top Chord

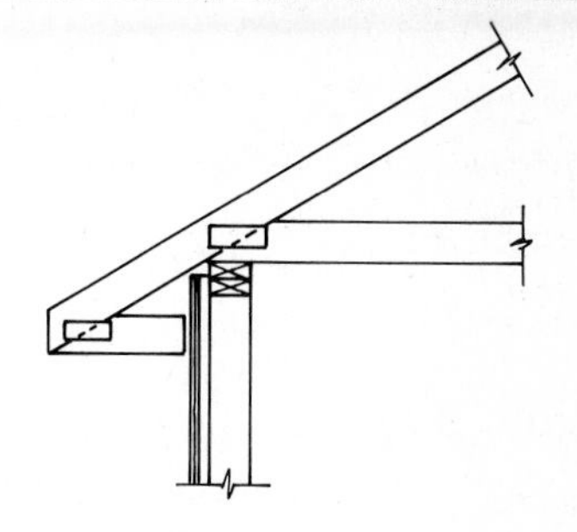

Boxed Soffit

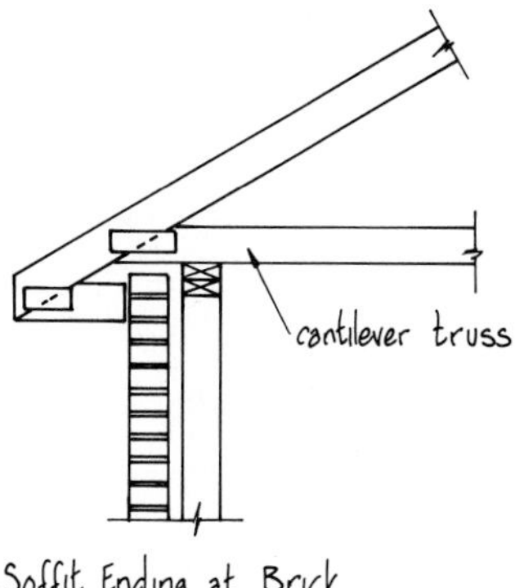

Soffit Ending at Brick

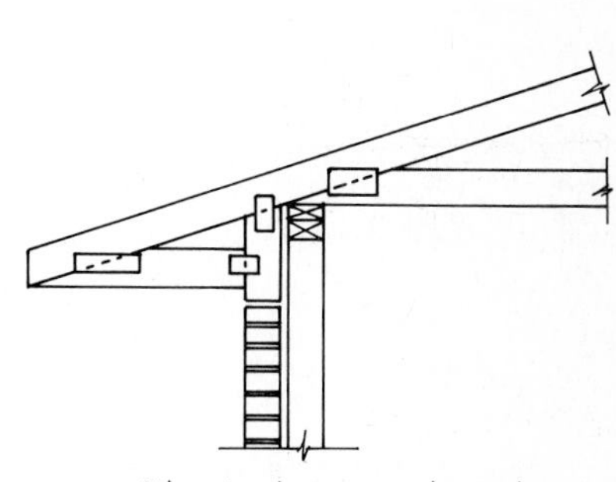

Soffit Extending Beyond Brick

HEEL CUTS

The heel or bearing of standard trusses as shown above are not deep enough to allow the installation of thick insolation over the wall plate and still maintain the required flow of ventilation air into the attic roof space. There are two ways to solve this problem:

- The heel of the truss can be raised to accommodate varing insulation thicknesses, as shown, or
- Ventilation channels can be tacked to the underside of the roof sheathing to compress the insolation and provide passage for ventilation air. The channels are generally molded plastic or fiber ducts of various shapes. See 10-20 for details of this procedure and suggested vent area sizes.

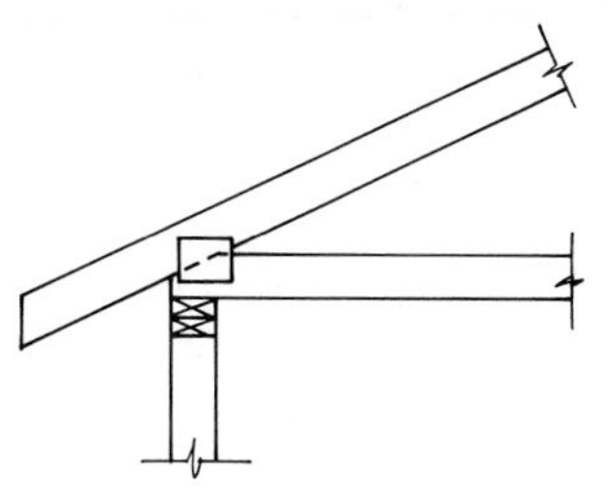

Modified Standard Truss

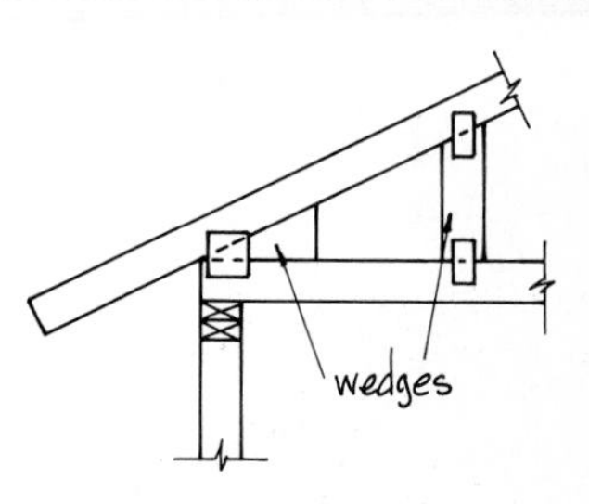

Modified Standard Truss

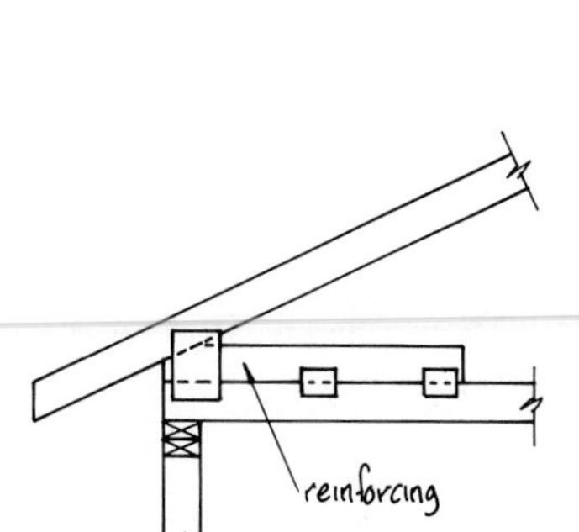

Reinforced Bottom Chord

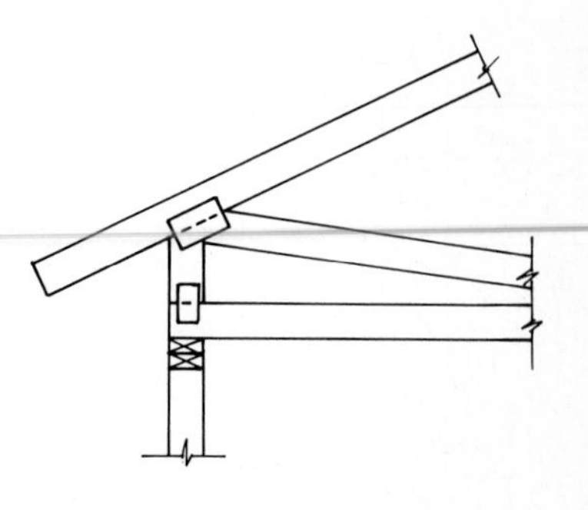

Bobtail Truss

RAFTER SYSTEMS

The design of raftered roof systems has been established over a considerable period of time and is generally referred to as "conventional" or "traditional" construction. Although truss systems are in widespread use, rafters are an equally valid choice for many buildings.

- In most roof types, sloping rafters are paired with horizontal ceiling joists to form a rigid roof frame.
- Rafters provide a base for the roofing material and support the roof loads.
- Ceiling joists support the ceiling finishes and act as ties between exterior walls. In some cases they also provide support for roof loads (see 8-24), or act as floors when attic space is used as a living area.
- Rafters are cut on-site and are erected one by one. Skilled carpenters are needed for all but the simplest roof types.
- Rafter and ceiling joist sizes are determined by span tables (see 8-19).

The types of rafters needed for a roof are determined by roof shape and style. The roof framing plan at the right and the matching pictorial drawing below indicate some of the various kinds of rafters and their typical names.

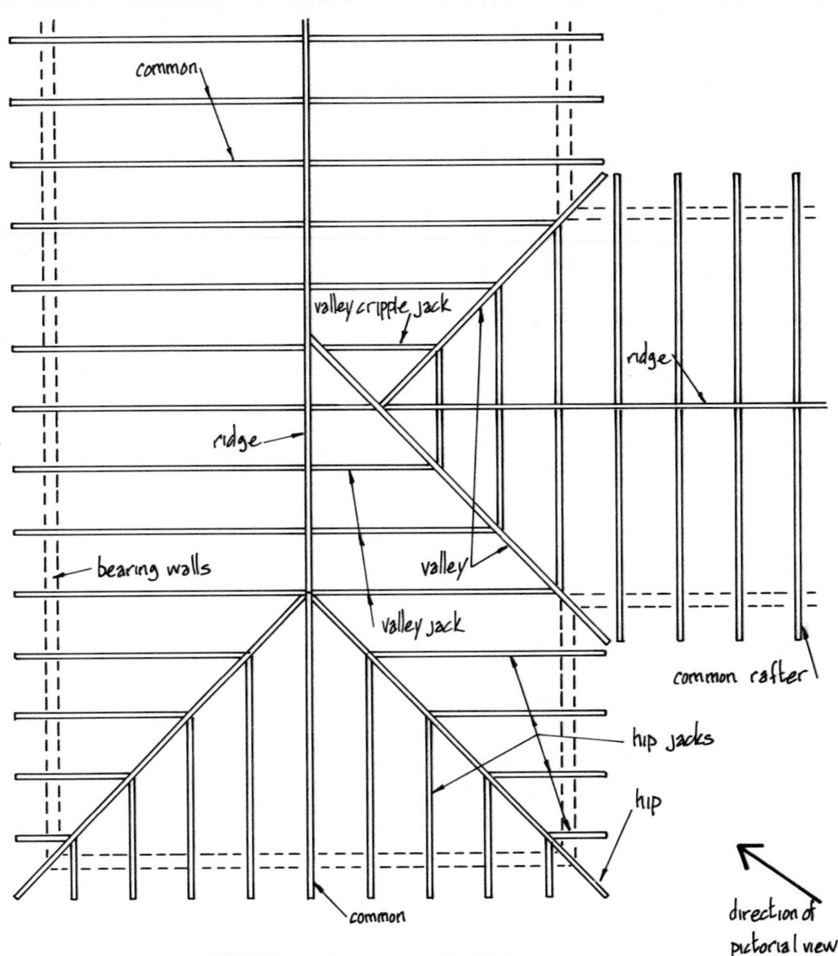

RAFTER LAYOUT – PLAN VIEW
Hip roof, T-intersection, with lower ridge line.

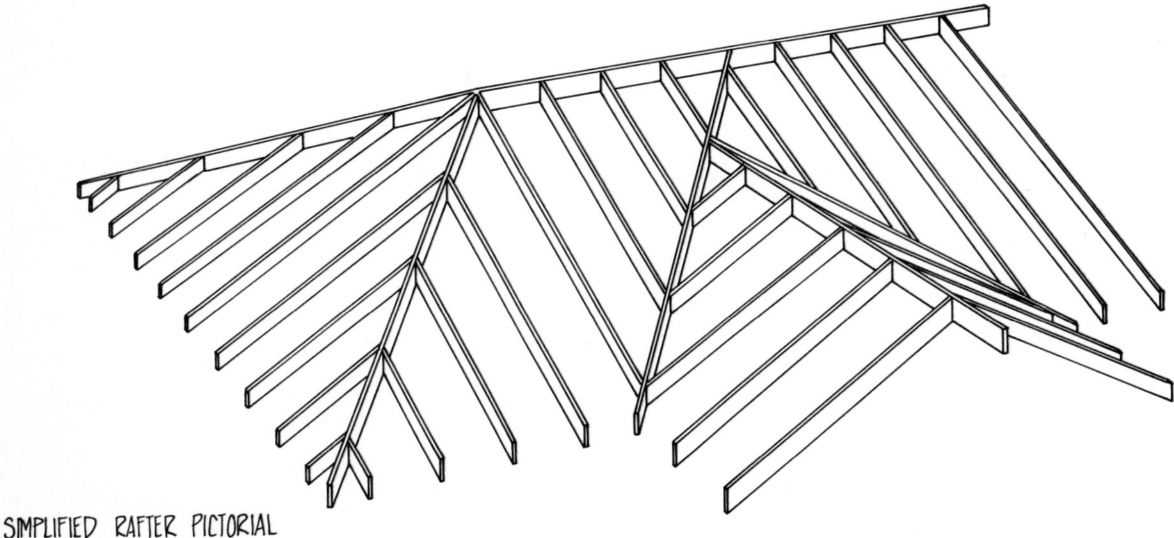

SIMPLIFIED RAFTER PICTORIAL
Ceiling joists and bearing walls omitted for clarity.

RAFTER CALCULATIONS AND CUTS

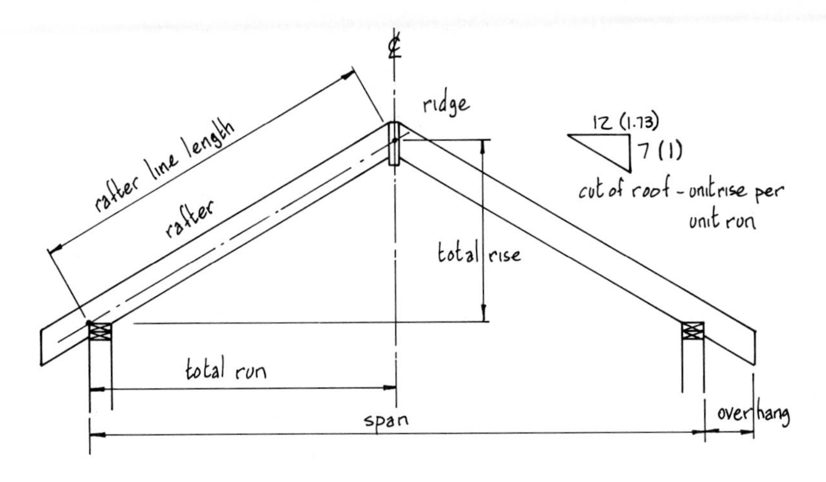

RAFTER FRAMING TERMS

Note Rafter line length does not take into account thickness or width of stock.

Calculation of rafter length is generally made by the framing carpenter and can be done in several ways:

- The quickest way is to use a carpenter's "framing square". This is an L-shaped flat metal measuring tool specially designed for setting-out rafter cuts. It can be used to step-off rafter length in increments relating to slope size or, using slope tables that are given on the face of the square, rafter length can be calculated directly based on the rafters span.
- By using the Pythagorean theorem of a right-angled triangle ($A^2 + B^2 = C^2$).
- By using trigonometry.
- See 8-3 for roof slope and pitch information.
- See 2-20 for metric/Customary comparisons.

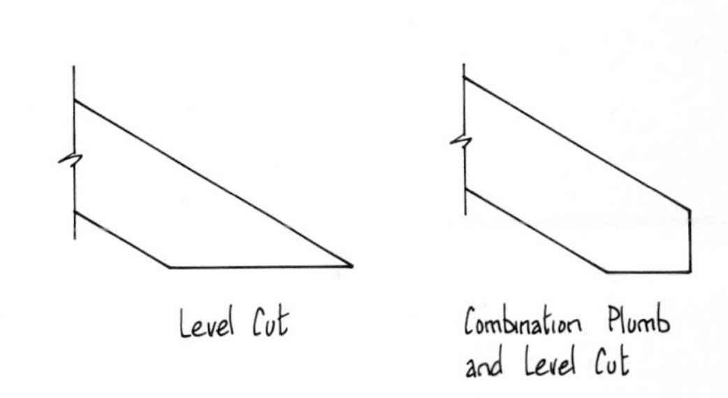

Ridge Bird's-mouth Plumb Cut

RAFTER CUTS

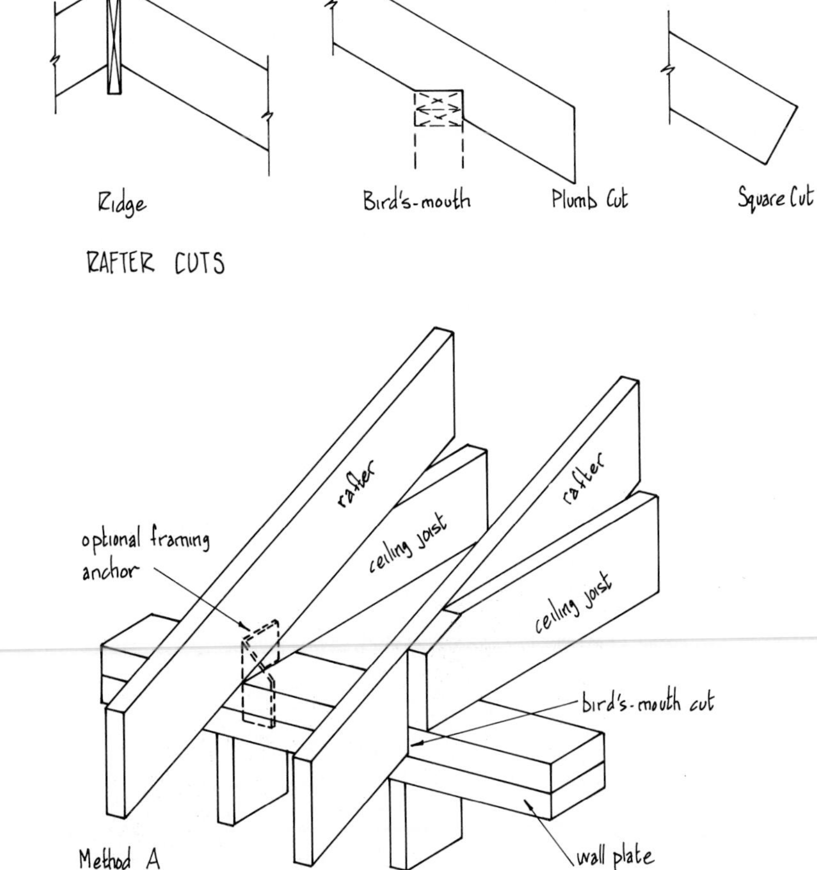

RAFTER-PLATE CONNECTION

The two standard methods of attaching rafters and ceiling joists to the wall plate are illustrated at the left.

- In method A, the ceiling joist is cut to suit the rafter slope while the rafter remains uncut. A metal framing anchor is recommended to increase the strength of the rafter attachment.
- In method B, the preferred method in most cases, the rafter is given a standard bird's-mouth cut and the ceiling joist is attached to the side of the rafter. Framing anchors are normally only required in high wind regions.

CEILING JOISTS

ONE and ONE-HALF STOREY CONSTRUCTION

At the left are two typical examples of 1½ storey roof framing. The left hand example is more common but has a reduced floor area due to the position of the knee wall. The right hand example retains the full floor area but has the disadvantage of an increased exterior wall height.

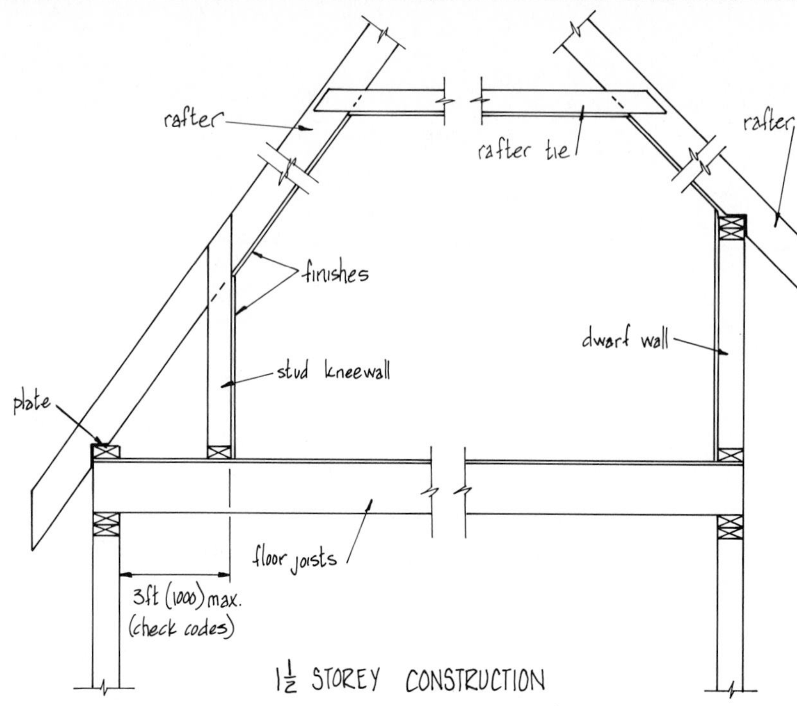

1½ STOREY CONSTRUCTION

Two alternate methods of rafter framing for 1½ storey structures.

Ceiling joists can span directly between rafter ends or be supported at midspan by a bearing wall. To reduce the joist size a center bearing wall is provided where possible, especially if the joist carries a portion of the roof load or a floor load.

After the wall frames are erected, the ceiling joists are nailed in place to prevent outward wall movement when the rafters are erected. Butting or lapping joists at the center bearing wall affects the rafter detail at the ridge as shown below.

Ridge boards are continuous and generally of 1-in (19) thick lumber although 2-in (38) stock can also be used.

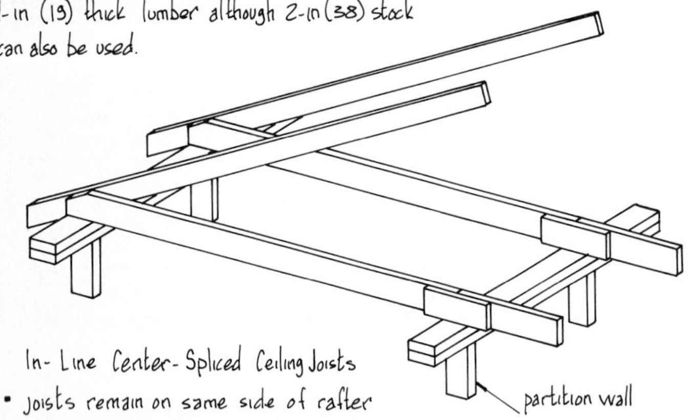

In-Line Center-Spliced Ceiling Joists

• joists remain on same side of rafter

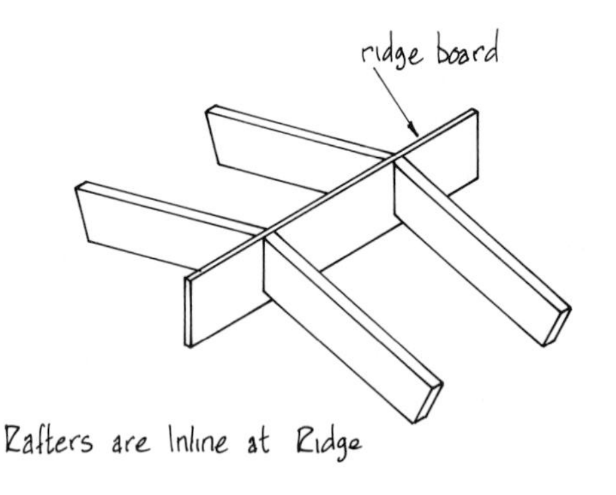

Rafters are Inline at Ridge

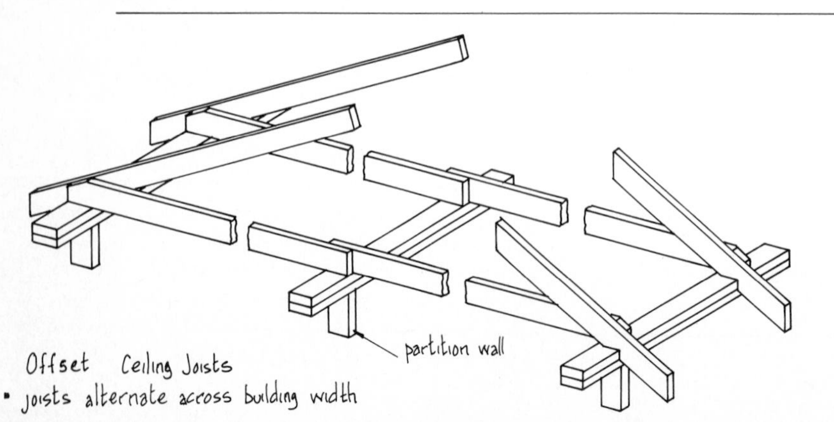

Offset Ceiling Joists

• joists alternate across building width

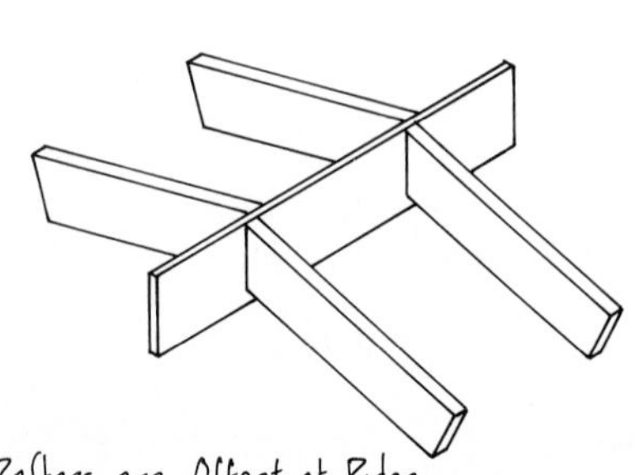

Rafters are Offset at Ridge

TYPICAL SPAN TABLES

RAFTERS – NOT SUPPORTING CEILING

GRADE	NOMINAL SIZE IN. ACTUAL SIZE (mm)	LIVE LOAD 30 PSF (1.5 kN/m²) RAFTER SPACING 16 in. (400) ft in (m)	LIVE LOAD 30 PSF (1.5 kN/m²) RAFTER SPACING 24 in. (600) ft in (m)	LIVE LOAD 20 PSF (1.0 kN/m²) RAFTER SPACING 16 in (400) ft in (m)	LIVE LOAD 20 PSF (1.0 kN/m²) RAFTER SPACING 24 in (600) ft in (m)
No. 1	2×4 (38×89)	8-1 (2.45)	6-9 (2.04)	9-3 (2.80)	8-0 (2.40)
	2×6 (38×140)	12-1 (3.66)	9-10 (2.99)	14-2 (4.30)	11-7 (3.51)
	2×8 (38×184)	15-11 (4.82)	13-0 (3.94)	18-9 (5.68)	15-3 (4.63)
	2×10 (38×235)	20-3 (6.15)	16-7 (5.02)	23-11 (7.24)	19-6 (5.91)
	2×12 (38×286)	24-8 (7.49)	20-2 (6.11)	29-1 (8.81)	23-9 (7.19)
No. 2	2×4 (38×89)	7-6 (2.26)	6-1 (1.85)	8-10 (2.67)	7-2 (2.18)
	2×6 (38×140)	10-11 (3.28)	8-11 (2.68)	12-10 (3.86)	10-6 (3.15)
	2×8 (38×184)	14-4 (4.33)	11-9 (3.53)	16-11 (5.09)	13-10 (4.16)
	2×10 (38×235)	18-4 (5.52)	15-0 (4.51)	21-7 (6.50)	17-8 (5.31)
	2×12 (38×286)	22-4 (6.72)	18-3 (5.48)	26-3 (7.90)	21-5 (6.45)

Notes to table:

- Spans given are for the horizontal span or run of a rafter and not line length.
- Table is based on rafters sloping 1 in 3 or greater. For lower slopes consult appropriate tables or provide intermediate rafter support (see 8-24).
- Table is for common rafters only. Generally, hip and valley rafters are 2 in (50) deeper to provide full contact with angle-cut jack rafters.

ROOF JOISTS – SUPPORTING CEILINGS

LIVE LOAD 30 PSF (1.5 kN/m²)

GRADE	NOMINAL SIZE in. ACTUAL SIZE (mm)	GYPSUM BOARD/PLASTER CEILING JOIST SPACING 16 in (400) ft in (m)	GYPSUM BOARD/PLASTER CEILING JOIST SPACING 24 in (600) ft in (m)	OTHER CEILINGS JOIST SPACING 16 in (400) ft in (m)	OTHER CEILINGS JOIST SPACING 24 in (600) ft in (m)
No. 1	2×4 (38×89)	6-5 (1.94)	5-7 (1.70)	7-4 (2.22)	6-5 (1.94)
	2×6 (38×140)	10-1 (3.05)	8-10 (2.67)	11-5 (3.47)	9-4 (2.83)
	2×8 (38×184)	13-4 (4.03)	11-7 (3.52)	15-1 (4.57)	12-4 (3.73)
	2×10 (38×235)	17-0 (5.14)	14-10 (4.49)	19-3 (5.84)	15-8 (4.77)
	2×12 (38×286)	20-8 (6.25)	18-1 (5.46)	23-5 (7.10)	19-1 (5.80)
No. 2	2×4 (38×89)	6-2 (1.88)	5-5 (1.64)	7-1 (2.15)	5-9 (1.75)
	2×6 (38×140)	9-9 (2.95)	8-5 (2.54)	10-4 (3.11)	8-5 (2.54)
	2×8 (38×184)	12-10 (3.89)	11-1 (3.35)	13-8 (4.10)	11-1 (3.35)
	2×10 (38×235)	16-5 (4.97)	14-2 (4.28)	17-5 (5.24)	14-2 (4.28)
	2×12 (38×286)	20.0 (6.04)	17-3 (5.20)	21-2 (6.37)	17-3 (5.20)

CEILING JOISTS – ATTIC NOT ACCESSIBLE BY A STAIRWAY

LIVE LOAD 30 PSF (1.5 kN/m²)

GRADE	NOMINAL SIZE in. ACTUAL SIZE (mm)	GYPSUM BOARD/PLASTER CEILING JOIST SPACING 16 in (400) ft in (m)	GYPSUM BOARD/PLASTER CEILING JOIST SPACING 24 in (600) ft in (m)	OTHER CEILINGS JOIST SPACING 16 in (400) ft in (m)	OTHER CEILINGS JOIST SPACING 24 in (600) ft in (m)
No. 1	2×4 (38×89)	9-3 (2.80)	8-1 (2.45)	10-7 (3.21)	9-3 (2.80)
	2×6 (38×140)	14-7 (4.41)	12-9 (3.85)	16-8 (5.05)	13-9 (4.18)
	2×8 (38×184)	19-3 (5.81)	16-9 (5.08)	22-0 (6.65)	18-2 (5.51)
	2×10 (38×235)	24-6 (7.41)	21-5 (6.48)	28-1 (8.49)	23-2 (7.03)
	2×12 (38×286)	29-10 (9.02)	26-1 (7.88)	34-2 (10.33)	28-2 (8.55)
No. 2	2×4 (38×89)	8-11 (2.71)	7-10 (2.37)	10-3 (3.10)	8-7 (2.59)
	2×6 (38×140)	14-1 (4.26)	12-4 (3.72)	15-3 (4.59)	12-5 (3.75)
	2×8 (38×184)	18-7 (5.62)	16-3 (4.91)	20-1 (6.05)	16-5 (4.94)
	2×10 (38×235)	23-9 (7.17)	20-9 (6.26)	25-8 (7.73)	21-0 (6.31)
	2×12 (38×286)	28-10 (8.72)	25-2 (7.62)	31-3 (9.40)	25-6 (7.67)

Notes to table:

- Table is for attics not accessible by a stairway.
- For joists supporting a floor or partial roof load, see the floor joist table on 6-3.

The span table examples on this page are typical of allowable spans for spruce lumber. Other species and grades have different properties. Local codes and available materials will govern exact requirements.
All metric sizes are soft conversions.

source: CMHC

GABLE AND SHED ROOFS

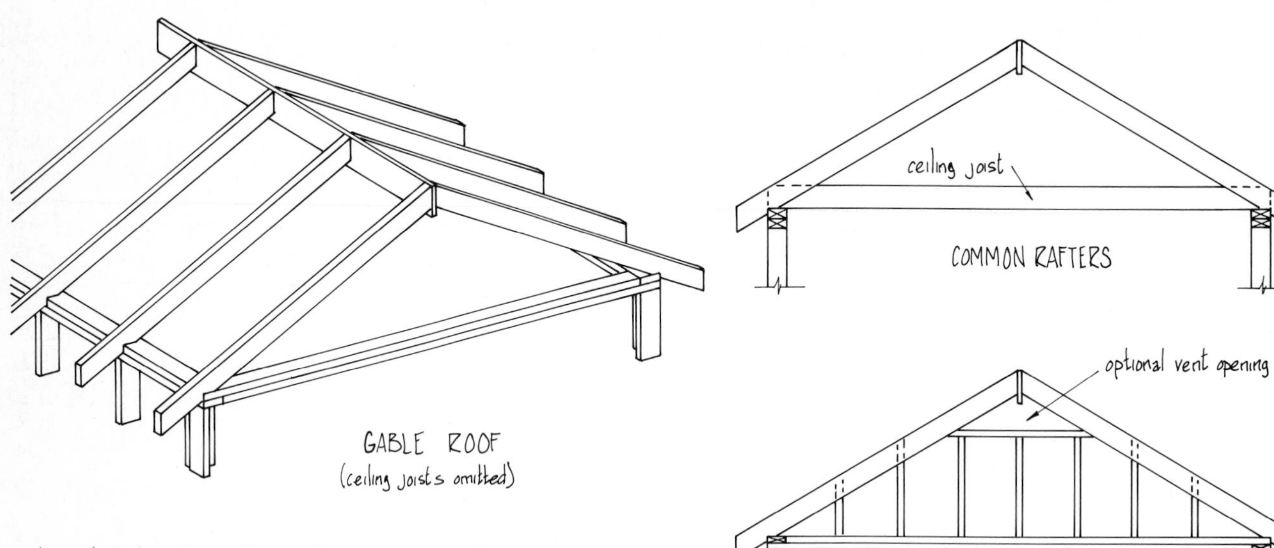

GABLE ROOF
(ceiling joists omitted)

COMMON RAFTERS

GABLE END RAFTERS with FRAMING

Gable and shed roofs are the simplest to design and construct.

- Each rafter is a common rafter, allowing quick repetition in fabrication.
- Shed rafters have two bird's-mouth cuts and two overhangs. Since each rafter is supported at both ends, a horizontal wall tie is not required, allowing ceilings to follow the rafter slope.
- The gable's vertical end is part of the exterior wall and is framed with studs as shown.
- See 8-14 for gable-end overhangs. Although the details on that page relate to trusses, the framing method is essentially the same. See, also, 7-8.

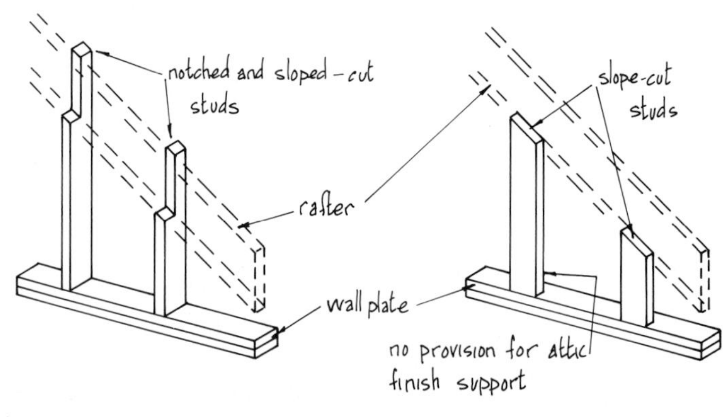

TWO METHODS of GABLE END STUD FRAMING

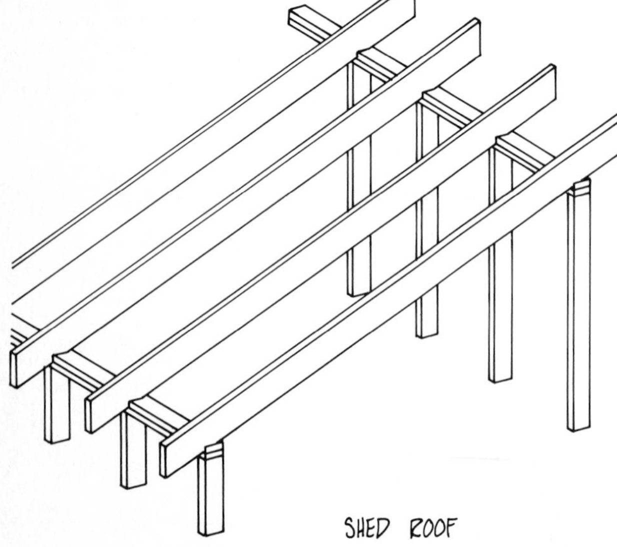

SHED ROOF

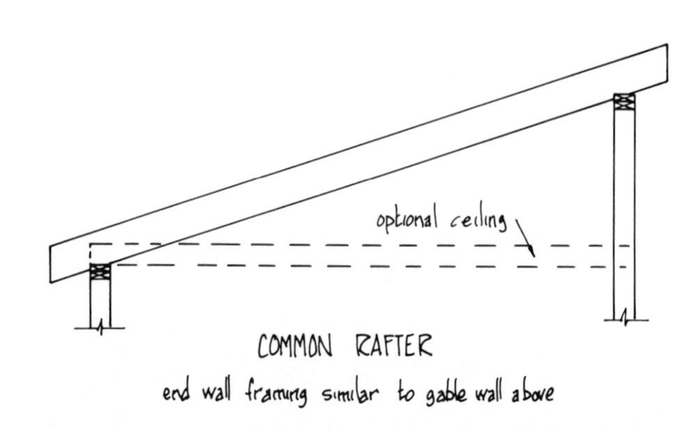

COMMON RAFTER
end wall framing similar to gable wall above

HIP ROOF

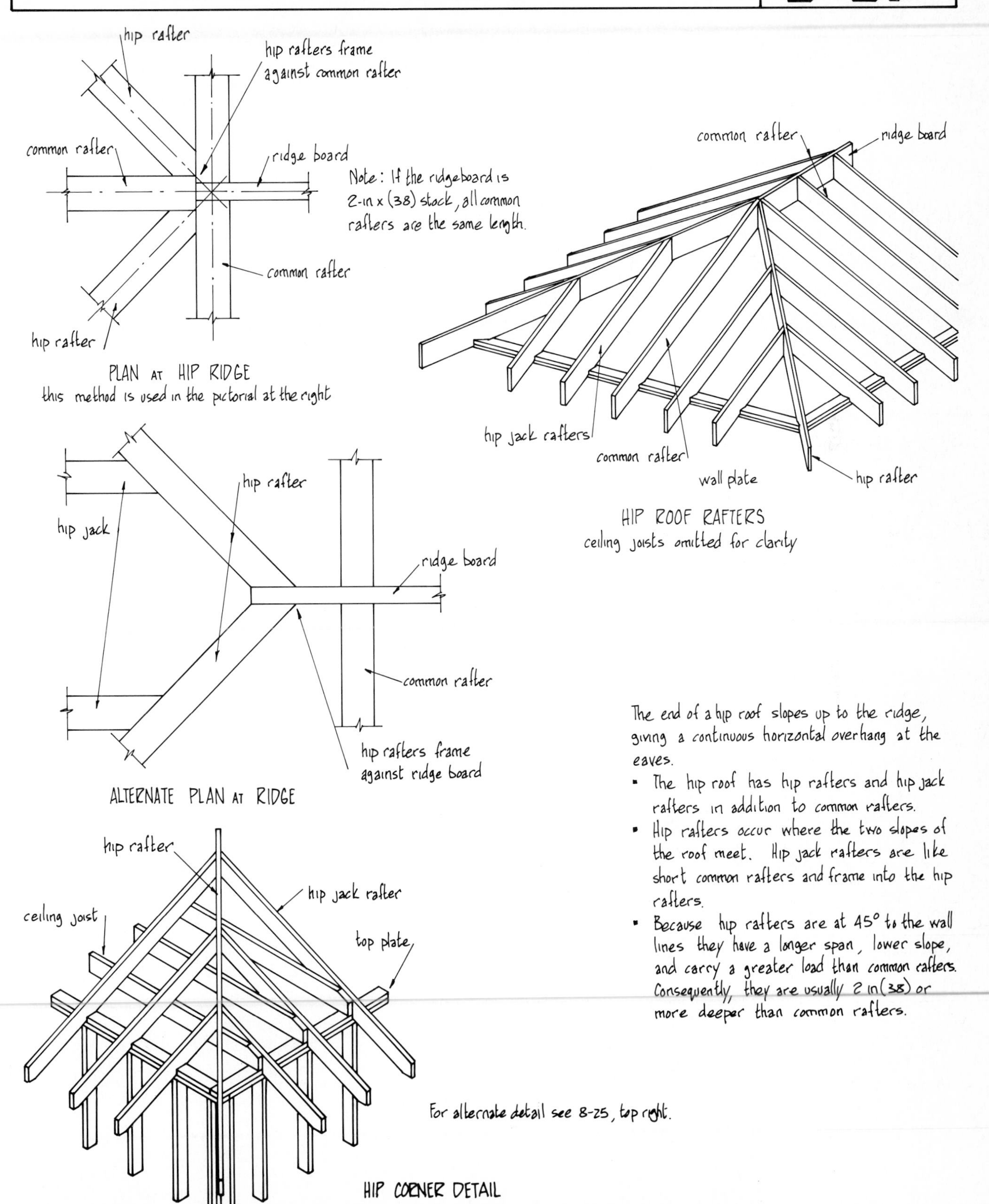

The end of a hip roof slopes up to the ridge, giving a continuous horizontal overhang at the eaves.

- The hip roof has hip rafters and hip jack rafters in addition to common rafters.
- Hip rafters occur where the two slopes of the roof meet. Hip jack rafters are like short common rafters and frame into the hip rafters.
- Because hip rafters are at 45° to the wall lines they have a longer span, lower slope, and carry a greater load than common rafters. Consequently, they are usually 2 in (38) or more deeper than common rafters.

For alternate detail see 8-25, top right.

GAMBREL AND FLAT ROOFS

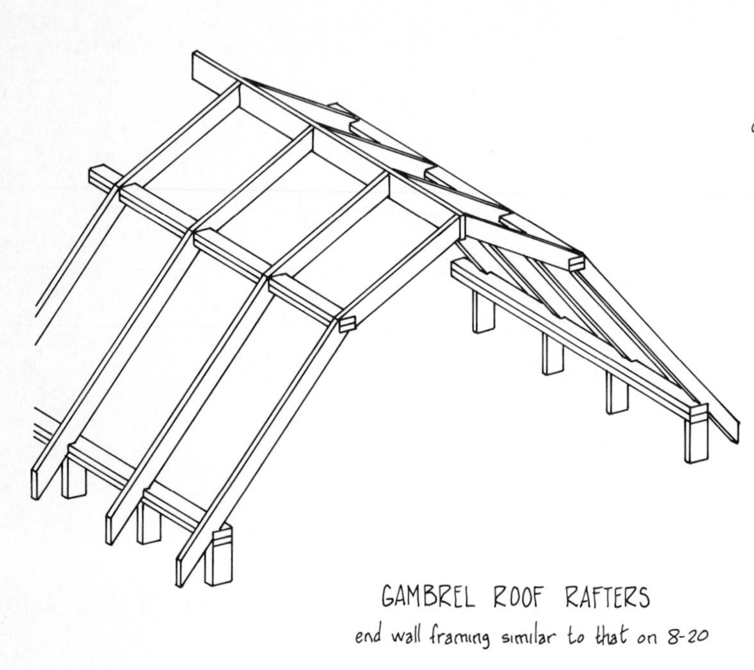

GAMBREL ROOF RAFTERS
end wall framing similar to that on 8-20

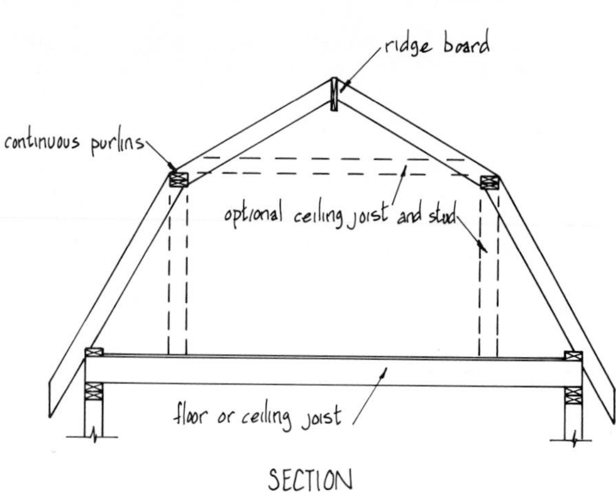

SECTION

GAMBREL ROOF

This roof is similar to a gable roof except that each slope is broken into two separate roof surfaces.

- The slope angle of the upper and lower roofs is generally 30 and 60° to the horizontal, respectively.
- The continuous purlins that receive both rafters can either act as a secondary ridge and be tied by collar beams across the roof or be the top plate of internal partition walls.

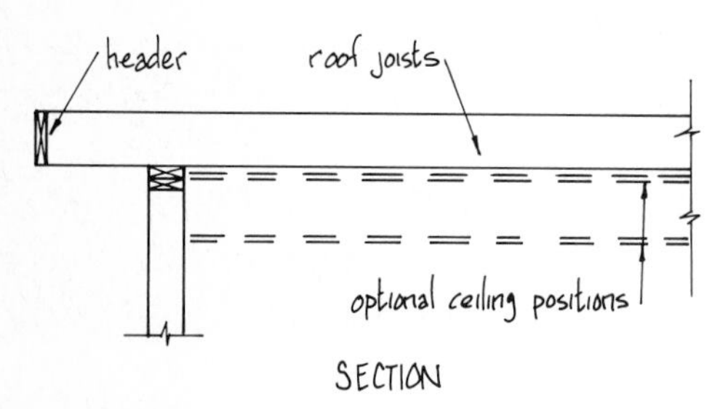

SECTION

FLAT ROOF

Flat roof construction is very similar to floor construction.

- Roof joists support both roof and ceiling loads and must be sized accordingly.
- Where an overhang is required on all sides, lookout rafters and the special corner detail shown at the right are necessary.
- To provide drainage the roof should be given a slight slope by tapering the top of each joist or adding a tapered strip.
- It is hard to adequately insulate this type of roof in cold climates and condensation buildup is an additional problem. See 10-21 for some solutions.

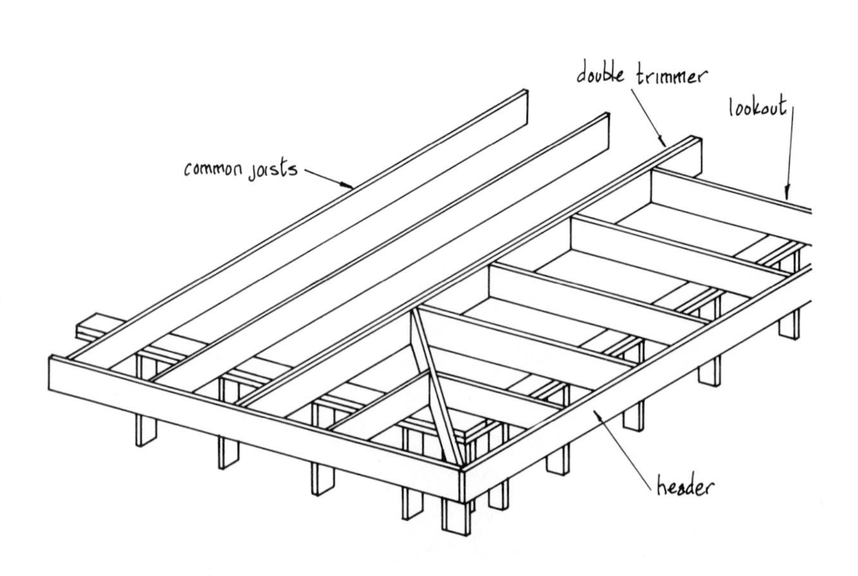

FLAT ROOF JOISTS

ROOF INTERSECTIONS

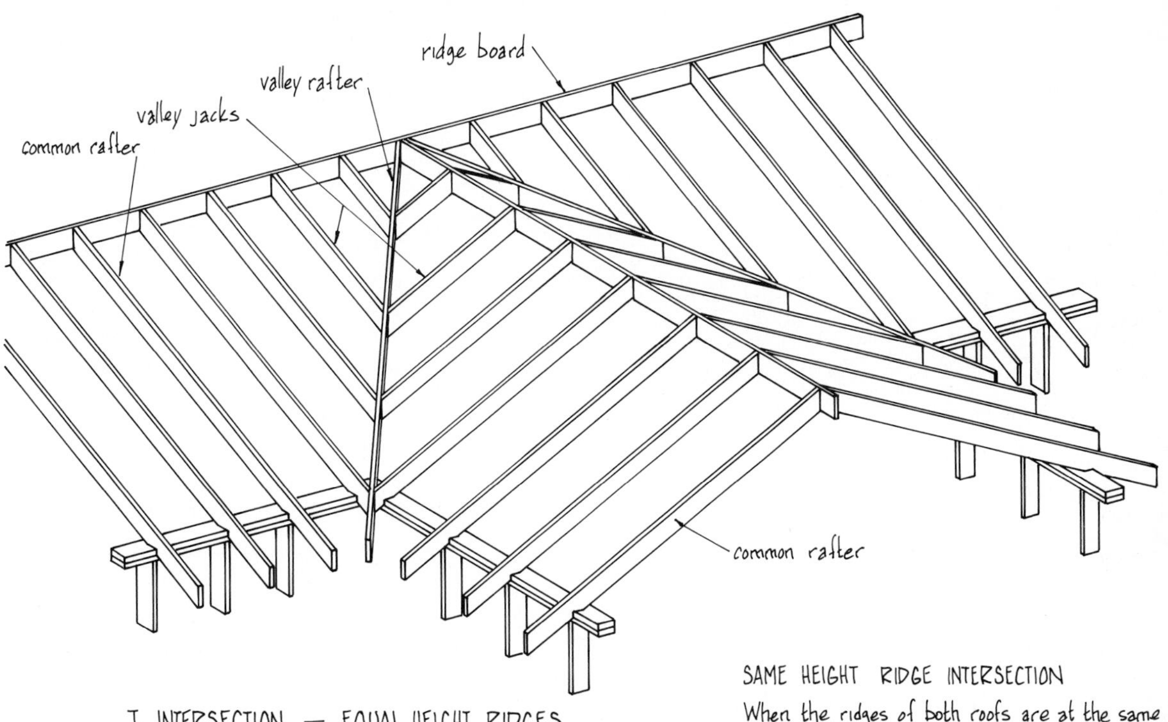

T- INTERSECTION — EQUAL HEIGHT RIDGES

for unequal height ridges see 8-16

SAME HEIGHT RIDGE INTERSECTION

When the ridges of both roofs are at the same level, two valleys are formed, each having a valley rafter that spans from wall line to ridge. Short valley jack rafters continue the roof slope between ridge and valley rafter.

DIFFERENT LEVEL RIDGES

To support the ridge of the lower roof, one of the valley rafters is extended to the higher ridge and carries both the ridge and second valley rafter. See 8-16 for a typical example of this type of roof.

hip rafter
hip jacks
common rafter
supporting wall
down
down
down
down
down
down
ridge
valley rafter
valley jacks
ridge

PLAN OF L- JUNCTION
equal height ridges

L- CORNER INTERSECTION

In this type of intersection, a valley is formed on the inside corner and a hip is formed on the outside corner.

INTERMEDIATE RAFTER SUPPORT

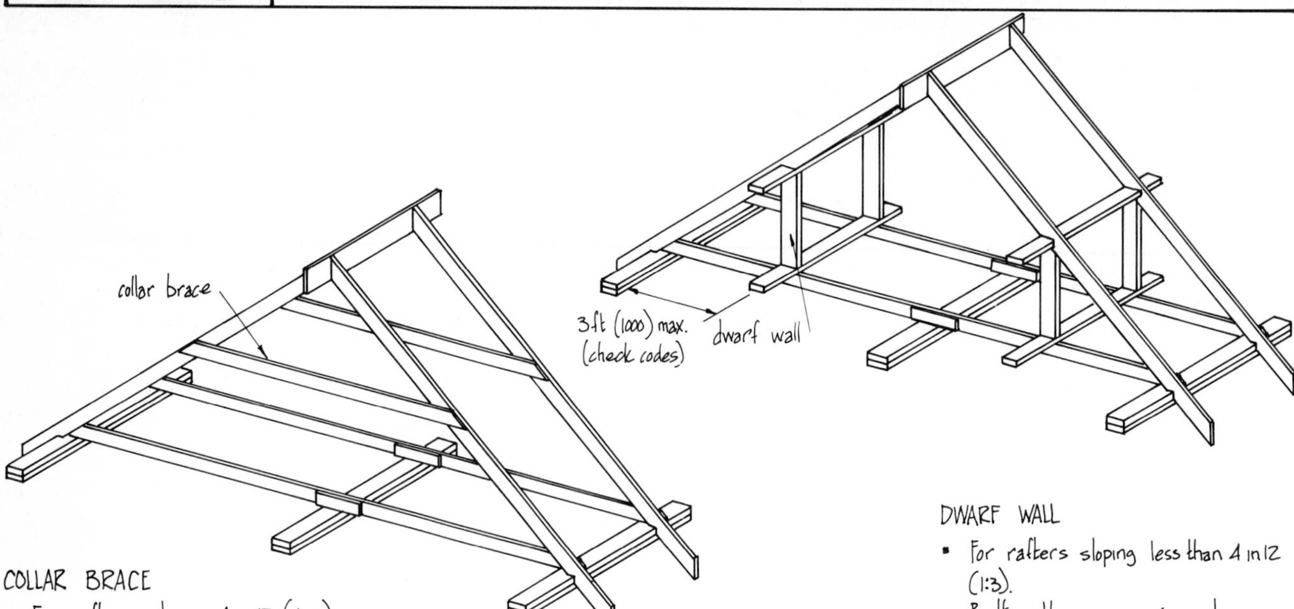

COLLAR BRACE

- For rafters sloping 4 in 12 (1:3) or greater.
- 2×4 in (38×89) brace nailed to the side of each rafter pair.
- The brace is in compression and if over 8 ft (2400) long must be laterally tied.

There are several methods of providing an intermediate rafter support that has the effect of shortening span and reducing rafter size. With the support in place, span is then taken from the intermediate point to the ridge or eave line.
Where possible, supports should connect with the ceiling joist at a bearing wall position; otherwise, an increase in ceiling joist size may be necessary.

DWARF WALL

- For rafters sloping less than 4 in 12 (1:3).
- Built in the same way as a bearing stud wall.
- Can have a single top plate provided that studs are directly under the rafters.

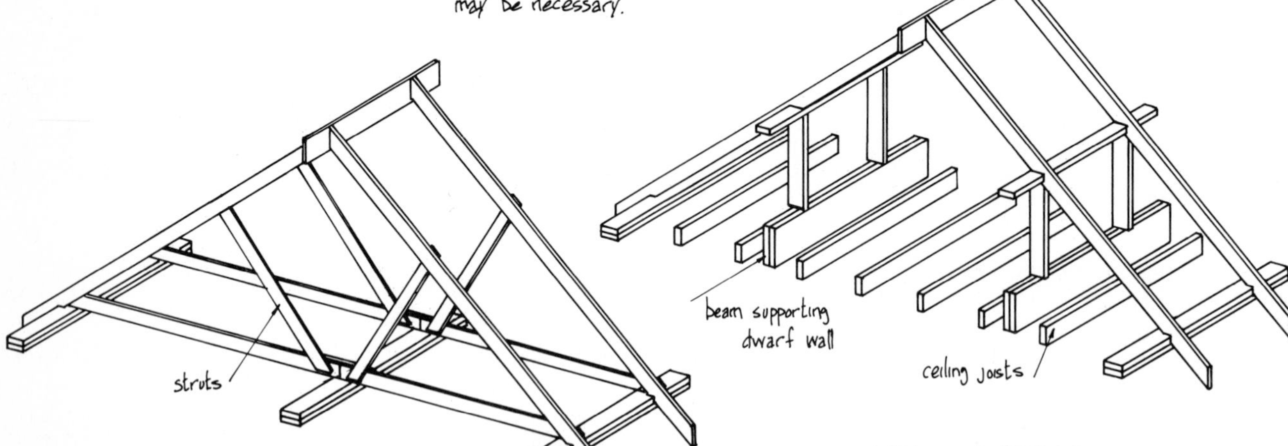

STRUTS

- Angle struts may be used in place of collar beams and dwarf walls.
- Supported by a load-bearing center wall.
- Angle of strut to the horizontal not less than 45°.

BEAM AND DWARF WALL

- Used when rafters run at right angles to the ceiling joists.
- Beam sits on a bearing partition but is raised a minimum 1 in (25) at bearing so that the beam's deflection will not damage the ceiling below.

RAFTER DETAILS

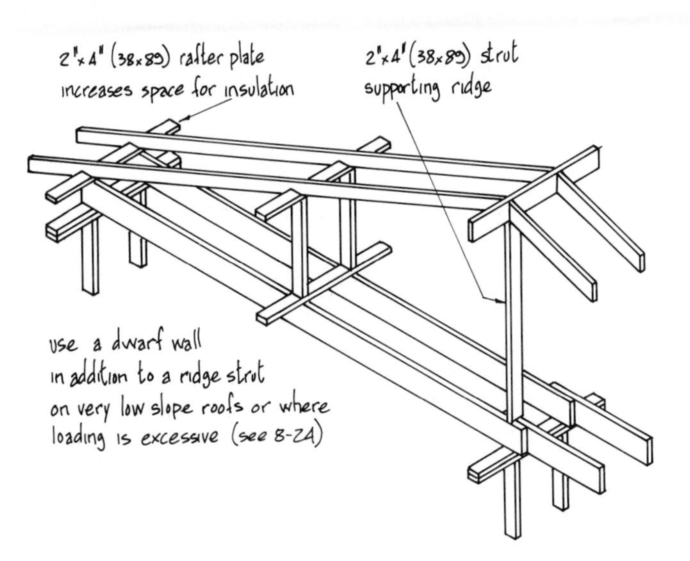

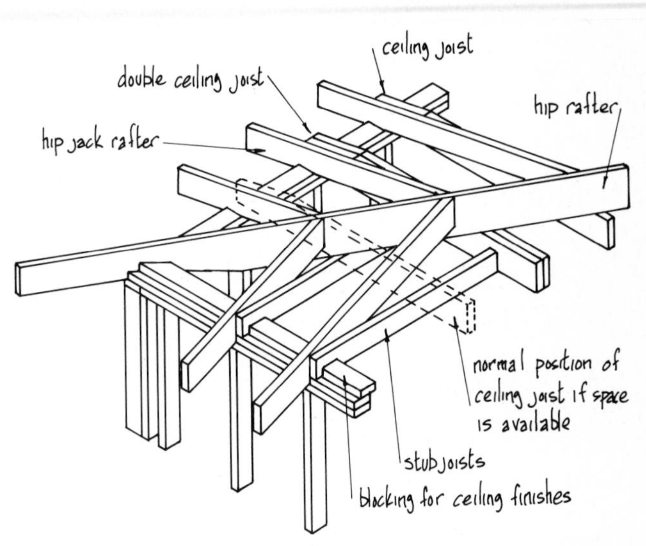

RIDGE BEAM

- For roofs with slopes of less than 4 in 12 (1:3), the ridge board or beam should be supported with a vertical strut spaced at 4 ft (1200) centers as shown above.
- A vertically-extended bearing wall can also be used.
- With this arrangement, the rafter is now supported at both ends, reducing outward thrust at the wall line, and eliminating the need for a ceiling-level horizontal wall tie.

HIP RAFTER AND CEILING

- On a low-slope roof with a deep hip rafter it may not be possible to install the last ceiling joist because of the hip rafter depth.
- The detail above solves this problem by adding stub joists at right angles to the ceiling joists.

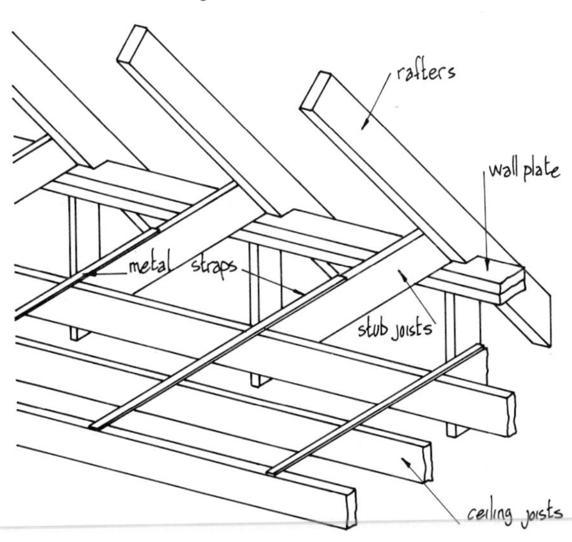

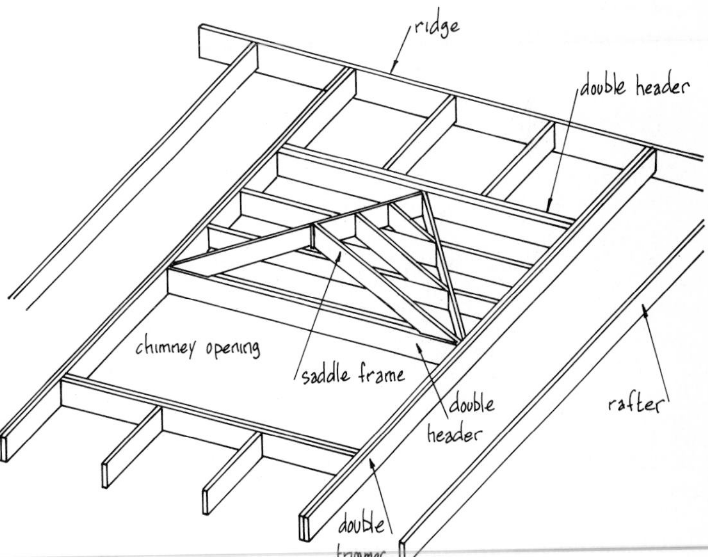

END WALLS

- At hip roof end walls where ceiling joists are parallel to the wall line, some form of roof frame reinforcing is necessary.
- Stub joists tied with metal straps extending over three ceiling joists are ideal.
- If a subfloor is applied over the ceiling joists, use metal framing anchors to tie the stub joists.

ROOF OPENINGS

- Small openings can be framed with single headers and trimmers.
- The opening shown above is for a chimney or other large roof penetration. A saddle frame is also shown (see 11-29).
- Openings in ceiling joists are framed as for floor joists (see 6-11).

DORMER FRAMING

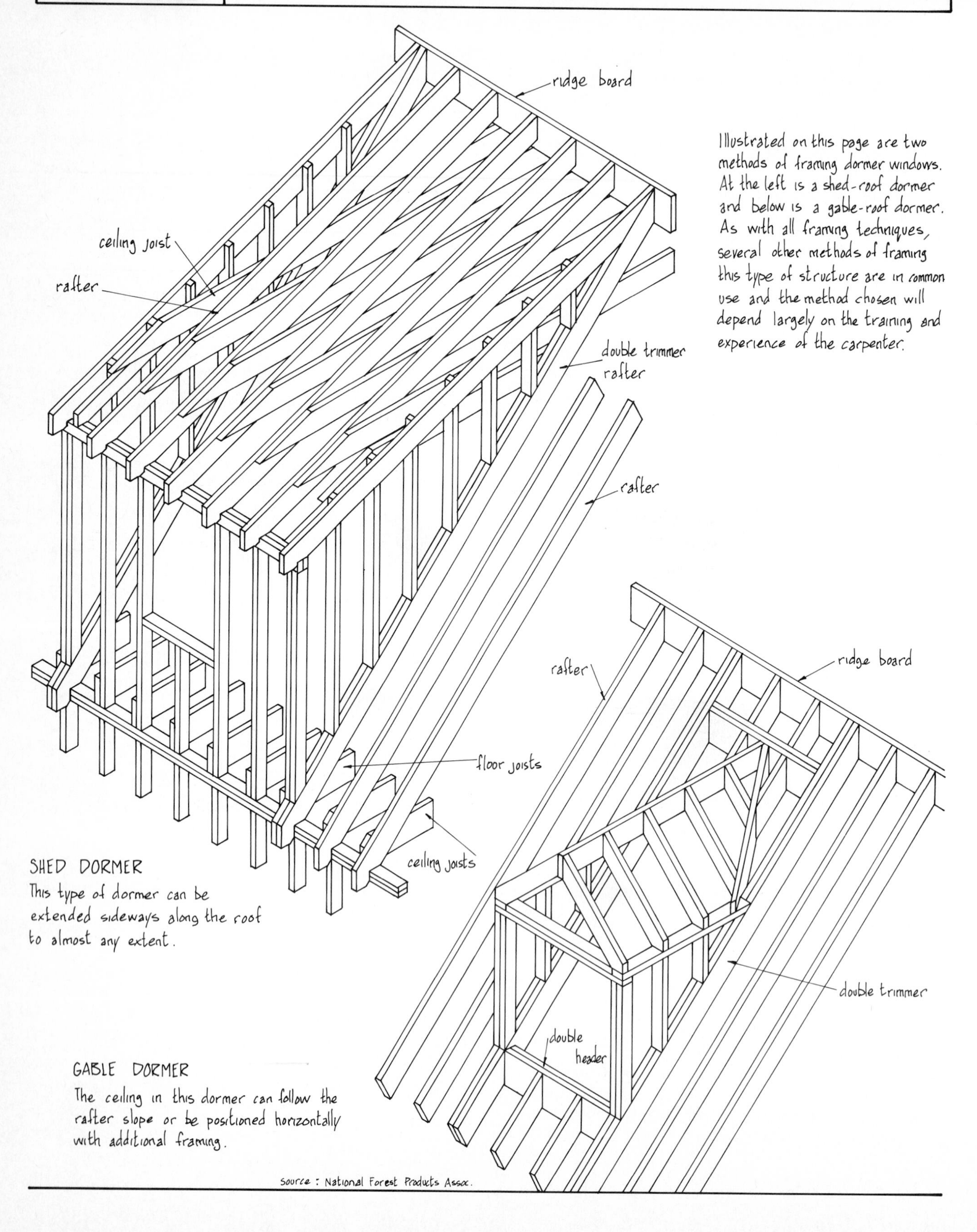

Illustrated on this page are two methods of framing dormer windows. At the left is a shed-roof dormer and below is a gable-roof dormer. As with all framing techniques, several other methods of framing this type of structure are in common use and the method chosen will depend largely on the training and experience of the carpenter.

SHED DORMER

This type of dormer can be extended sideways along the roof to almost any extent.

GABLE DORMER

The ceiling in this dormer can follow the rafter slope or be positioned horizontally with additional framing.

Source: National Forest Products Assoc.

WALL SETBACKS

Where a wall line is recessed, an alternate method of supporting the rafter ends must be used. The illustration at the right shows a vertical extension of the bearing wall that provides an intermediate bearing for the rafters. The rafters then overhang the wall line and receive additional support from the ceiling joists that act as a cantilever.

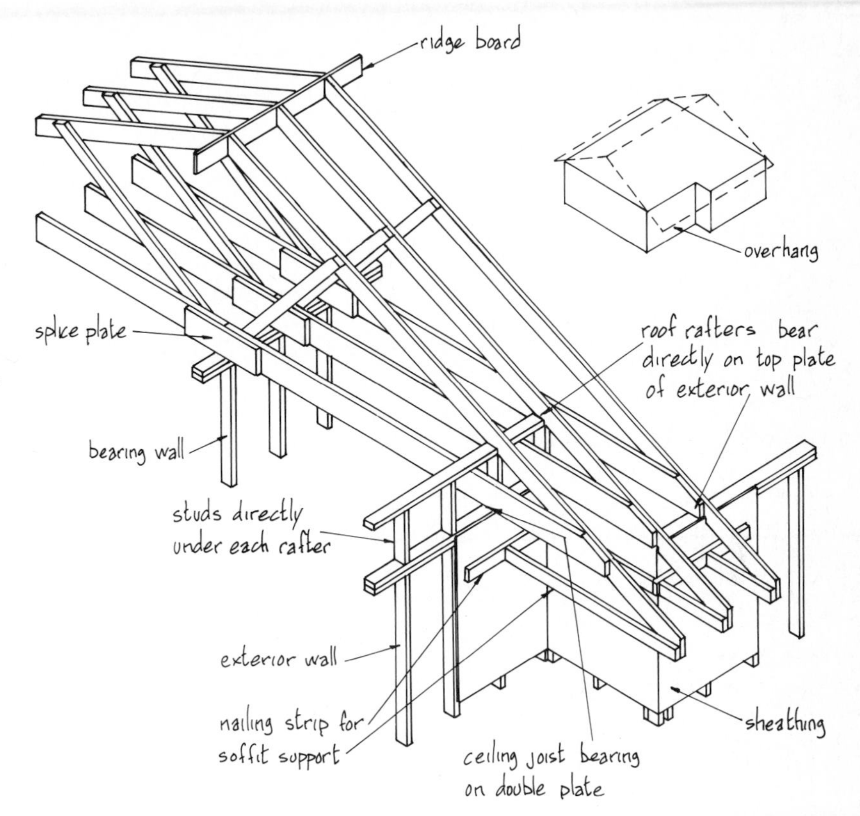

BEAM AND COLUMN SUPPORT

In this illustration, a beam is used to support the rafter ends. A steel beam may be necessary if the clearance between rafter and soffit is limited.
Larger loads and wider overhangs can be obtained with this method.

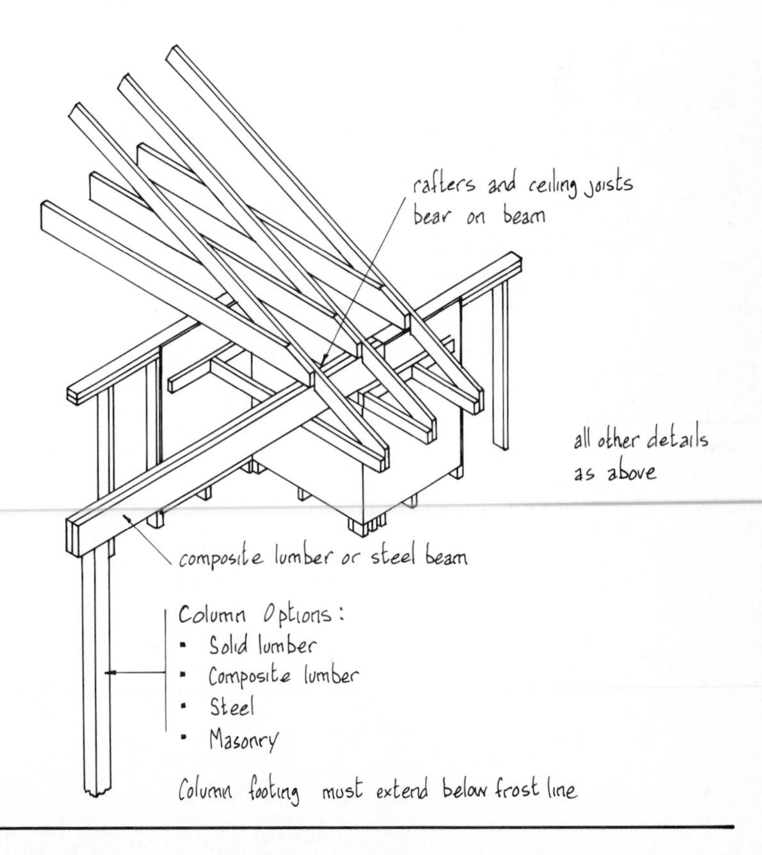

source: CMHC

ROOF SHEATHING

Roof sheathing provides a nailing base for the roof finishes and a structural tie for the trusses or rafters. Sheathing materials in common use are plywood, particleboard, waferboard, and for some applications, lumber boards. All are subject to span and loading constraints.

PLYWOOD

Sheathing grade (unsanded) is generally used for roof applications.

- Panels are placed so that the long dimension spans across supports and each panel must span over at least two supports.
- Panel ends must be staggered and bear on the supports.
- To accommodate expansion and contraction leave a 1/16 in (2) gap at panel ends and a 1/4 in (6) gap at the edges.
- Panels can have edges either supported or unsupported. Unsupported panels have a lower span rating.
- Edges are commonly supported with H-clips that connect adjoining panels. Tongued and grooved edges or blocking nailed between supports can also be used.
- Plywood thicknesses shown at the right are minimum sizes only. Check local codes for exact sizes, especially in regions with heavy snow loads.

Follow the general information above for particleboard, waferboard, or other suitable composite materials.

JOIST OR RAFTER SPACING	MINIMUM PLYWOOD THICKNESS		MINIMUM THICKNESS PARTICLE/WAFERBOARD
	EDGES SUPPORTED	EDGES UNSUPPORTED	EDGES SUPPORTED
IN (MM)	IN (MM)	IN (MM)	IN (MM)
12 (300)	5/16 (7.5)	3/8 (9.5)	3/8 (9.5)
16 (400)	5/16 (7.5)	3/8 (9.5)	3/8 (9.5)
20 (500)	3/8 (9.5)	1/2 (12.5)	7/16 (11.1)
24 (600)	3/8 (9.5)	1/2 (12.5)	7/16 (11.1)

MINIMUM ROOF SHEATHING THICKNESS

Check local codes for exact requirements.
It is recommended that 1/2-in (12.5) supported edge plywood be used on 24-in (600) spaced trusses or rafters.

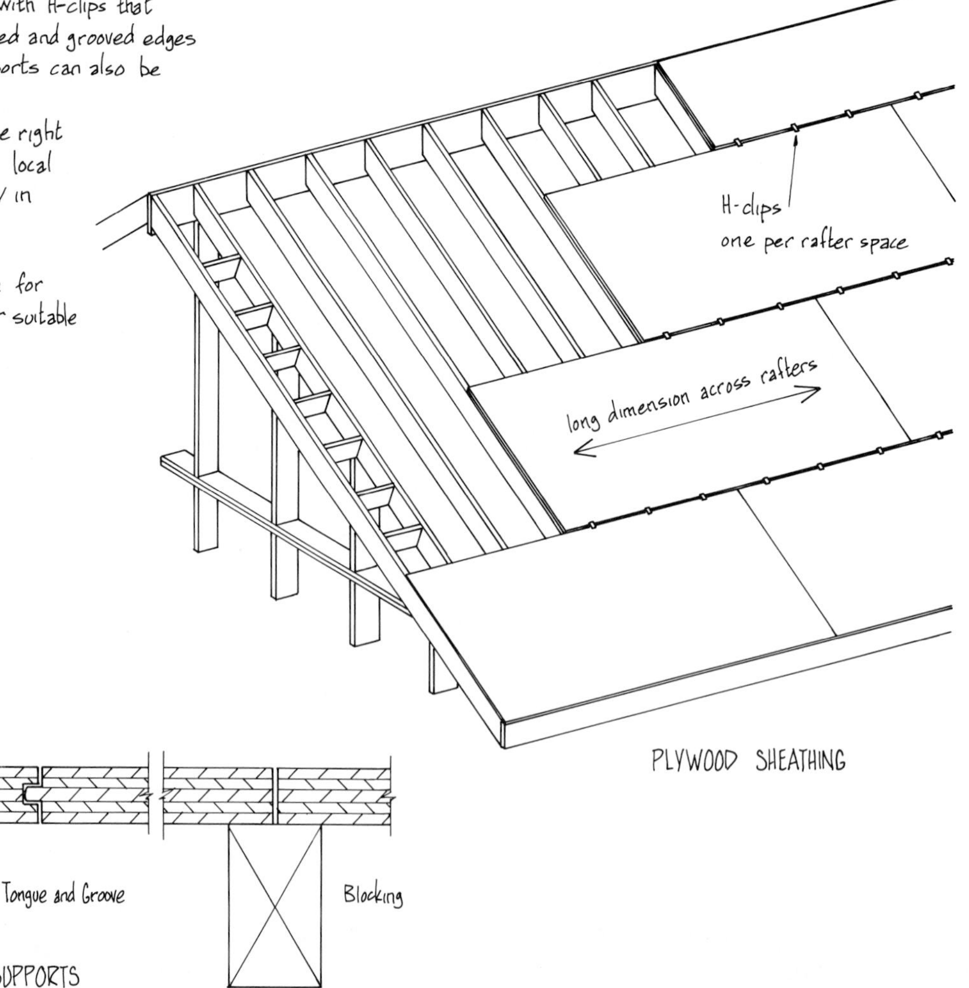

PLYWOOD SHEATHING

H-Clip

Tongue and Groove

Blocking

SHEATHING EDGE SUPPORTS

ROOF SHEATHING

TRUSS OR RAFTER SPACING IN (MM)	BOARD THICKNESS IN (MM)
up to 16 (400)	11/16 (17)
16-24 (400-600)	3/4 (19)

MINIMUM BOARD SHEATHING THICKNESS

- Lumber boards may be up to 12 in (286) wide for closed sheathing.
- Boards for spaced sheathing are generally 4 in (89) or 6 in (140) wide and are spaced at the same centers as the shingle exposure dimension.

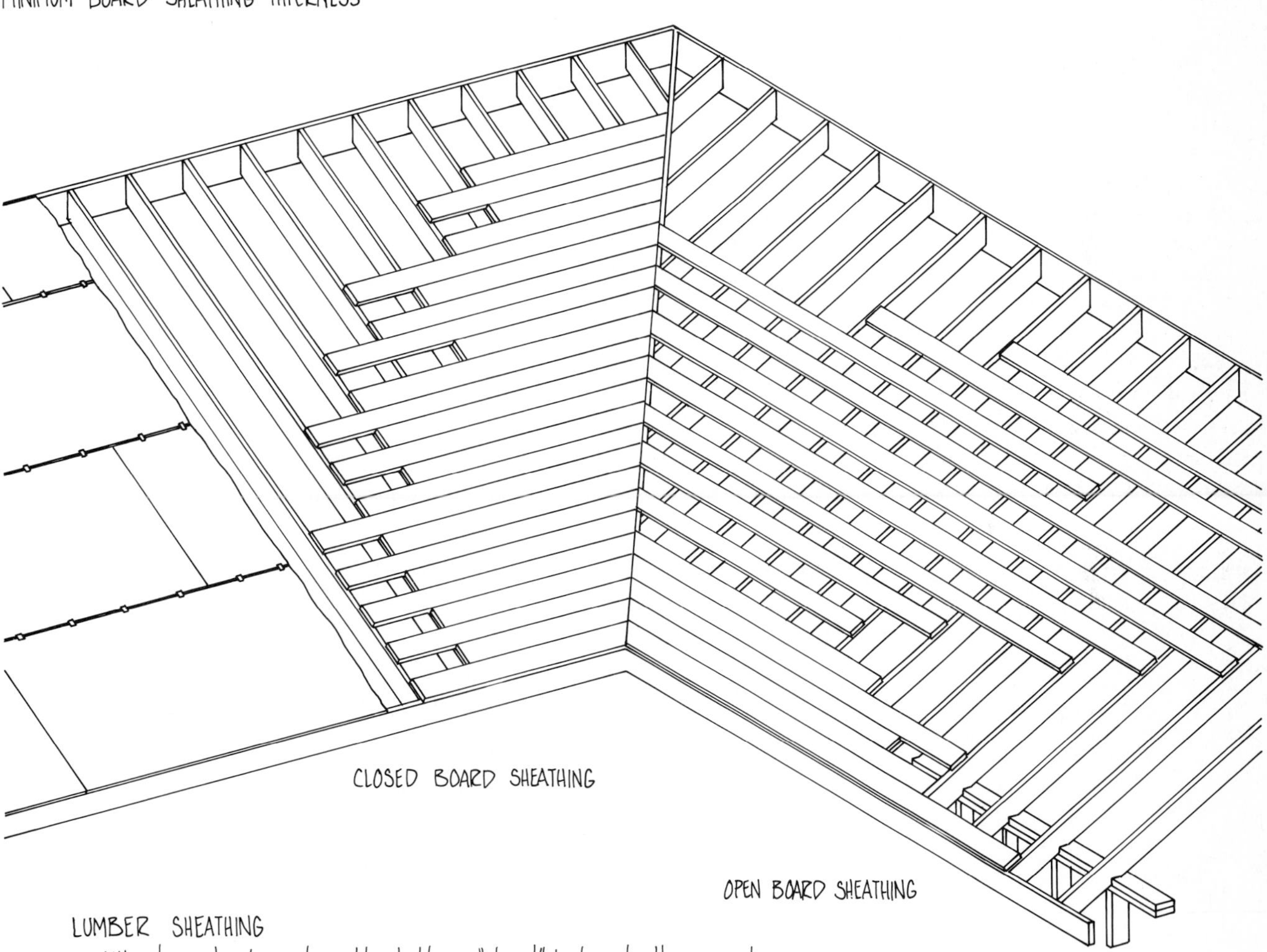

LUMBER SHEATHING

- Although used extensively in older buildings, "closed" lumber sheathing is not generally used in modern construction because of material and labor costs.
- "Open" lumber sheathing, however, is usually used for supporting wood shingles or shakes. Open boards reduce the possibility of decay by allowing the roof materials to dry faster.
- The joints of square-edged boards should be over a rafter and be staggered. Joints of t&g boards need not be on a rafter.
- All boards should span continuously over at least two rafters.
- See 11-30 to 11-32 for shingles and shakes.

See 11-24 onward for roofing materials.
See 8-38 and 8-39 for commercial roofing materials.

ROOF FUNCTIONS AND DRAINAGE

Besides serving as a weatherproof covering for the building, commercial and industrial roofs are often designed to fulfill additional functions, depending on the intended use of the building:

- Most roofs serve as a platform for a variety of mechanical equipment, including air conditioning/heating units, fans, vents, and hoists.
- Additional glazing in the form of northlights or skylights is often added.
- Roof gardens, exercise areas, and restaurants are common on commercial buildings.
- Below the roof a large variety of equipment can be suspended from the roof structure. Most mechanical and electrical services are run through the roof/ceiling space and they can be left visible or hidden above the ceiling finish.
- Industrial buildings are often subject to heavy point or impact loads from hoisting or other equipment suspended from the roof structure. Specially designed structures are essential for these conditions.

In many buildings, particularly multistoried with flat roofs, it is common for the roof system to parallel the floor system in both design and construction. This makes both economic and design sense since the same basic materials are used for floor and roof after allowances are made for loading, finishes and insulation values. See 8-38, 8-39 and 11-34 for some typical roof finishes.

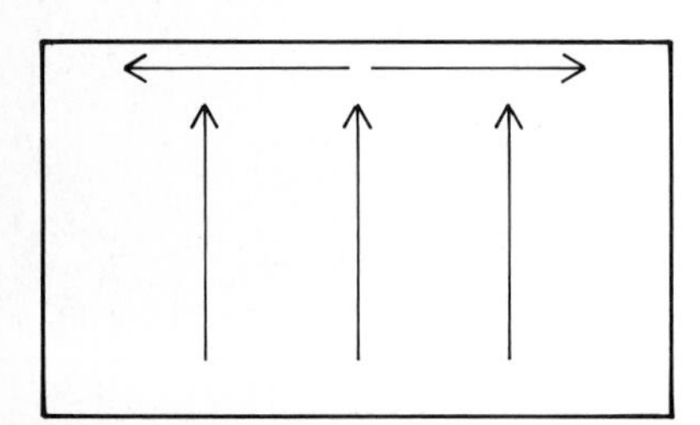

Flat or Monopitch

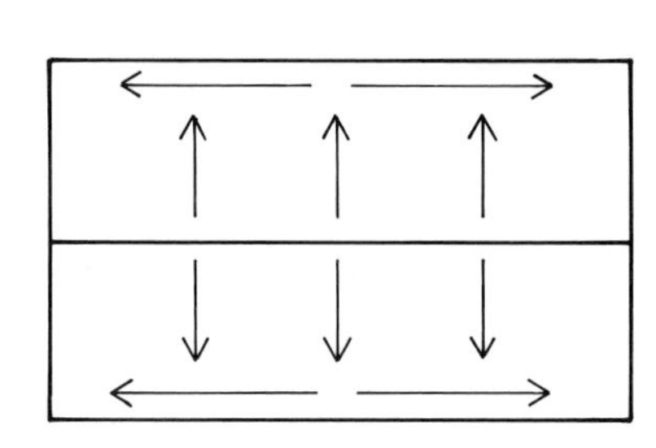

Flat or Double Pitch with or without a ridge line

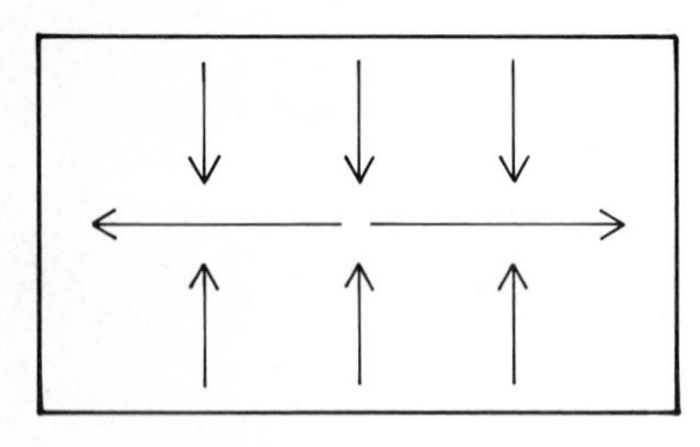

Butterfly Flat or Pitched

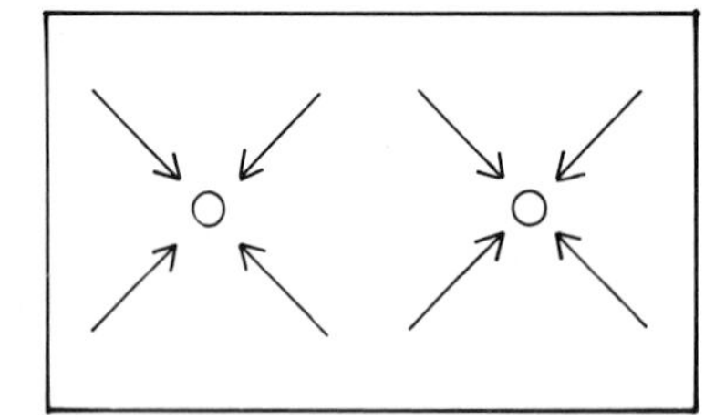

Flat with Center Roof Drains

ROOF PLANS SHOWING DRAINAGE PATTERNS
arrows indicate water flow

ROOF DRAINAGE

Efficient roof drainage is extremely important if the roof is to remain weathertight for the extended number of years expected of a commercial or industrial roof.

PITCHED ROOFS

- Generally, mono (single) or double pitched.
- Water drains to eave line with gutters at roof edge or behind a parapet wall.
- Roof pitch and structure type usually determine type of roof finish.

FLAT ROOFS

- Flat roofs are rarely totally flat unless designed to pond water for air conditioning purposes. Ponding is dangerous on roofs not so designed.
- Minimum slope of 1/4 in per foot (1:50) is recommended for adequate drainage.
- Water drains to edge gutters or to center roof drains.

Collected water can be discharged at ground level, into ponds, or streams, to a storm sewer, or in some regions, to a sanitary sewer. Consult local codes for discharge restrictions. See 16-18 and 19 for more information.

LAMINATED WOOD STRUCTURAL SYSTEMS

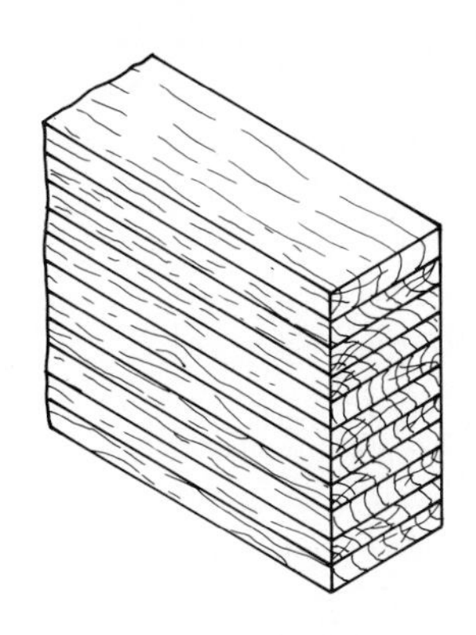

TYPICAL LAMINATED TIMBER BEAM
Laminations are glued together under pressure. Mechanical fasteners are not required.

GLUE-LAMINATED WOOD BEAMS

These are manufactured beams made from standard lumber sizes that are glued together under pressure. When dry the beam is sanded smooth and finished in a variety of ways.

- A laminated beam is generally one-third stronger than a comparable sized solid wood member.
- Type of glue used depends on strength requirements and exposure conditions.
- Manufacturers carry stock sizes of laminated beams or will custom fabricate.
- The beams can be formed into curved shapes during fabrication, allowing wide design freedom.
- Laminated beams are generally used where strength and aesthetic appeal are required.

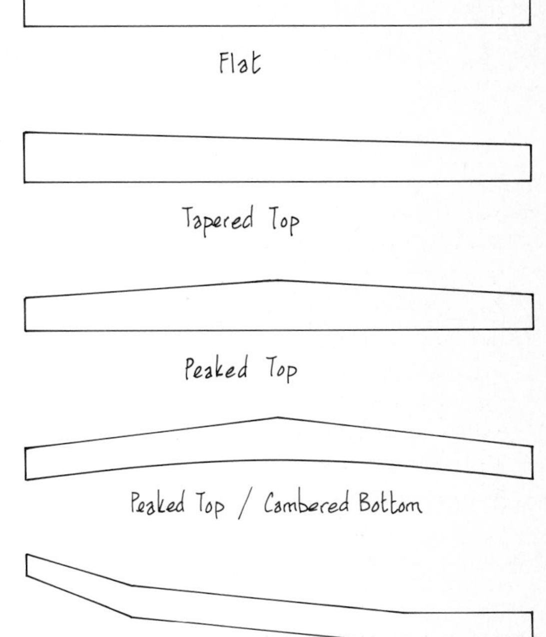

The beam shapes shown above are typical of the stock types produced by fabricators. Spans up to 50 ft (15m) are common when used as main beams. In smaller sizes they can also be used as purlins spanning between main beams.

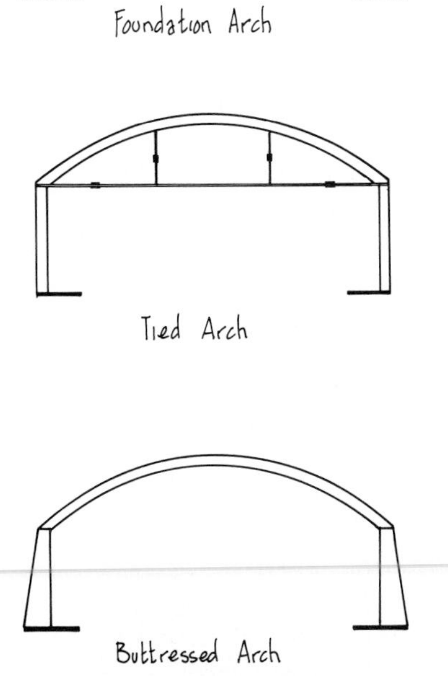

Foundation Arch

Tied Arch

Buttressed Arch

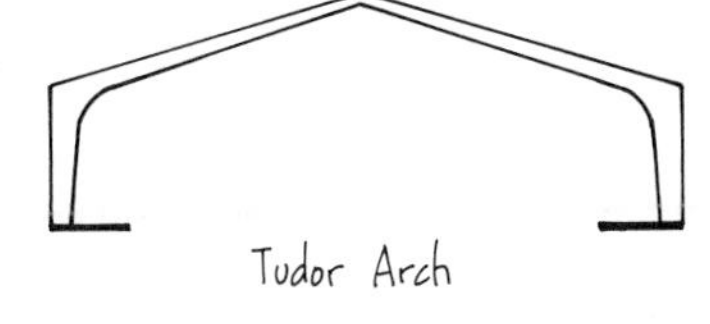

Tudor Arch

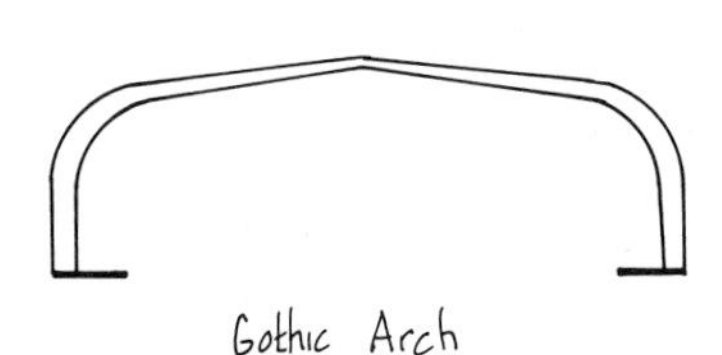

Gothic Arch

Continuous Arch

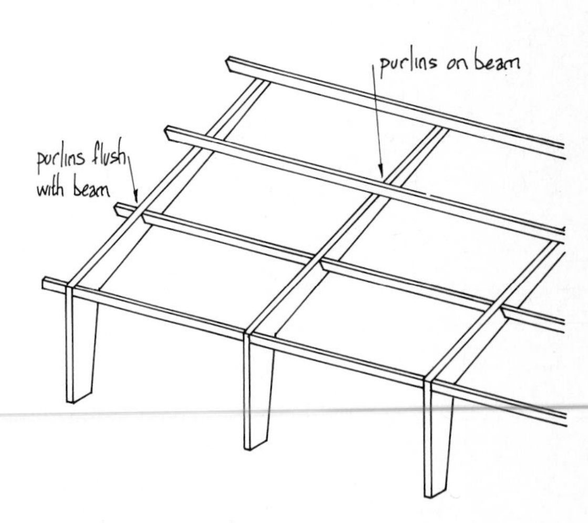

Laminated beams and columns can also be used in a standard beam and column type of structure. Structural design parallels that of a steel-framed building.

TWO-HINGED BARREL ARCHES

- A continuous span arch requiring support for lateral thrust at the bearings. The arches above show three types of thrust support.
- Spans of 250 ft (76 m) and above are possible.

THREE-HINGED ARCHES

- There are three main types, as shown above.
- Spans up to 100 ft (30m) are usual.
- Used where graceful, aesthetic effects are required.

HEAVY WOOD TRUSSES

TYPICAL TRUSS TYPES

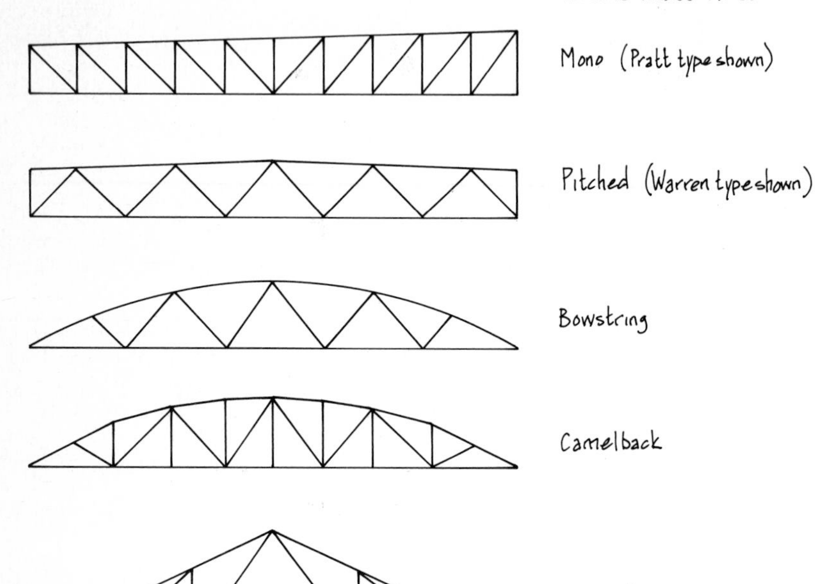

For many commercial and industrial applications the standard prefabricated lightweight trusses detailed on 6-20 and 8-8 are quite adequate. For longer spans and heavier loads, trusses must be carefully engineered to meet the service conditions.

- Heavy trusses are generally spaced 8-20 ft (2.5-6 m) apart.
- Purlins spanning between trusses are used to support the roof covering. Typical purlin connections are shown below.
- Web-to-chord connections will vary considerably depending on truss construction and design.

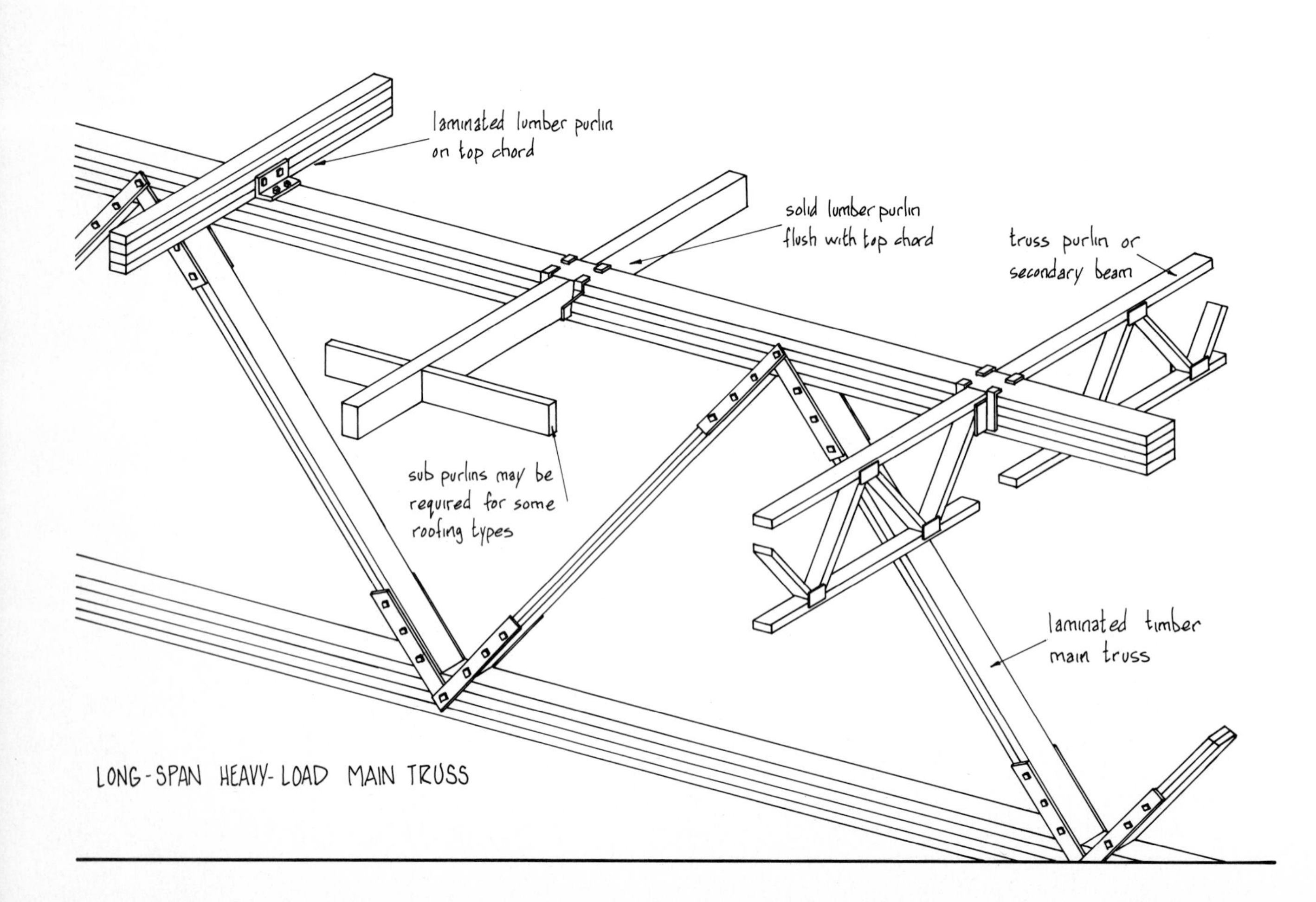

LONG-SPAN HEAVY-LOAD MAIN TRUSS

HEAVY STEEL TRUSSES

For small and medium span buildings the same type of lightweight trusses that were used for floors can be applied to roofs (see 6-24 for details). For larger spans and heavier loads, other truss types and materials must be used.

- The truss types shown on the previous page apply equally to steel trusses. Other truss types and arrangements for industrial buildings are shown at the right.
- Typical purlin types and connections are shown below on a welded double-angle and gusset plate, heavy parallel chord truss.

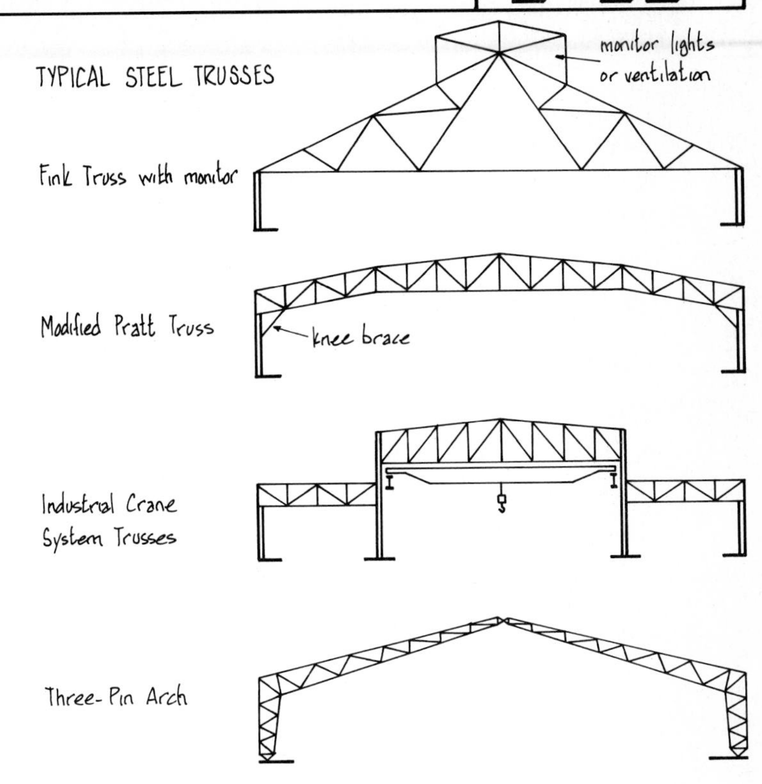

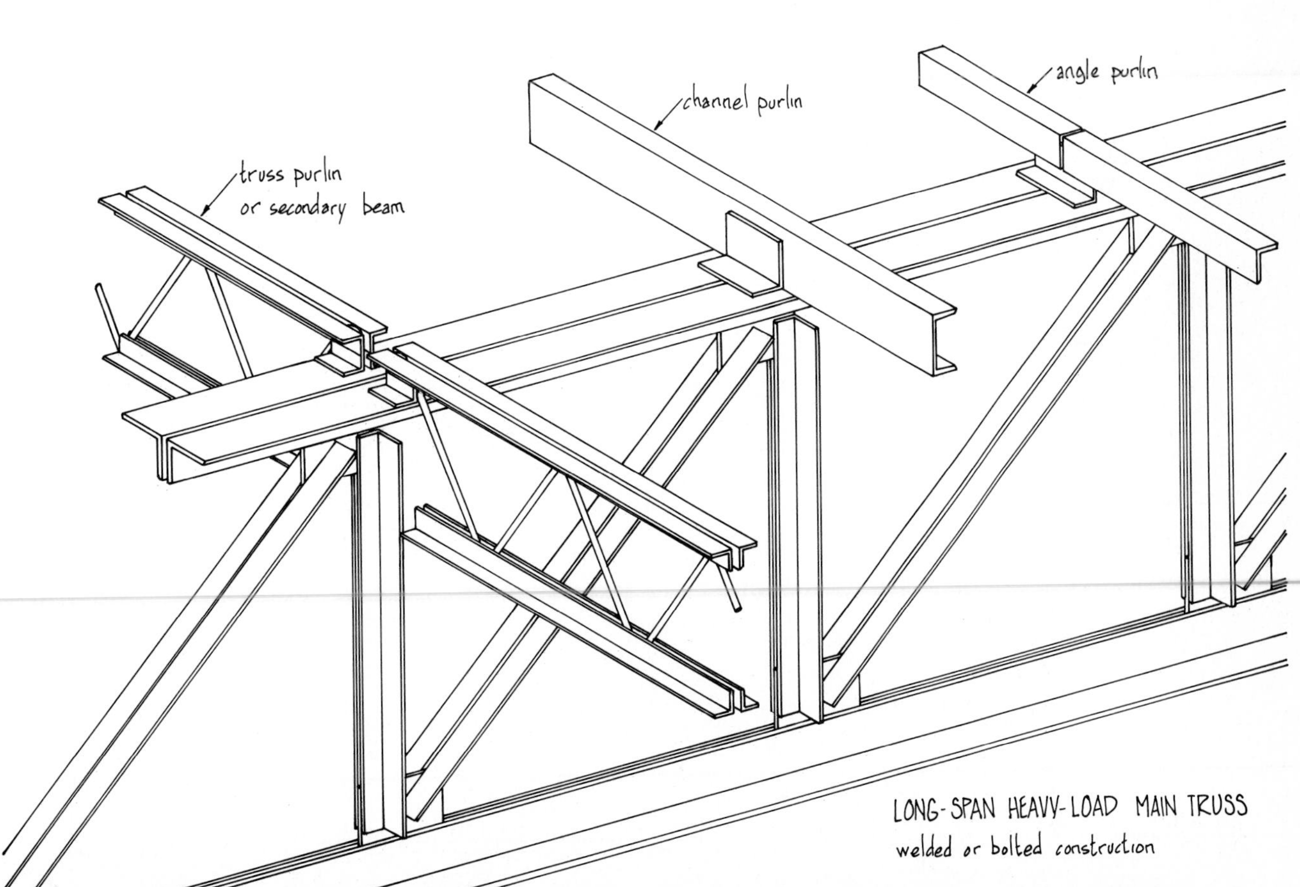

STRUCTURAL STEEL

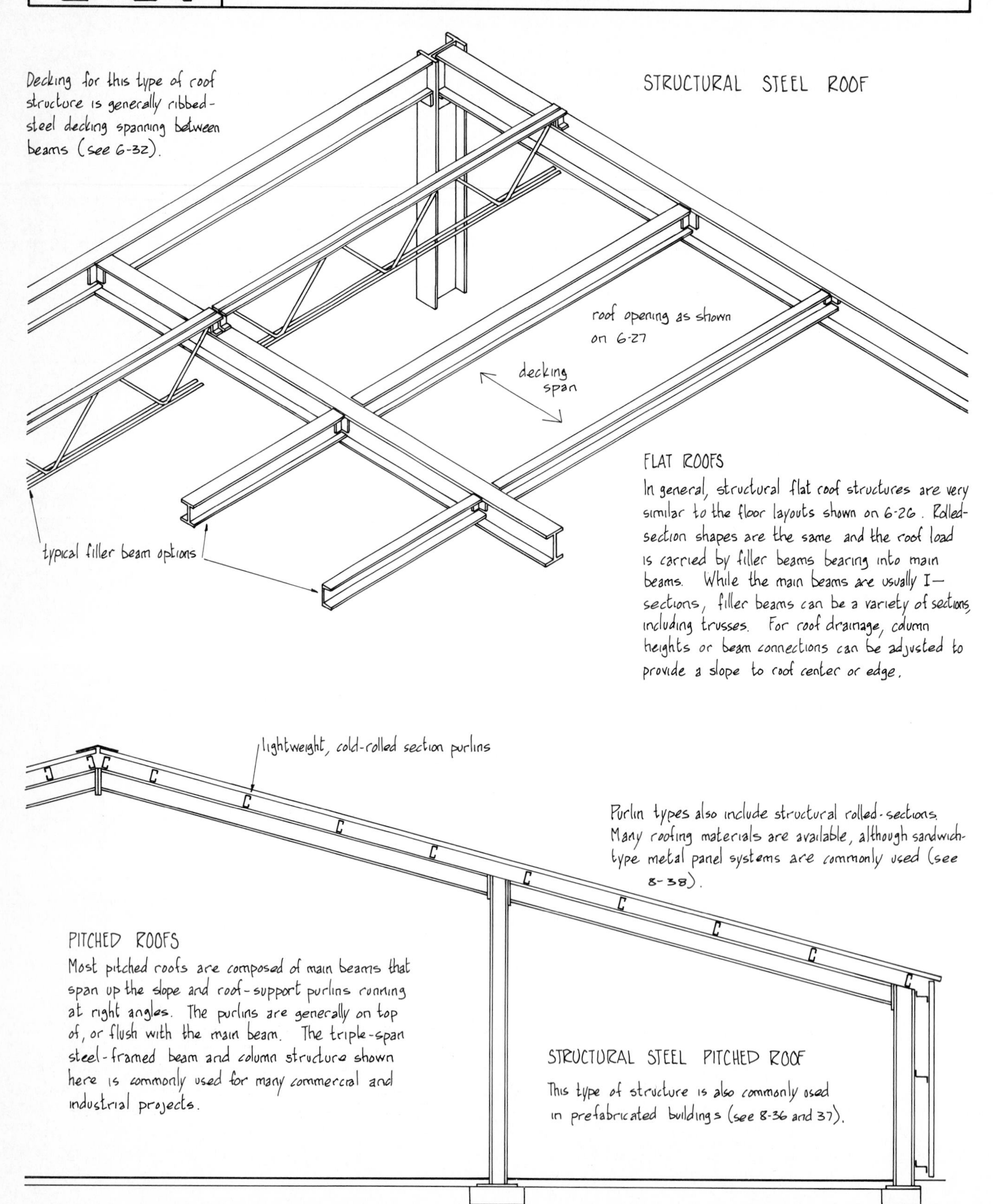

PRECAST CONCRETE

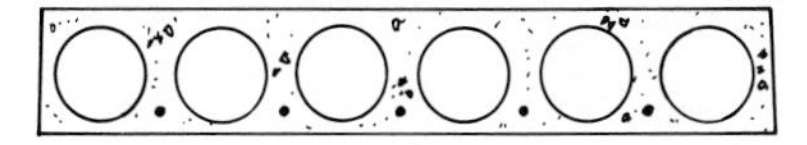

PRECAST HOLLOW CORE SLAB (see 6-28)

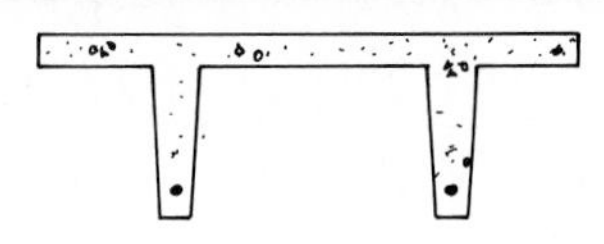

PRECAST T-BEAM (see 6-28)

For precast or monolithic (site poured) concrete flat roofs, the floor units and floor types detailed on 6-28 to 6-31 are equally applicable to roofs. These structural members work very well with standard built-up roofing systems and are commonly seen in modern buildings. For pitched-roof concrete beam systems see below.

FLAT ROOF CONCRETE SYSTEMS

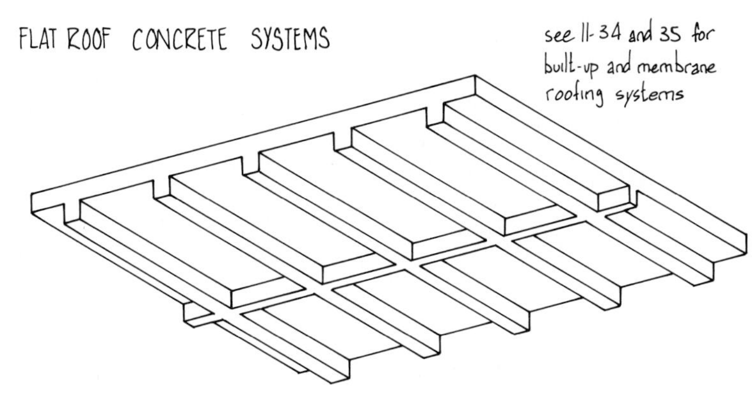

CAST-IN-PLACE TWO-WAY RIBBED SLAB (see 6-30)

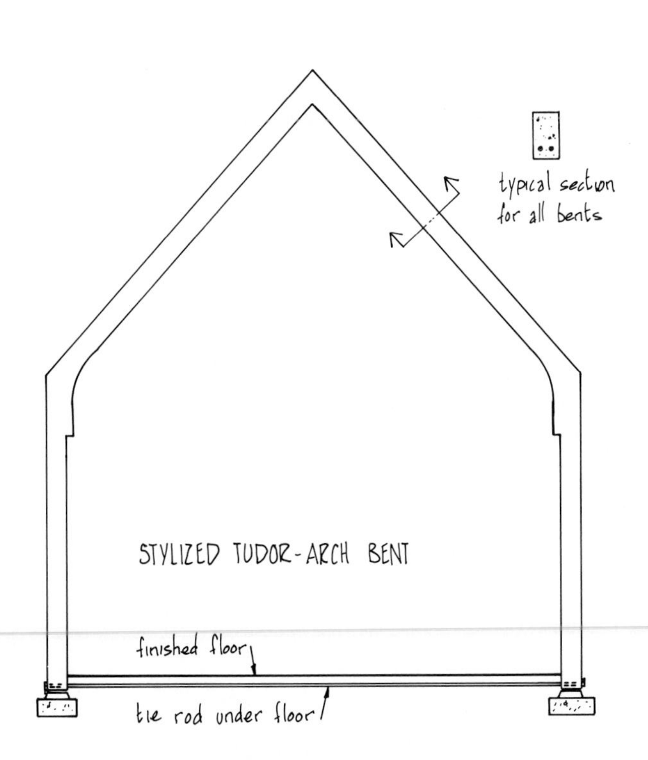

STYLIZED TUDOR-ARCH BENT

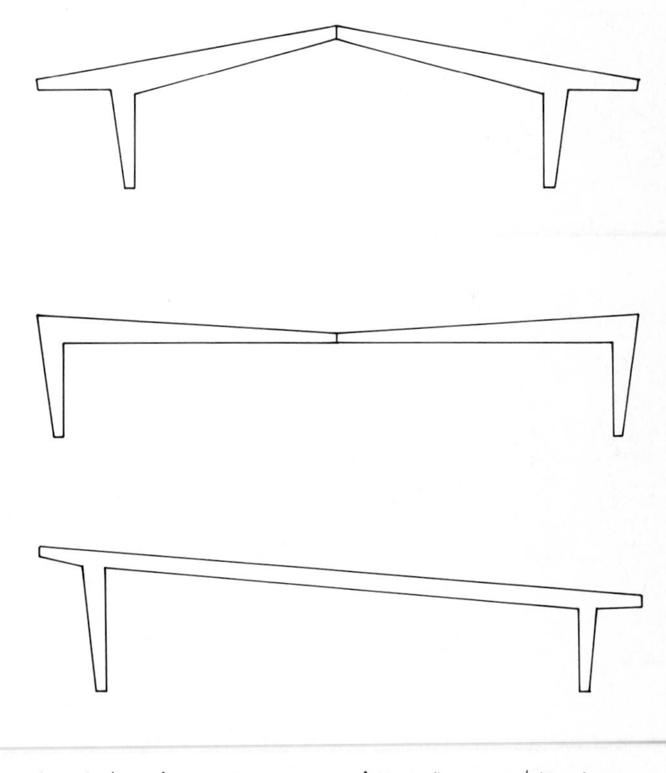

Units are bolted to foundation beams and site-connected at the peak. Note the tie rod underfloor at the base to counteract tension loads at the foundation connection.

A wide selection of precast concrete roof "bents" are available from manufacturers. The stock shapes shown above are typical of these decorative, fireproof units. Spans up to 50 ft (15 m) and above are practical and most types of long-span concrete, wood, or compressed fiber roof deck systems are compatible with this roof beam system.

PREFABRICATED METAL BUILDINGS

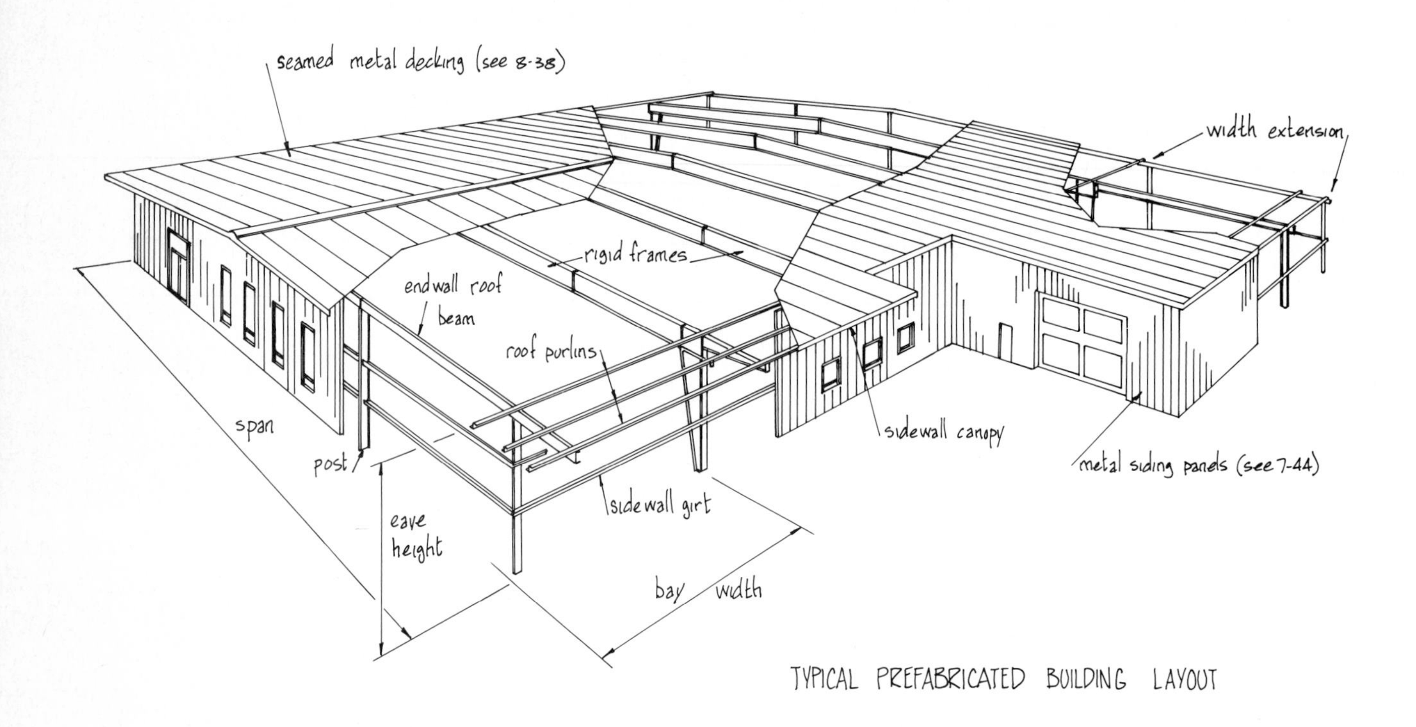

TYPICAL PREFABRICATED BUILDING LAYOUT

This type of prefabricated rigid steel frame building integrates both roof and wall support structures into one continuous system. They are used extensively for manufacturing plants, storage facilities, sports arenas, and other operations where a wide clear span is required.

- In most cases the frame manufacturer designs, fabricates, and erects the entire structure, including siding and roofing. The systems are lightweight, efficient, economical, and are erected quickly.
- There are a variety of frame styles to suit most span and load conditions. Most manufacturers have a selection of designs that can be ordered "off the shelf" with minimal additional engineering, or buildings can be totally or partially custom designed.
- When all elements of the building are totally supported by the frame, a full foundation system is normally required only under the frame's legs. However, if masonry exterior walls are specified, full perimeter foundation walls are required.
- One unique aspect of this type of building is that it can be dismantled and re-erected at another location should this be necessary.

source: Butler Manufacturing Co.

PREFABRICATED METAL BUILDINGS

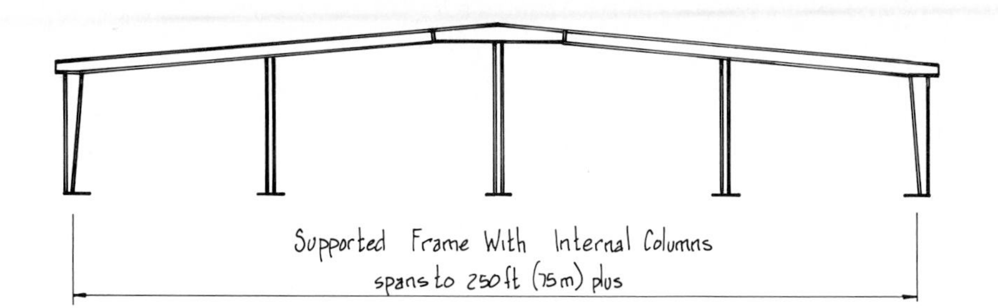

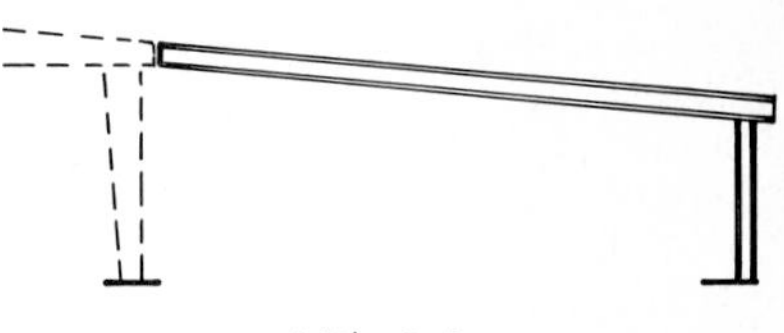

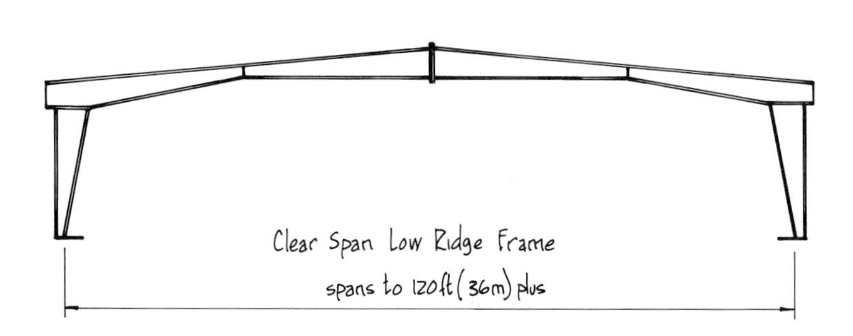

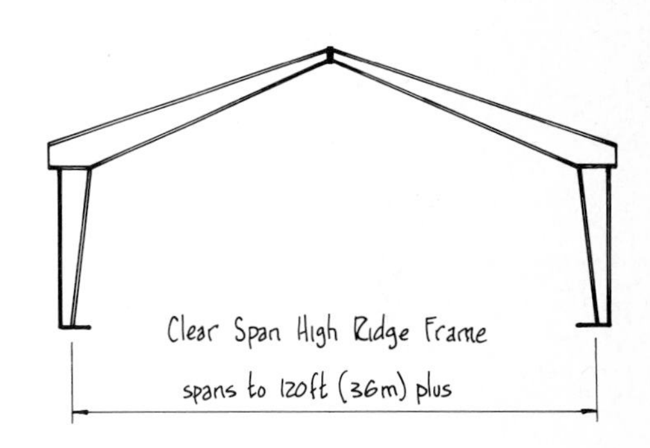

TYPICAL FRAME PROFILES

The frame styles shown above are typical of those offered by most manufacturers. Other styles and sizes are generally available to suit most design conditions. See 8-38 for roofing and 7-44 for siding.

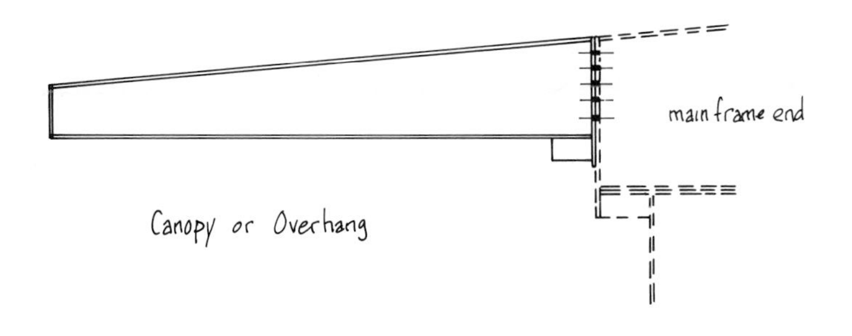

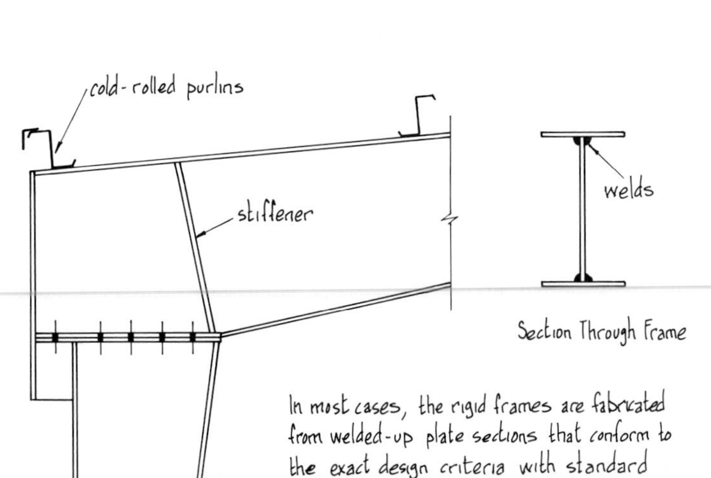

In most cases, the rigid frames are fabricated from welded-up plate sections that conform to the exact design criteria with standard hot-rolled structural steel sections being rarely used.

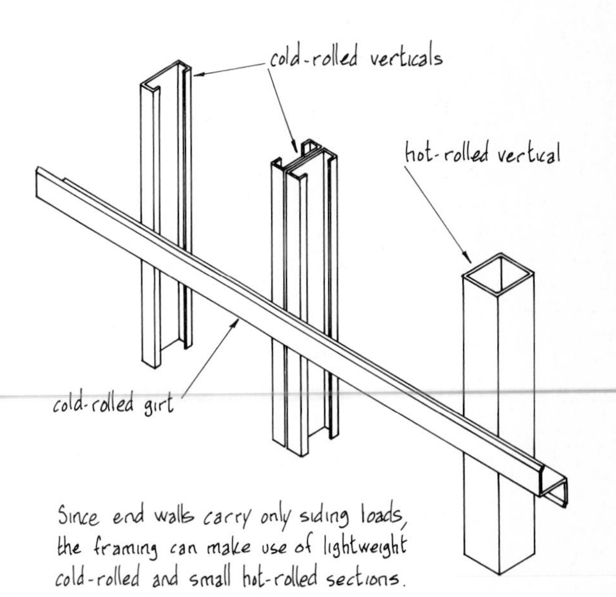

Since end walls carry only siding loads, the framing can make use of lightweight cold-rolled and small hot-rolled sections.

source: Butler Manufacturing Co.

ROOF DECKING

The decking systems shown here are typical of those used in modern construction. Many other roofing systems are available, each having specific advantages and applications. Each is chosen with the following points in mind:

- The deck chosen generally reflects the type of roof structural members, i.e., steel deck for steel-framed buildings, plywood or pressed wood fiber for wood-beamed buildings, etc.
- Any deck must be able to carry the designed loads over the intended spans without undue deflection.
- If the deck is to act as a finished ceiling, it must be visually acceptable as determined by the intended use of the building.
- The roof finish, drainage method (see 8-30) and pitch have an effect on decking choice.
- Fast building close-in is normally a design requirement, and most decking systems allow for rapid erection.

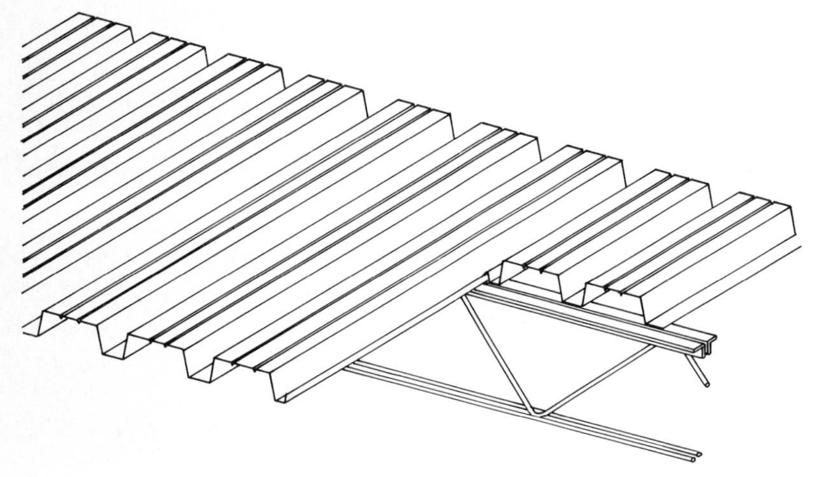

RIBBED STEEL DECKING

RIBBED STEEL DECK

- See 6-32 for examples of typical deck sections.
- Deck has interlocking side seams and is generally welded to the steel roof beams or trusses.
- To increase roof strength and provide a flat surface for the roof covering, the deck can be covered with reinforced poured concrete or gypsum concrete. Insulation and finishes are then applied (see 11-34).
- Where extra strength is not required, insulation boards and roof finish are applied directly to the deck.
- A variety of span and load conditions can be met with steel decking.

See 11-33 for other types of metal roofing.

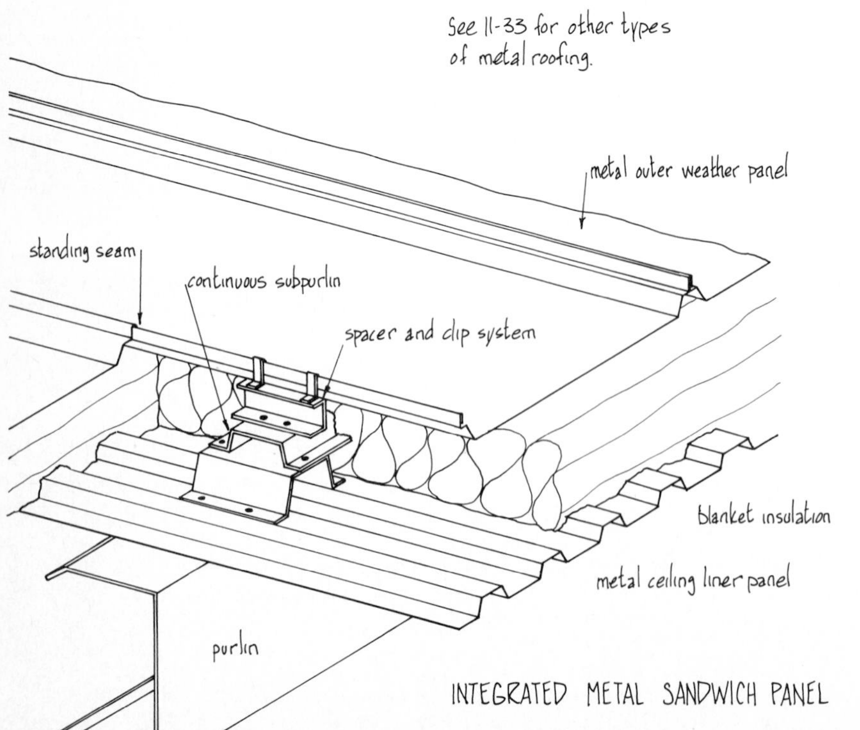

INTEGRATED METAL SANDWICH PANEL

METAL SANDWICH PANELS

- Used mostly on prefabricated metal buildings (see 8-36).
- The principal advantage is that the system is totally integrated and can provide high insulation values with up to R-50 (RSI 8.8) commonly available.
- Most systems consist of an inner ceiling liner, outer panel support system, fiberglass batt insulation, and an outer weather panel with waterproof side seams.

ROOF DECKING

CONCRETE PLANKS

- Precast concrete planks have tongued-and-grooved edges and wire mesh reinforcing. They are generally reversible, allowing either side to be used as the finished ceiling.
- Most can be nailed to the supporting structure or tied with metal clips.
- Most are not designed for long spans, with 8 ft (2400) being the general maximum.

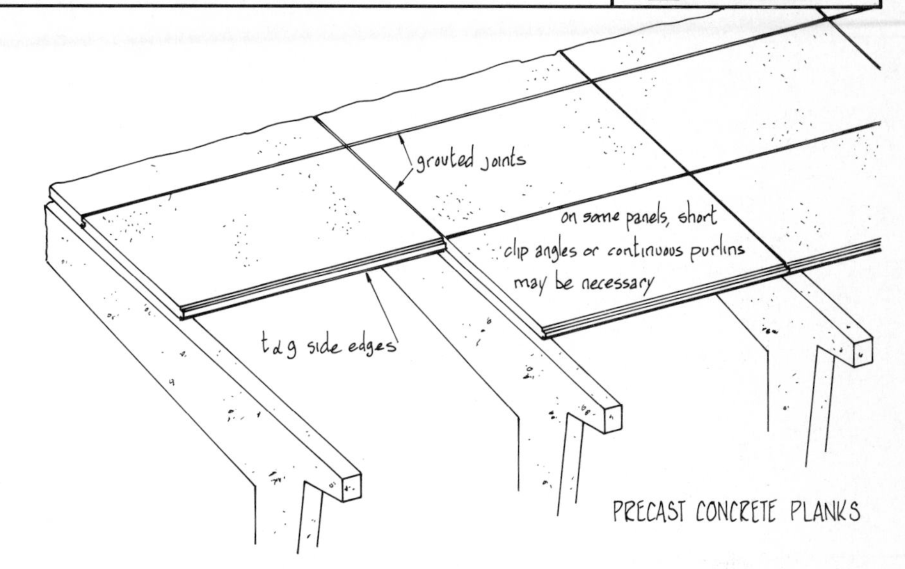

PRECAST CONCRETE PLANKS

PREFRAMED PLYWOOD ROOF PANELS

- Individual panels are prefabricated before erection using unsanded PS1 plywood panels and lumber stiffeners.
- Spans to 12 ft (3600) are common but can be 30 ft (9 m) or longer.
- Stiffeners are attached at 16- or 24-in (400 or 600) centers depending on load and span.
- Panels are strong and lightweight and are compatible with all types of wood structural systems.

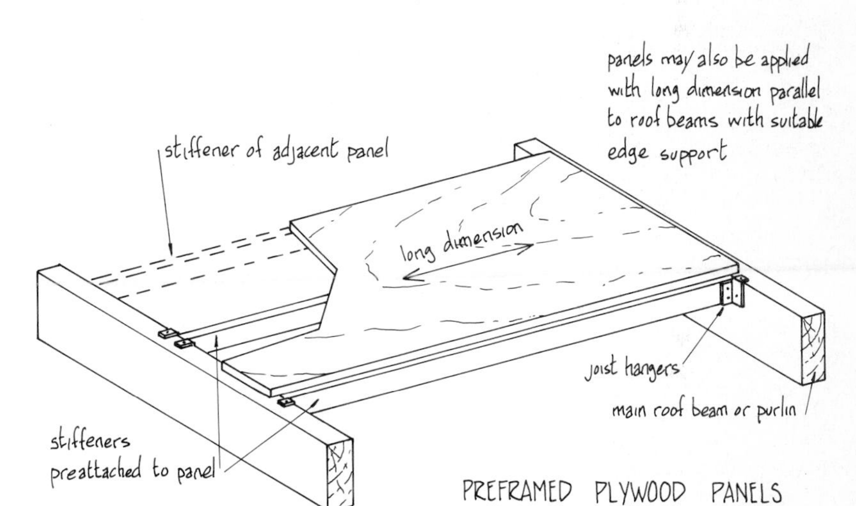

PREFRAMED PLYWOOD PANELS

PRESSED WOOD FIBER PANELS

- Panels are rigid boards made from pressed and glued wood fibers.
- They can be used with virtually all types of roof structures and provide an attractive finished ceiling.
- Panels are available with tongued-and-grooved edges to span without edge support or with square edges for use with a compatible edge support system.
- Fastening is by metal clips, direct nailing, or an edge support and grout system.

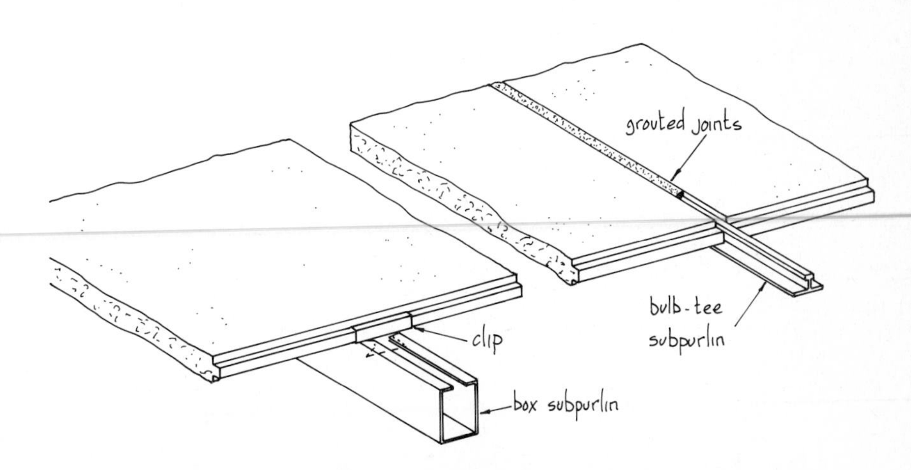

PRESSED WOOD FIBER PANELS

CHAPTER PAGE TITLES

TRANSVERSE BEAMS

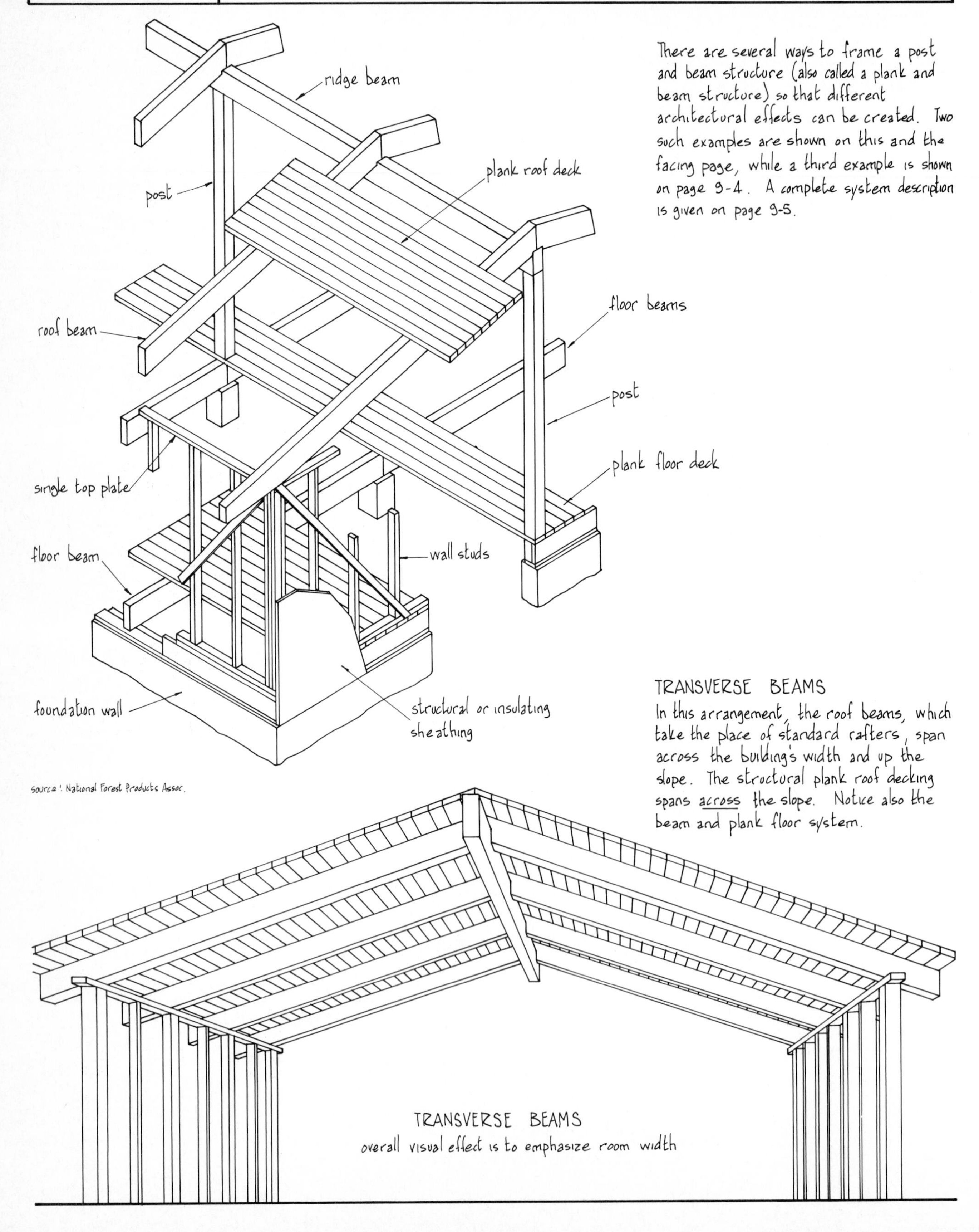

There are several ways to frame a post and beam structure (also called a plank and beam structure) so that different architectural effects can be created. Two such examples are shown on this and the facing page, while a third example is shown on page 9-4. A complete system description is given on page 9-5.

TRANSVERSE BEAMS

In this arrangement, the roof beams, which take the place of standard rafters, span across the building's width and up the slope. The structural plank roof decking spans across the slope. Notice also the beam and plank floor system.

TRANSVERSE BEAMS
overall visual effect is to emphasize room width

LONGITUDINAL BEAMS

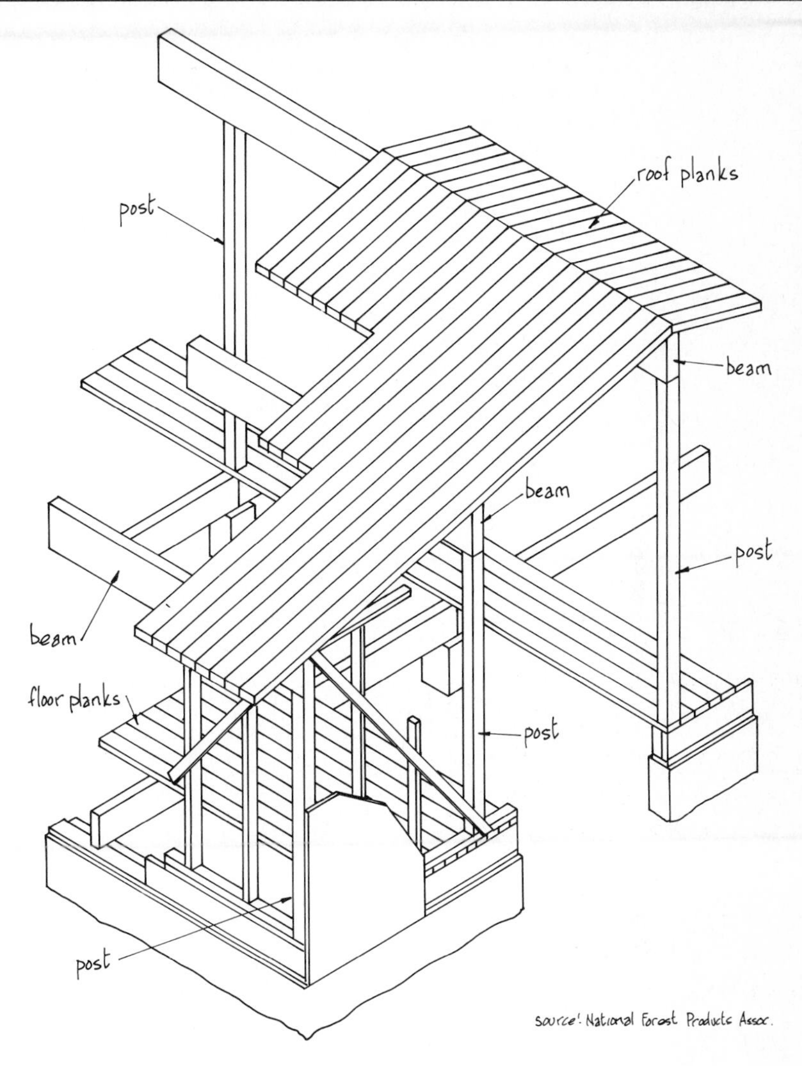

LONGITUDINAL BEAMS

In this arrangement the structural decking spans with the slope and is supported by beams running parallel to the side walls. Longitudinal beams are generally larger than those placed transversely because of the longer span. Floor beams and decking remain the same.

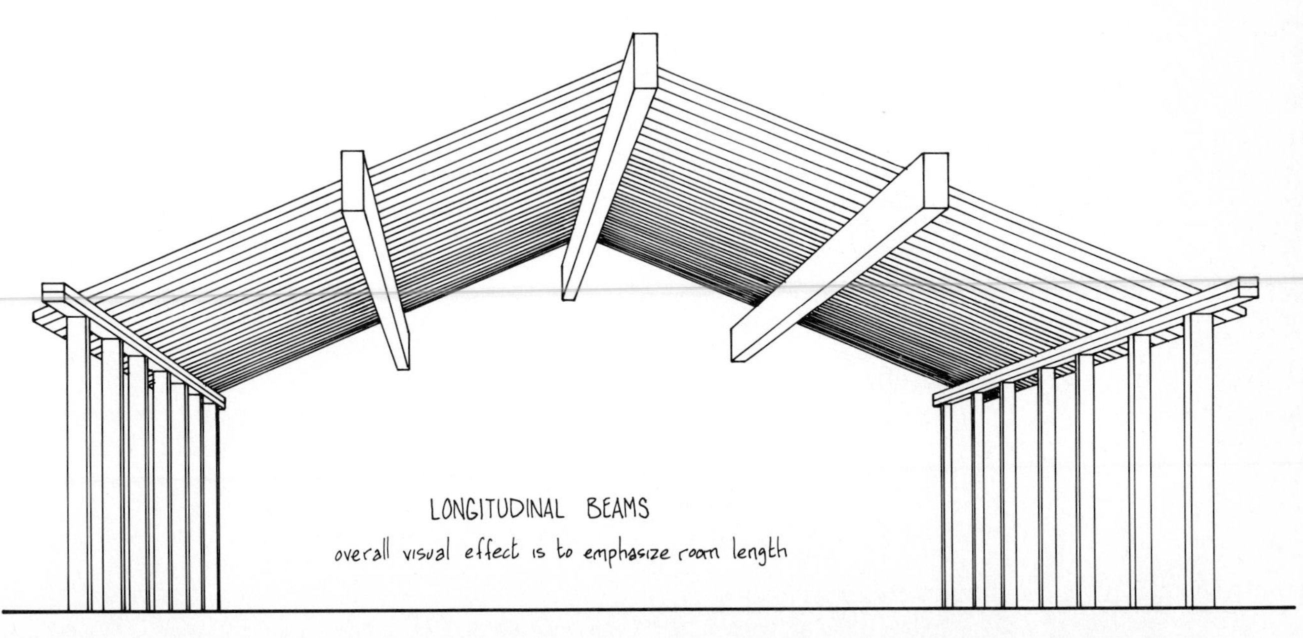

LONGITUDINAL BEAMS

overall visual effect is to emphasize room length

COMBINED FRAMING

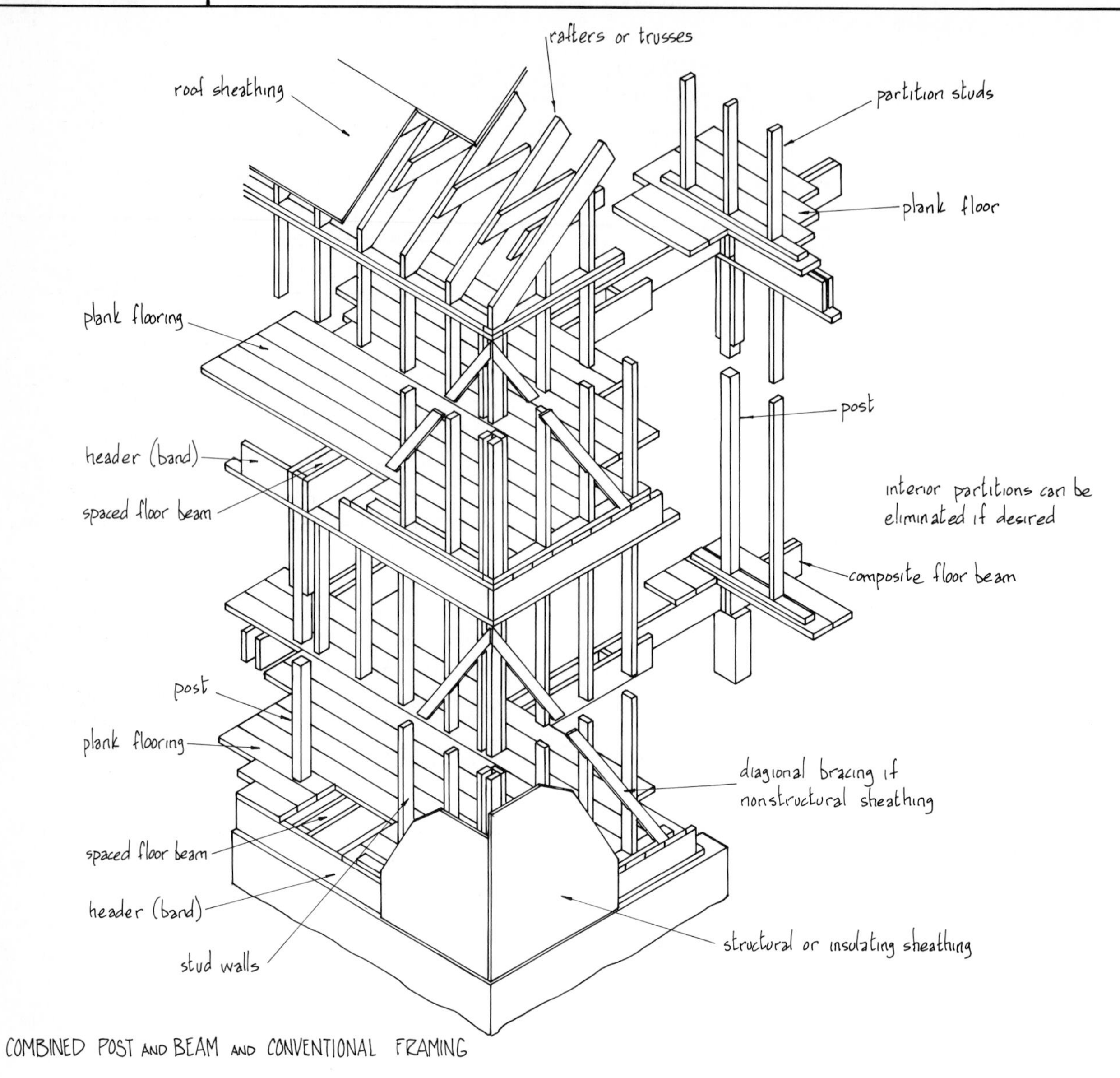

COMBINED POST AND BEAM AND CONVENTIONAL FRAMING

Post and beam techniques are often combined with conventional framing methods. This can be done in many ways, but the general approach is to capture the visual effects of post and beam while retaining the structural advantages of conventional framing.

The illustration on this page shows a two storey structure that has plank and beam floors but conventional platform wall and roof framing.
Solid floor beams can be used but, as in the example above, spaced beams (with casing, see 9-6) and composite beams are also commonly used. The rafter system (or trusses), with plasterboard and 1-in (19)-thick boards attached to the underside as a ceiling finish, allow high insolation values.

Source: National Forest Products Assoc.

STRENGTHS AND WEAKNESSES

As can be seen on the preceding three pages, a post and beam structure differs from conventional framing in the following ways:

- The structure is a series of in-line beams and posts spaced at 4- to 8-ft (1200-2400) centers.
- The beams support plank decking that functions as the structural floor or roof. The decking also functions as the finished floor or ceiling after suitable treatment.
- All (or most) of the structural members of the frame are intended to be visible. This is quite opposite to conventional framing, where all structural members are hidden behind the finishes.

Post and beam framing, like all other framing systems, has strengths and weaknesses. Some of the more important points are:

STRENGTHS

Visual Effects

- Post and beam structures have wide design flexibility. Strong, creative architectural effects are possible when advantage is taken of spacious interiors and wood textures.
- Partitions within the structure do not normally carry loads and can be eliminated or placed to best advantage with only minimal floor support (see 9-10).
- Since headers are not required between posts, large areas of glass inside and out are possible. It should be noted, though, that wind bracing, especially on walls, is essential. This is normally achieved by inserting conventionally framed and cross-braced panels between some posts.
- The timber structure can be finished in a variety of ways to reflect intended design effects. Staining, varnishing, or painting is standard.

Erection

- Post and beam structures are generally faster to erect since there are considerably fewer members in the structure than there are in conventional framing.
- The system also lends itself to prefabrication techniques. With careful design, most members can be shop- or site-fabricated, allowing erection to proceed smoothly and quickly.

Other Factors

- Foundations for post and beam structures can be single piers located under each post (see 5-21) or continuous foundation walls (see 5-10 to 5-16).
- Wall overhangs are easily obtained by extending the main beams past the wall line.
- As can be seen on the preceding page, integration of this system with conventional framing is very easy, allowing the strengths of both systems to be used to best advantage.

WEAKNESSES

Any weaknesses in the system should be considered only as limitations, most of which can be overcome with careful planning and design.

- For most regions of the country heat loss or heat gain through the structure is a prime consideration. In the past, roofing was generally applied directly to the plank roof decking without additional insulation material. Today's power costs for heating and cooling demand that additional steps be taken to reduce fuel bills. For roofs this means adding insulation above the decking if the decking is to be the ceiling finish (see 10-21) while the walls may need more conventionally framed and insulated panels between posts.
- There is less space in post and beam framing for electrical and mechanical services since the structure is visible, solid, and widely spaced. Careful planning, together with spaced beams and posts plus stud partitions, may be necessary to conceal services.
- The plank floors are designed for moderate uniform loads. Heavy or concentrated loads from bearing walls, bathtubs, appliances, or other items, will require extra framing where such loads occur.
- Solid timber beams in large sizes or long lengths may be either hard to obtain or prohibitively expensive in some regions, forcing the use of made-up members.
- Greater care than is normal must be taken by all trades to avoid unnecessary damage to structural members during erection. The final visual effects are most important and most of the carpentry work must be regarded as "finish work."

MATERIAL TYPES

BEAMS | POSTS

SOLID SAWN TIMBER

Easiest to work with but may be hardest to obtain. Lengths and sizes available will vary with species and job location but can be specially ordered from most suppliers. Can be used either rough sawn or dressed, depending on desired visual effects. Must be properly seasoned to avoid twisting and checking.

LAMINATED SOLID LUMBER

Glue-laminated members are stronger than sawn solid timber and can be supplied as a completely finished item. They are dimensionally stable, can be straight, curved, or tapered, and fabricated in any size or length. Most laminated members are custom fabricated but some types are available off the shelf.

BUILT-UP LUMBER

post as above

Used when solid lumber is uneconomical or unobtainable. Made up of standard dimensional lumber well nailed together. These beams and columns are generally "cased" with trim lumber to improve appearance (see below).

SPACED BEAMS AND POSTS

Spaced beams are lightweight and easy to handle. Made from dimensional lumber, they can be sized to meet most medium span and load conditions. When cased, these beams and posts are ideal for concealing electrical and mechanical services.

PLYWOOD BOXED-BEAMS

Generally fabricated from a "ladder"-type dimensional lumber truss, boxed on both sides with plywood panels. These beams are lightweight and strong, and can be fabricated in almost any shape and size. Standard wood trusses (see 6-20) or wood I-beams (see 6-18) can also be used as the core beam.

BEAM CASING AND FINISHING

The undersides of spaced, built-up, or other nonsolid beams need to be finished for visual effect. Several methods of casing are available, two of which are shown here. Posts can be finished in the same manner. Solid beams, if not dressed, may need chamfered edges as shown to improve appearance.

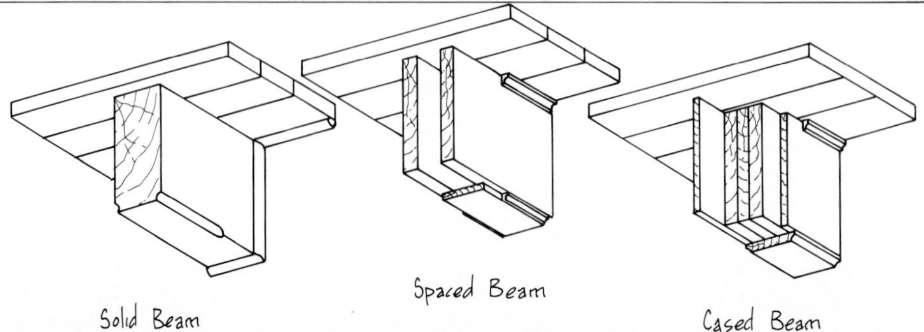

PLANKS - MATERIALS, SPANS, FINISHES

In post and beam construction the floor and roof decking is almost always left exposed as a finished ceiling and often used as a finished floor. Decking lumber must therefore be of good quality, correctly seasoned to prevent cracking and separation, and carefully installed to avoid damage. Planks are produced in many sizes and styles, four examples of which are shown at the right.

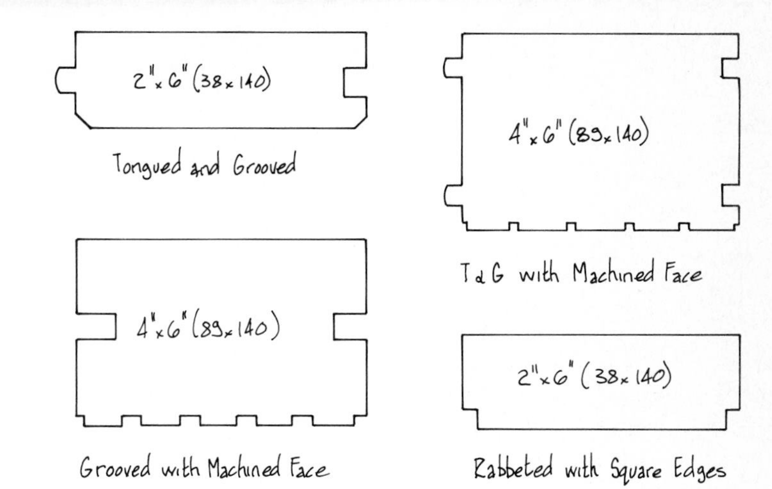

The planks can be designed to span across supports in several ways and the method chosen has an impact on the strength and deflection of the deck. Basic span choices are:

- Simple span, where planks span only between each support. This method has the least stiffness.
- Two-span continuous is the most efficient of the fixed-span methods and has the least deflection.
- Controlled random pattern allows variable-length planks to be used provided that end joints are staggered. With this method nonstandard or variable support spacing can be used. Has a medium-value stiffness.

Size, species, and type of decking chosen depends on local codes, loading conditions, spans, and desired finished appearance. Check manufacturers' span tables for sizing information and plank availability.

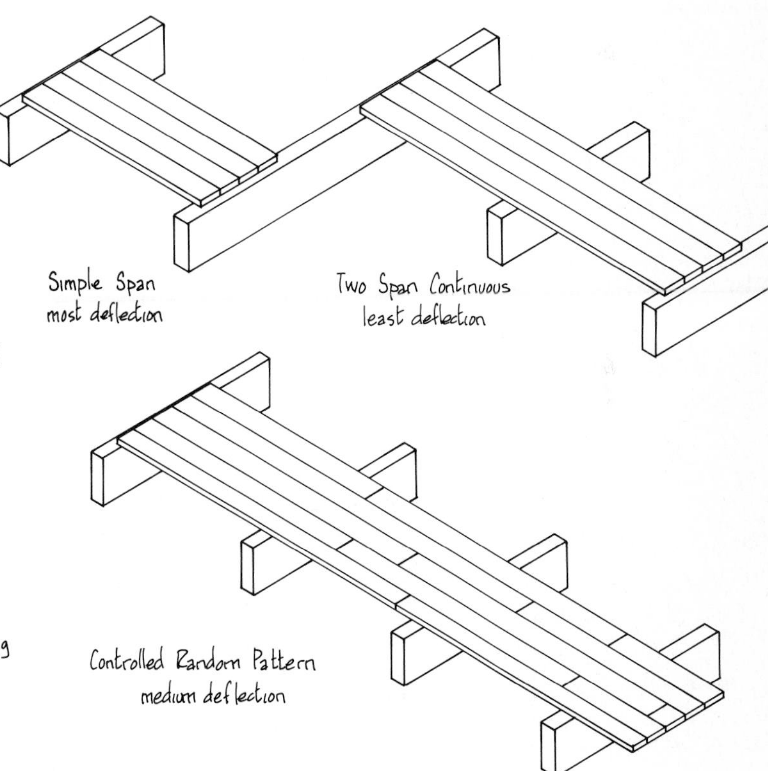

The visual effects for ceilings created by different decking types are many and varied. Most effects derive from the treatment of plank joints and four typical examples are shown below. Provided that the lateral distribution of deck load is not compromised, many other plank joints are possible.

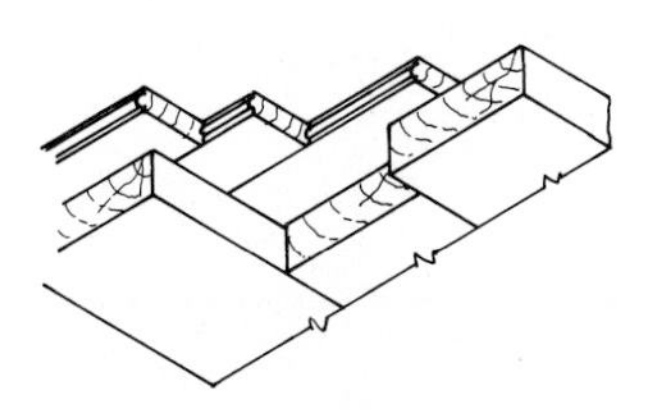
Square Edge

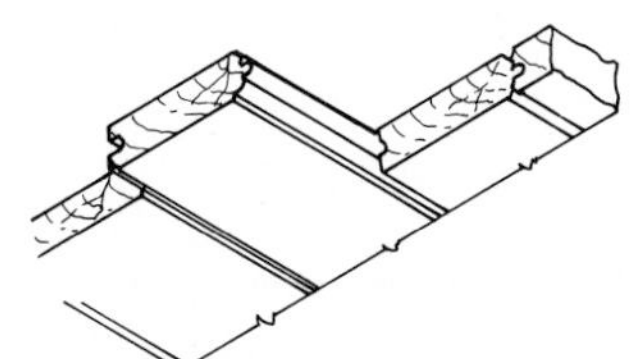
Tongued and Grooved

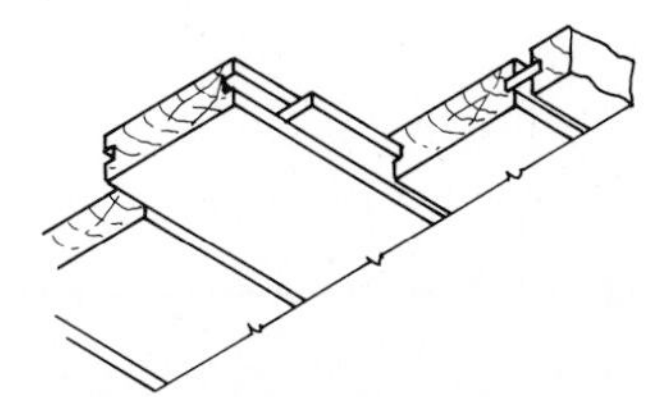
Grooved with Spine

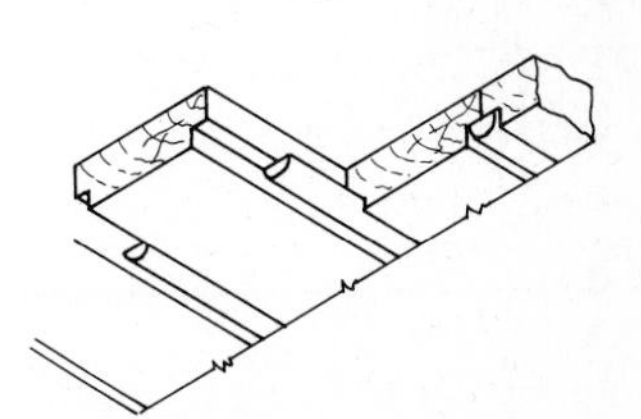
Rabbeted

CONNECTION DETAILS

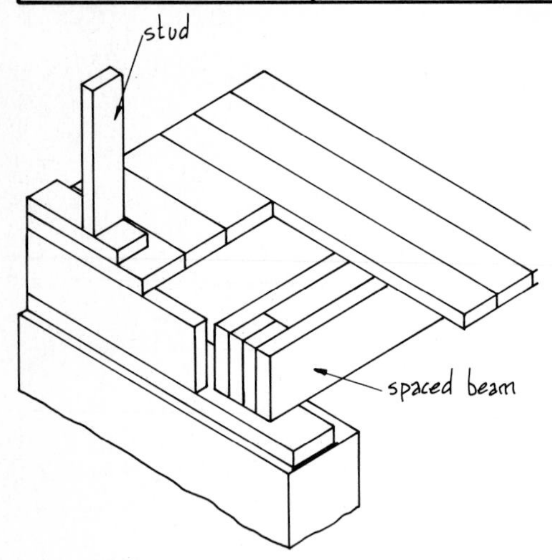

First Floor - Exterior Wall - Spaced Beam On Sill

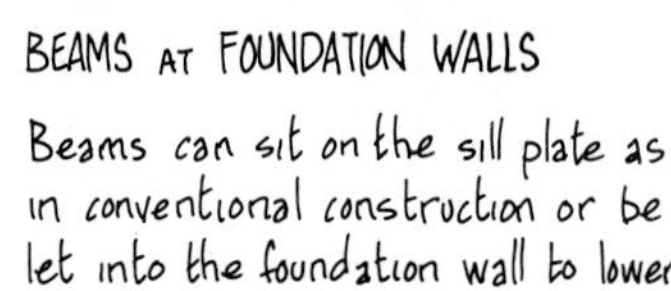

BEAMS AT FOUNDATION WALLS

Beams can sit on the sill plate as in conventional construction or be let into the foundation wall to lower the floor level in relation to outside grade.

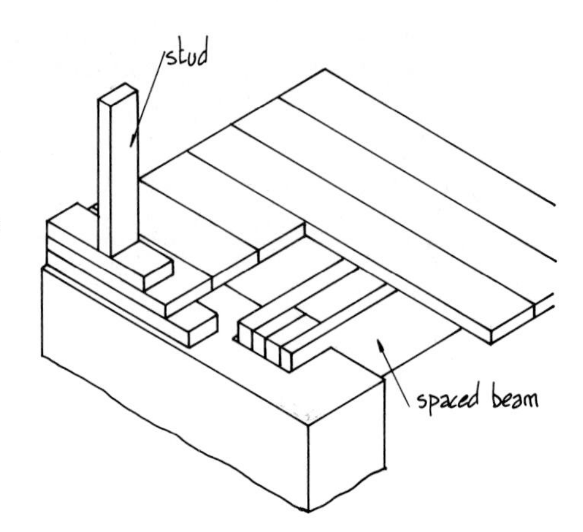

First Floor - Exterior Wall - Spaced Beam in Pocket

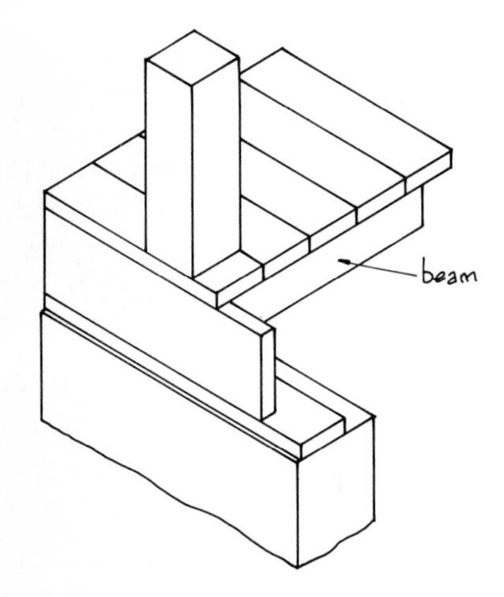

Post on Plank Floor at Beam Junction

See pages 9-12 and 9-13 for typical fasteners.

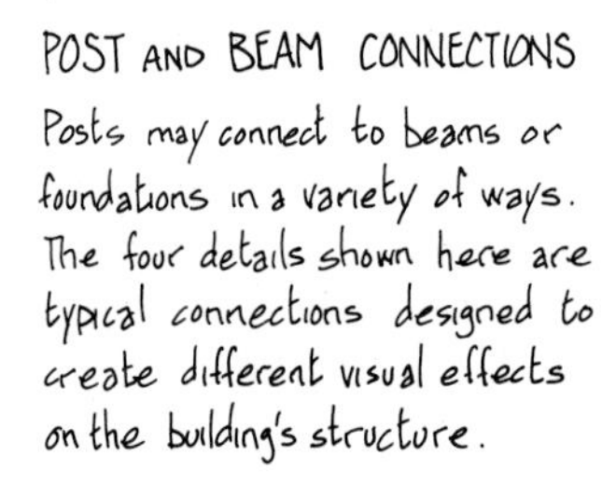

POST AND BEAM CONNECTIONS

Posts may connect to beams or foundations in a variety of ways. The four details shown here are typical connections designed to create different visual effects on the building's structure.

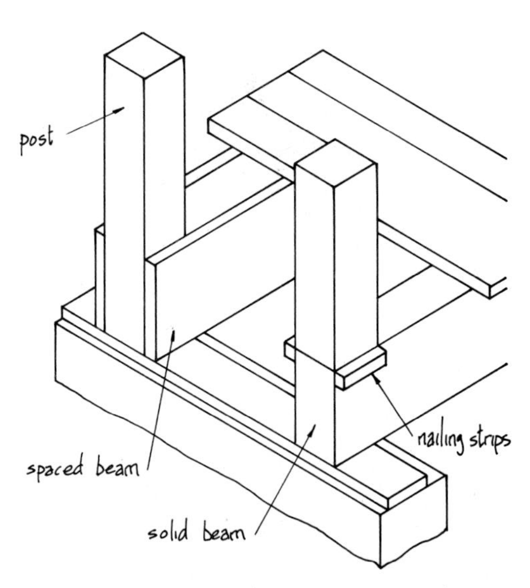

Posts at Wall Line and Beam Ends

Cantilever Beams with Post at Wall Line

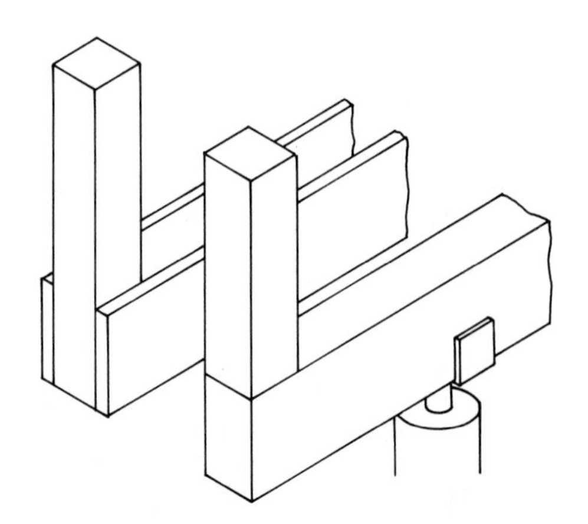

Posts at Cantilever Ends

CONNECTION DETAILS

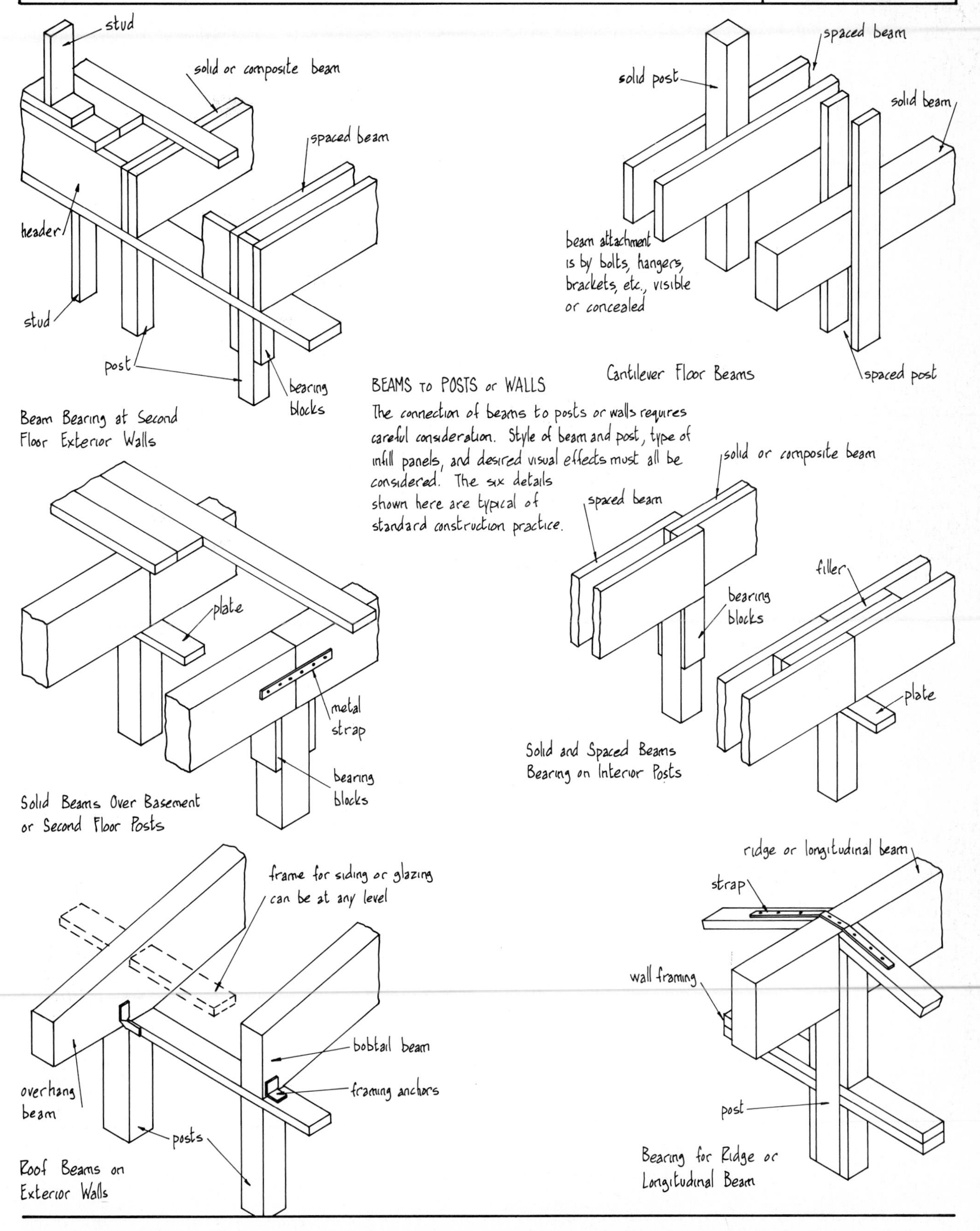

BEAMS TO POSTS or WALLS

The connection of beams to posts or walls requires careful consideration. Style of beam and post, type of infill panels, and desired visual effects must all be considered. The six details shown here are typical of standard construction practice.

PARTITIONS AND RAFTERS

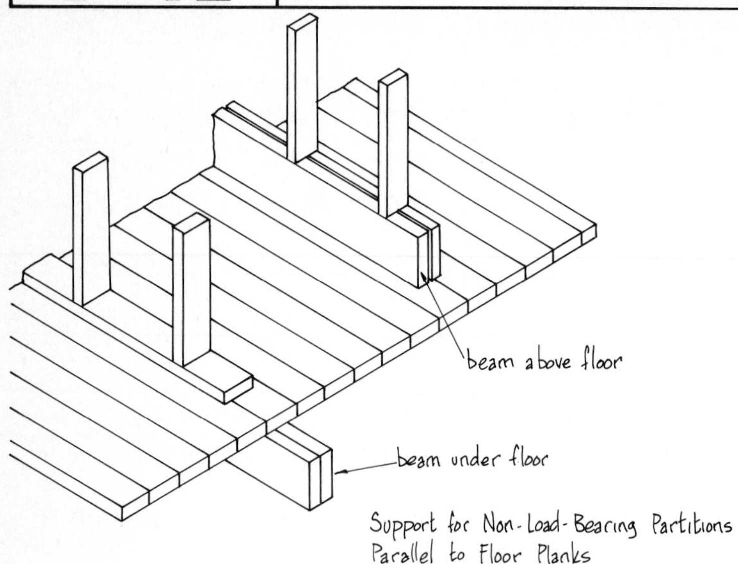

Support for Non-Load-Bearing Partitions Parallel to Floor Planks

PARTITION SUPPORT

Non-load-bearing partitions that run parallel to floor planks need extra support. The two methods shown at the right are generally the most practical. Non-load-bearing partitions that run at right angles to the planks do not normally need extra support.
For load-bearing partitions a full-sized beam or other adequate support system will be required.

See 9-12 and 9-13 for typical fasteners.

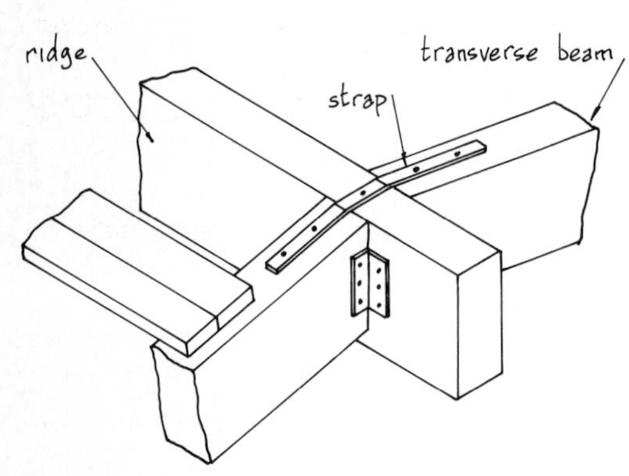

Beams Into Ridge

RAFTER SUPPORT

On all rafter systems there must be some means of absorbing horizontal thrust developed at the wall line.
Ideally, the rafters will be connected to a supporting ridge beam and each other so that horizontal thrust does not develop. Typical connections are shown in the two middle details.
Where a ridge beam is either impossible or undesirable, a horizontal tie must connect the beams at wall plate level. The two lower details show possible approaches to this problem.

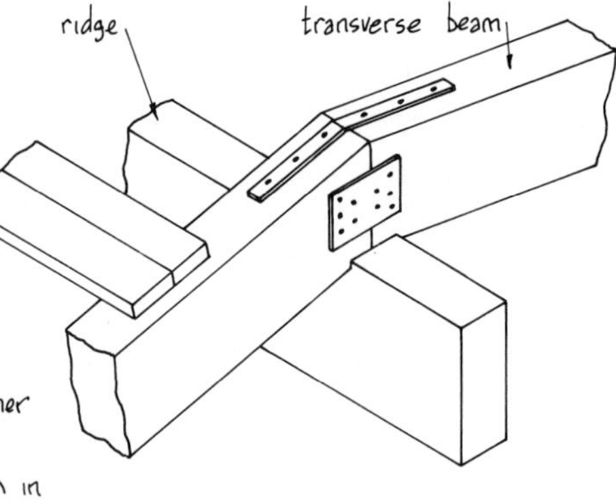

Beams On Ridge

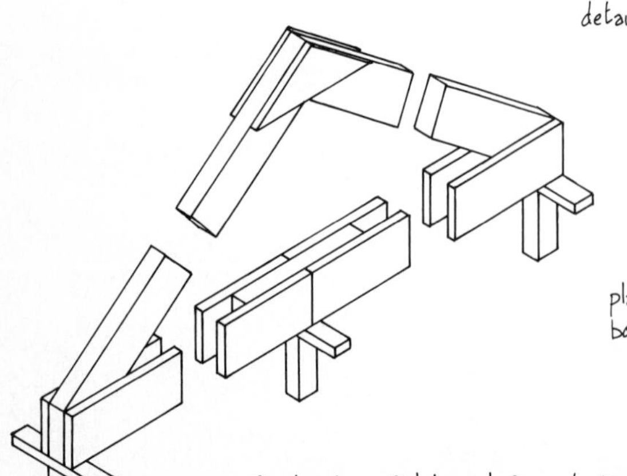

plated, nailed, or bolted connections

Combination Solid- and Spaced-Beam Truss

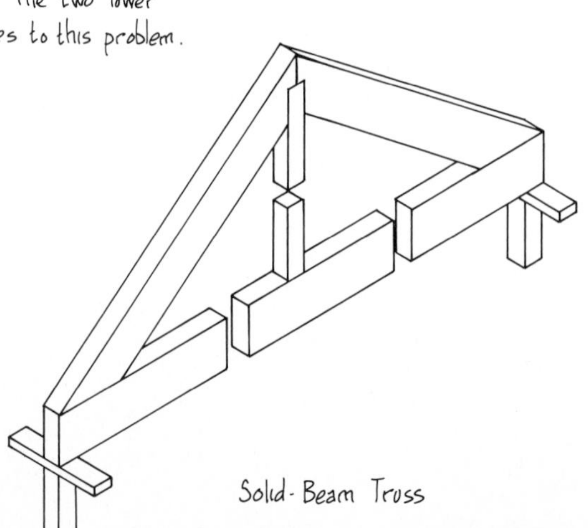

Solid-Beam Truss

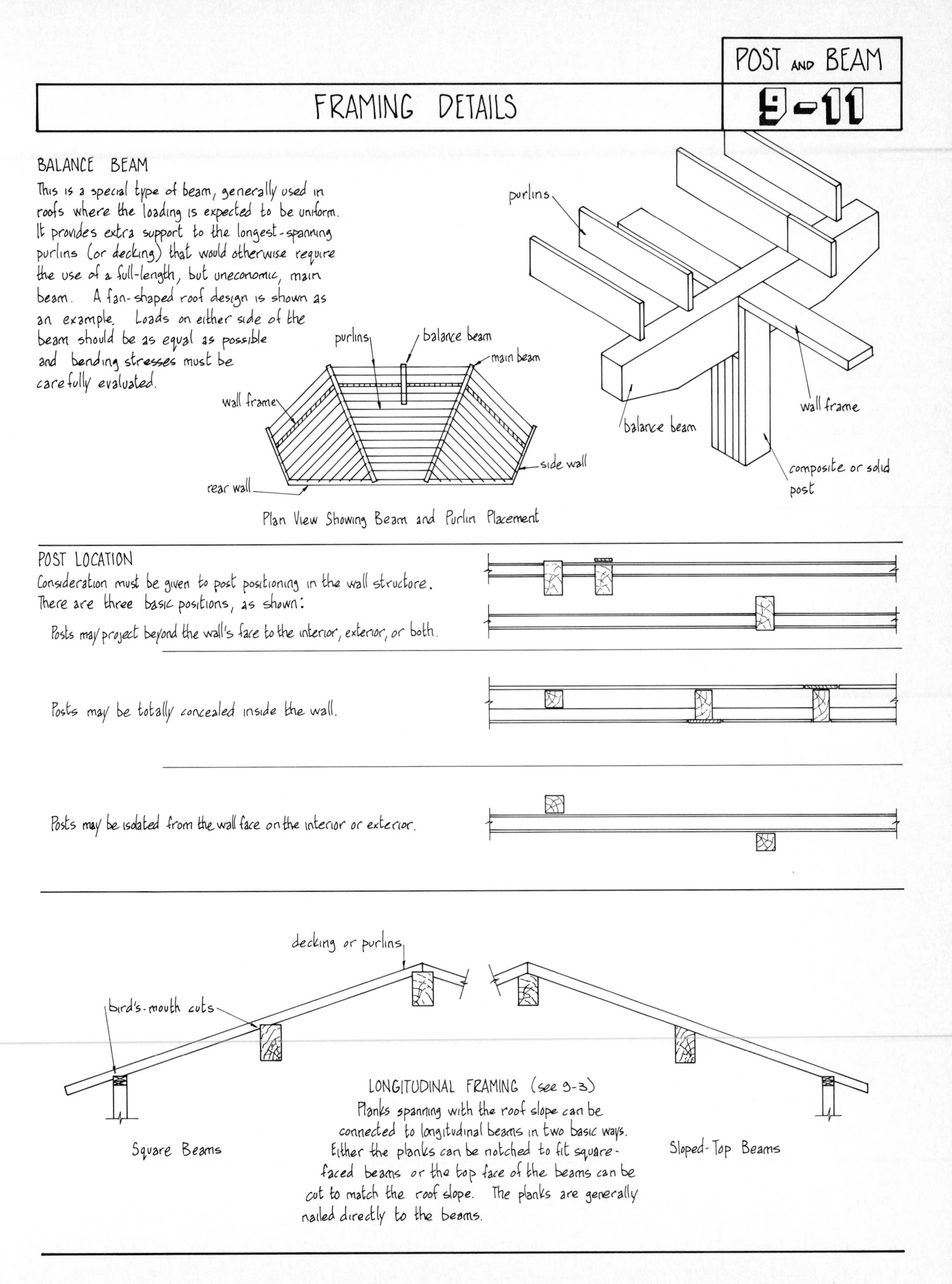
POST AND BEAM
FRAMING DETAILS
9-11
BALANCE BEAM
This is a special type of beam, generally used in roofs where the loading is expected to be uniform. It provides extra support to the longest-spanning purlins (or decking) that would otherwise require the use of a full-length, but uneconomic, main beam. A fan-shaped roof design is shown as an example. Loads on either side of the beam should be as equal as possible and bending stresses must be carefully evaluated.
purlins
balance beam
main beam
wall frame
side wall
rear wall
Plan View Showing Beam and Purlin Placement
purlins
balance beam
wall frame
composite or solid post
POST LOCATION
Consideration must be given to post positioning in the wall structure. There are three basic positions, as shown:
Posts may project beyond the wall's face to the interior, exterior, or both.
Posts may be totally concealed inside the wall.
Posts may be isolated from the wall face on the interior or exterior.
decking or purlins
bird's-mouth cuts
LONGITUDINAL FRAMING (see 9-3)
Planks spanning with the roof slope can be connected to longitudinal beams in two basic ways. Either the planks can be notched to fit square-faced beams or the top face of the beams can be cut to match the roof slope. The planks are generally nailed directly to the beams.
Square Beams
Sloped-Top Beams

FASTENINGS

Unlike conventional construction, post and beam structures have a limited number of joints. This places a greater load on each joint, and care must be taken to adequately reinforce connections between members. The details on this page and the next show some typical connections, both visible and concealed.

POSTS to FOUNDATIONS or FLOORS

- Since all wall, upper floor, and roof loads are carried to the foundations by the posts, carefull consideration must be given to this joint.
- Load should be spread over the post's cross-sectional area and the post must be restrained from any lateral movement.
- Whether the post connection is visible or concealed will, to a large extent, determine the type of fastening used.
- Fasteners include plates, straps, hangers, bolts, lag screws, dowels, and pegs.
- Posts exposed to weather should not be in contact with foundations, and in wet regions the use of a wood preservative is advisable.

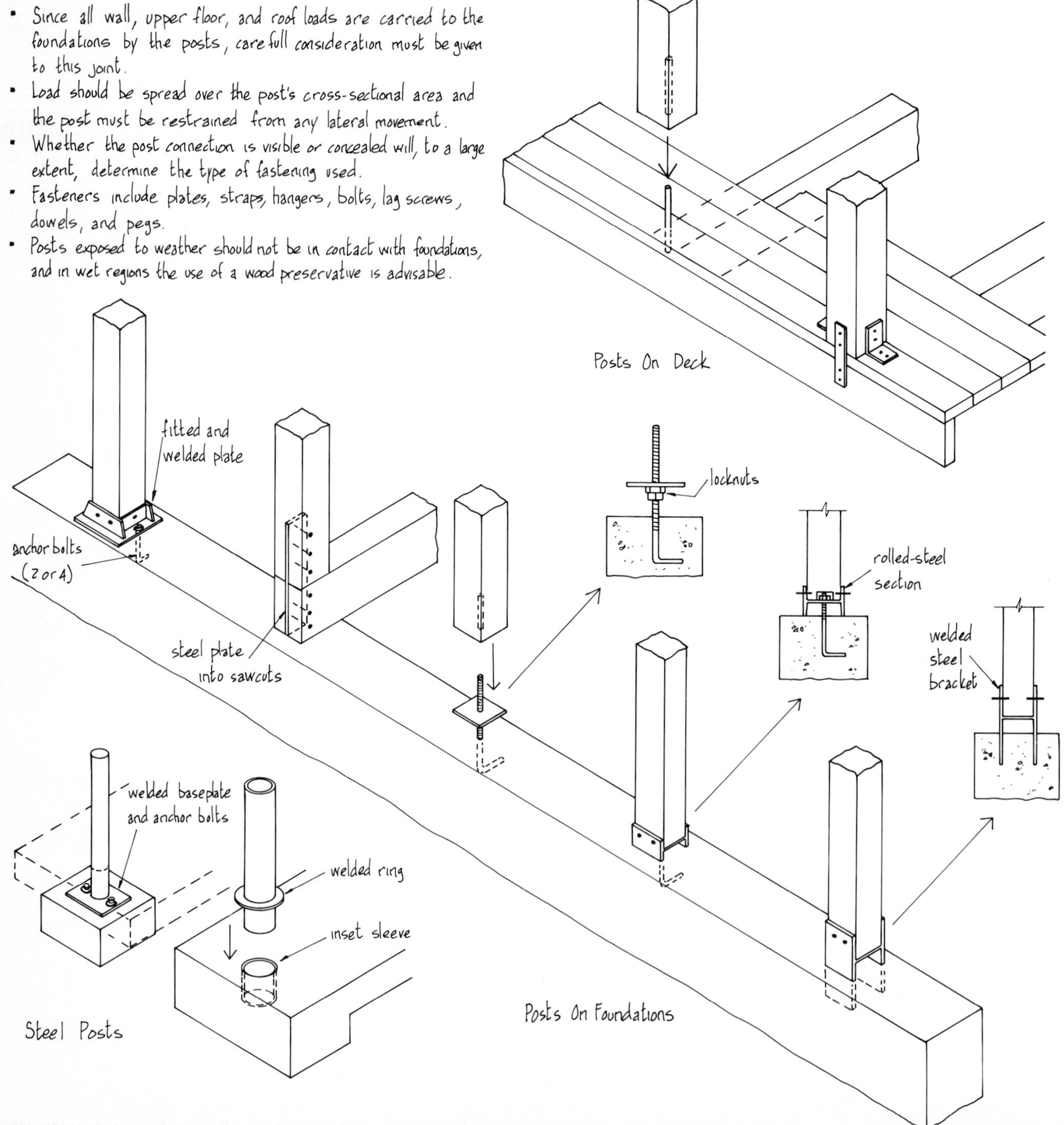

FASTENERS

POST to BEAM - BEAM to BEAM

- As with post to foundation fastening, loads must be transmitted adequately and lateral movement restrained.
- Where visible fastenings will detract from finished appearance, use notches, pins, dowels, or other concealed fastenings. Bolts or lag screws can be set in countersunk and plugged holes.
- To prevent roof uplift in regions where high wind loads are common, special fastenings and connections may be specified by local codes.

All mechanically fastened connections should be screwed or bolted rather than nailed.

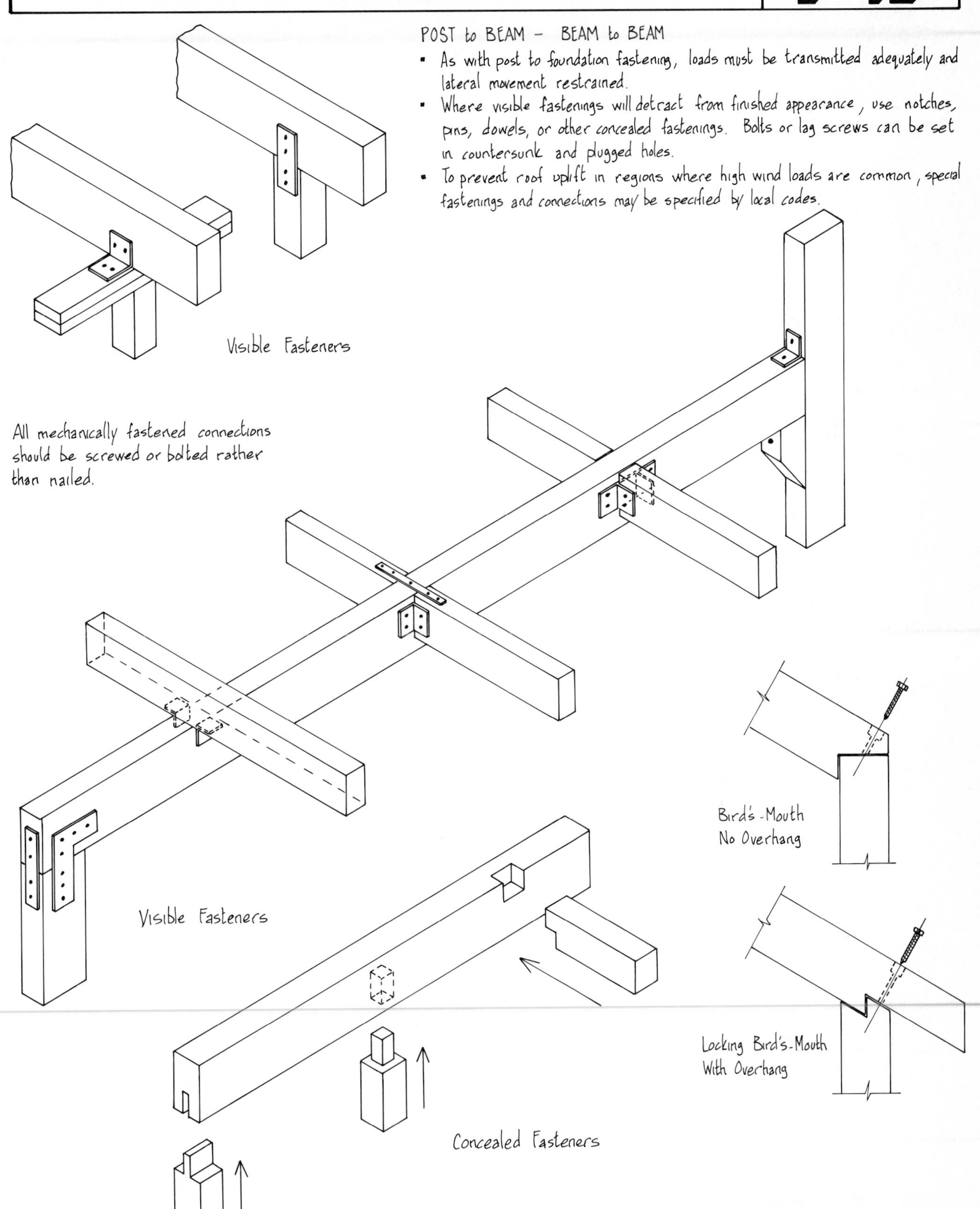

SOLID TIMBER FRAME

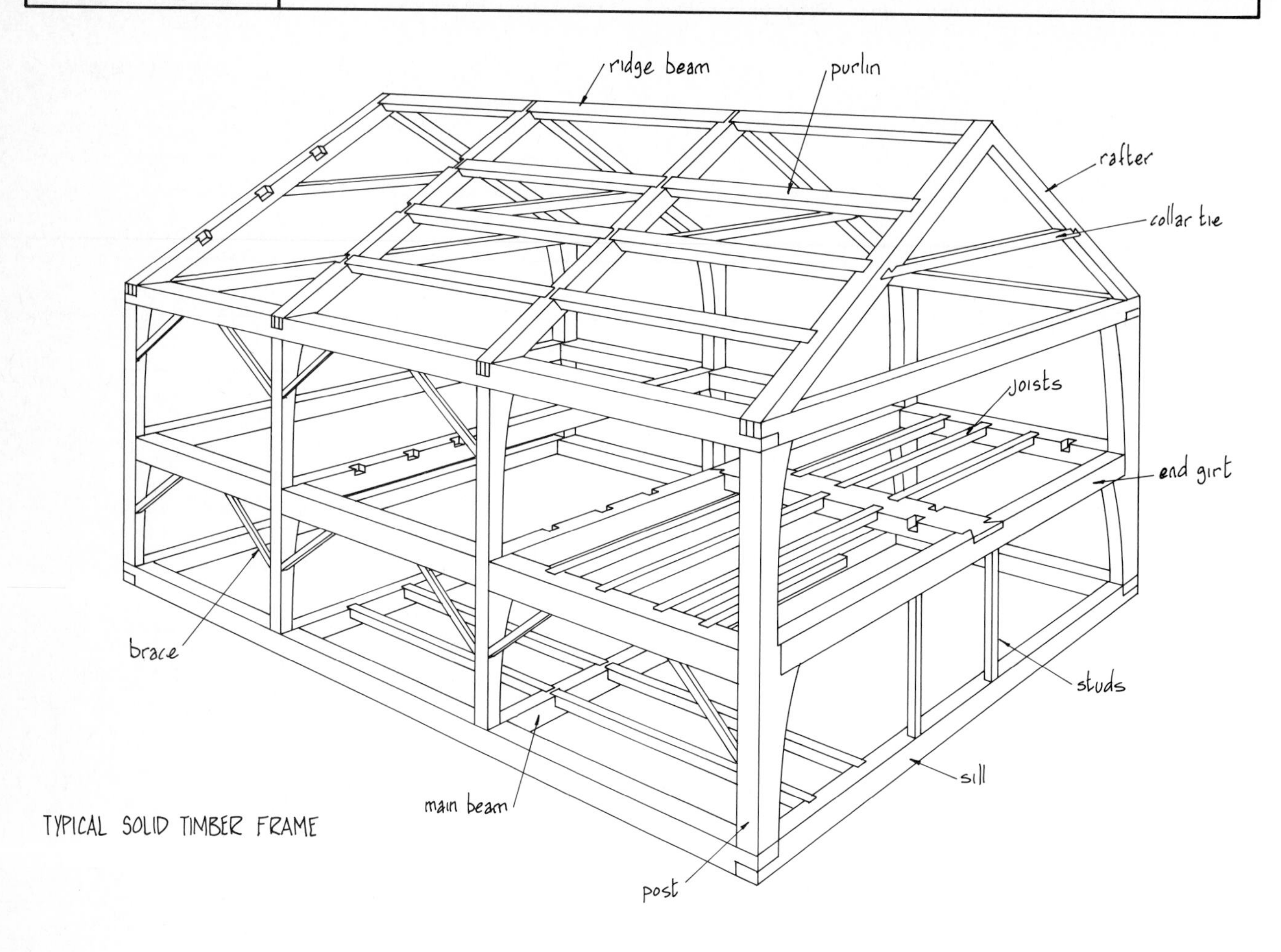

TYPICAL SOLID TIMBER FRAME

The information on this page and the next is presented to show the type of structure that is available to homeowners who desire what could be called the ultimate in timber framing and the exact opposite of conventional "mass-produced" housing.

The heavy, solid timber frame shown above is similar to the type of framing system employed by New England colonists in the mid-1700s and follows the same basic traditional proportions, structural details, ceiling heights, and roof pitches. Although this type of structure is often designed, fabricated, and erected by specialist contractors, many are built by individual owner-builders as a personal statement of good design and craftsmanship.

Some pertinent facts are:

- The chief advantages of this type of frame is its aesthetic beauty, great strength, and extended longevity.
- To be relatively economical, most are built in regions where large timbers are readily available. "Green" pine is generally used since seasoned timber in the sizes needed is unobtainable. The advice of local saw mills should be sought before design begins.
- All members of the frame must be individually detailed and precut before erection takes place. This requires very careful planning if erection is to proceed smoothly.
- The craft of joinery is employed to shape the ends or sides of timbers into interlocking units. Although some joint notches appear quite large, provided that the wood removed from one timber is totally replaced by the wood of the connecting timber, the full strength of the original timber will not be significantly reduced.
- Very few joints, if any, are connected with metal fasteners. Most are connected with hardwood pegs that tie the self-locking joints together. Peg size varies with the size and thickness of joints.

Shown on the next page are a small selection of typical joints used in this type of frame.

TIMBER JOINTS

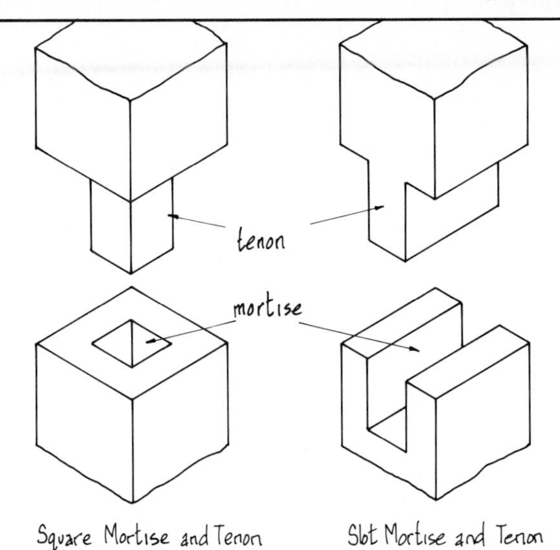

SIMPLE MORTISE AND TENON

- The mortise is the hole or slot in the receiving timber.
- The tenon is the projection that fits into the mortise.

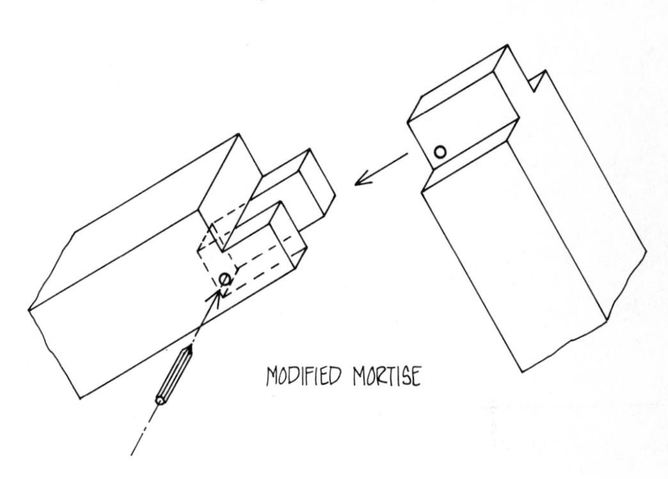

MODIFIED MORTISE

- This particular joint is used at the rafter peak to connect each pair of rafters to the ridge beam.

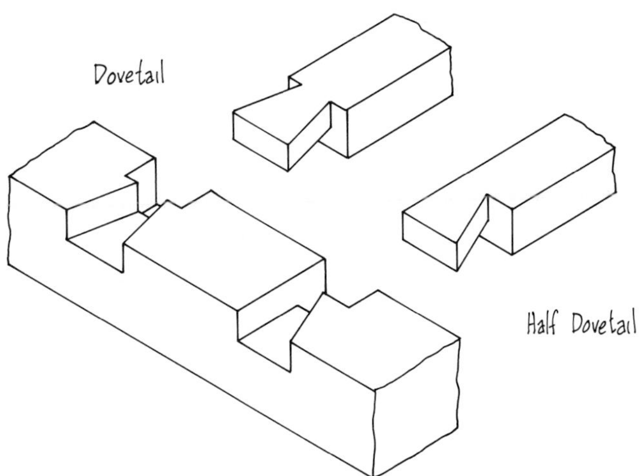

SCARF

- Used to join two timbers end to end. The joint is interlocking and needs no post beneath. Other types of scarf joints may need a supporting post.

DOVETAIL

- A full dovetail is the strongest joint for connecting two timbers at right angles.
- The half dovetail is used to connect collar ties to rafters, sill corners, and corner bracing to posts or girts.

Note that many other types and variations of joints are employed than are shown here. Consult specialized handbooks for more information.

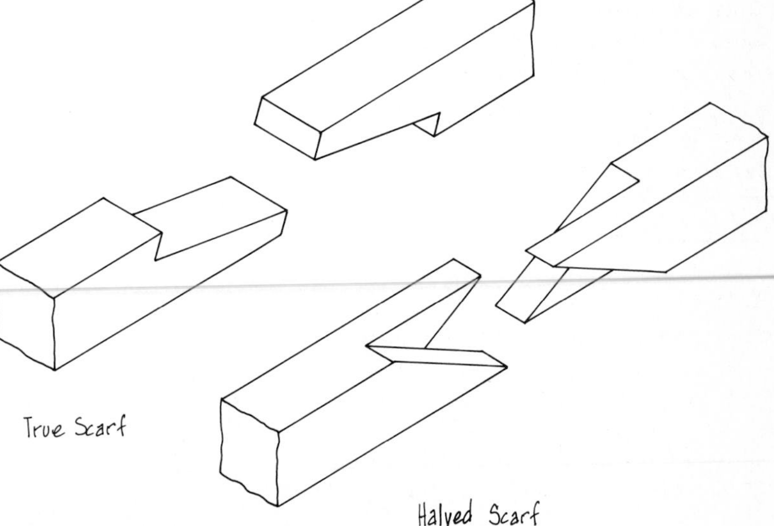

CHAPTER PAGE TITLES

Note to Instructors:
A complete course outline entitled "Building for Energy Efficiency" is presented in the Instructors Manual for this text. ISBN reference 0-13-453243-0

ENERGY CONSERVATION

In recent years energy conservation has become an extremely important part of the building design process. While the ultimate goal of the designer is to achieve maximum human comfort, the total cost of that comfort must also be considered. The designer must be able to balance the many factors that will influence the final design in a way that will ensure acceptable comfort at reasonable cost.

From the point of view of the building's owner the important factors are:

- Energy costs have been increasing faster than the rate of inflation – a situation that is likely to continue.
- Mortgage costs are relatively stable and in terms of real dollars, tend to decline over the long term.
- In view of these factors, it makes good economic sense to "invest" in energy conservation measures that will yield increasing benefits over the life of the building. The added mortgage cost of increased conservation is offset to a large degree by energy savings.
- The two charts at the right compare the simplified energy costs of a standard and an upgraded structure.

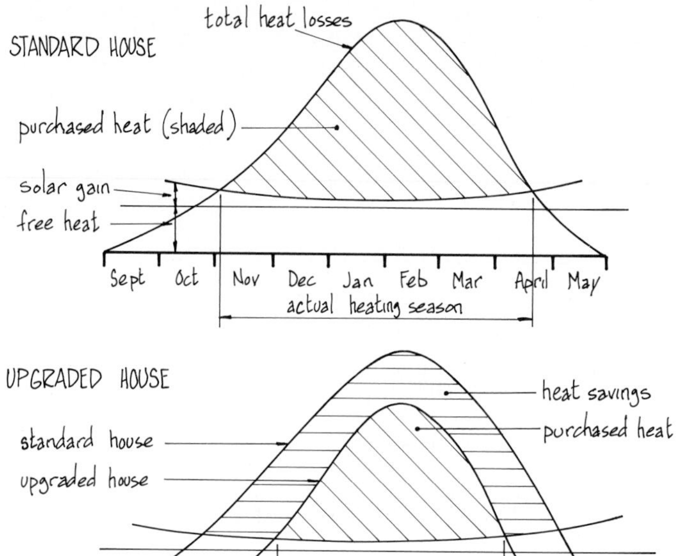

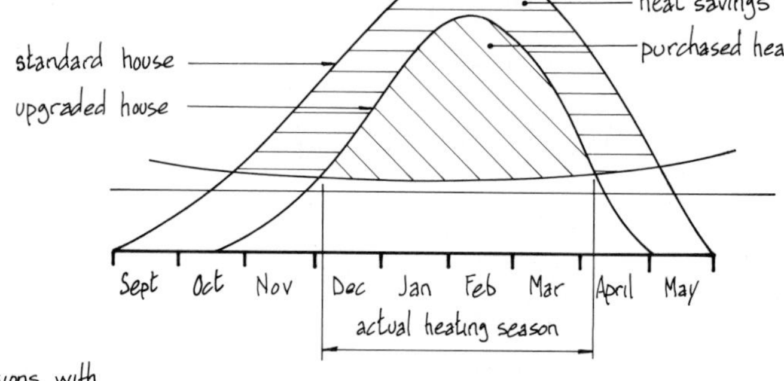

Note: Although the details in this chapter are intended for regions with heating as the primary energy cost, they apply equally, with only minor modification, to hot and humid regions requiring extensive air-conditioning.

From the point of view of the designer the factors to be considered are shown at the right. As can be seen, energy efficiency is not simply a matter of adding more insulation to a structure. All the factors must be balanced with each other in terms of their effect on building function as well as their impact on building cost.

This chapter concentrates on illustrating the two major areas of building construction: air tightness and insulation values. The third area, cost factors, is influenced by the chosen design features and intended marketing area and is therefore the ultimate responsibility of the designer.

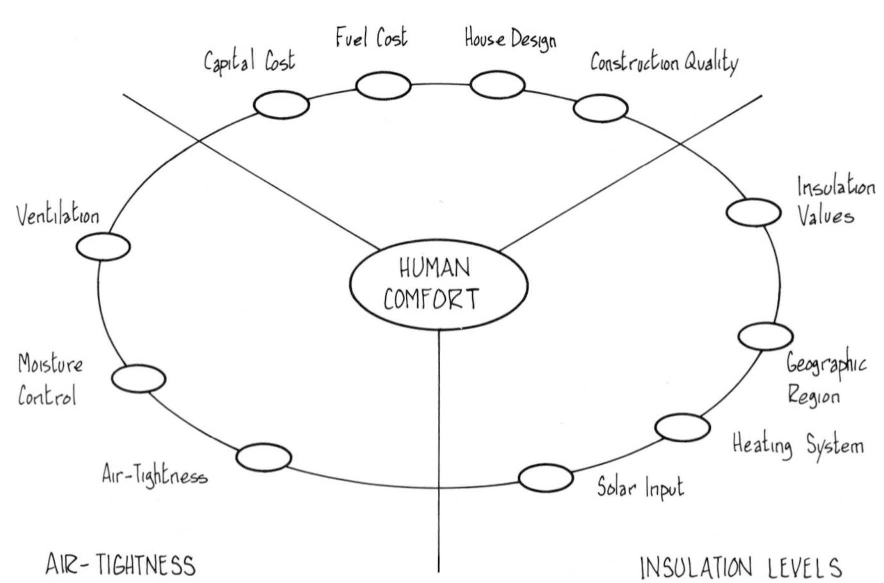

Diagram represents the design factors necessary to achieve human comfort and energy efficiency.

source: Canadian Home Builders' Assoc.

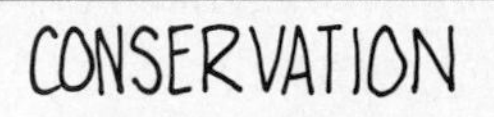

ENERGY BALANCE

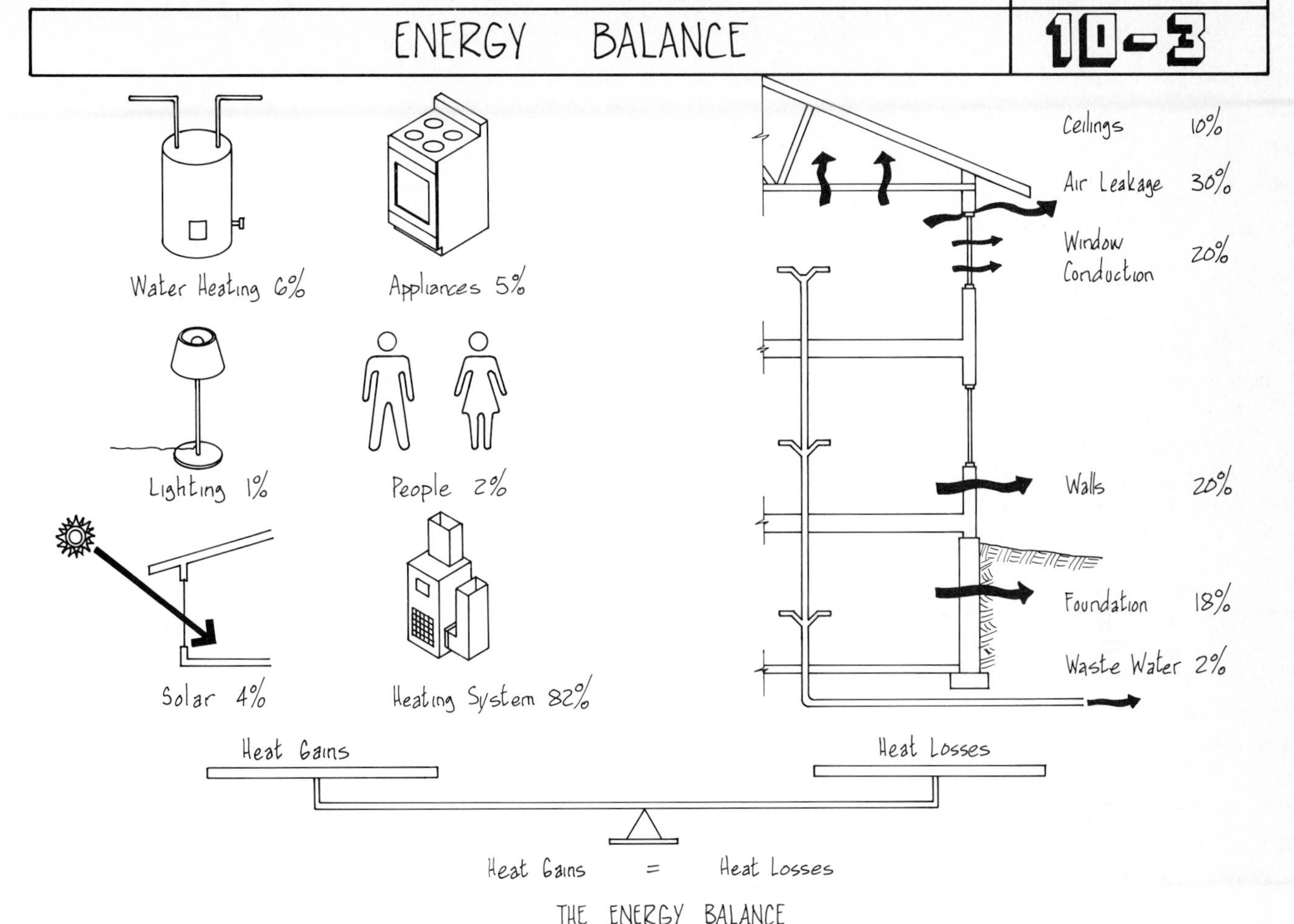

THE ENERGY BALANCE

STANDARD STRUCTURES

For any structure there is always an energy balance where heat losses equal heat gains. Although the percentages will vary depending on actual operating conditions, a typical balance for an average house can be illustrated as shown above. Assume that the house is at latitude 40° N on a cold winter day and note the following:

- The heating equipment supplies only 82% of the heat gains with the balance from other sources. On warmer days this percentage will be considerably reduced, while the other sources will increase in relative value.
- In early fall and late spring the heating equipment may not operate at all because of the "free heat" from the other sources (see the preceding page, first chart).
- Over the entire heating season the heating equipment may supply 60-70% of the total heating load.
- The largest single heat loss in the structure is air leakage. This accounts for air both infiltrating and exfiltrating the structure (see 10-8).

UPGRADED STRUCTURES

When extra measures are taken to increase energy efficiency over and above minimum code requirements, the building remains in balance but the relative proportions of the heat sources will change. The two most direct methods of increasing efficiency is to upgrade insulation values and to reduce air leakage. A substantial added bonus can be achieved by designing the structure to accept greater solar gains through passive solar techniques (see 3-6 to 3-21).

After upgrading, the following results can be expected:

- Total heating costs will be reduced. Reductions of 50-75% and greater are entirely possible if all reasonable techniques are employed.
- The heating equipment's contribution decreases, while the free heat value increases proportionally (see the preceding page, second chart).
- The length of the heating season will be shortened and the size of the heating equipment can be reduced.

It should be noted, though, that while increasing air tightness has a very positive effect on energy efficiency, it also creates some potential problems, especially with regard to humidity and moisture control. These problems and their solutions are discussed on 10-24.

Also note that any measures taken to increase efficiency should have a "payback period" where, over a reasonable period of time, the extra cost of those measures are paid back by actual energy savings.

source: Canadian Home Builder's Assoc.

HEAT LOSS IN STRUCTURES

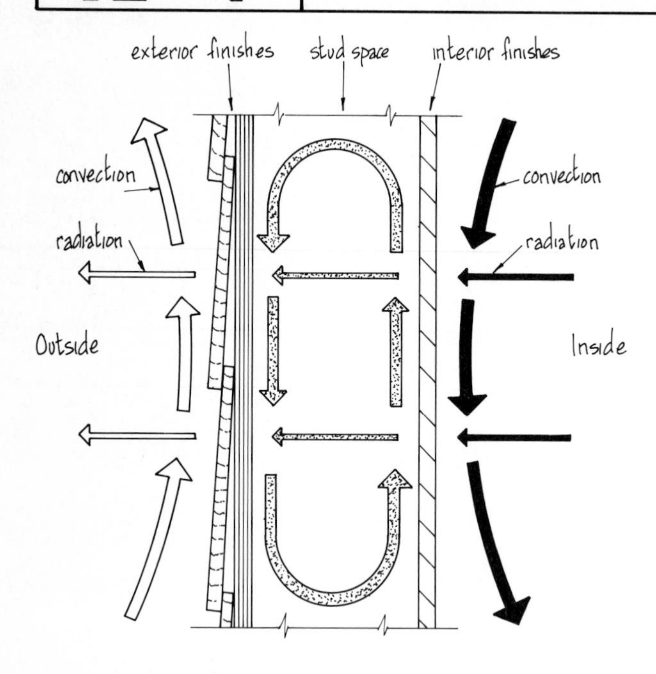

SECTION THROUGH UNINSULATED STUD CAVITY

In addition to convection and radiation, heat is also lost by conduction through the studs and finishes, and by direct air flow across the cavity between interior and exterior.

HEAT FLOW

The heat that is contained within a structure must eventually be lost to the outside. In an uninsulated building heat is lost through a wall or other part of the structure in several ways:

- Conduction is heat flowing through a solid material, in this case the wall components.
- Convection is heat transported by moving air. Air heated by the warm interior surface rises in a natural convection cell and is cooled by the exterior surface. Large quantities of heat are lost by convection.
- Radiation is heat flowing across an open space in the same way that the sun heats the earth. Even if the wall components were separated by a vacuum, heat would still be lost by radiation.
- Airflow through holes and cracks in the structure itself. Air leakage is often the highest heat loss factor in a building but also the one most likely to be neglected. See 10-8 for airflow causes.

The correct use of insulation and an air/vapor barrier is therefore very important if heat loss is to be reduced to reasonable levels.

HEAT LOSSES BELOW GRADE

Significant factors are:

- The soil against the foundation tends to be warmer than the air above in fall and early winter and cooler than the air above in spring and early summer.
- Dry soil acts as an insulation material and slows heat loss.
- Wet soil acts as a heat drain and will rapidly conduct heat away from the foundations.
- Hollow block walls have an added convection effect in the block's cells.
- Snow cover on the soil's surface has an insulating effect under certain conditions.

Heat loss through the foundations of a building is complex and difficult to measure because of the wide range of conditions that can exist. In most cases, however, such heat loss is entirely controllable with standard insulating methods.

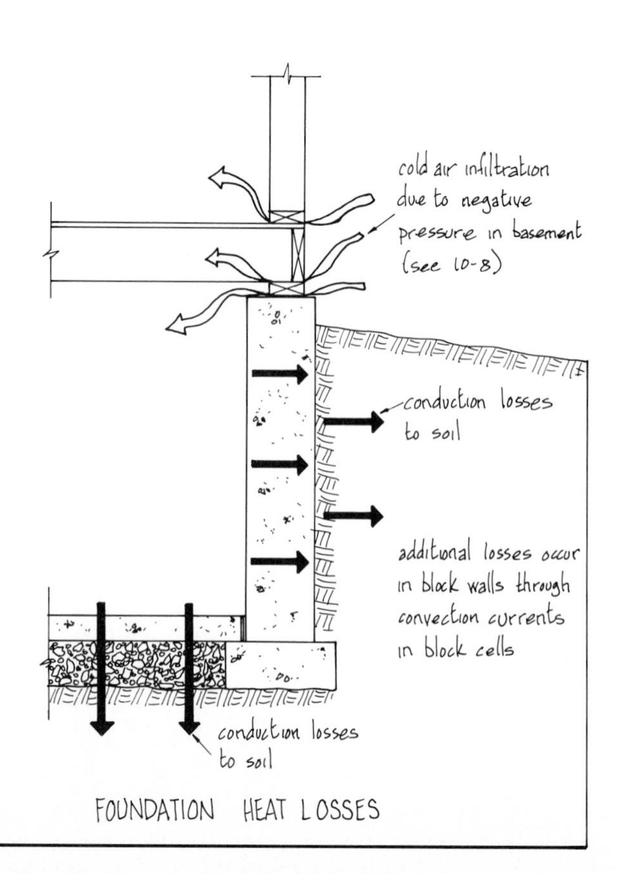

FOUNDATION HEAT LOSSES

HOUSE STYLES AND HEAT LOSS

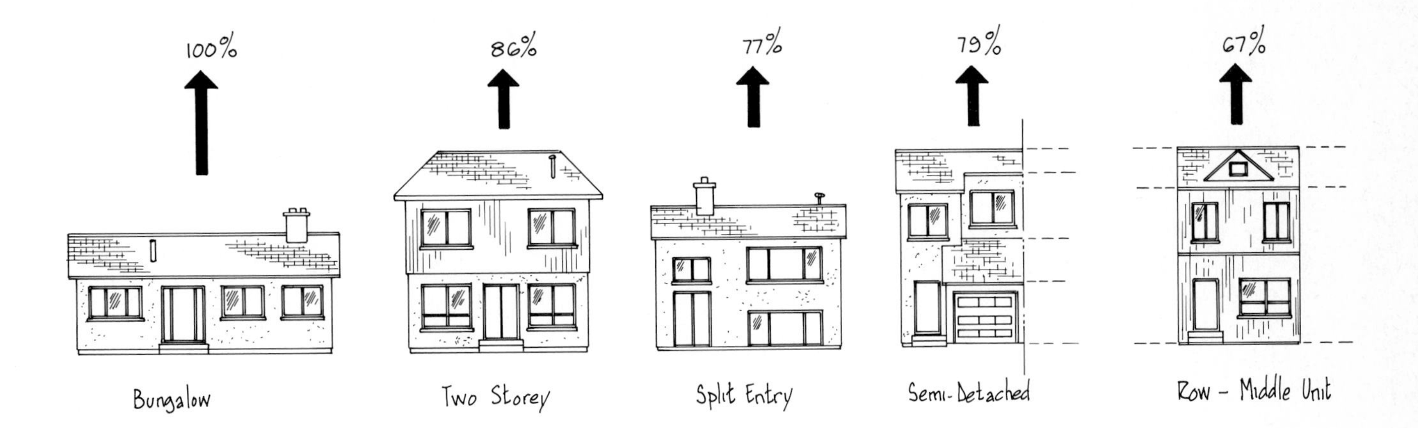

HOUSE STYLE COMPARISON

The illustration above compares the heat loss characteristics of five basic house styles. For measurement purposes each has the same heated floor area and insulation values and each is compared to the bungalow since that style has the largest heat loss.

Basically, the ratio of exposed surface area to internal volume causes the variations in heat loss. For example, the row house (middle unit) is the most efficient because it has units on either side, resulting in very little heat loss through its long side walls.

It follows that for maximum efficiency a house shape should be as compact and simple as possible. Although this approach is probably unacceptable to most buyers, attractive design features can be added that soften the building's lines. Alternatively, if high-energy-efficiency techniques are employed, other house styles can be built without undue penalty. In general it is best to avoid projecting into or out of the house envelope. For example, interior garages and exterior dormers increase the heat-losing area of the structure and should be eliminated where possible.

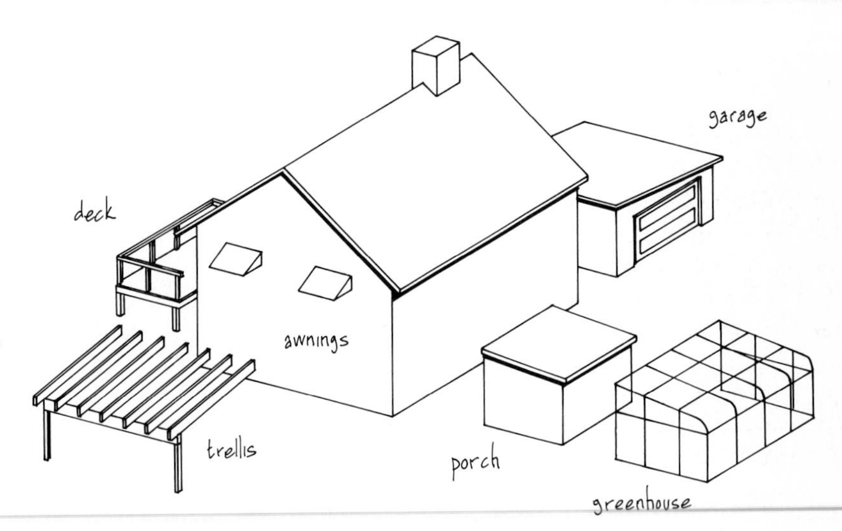

ADDED UNHEATED VISUAL ELEMENTS

Some elements, notably garages, greenhouses, and porches, have the additional value of sheltering and increasing the insulation value of the walls to which they are attached. Ideally, place garages and porches on the north side.

source: Canadian Home Builder's Assoc.

GENERAL DESIGN

OVERALL DESIGN FEATURES

Although few houses ,are specifically designed as "solar" structures, many of the features used in passive solar design will contribute greatly to overall energy efficiency and can easily be included in many house styles.

- Rooms occupied the most should be positioned in the south to receive the most sunlight. Rooms with limited occupation or those that act as transition spaces should be positioned on the north side of the dwelling.
- Glazing should be concentrated on the south side when overall design will permit. Glazing on north, east, and west should be kept to a minimum.
- For summer cooling, make use of overhangs, shutters, and shades on south, east, and west windows. Overhang size depends on window height and latitude of dwelling.
- Where practical, site the house and design landscaping to assist efficiency. Plant deciduous trees to the south and coniferous trees to the north and west. Use existing topography or add berms to provide protection from wind.

For greater detail on passive solar design, see 3-6 to 3-21.

In addition to specific design details there are other factors that must be considered :

- Local building codes will specify the minimum insulation and airtightness levels required in any structure. However, building codes react slowly to market pressures, and the efficiency levels specified are generally too low in relation to current and projected energy costs. Designers and contractors should therefore be selecting efficiency levels often considerably higher than code requirements.
- There must be a balance between the extra cost of increased efficiency and the selling price of a house in its intended market area. Buyers must be assured that their purchase is cost effective if energy efficiency is a selling feature.
- There are a large number of efficiency options to choose from. Designers and contractors must understand the effects and implications of each feature and aim for total compatibility of options.

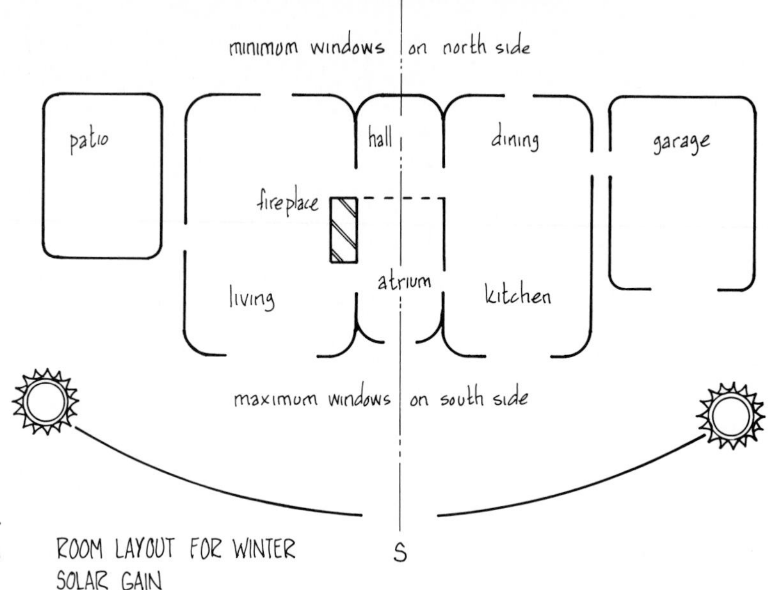

ROOM LAYOUT FOR WINTER SOLAR GAIN

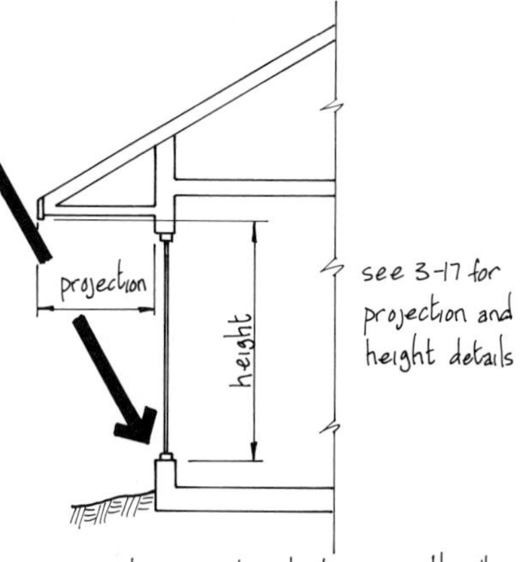

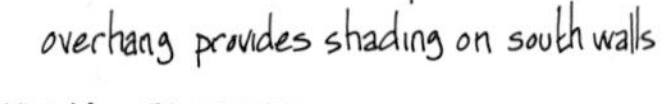

overhang provides shading on south walls

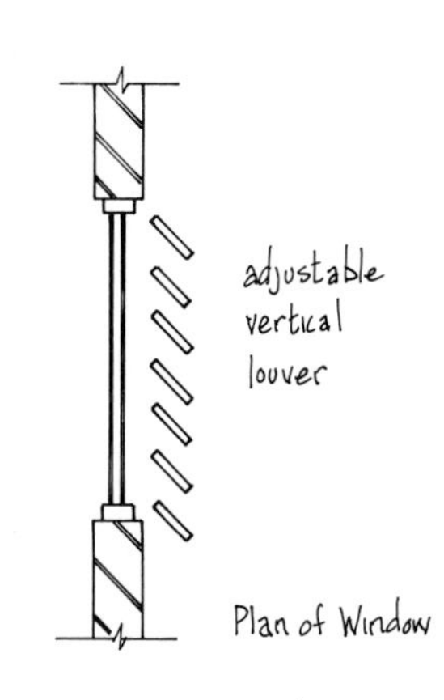

vertical adjustable shutters provide shade on east and west walls

SHADING FACTORS

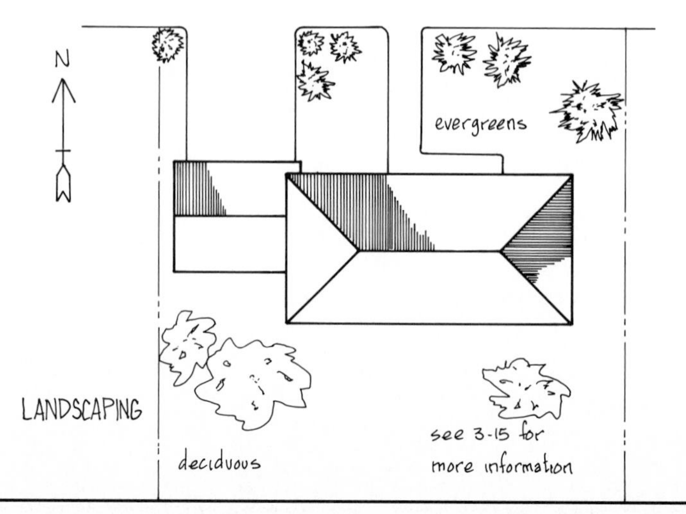

LANDSCAPING

PROBLEMS AND SOLUTIONS

As mentioned previously, the two main sources of heat loss are air leakage and low insulation levels. Modern technology and appropriate construction methods can reduce heat losses from these sources to negligible quantities, but, in doing so, other problems are created that must, in turn, be considered.

Every building must be regarded as a self-contained environment wherein all factors are carefully balanced to maintain human comfort. The basic sequence of problem, solution, and problem is illustrated in the flowchart below.

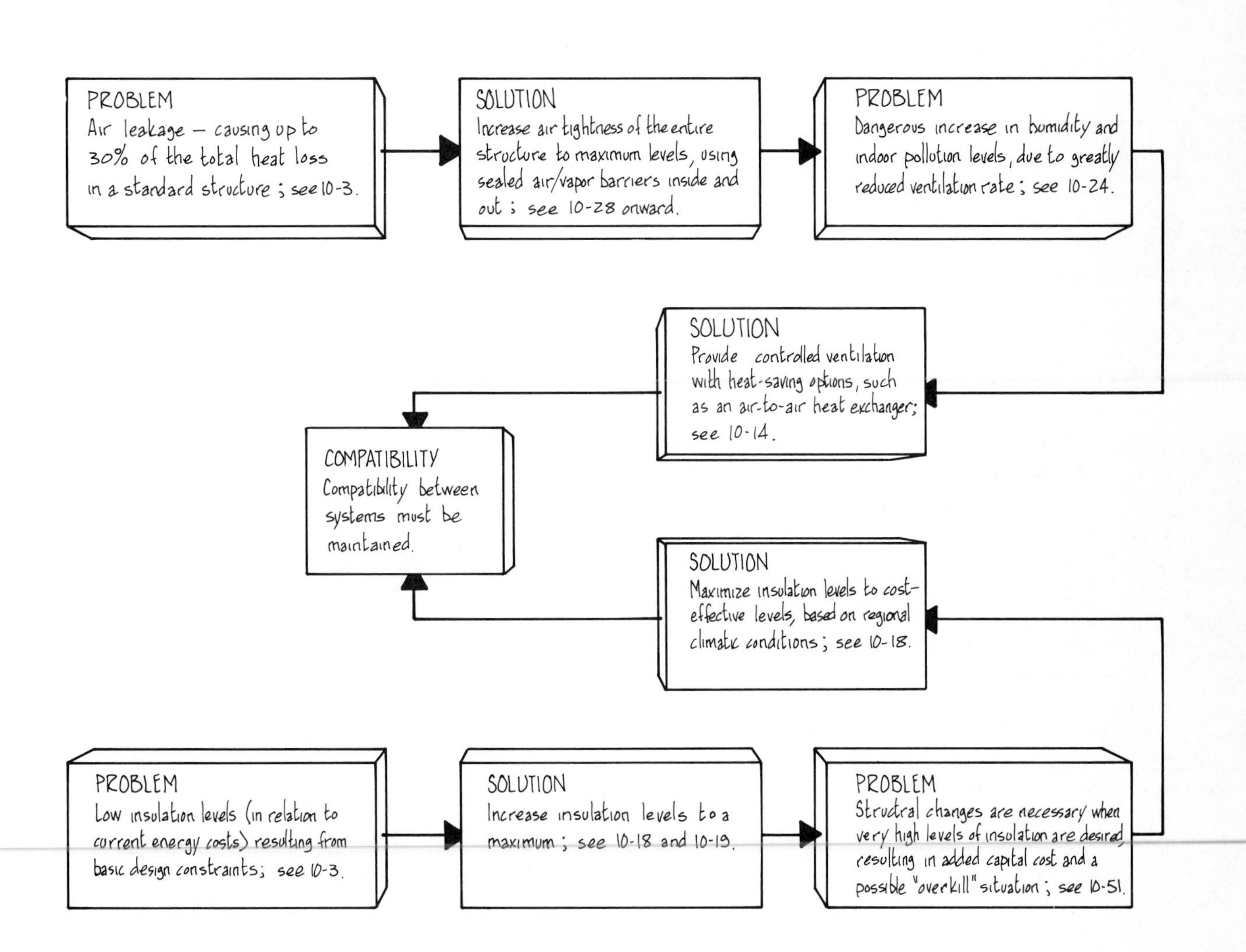

On the following pages, each problem and its recommended solution is discussed in turn.

CONSERVATION

10-8

AIR LEAKAGE CAUSES

Air leakage into (infiltration) and out of (exfiltration) a dwelling will occur through even the smallest hole or crack in the structure. The forces that drive this leakage can be defined in three ways:

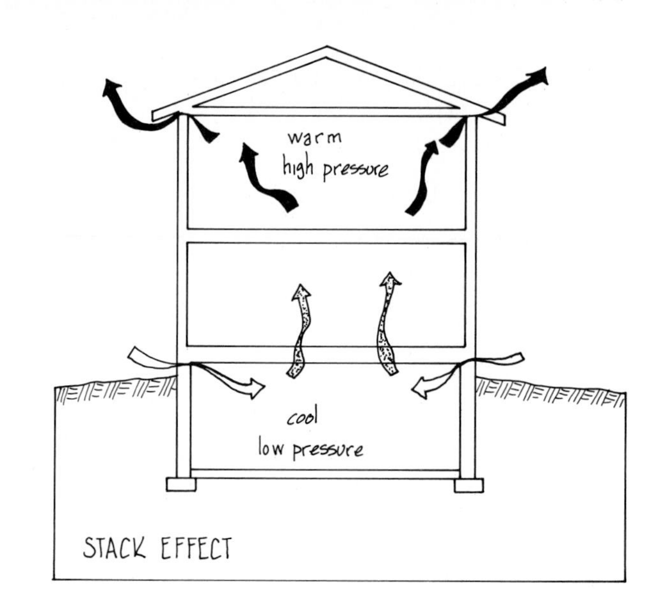

- Stack Effect. Any structure is subject to internal stratification of warm and cold air. Warm air rises, creating a positive pressure that forces exfiltration through the upper structure. Negative pressure causes cool air to infiltrate into the lower structure to maintain the cycle. Tall structures or large inside/outside temperature differences increase the stack effect.

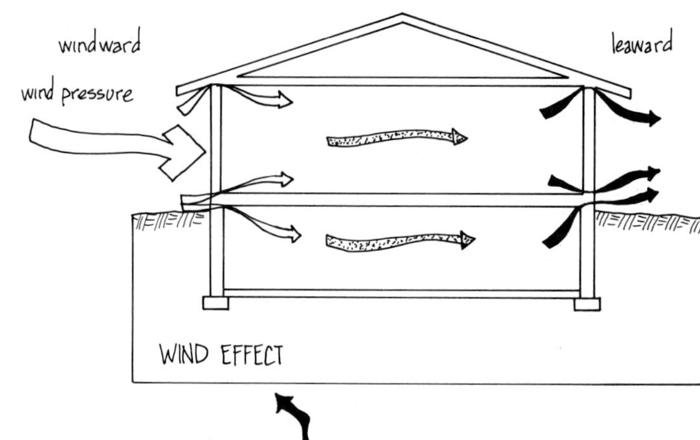

- Wind Effect. Outside air pressure increases to windward of the structure forcing cold air to infiltrate on that side. This in turn increases internal air pressure and forces warm air exfiltration on the lower-pressure leaward side. This effect increases with wind velocity to the point where cold drafts or movement of window drapes can be observed in poorly sealed houses.

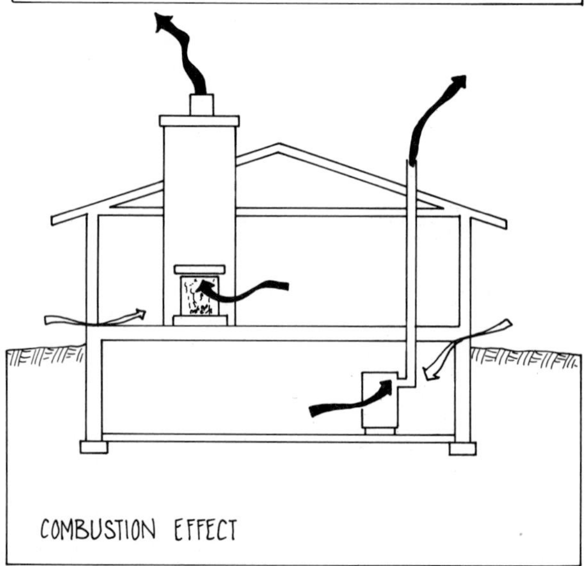

- Combustion Effect. Dwellings equipped with combustion heating systems (gas, oil, propane, wood, etc.) and which use internal air for combustion create a negative pressure that is normally relieved by cold air infiltration. Even when not operating, the undampered flue of the equipment has a continuous stack effect of its own.

Stack and wind effects can virtually be eliminated with good air-sealing techniques. The combustion effect can be countered by all-electric heating or combustion equipment with closed-loop external-air-feed systems (see 14-4, continuous condensing furnace).

NEUTRAL PLANE POSITIONS

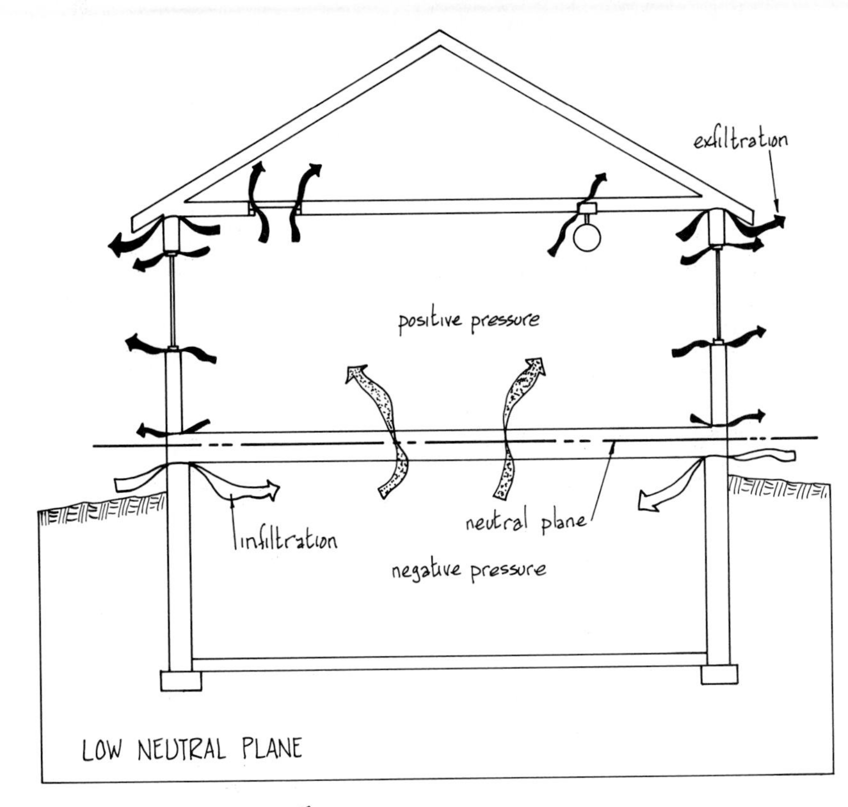

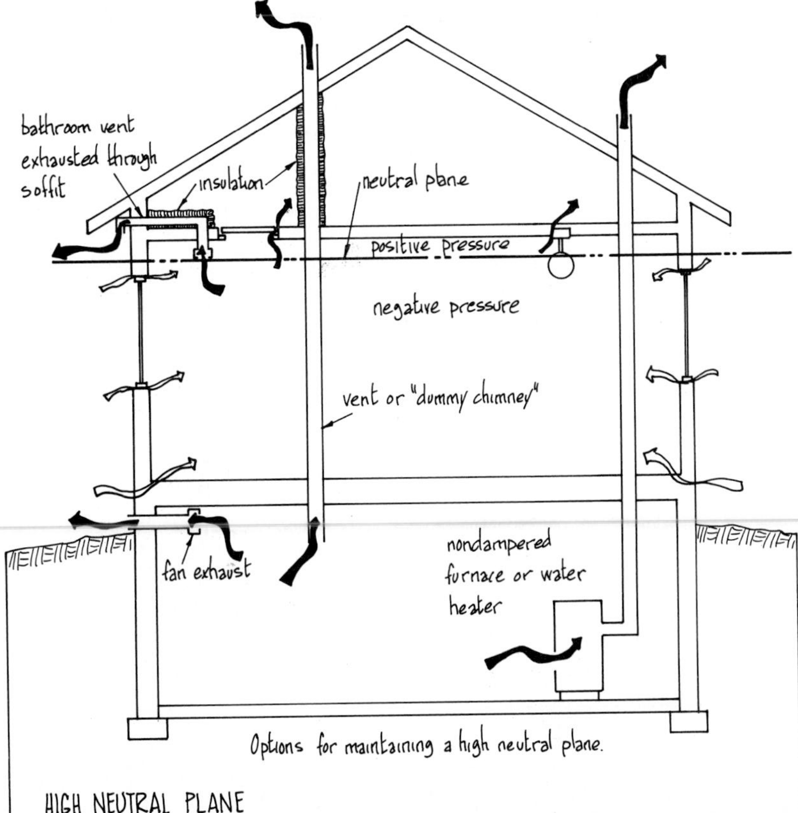

In all structures a certain quantity of air will always infiltrate and exfiltrate because of the positive and negative pressure effects. At a certain location in the building, however, internal pressure will be equalized with the external pressure and air will flow neither in nor out. This zone is termed the neutral plane (also called neutral pressure point or zero pressure point) and its position in any structure is subject to the sum of the air leakage effects described on the preceding page. The two illustrations at the left show air movement effects in a low or high neutral plane situation.

Some additional factors:

- If air infiltrates into the lower part of the structure more easily than air exfiltrates at the top, the upper half will be slightly pressurized and the neutral plane will be low.
- If air exfiltrates in the upper regions faster than it can be replaced by infiltration, a partial vacuum is created at the top and the neutral plane will be high.
- The neutral plane position is not dependent on the quantity of air infiltrating and exfiltrating—what matters is the relative ease of its entry and exit that creates the pressure differential.

In all structures, efforts should be made to keep the neutral plane as high as possible. This ensures that most of the airflow through the structure is in the form of infiltration, which provides a drying mechanism for the cold insulated cavities where condensation may form (see next page).

Alternate methods of achieving a high neutral plane are illustrated in the lower illustration. Options include:

- An open vent or "dummy chimney" extending from the basement (or lowest room) to and through the roof. A standard 4 in (100) plumbing vent stack is ideal. A damper may be used if desired.
- The undampered flue of a combustion furnace or domestic water heater that extends through the roof.
- A bathroom exhaust vent that discharges through the exterior soffit. Depending on the structures' air-tightness, a continuously running fan may be necessary.
- A powered exhaust system discharging through the basement wall. This may be part of a heat exchange or makeup air system (see 14-16 to 14-23).

In a tight structure, the quantity of infiltrating air is very limited and the cost of the heat loss generated is small compared to the cost of the structural damage caused by condensation in insulated spaces.

HUMIDITY

IDEAL INDOOR RELATIVE HUMIDITY LEVELS

OUTSIDE TEMPERATURE °F (°C)	60 (15)	50 (10)	40 (5)	32 (0)	25 (-5)	15 (-10)	5 (-15)	-5 (-20)	-15 (-25)	-20 (-30)	-30 (-35)	-40 (-40)
RH %	85	70	60	55	45	40	35	30	25	20	15	5

Based on 70°(21°C) indoor temperature and double glazing.

All air contains a certain amount of water vapor, refered to as humidity. Air that contains all the water vapor that it can hold has a relative humidity (RH) of 100%, while air containing only half the possible total of water has a 50% RH. The total water vapor capacity of air is directly proportional to its temperature; the warmer the temperature, the greater the capacity.

From a dwelling's point of view there are limits to the level of RH that can be sustained without causing condensation problems. Condensation occurs when warm damp air is cooled to its dew point and liquid water drops are formed. Condensation usually occurs on windows, wall interiors, and in attic spaces.

The table above gives maximum RH levels that can reasonably be sustained in a dwelling without causing undue condensation on interior window surfaces. Note that humidity levels are based on outside temperatures and that levels should be at, or less than, the indicated percentages.

INDOOR HUMIDITY SOURCES

ACTIVITY	lbs	(kg)
cooking (3 meals per day)	2	(0.90)
dishwashing (3 meals per day)	1	(0.45)
bathing - shower or tub	1	(0.45)
clothes washing (per week)	4	(1.80)
clothes drying indoors or with unvented dryer (per week)	26	(11.80)
floor mopping (per 100sf/10 m²)	3	(1.36)
occupants (family of 4 per day)	12	(5.40)

Listed above are the principal daily sources of water vapor in a dwelling, based on a family of four. In a tightly sealed structure provision must be made to vent this daily water vapor load to the outside if serious condensations are to be avoided.

Humidity exits the building shell in two ways:

- Vapor pressure forces water vapor to diffuse through the construction materials. This is a relatively slow process in most materials.
- Air exfiltration (see neutral plane concept, preceding page) carries large quantities of water vapor through the structure very quickly. Air leakage is therefore considerably more important than diffusion in most structures.

Water vapor, reaching the cold outer structure, will condense into water droplets and accumulate, often in close proximity to the leakage source. Provided that this water is not excessive and can dissipate over the summer months, little harm will occur. However, water that does not dissipate can cause significant structural and other damage:

- Dry rot can seriously weaken the structural frame.
- Insulation will absorb water and lose all thermal resistance.
- Plaster and gypsum board will crumble and paint finishes inside and out will fail and peel.

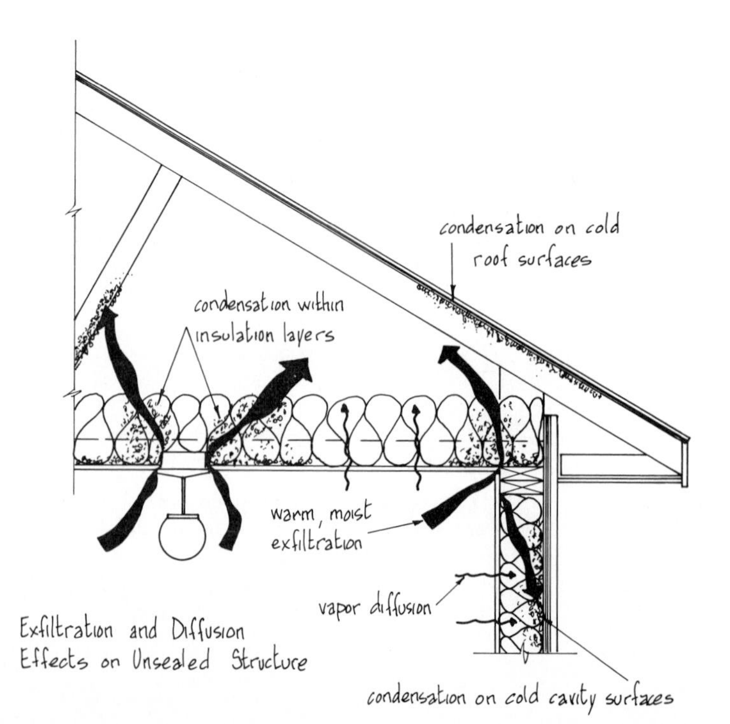

Exfiltration and Diffusion Effects on Unsealed Structure

AIR / VAPOR BARRIERS

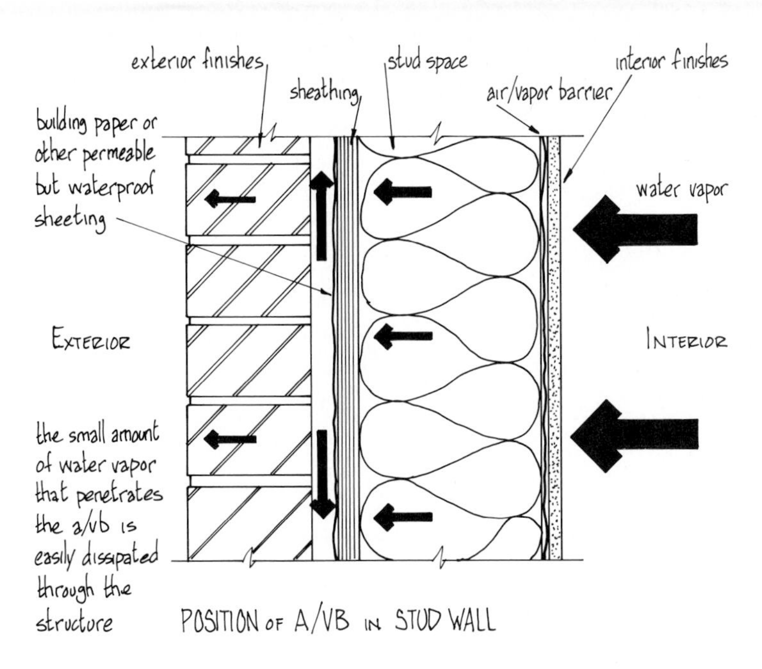

POSITION OF A/VB IN STUD WALL

To stop the movement of heat and humidity through cracks and holes in the structure and to eliminate water vapor diffusion through the building materials, the installation of an air/vapor barrier is necessary. To be effective, an air/vapor barrier must have the following properties:

- Low permeability, i.e., vapor diffusion through the barrier is very low.
- Available in large enough sheets so that joints are kept to a minimum.
- Flexible enough to be molded around such obstructions as electrical fixtures, plumbing, windows, etc.
- Strong enough to resist damage during construction.

Various materials are employed as air/vapor barriers, but the one in general use is polyethylene film. If carefully installed, this material can provide an almost continuous airtight barrier and is most effective. Ideally, a 6 mil (0.15) thickness should be used to reduce construction damage. Other material types are detailed on 10-12 and 10-13.

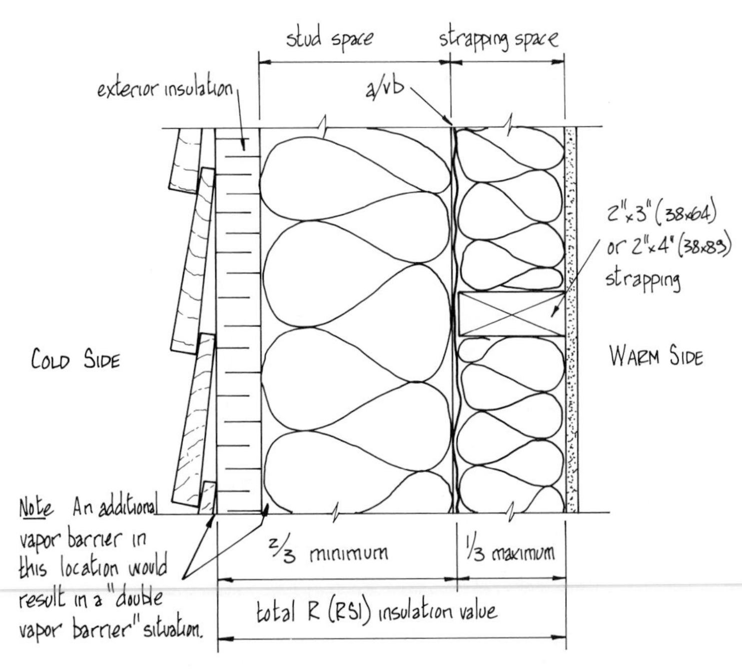

ILLUSTRATION OF THE ONE-THIRD/TWO-THIRDS RULE

See 10-18 and 10-19 for more information on this type of construction.

THE ONE-THIRD/TWO-THIRDS RULE

In conventional buildings the air/vapor barrier is placed between the structural members and the interior finish as shown in the upper illustration. In some upgraded designs, however, the barrier is located inside the structure between insulation layers as shown in the lower illustration. This is quite acceptable provided that the barrier is no more than one-third of the total insulation value in from the warm side (at least two-thirds of the effective insulation value is on the cold side).

CAUTION – DOUBLE VAPOR BARRIERS

Because water vapor will condense on a cold surface, air/vapor barriers are always placed on the warm side of an insulated space. The barrier is then at or near room temperature, and condensation cannot form on its surface.

However, great care must be taken to ensure that a nonpermeable vapor barrier is not placed on the cold side of an insulated space. A cold barrier traps vapor and promotes condensation in the insulated space with the results described on the preceding page. The severity of this problem increases with colder temperatures. A wide range of materials, including some insulation products, are nonpermeable and require careful detailing if used on the cold side of a structure.

Note: In extremely cold regions the 1/3 – 2/3 ratio should be replaced by a 1/4 – 3/4 ratio to ensure an effective a/vb.

AIR / VAPOR BARRIERS – MATERIALS

The primary function of an air/vapor barrier (a/vb) is to stop all air infiltration and exfiltration through the structure. Its secondary function is to stop vapor diffusion through the building materials. The a/vbs shown here fulfill these requirements to varying degrees – the designer must choose which is most appropriate after balancing efficiency and cost factors. The a/vb must be carefully sealed at all ceiling/wall/floor junctions, window and door openings, and holes where plumbing, electrical, or other services penetrate (see 10-25 onwards for more details).

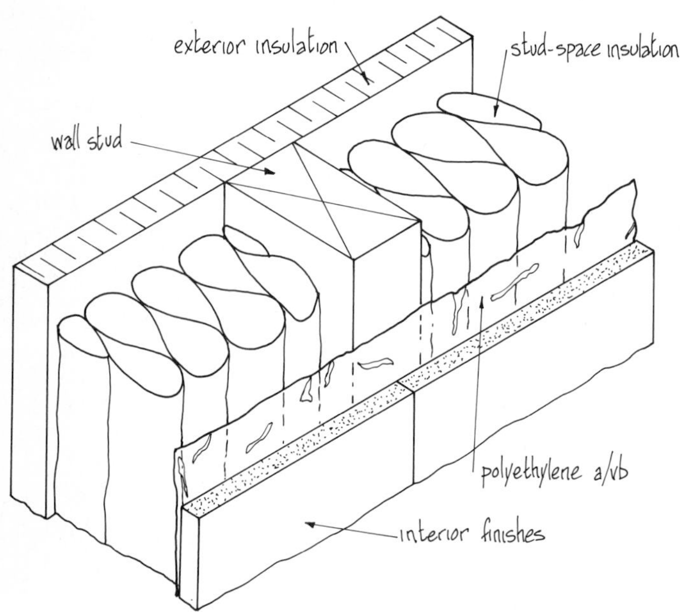

SHEET POLYETHYLENE

- This is the usual material of choice since it is extremely flexible and easy to handle.
- While a 4 mil (0.10) thickness meets permeability requirements, use of the stronger 6 mil (0.15) is preferred to reduce on-site damage.
- Use large, full-wall-height poly rolls to minimize joints.
- All joints should be made over a structural member for support and sealed with caulking or tape after lapping on framing space (see 10-25).

TYVEK HOUSEWRAP

This is a speciality product manufactured by DuPont that is extremely effective in controlling airflow through a structure. However, this material is highly permeable and is not a vapor barrier. Its use in a structure must therefore be carefully controlled.

- Because it is not a vapor barrier, Tyvek can be "wrapped" around the exterior walls, over insulation or sheathing, to stop air passing through the wall cavity. Used in this way on conventional construction it reduces airflow into the building by 30-40%.
- It provides an ideal air seal at floor and wall junctions by wrapping around floor headers as shown at the right. (Note: Polyethylene can also be used in this way, but care must be taken to ensure that the one-third/two-thirds rule is observed. See 10-28 for a typical detail.)
- Although not a vapor barrier, Tyvek is highly liquid-water resistant.
- Made from spunbonded olefin (synthetic fibers) it will not shrink or rot and is tear and puncture resistant.
- Two sizes are available: 9-ft (3m) and 3-ft (1m) wide rolls each 195 ft (60m) long.
- When used as a housewrap, it should be stapled to every second stud and caulked or taped to the structure top and bottom. Vertical joints should be lapped and sealed.

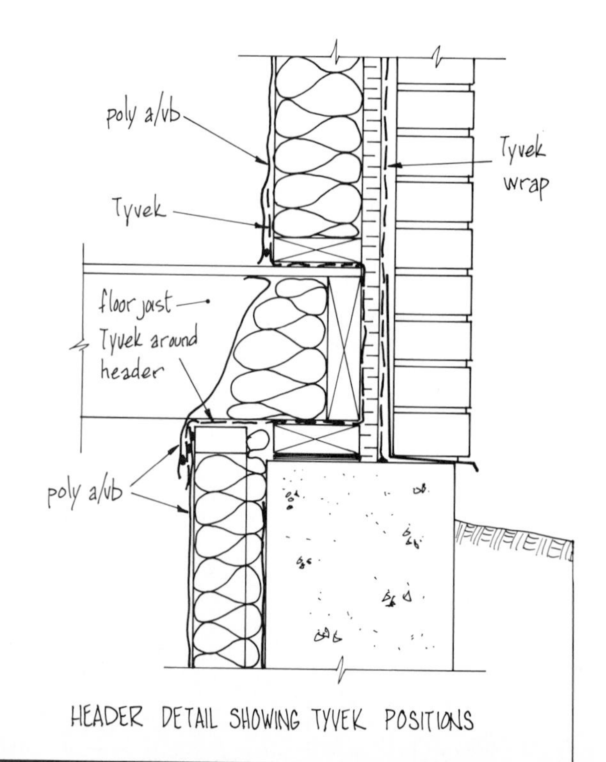

HEADER DETAIL SHOWING TYVEK POSITIONS

AIR/VAPOR BARRIERS - MATERIALS

RIGID FOAM INSULATION

- Aluminum-foil-faced rigid foam insulation boards are used on the interior (warm side) of the structure.
- All panel joints and penetrations are tightly sealed with aluminized tape.
- Floor/wall/ceiling junctions are taped or caulked.
- Interior finishes (gypsum board, plaster, etc.) are overlaid with staggered joints.
- CAUTION. When used on the cold side of a structure, this material acts as a double vapor barrier. Without an effective a/vb on the warm side, condensation problems can occur, especially in cold climates.

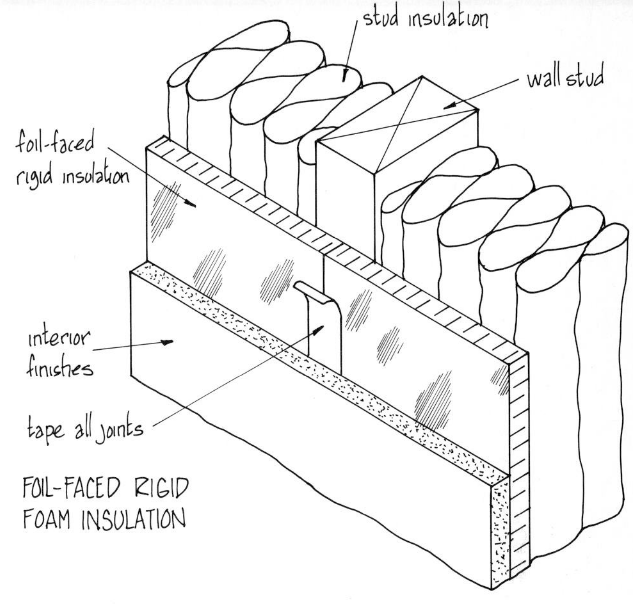

FOIL-FACED RIGID FOAM INSULATION

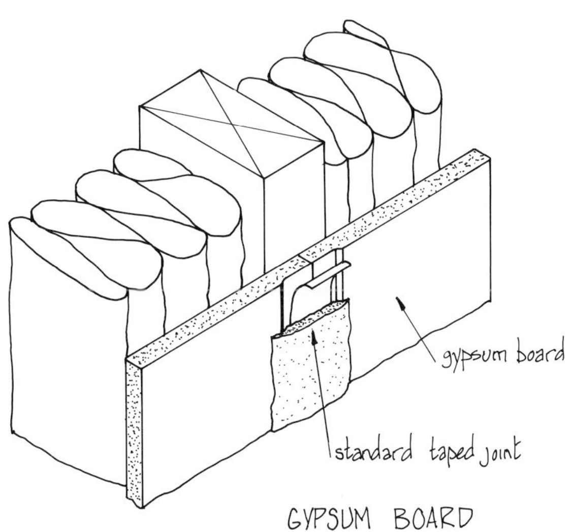

GYPSUM BOARD

GYPSUM BOARD

- Standard drywall sheets or foil-faced one-side sheets are used.
- Caulking or flexible gaskets are required at floor/wall/ceiling junctions.
- Joints are taped and filled as normal, but a specially formulated vapor-barrier paint system is required as a primer sealer.

FACED INSULATION BATTS

- Fiberglass batts of various sizes with a layer of brown kraft paper or aluminum foil attached.
- The paper or foil lips are stapled to each framing member, overlapping each other.
- While the paper or foil is an effective vapor barrier, the method of attachment makes this material useless as an air barrier.
- Consequently, this material is not recommended as an a/vb, and in many regions this material is now unavailable.

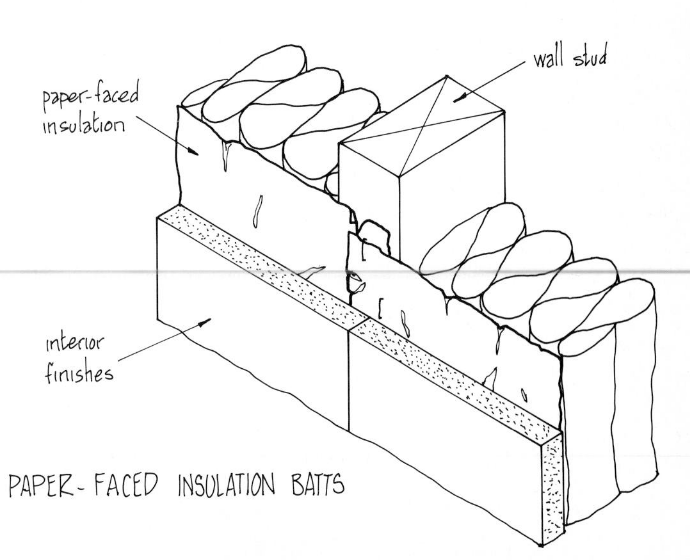

PAPER-FACED INSULATION BATTS

INSULATION

INSULATION - DEFINITION

Insulation is essentially a material that creates a space where air is trapped and cannot move around. It is this dead air that resists heat movement by convection, conduction, and radiation through the material (see 10-4). Insulation can be made from a variety of materials, including glass fibers, mineral fibers, plastic cells, or even ground-up newspapers, with each material intended for use in general or specific situations.

How much insulation should be installed is discussed on page 10-23 and the various types of insulation are catalogued on 10-15 to 10-17.

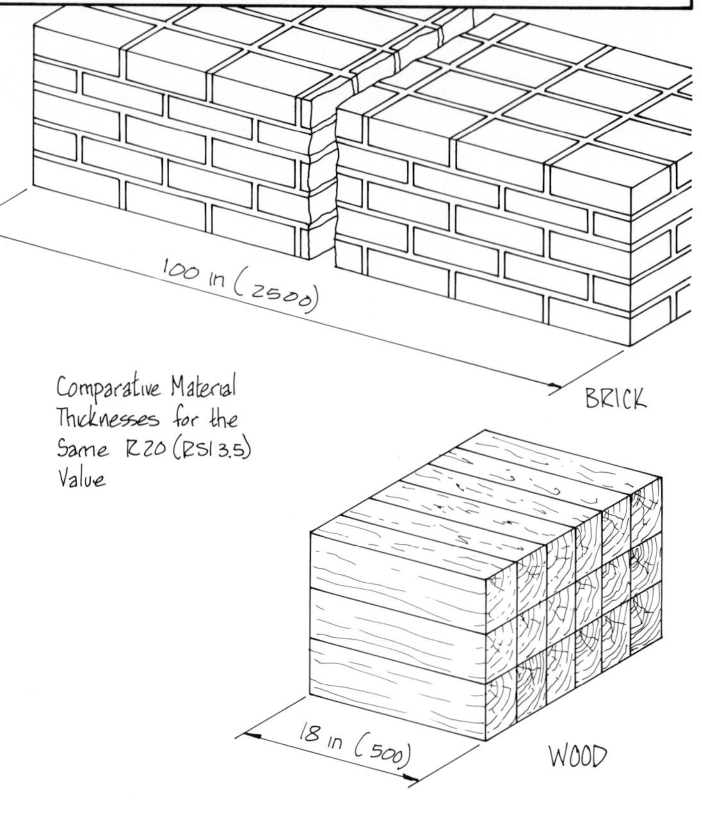

Comparative Material Thicknesses for the Same R20 (RSI 3.5) Value

THERMAL RESISTANCE

- The thermal resistance of a material is expressed as an "R" value (R12, R20 etc.) indicating the resistance of a material to heat flow. In the metric system the symbol is RSI and is measured in different units (see 2-17 for details).
- Resistance value can be stated as a total for the entire thickness of a material or as the resistance per inch (mm) of the material.
- Even materials not classified as insulation have a certain amount of thermal resistance. However, this resistance is so low in some materials that they are considered to be a contributor to heat loss (or heat gain).
- R values for a small selection of building materials are given at the right with a graphic illustration of comparative material thicknesses for a given R (RSI) value.
- The type of insulation used at any particular location within a building is influenced by R value, cost, and suitability. Insulation types are compared starting on the next page.

Building Material Insulation Values

MATERIAL	R per INCH	RSI per 25 MM
asbestos-cement	0.25	0.05
gypsum board	0.90	0.16
plywood	1.25	0.22
fiberboard	2.38	0.42
soft woods (average)	1.25	0.22
hard woods (average)	0.90	0.16
cedar	1.50	0.26
concrete (lightweight)	0.40	0.07
brick (common)	0.20	0.04
brick (face)	0.11	0.02
concrete block (average)	1.50	0.26
metals	negligible	
building paper, poly, etc.	negligible	
air space in wall - 3/4"-4" (19-100)	0.95	0.17

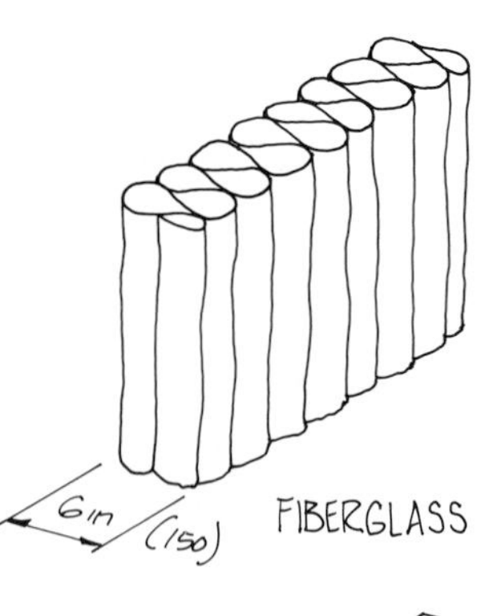

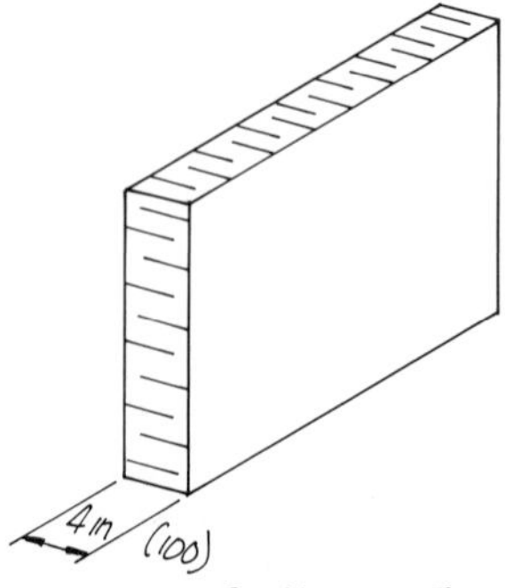

CODE REQUIREMENTS

Local building codes specify the minimum levels of insulation that a structure must have in one of two ways:

- The R value of the insulation alone that must be included in the structure.
- The total R value of the entire building assembly, i.e., the insulation plus the various building materials, air spaces, and air films, etc.

In each case the R value is taken through the cavity and does not include the heat loss effects of the framing, which is often not insulated. An energy-efficient design should include a means of insulating the frame as well as the cavities (see 10-18 and 10-19).

INSULATION MATERIALS

The insulation products shown in the following tables are typical of those currently available. Note, though, that new and different products are marketed constantly, and while you should keep abreast of manufacturers' offerings, some caution in their use is advised since new products are generally unproven products. This factor is particularly important in buildings where insulation is inaccessible after construction and must perform efficiently for the life of the building.

NOTES TO CHARTS

- The R per inch (RSI/mm) values are average for that type of material. Similar products from different manufacturers vary slightly in value.
- Cost of material is stated as high, medium, or low and should only be used as a general comparison. A more accurate cost analysis calculation is given on 10-17.
- The drying-ability column indicates how well the material retains its insulating properties after being wetted and then dried. This is particularly important since some materials are very adversely affected during and after wetting. Obviously, such materials should not be used where wetting may occur.

INSULATION MATERIALS

* indicates soft metric (see chapter 2)

MATERIAL TYPE	R per in. (RSI) per mm.	COST	DRYING ABILITY	COMMENTS
BATT or BLANKET				
Glass Fiber Material – loosely woven long fibers of spun glass length – batts 4 ft (1.22* m) – blankets, up to 80 ft (23 m*) width – friction fit for 16- and 24-in (400, 600) stud spacing. thickness – 3 to 12 in (76-304*)	3.2 (0.022)	low	good	Uses: all accessible walls, floors, ceilings, heating ducts, etc. Comments: easy to install in open cavities and the most economical type for this application. Batts are friction fit between structure members. Fire, fungus, and vermin resistant. This is the most used type of insulation. Limitations: skin irritant only during installation; wear simple protective clothing. Thick batts must be installed carefully to avoid cavity spaces (see 10-27). Cannot be installed in closed cavities and difficult in uneven spaces. Avoid wetting. Has little airflow resistance. Paper- or foil-faced batts are available in some areas but should not be used as an a/vb (see 10-13).
Mineral Wool Material – recycled mineral waste sizes – comparable to fiberglass	3.7 (0.024)	low	good	Similar in most respects to fiberglass. Excellent fire resistance.
LOOSE FILL — BLOWN-IN Specialized blowing equipment is required to install these materials. Care must be taken to install evenly to the correct depth for the R value required, and to avoid blocking eave ventilation openings. Some form of eave blocking is required to prevent displacement in high wind conditions.				
Glass Fiber Material – similar to batts but chopped for blowing. Installation – should be blown to a density of 1.5 to 1.7 lbs/cu ft (24 to 28 kg/m³).	2.9 (0.02)	low to medium	good	Easily fills irregular spaces, quick to install, and light weight. Fairly noncombustible. May settle slightly. Limitations: avoid high moisture levels or high temperature sources (e.g., light fixtures). Relatively low R for installed thickness. Skin irritant during installation.
Mineral Wool Material – similar to batts but chopped. Installation – should be blown to a density of 2 to 3.5 lbs/cu ft (32 to 56 kg/m³).	2.7 (0.019)	low to medium	good	Similar to glass fiber. Highly fire resistant.
Cellulose Material – finely shredded newspapers, treated with antifungal and fire resistant chemicals. Installation – should be blown to a density of 3.5 to 4.2 lbs/cu ft (56 to 64 kg/m³).	3.6 (0.025)	low	fair to poor	Good R-values, fills confined spaces easily, and can be a good air infiltration barrier when properly installed. Limitations: normal water vapor quantities have little effect but it absorbs liquid water easily with loss of R value. Chemical additives may bleed. Very dusty during installation. Likely to settle if not blown to correct density. May not be an acceptable material in some regions.
LOOSE FILL — POURED Similar in application to blown insulation except that specialized equipment is not required. May be installed in vertical cavities if they are open from the top. More settling can be expected, requiring additional material to achieve rated R values.				
Glass Fiber Material and installation – similar to blown material.	3.0 (0.021)	low to medium	good	Similar to blown variety. Low resistance to air flow.

INSULATION MATERIALS

MATERIAL TYPE	R per in (RSI/mm)	COST	DRYING ABILITY	COMMENTS
LOOSE FILL-POURED (continued) Mineral Wool Material and installation - similar to blown variety.	3.00 (0.021)	low to medium	good	Similar to blown variety.
Cellulose Material and installation - similar to blown variety.	3.6 (0.025)	low	fair to poor	Similar to blown variety.
Vermiculite Material - expanded mica. Installation - pour to a density of 7 lb/cuft (112 kg/m³).	2.3 (0.016)	medium	poor	Pours very easily, lightweight and fire resistant. Fills irregular spaces well. Good fungus and vermin resistance. A finely ground and coated product is available for inside masonry walls. Limitations: poor drying ability, not a good air barrier, and settles.
Polystyrene Beads. Material - available in shredded form from extruded poly, or beads from expanded poly. Installation - pour to density of 1 lb/cuft (16 kg/m³)	2.8 (0.02)	low	good	Pours easily although highly susceptable to static electricity. High resistance to moisture, and lightweight. Limitations: highly flammable and gives off toxic fumes when burning. Easily displaced by wind. Low air flow resistance. Recommended installation only in masonry walls.
Wood Shavings. Material - recycled wood waste often with chemical additives. Installation - pour to a density of 4 lb/cuft (64 kg/m³)	2.4 (0.017)	low	poor	The only advantage to this material is its low cost, especially if produced locally. Limitations: a fire hazard, subject to fungus and vermin if not carefully treated. Readily absorbs water and is slow to dry. Extensive settling, especially in walls.
RIGID BOARD Manufactured from glass fiber or foamed plastic, rigid boards tend to have high R values but also tend to be relatively expensive. Ideal for below-grade applications or as an insulated facing for stud walls (but is not a structural sheathing). Boards are very effective for large flat areas, but are difficult to fit into irregular spaces. Some boards can create a double vapor barrier (see 10-13). Foam boards are flammable and give off toxic fumes in a fire situation.				
Glass Fiber - Semirigid Material - denser fiber than batts, usually with a highly permeable synthetic-paper face (usually Tyvek see 10-12). Available in building-sheet sizes. Installation - nailed to structure with wide-headed fasteners. Joints should be taped for a tight air seal.	4.2 (0.029)	medium	excellent	A very useful product. Used on walls, ceilings, floors, interior foundation walls, ducts, and pipes. Dimensionally stable, dries exceptionally well, easy to handle. Limitations: crushable; care must be taken when applying finishes. Skin irritant during installation. Paper facing may strip or nails may pop under high wind conditions before close-in.
Glass Fiber - Rigid. Material - high density fibers, usually without facing. Installation - as semi-rigid but without taped joints.	4.4 (0.030)	medium	excellent	Can be used in place of semirigid boards. Some products are specially designed for exterior below-grade foundations where it absorbs lateral earth forces and conducts subsurface water to weeping tiles (see 10-32).
Extruded Polystyrene Material - high-density foam plastic with fine, closed air cells. Often blue in color. Sizes: 16 and 24 in (400, 600) wide and 4 or 8 ft (1200, 2400) long. Thicknesses 1 to 4 in (25-100). Installation - mechanically fastened or glued.	4.6 (0.032)	high	excellent	Very versatile material; can be used anywhere in a structure with proper application methods. Highly moisture resistant. Ideal for below-grade applications. Can be glued to interior concrete walls and used as an a/vb under some conditions. Limitations: must be protected from fire and sunlight and high temperatures. Easily damaged. May cause double vapor barrier on exterior; panels must not be jointed or sealed at edges.
Expanded Polystyrene (beadboard). Material - low-density plastic foam with large closed cells or beads. A higher-density material is also available. Sizes as extruded poly plus 4 ft (1200) wide and thickness to 6 in (150). Installation - similar to extruded polystyrene.	3.6 (0.025)	medium	fair	Lowest cost per R of all foam boards. Not a vapor barrier; suitable for exterior applications. Can be glued to concrete or wood. Limitations: flammable and gives off toxic fumes; must be fire protected. Not recommended for below-grade use. Subject to some shrinkage and degrades with sunlight and water exposure. Easily damaged.
Phenolic Foam. Material - expanded phenolic plastic usually faced both sides with stiff paper. Sizes: boards 4 ft (1200) wide by 8- or 9-ft (2400, 2700) long. Various thicknesses. Installation - nailed to structure with wide-headed nails. Joints may be taped. (Continued)	4.1 (0.028)	high	fair	Strong, easy to handle, and relatively fire resistant. Not a vapor barrier, making it ideal for exterior wall use. Limitations: Not recommended for below-grade use; subject to moisture damage. Fairly brittle and degrades with sun exposure. Corrosion can occur if galvanized fasteners not used.

INSULATION MATERIALS

MATERIAL TYPE	R per in (RSI) per mm	COST	DRYING ABILITY	COMMENTS
RIGID BOARDS (continued) Polyisocyanurate Foam Material – rigid isocyanurate foam boards with aluminum-foil facing both sides. Sizes – as for phenolic foam. Installation – nailed to structure with galvanized roofing nails. Long edges must not be tightly fitted.	7.0 (0.052)	high	medium	Very high R values and good rigidity due to the foil facings. May be used as an interior a/vb with taped joints (see 10-13). Limitations: high cost. Low permeability can pose problems with exterior usage if care is not taken to provide stud space ventilation and an efficient warm-side vapor barrier. Low fire resistance; must be suitably protected.
FOAMED-IN-PLACE Polyurethane Material – closed-cell, foamed with refrigerant gas. Installation – special equipment required.	6.0 (0.042)	high	fair	High R value and can be applied to irregular surfaces. Provides a complete air seal. Applied in 2-in (50) layers. Limitations: must be covered; flammable with toxic fumes. Expands rapidly on application and slowly with age. Should not be used in an enclosed cavity.
Urea Formaldehyde Material – mixed-on-site foam, special equipment. Installation – similar to polyurethane but requires careful mixing under correct conditions.	2.5 (0.018)	high	poor	Properties similar to polyurethane but low R value. Limitations: will shrink after application with loss of R value. Under certain conditions will break down and outgas formaldehyde, a serious health hazard. Use is banned in some states and all of Canada.
ALUMINUM FOIL Material – single or multiple layers of foil attached to board material or separately in accordian-type sheets inserted in stud spaces.	varies	medium	good	Excellent protection against radiation heat losses. Excellent a/vb material when attached to rigid boards and with joints taped. Nonflammable. Limitations: little effect on convection or conduction heat losses. Expensive and damages easily. Requires a minimum reflective air space of 1/2 in (13) to function properly. Dust or condensation on foil surface reduces efficiency.
INSULATING MASONRY BLOCK Material – lightweight concrete blocks with insulating properties. Some types also have insulated cores. Installation – dry stacked and face bonded.	1-2 (0.007-0.014)	medium	excellent	An efficient method of preinsulating block walls in terms of reduced labor costs, excellent structural strength, and good R values. The units can be substituted for standard blocks under most conditions. Limitations: more expensive than regular units but will reduce or eliminate the need for additional interior or exterior insulation. See 10-38 for more information.

COMPARATIVE COST CALCULATION

The best method of comparing insulation cost is to calculate the cost of each R per square foot of material ($/RSI/m²). This approach avoids the difficult comparison of R and material thickness, which varies greatly.

The formula is as follows:

$$\frac{(\text{cost}/\text{area})}{\text{total R}} = \$/\text{sf}/\text{R}$$

where: cost is actual cost of material
area is face area of material
total R is the full R value of that material

Example: Extruded polystyrene sheet 2 in thick, R10 value, 2 x 8 ft face size, and $15.00 cost.

$$\frac{(15/16)}{10} = \$\,0.09375/\text{SF}/\text{R}$$

or, expressed in cents, 9.375 cents/sf/R. The same calculation can be performed for metric materials simply by substituting m² for area and RSI for R.

When this base cost is known, the total insulation cost of a building can be calculated. Multiply each base cost by the total R and the total square footage (m²) of the assembly.

UPGRADED WALL TYPES

There are four basic types of upgraded walls. Each is shown here together with the maximum R values reasonably obtainable. Internal and external finishes have been omitted for clarity, and in each illustration the wall is viewed from the inside. Note that all walls are solidly filled with insulation—there should be no air spaces between insulation layers.

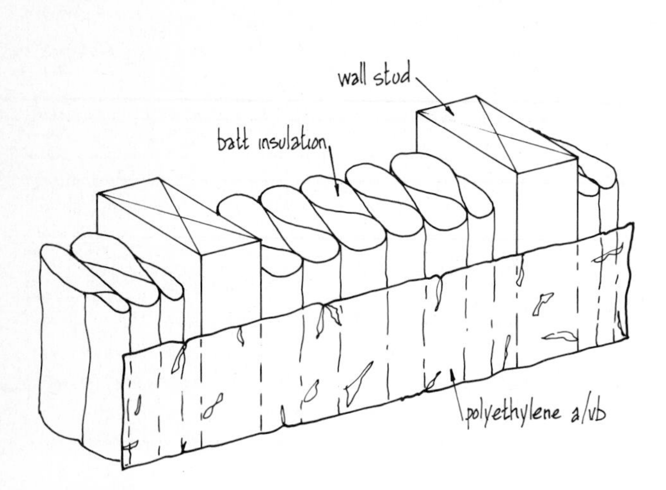

SINGLE STUD

- Conventional construction using 4-in (89) or 6-in (140) studs.
- 8 in (184) studs are generally uneconomical in most regions.
- R12 (2.1) with 4-in (89) or R19 (3.34) with 6-in (140) studs.
- A/vb on inside wall face.
- Use 24-in (600) spacing for 6-in (140) studs where allowed by codes.

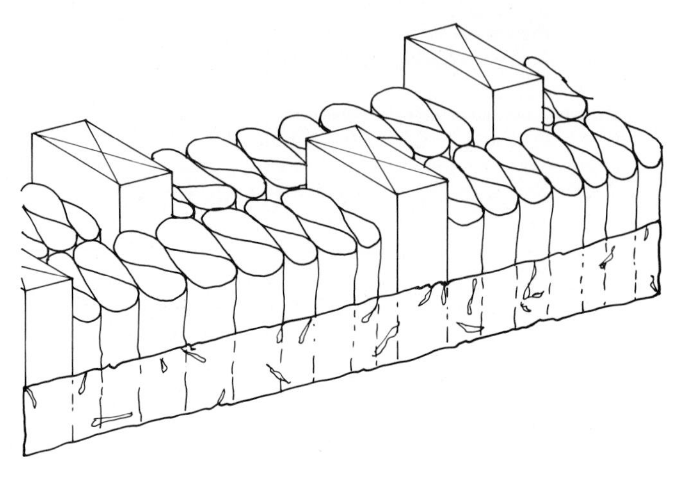

STAGGERED STUDS

- Increases wall thickness without major changes.
- With no across-wall contact, studs are thermally broken.
- Either 8-in (184) or 10-in (235) top and bottom plates are used.
- A/vb is still on the interior stud face.
- Up to R30 (5.25) is standard.

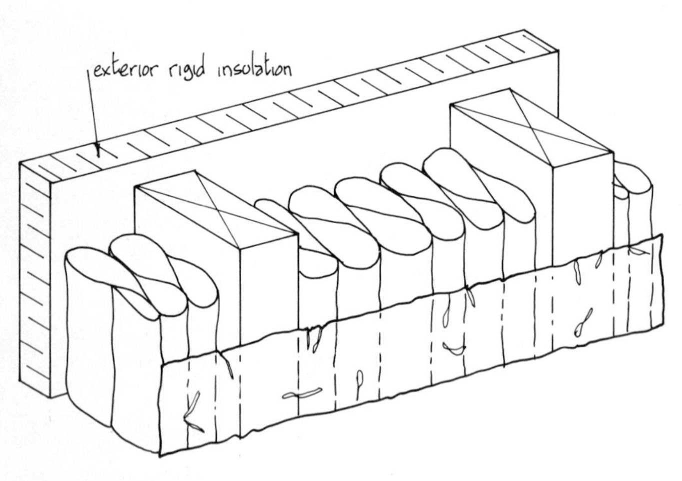

SINGLE STUD WITH INSULATED SHEATHING

- Rigid insulation is added to the exterior or interior face of the stud to reduce the thermal bridging effect through the studs.
- If placed on the interior, the a/vb may be omitted if the insulation acts in its place (see 10-13).
- Up to R35 (6.16) is obtainable with 6-in (140) studs and high-density rigid insulation.

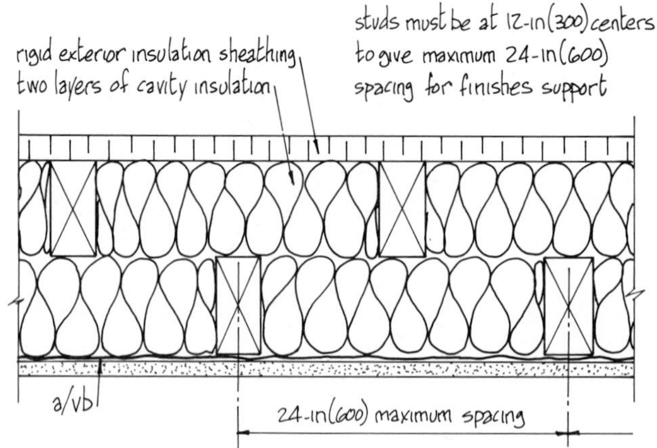

STAGGERED STUDS WITH INSULATED SHEATHING

- Same advantages as for single stud at the left.
- Up to R40 (7.04) is obtainable.
- Note the disadvantage of a maximum 12-in (300) stud spacing that is required for finishes support.

UPGRADED WALL TYPES

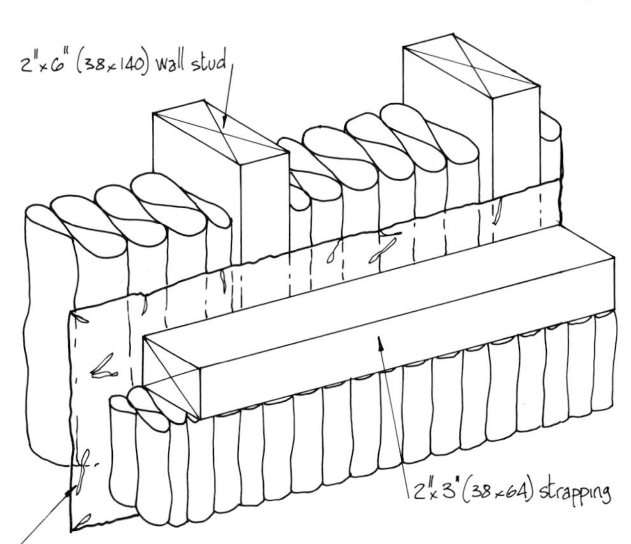

STRAPPED WALLS

- Thermal bridging is reduced and insulation can be increased.
- The a/vb can be located between the strapping and the stud for protection from damage and penetration by services.
- 2-, 3-, or 4-in (38, 64, 89) strapping is commonly used.
- Note that the one-third/two-thirds rule must be observed (see 10-11).
- Up to R32 (5.36) with 6-in (140) studs and 4-in (89) strapping.

DOUBLE STUD WALL

- Two separate stud walls with space between. Inner or outer walls can be load bearing.
- A/vb usually attached to outer face of inner wall for total air tightness as shown. Separately erected walls allow a variety of a/vb and insulation placement methods.
- R40 (7.04) with a 12-in (300)-thick wall.
- One-third/two-thirds rule applies.

STRAPPED WALL WITH INSULATED SHEATHING

- Exterior insulated sheathing is often added so that the one-third/two-thirds rule can be observed when using wide strapping.
- Up to R45 (7.92) is obtainable, with the added advantage of a tighter a/vb.

exterior insulation
2"x6" (38x140) stud
a/vb
2"x4" (38x89) strapping
minimum 2/3
maximum 1/3
R values
Section

DOUBLE STUD WITH INSULATED SHEATHING

- Up to R55 (9.68) or higher is possible, but this type of wall is unlikely to be economic in all but the coldest regions.

stud space
a/vb
stud space
minimum 2/3
maximum 1/3
R values
Section

As can be seen from the first two wall types on the preceding page, higher insulation levels are achievable with only minor changes in conventional construction techniques. To achieve very high levels, or to give added protection to the a/vb, different and more expensive techniques are necessary, as shown on this page.
The choice of wall type is dependent on many factors, not the least of which is cost-effectiveness. Some of these factors and suggested insulation levels are given on 10-22.

ROOF VENTILATION

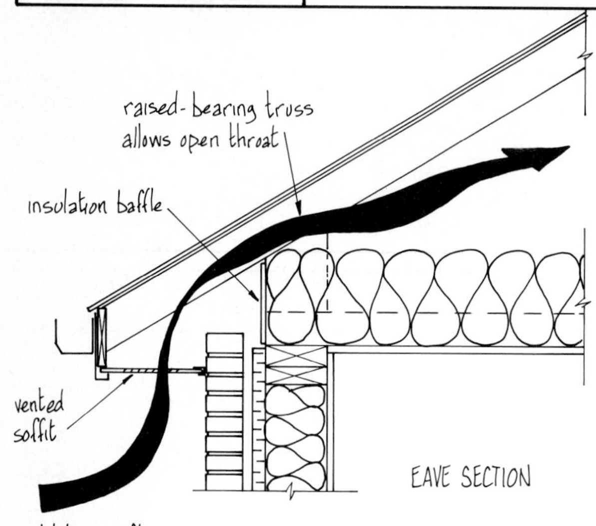

ATTIC ROOFS

- In an attic roof, values of R 60 (10.56) or higher are obtainable simply by adding batt or loose-poured insulation.
- To maintain attic ventilation (see below) a raised bearing truss, as shown at the left, is ideal (see 8-15 for details).
- Where rafters or standard-heel trusses are used, the ceiling insulation, which should be carried over the wall plate, must not be allowed to obstruct the throat opening. Three methods of doing this are shown below.
- With deep insulation, a continuous baffle attached to the wall plate or truss end will stop insulation spilling into the soffit vent or being displaced by the wind.

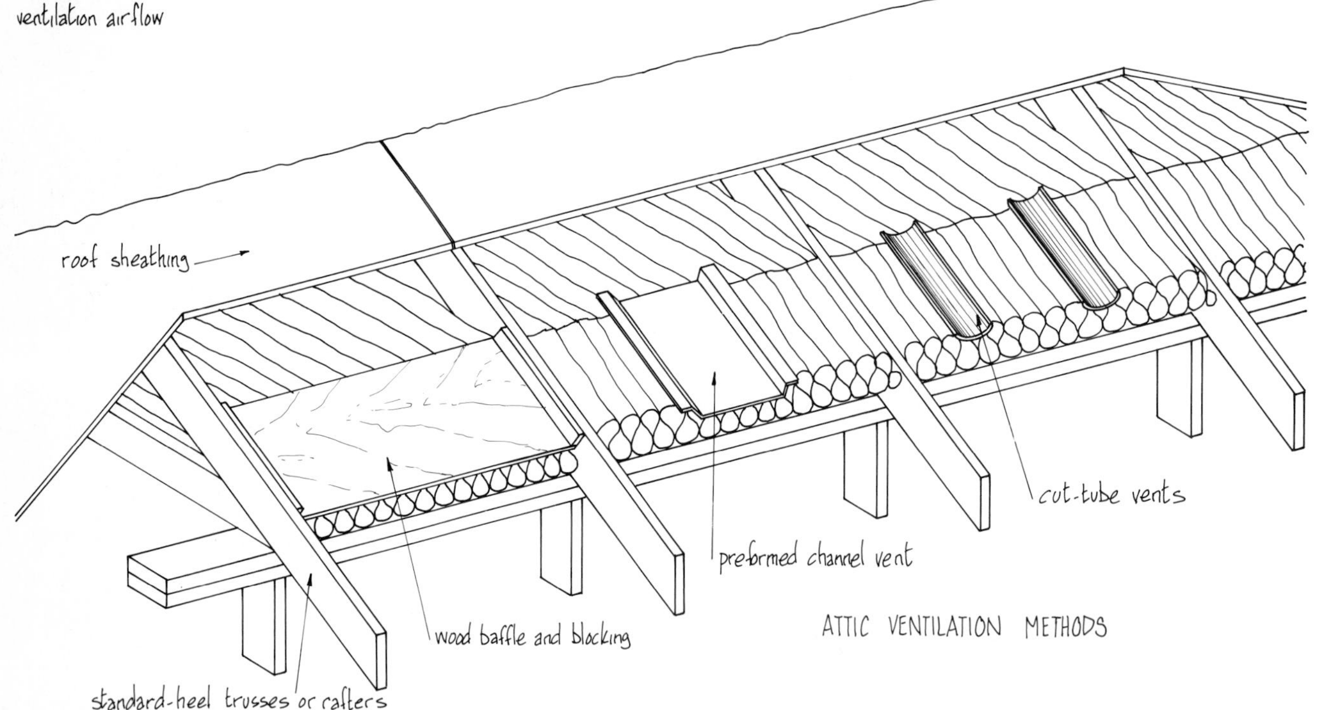

ATTIC VENTILATION METHODS

ridge vents
roof vents
gable vents
soffit vents

ATTIC VENTILATION

ROOF VENTILATION

- Even in tightly sealed houses, some condensation and frost will occur in open attic roofs. Attic ventilation is therefore essential if problems are to be avoided.
- Local codes specify minimum requirements but generally call for a free vent area of 1/300 of ceiling area for standard roofs and 1/150 for low-slope roofs.
- Ideally, half the vent area should be at the soffit level and half at a higher level in the form of ridge, gable, or roof vents.
- Flat roofs are something of a problem in cold regions and considerable research is being done in this area. Natural convection will not occur without a height differential and open venting serves little purpose. It is perhaps best to totally fill the ceiling space with insulation and tightly seal all openings to the roof to prevent warm moist air entering the roof space. Check with local authorities for current building practice.

UPGRADED ROOF TYPES

CATHEDRAL CEILINGS

Trusses

- High insulation levels in a cathedral ceiling are best obtained by using parallel chord trusses. These are economic in terms of material and erection time and allow cross-ventilation between truss spaces.
- Although the truss can be made deep enough to accept any R value, the added roof depth will increase wall heights and change visual effects. If this is unacceptable, use a scissor truss (see 8-9).

Rafters

- In a raftered roof the addition of cross purlins allows insulation to be increased but restricts the natural ventilation flow. For a better solution, see 10-53.
- Insulation values are limited by the size of the rafters and code requirements for the ventilation air space above the insulation.

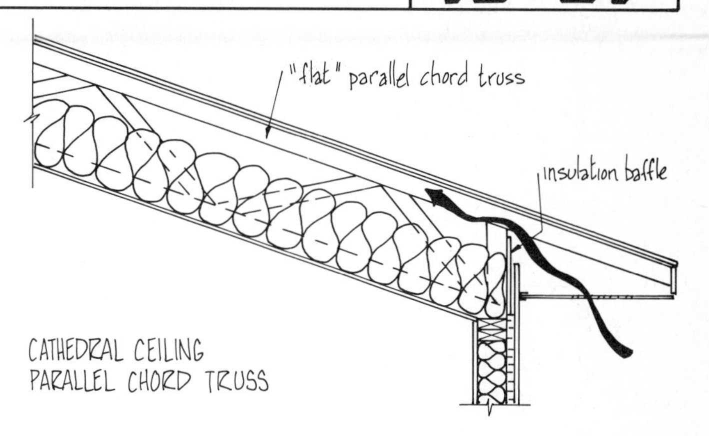

CATHEDRAL CEILING
PARALLEL CHORD TRUSS

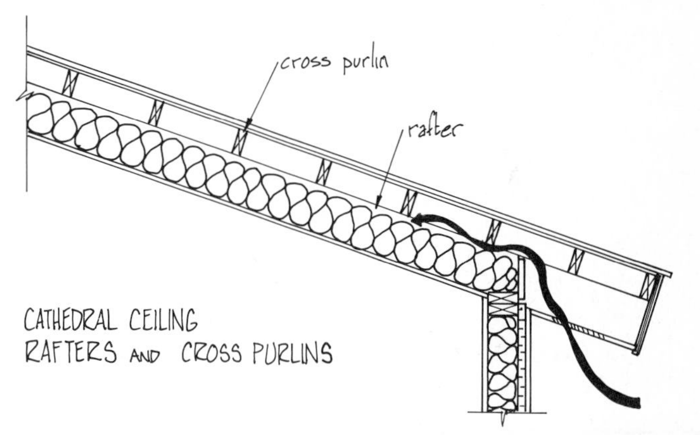

CATHEDRAL CEILING
RAFTERS AND CROSS PURLINS

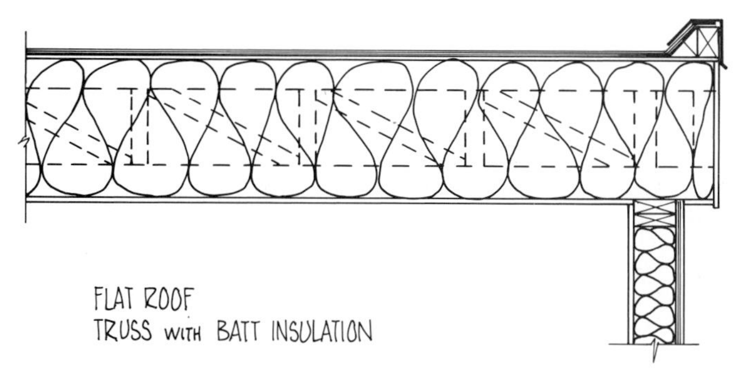

FLAT ROOF
TRUSS WITH BATT INSULATION

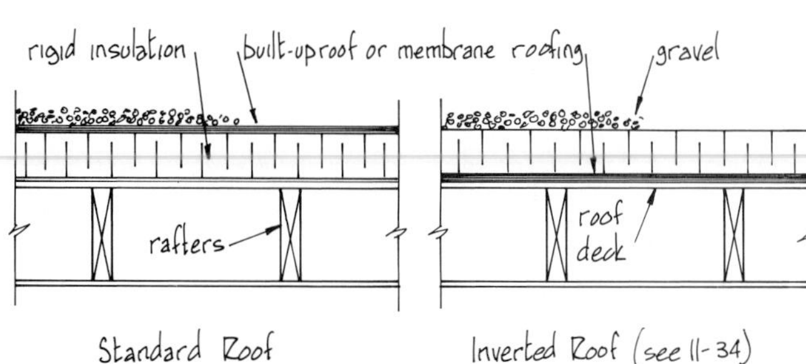

Standard Roof

Inverted Roof (see 11-34)

FLAT ROOF
RAFTERS WITH RIGID INSULATION

FLAT ROOFS

Batt or Loose Insulation

- Parallel chord trusses are again the best choice. Truss depth can be less if a ventilation airspace is not required (see the preceding page, bottom).
- Roof beams with cross-purlins are also in general use, and if the ceiling space is to be sealed and filled with insulation, greater R values can be obtained by increasing purlin size.

Rigid Insulation

- Rigid insulation can be placed on the top of the roof sheathing or decking (flat or pitched roofs), eliminating a cold ceiling cavity— an ideal method for post and beam structures (see chapter 9). However, rigid insulation is expensive and will require several layers to meet even minimum requirements.
- Rigid insulation can be placed either over or under the roof membrane. When placed over it is termed an "inverted roof" and protects the membrane from temperature variations and ultraviolet radiation. To avoid flotation of the insulation, good drainage is essential for inverted roofs.

INSULATION LEVELS

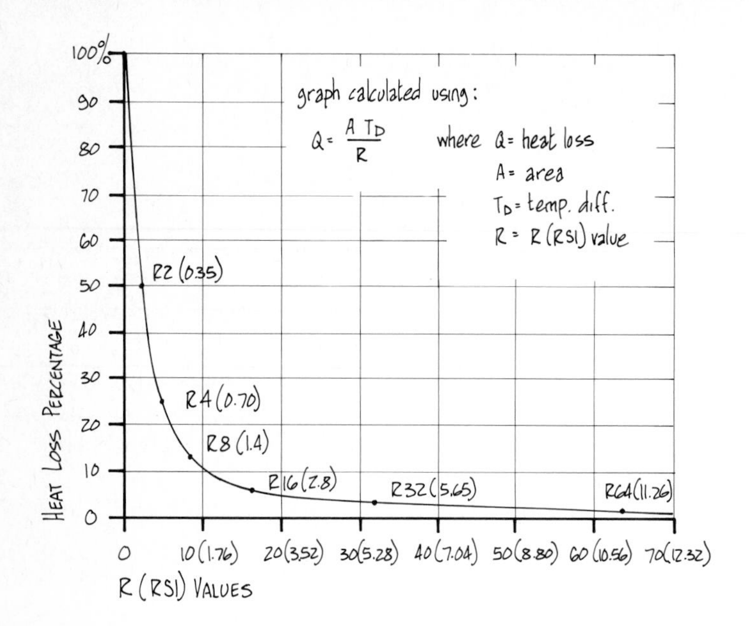

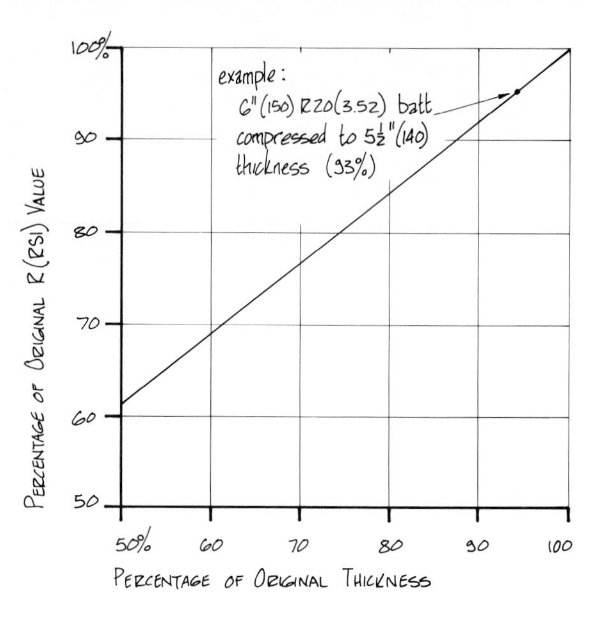

R VALUES AND HEAT LOSS

The graph above is very significant. It shows the relationship between R value and heat loss in any part of a structure under simplistic, straight-line conditions. Note the following:

- At zero R, heat loss is presumed to be 100% for comparison purposes.
- When the R value is doubled from its previous value, heat loss is reduced by 50% from its previous value.
- Note that at some level of R value there is a break-even point where adding more insulation has very little effect on reducing heat loss. On this graph that point is around R16 (2.8).
- However, the graph does not take into account inputs such as climate, present and future fuel costs, and construction type. Adding these factors moves the graph line up and to the right, giving a larger break-even R value. Even so, extremely high levels of insulation are not generally cost-effective except in the very coldest regions.

INSULATION COMPRESSION

The graph above shows how R value is reduced when insulation is compressed. This applies mostly to the non-rigid types of insulation such as fiberglass and related products.

Eg. a 6-in (150) R20 (3.52) fiberglass batt when installed in a 5½-in (140) stud space is compressed to 93% of its original thickness but retains 96% of its rated R value. Minor compression is therefore acceptable, but major compression will seriously reduce effective R values.

CHOOSING INSULATION VALUES

The choice of suitable insulation values for a structure is subject to several factors:

- Minimum levels are set out in local codes, but these are generally too low and are not revised fast enough to reflect current conditions.
- Various government and private organizations such as HUD in the United States and the NRC in Canada publish suggested levels more in line with today's conditions.
- Climate has an obvious effect on the levels chosen. Very cold and very hot regions need higher levels than temperate regions.
- Cost-effectiveness is very important. There must be an acceptable payback period wherein all extra costs are recovered through energy savings.

RECOMMENDED INSULATION VALUES

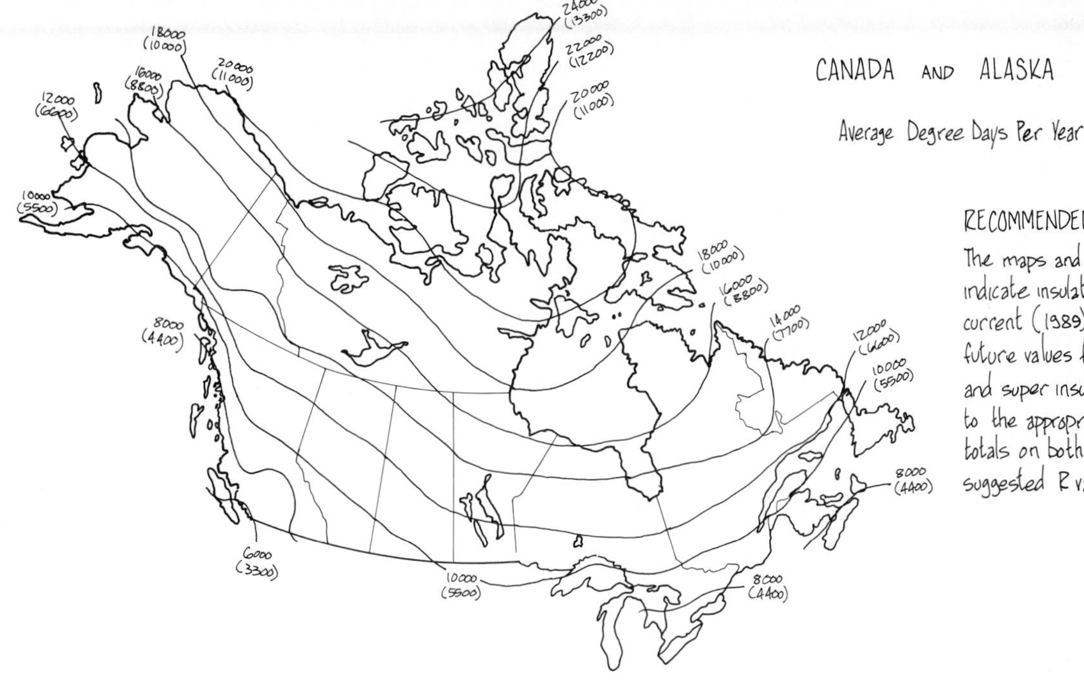

RECOMMENDED R-VALUES

The maps and table on this page indicate insulation levels that reflect current (1989) and reasonable projected future values for both energy efficient and super insulated buildings. Refer to the appropriate degree-day (DD) totals on both map and table for the suggested R values.

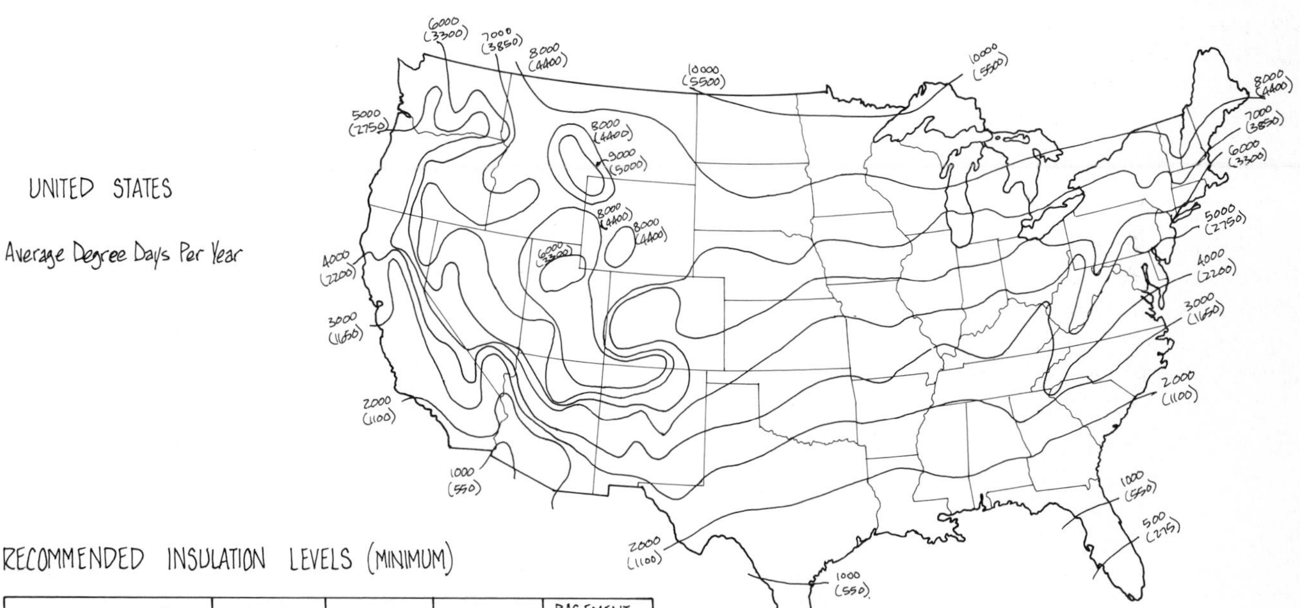

RECOMMENDED INSULATION LEVELS (MINIMUM)

DEGREE DAYS °F (°C)	CEILINGS R (RSI)	WALLS R (RSI)	FLOORS R (RSI)	BASEMENT WALLS R (RSI)
below 4000 (2200) temperate	20 (3.52)	12 (2.11)	12 (2.11)	5 (0.88)
below 4000 (2200) hot – humid/dry	35 (6.16)	20 (3.52)	20 (3.52)	12 (2.11)
4000 - 8000 (2200 - 4400)	30 (5.28)	20 (3.52)	20 (3.52)	12 (2.11)
8000 - 12000 (4400 - 6600)	40 (7.04)	25 (4.4)	25 (4.4)	20 (3.52)
above 12000 (6600)	50+ (8.8+)	35+ (6.16+)	35+ (6.16+)	30 (5.28)
super insulated all regions	50 - 60 (8.8 - 10.6)	35 - 45 (6.2 - 7.9)	35+ (6.16+)	25+ (4.4)

NOTES TO TABLE

- The degree-days measurement is the total daily accumulation of degrees during which the average outside temperature falls below 65°F (18°C). This measurement can have significant local variations (especially in mountain regions) within the broad curve shown on the maps. Refer to local weather statistics for exact DD values.
- Hot humid or dry regions (with low DDs) require high levels of insulation to reduce cooling costs.
- The tightness of the a/vb should be increased proportionally with the DD total. Above 8000 (4400) DD, airtightness should be as total as possible.

10-24 LIVING SPACE VENTILATION

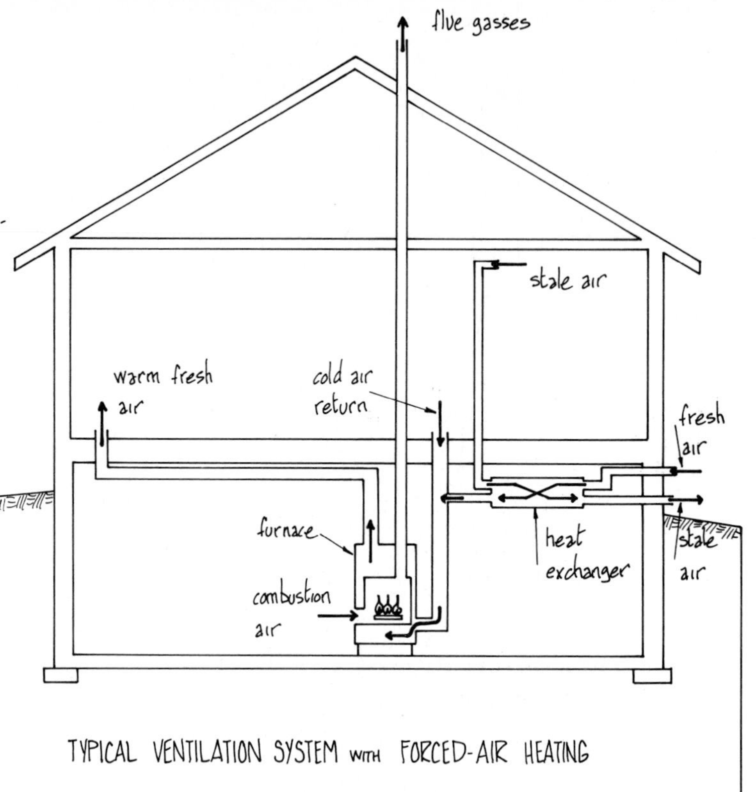

TYPICAL VENTILATION SYSTEM WITH FORCED-AIR HEATING

NEED FOR VENTILATION

In conventionally built housing, uncontrolled ventilation occurs naturally and in large quantities through holes and cracks in the structure. In upgraded airtight housing with minimal leakage, natural ventilation does not occur and must be replaced by a forced-ventilation system. Ventilation is absolutely essential for three basic reasons:

- Removal of stale odors from cooking, toxic fumes from combustion furnaces and hot water heaters, and outgassing from building materials.
- Excess humidity from various sources (see 10-10) must be removed if visible surface and hidden insulated space condensation is to be reduced or eliminated.
- Air used by heating equipment for combustion must be replaced (makeup air).

In addition, controlled ventilation is used to maintain a high neutral plane (see 10-9) so that the minimal air leakage remaining is directed inward to dry insulated cavities.

VENTILATION QUANTITIES

The amount of ventilation in a building can be expressed in two ways:

- Air changes per hour (ac/h) which indicates the number of times each hour the total air volume of the building is replaced by incoming fresh outside air. Conventional houses have rates of 1 to 4 ac/h, compared to airtight houses with 0.1 to 0.5 ac/h.
- Airflow rates, measured in cubic feet per minute (cfm) or, in metric, liters per second (L/s). This method is the most practical since it can be calculated on a demand basis based on the number of occupants, room usage, and makeup air requirements, all of which are independent of building size.

VENTILATION SYSTEMS

In energy-efficient housing the ventilation system is extremely important if acceptable comfort conditions are to be maintained. Each system must be designed to function within a specific structure and should be integrated with the heating or cooling equipment installed therein. In addition, most systems should be equipped with heat-recovery devices that increase ventilation efficiency, especially in cold regions.

Because ventilation systems are so closely tied to the building's heating and cooling equipment, discussion of this subject is presented in Chapter 14, beginning on 14-16. Please be sure to refer to this information.

A/VB INSTALLATION

The pages from here to the end of the chapter detail the installation of air/vapor barriers and insulation for most types of structures. Some of the techniques are somewhat different from long established "conventional" construction. Designers and contractors therefore need to revise their standard methods in both detailing and on-site work if economy and efficiency are to be achieved.

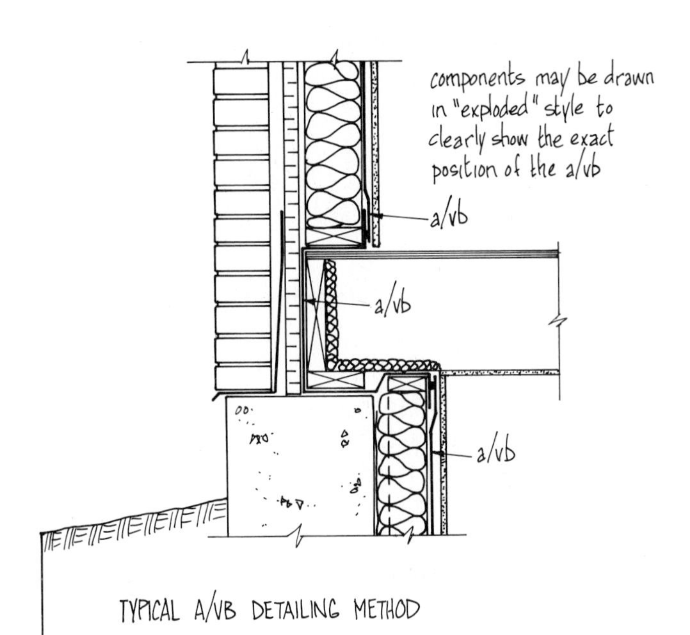

TYPICAL A/VB DETAILING METHOD

The illustration above shows one method of detailing the position of the a/vb on a construction drawing. Designers should note that the construction trades need to know the position, placement sequence, and type of a/vb for correct installation. This is especially so for critical junctions in the structure, such as at headers, wall/floor/ceiling joints, and corners, etc., where most air leakage is likely to occur.

CAULKING

The type of material chosen for sealing the a/vb (as distinct from exterior weather caulking) is very important. Ideally, the sealant should remain flexible and crack free throughout the building's life. Suitable types include acoustical sealant which remains tacky, and butyl or rubber caulking, which cures but remains flexible. It is the designer's and contractor's responsibility to specify suitable sealants. Check manufacturers' recommendations.
It is important to note that the sealant acts only as a gasket and will strip off the poly under wind and mechanical loads unless applied over a solid backing and compressed by the interior finishes.

INSTALLATION PRACTICE — POLYETHYLENE

- The a/vb should be installed to create a completely sealed envelope that is as airtight as possible. The a/vb must also be sealed to doors, windows, and vents or other obstructions that penetrate the envelope.
- Joints between sheets should be made over a solid backing (studs, plates etc.) with a continuous run of sealant applied between sheets. Both sheets are then stapled to the backing after being lightly pressed together. Stapling is necessary since the sealant is not an adhesive.
- Sheet overlap where sealant is not used should be at least one stud space, with staples at 6-in (150) centers.
- At corners and intersections allow spare material to avoid sheet stretching and tearing during plasterboard installation.
- Poly sheets that are placed before the main a/vb installation (such as at headers and other wall/floor junctions) should be protected with temporary covers. This is particularly important at door openings, where traffic is heavy.
- After installation, but before interior finishes are applied, the a/vb should be checked for even the smallest damage. While small holes can be patched with suitable tape, large holes require a large patch of poly sealed and stapled to the nearest solid backing.

It is obvious from the above that this type of careful a/vb installation is quite different from established conventional techniques. It is important that contractors and designers promote trade awareness of these different methods, particularly since the a/vb requires considerable preparatory work during all phases of the framing. Finishing trades, especially, must be cautioned not to damage the seal but also to report any holes that are made.

JOINTING POLY A/VBs

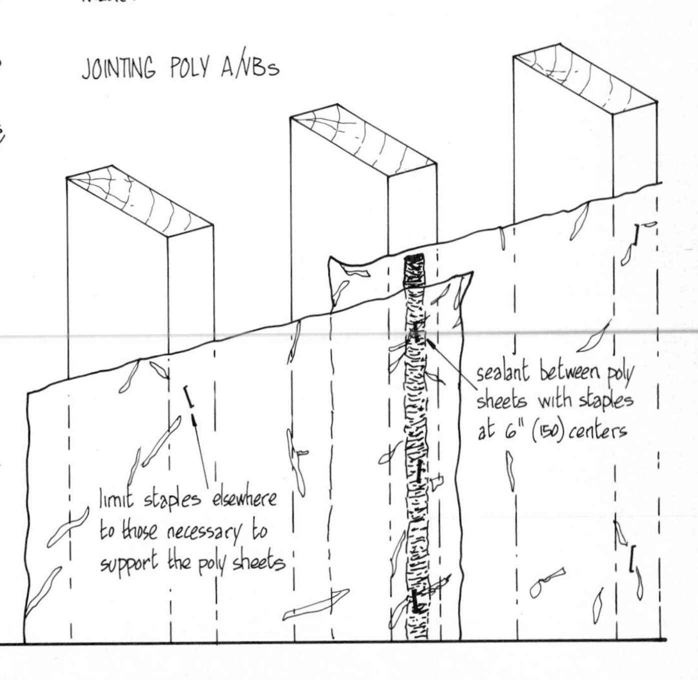

A/VB POSITIONS

A/VB POSITIONS

These two illustrations show the general position of the a/vb in a structure. Exact a/vb positions in specific structure types are detailed on the following pages.

- All a/vbs are on the warm side, with the exception of basement and crawl space floors, where the second vb's function is to stop water vapor infiltration from groundwater.
- Note that a moisture barrier, not a vapor barrier, is placed against the inside face of basement walls to control moisture seepage into the insulation.

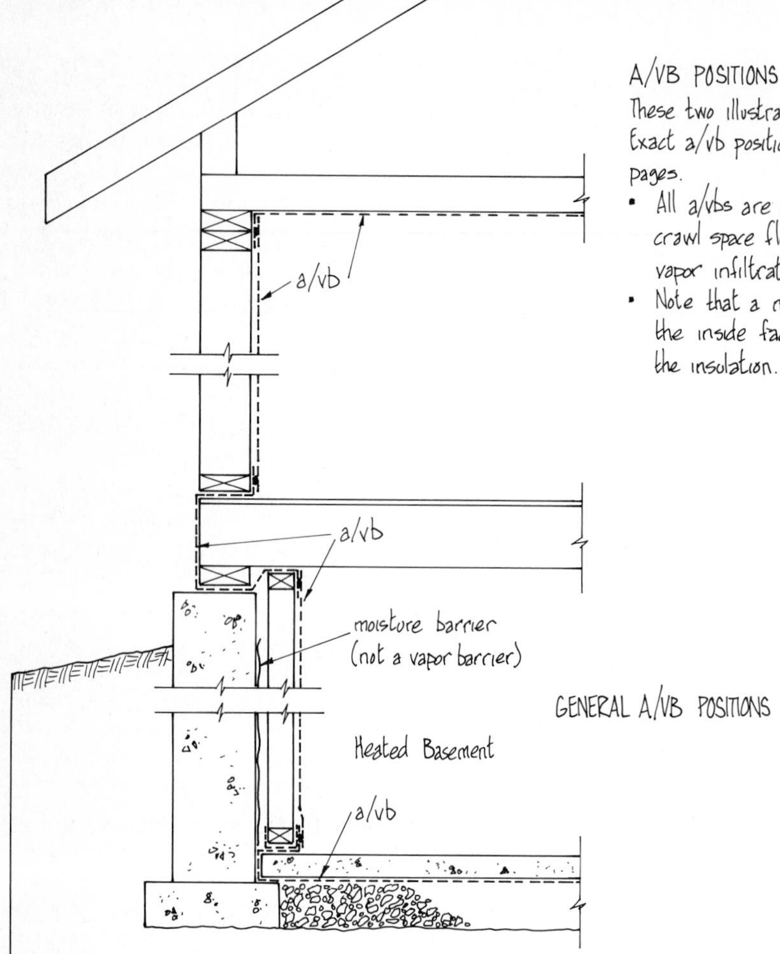

GENERAL A/VB POSITIONS

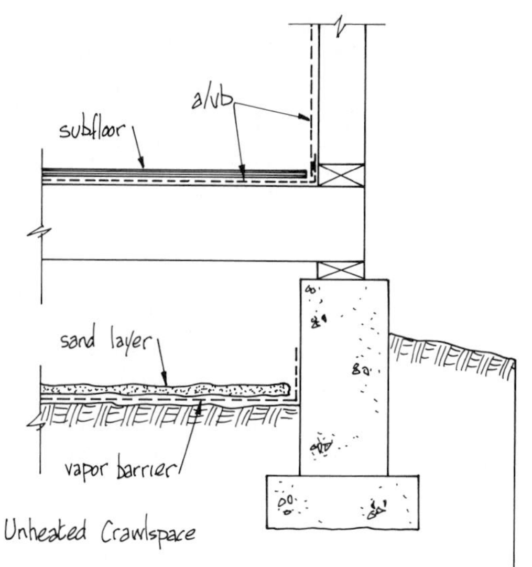

EXTERIOR WALLS

- Use full-width sheets that overlap top and bottom plates.
- Staple to top and bottom plates, and to studs only if necessary for stability. Interior finishes will provide final support.
- Cover wall completely and cut openings later.
- Windows and doors should be provided with a preapplied a/vb (see 10-60). Seal this to the main wall sheet after openings are cut.
- Apply the same procedures to ceilings and other areas.

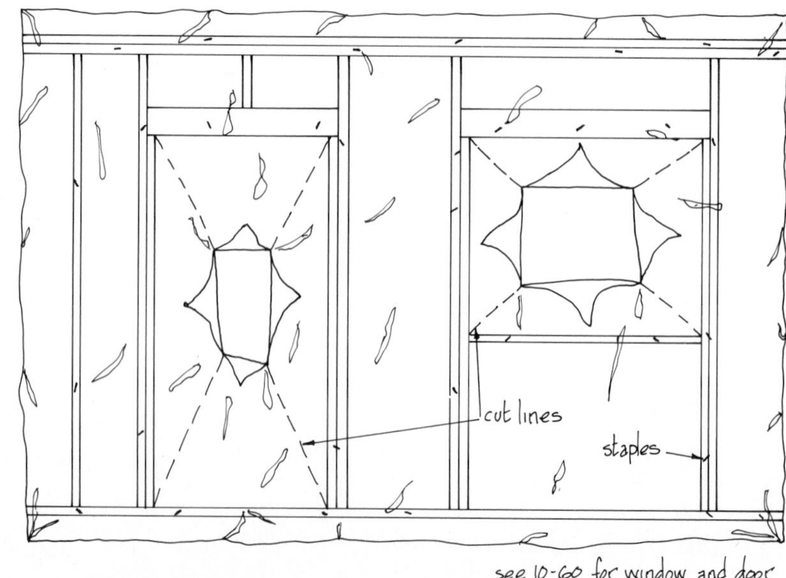

WALL ELEVATION

see 10-60 for window and door a/vb application techniques

INSULATION INSTALLATION

Each type of insulation should be installed in specific ways to ensure full efficiency and to avoid potential double-vb problems. The illustrations on this page give general procedures for four common insulation types. Information for other types appears on the following pages, but in all cases check manufacturers' recommended procedures.

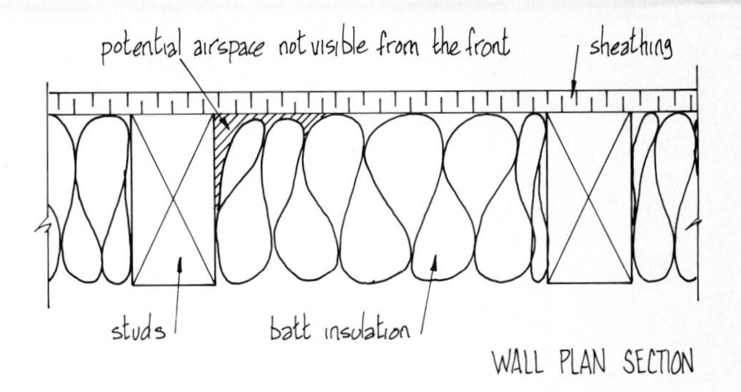

WALL PLAN SECTION

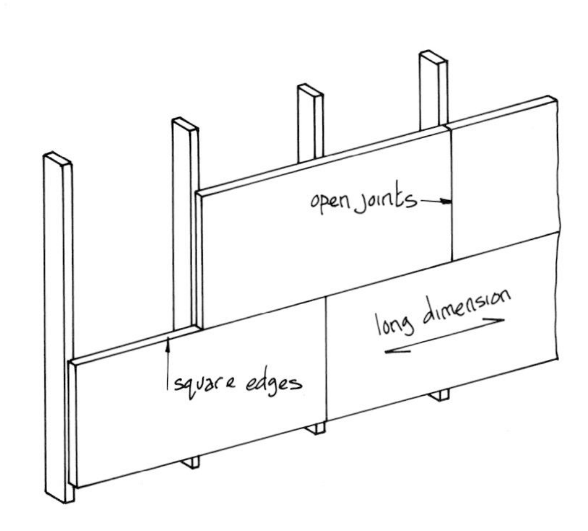

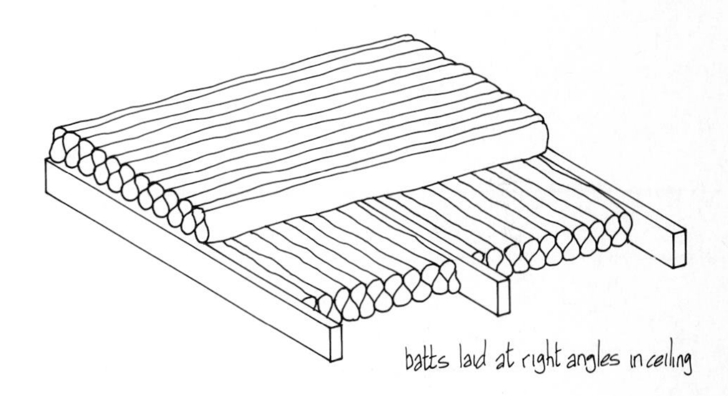
batts laid at right angles in ceiling

RIGID PLASTIC-FOAM BOARDS

- When installed as exterior sheathing care must be taken to avoid creating a double vapor barrier with any product that is nonpermeable. Wall cavities must be vented in some way to reduce condensation buildup. E.g., polystyrene sheathing should be installed horizontally in maximum 2-ft (600)-wide sheets with square (not jointed) edges. Joints must not be taped. It is important to refer to manufacturers' instructions with these products.

FIBERGLASS BATTS

- When installed in a wall cavity (upper drawing), the batt must completely fill the cavity with no internal airspaces. Even the smallest airspace will allow convection currents to occur, especially if a crack extends to the warm inner surface. This problem increases with batt thickness. For correct installation, the batt must be pushed hard into the cavity, especially at the edges, and then pulled out flush with the stud faces.
- For ceilings it is best to use two thinner batts rather than one thick batt. The two layers can then be laid at right angles to completely cover the ceiling frame and the side joints of the lower batts. This procedure also applies to walls with multiple batt layers.

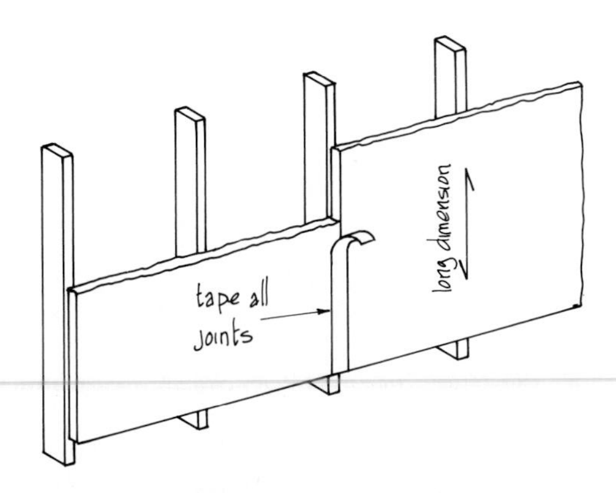

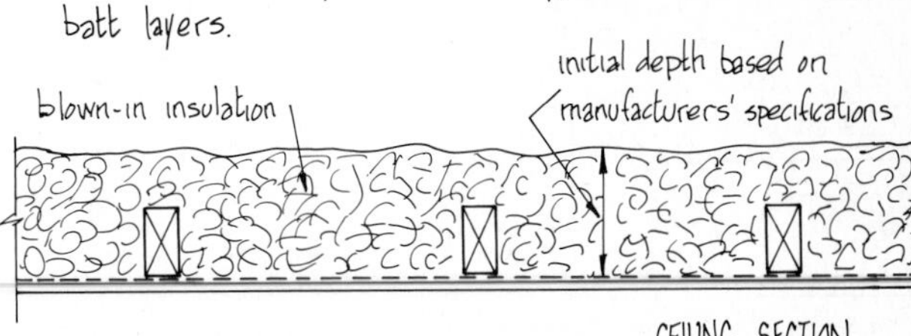

CEILING SECTION

RIGID FIBERGLASS SHEATHING

- This product may be installed in full-sized sheets with taped joints provided that the facing is of Tyvek or similar permeable material (see 10-20). Foil-faced sheathing may not be taped.

LOOSE-FILL INSULATION

- This material is normally blown in place with special equipment. Since some products will settle after installation (reducing the effective R value), they are initially installed on the basis of a maximum coverage per bag of material or on a minimum weight per unit area. After settlement the material should then be at design R value.
- Care must be taken not to block eave ventilation with these products.

HEADER DETAILS - CONVENTIONAL

The header area, where floor joists meet supporting walls, is often the largest source of air infiltration and frame heat loss in a conventional structure. While the exact construction at a header (band) is influenced by the type of wall, floor, and finishing materials, the steps taken to achieve energy efficiency are basically the same and can be modified to suit most situations. Specific header solutions are given on the next three pages; other solutions appear with wall and floor details.

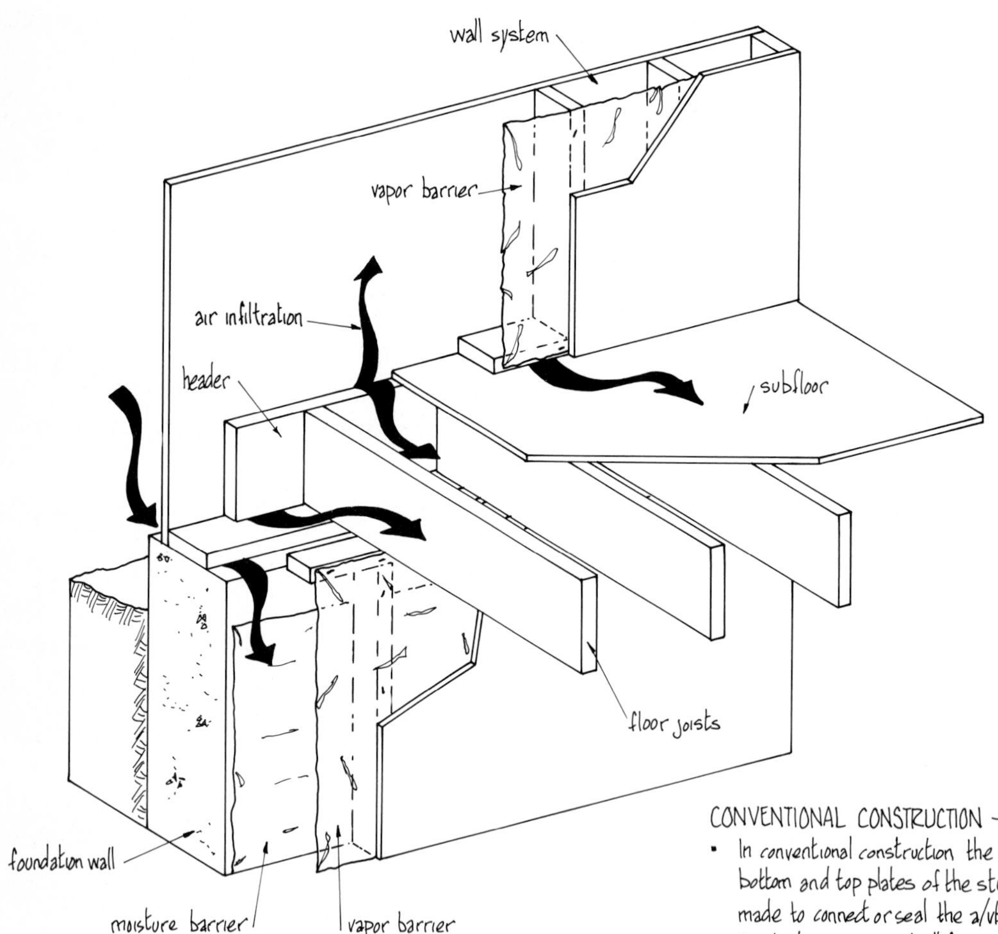

TYPICAL AIR INFILTRATION IN CONVENTIONAL CONSTRUCTION

CONVENTIONAL CONSTRUCTION – AIR LEAKAGE

- In conventional construction the a/vb is simply tacked to the bottom and top plates of the stud walls, and no attempt is made to connect or seal the a/vbs.
- Air leakage occurs at all frame junctions and will infiltrate or exfiltrate depending on pressure conditions.
- Tightly packing insulation materials into the spaces formed by the structure does little to stop air movement.
- This problem is the same whatever wall types are used and also occurs at second-floor headers.

HEADER DETAILS - UPGRADED

A/VB CONTINUATION WITH POLYETHYLENE SHEET

- A separate sheet of polyethylene is continued around sill and header and sealed to both upper and lower a/vbs. The poly-strip must be attached before floor framing starts.
- In this detail the foundation wall is insulated on the outside. This keeps the sill plate relatively warm and the poly can be placed under the sill. For interior insulated (batt or rigid) foundations the poly should be placed above the sill (see 10-34).
- Note that the one-third/two-thirds a/vb placement rule must be observed. The insulation value outside the a/vb must be at least two-thirds of the total header insulation. There is a temptation to fill the header space between floor joists with high levels of insulation. This must be resisted or the a/vb will be placed on the cold side and condensation will result.
- With this method the maximum level of insulation at the header is relatively low.

polyethylene a/vb extended around header

sealant

limited insulation 1/3 - 2/3 rule applies

rigid insulation

floor joist

sill plate

sealant

The two details shown will apply equally to other floor systems such as trusses or wood I-beams

POLYETHYLENE - SEALED HEADER

AIR BARRIER CONTINUATION WITH TYVEK SHEET (see 10-12)

- Tyvek sheet can be used to continue the air barrier around the header as shown. It is sealed to the upper and lower poly a/vb in the normal way.
- Note, however, that Tyvek is not a vapor barrier and the interior basement wall poly a/vb must be extended and sealed to the subfloor sheathing and floor joist sides. If this is not done, water will condense on the cold header, joist ends, and sill.
- The advantage of this approach is that any amount of insulation can be added to the header area since the one third/two-thirds rule does not apply to Tyvek sheet.

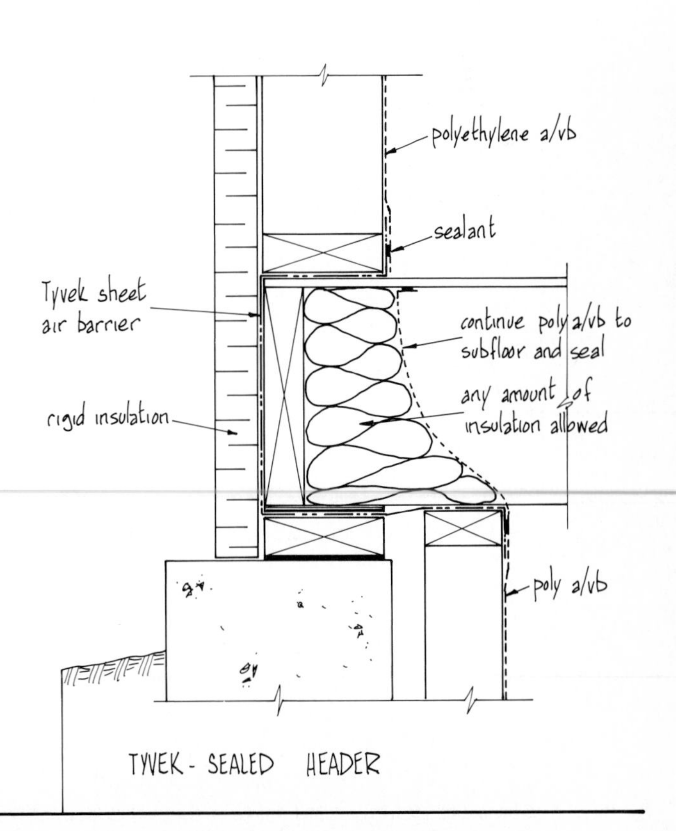

TYVEK - SEALED HEADER

HEADER DETAILS - UPGRADED

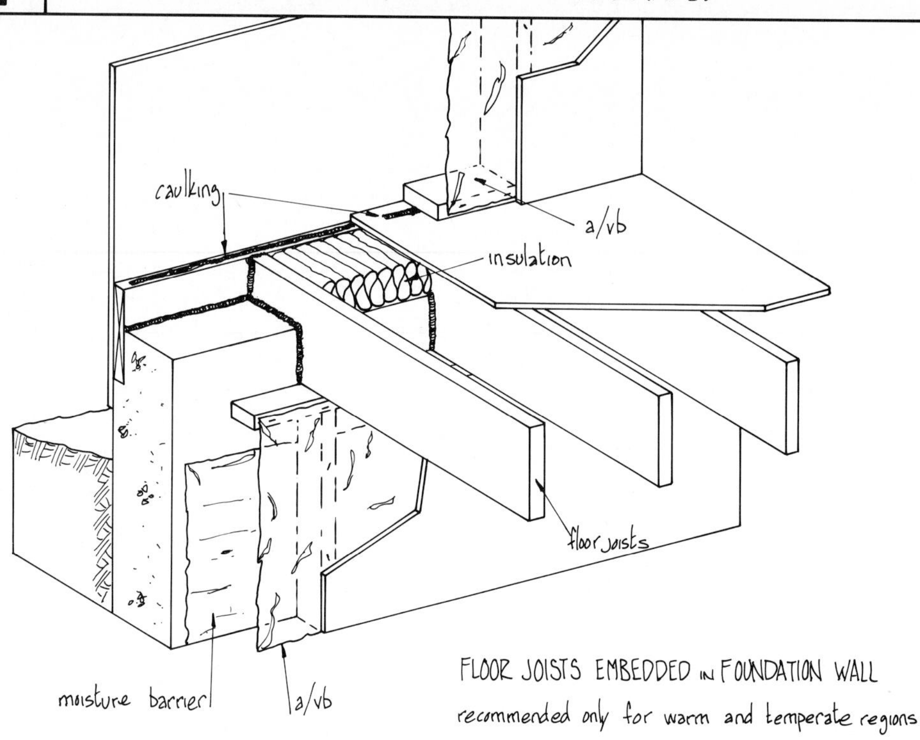

FLOOR JOISTS EMBEDDED IN FOUNDATION WALL
recommended only for warm and temperate regions

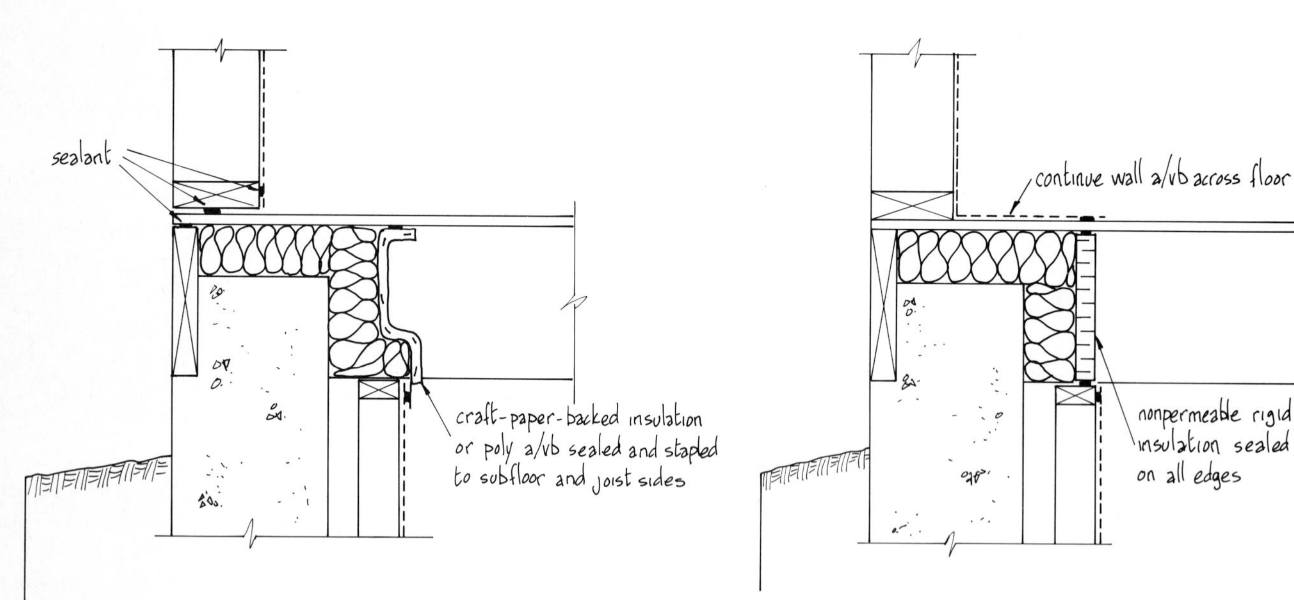

This type of framing, where the floor joists are either embedded or placed in pockets in the foundation wall, poses special problems for which there are three basic solutions:

- All cracks and joints can be caulked as shown above, with insulation placed behind the header. However, there is no vapor barrier on the warm side, and structural damage may occur.
- A better solution is either to seal the interior a/vb to subfloor and joists (middle left) or to tightly fit and seal individual pieces of nonpermeable rigid insulation between floor joists (middle right).

None of these solutions are totally practical. Each requires considerable labor to produce an effective seal and may result in structural deterioration.

It is strongly recommended that this type of construction not be used for energy-efficient structures.

HEADER DETAILS – UPGRADED

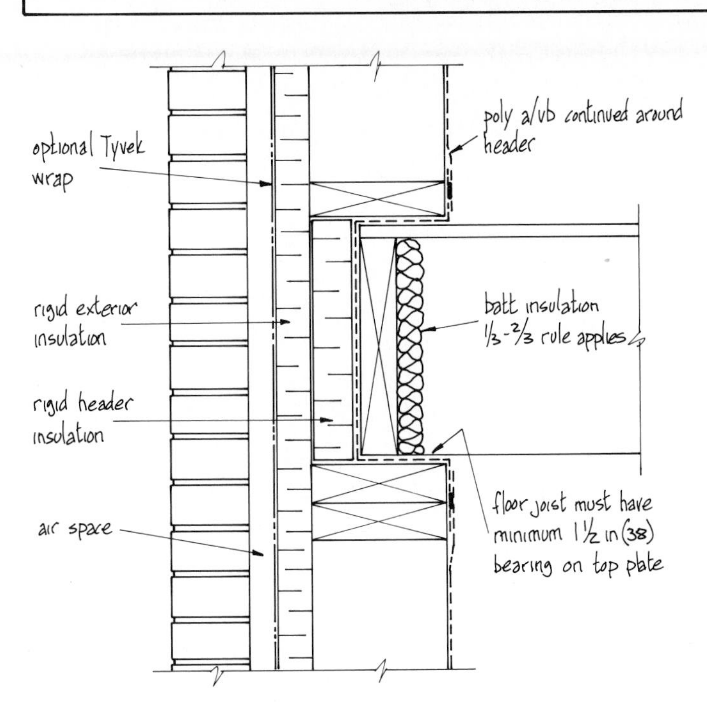

SECOND-FLOOR HEADER DETAIL

A/VB AT SECOND-FLOOR HEADERS

- This junction can be treated in much the same way as sill headers. A separate strip of poly is attached to the lower wall top plate before the floor is framed.
- In the example shown, the header is moved in from the wall line, allowing extra insulation outside the poly a/vb to meet the one-third/two-thirds rule. Note that the floor joists must have a minimum 1½-in (38) bearing on the top plate and that the upper stud wall must have adequate support at the bottom plate.
- Additional insulation value, or a larger bearing area, could be achieved by fitting the header between the floor joists, but this is a more expensive option.
- As with all details of this type, the connecting strip of poly must be installed before the floor framing starts, and protected during further construction.

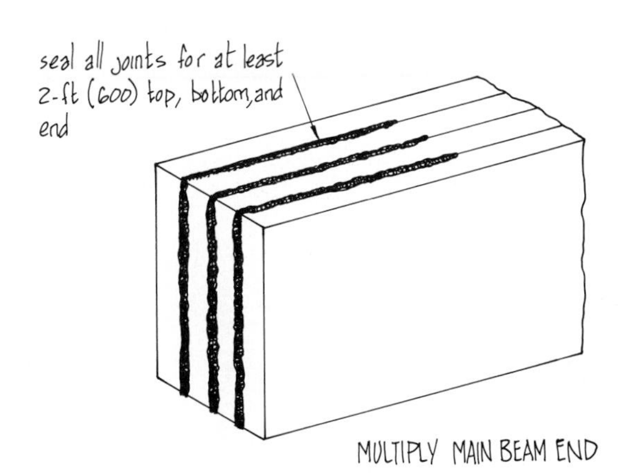

MULTIPLY MAIN BEAM END

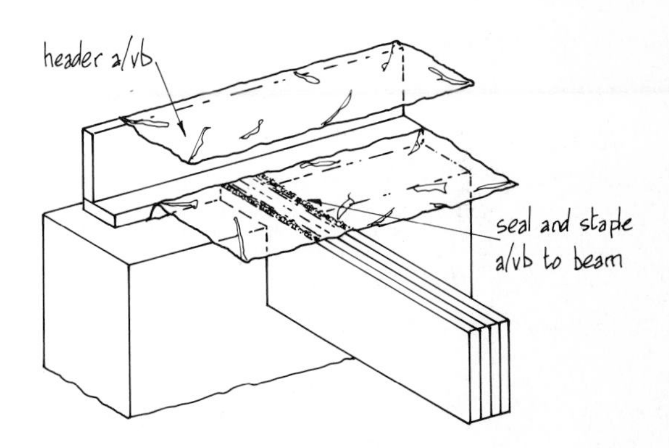

Main supporting beams present a double problem at foundation walls; the header a/vb must be continued around the beam and the plys of the beam allow air infiltration to bypass the a/vb.

- The first step is to seal each ply joint before erection (above). Wrapping in poly is an alternative but is subject to construction damage.
- The header a/vb is then sealed to the beam top and protected during construction (top right).
- The header a/vb is then cut alongside the beam and folded down to seal to the stud frame a/vb below (lower right).

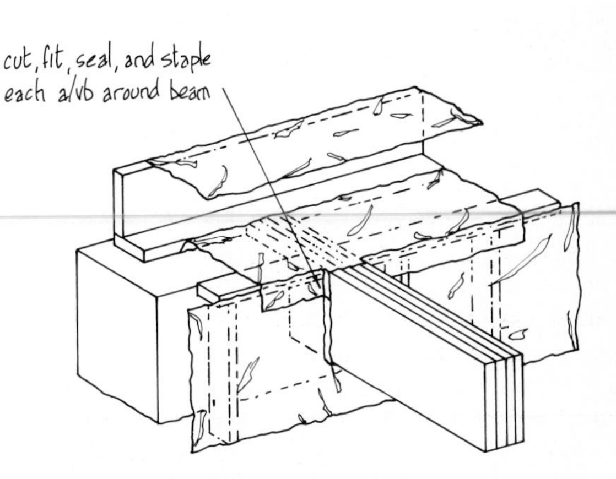

BASEMENT WALL INSULATION

When possible it is generally better to insulate a basement wall on the outside, for the following reasons:

- The wall is protected from temperature extremes and is less likely to be damaged by cracking.
- The mass of the wall acts as a heat storage medium that reduces daily internal temperature swings (see 3-11).
- The insulation helps protect the dampproofing membrane on the wall's exterior and may reduce leakage through the wall.

Exposed insulation must be protected from mechanical damage and ultraviolet radiation to a minimum below-grade depth of 12 in (300). Standard protection methods are:

- Cement parging on wire mesh or expanded metal (see 11-20).
- Pressure-treated plywood sheathing.
- Asbestos board.

EXTERIOR INSULATION

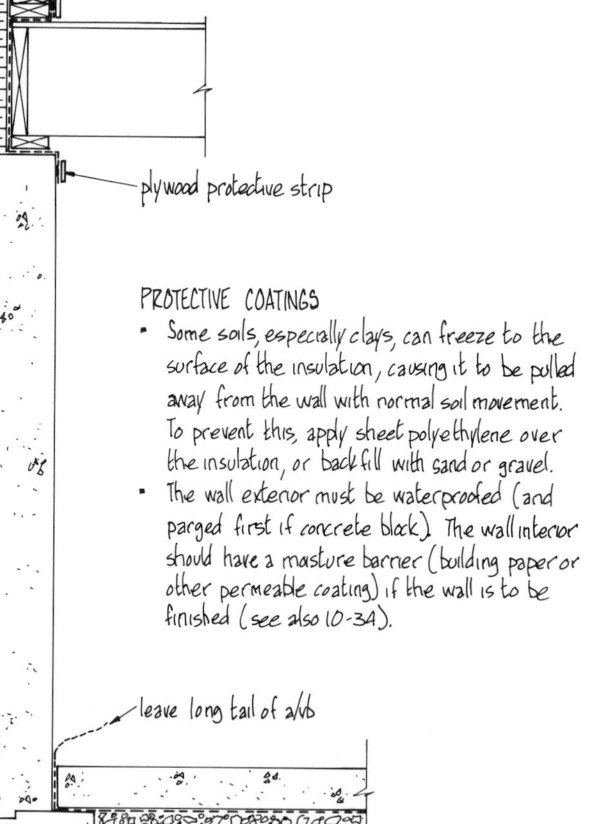

INSULATION TYPES

There are two insulation types in general use below grade:

- Extruded Polystyrene. This is the only type of polystyrene recommended for long-term below-grade applications (see 10-16). It will not degrade (unless exposed to sunlight) and absorbs very little water. Install in full-sized panels with or without interlocking joints. Expanded poly is sometimes used, but is subject to water saturation and physical breakdown over time and is not recommended.
- Rigid Fiberglass (below-grade type). This material is specially designed for below-grade applications and has certain advantages. Although the material readily absorbs water, the outer layer of fibers will conduct all moisture down to the weeping tile for drainage. This material must therefore be installed down to, and make good contact with, the weeping tile. It must not be stopped half-way or a water buildup will occur above the footings.

Installation note: insulation panels are normally installed vertically and are held in place by the earth or gravel backfill. In some cases, to provide stability during installation and backfilling, adhesive or mechanical fasteners are recommended for the above-grade portion of the panels.

PROTECTIVE COATINGS

- Some soils, especially clays, can freeze to the surface of the insulation, causing it to be pulled away from the wall with normal soil movement. To prevent this, apply sheet polyethylene over the insulation, or backfill with sand or gravel.
- The wall exterior must be waterproofed (and parged first if concrete block). The wall interior should have a moisture barrier (building paper or other permeable coating) if the wall is to be finished (see also 10-34).

To stop convection currents in concrete block foundation walls, use solid blocks for the top and at least one intermediate course (see 5-13).

The underslab a/vb can be terminated at floor level in two ways:

- If the basement interior will not be finished, the a/vb can be terminated and sealed at floor level.
- For finished basements leave a minimum 12-in (300) tail for connection to the finishes a/vb.

FOUNDATION WALLS - EXTERIOR INSULATION

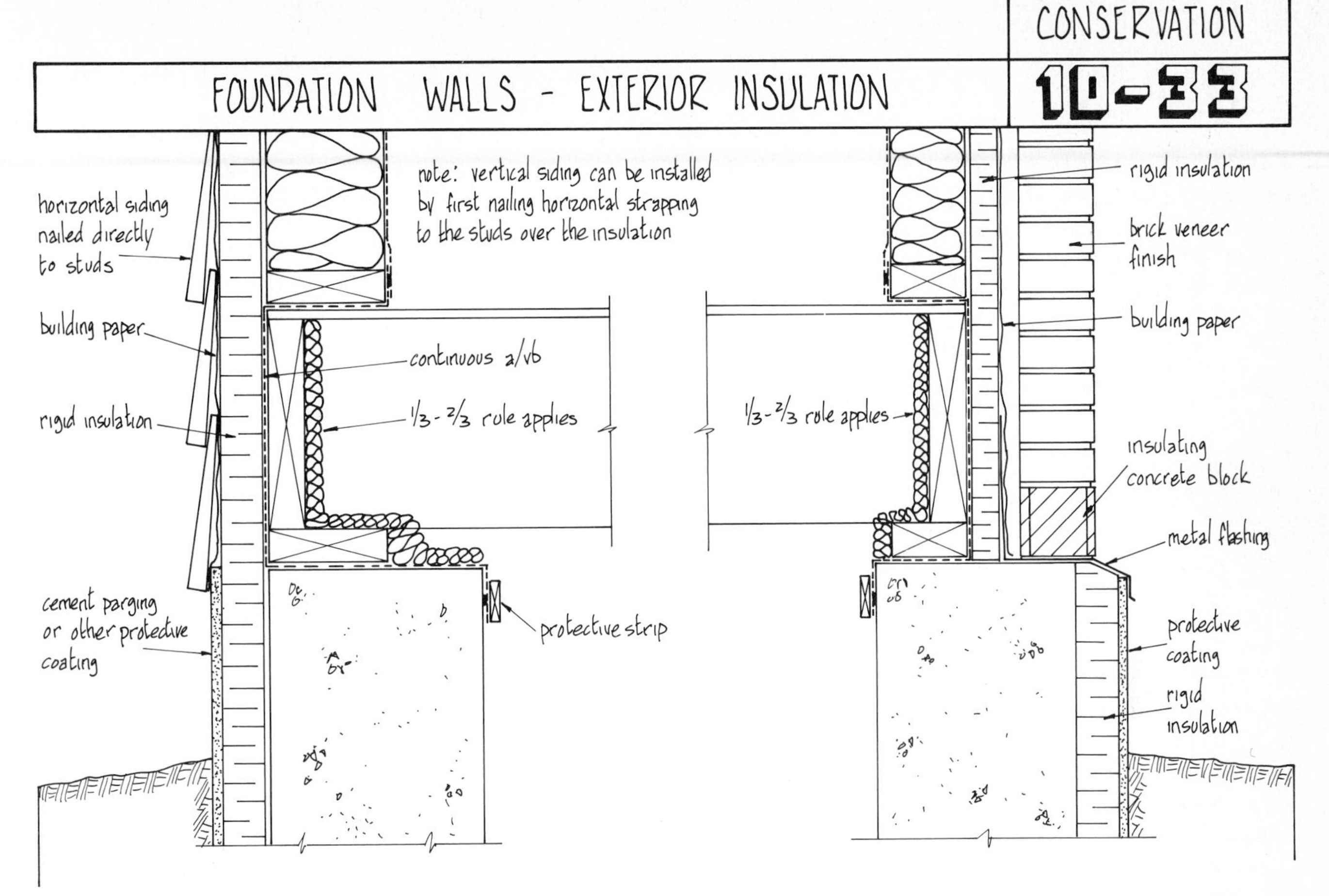

The section above shows one typical application where siding is used as the exterior finish. Exact details will depend on type of siding, insulation material thickness, and method of protection for below-grade insulation.

The brick siding option above uses a course of insulating block (see 10-38) to eliminate the cold area where the brick finish is normally supported on the foundation wall.

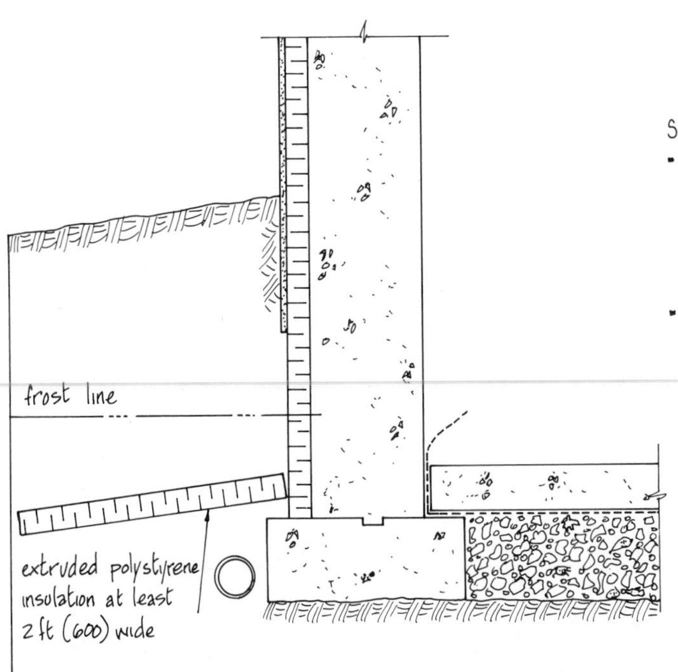

SHALLOW BASEMENTS

- Even though the footing must be below the frost line (see 5-8), it is advisable for this type of basement to extend the below-grade insulation out over the weeping tile. With a fully insulated basement wall the shallow footing area is relatively cold and this procedure helps keep the footing frost-free.
- An additional efficiency measure is to make the above-grade portion of the foundation wall as small as possible and to continue the basement wall up to first-floor level with fully insulated stud construction.

BASEMENT WALLS - INTERIOR INSULATION

Basement walls are often insulated on the inside, especially when the basement is to be finished. This approach has the advantage of providing a flat, vertical base for the finishes and containment for the insulation.

In this detail, because the sill plate is relatively cold, the poly a/vb is placed over the sill as shown, to exclude the moist interior air and stop condensation forming. The a/vb should then be protected from construction damage with a thin strip of plywood.

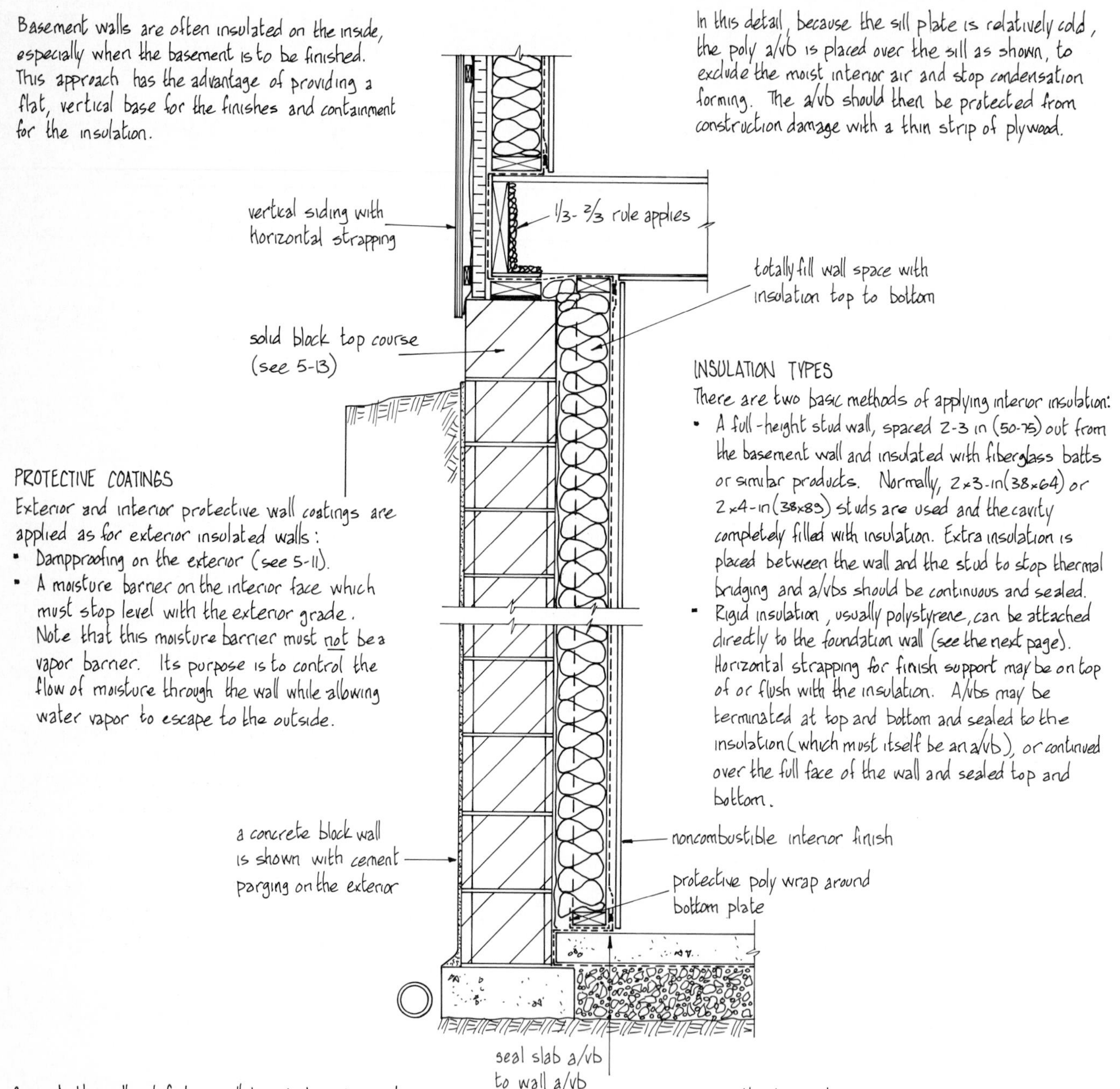

PROTECTIVE COATINGS

Exterior and interior protective wall coatings are applied as for exterior insulated walls:

- Dampproofing on the exterior (see 5-11).
- A moisture barrier on the interior face which must stop level with the exterior grade. Note that this moisture barrier must not be a vapor barrier. Its purpose is to control the flow of moisture through the wall while allowing water vapor to escape to the outside.

INSULATION TYPES

There are two basic methods of applying interior insulation:

- A full-height stud wall, spaced 2-3 in (50-75) out from the basement wall and insulated with fiberglass batts or similar products. Normally, 2×3-in (38×64) or 2×4-in (38×89) studs are used and the cavity completely filled with insulation. Extra insulation is placed between the wall and the stud to stop thermal bridging and a/vbs should be continuous and sealed.
- Rigid insulation, usually polystyrene, can be attached directly to the foundation wall (see the next page). Horizontal strapping for finish support may be on top of or flush with the insulation. A/vbs may be terminated at top and bottom and sealed to the insulation (which must itself be an a/vb), or continued over the full face of the wall and sealed top and bottom.

Since both wall and footing will be at low temperatures, good drainage is essential. Backfill material should be carefully selected to exclude poor-drainage or clay-type soils. Use granular backfill for extremely wet conditions.

Some authorities advise leaving a 6- to 12-in (150-300) insulation gap at the wall base that allows a heat flow to the footings and reduces the danger of freezing. This option is not necessary for new construction provided good drainage can be assured, but may be required in some problem localities. Check with local authorities.

BASEMENT WALLS - INTERIOR INSULATION

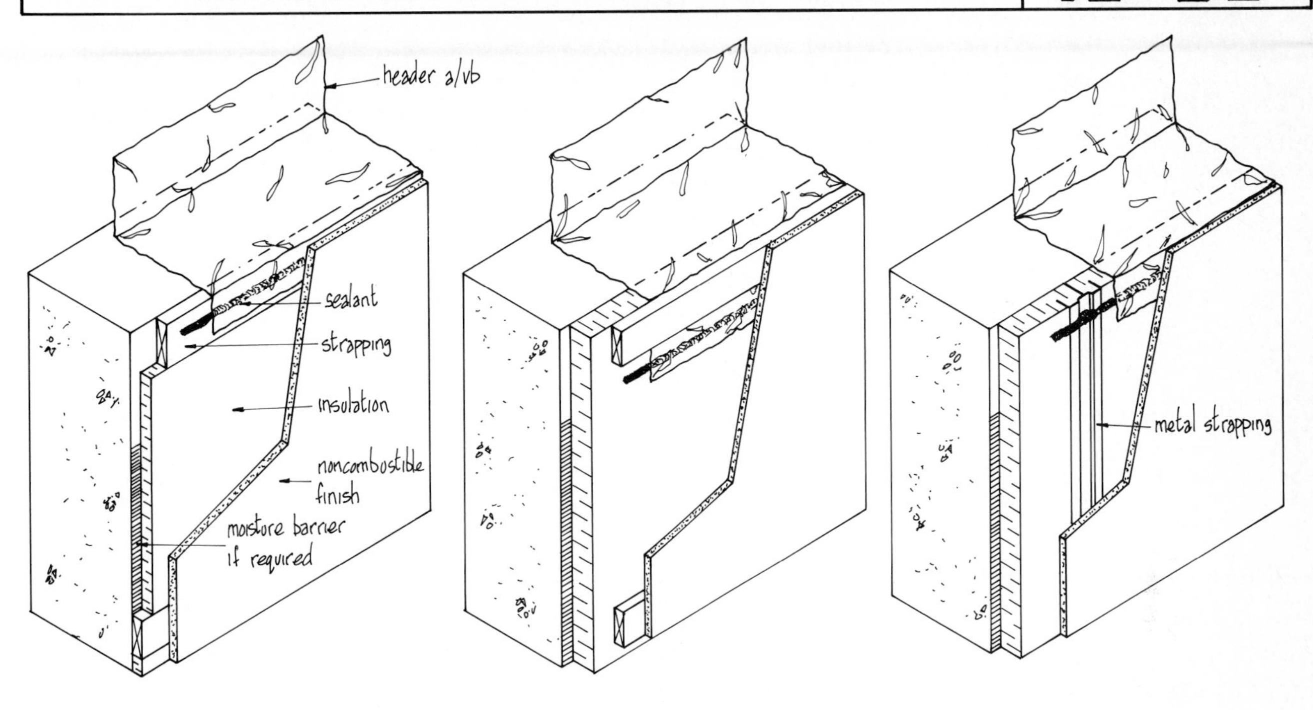

RIGID INSULATION OPTIONS

Shown above are three typical rigid insulation applications. Note the following:

- Insulation must be as tight to the wall face as possible to stop convection currents behind the insulation.
- Where a painted moisture barrier is used, insulation may be glued to the wall. Where a building paper moisture barrier is used, insulation should be mechanically fastened to the wall.
- Strapping must always be nailed or screwed to the wall.
- The entire system must be covered with noncombustible material (eg., plasterboard), which may then be used as the final finish or as a base for wood paneling or similar products. The plasterboard may be glued to the insulation but must be mechanically fastened to the strapping. The joints need not be taped if the board is used as a base for other finishes.
- There are two main disadvantages to these systems. First, there is thermal bridging through the mechanical fasteners (and the strapping in the first detail), and second, it may be impossible to install a hidden electrical system, forcing the use of surface-mounted wiring and fixtures.

CONCRETE CURING

- A newly poured concrete foundation wall should be allowed to cure and dry out for at least 6 months before being sealed behind an airtight a/vb.
- Where a basement is to be left unfinished for a year or more, the header can be made airtight by using the detail shown at the right. Connection of the interior wall a/vb to the wood strip can be made later.

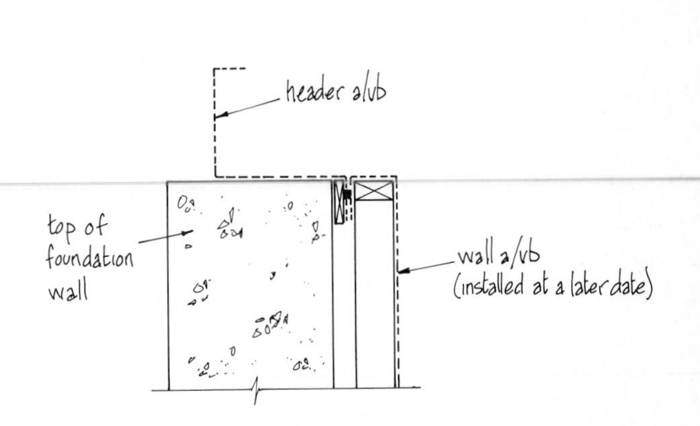

PRESERVED WOOD FOUNDATIONS

The top plate of the foundation wall is normally an area of large heat loss. To reduce heat flow, extend the upper wall exterior insulation down to cover both header and top plate.

Preserved wood foundations are relatively easy to insulate and seal since they are basically extensions of the upper walls. One significant advantage of this type of construction, besides speed of erection, is that electrical and plumbing services are easily hidden in the stud wall.

Standard PWF construction details remain the same as shown on 5-14. A preserved plywood strip protects the 6-mil (0.15) poly moisture barrier, which terminates at grade level.

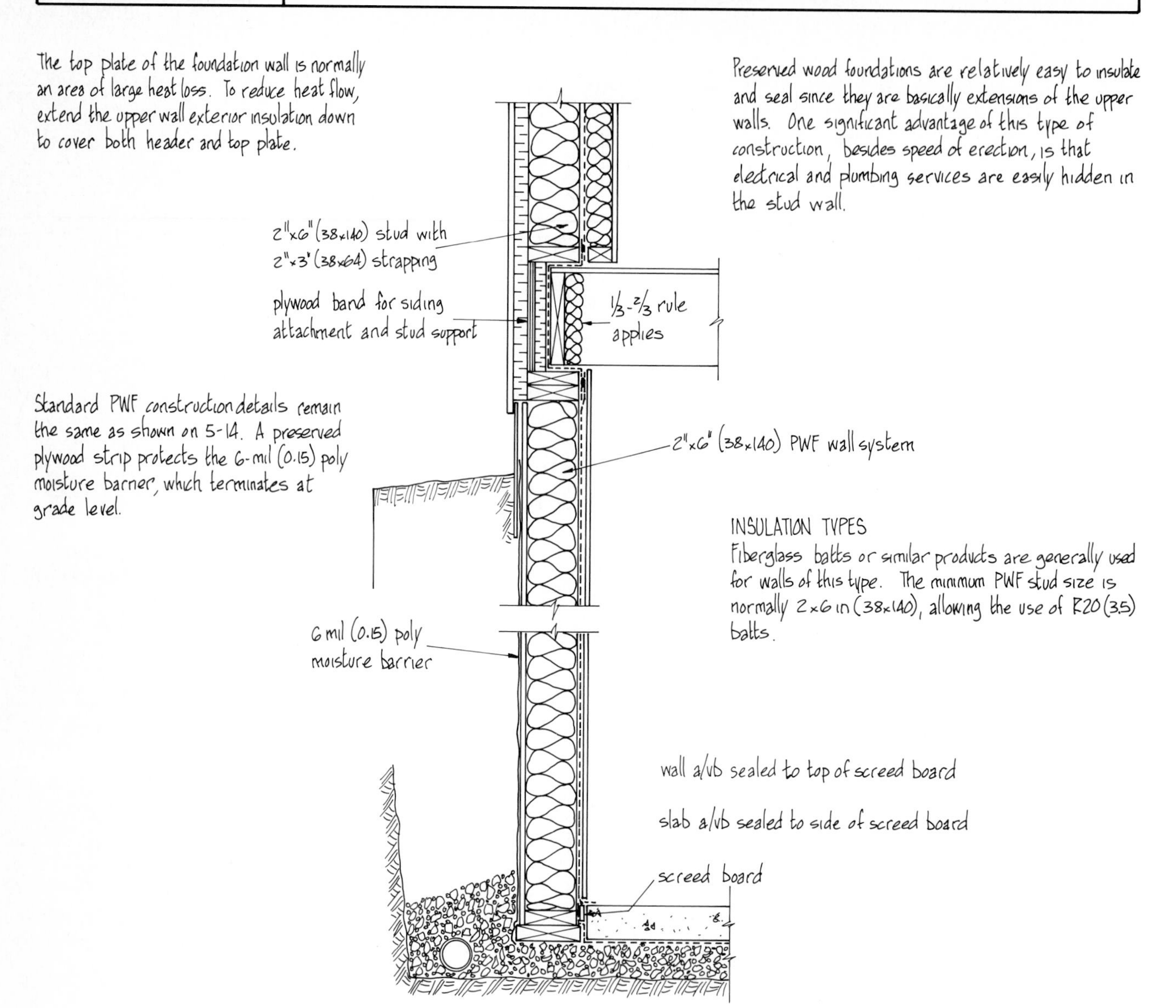

INSULATION TYPES

Fiberglass batts or similar products are generally used for walls of this type. The minimum PWF stud size is normally 2x6 in (38x140), allowing the use of R20 (3.5) batts.

see 5-15 for alternative footing detail
see 10-41 for wood floor information

A/VB CONNECTION

- In this detail the screed board is the only horizontal wood member that can provide a solid backing for the a/vb connection.
- The underfloor a/vb is terminated at the screed board because the exposed edge of poly would be subject to considerable damage during concrete pouring and finishing.
- See the next page for an alternative option.

PRESERVED WOOD FOUNDATIONS

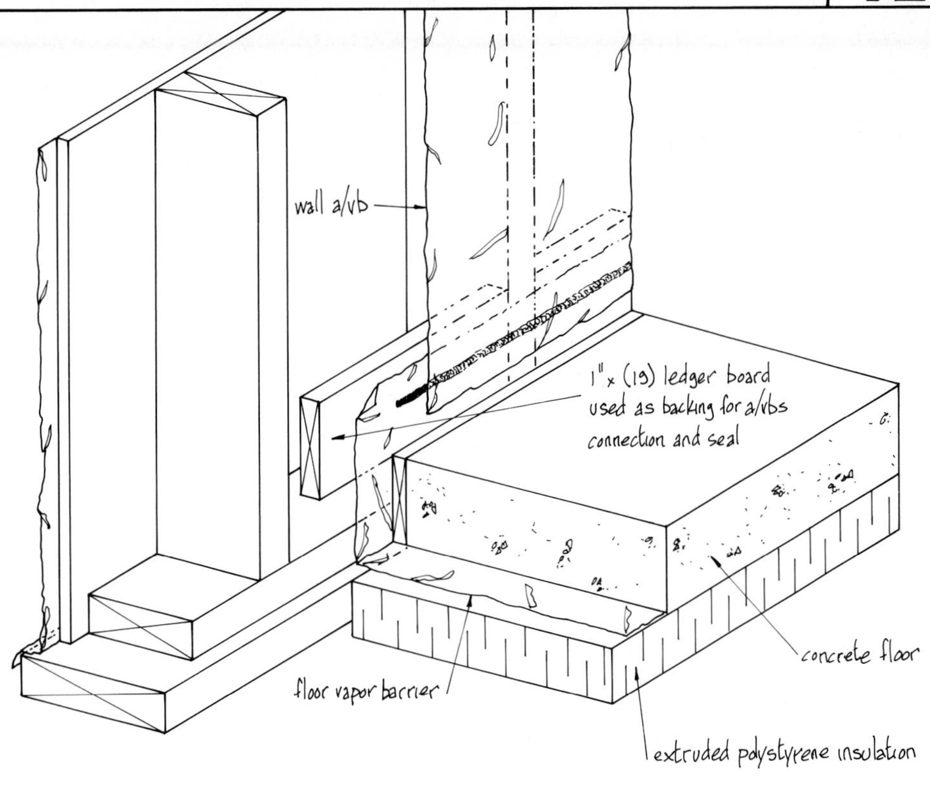

PWF FOOTING WITH FLOOR AND WALL A/VB CONNECTION

BASE OPTION

- The detail above, which provides a direct connection between wall and floor a/vbs, should be used only if great care is taken not to damage the floor a/vb during concrete pouring and finishing.
- A 1-in x (19) ledger board is inserted between the studs to provide the necessary solid backing for the a/vb connection.

GENERAL NOTES

- Exterior wall insulation is not normally applied to PWFs since the insulation must be mechanically fastened to the outer sheathing. To do this would destroy the watertightness of the exterior poly moisture barrier.
- To increase the R value of the wall, a better solution is to add insulation to the inner face. Strapping the wall as detailed on 10-19 is ideal and has the added benefit of a protected a/vb.

CONSERVATION

10-38

INSULATING CONCRETE BLOCKS

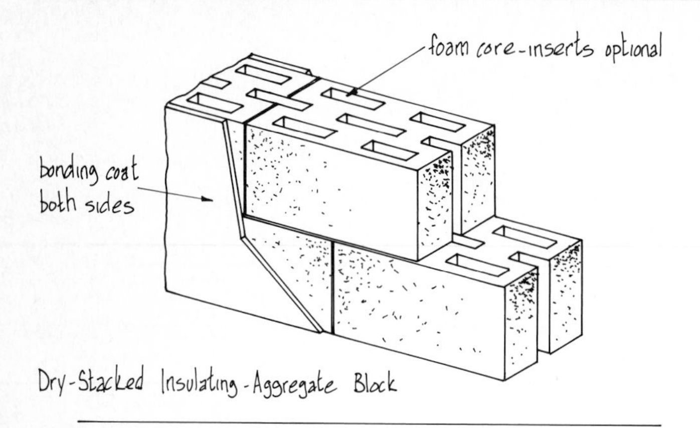

Dry-Stacked Insulating-Aggregate Block

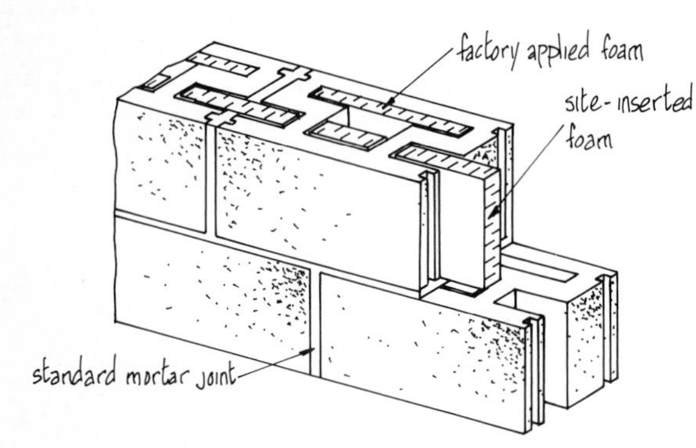

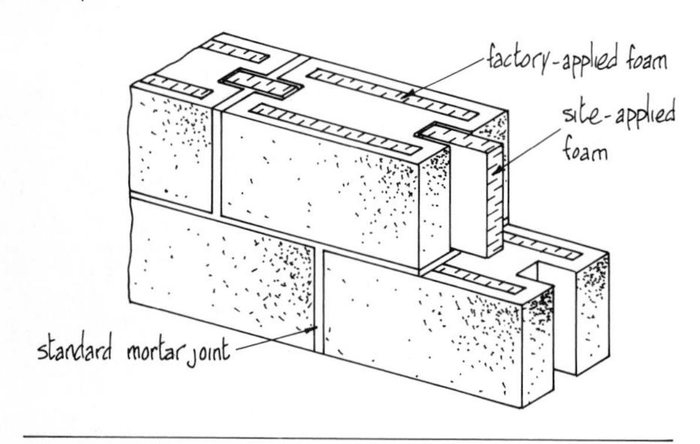

Two Examples of Standard Blocks with Foam Inserts

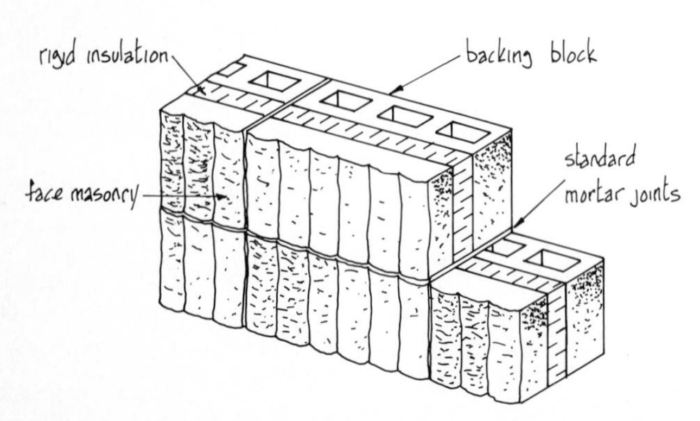

Bonded Insulated Unit

INSULATING MASONRY UNITS

These are special types of concrete modular blocks that provide an insulated wall system for above or below grade use. The systems are designed as load-bearing masonry walls with structural properties similar, or equal to, those of standard concrete block construction.

There are two basic types of insulating concrete blocks:

- Lightweight units containing an aggregate that has insulating properties (such as polystyrene beads) and which may also have insulated cavities. Reasonably high R values are obtained with these units, up to R30 (RSI 5.28) for a 12-in (300) thick block, but the cost is relatively high.
- Standard or lightweight units that rely on cavity insulation and specially designed shapes to increase R value. Overall insulation values are lower with this type of unit but so are the costs.
- Composite units having factory-bonded block, insulation, and face masonry in a single unit. Cost is high, but these units provide a finished exterior wall in one installation sequence.

Erection procedures vary with the type of block:

- Lightweight, insulating aggregate units are generally dry-stacked (without mortar) and are bonded with a surface coat of high-strength parging on both faces of the wall.
- Standard-weight units are generally laid-up with mortar joints in the traditional blocklaying procedure.
- In either case, the option of using vertical steel reinforcing rods to increase the wall strength is not affected.

The type and installation methods of the core insulation also varies. Some types have factory-installed foam only, some have a combination of factory-installed and site-installed inserts, while others rely on site inserts only. Several examples of typical block types are shown on this page.

Note: The long-established practice of pouring loose-fill insulation material (expanded mica, polystyrene beads) into the core cavities to increase insulation values is not included in this section.

INSULATING BLOCK WALLS

DRY STACKING

- Blocks are manufactured in standard modular sizes (see 7-22) and preground for accuracy.
- The first course is laid in a bed of mortar and carefully leveled and aligned.
- All other courses are laid without mortar in running bond (see 7-27), with a minimum 6-in (150) overlap.
- After stacking, the wall is coated both sides with a special surface bonding cement that structurally bonds the blocks together.
- Generally, blocks stacked higher than 10-12 times the block thickness should be braced until surface bonded.

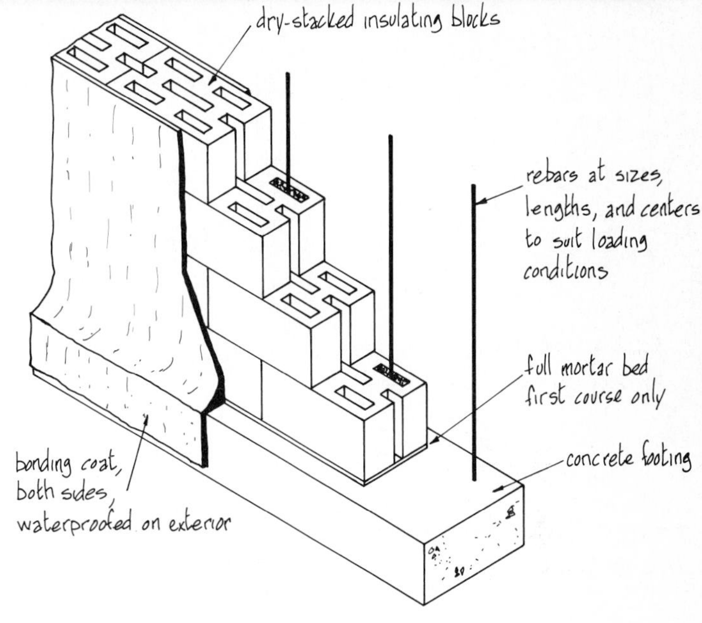

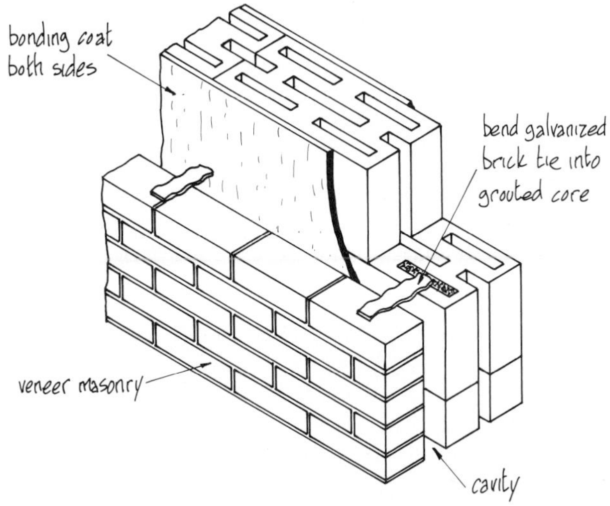

VENEER CONSTRUCTION

Brick veneer walls may be tied to the dry-stacked blocks by inserting galvanized metal ties between the blocks and bending them down into a grouted core before the surface bond cement is applied.

BASEMENT WALLS

- Vertical reinforcing bars grouted in the core space at every second block may be required, depending on wall, block type, height, and block width. Refer to manufacturers' specifications.
- Surface bonding cement should form a cove at the junction of the wall and footing and continue at least halfway down the footing depth. The bonding coat replaces the parge coat normally used on standard concrete block. The below grade portion of the wall is waterproofed as normal.
- With most systems backfilling must be done with caution. Use only granular backfill (sand or gravel) and backfill gradually without sudden impact against the basement wall. Before backfilling, the first-floor subfloor should be in place (as is usual), or alternatively, the wall should be securely braced on the inside.

STRAPPED WALLS

- Strapping, nailed to the bonded wall, is used to support exterior siding or interior finishes.
- Electrical cables can be run through the block cores, recessed in the wall surface, or run in the space made by the strapping. Check local electrical codes for allowable cable types.

INSULATED FLOOR SLABS

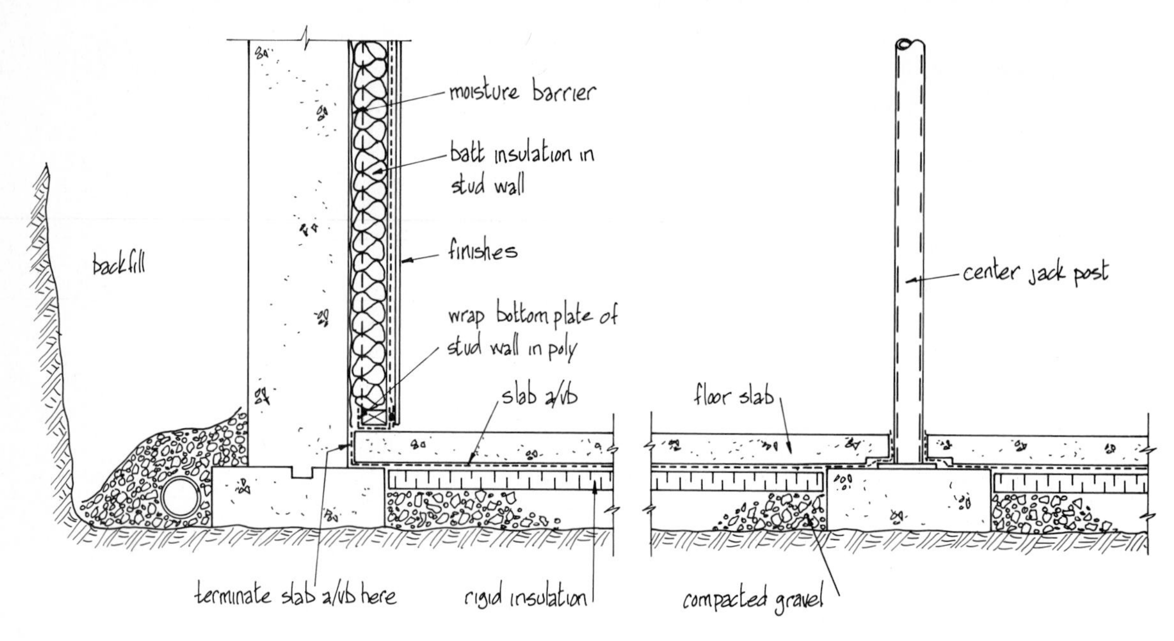

INSULATED BASEMENT FLOOR SLAB

The illustration above is typical of insulated concrete basement floors in energy-efficient housing. There are several options to choose from:

A/VBs

- The a/vb should be continuous under the entire slab. Its primary purpose is to act as a moisture barrier to prevent water movement from the ground into the house.
- Since air moves quite easily through the soil, it also acts to stop air leakage through cracks in the foundations or slab.
- Reducing airflow is important. Radon gas, generated by naturally occuring radioactive materials in the soil, can accumulate in the basements of otherwise well-sealed but poorly ventilated houses (see 14-16 onward).
- Ideally, the underslab a/vb should be extended to connect with the wall a/vb above. This can only be done if extra care is taken not to damage the a/vb during slab pouring and finishing operations. Failing this, the a/vb should be trimmed at the top of the slab level as shown.

INSULATION

- Although slab heat loss is generally low in dry soils, insulation will still provide some savings and result in warmer floors. In wet soils or cold regions, heat flow can be substantial and large savings will be realized.
- Extruded polystyrene is the recommended insulation choice. It is normally laid directly on the compacted gravel and covered with the a/vb as shown.
- With this method there is still a heat loss through the uninsulated footings, but this is usually considered an advantage since it warms the footing area. For completely insulated footings, see the option on the next page.

CONCRETE CURING

- Placing an a/vb below the slab tends to slow the curing time of the concrete. If the contractor considers this to be a potential problem, one solution is to place a layer of sand over the poly a/vb before pouring the concrete to act as a water reservoir.

INSULATED BASEMENT FLOORS

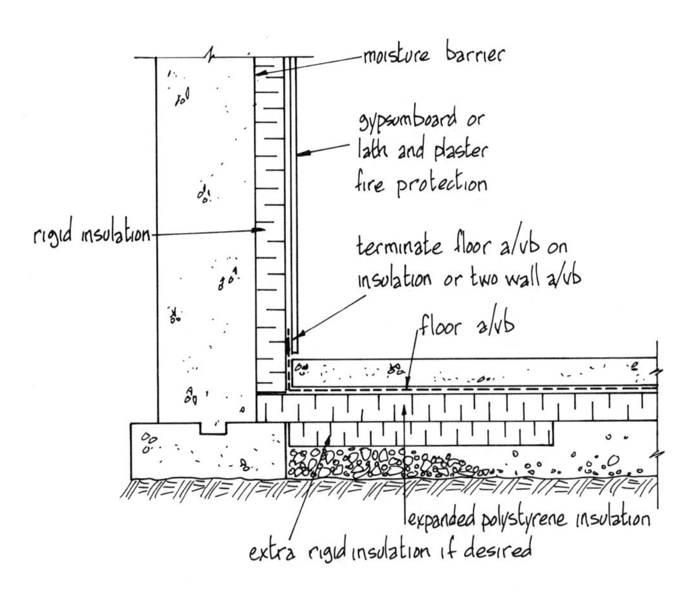

CONTINUOUSLY INSULATED WALL AND FLOOR

FOOTING INSULATION OPTION

- The detail at the left shows a completely insulated footing area. Floor and basement wall insulation is continuous and the under-slab a/vb is sealed to the wall insulation at the wall base or connected to a full wall a/vb. Extreme care must be exercised when pouring and finishing the concrete floor not to damage the floor a/vb.
- As an additional option, an extra layer of insulation is placed inside all footings since this is the area of greatest heat loss.
- Using this detail, frost damage may occur at the footing level unless the footing is below the frost line and good drainage is assured.

PWF INSULATED WOOD FLOOR

The illustration at the right shows typical details of a wood basement floor used with the PWF system (see 5-14).

- Insulation is fiberglass or mineral wool batts installed between the floor joists.
- The continuous a/vb is located between the subfloor and joists and is sealed to the wall a/vb above.
- A moisture barrier, such as building paper with overlapping joints, must be placed over the gravel layer to protect the insulation.
- Under certain conditions this type of floor is subject to upward buckling caused by lateral earth pressures at the foundation walls. Check with local authorities and trade organizations such as the American Plywood Association or Canadian Wood Council for exact design details.
- For PWF and insulated concrete floors, see 10-36.

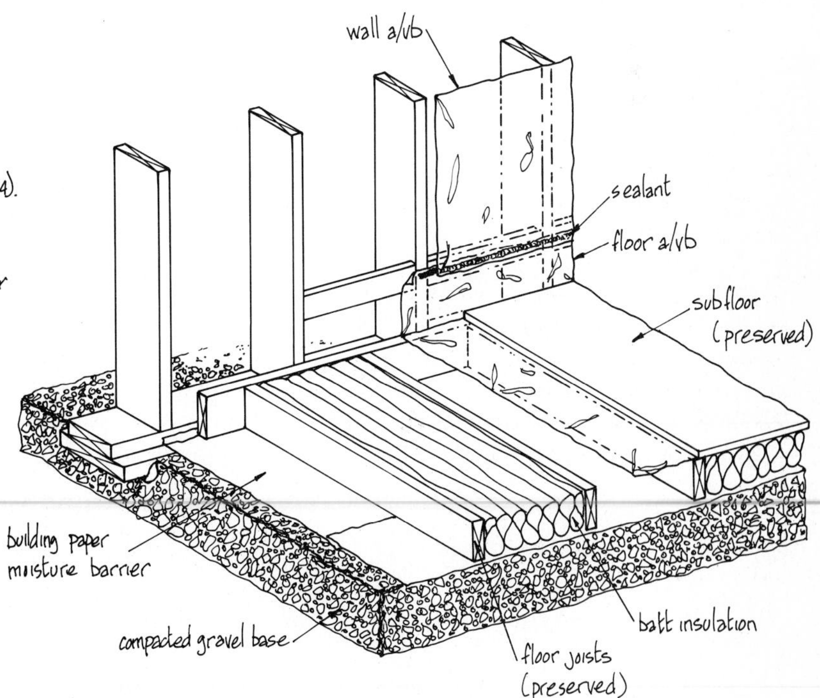

INSULATED PWF BASEMENT FLOOR

SLAB - ON - GRADE

SLAB - ON - GRADE

- This type of foundation is essentially the same as a basement slab and the procedures for insulating and air sealing are therefore similar. However, this type of slab has greater potential for heat loss than a deep basement because much of the heat-flow buffering effect of the earth is lost.
- In all cases the best procedure is to insulate on the outside with rigid polystyrene or fiberglass. Routing of the a/vb will depend on the combined details of walls and floor. In cold regions with a deep frost line, heavy insulation of the wall or grade beam and floor is essential to avoid excessive heat loss.
- In cold and very cold regions where a basement is not required, it is more practical to use crawl space construction rather than slab-on-grade.

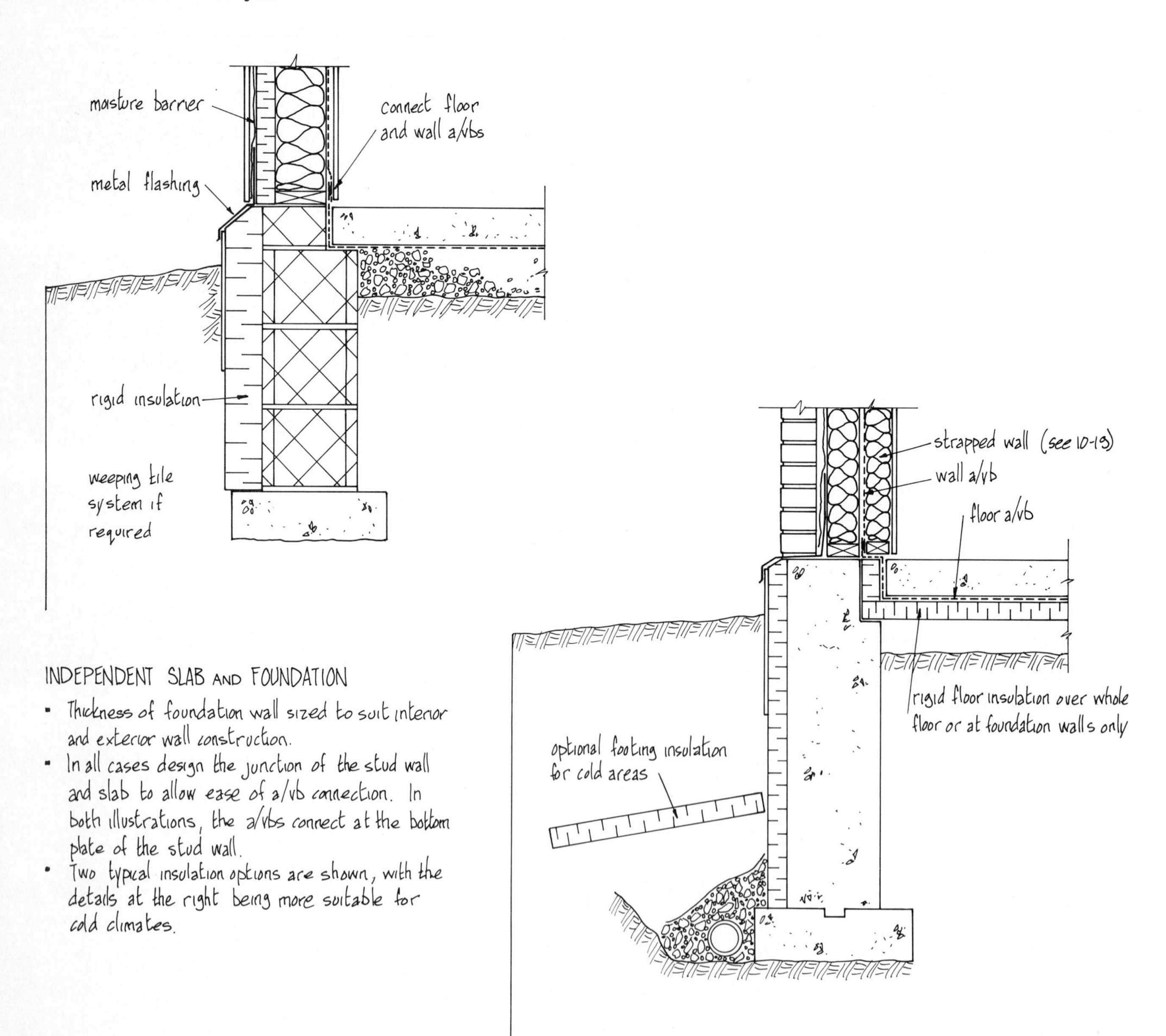

INDEPENDENT SLAB AND FOUNDATION

- Thickness of foundation wall sized to suit interior and exterior wall construction.
- In all cases design the junction of the stud wall and slab to allow ease of a/vb connection. In both illustrations, the a/vbs connect at the bottom plate of the stud wall.
- Two typical insulation options are shown, with the details at the right being more suitable for cold climates.

SLAB-ON-GRADE

COMBINED GRADE BEAM AND SLAB

- The grade beam can be designed to accommodate any thickness of wall.
- The a/vb can be separate and unconnected as shown at the right or be continuous around the footing as shown below.
- The a/vb may also be positioned on top of the slab, depending on the final flooring material. In this case, the underslab a/vb must be omitted.

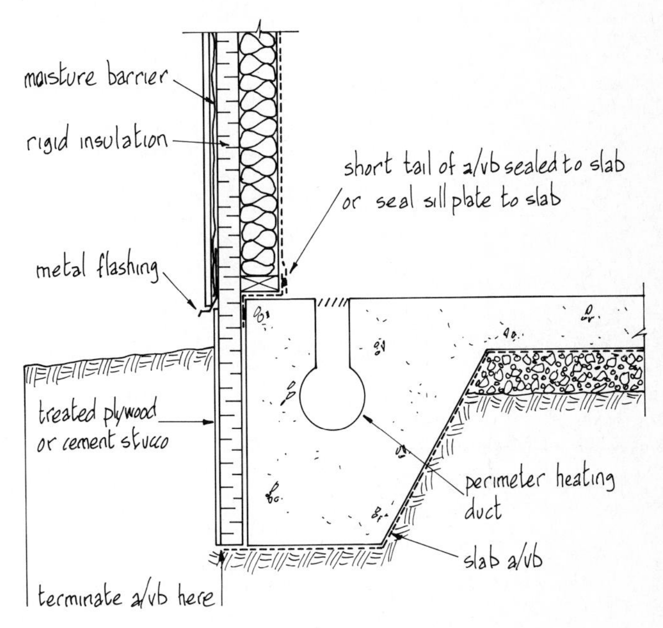

UNCONNECTED A/VB AND SIDEWALL JUNCTION

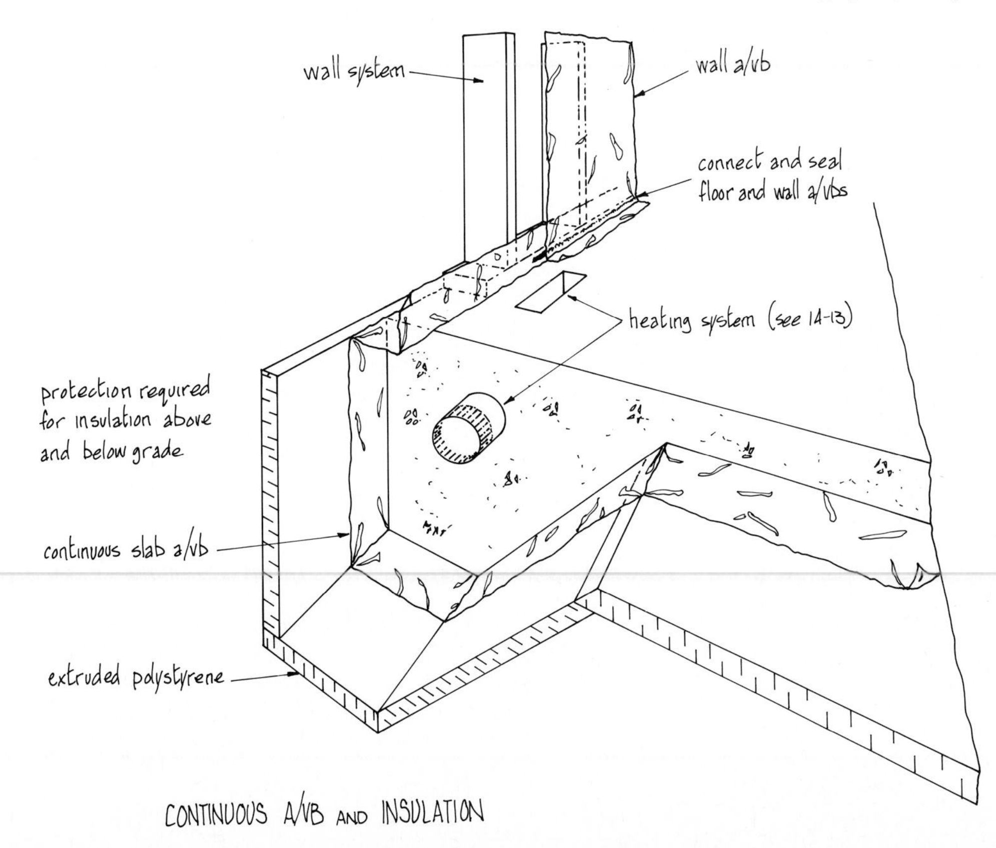

CONTINUOUS A/VB AND INSULATION

CRAWL SPACES

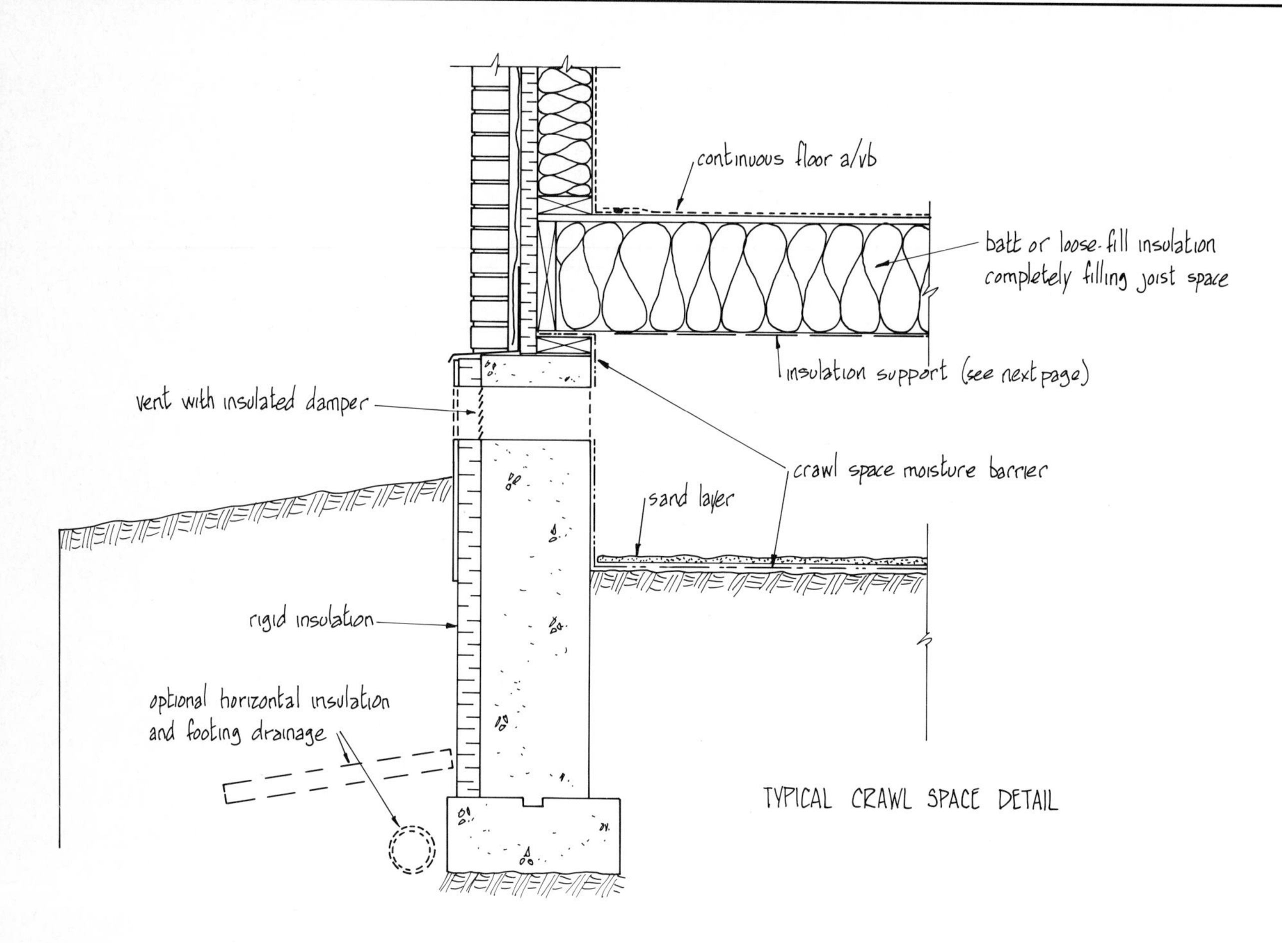

TYPICAL CRAWL SPACE DETAIL

UNHEATED CRAWL SPACES

- With this type of construction the insulation layer must be between the floor joists since the crawl space is cold.
- The insulation, in batt or loose form, should be supported and protected with mesh or rigid boards, as illustrated on the next page.
- The insulation should completely fill the floor joist cavity to maximize R values.
- The a/vb is positioned under the subfloor and all overlaps are sealed. Connection with the wall a/vb can be made on the floor as shown or on the stud wall bottom plate.
- Penetrations of the a/vb for plumbing or heating services, etc. should be carefully sealed. All electrical services should be designed to run in the walls above without extending through the floor. Plumbing pipes or heating ducts under the floor must be fully insulated.
- Winter temperatures in the crawl space should be kept above freezing to avoid frost heave or other damage. A small dampered duct, supplying heat from the main system, may be required in cold regions. Perimeter wall insulation, installed as for basement walls, helps retain most of the heat.
- A moisture barrier is required on the earth floor to limit the amount of water vapor entering the crawl space. This must be continuous over the whole floor and should extend up the walls to at least floor header height. The moisture barrier, which may be polyethylene, does not need to be sealed at overlaps or header termination.
- The sand layer acts as both ballast and protection for the a/vb.
- Crawl space venting that can be opened in summer and closed in winter is essential for the dissipation of accumulated moisture. Vent closures should be weather-stripped and insulated. Check local codes for current net vent area requirements.

CRAWL SPACES

FLOOR INSULATION SUPPORT

- For batt insulation attach wire mesh or expanded metal to the underside of the floor joists. The mesh should be small enough and strong enough to resist rodent attack. The batts can then be installed from above.
- Loose-fill insulation must be supported by a solid board. In the detail shown, plywood strips are nailed to the underside of the floor joists and support thin ply or waferboard that is dropped in from above. This method has the added advantage of reducing airflow through the insulation.

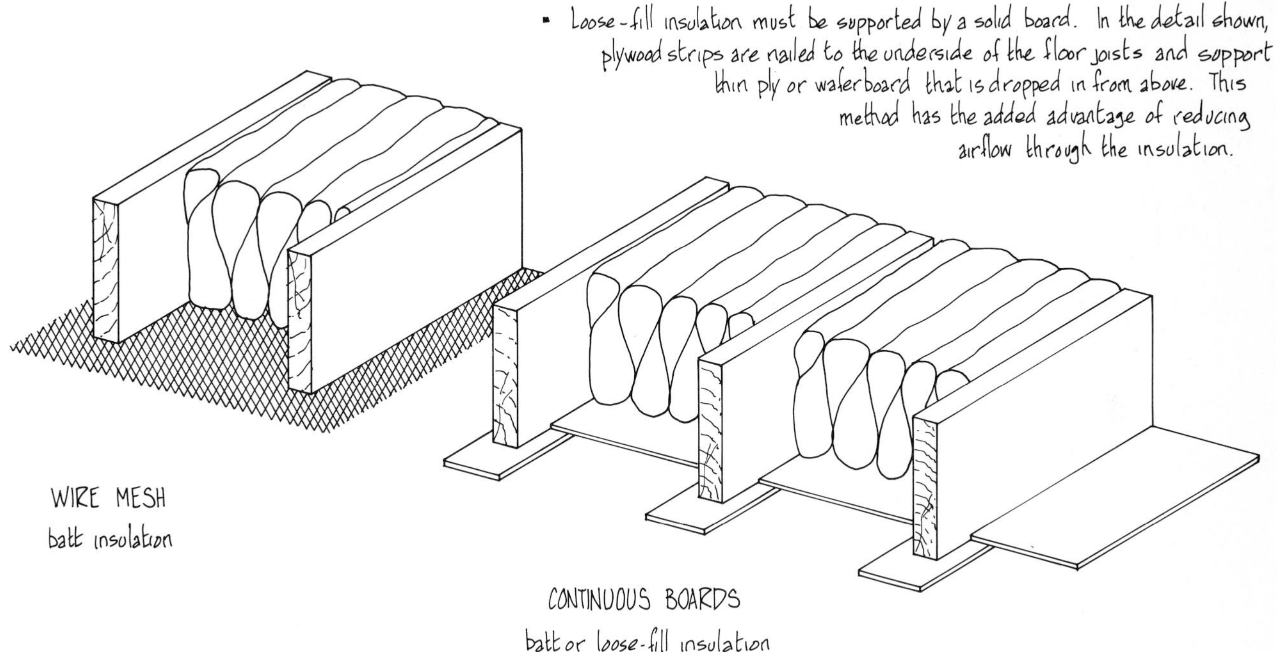

WARM CRAWL SPACES

A crawl space can be treated as a heated or unheated but warm basement space. The illustration at the right shows suggested details, and the following points should be noted:

- Interior insulation is shown as an option. Exterior insulation as on the preceding page is also appropriate.
- Insulation extends across the crawl space floor to reduce ground heat loss. Depending on climate, this insulation can be applied only to the wall perimeter or be extended across the entire floor. No insulation is required between the floor joists.
- The a/vb must now make an airtight joint at the floor header and extend to the wall a/vb.
- If the crawl space is to be used as a hot air plenum for heating and cooling purposes (see 14-12), venting is not required.
- In temperate to hot climates, the vapor barrier may be placed below the insulation on the crawl space floor.

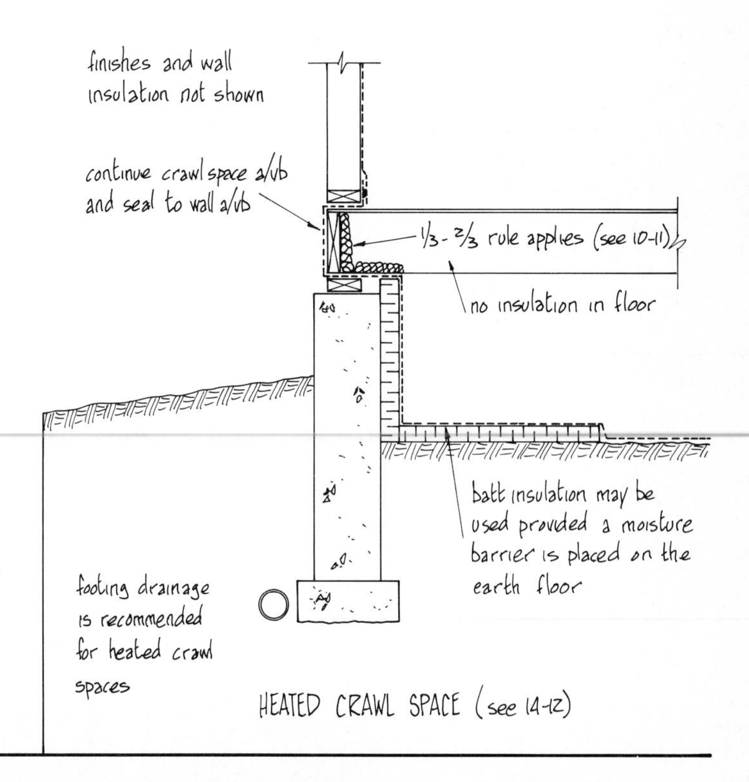

HEATED CRAWL SPACE (see 14-12)

CONSERVATION

10-46

WALLS - SINGLE STUD

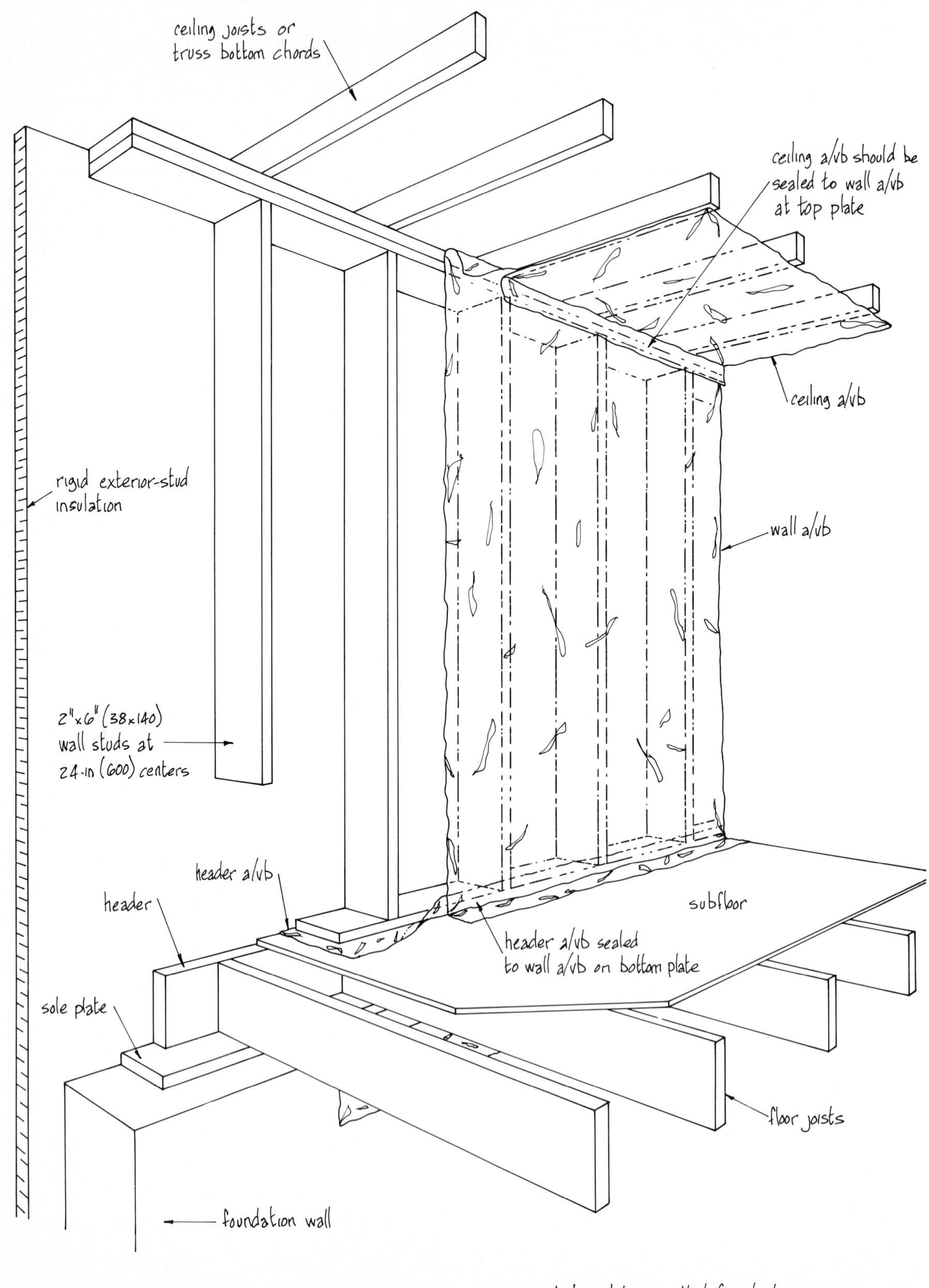

stud insulation omitted for clarity

WALLS – SINGLE STUD

SINGLE STUD WALLS

This is the most common type of wall system in use today. Although it has been well proven over the years, there are limits to the energy efficiency levels obtainable with this method.

- Maximum R values obtainable are detailed on 10-18. The illustration at the left uses 1½-in (38) exterior insulation sheathing to increase R value and reduce thermal bridging through the studs. See 10-15 for suitable rigid insulation materials.
- The a/vb must be placed on the inside face of the wall, where it will be penetrated by electrical and plumbing services. Careful sealing techniques will be necessary at these points.
- A 2 x 6-in (38 x 140) stud size is standard for these walls – larger studs tend to be uneconomical.
- A staggered stud wall (see 10-18) follows the same basic details. While a higher R value is obtainable with this type of wall, the same disadvantages apply, notably the unprotected position of the a/vb.
- For window and door sealing techniques, see 10-60.

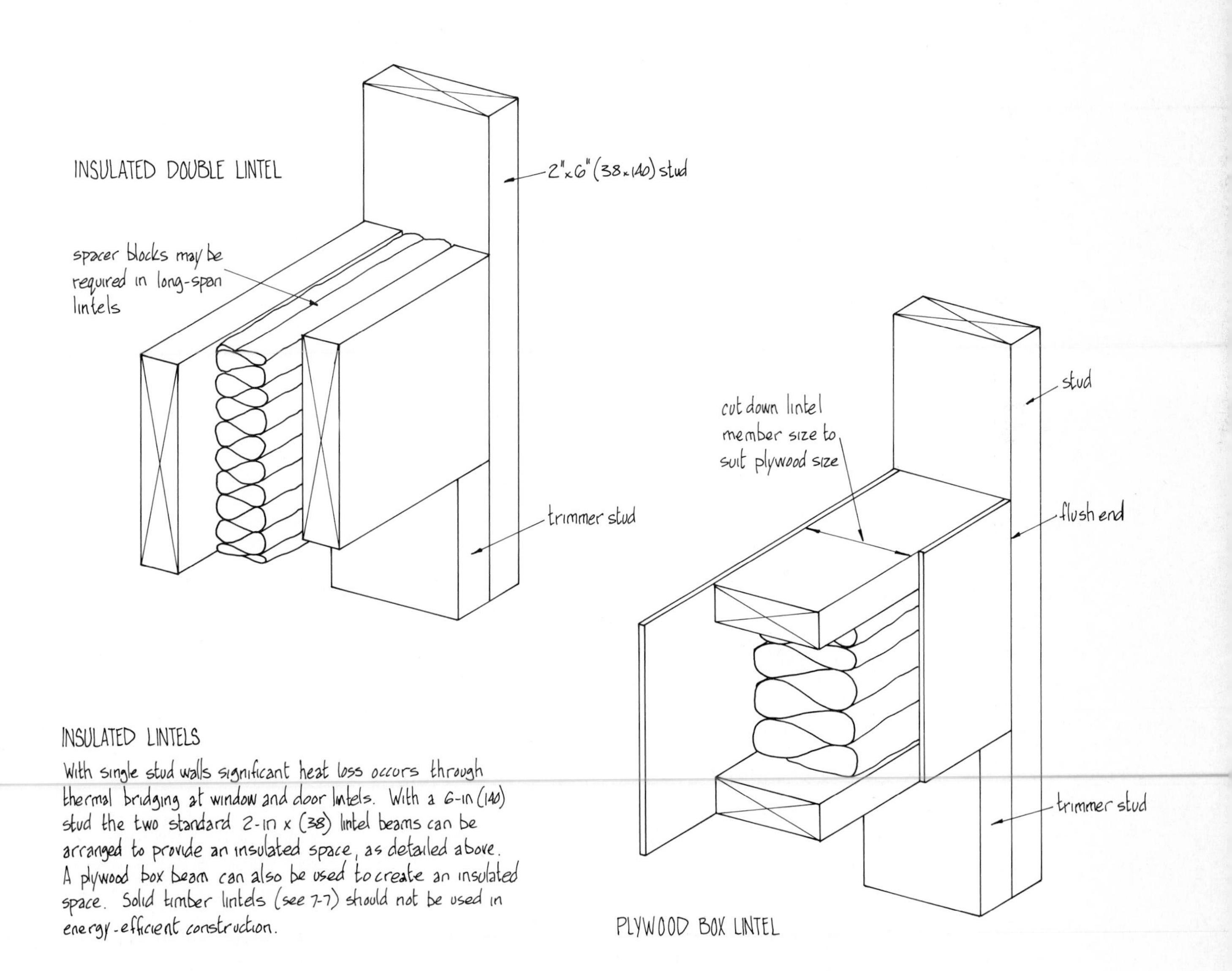

INSULATED LINTELS

With single stud walls significant heat loss occurs through thermal bridging at window and door lintels. With a 6-in (140) stud the two standard 2-in x (38) lintel beams can be arranged to provide an insulated space, as detailed above. A plywood box beam can also be used to create an insulated space. Solid timber lintels (see 7-7) should not be used in energy-efficient construction.

WALLS - STRAPPED SINGLE STUD

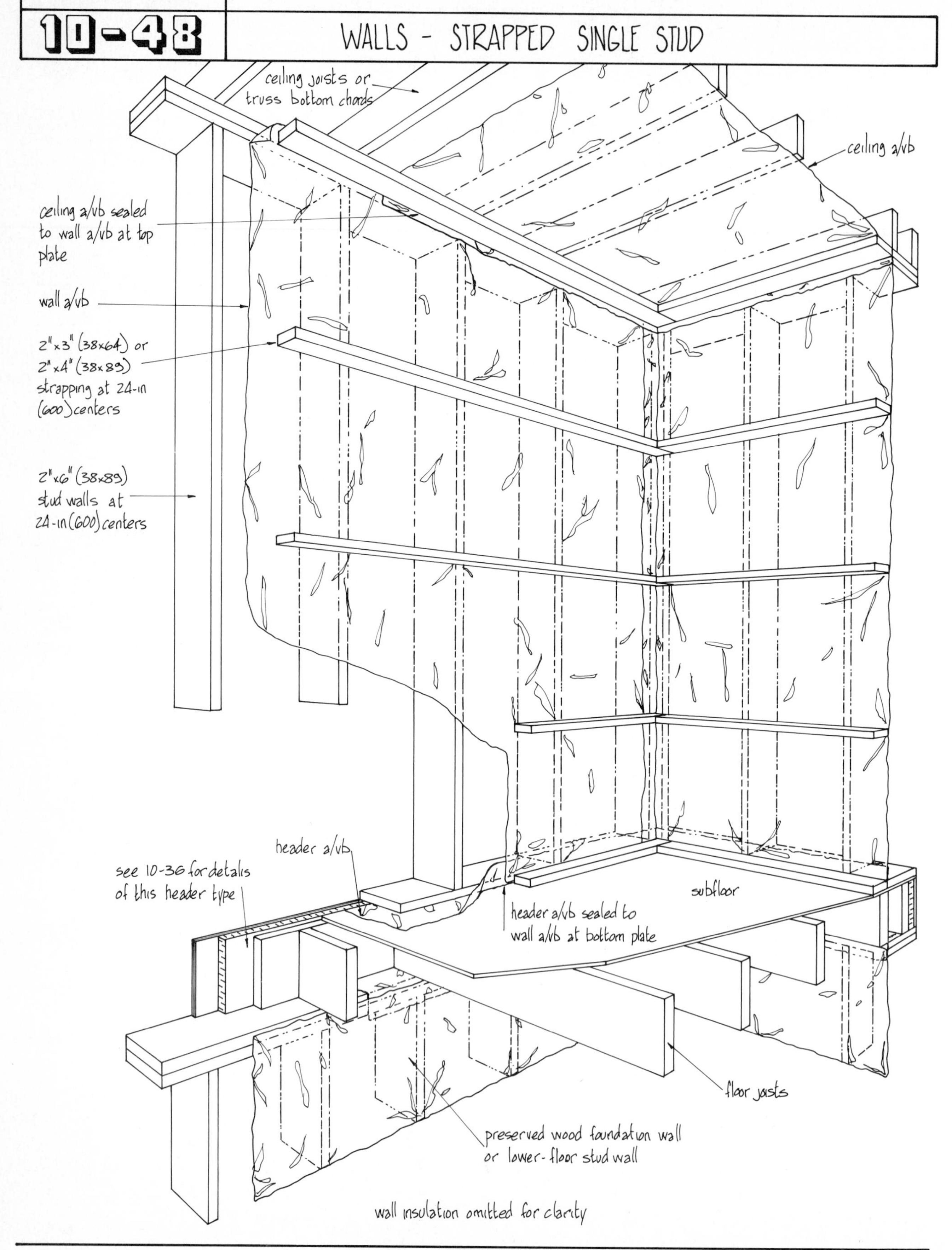

WALLS - STRAPPED SINGLE STUD

STRAPPED WALL

This method, which enhances the traditional single stud system, is a good compromise when comparing single stud with double stud walls.

- Horizontal strapping is applied to standard single stud walls as shown at the left below.
- The a/vb is applied to the face of the stud before the strapping is added. This protects the a/vb and allows the electrical and plumbing systems to be run in the cavity provided by the strapping. With careful planning, penetration of the a/vb can be kept to an absolute minimum.
- The a/vb must be placed in accordance with the one-third/two-thirds rule (see 10-11).
- Thermal bridging through the studs is substantially reduced, especially with the addition of exterior insulated sheathing.
- Interior finishes such as plasterboard are generally installed horizontally on the strapping.
- See 10-19 for maximum R values.

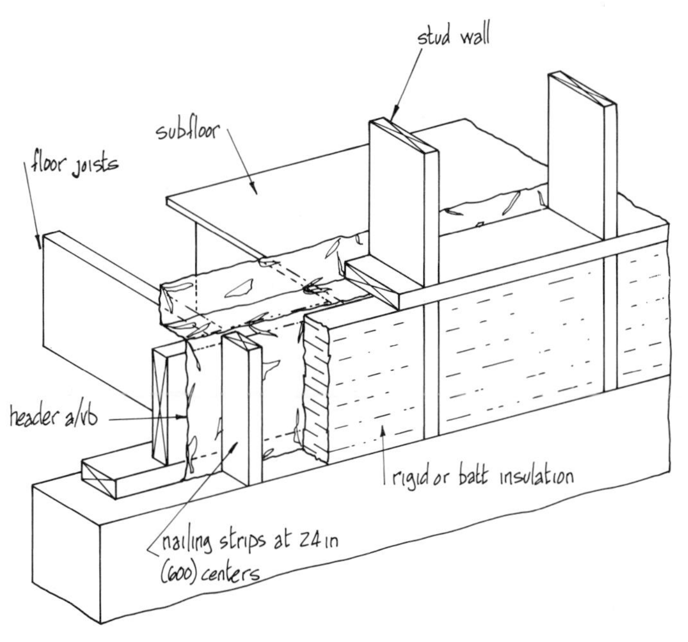

BLOCKING AT HEADERS

Where headers are set back in the wall and have more than 1½ in (38) of rigid insulation thickness on the outside to comply with the one-third/two-thirds rule, it is often difficult to attach the exterior finishes to the header. To provide backing for the finishes, install vertical blocking (or horizontal blocking if using vertical siding) at 24-in (600) centers on the header. Fit the exterior insulation between the blocking.

STUD VENTILATION

A/vbs are rarely perfectly sealed, and some moisture accumulation is likely to occur in the stud cavity. When nonpermeable insulation is used, e.g., extruded polystyrene or foil-faced isocyanurate, vapor movement out of the cavity is restricted and water accumulation can be a problem. Provision of a breathing opening as shown at the right is recommended. Stop the insulation 1 in (25) short of the top (or bottom) plate. Alternatively, use an insulation that does not act as a vapor barrier, such as rigid fiberglass.

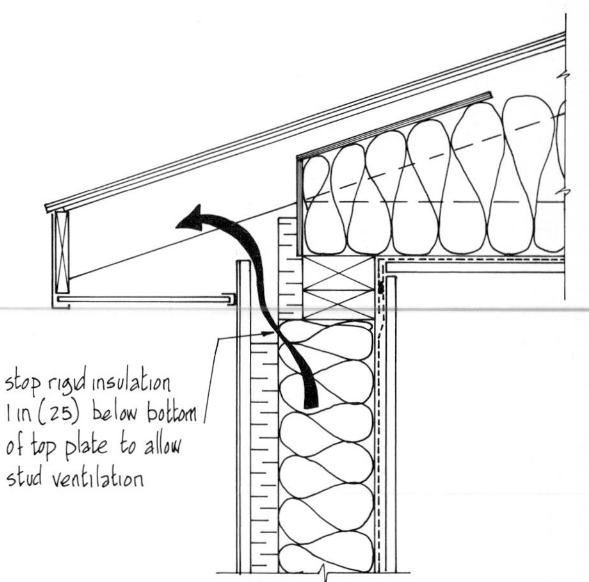

WALLS - DOUBLE STUD

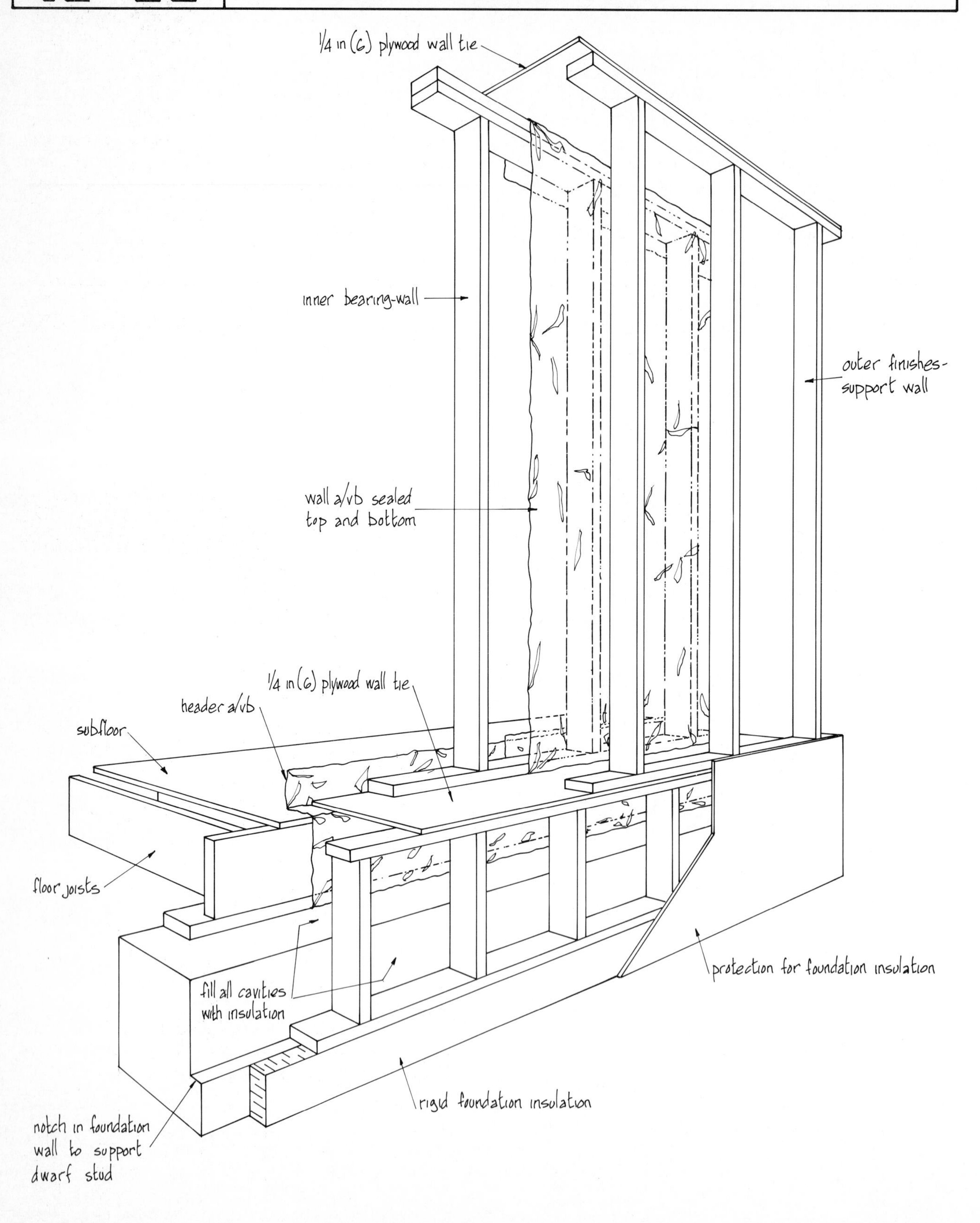

WALLS - DOUBLE STUD

DOUBLE WALL STUDS

As described on 10-19, this type of wall can be designed to provide very high insulation levels. It is used for "super insulated" homes, especially in cold and very cold regions. If the entire structure is designed along the same lines, total heat loss can be reduced to the point where the "free heat" of a dwelling (see 10-3) is sufficient to support the entire heating load in all but the most extreme weather conditions.

- Although there are several variations of this type of wall, each is framed in one of two basic ways: as one complete wall or as two separate walls tied together after erection. In each case the wall is framed horizontally on the floor deck, as is standard practice with platform construction. Plywood spacer plates tie the walls together as shown at the left.
- The inner wall is normally the bearing wall, with the outer wall supporting the outside finishes.
- The a/vb is positioned on the outer face of the inner wall for maximum protection. Electrical and plumbing services run in the cavity provided.
- The total R value will depend on the overall thickness of the wall, which must be completely filled with insulation (see 10-19 for maximum values).
- Because of wall thickness, window and door frames will need extra large extension pieces or special triming and finishing techniques (see 12-8, bottom).
- One variation of this type of framing, the trussed wall, is shown at the left, below.

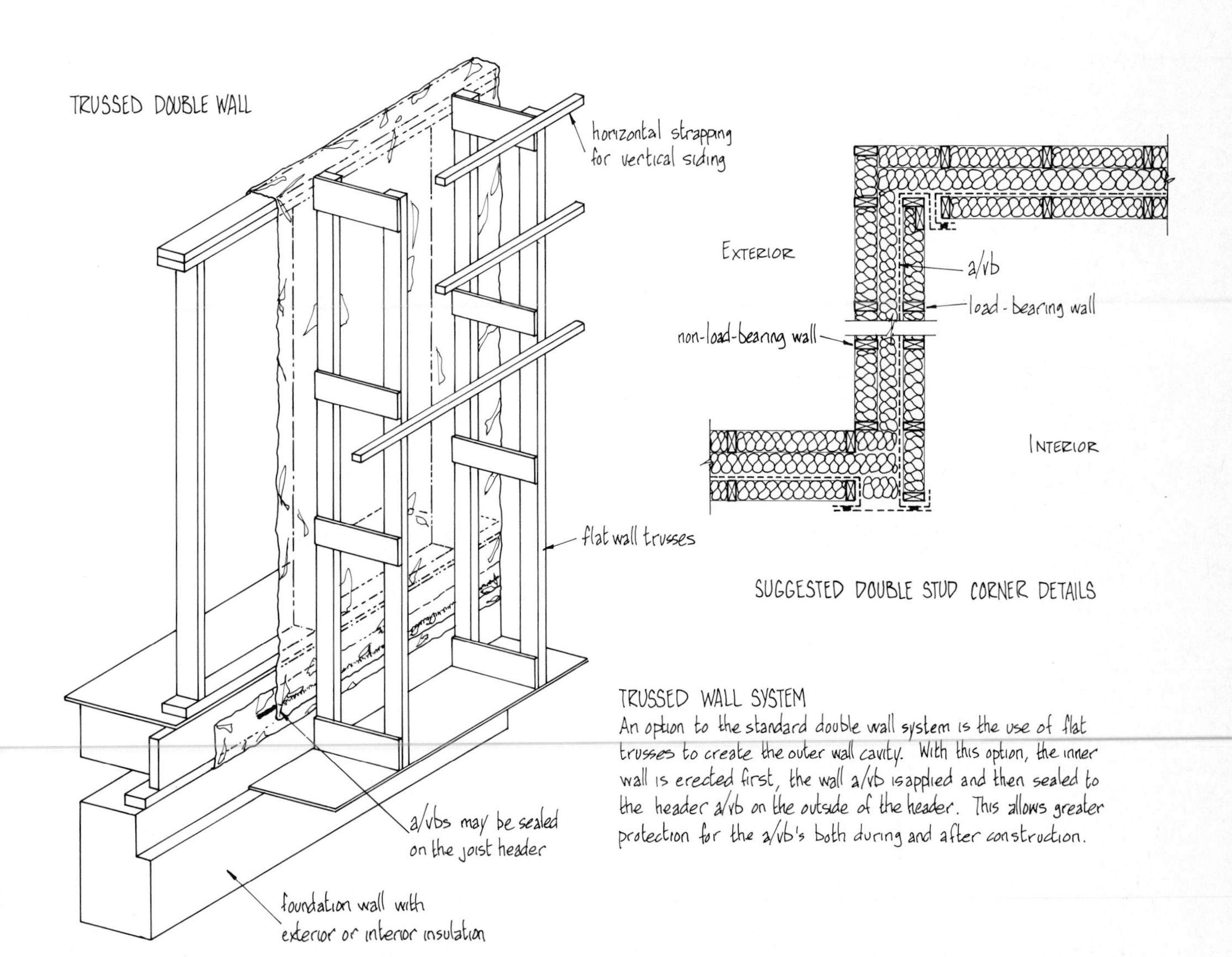

TRUSSED WALL SYSTEM

An option to the standard double wall system is the use of flat trusses to create the outer wall cavity. With this option, the inner wall is erected first, the wall a/vb is applied and then sealed to the header a/vb on the outside of the header. This allows greater protection for the a/vb's both during and after construction.

FLAT CEILINGS

Refer to 10-20 for general flat-ceiling roof details and to 8-8 for prefabricated truss information. Treatment of electrical boxes in ceilings is discussed on 10-59.

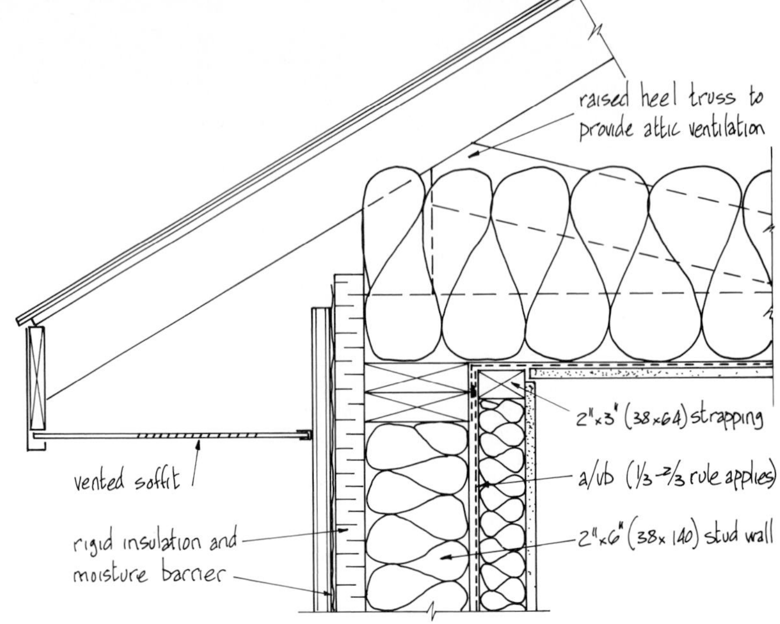

CEILING/WALL JUNCTION WITH STRAPPING AND RAISED TRUSS

CEILING INSULATION

Either batt or poured fiberglass is the usual choice for ceiling insulation. Batts have long term stability but are not easy to fit around obstructions and may result in air gaps and some heat loss. Poured insulation solves this problem but may settle over time, reducing R value.

In either case the weight of deep insulation can cause the ceiling plasterboard to sag under certain conditions:

- High humidity caused by concrete coring and some finishing operations can weaken the plasterboard.
- Sprayed textured-plaster ceiling finishes will accentuate the problem by wetting the plasterboard.

When the ceiling insulation exceeds 10 in (250) in thickness, the following procedures are recommended:

- When trusses are spaced at 24-in (600) centers use 5/8-in (15.9) plasterboard in place of 1/2-in (12.7); or
- The ceiling can be strapped with 1x2-in (19x38) stock at 24-in (600) centers and at right angles to the trusses, as shown below. This method has the advantage of providing a space for electrical wiring installation without damaging the a/vb.

WALL/CEILING JUNCTIONS

Termination and sealing of the a/vbs depends on the design of the wall/ceiling junction.

- The detail above shows a standard flat-ceiling truss and stud walls with 2x3-in (38x64) strapping. Connection of the a/vbs is made on the stud's top plate before application of the strapping.
- For single studs without strapping the connection is made on the inside face of the top plate.
- On a double stud wall the a/vb is attached to the outside face of the inner stud (see 10-50). The a/vb is continued over the top plate to the room interior and connection with the ceiling a/vb is made on the face of the top plate.

STRAPPED CEILING

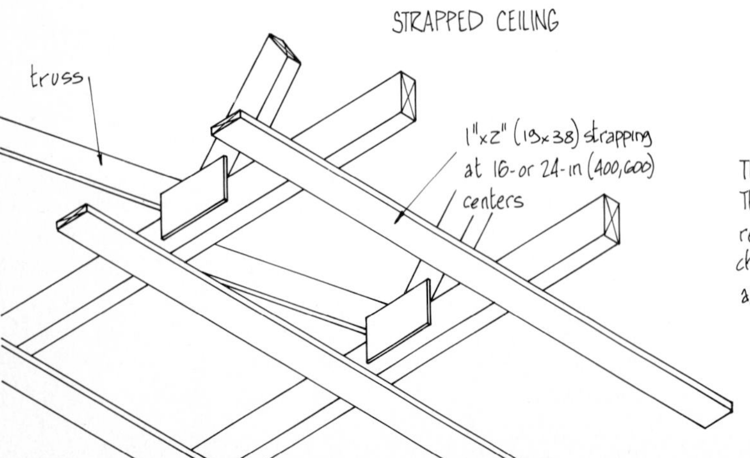

TRUSS UPLIFT

The flat ceiling chord of trusses can be affected by what is referred to as "seasonal truss uplift", which forces the bottom chord upward, cracking plasterboard joints and damaging the a/vb. This problem is discussed in detail on 10-54.

SLOPING CEILINGS

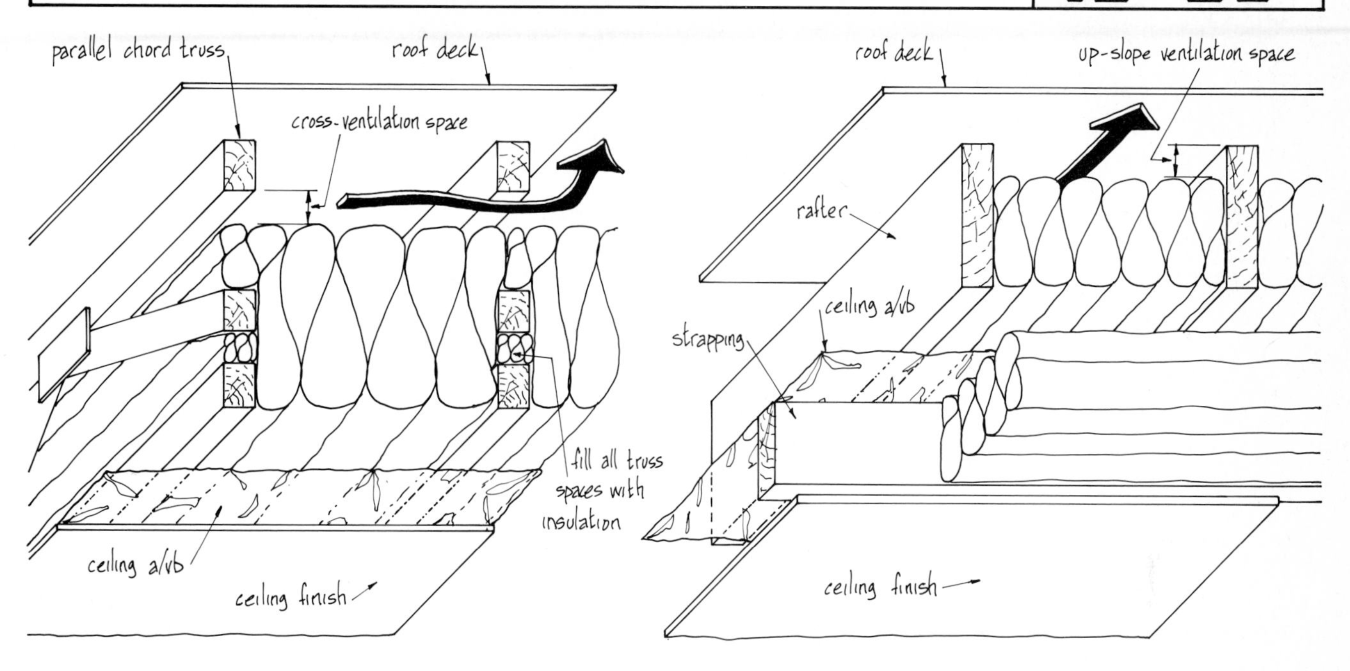

SECTION THROUGH TRUSS ROOF SPACE

RAFTERED AND STRAPPED ROOF

SLOPED-CEILING ROOF VENTILATION

- When parallel chord or scissor trusses are used, ventilation across the trusses is achieved by keeping the insulation below the top chord. Individual roof vents can then be used at suitable locations near the roof peak.
- With rafters strapped on the underside, insulation levels can be increased and ventilation air will flow naturally up the slope. However, since each rafter space is closed, continuous ridge vents must be provided. Continuous soffit vents may also be required, depending on soffit details. The a/vb is placed between strapping and rafter in compliance with the one-third/two-thirds rule, and all electrical services are located in the strapping space.
- Wood I-beams (see 6-18) may also be used as rafters. See below and the right.

STRAPPED RAFTERS

On a raftered roof, adding strapping to the underside of the rafters increases insulation levels and allows an effective ventilation flow up the slope.

- The a/vb is placed between the strapping and the rafter and must conform to the one-third/two-thirds rule.
- All electrical services are located in the strapping cavity and do not penetrate the a/vb.
- Check with local codes for the required ventilation airspace.

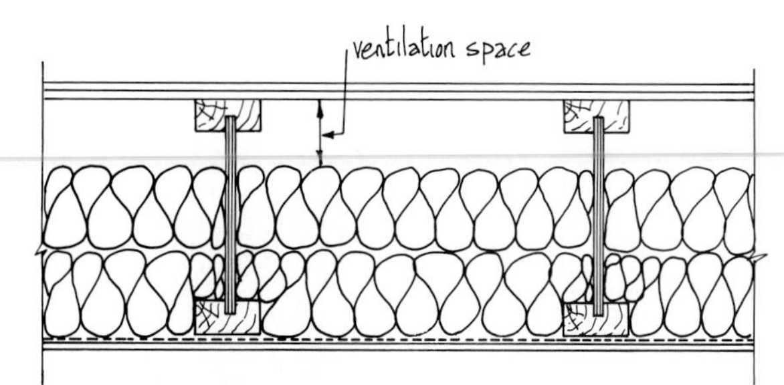

WOOD I-BEAM ROOF SECTION

It is generally best to install thick insulation batts in two separate layers.

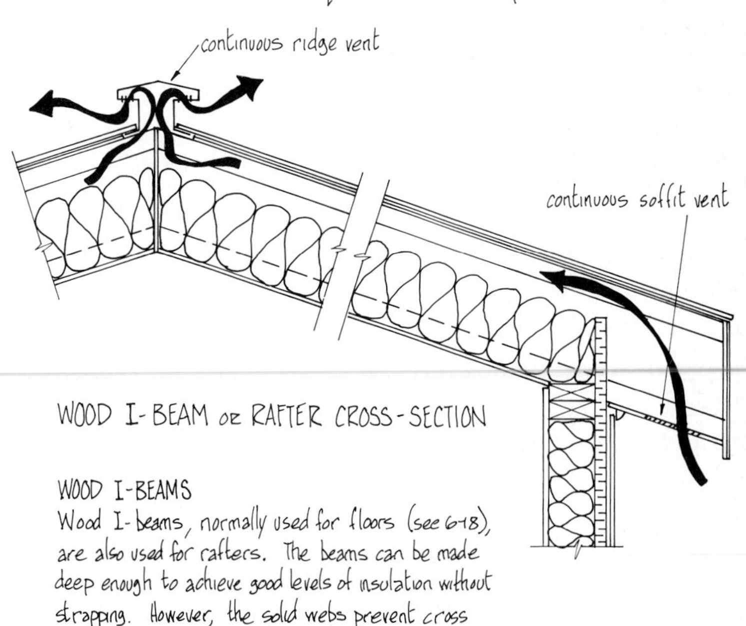

WOOD I-BEAM OR RAFTER CROSS-SECTION

WOOD I-BEAMS

Wood I-beams, normally used for floors (see 6-18), are also used for rafters. The beams can be made deep enough to achieve good levels of insulation without strapping. However, the solid webs prevent cross ventilation, and the details above should be noted.

PARTITION WALLS

INTERIOR PARTITIONS – TRADITIONAL ERECTION SEQUENCE

- In the traditional framing sequence interior partitions and exterior walls are constructed together in one continuous operation. Trusses (or ceiling joists) are added after the wall framing is finished.
- With this sequence the a/vb must be fitted between each partition at ceiling and exterior walls, creating many joints and air leakage points. The basic installation details of this method are shown on the next two pages.

ENERGY-EFFICIENT ERECTION SEQUENCE

- Roof trusses are designed to span between exterior walls without partition wall support.
- To install the minimum number of a/vb joints in an airtight structure, frame all interior partitions after erection of the roof trusses and the installation of exterior wall and ceiling a/vbs. With this procedure the a/vb can be installed on the walls and ceilings of the building shell in one continuous operation without interference from partitions.
- In addition, it is often advantageous to install the plasterboard on the outer shell before erection of the partitions. This helps protect the a/vb during future building operations. Note that the trusses are quite capable of carrying the plaster board without assistance from the partitions.
- This technique is especially recommended in regions where truss uplift is a potential problem.

truss or rafters
a/vb
ceiling finish
partition – lift up into place
floor

shims
3/4 in (19)

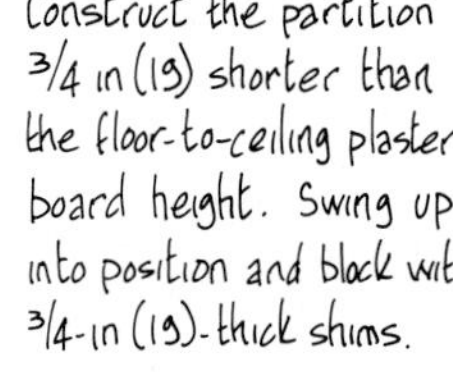
Construct the partition 3/4 in (19) shorter than the floor-to-ceiling plaster-board height. Swing up into position and block with 3/4-in (19)-thick shims.

SEASONAL TRUSS UPLIFT

This phenomenon, which occasionally occurs in cold and very cold regions, is caused by different moisture contents in the top and bottom chords of a truss.

- The bottom chord, being covered in insulation, is relatively warm and dry.
- The top chords, exposed to the atmosphere, are colder and have a higher moisture content.
- This differential causes the top chords to expand, lifting the bottom chord in the center (see the diagram at the right).
- The lifting effect causes cracking at ceiling and partition joints. It also tends to break the a/vb seal at these joints if the a/vb was installed after partition framing.
- To minimize the effects of this potential problem, frame the partitions after plasterboard installation, as described above. Although cracks will still form, the continuous a/vb will rise with the truss and should remain relatively intact.
- The uplift effect is under study by several trade organizations. Check with local authorities for updated information.

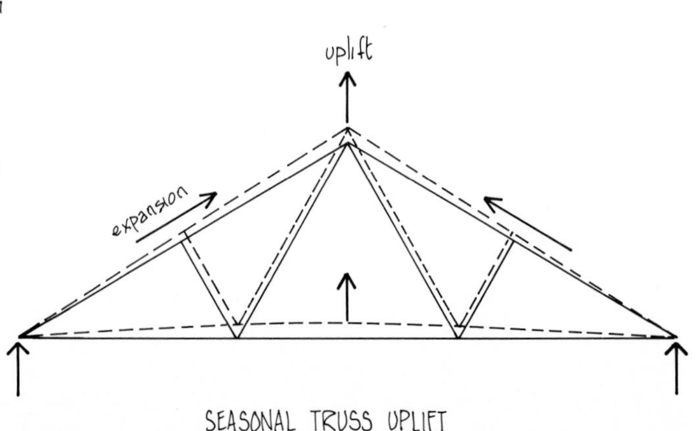

SEASONAL TRUSS UPLIFT

PARTITIONS

The details on this page apply to partitions which will be framed before truss erection.

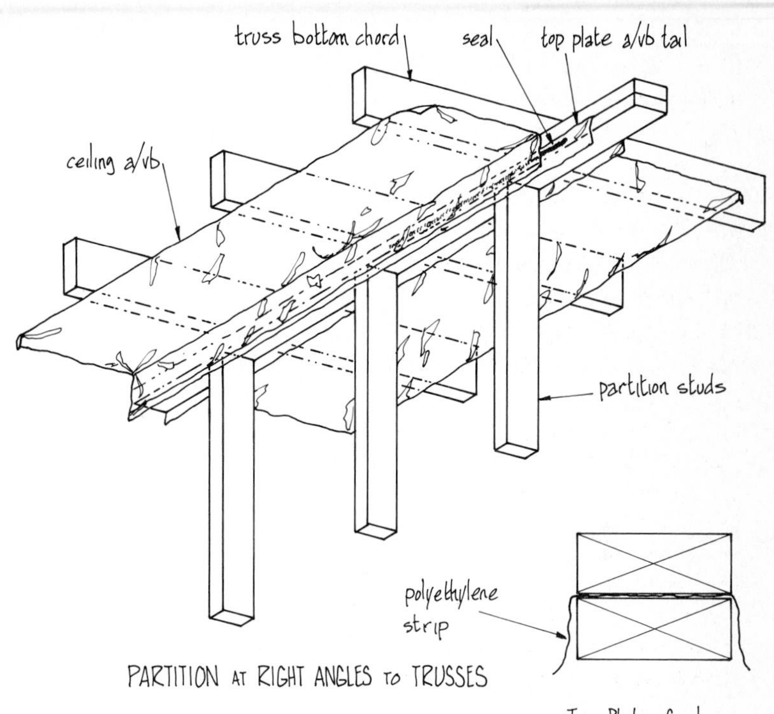

PARTITION AT RIGHT ANGLES TO TRUSSES

Top Plate Section

PARTITION AT RIGHT ANGLES TO TRUSSES

- Use a double top plate with a strip of poly sandwiched between the top plates.
- Seal the ceiling a/vb to the partition a/vb strip.
- This detail has several advantages: a/vb continuity is maintained, the poly is protected during further construction, and the second top plate provides a nonslip surface when erecting the trusses.

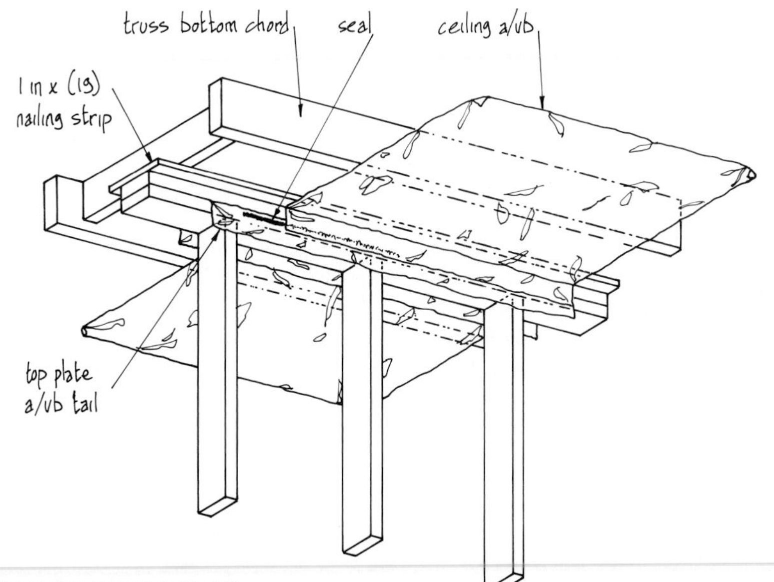

PARTITION PARALLEL TO TRUSSES

- The same procedure as above is followed, with the addition of a nailing strip for the plaster-board.
- Blocking to tie the head of the partition to the trusses must also be provided.
- A single top plate could be used, but longer studs are required for this type of wall. It is generally easier for both wall types to follow the same framing methods and stud sizes.

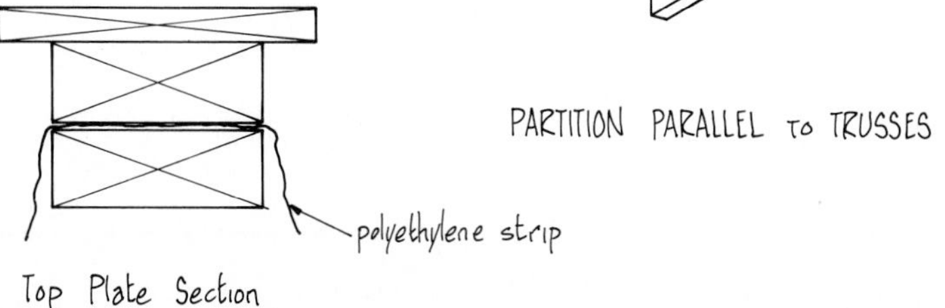

PARTITION PARALLEL TO TRUSSES

Top Plate Section

PARTITIONS

The details on this page apply to partitions which will be framed before truss erection.

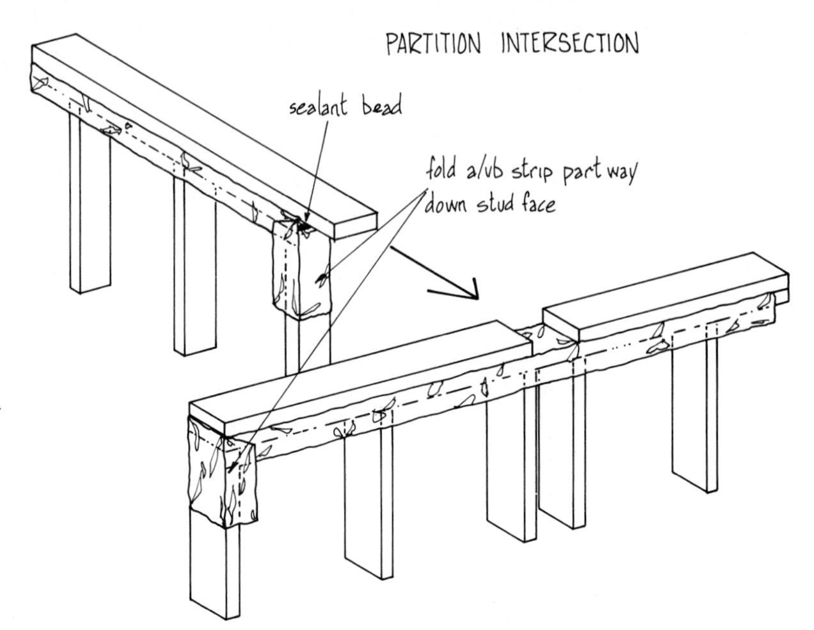

A/VB JOINT BETWEEN TWO PARTITION WALLS

- The illustration at the right shows the preparation of each partition wall before connection.
- At the free ends of the partitions, continue the strip of poly, sandwiched between the top plates, down the end face and fold backward over the end stud.
- Apply a bead of sealant at the point of contact before the partitions are fastened together.
- Wall and ceiling a/vbs can be fitted in the usual way.

PARTITIONS AT EXTERIOR WALLS

- The same basic procedure as above is followed except that the sandwiched poly strip is continued down to the bottom of the end stud.
- With the partition in place, the partition a/vb is sealed to the header a/vb.
- If the floor is over an unheated crawl space, a strip of poly should be positioned on the floor before erection of the partition.
- Use horizontal blocking (or other suitable detail) to provide support for the partition and plasterboard.
- Wall and ceiling a/vbs can now be fitted and sealed.

exterior wall before a/vb application

horizontal blocking

floor a/vb tail if over an unheated crawl space

seal junction

header a/vb

EXTERIOR WALL AND PARTITION JUNCTION

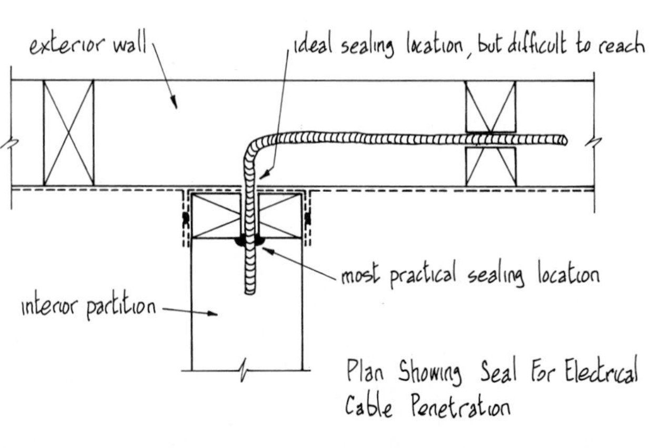

Plan Showing Seal For Electrical Cable Penetration

OVERHANGS AND SPLIT LEVELS

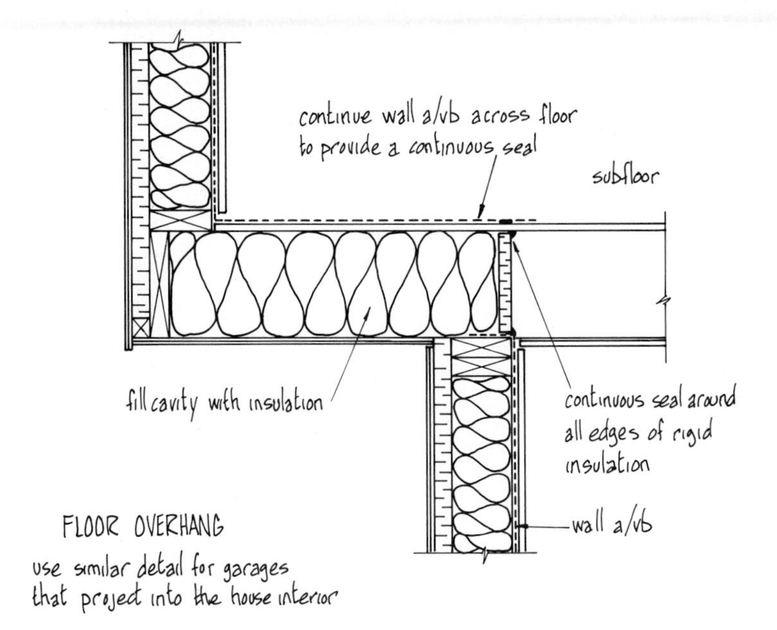

FLOOR OVERHANGS AND INTERIOR GARAGES

- The illustration at the left shows a floor overhang. Interior garages create an almost identical problem since they, too, intrude into the heated building space.
- For good insulation practice and a/vb continuity the floor over the cold space must be fully insulated and the a/vb must be on the warm side.
- To bridge the a/vb gap between the floor and the lower wall use nonpermeable rigid insulation board fitted tightly between the floor joists. Seal continuously on all edges.
- This procedure is obviously time-consuming and relatively expensive. It is recommended that overhangs or interior garages not be included on an energy-efficient dwelling if at all possible.

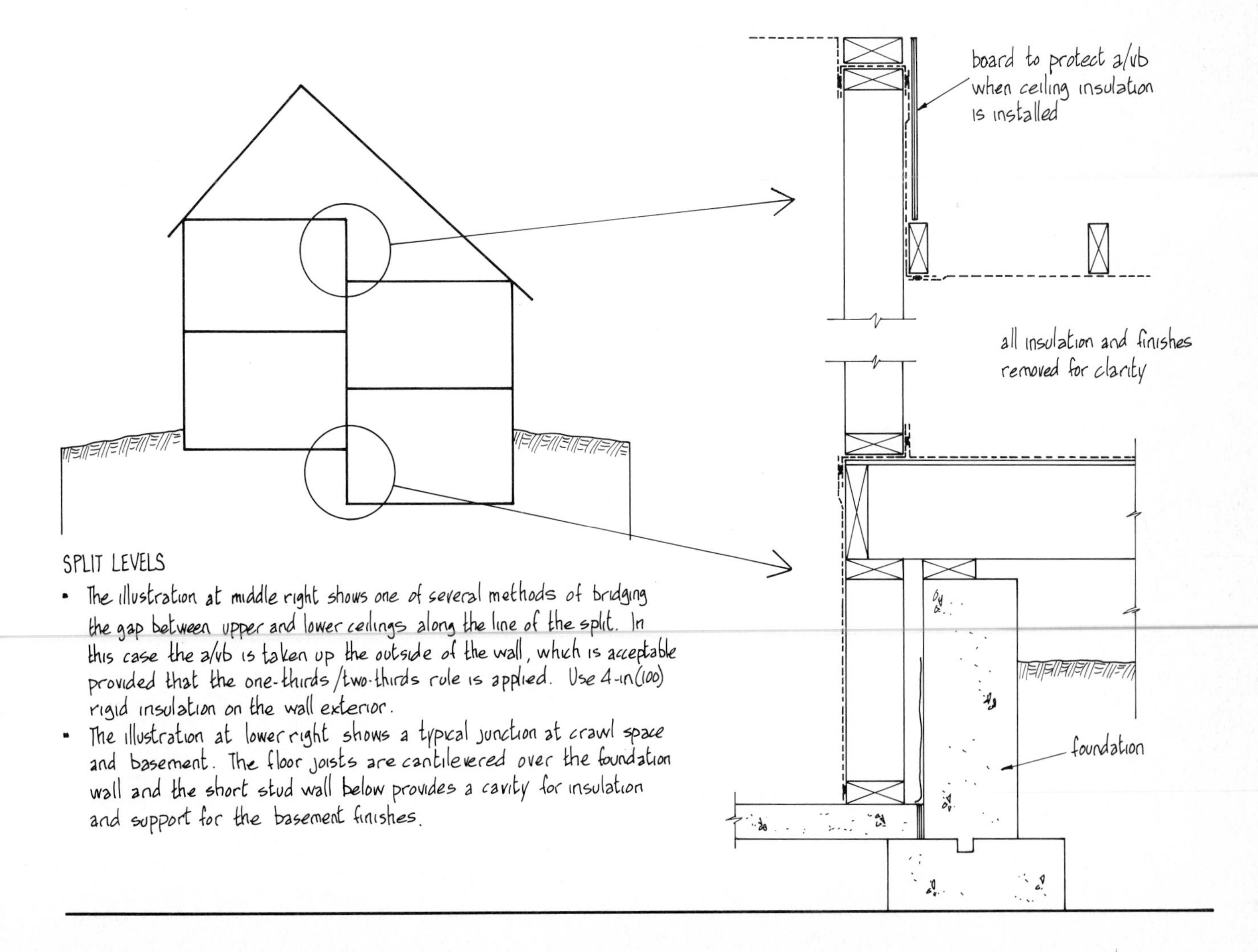

SPLIT LEVELS

- The illustration at middle right shows one of several methods of bridging the gap between upper and lower ceilings along the line of the split. In this case the a/vb is taken up the outside of the wall, which is acceptable provided that the one-thirds/two-thirds rule is applied. Use 4-in (100) rigid insulation on the wall exterior.
- The illustration at lower right shows a typical junction at crawl space and basement. The floor joists are cantilevered over the foundation wall and the short stud wall below provides a cavity for insulation and support for the basement finishes.

MISCELLANEOUS A/VB DETAILS

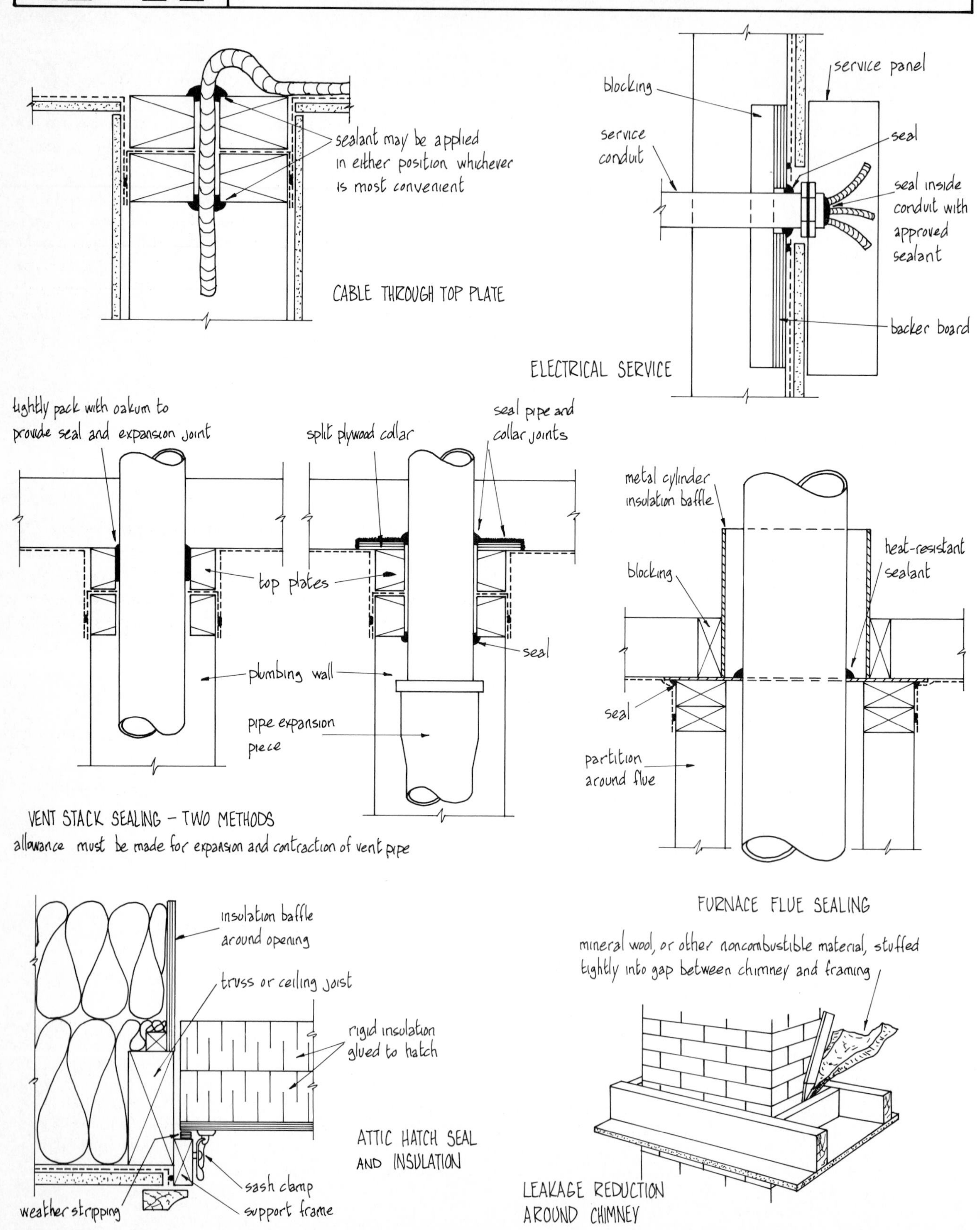

SEALING ELECTRICAL BOXES

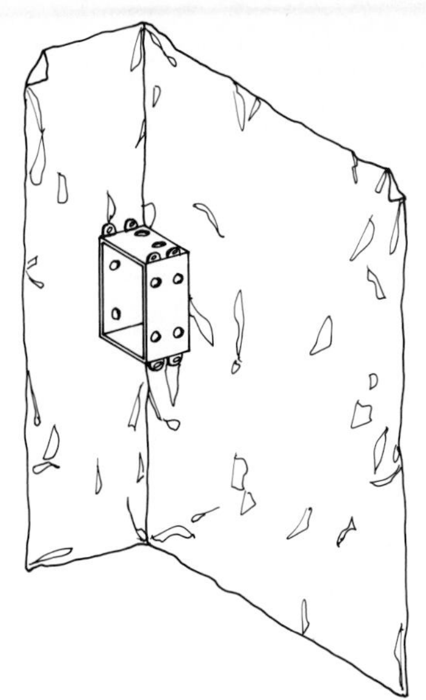

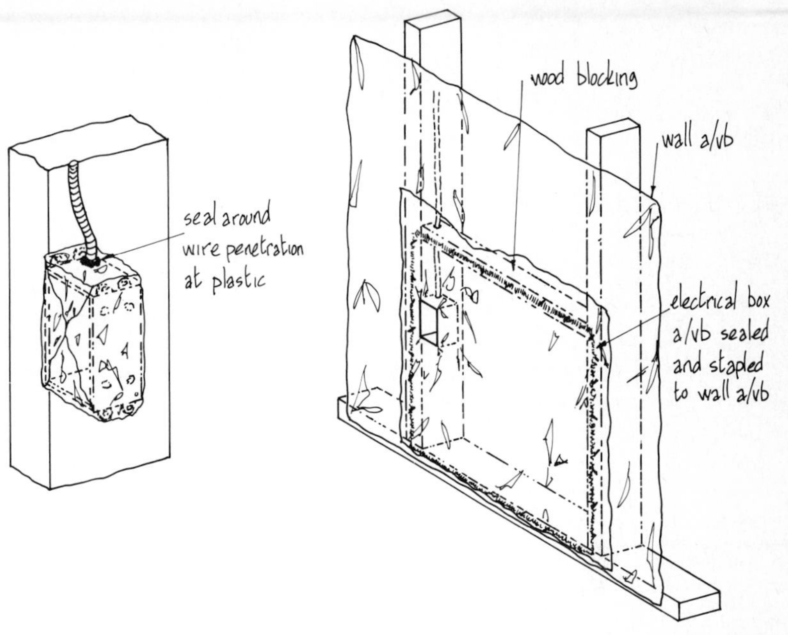

CONTINUITY OF THE A/VB AT ELECTRICAL BOXES

Step One
Cut a rectangle of 6 mil (0.15) polyethylene large enough to overlap the stud space, bottom plate, and blocking. Allow plenty of spare material. Position plastic behind the box.

Step Two
Attach the box to the stud and wrap the box loosely with the plastic. Run the electrical cable through the plastic and into the box, being careful to minimize damage to the plastic. Seal around the cable where it penetrates the plastic. Push all loose material into the box cavity before the wall a/vb is installed.

Step Three
Install the wall insulation and the wall a/vb. Carefully slit the wall a/vb at the box opening and pull out the box a/vb. Spread the box a/vb over the face of the wall a/vb, sealing and stapling together over the solid backing of the studs, blocking and bottom plate.

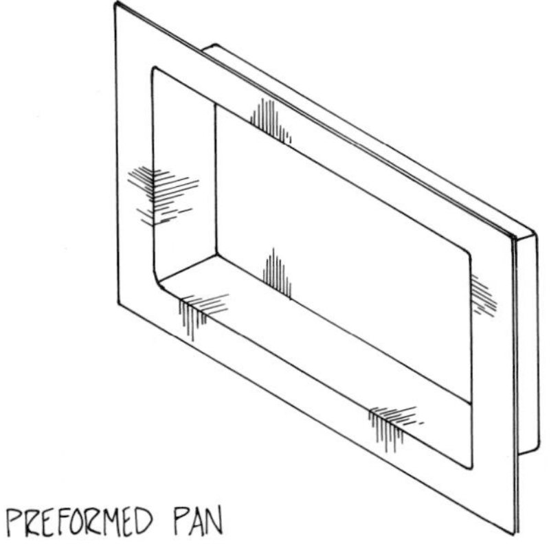

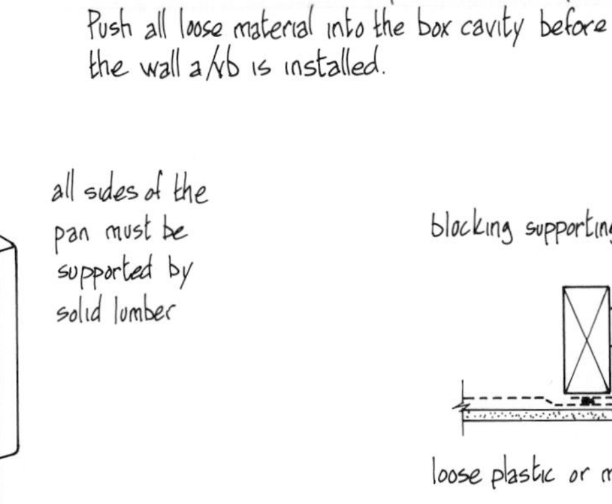

PREFORMED PAN

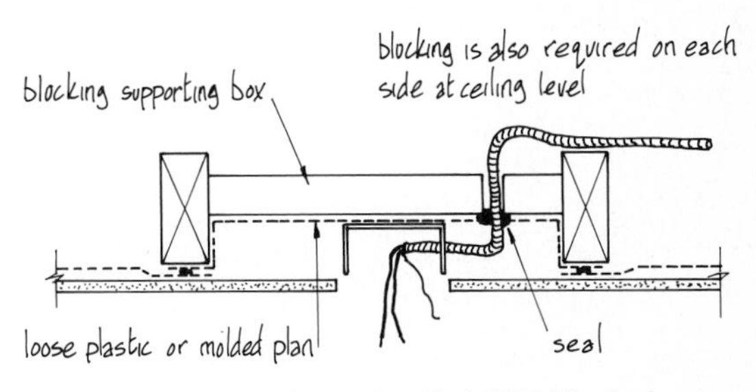

SECTION AT CEILING ELECTRICAL BOX

Several commercial products are available that provide a continuous vapor barrier around electrical boxes. The type shown above, a shallow rectangular box, molded in polyethylene, is typical. It is designed to fit between two studs with the horizontal flanges supported on blocking. The box is fastened to the stud through the side of the pan, the cable is installed and sealed where it penetrates the pan. High R-value insulation should be installed behind the pan to avoid condensation problems, while standard batt insulation can fill the pan depth. Apply sealant to the pan flanges before the wall a/vb is installed.

Although details on this page show sealing of electrical boxes in exterior walls and ceilings, an attempt should be made to locate all receptacles on interior walls and to use wall-mounted lighting. Doing this will eliminate a large number of a/vb penetrations.
All types of recessed light fixtures that extend into an insulated space should be eliminated since most fire codes do not allow these fixtures to be either insulated or sealed.

WINDOWS

All windows are a source of heat loss, even those that face south and are used as solar collectors (see 3-14). In an energy-efficient dwelling, the type of windows chosen and the method of their installation is therefore an important consideration.

There are three principal areas of heat loss in every window installation:

- Air infiltration (or exfiltration) between the window frame and the rough opening in the stud wall.
- Similar air leakage through joints between the window components.
- Conduction and convection through the glass and frame.

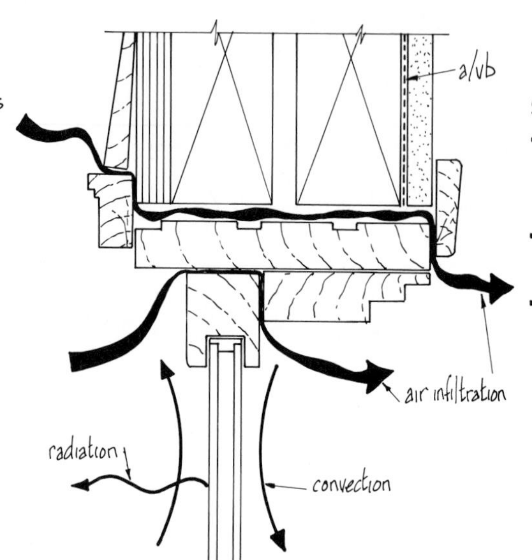

WINDOW-HEAD SECTION
showing heat loss through infiltration, convection, and radiation

SOLUTIONS

- Ensure that the a/vb is continuous between window frame and stud wall framing (see below).
- Use window types designed for low air leakage (see the next page).
- Use windows with the highest thermal resistance through both glass and frame (see the next page).

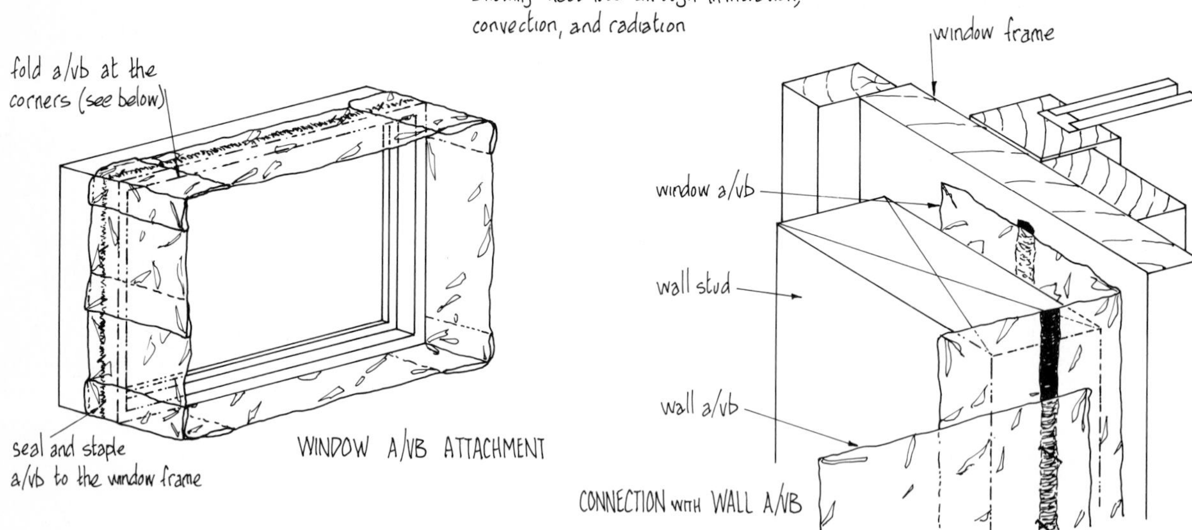

WINDOW A/VB ATTACHMENT

CONNECTION WITH WALL A/VB

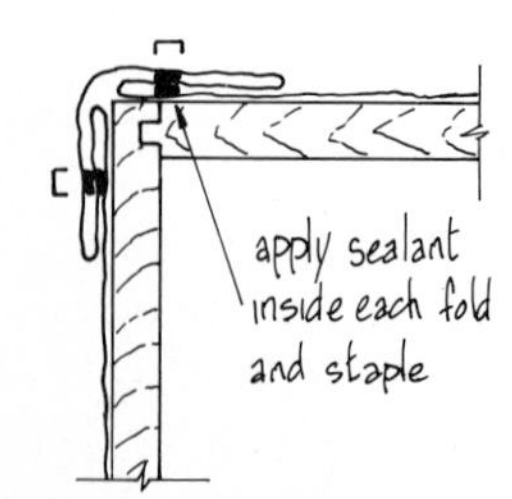

CORNER SECTION
showing spare-material folds

WINDOW INSTALLATION

- Prepare the window by stapling and sealing a strip of poly to the frame as shown above left. Allow plenty of loose material at the corners as indicated in the small section at the left.
- Install the window in the rough opening following the manufacturers' instructions. Insulate the gap around the window with loose fiberglass (do not pack tightly) or with polyurethane foam.
- Fold the window a/vb back onto the stud frame and seal to the wall a/vb already in place. Avoid cutting and patching the window if possible.
- These procedures apply to single stud and strapped walls. For double stud walls the wall a/vb must be folded around the inner face of the inner stud for sealing to the window a/vb.

WINDOWS AND DOORS

CASEMENT

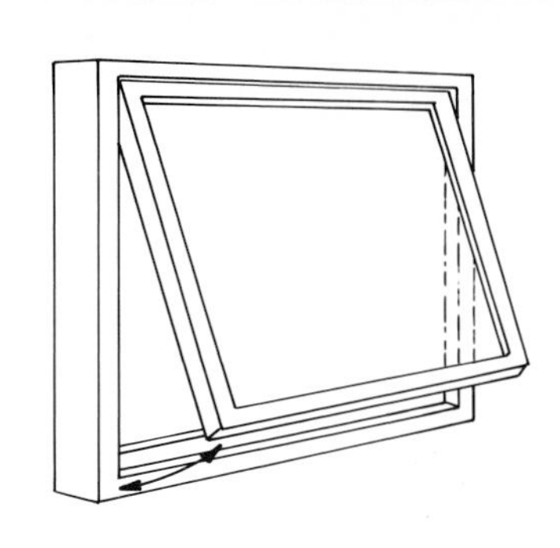

AWNING
reverse for hopper

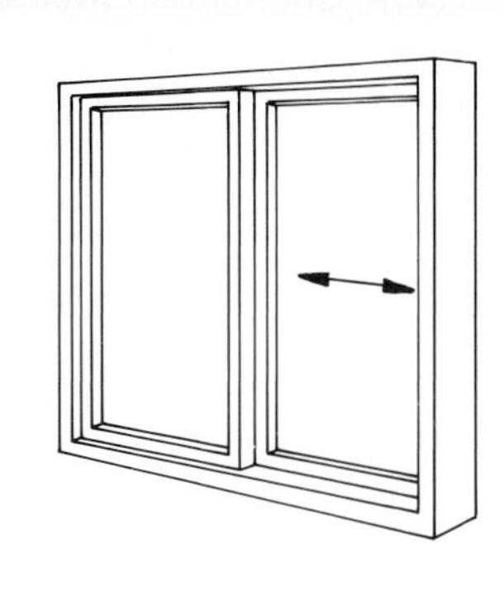

HORIZONTAL SLIDER

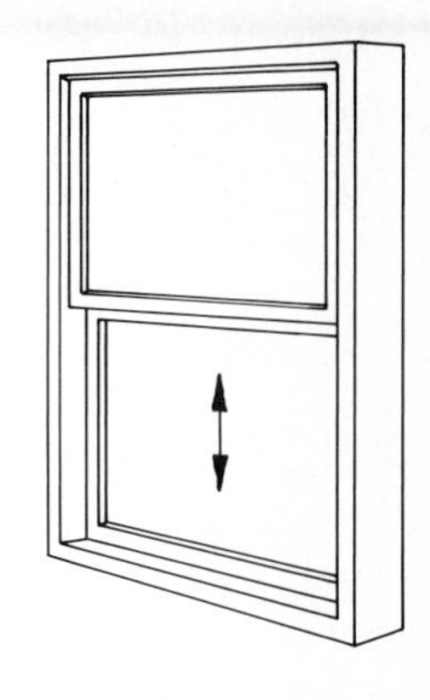

VERTICAL SLIDER

WINDOW AIRTIGHTNESS

- The type and construction of a window determines its resistance to air leakage. Shown above are the four main window types in general residential use.
- Generally, windows that are hinged (casement, awning, and hopper) have the best resistance. They have better weatherstripping and fewer joints.
- Sliders, either horizontal or vertical, have the least resistance since it is very hard to adequately seal sliding joints.
- When choosing windows, obtain performance information from manufacturers for comparison purposes.

THERMAL RESISTANCE

- The number of glazing layers in a window determines the effective R-value for that window provided that the airspace between the panes is of the correct size. Optimum space between panes is 5/8 to 3/4 in (16-19). R value decreases sharply below this spacing size.
- Double glazing is standard in most regions (and may be the specified code minimum). Triple and quadrupal glazing is also available, but the additional cost must be balanced against projected energy savings.
- Wood frames offer the most resistance. Metal (and some plastic frames) offer the least resistance and are often subject to interior condensation and frosting even when thermally broken to reduce conduction.

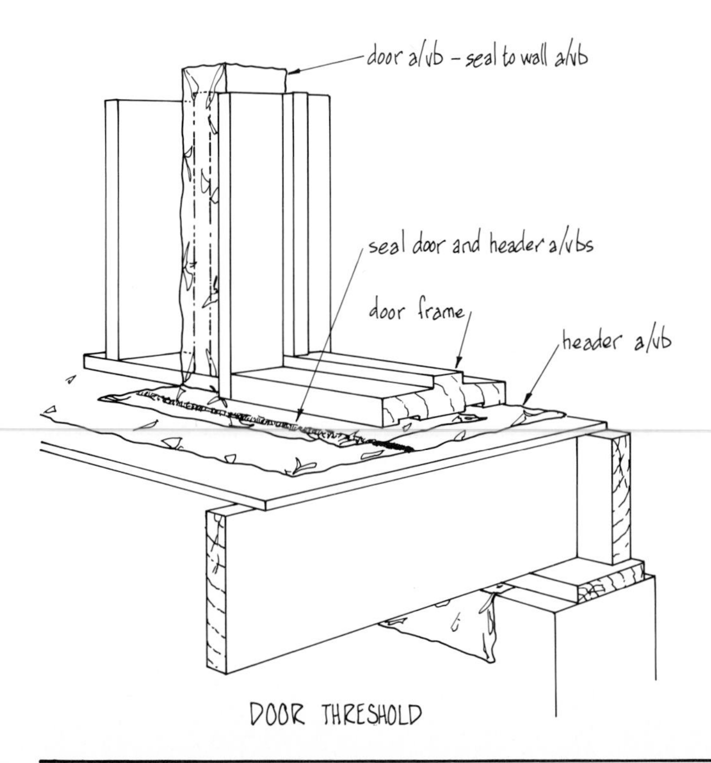

DOOR THRESHOLD

DOORS

Doors are similar to windows in terms of air leakage and heat loss.

- Follow the same attachment procedure as for windows except that at the sill the poly is sealed to the header a/vb as shown at the left. The exposed plastic on the subfloor must be protected from construction damage.
- The door must be as tightly weather-stripped as possible. Owners should be made aware that weather stripping may have to be replaced after a number of years to maintain peak efficiency. Glazed patio doors should be regarded as large slider windows.
- The insulation value of a door varies widely with type of construction. Solid wood doors have R1 to R2 (RSI 0.18-0.35) values, while some insolated steel doors provide R15 (RSI 2.6). In addition, the steel doors have magnetic weather stripping attached to the wood frames, which effectively seals the doors against air leakage.

CHAPTER PAGE TITLES

GYPSUM BOARD (DRYWALL)

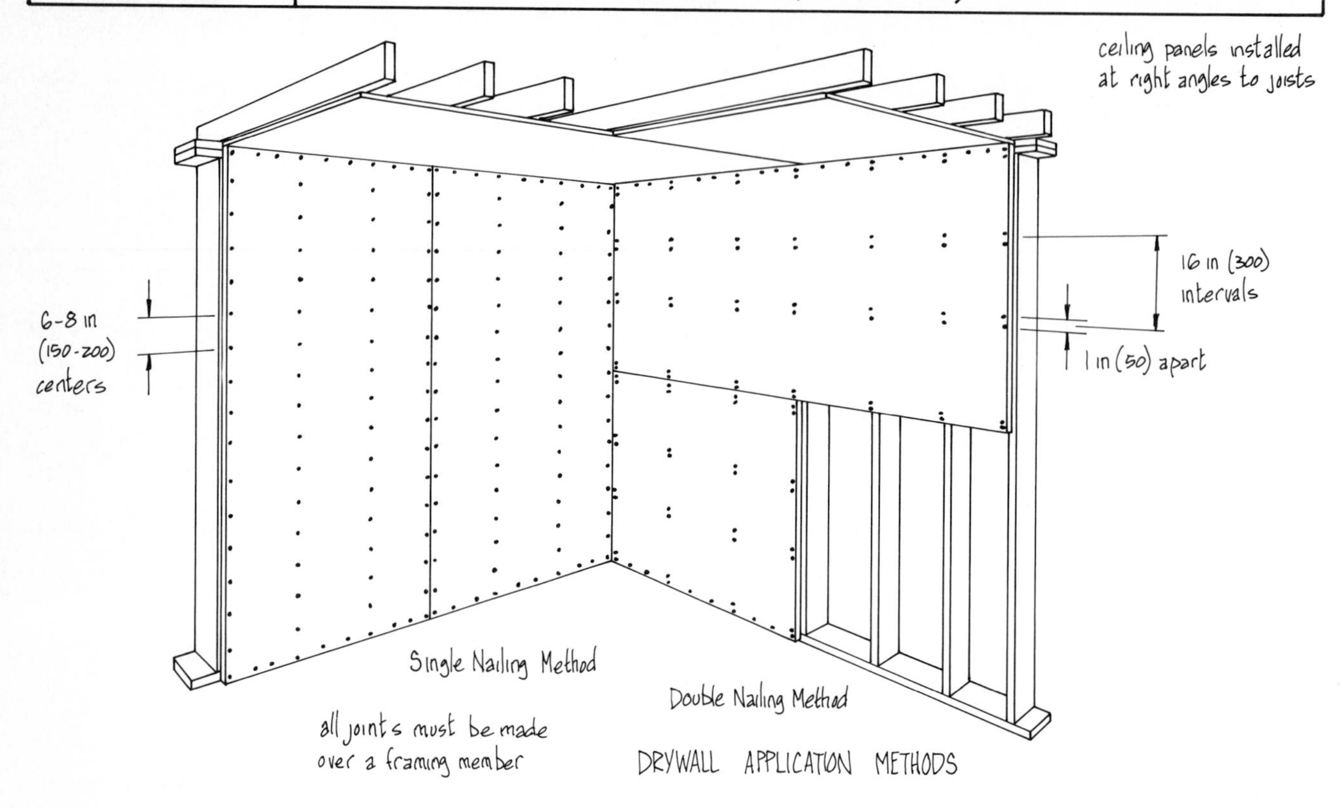

DRYWALL APPLICATION METHODS

Gypsum board is the most used interior wall and ceiling finish in residential and commercial construction because of its time- and cost-saving advantages. Gypsum panels are composed of a fire-resistant gypsum core encased in heavy paper on the face and back sides. The paper face is folded around the long edges to reinforce and protect the core, and the ends are square cut. There are several types of board available:

- The standard gypsum board as described above. Has tapered long edges and square short edges. Good fire rating.
- Fire-code panels, with a specially formulated core for increased resistance to fire exposure.
- Water-resistant panels, for use in bathrooms or other wet areas. Both core and paper faces are water resistant.
- Prefinished panels with a variety of plastic, paper, or other finishes factory applied to the outer face. This type is used extensively in commercial construction.
- Aluminum-foil-backed panels for use in a combined drywall-vapor barrier system (see 10-13).

Gypsum panels are usually applied either vertically or horizontally in single sheets directly to the framing members as illustrated above. Horizontal application is generally recommended since this makes the joints less visible. Nails or screws may be driven in a single- or double-nailed pattern at the spacings shown above.

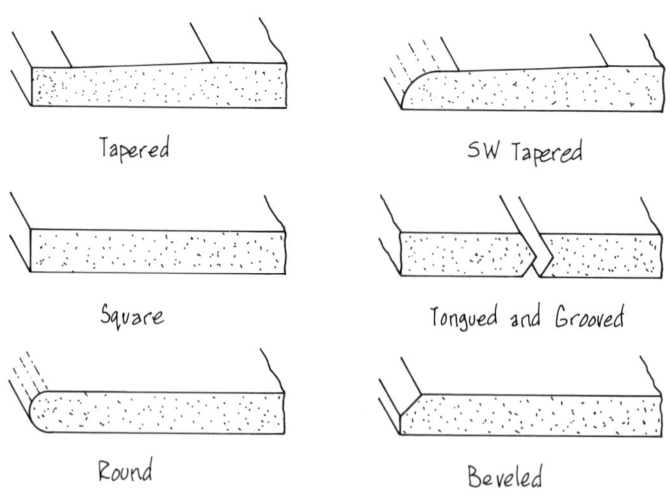

There are several edge types available, as shown above, although each is generally related to a specific type of panel. The tapered edge is the most common and is used for all standard-type panels. Note that only the long edges of a panel are shaped — short edges are always square.
Most panels are made in a standard 4-ft (1200) width, with varying lengths between 8 ft (2400) and 16 ft (4800) maximum.

GYPSUM BOARD (DRYWALL)

MAXIMUM FRAME SPACING FOR GYPSUM PANELS - SINGLE LAYER

PANEL THICKNESS in (mm)	LOCATION	APPLICATION METHOD [1]	MAX. FRAME SPACING in (mm)
3/8 (9.5)	ceilings [2]	perpendicular [3]	16 (400)
	sidewalls	parallel or perpendicular	16 (400)
1/2 (12.7)	ceilings	parallel [3]	16 (400)
		perpendicular [4]	24 (600)
	sidewalls	parallel or perpendicular	24 (600)
5/8 (15.9)	ceilings	parallel [3]	16 (400)
		perpendicular	24 (600)
	sidewalls	parallel or perpendicular	24 (600)

Notes:
1. Long edge position relative to framing.
2. Not recommended below unheated spaces.
3. Not recommended with water-based spray-textured finishes.
4. Maximum spacing 16 in (400) if used with finishes as in note 3.

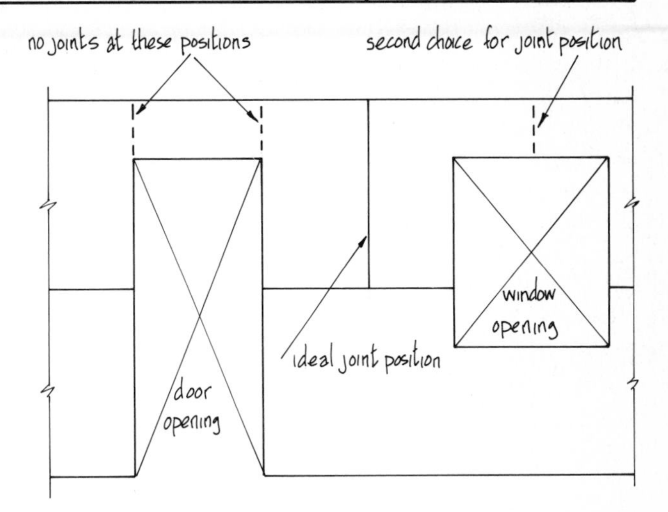

JOINTS

Ideally, all holes in panels for windows, doors, etc., should be made within the length of a sheet as illustrated above. Where panel ends must be joined at an opening it is best to do so near the center of the opening. Joints occurring over the opening's sides should be avoided.

RECOMMENDED PANEL SPANS

Gypsum panels, depending on type, are available in thicknesses ranging from 1/4 to 1 in (6-25.4*). The table above gives recommended thicknesses for standard panels in relation to panel application and position. For other panel types, refer to manufacturer's instructions.

ringed drywall nail

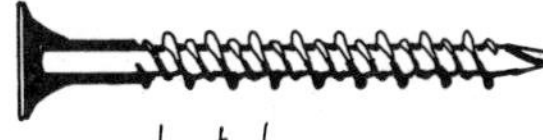

wood-stud screw

metal-stud screw

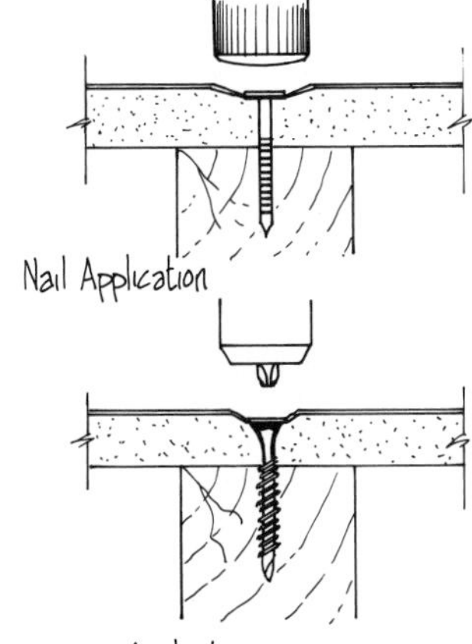

Nail Application

Screw Application

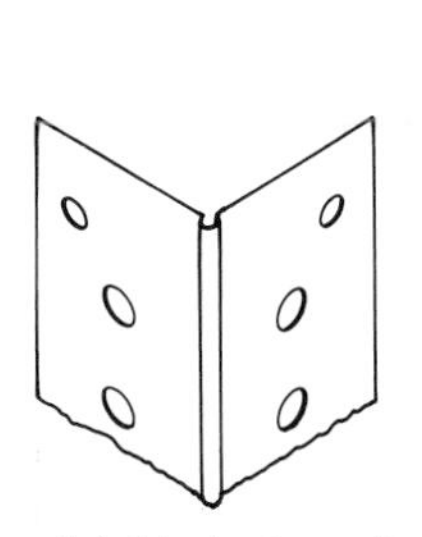

Metal Outside-Corner Bead

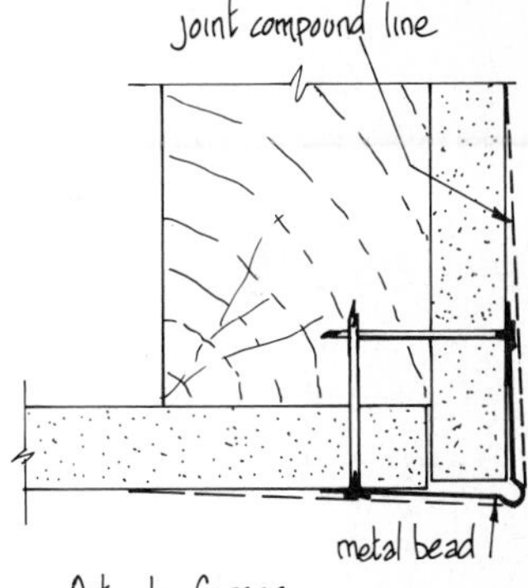

Outside Corner

paper tape

joint compound

Inside Corner

CORNERS

Outside corners must be reinforced with metal corner beads. The bead is nailed or screwed through the panels to the framing and finished with joint compound. Interior corners have lapped panels reinforced with paper or fabric tape (see page 11-5) and then finished with joint compound.

NAILS AND SCREWS

The nail and screw types shown above are typical of those used for standard gypsum panel applications. The nails are annular ringed with a 1/4-in (6) head and are driven sufficiently to "dimple", but not break, the paper face. The screws are for wood or steel stud framing and are driven by a special drywall "screwgun" equipped with a clutch that stops driving when the screw is at the correct depth. Length of nails or screws depends on panel thickness. Many other types of nails and screws are manufactured for specific applications.

chart source: Canadian Gypsum Co.

GYPSUM BOARD (DRYWALL)

It is important that gypsum panels be tight against the framing and that the framing members are straight and in line with each other. Studs that are bowed can be straightened by making saw cuts on the concave side and then forcing the stud into alignment. Cleats must be nailed on each side to maintain the alignment. Ceiling joists that are out of alignment must not be cut. The best method is to strap the ceiling with 1 x 3-in (19x64) lumber and make alignment adjustments with wedges inserted between the strapping and the joists.

wedges

1"x3" (19x64) strapping

CEILING STRAPPING

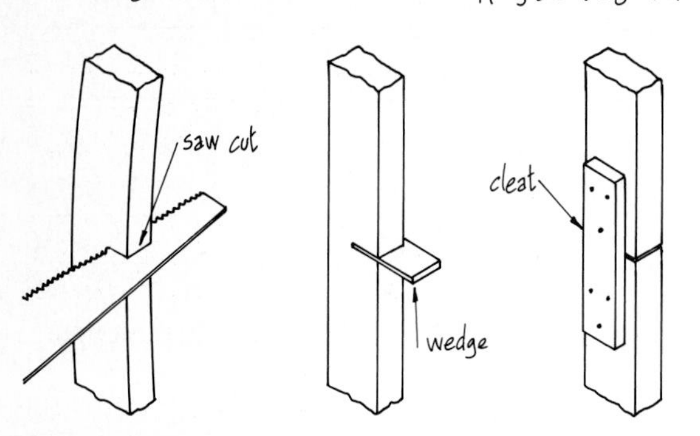

STRAIGHTENING A BOWED STUD
more than one cut may be necessary for a badly bowed stud

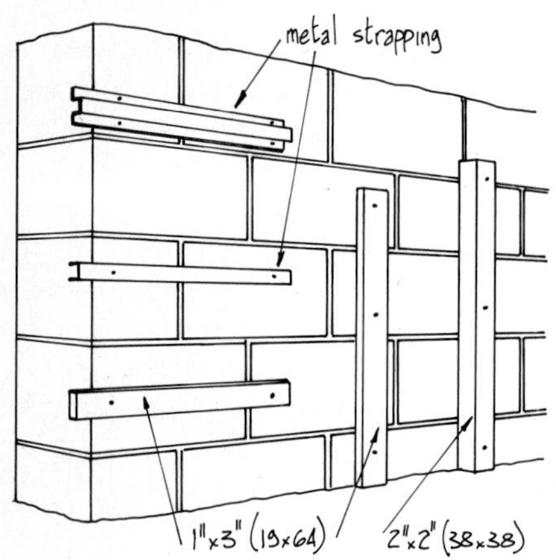

MASONRY WALLS
Gypsum panels may be nailed or screwed to wood or metal furring strips attached to masonry exterior walls. Panels may also be glued directly to an <u>interior</u> wall provided that it is straight and true. Rigid foam or fiberglass batt insulation is fitted between the furring strips. Some rigid insulation systems have metal furring strips set into the face of the insulation, as shown at the right. Although the strip is screwed or bolted to the wall, it is not in direct contact with the wall face, thus greatly reducing thermal bridging (see 10-35).

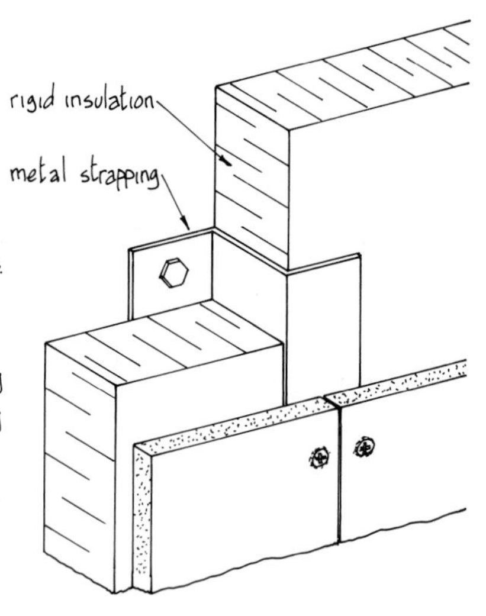

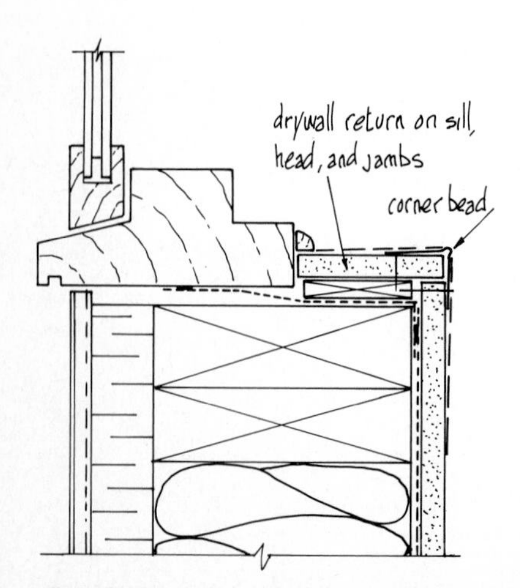

WINDOW, DOOR, AND BATHTUB TRIM
The cut edges of panels must be hidden by casing trim around windows and doors as shown at the left. Window returns can be continued in gypsum board using corner beads, but this is not generally recommended because of potential damage to panel surfaces. A typical detail at the edge of a bathtub is shown at the right.

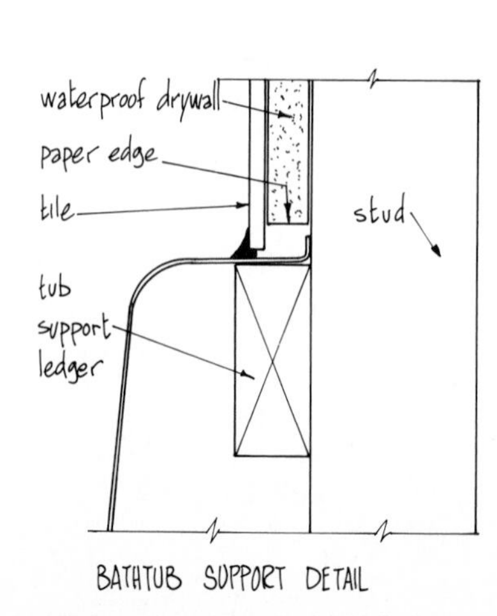

BATHTUB SUPPORT DETAIL

JOINT FINISHING

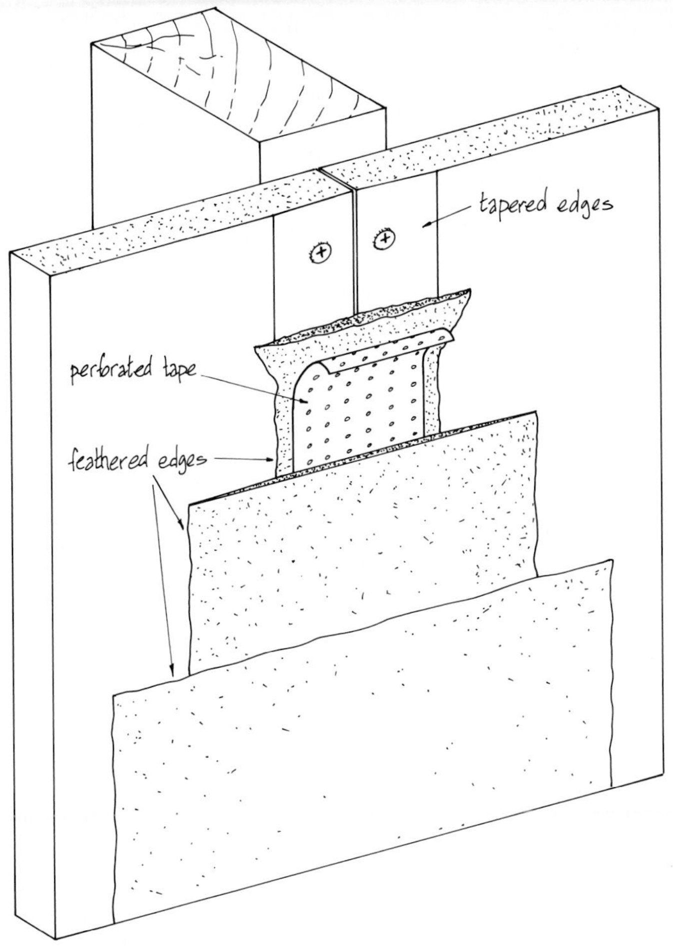

JOINT FINISHING

The following procedure is the standard joint finishing method:

- Joints between panels greater than 1/8 in (3) should be filled with joint cement and allowed to dry before taping starts.
- A band of cement about 5 in (125) wide is applied along the length of the joint. Continuous paper tape is then pressed into the fresh cement with a trowel or putty knife, ensuring that the tape is smooth and the cement edges are feathered to zero thickness.
- After the first layer of cement is set, a second layer about 8 in (200) wide is applied, then smoothed and feathered with a putty knife. This layer should be about 10 in (250) wide over the square end joints.
- A third cement layer is applied after the second layer is dry. It should be 10-12 in (250-300) wide over tapered joints and 16 in (400) wide over square joints. This final layer should be smoothed carefully to eliminate bulges and rough spots and to completely feather the edges. When this layer is dry, it is lightly sanded to finish the smoothing process.
- Nail or screw holes elsewhere in the panel are cement filled and feathered two or three times using a putty knife and lightly sanded for a smooth finish.

Drywall contractors use a variety of special tools to speed the taping and cementing process. Other types of tape are also available, one of which is an adhesive-coated open-weave glass-fiber tape that is attached directly to the panel faces by hand pressure and finished with two coats of quick-drying cement.

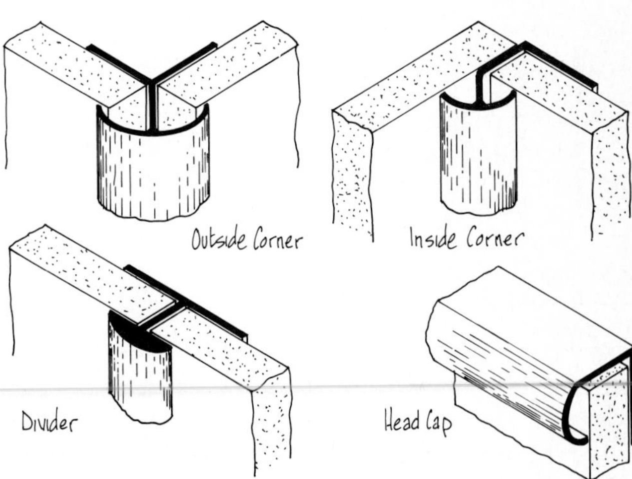

DOUBLE-LAYER APPLICATION

Double layers of gypsum board are often installed for fireproofing purposes. The first layer is applied vertically and fastened to the framing as usual. The second layer is applied horizontally and glued to the first using double-headed nails as temporary fasteners. When the glue is dry, all nails or other supports are removed and the joints finished as described above.

DECORATIVE MOLDINGS

Depending on type, most prefinished gypsum panel edges are concealed by matching vinyl or metal moldings such as those illustrated here. Panels are attached to their supports by hidden nails or screws, clips, or brackets.

INTERIOR PLASTER

Plaster is applied in two or three coats to a rigid base that provides both support and a good bond or key for the plaster. The two common types of plaster base are:

- Gypsum board lath, similar in manufacture and installation to standard drywall (see the preceding pages). Standard size is 16 x 48 in (400 x 1200), but other sizes are also available. Some types are perforated to strengthen bonding and increase fire resistance. Use 3/8-in (9.5) thickness for stud or joist spacings of 16-in (400) and 1/2-in (12.7) thickness for 24-in (600) spacings. Gypsum lath is nailed or stapled to wood framing and screwed to metal framing.
- Expanded-metal lath, made from slit and expanded sheet metal with a rust-resistant finish. Available in several sheet sizes, it is applied with ends and sides overlapping and is nailed, stapled, or screwed to the framing as for gypsum lath. No extra reinforcing is necessary except for corner beads at outside corners (see below). Used extensively in commercial construction, but generally only for shower and bathroom walls and ceilings in residences.

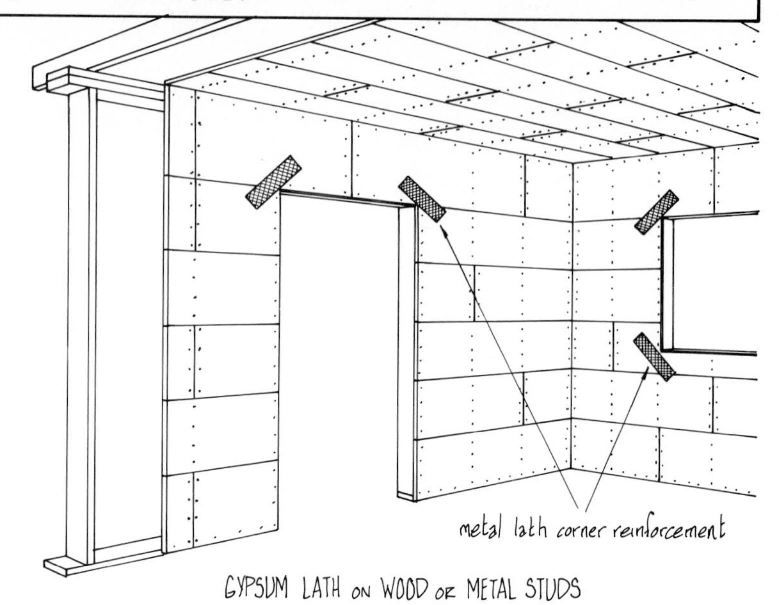

GYPSUM LATH ON WOOD OR METAL STUDS

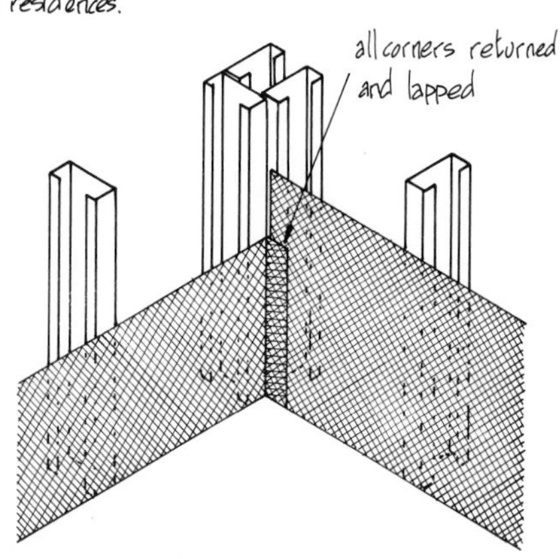

EXPANDED-METAL LATH

At the left, expanded-metal lath is shown attached to metal studs. When expanded-metal is used over wood framing, asphalt-saturated building paper is first applied to the framing face to control moisture absorbtion or to protect a vapor barrier.

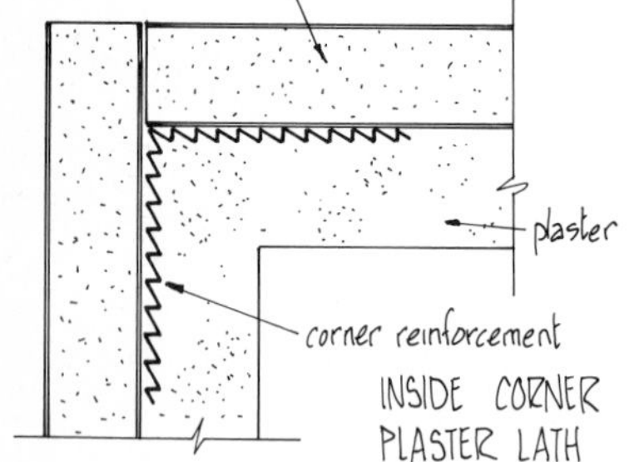

INSIDE CORNER PLASTER LATH

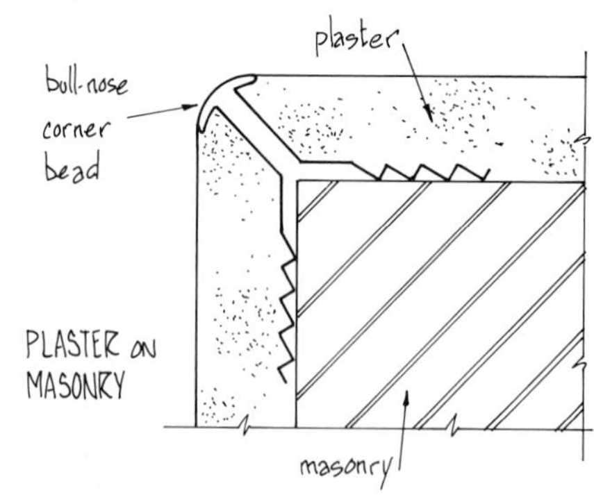

PLASTER ON MASONRY

REINFORCING

Shown here are four reinforcing details for inside and outside corners and over structural members. Reinforcing for gypsum lath at openings is shown above. It is important that corner beads be plumb and straight and that the bead is flush with the finished face of the plaster.

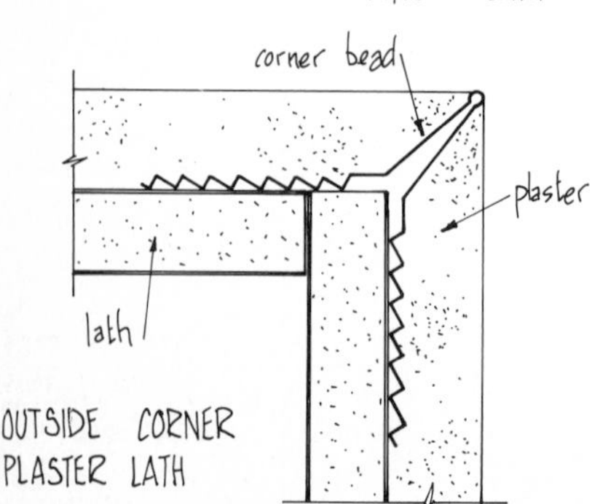

OUTSIDE CORNER PLASTER LATH

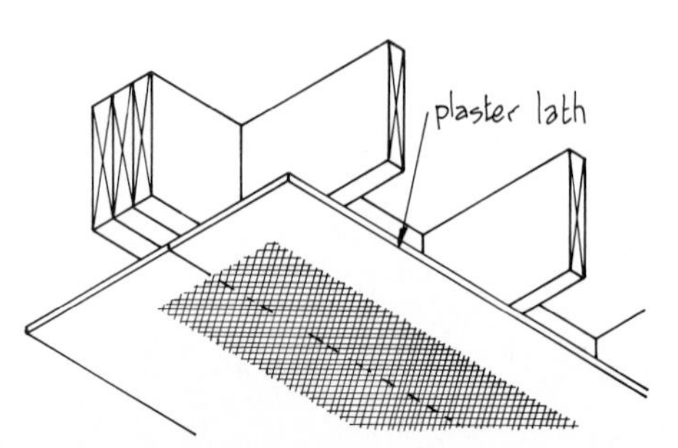

METAL LATH REINFORCING AT FLUSH MAIN BEAM

INTERIOR PLASTER

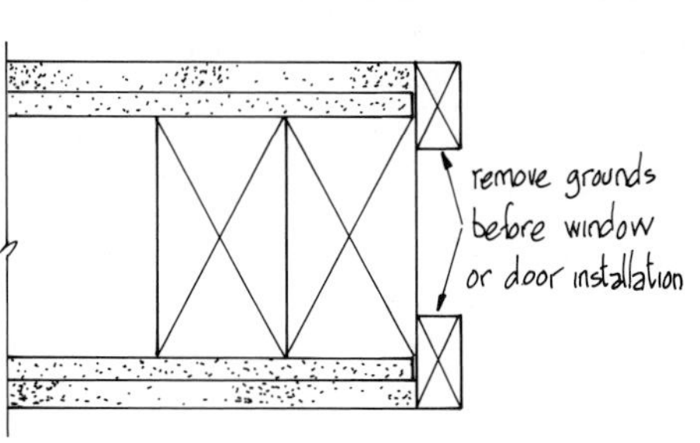

TEMPORARY GROUNDS

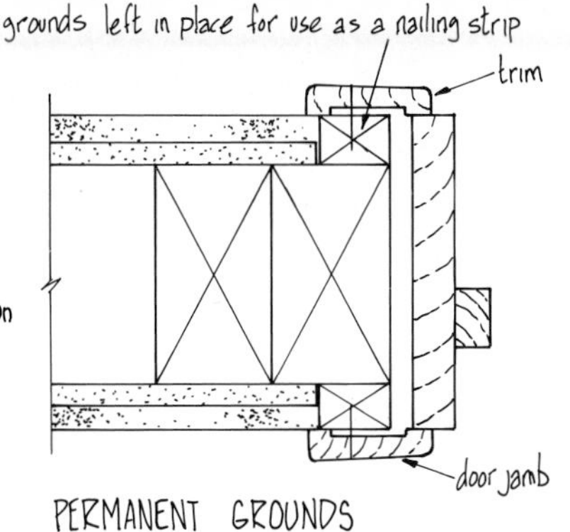

PERMANENT GROUNDS

PLASTER GROUNDS

To help control plaster thickness, plaster grounds, sized to match the total thickness of lath and plaster, are attached to all openings and wall bases as shown in the illustrations at the left. The grounds may be wood strips as shown or formed metal sections that are available in a variety of shapes and sizes. The grounds are either designed for permanent installation and used as a nailing base for trim or designed to be removed after plastering. Some windows, doors, electrical fixtures, etc., are designed to provide their own plaster grounds, as indicated by the hollow metal door jamb at the left.

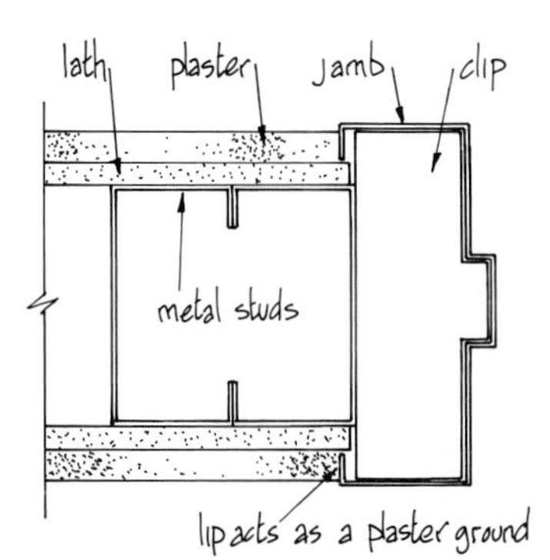

HOLLOW METAL DOOR FRAMES

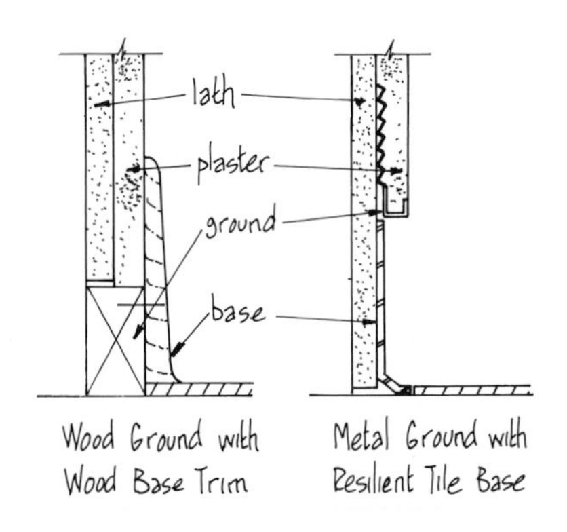

WALL BASE EXAMPLES

PLASTER COATS over METAL LATH

Plaster is applied to metal lath in a three-step process:

- The first or "scratch coat" is applied directly to the lath. It is allowed to stiffen (called "taken up") and the top surface is scratched or raked to provide a good bond for the second coat. The usual plaster mix is one part gypsum plaster to two parts sand.
- The second coat (called the "brown" or "leveling" coat) is then applied and carefully leveled to the thickness of the plaster grounds. Usual mix for this coat is one part gypsum plaster to three parts sand.
- The final finishing coat or "putty coat" is applied in a thin layer after the second coat is set and troweled to a smooth, hard finish. The usual mix is one part gypsum plaster to three parts lime.

PLASTER COATS over GYPSUM LATH

Plaster is applied to gypsum lath in a two-coat process:

- After the scratch coat has been applied, it is followed almost immediately by the brown coat, which is leveled to the thickness of the plaster grounds. It is not necessary to scratch the first coat since it is allowed to dry only long enough to prevent sagging. The usual mix for both coats is one part gypsum plaster to two and a half parts of sand.
- The finish coat is applied as in the three-coat process.

PLASTER THICKNESS

Metal lath – minimum 3/4 in (19) measured from the back of the lath.
Gypsum lath – minimum 1/2 in (12.7).
Masonry, tile, or brick — minimum 5/8 in (16).

INTERIOR PANELING - MATERIALS

Paneling for interior use is available in a wide range of material types and finishes. Aside from aesthetics, cost is a major factor in choosing a particular type of product. For example, plywood panels, faced with selected hardwood veneers, cost many times that of popular vinyl-faced plywood or hardboard panels. The information below outlines the main product groups in general terms. Always refer to manufacturers or suppliers for current product offerings.

PLYWOOD

Available in a wide range of hardwoods and softwoods, with plain, textured, grooved, or prefinished finishes. Panel sizes are 4 ft (1200) wide and up to 10 ft (3000) long. Thicknesses range from 3/16 to 1/2 in (5-12.7*) with 1/4 in (6) being most readily available. Most common, and least expensive, are printed wood grain vinyl-paper finished softwood panels. The panel edges are available in square, beveled, or shiplapped, with a variety of plastic or wood moldings to cover joints or provide accent strips.

HARDBOARD

A dark-colored composition board, made from glued softwood pulp. On unfinished panels, one face has a smooth surface for painting. Hardboard is also available in a wide range of prefinishes including tileboard, which has a tough, baked-on plastic finish that can be used in bathrooms and kitchens. Panel sizes are similar to plywood. Panels less than 1/4-in (6) thick must be installed on a solid backing.

FIBERBOARD

A soft composition material available in 4 by 8 ft (1200 x 2400) panels or in boards of various width. Thickness is usually 1/2 or 3/4 in (12, 19*). A wide variety of prefinishes are available. Because this material is soft, it should not be used where mechanical damage is likely to occur.

PARTICLEBOARD AND CHIPBOARD

A fairly brittle composition board available in sizes similar to plywood, that should be applied on a solid backing if less than 1/2 in (12*) thick. Usually prefinished with veneers or plastic laminates.

PLASTIC LAMINATE

Although most often used on horizontal surfaces in kitchens or bathrooms, laminates are also used as a face veneer for panels. Durable and washable, and available in an extremely wide range of colors, textures, and patterns. Glue-laminated to a rigid backing board with trimmed or capped edges (see 13-18).

SOLID LUMBER BOARDS

Several species of wood are used for panelling, ranging from redwood, cypress, and cedar which give darker tones, through ponderosa pine, fir, and hemlock giving medium tones, to white pine and spruce which give lighter tones. Boards are usually nominally 3/8 or 3/4 in (10, 19*) thick and have square, tongued and grooved, or shiplapped edges. They are applied vertically, horizontally, or diagonally and are available in several widths. Boards are usually finished after installation but can be purchased prefinished.

INTERIOR WOOD - INSTALLATION

The wall conditions illustrated here are typical for sheet or solid lumber paneling installed vertically. For horizontal paneling, either the vertical studs are used or vertical furring is applied to the wall face.

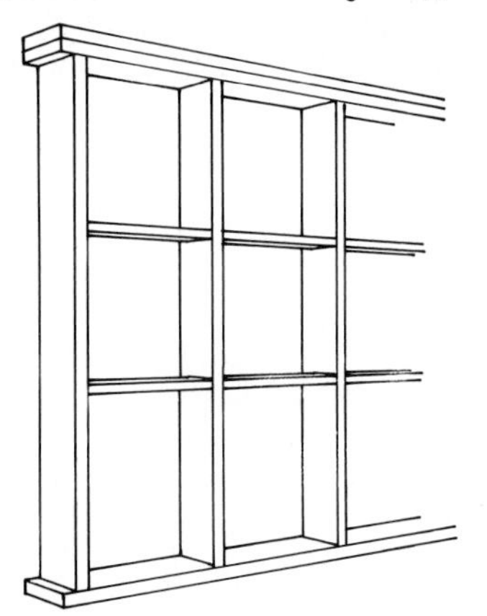

STUDS AND BLOCKING
Horizontal blocking is fitted between studs at centers to suit the type of panel. Blocking must be fitted before the vapor barrier is installed if on an outside wall.

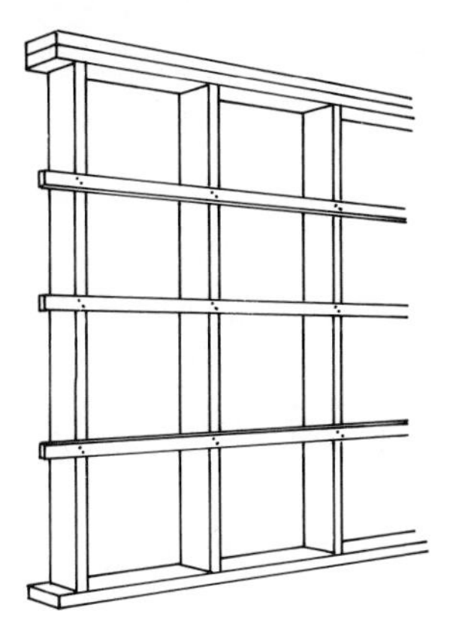

FURRING STRIPS
Furring, usually 1x2 in (19x38) or 2x2 in (38x38), applied at suitable centers. Faster and cheaper than blocking and can be shimmed to provide a flat surface (see also 10-49).

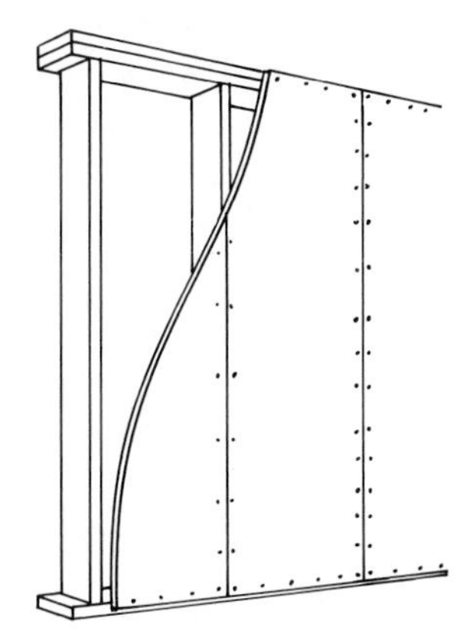

GYPSUM BOARD
A standard application of gypsum board but without taped joints provides a fire-rated flat base. Paneling is usually glued to the drywall and is the only approved panel installation method in some regions (check local codes).

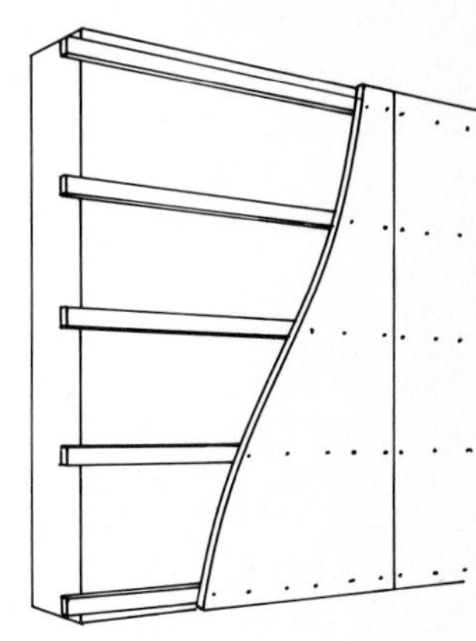

MASONRY WALLS
Wood or metal furring strips mechanically fastened to the wall. For exterior insulated walls, 2-in x (38) furring or a free-standing stud wall is common (see 11-4 or 10-34).

THREE EXAMPLES OF PLYWOOD EDGE TYPES

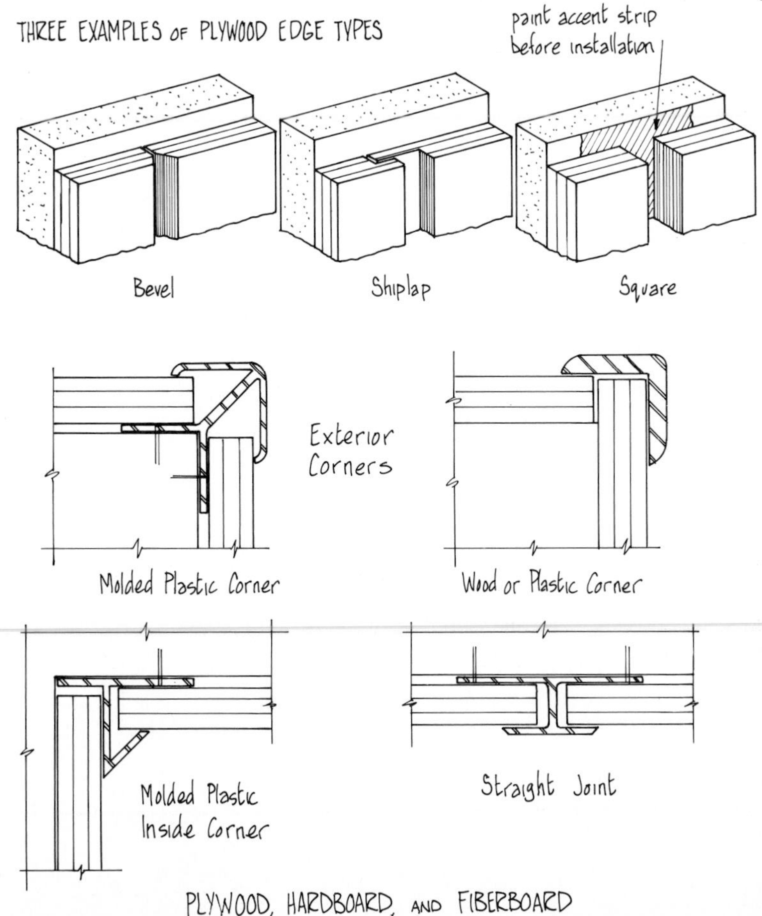

PLYWOOD, HARDBOARD, AND FIBERBOARD

TYPICAL SOLID WOOD BOARD PROFILES

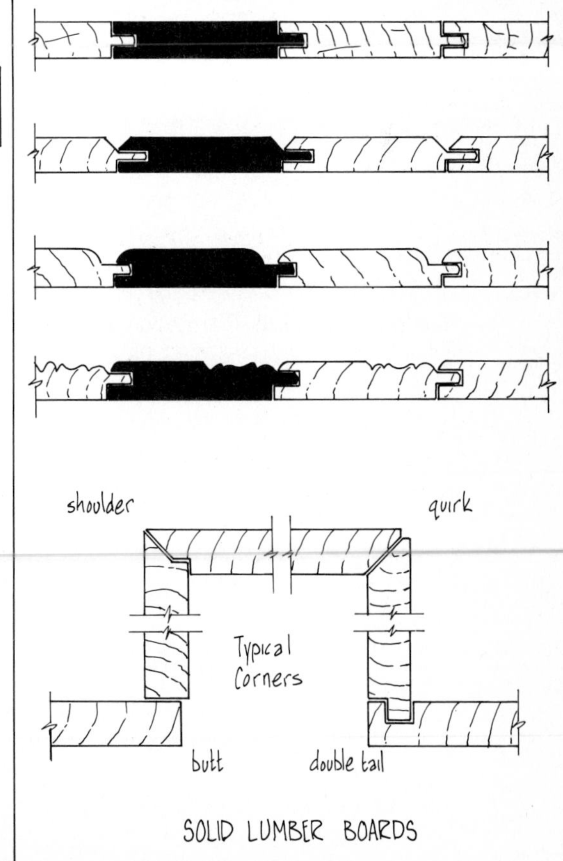

SOLID LUMBER BOARDS

FINISHES

11-10 CERAMIC TILE

Ceramic tile is commonly used in shower stalls, around bathtubs, on kitchen walls and countertops, and any floor surface where hard wearing or decorative qualities are required.

TILE TYPES

Designers are able to choose from a vast selection of tile shapes, sizes, types, and finishes. Most tiles fall into three production groups and three usage groups:

PRODUCTION

- Glazed tile has a glaze applied to the surface of the clay body before final baking. The finished surface can be produced in many textures from high gloss to a dull, pebbly finish.
- Unglazed tile has no surface glaze and the color runs throughout the tile body.
- The vitreous (glasslike) rating determines how well a tile absorbs water. Porcelain absorbs the least, while common glazed wall tile absorbs the most.

USAGE

- Wall Tiles. Light-bodied glazed tiles designed for vertical applications, countertops, and some light-traffic floor areas. Many types have matching trim pieces as illustrated at the right and include accessories such as glazed ceramic soap dishes, towel bars, glass and brush holders, etc.
- Floor Tiles. Thicker, heavier-bodied glazed or unglazed tiles. Care must be taken to choose a type that is nonslip when wet. Many exotic shapes are available as illustrated, and some types come with trim pieces.
- Ceramic Mosaics. Small tiles in various thicknesses and shapes, generally mounted in a group on thread mesh, paper, or silicone rubber backing sheets. The backing sheet enables large areas of tile to be applied quickly and allows the tile to follow the contours of columns, swimming pools, etc., if desired.

INSTALLATION

In all cases the tile base must be firm, flat, and rigid. Almost all building surfaces can act as a base provided that all holes are filled and smoothed. Wood joist floors, in particular, will likely need additional stiffening to reduce flexibility and may also need an underlayment over the subfloor (see 6-16). There are two basic methods of attaching tile to the backing:

- Thin-Set Adhesives. Can be used directly on almost any surface and are of three basic types: organic or mastics, cement based, and epoxy. The choice of adhesive depends on type and position of base and the tile to be used. All are applied in a thin layer with a comb-type applicator.
- Mortar Bed. A layer of cement mortar is first applied to the surface to be tiled. Reinforcing is generally added to floor mortars, while wall mortars are applied in a manner similar to a plastered wall. Tiles are attached to the dry mortar bed with cement or dry-set mortar.

FINISHING

After the tiles are installed and the mortar has set, the joints between the tiles are filled with a grout that highlights the tile pattern and keeps out dirt and liquids. Grouts are of three types: cement based, silicone rubber, and epoxy. The grout used depends on tile type, location, joint width, and adhesive. Grout color can be matched to tile color or used to provide a color contrast.

TYPICAL TILE SHAPES

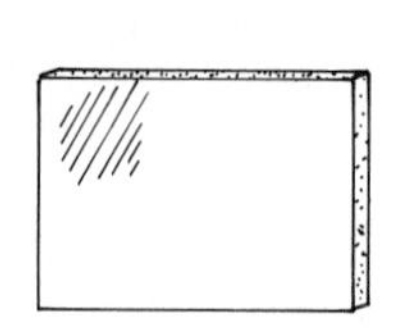
Rectangular

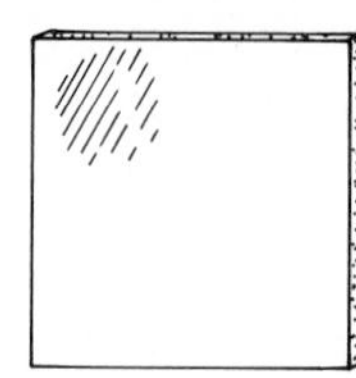
Square

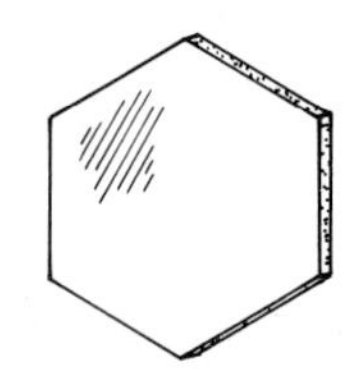
Hexagon

Sizes range from 4x4 in (100x100) to 12x12 in (300x300) in floor and wall tiles. Refer to suppliers for available products.

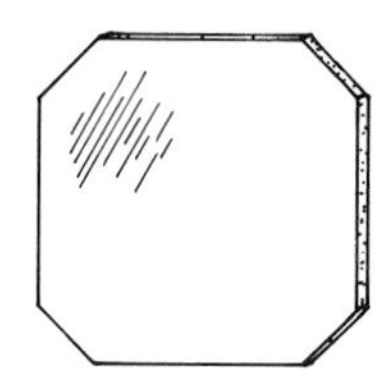
Octagon

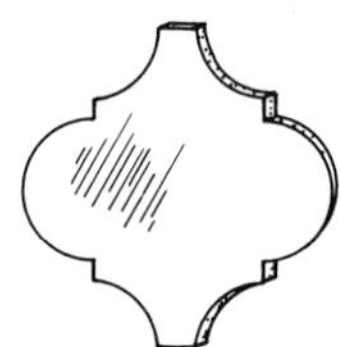
Spanish

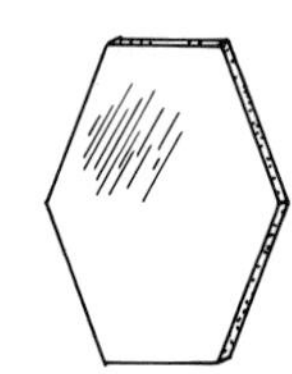
Elongated Hexagon

Mosaic tile, while available in the same profiles, range in size from 1x1 in (25x25) to 3x3 in (150x150).

EXAMPLES OF TRIM PIECES

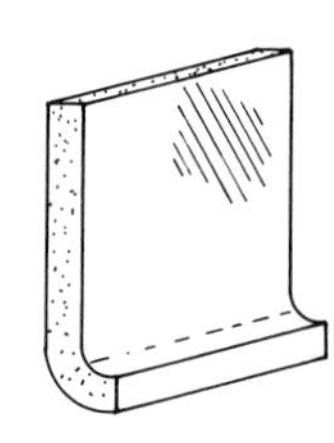

Cove

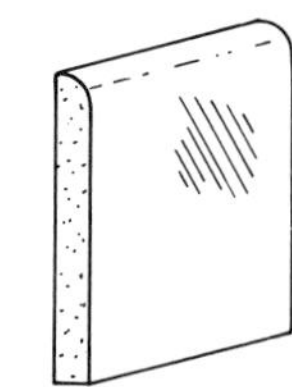
Surface Bullnose

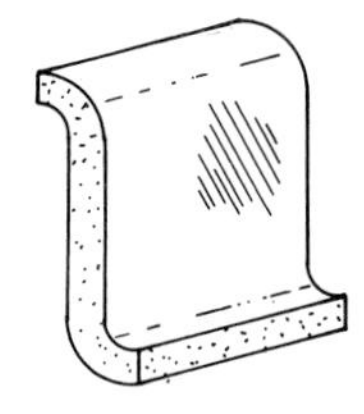
Base

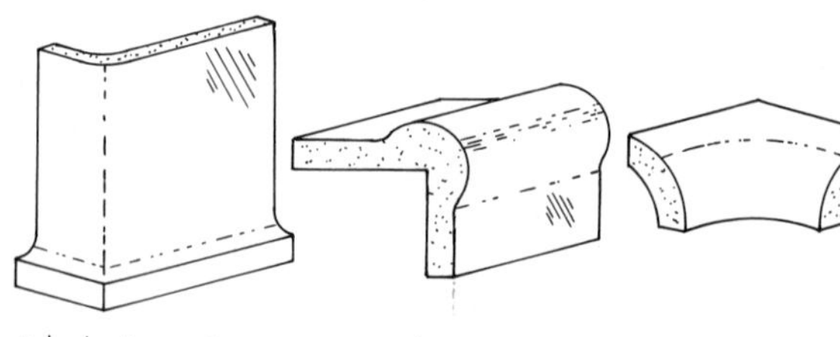
Outside Corner Cove Sink Cap Inside Corner

Selected trim pieces are available for most wall tile in a variety of shapes to suit installation conditions. It is important to check availability with suppliers before design and installation work begins.

WOOD TILE

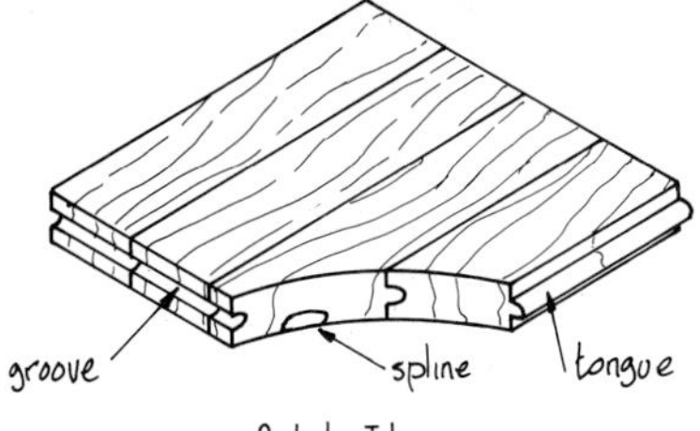

Solid Tile

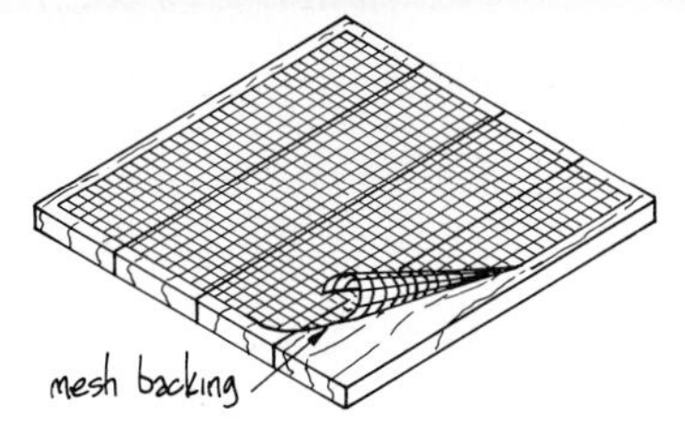

Solid blocks (above) are splined together or come with a cotton or plastic mesh to hold the pieces in position until laid. The layers of laminated blocks (left) are bonded with waterproof glue under high temperature and pressure.

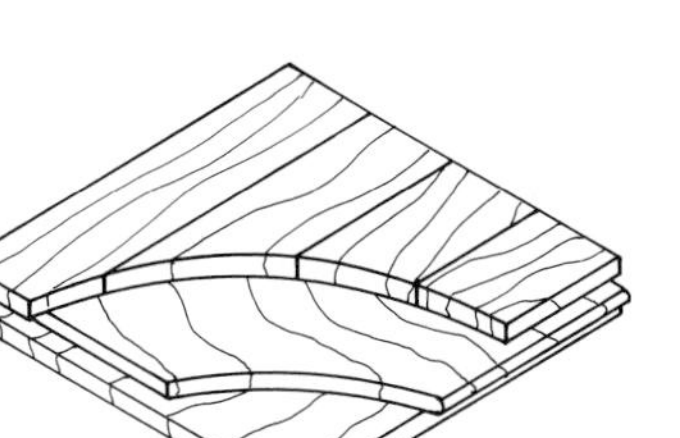

Laminated Tile

WOOD BLOCKS

Wood block tile are made from precision-cut strips of hardwood, factory assembled into square or rectangular blocks. They are of two types, solid or laminated, as illustrated at the left, and may be factory prefinished or unfinished. A wide variety of face patterns and textures are available in several different hardwood species, and all are generally designed to create the impression of an individually laid wood-strip parquet floor.

APPLICATION

In new construction, the blocks are usually laid over subfloors of plywood, concrete, or wood board. In each case the floor must be flat and smooth with all holes filled. Concrete floors must be completely dry (and remain so), while plywood and board subfloors may need a plywood underlayment (see 6-16). Thinner blocks are glued to the subfloor with a suitable thin-set adhesive, with care being taken to maintain even and straight patterns. Some blocks have a self-stick adhesive on the back. Thicker blocks are usually blind-nailed to wood subfloors.

OTHER SURFACES

Wood blocks can be applied to walls, countertops, and doors to match or contrast with other room finishes.

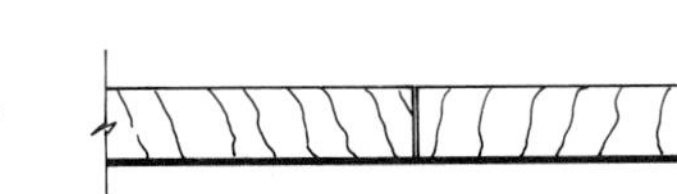

Tiles Toe-Nailed to Subfloor

Tiles Glued to Subfloor

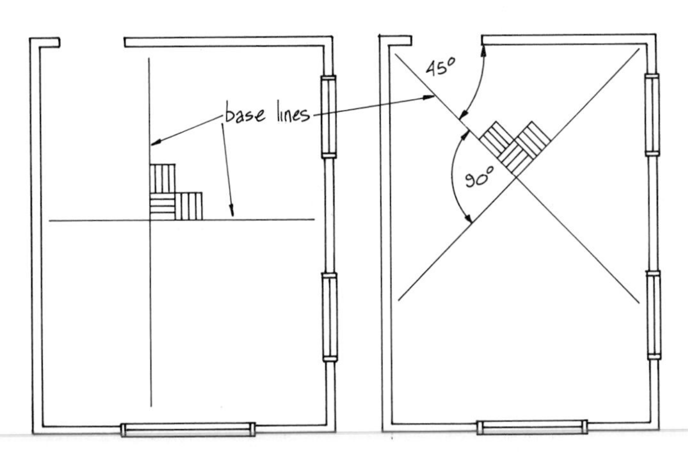

Square Pattern Layout

Diagonal Pattern Layout

Tiles may be laid square with the walls or at a 45° diagonal as shown above. In either case, care must be taken to position the tiles so that an equal space is obtained at each opposing wall.

FOUR EXAMPLES OF TILE PATTERNS

Stone

Finger Block

Haddon Hall

Canterbury

WOOD STRIP FLOORING

Most strip flooring is made from hardwoods: usually white or red oak, although hard maple, birch, beech, and other hardwoods are occasionally used. Softwoods such as fir, southern pine, or hemlock are used for economy or where special effects are required. Strip flooring is available in three types:

- Strip flooring, which has tongued-and-grooved sides and ends that have been accurately machined for tight-fitting joints. It is supplied in random lengths and in either unfinished or finished form. Unfinished strips have square face corners and are sanded after installation to remove unevenness. Finished strips cannot be sanded after installation and so have slightly chamfered face edges to hide unevenness.
- Plank flooring, is wider and generally thicker than strip flooring and has t&g sides but not ends and is also supplied in random lengths. Planks are often installed with plugged screws as illustrated below to simulate the wood peg fasteners of antique plank flooring. For the type of flooring used in post and beam construction, see 9-7.
- Parquet flooring, is similar to strip flooring but is precision cut to exact lengths and widths so that patterns can be made with the strips. Parquet is generally laid with a mastic adhesive, although blind nailing can also be used. Patterns such as those shown on 11-11 or 11-13 are common.

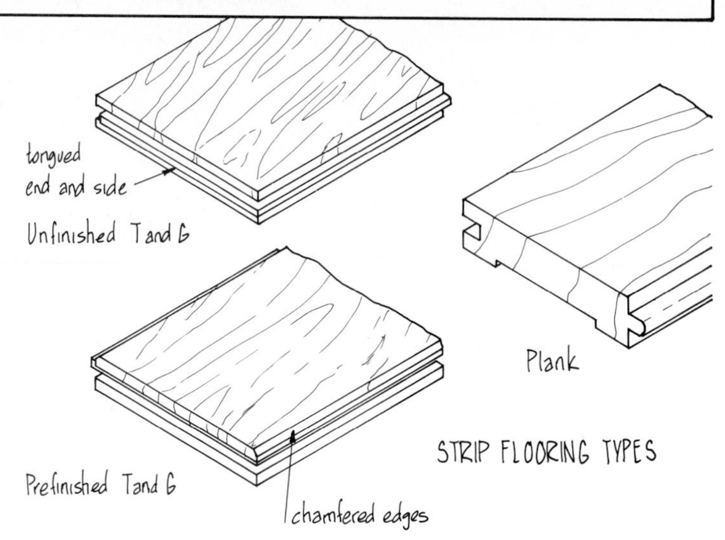

STRIP FLOORING TYPES

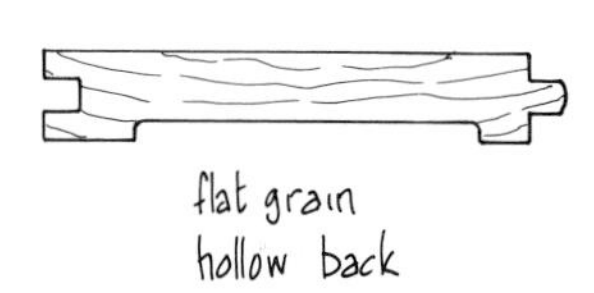

flat grain
hollow back

edge or vertical grain
scratch back

Depending on species and flooring type, either flat or edge grain may be specified to provide a more uniform face presentation. The underside of the strips may be machined with a hollow or scratch undercut to reduce warping and provide a more even installation.

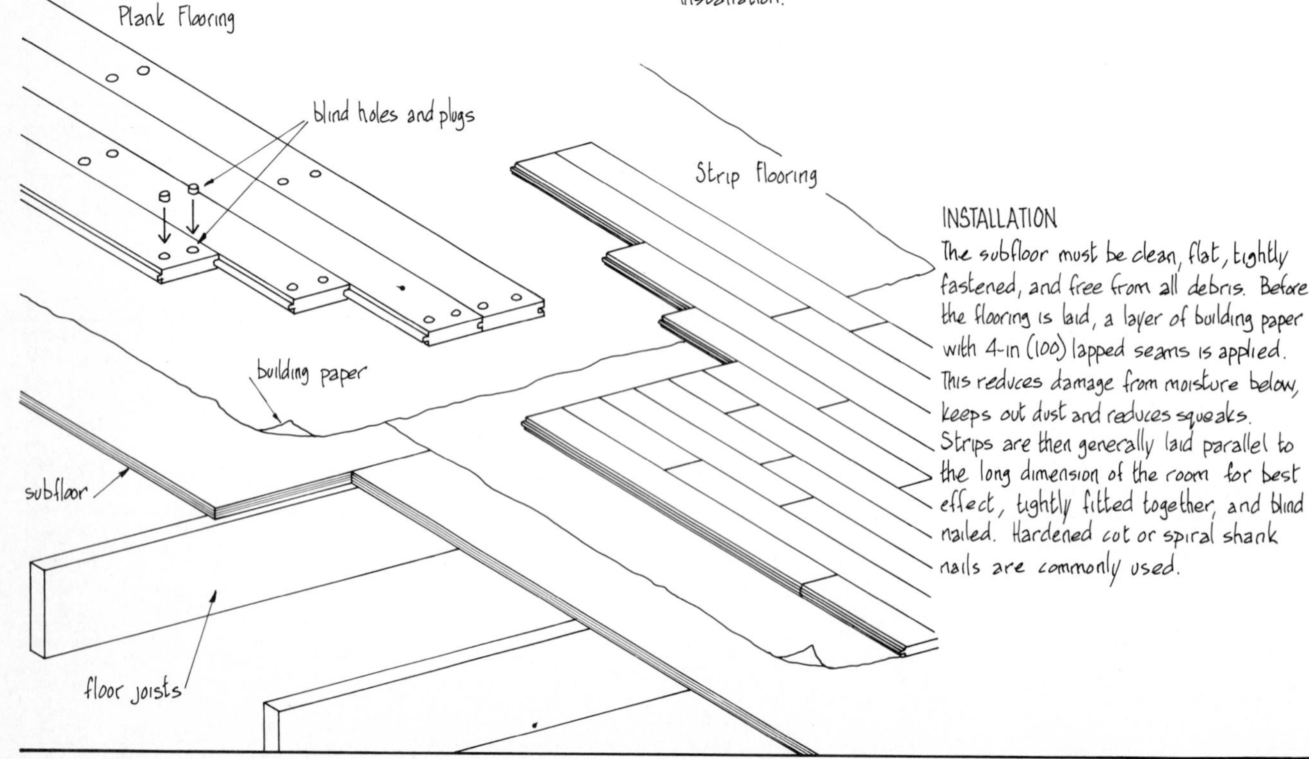

INSTALLATION

The subfloor must be clean, flat, tightly fastened, and free from all debris. Before the flooring is laid, a layer of building paper with 4-in (100) lapped seams is applied. This reduces damage from moisture below, keeps out dust and reduces squeaks. Strips are then generally laid parallel to the long dimension of the room for best effect, tightly fitted together, and blind nailed. Hardened cut or spiral shank nails are commonly used.

WOOD STRIP FLOORING

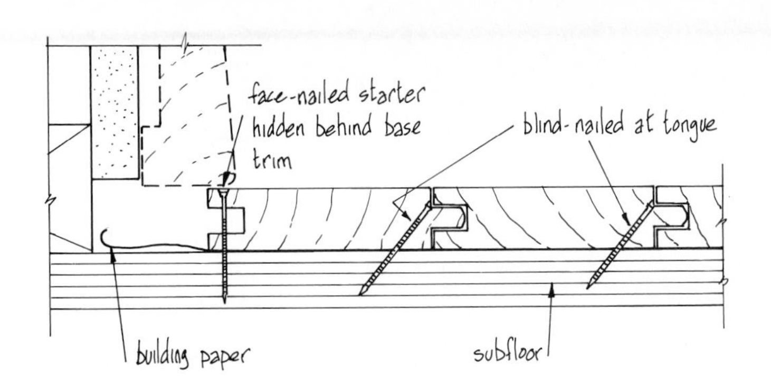

BLIND NAILING

The starter strip at the wall must be face nailed and filled. Leave a 1/2-in (13) clearance at the wall for floor expansion. All subsequent courses are blind nailed through the tongue at an angle of approximately 50°.

STRIP FLOORING PATTERN EXAMPLES

Herringbone

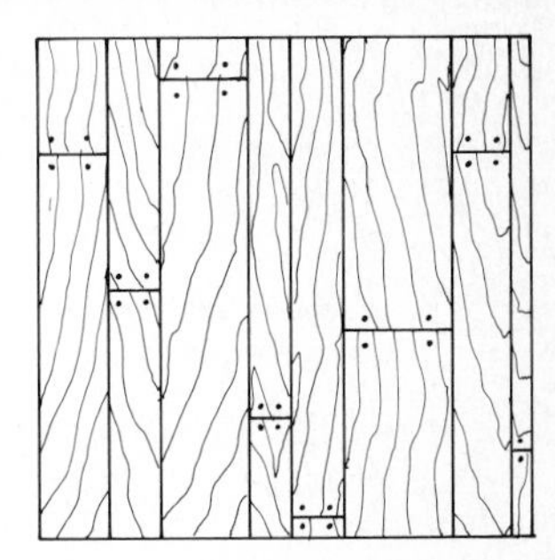

Random Width - Face Nailed

Butt Herringbone

Basketweave

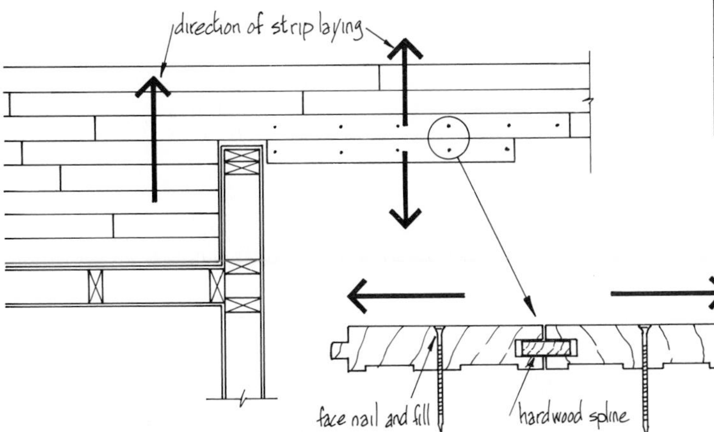

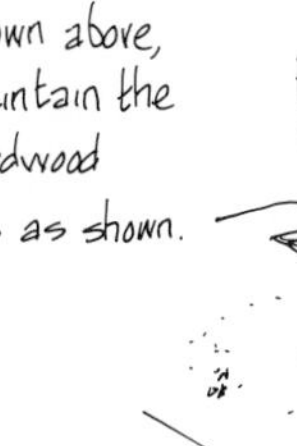

DIRECTION CHANGES

When a floor returns around a corner such as that shown above, the direction of strip laying must be reversed to maintain the blind-nailing sequence. To reverse the strips, a hardwood spline is inserted in the grooves of adjacent strips as shown.

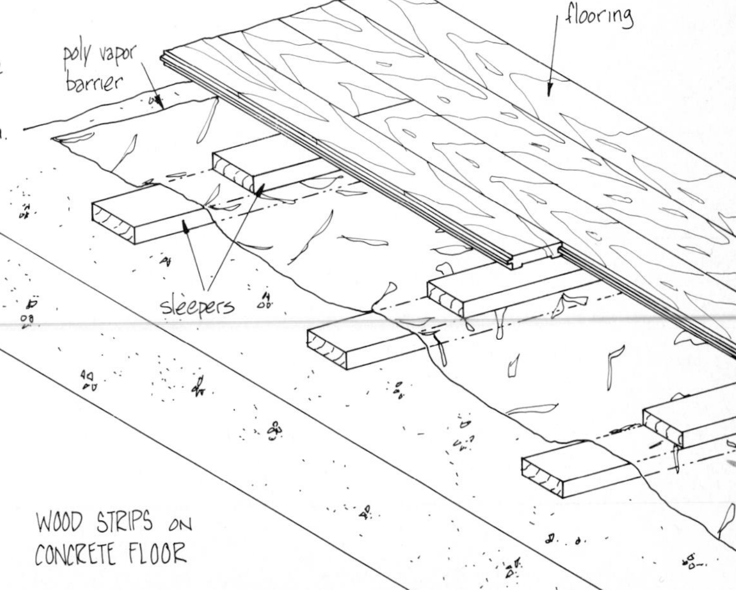

WOOD STRIPS ON CONCRETE FLOOR

CONCRETE FLOORS

Wood floors cannot be laid directly on concrete floors that are on or below grade because of moisture problems. One method of providing a moisture barrier is illustrated at the right. A 1x2 in (19x38) sleeper at 12- or 16-in (300-400) centers is set in mastic and nailed to the concrete floor. A layer of 4-mil (0.10) polyethylene sheeting is applied and secured with a second sleeper. The strip flooring is then nailed to each sleeper.

RESILIENT FLOORING

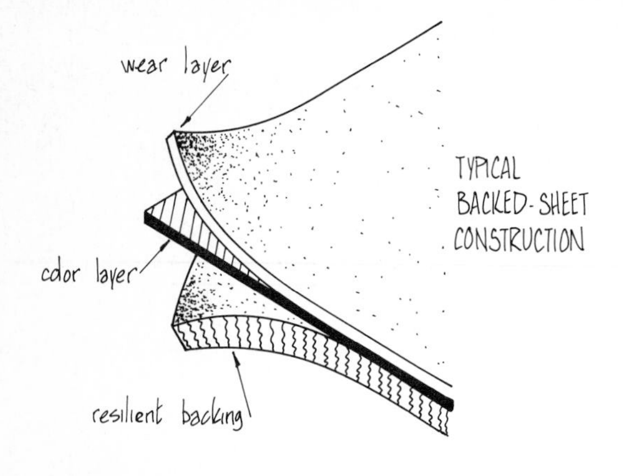

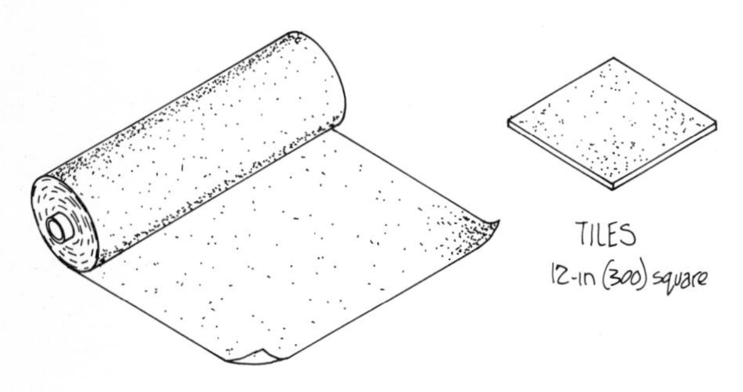

ROLLS 6 ft to 15 ft (2-5 m) wide

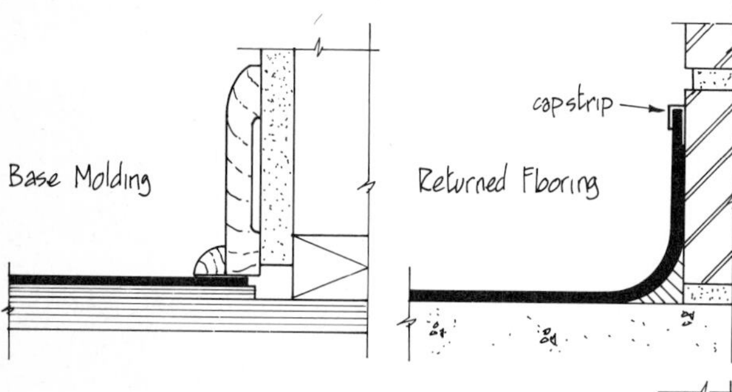

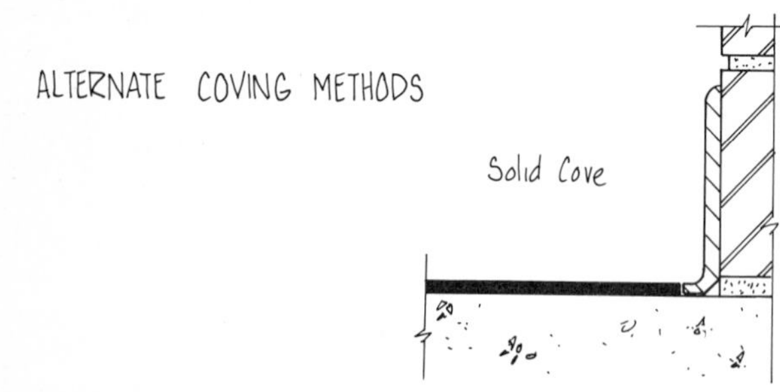

ALTERNATE COVING METHODS

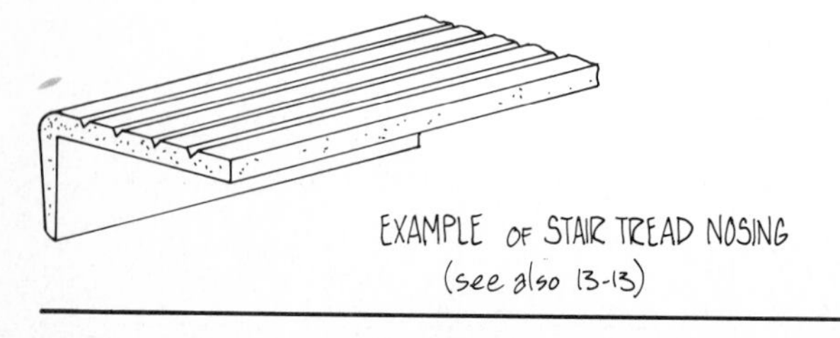

EXAMPLE OF STAIR TREAD NOSING
(see also 13-13)

Vinyl is the predominant material used in resilient flooring production. Other materials include cork and rubber, particularly for items such as stair treads, coves, and other fitments. Vinyl flooring is made in two ways:

- Backed sheets, with two or more layers of sheeting bonded together. There are two types of backed sheet: "inlaid," where vinyl particles are compressed into a thin layer at high temperatures and bonded to a backing sheet, and "rotogravure" where the color and pattern are printed on a flooring sheet, covered with a vinyl wear-layer, and bonded to a backing sheet. Inlaid sheets are very durable, but the hard surface makes the flooring stiff and difficult to install. Rotogravure sheets are generally very flexible and easy to handle and are the material of choice for most average installations.
- Solid sheets, which are a nonlayered, nonbacked material made from solid vinyl, or vinyl with the addition of fillers and pigments. The color and pattern extends from front to back. The material is very stiff and heavy and is fairly brittle. It is used mostly where very heavy wear is expected or in specialty rooms where a clean or sterile environment is required.

Depending on type, flooring is available in rolls of 6-, 9-, 12-, or 15-ft (2, 3, 4, 5 m) width and/or in 12-in (300) square tiles.

INSTALLATION
Actual installation method depends entirely on subfloor type, flooring material, and intended usage. It is important that the manufacturer's instructions be followed carefully and that the subfloor be stable, smooth, clean, and clear of even the smallest debris. There are three basic installation options:

- Full-spread adhesive, where the flooring is cemented to adhesive spread over the whole floor.
- Self-stick flooring tiles with a factory-applied contact adhesive.
- Perimeter only attachment with staples or adhesives. No adhesive is required on the body of the floor. Only certain types of flooring can be installed in this way.

Sheet flooring seams can be glued, chemically bonded, or heat welded, depending on type of flooring and expected service conditions. Coves at floor/wall junctions can be handled in three ways:

- Flooring can stop at the wall line and be covered with a base molding.
- Flooring can be extended up the wall as a "flash cove" and the edge covered with trim.
- A preformed matching or contrasting cove can be attached to the wall and the flooring butted against the cove base.

CEILING TILE AND SUSPENDED CEILINGS

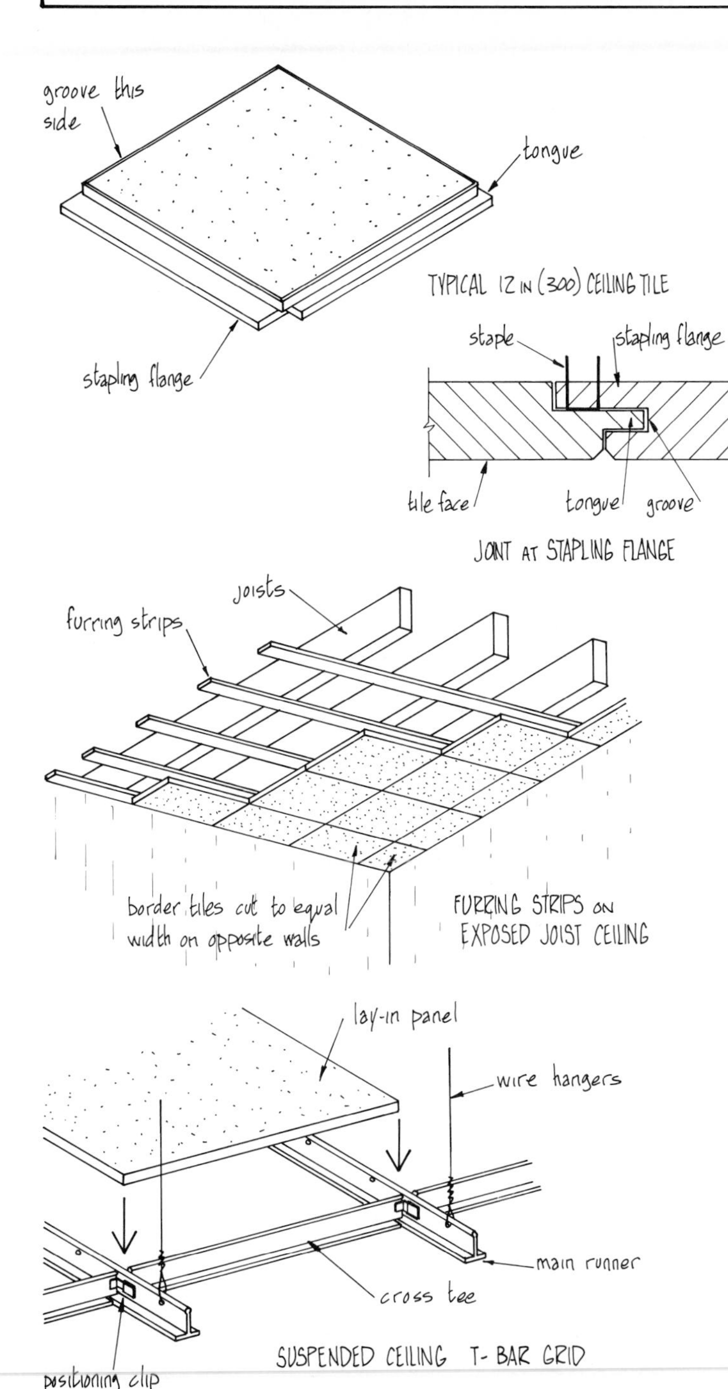

CEILING TILE

Most ceiling tiles are made from wood or vegetable fibers, mineral fibers, or rigid fiberglass and have tongued-and-grooved edges for self-alignment and a nailing strip for concealed fastening. Standard size is 12×12 in (300×300) and 1/2- to 3/4-in (12.5 - 19) thickness. Other sizes are also available. The tile face is factory prefinished and is either plain or molded with fissures, perforations, or embossings to increase acoustical properties and create visual effects.

INSTALLATION

The ceiling tiles are generally stapled or glued to either a flat ceiling surface such as plaster board, plaster, and fiberboard, or attached to wood or metal furring strips nailed at right angles to floor or ceiling joists. In either case, the support structure must be as flat and even as possible and the firring strips should be shimmed as required (see 11-4).

In most installations, the tiles are set out on a line equidistant from the long walls, with the border tiles butting against each wall cut to equal width as shown at the left. Installation must start from a corner and proceed in the tongue and groove direction.

LAY-IN PANELS — T-BAR CEILINGS

Suspended T-bar ceilings with lay-in panels are used extensively for ceilings where mechanical or other services are suspended beneath exposed ceiling joists, as is standard practice in most commercial and some residential buildings. The metal or plastic grid system can be suspended at any required depth, while the panels are easily removable for ceiling space access. It is also common practice for the enclosed ceiling space to be used as the warm-air return duct in an air-conditioning system.

The panels are made and finished as for the ceiling tiles above but in a wider range of colors and patterns. Panel edges are usually square, recessed, or have grooves that allow the grid system to be hidden. Standard sizes are 24 × 48 in (600×1200) or 24×24 in (600×600). A variety of light fixtures are designed to fit directly into the grid sizes or, as in a "luminous" ceiling, the lights are installed above the grid and the fiber panels are replaced with translucent or open egg-crate acrylic panels.

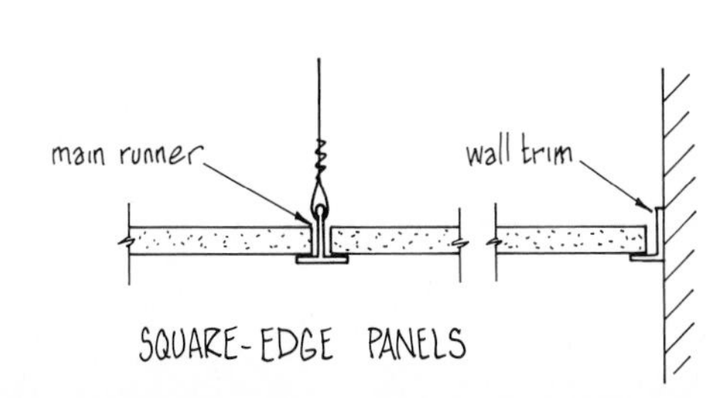

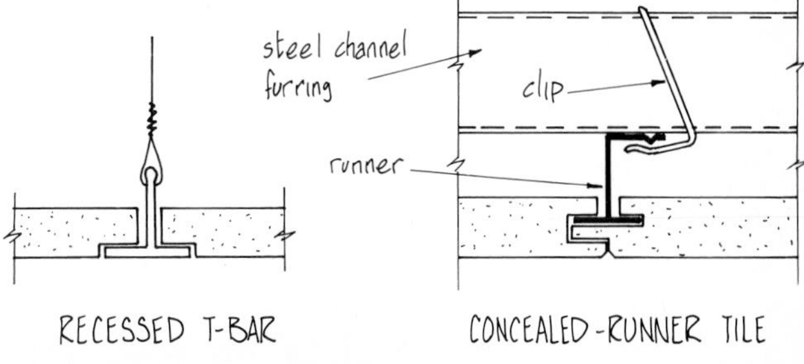

WALL SHEATHING

Exterior wall sheathing serves the double purpose of closing the wall cavity and, if required, providing a base for the exterior finishes. Sheathing is classed as either structural or nonstructural with regard to its nailing base and wind-bracing abilities. Sheathing materials include plywood, fiberboard, composition board, gypsum board and a variety of insulation materials. Lumber boards, usually 1 in (38) x 6 or 8 in (140-184), are rarely used as sheathing today. The three major sheathing material groups are detailed below. Refer to codes and manufacturers' recommendations for exact installation procedures.

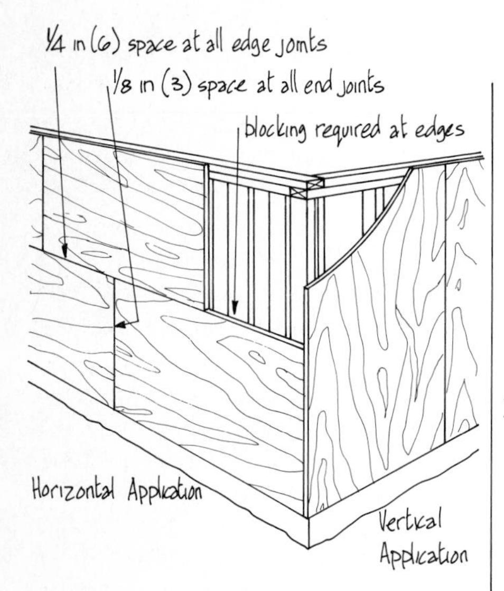

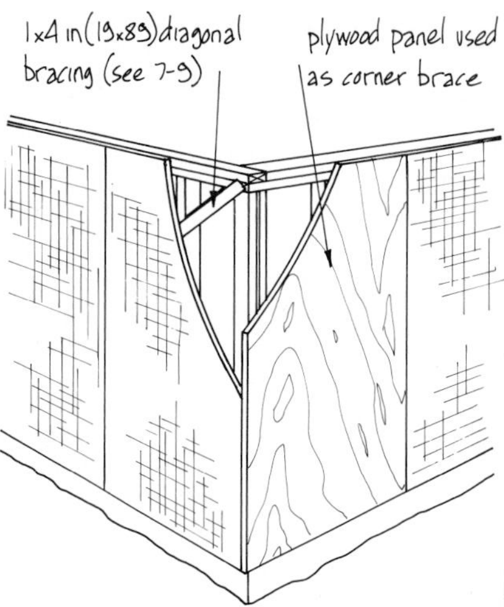

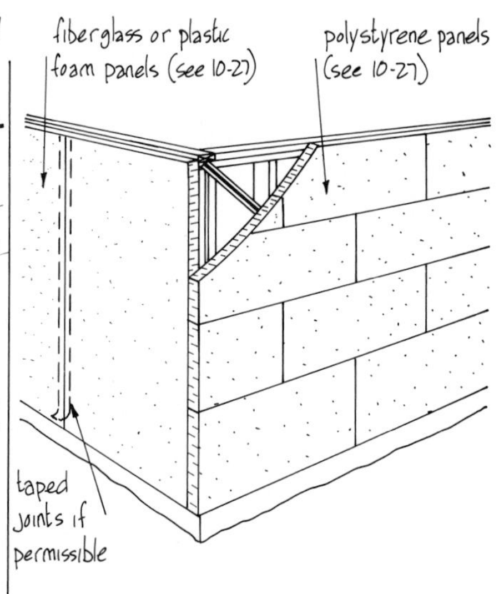

PLYWOOD

- Strongest of the sheathing boards.
- Installed vertically or horizontally and eliminates the need for diagonal bracing (see 7-9).
- Minimum thickness 5/16 in (8) for studs at 16 in (400) on center and 3/8 in (10) for 24-in (600) centers. Thicker panels, usually 1/2 in (12.7), are required for some types of siding and for stucco finishes.
- Siding can generally be nailed directly to ply sheathing.
- Single-ply wood panels, installed vertically on both sides of a corner, are used in combination with other sheathing to eliminate diagonal corner bracing (see center illustration).

FIBER/COMPOSITION/GYPSUM BOARD

- Fiberboard is relatively soft and usually has t & g edges and asphalt-coated faces for water resistance. Standard thickness is 1/2 or 25/32 in (12.7 or 20). Some types, when installed vertically, may substitute for diagonal wall bracing. Siding must be nailed directly to the framing.
- Composition board is usually of the wafer- or particleboard type. Performance is similar to plywood when used in a greater thickness. Some boards must be covered with water-resistant building paper before finishes are applied.
- Gypsum board, with a water-resistant paper face, is used when a higher level of fire resistance is required. Gypsum boards cannot substitute for diagonal bracing, and siding must be nailed to the frame.

INSULATION BOARD

- There are two basic types of insulation board: rigid foamed plastic and semi-rigid, crushable fiberglass (see 10-16).
- Plastic foam, with or without foil faces, must be installed so as not to create a second vapor barrier (see 10-11). Some types allow siding to be nailed directly through to the framing without intermediate support.
- Fiberglass types do not create double vapor barriers but will require the installation of intermediate supports for nailable siding.
- See 10-27 onward for installation details.

VERTICAL OR HORIZONTAL PANEL SIDING

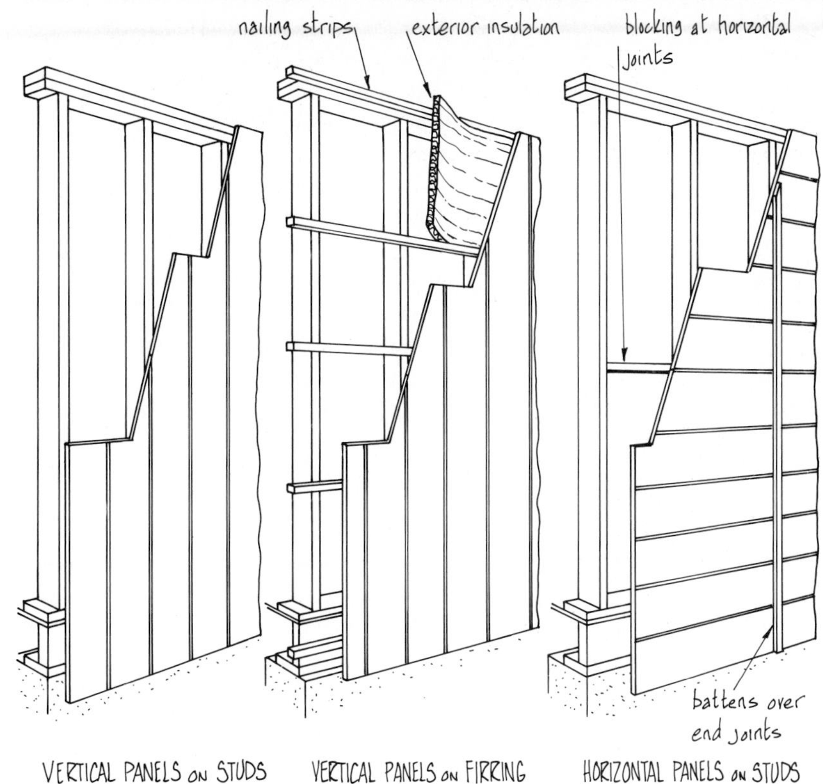

VERTICAL PANELS ON STUDS
VERTICAL PANELS ON FIRRING
HORIZONTAL PANELS ON STUDS

PANEL SIDING

MATERIALS

- Plywood is used extensively for panel siding. It is available in a wide variety of materials, including redwood, Douglas fir, and cedar and in an equally wide range of finishes from smooth to roughsawn to grooved. Panels are produced in several exterior-type grades and may be supplied plain sanded or prefinished with stain, sealer, or other special finishes. Panels are generally 4 ft (1200) wide and up to 10 ft (3000) long. Thickness depends on panel type and intended application.
- Hardboard panels are also available in a wide variety of styles and finishes, including simulations of wood shingles, rough cedar, knotty pine, and stucco.

INSTALLATION

Installation methods depend on the type of wall structure, sheathing material, and vertical or horizontal panel application. Although reference to the manufacturer's instructions should always be made, some general points can be mentioned:

- Panels can be nailed directly to the framing as shown at the left, or over structural or nonstructural sheathing (see preceding page). Panels installed horizontally on framing may need blocking at the panel edges.
- Panels over semirigid insulation may need firring strips, nailed to the framing at 2 ft (600) vertical centers, to avoid compressing the insulation at the nailing points.
- Building paper may be needed over some types of sheathing or under some types of panels to waterproof the wall structure.
- Panel joints are treated in a variety of ways to waterproof and/or to create visual effects. Some typical joint details are illustrated on this page.
- Refer to manufacturers' recommendations regarding clearance between panel edges and ends. A gap of 1/8 in (3) is standard.

VERTICAL JOINTS

Butt and Caulk
Shiplap Channel
Batten

HORIZONTAL JOINTS

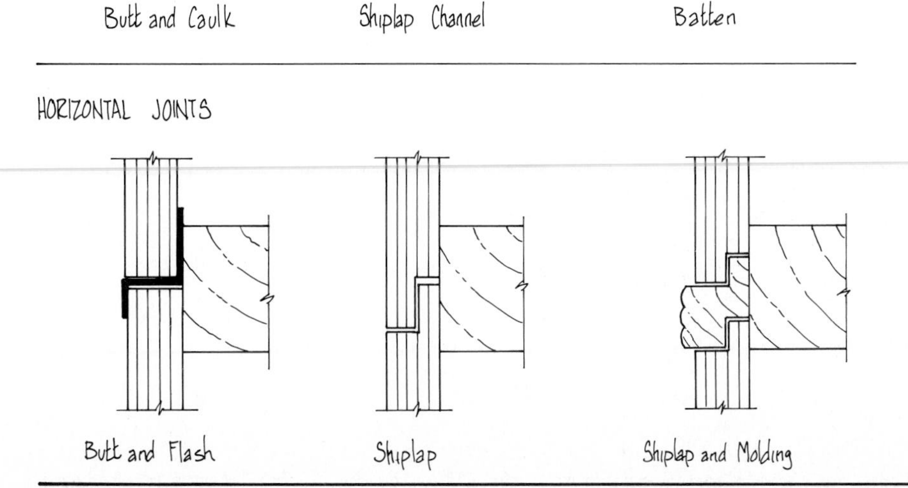

Butt and Flash
Shiplap
Shiplap and Molding

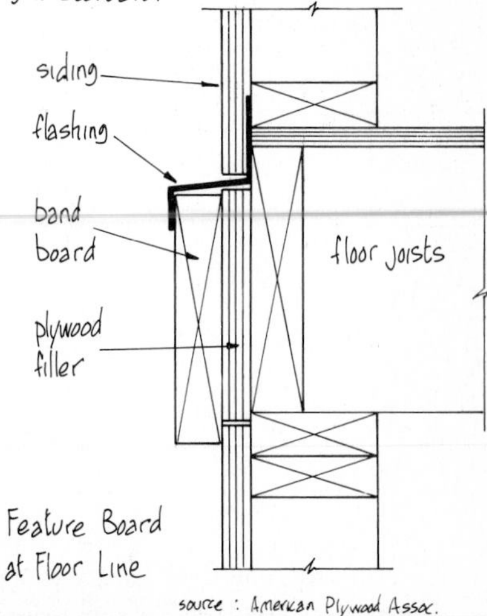

Feature Board at Floor Line

source: American Plywood Assoc.

LAP SIDING – MATERIALS

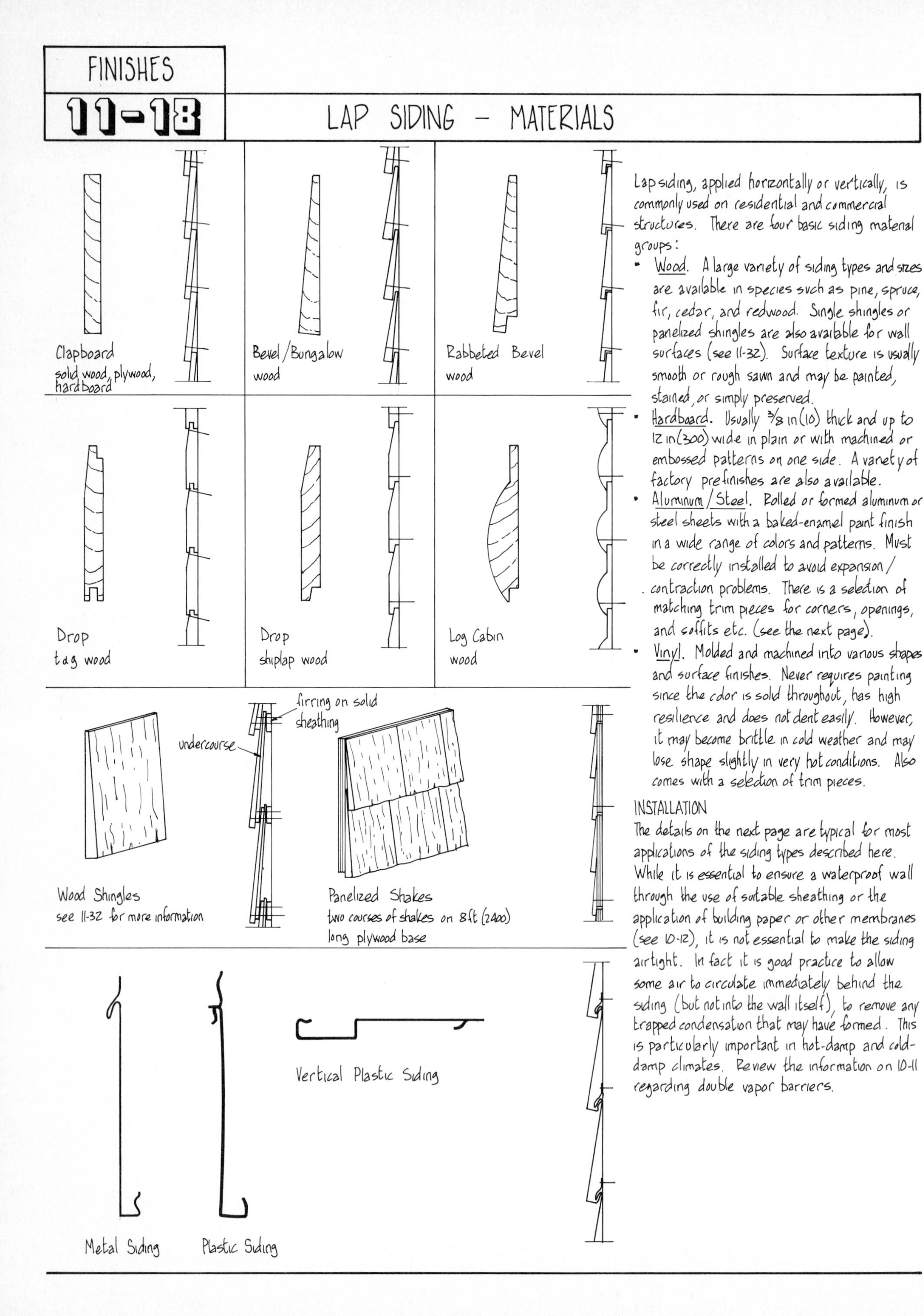

Lap siding, applied horizontally or vertically, is commonly used on residential and commercial structures. There are four basic siding material groups:

- Wood. A large variety of siding types and sizes are available in species such as pine, spruce, fir, cedar, and redwood. Single shingles or panelized shingles are also available for wall surfaces (see 11-32). Surface texture is usually smooth or rough sawn and may be painted, stained, or simply preserved.
- Hardboard. Usually 3/8 in (10) thick and up to 12 in (300) wide in plain or with machined or embossed patterns on one side. A variety of factory prefinishes are also available.
- Aluminum / Steel. Rolled or formed aluminum or steel sheets with a baked-enamel paint finish in a wide range of colors and patterns. Must be correctly installed to avoid expansion / contraction problems. There is a selection of matching trim pieces for corners, openings, and soffits etc. (see the next page).
- Vinyl. Molded and machined into various shapes and surface finishes. Never requires painting since the color is solid throughout, has high resilience and does not dent easily. However, it may become brittle in cold weather and may lose shape slightly in very hot conditions. Also comes with a selection of trim pieces.

INSTALLATION

The details on the next page are typical for most applications of the siding types described here. While it is essential to ensure a waterproof wall through the use of suitable sheathing or the application of building paper or other membranes (see 10-12), it is not essential to make the siding airtight. In fact it is good practice to allow some air to circulate immediately behind the siding (but not into the wall itself), to remove any trapped condensation that may have formed. This is particularly important in hot-damp and cold-damp climates. Review the information on 10-11 regarding double vapor barriers.

LAP SIDING – INSTALLATION

The installation details shown on this page are typical for many types of siding. Refer to manufacturer's instructions for exact details.

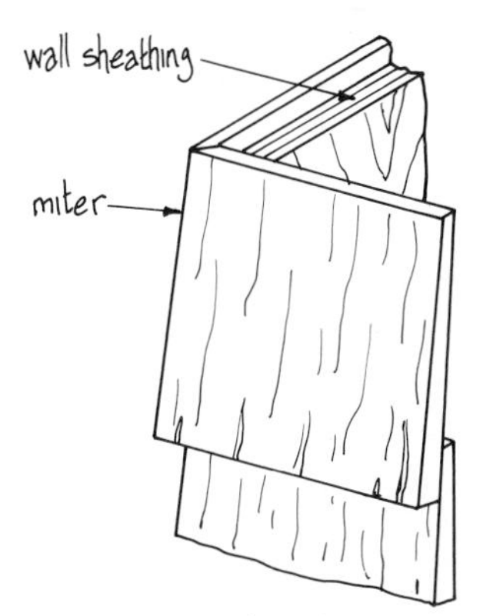

Mitered Outside Corner

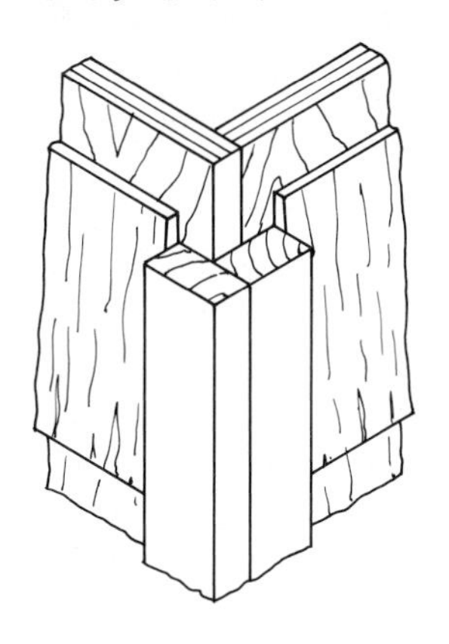

Butt Outside Corner Boards

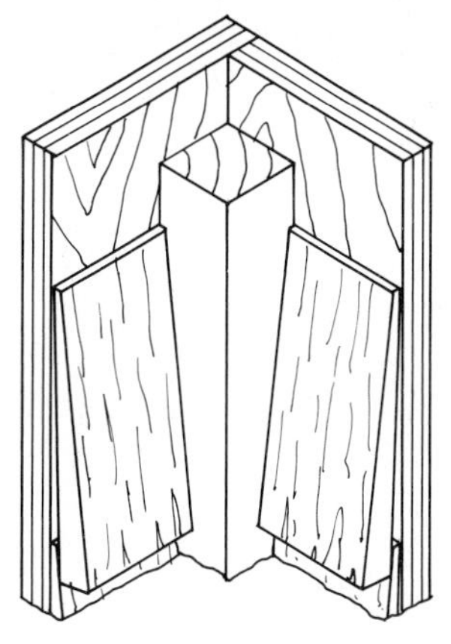

Inside Corner Wood Strip

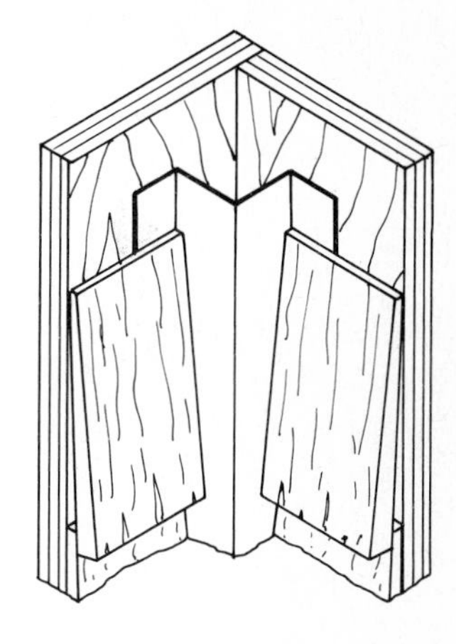

Inside Corner Metal or Plastic Post

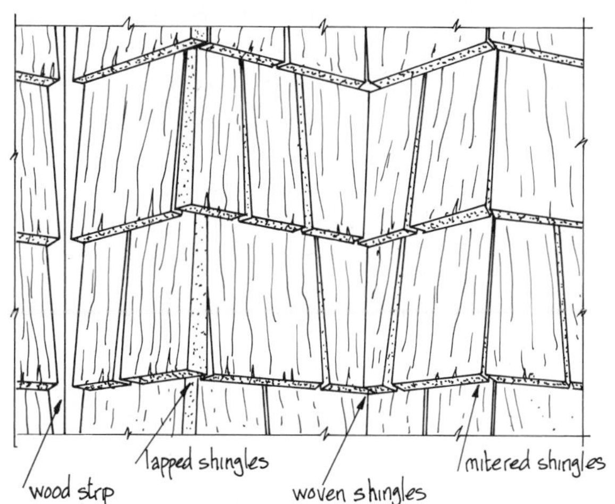

Shingle Corners

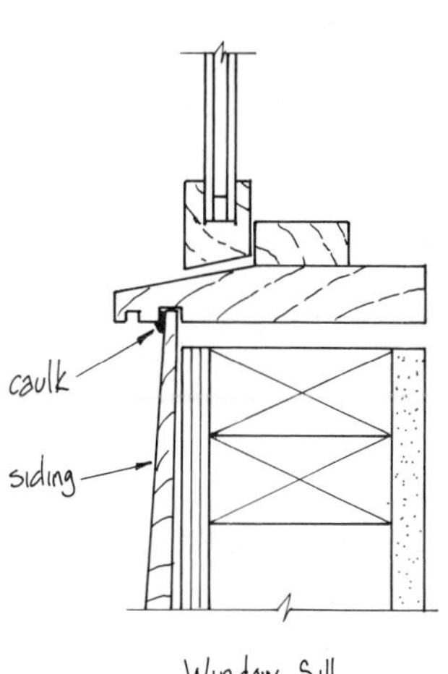

Window Sill

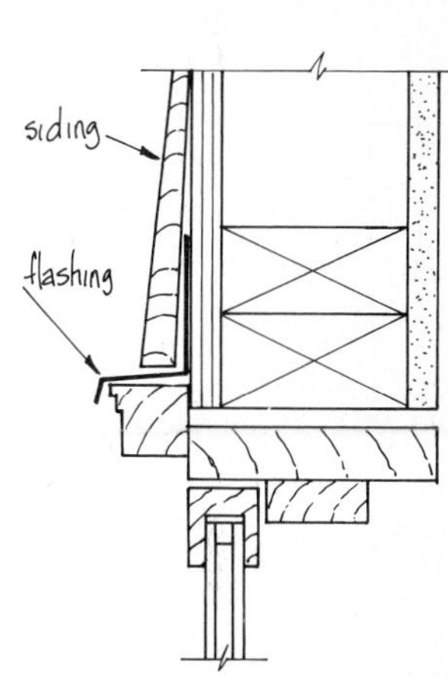

Window/Door Head

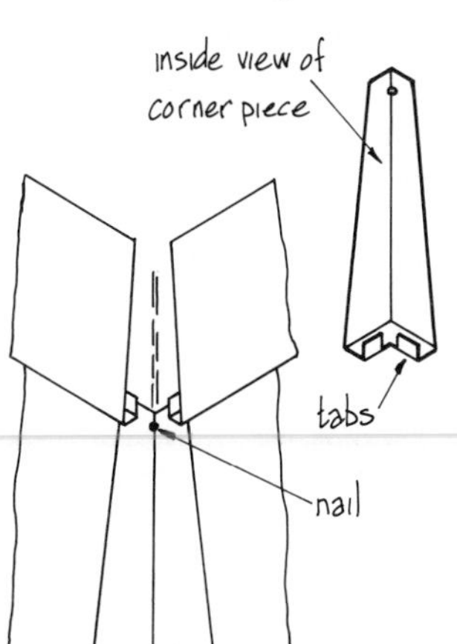

Metal Corner Pieces

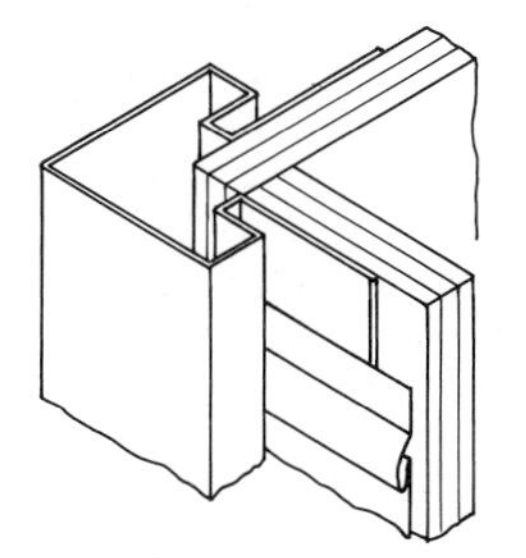

Metal/Plastic Outside Corner

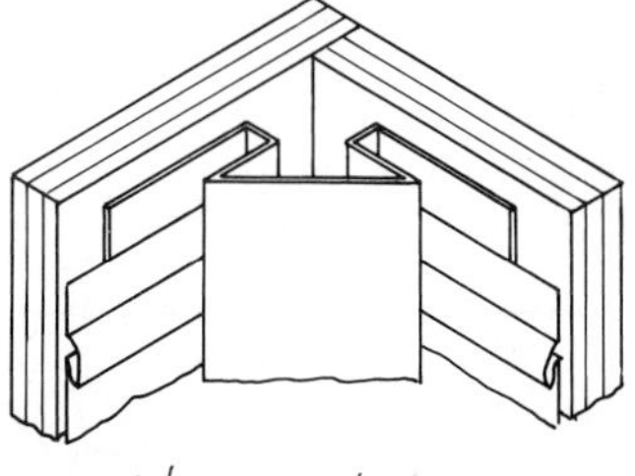

Metal/Plastic Inside Corner

Metal/Plastic Siding Accessories

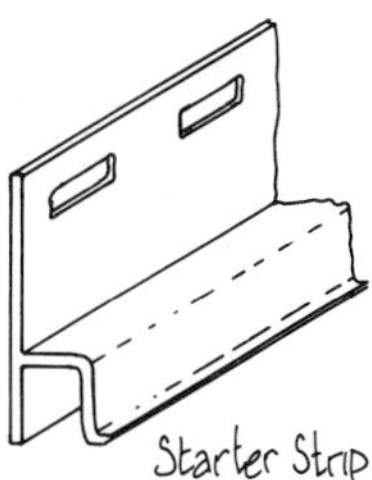

Starter Strip

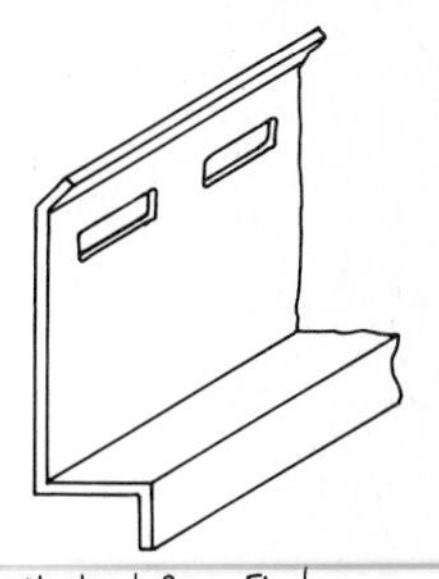

Vertical Base Flashing

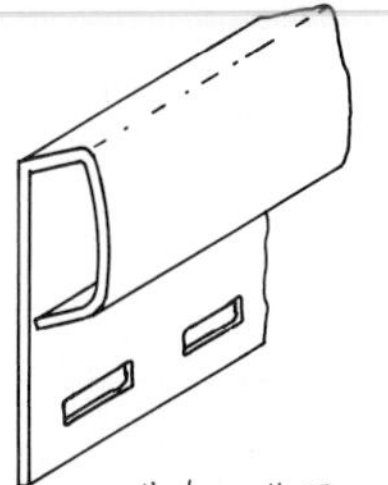

Undersill Trim

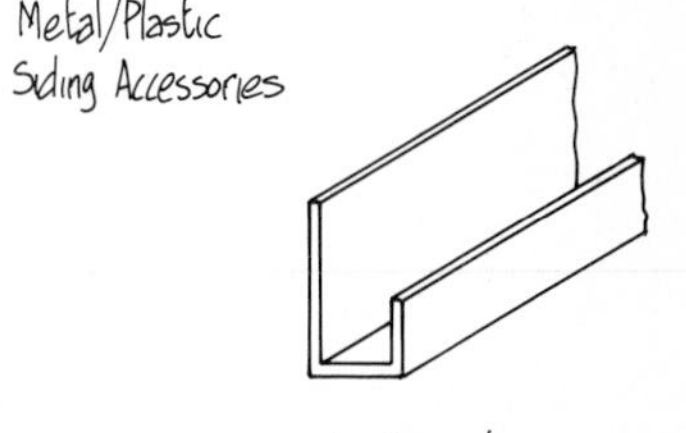

J-Channel

STUCCO

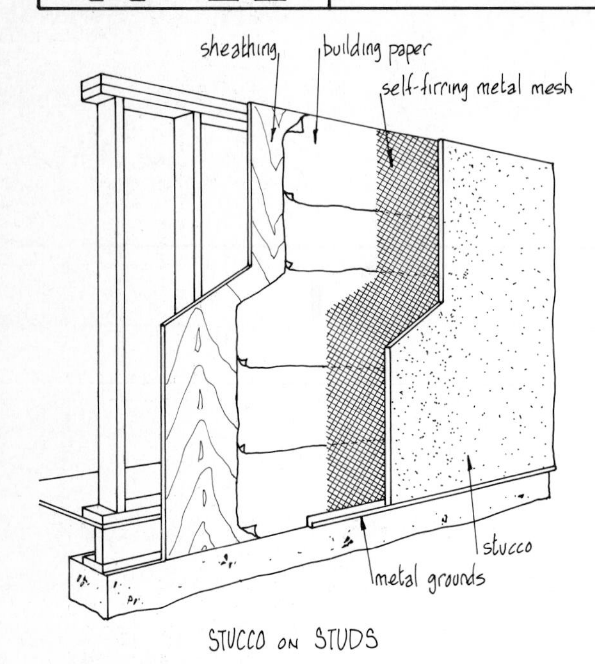

STUCCO ON STUDS

Stucco is the exterior version of a plastered wall and follows the same two-or three-coat application method. A mix of Portland cement, sand aggregate and lime is generally used although epoxy mixes are also available. Color pigments can be added to the finish coat, the finished wall can be painted or stones can be embedded in the surface.

STUD-WALL BASE

- Heavy-gage self-furring welded or woven stucco reinforcing mesh is nailed horizontally to the studs over asphalt sheathing paper and sheathing. The mesh must be firmly attached with all joints and corners lapped at least 2 in (50). Both mesh and fasteners must be heavily galvanized.
- To reduce structural shrinkage and stucco damage the use of balloon framing is recommended for walls over one story high (see 6-5).
- A minimum 3/4 in (19) thick three-coat application is standard (see 11-7) unless a "stone-dash" finish is required, where mineral chips are hand-thrown into the fresh mortar of the second coat.

CONCRETE / MASONRY WALL

- A minimum 5/8 in (16) thick two-coat application is standard.
- The wall surface must be clean, fairly absorbent, and rough enough to provide a good key for the stucco.
- Expansion joints, located at the wall expansion joints, are essential to prevent cracking. Typical stucco expansion joints are illustrated below.

The use of plaster grounds to help establish the correct thickness of stucco is recommended at all openings, bases, and tops of walls.

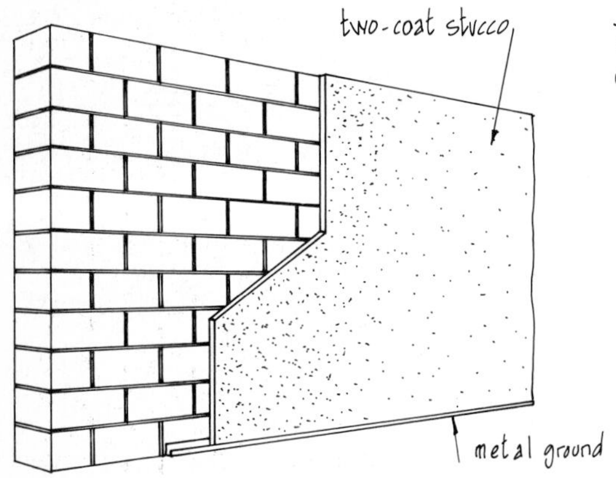

STUCCO ON MASONRY

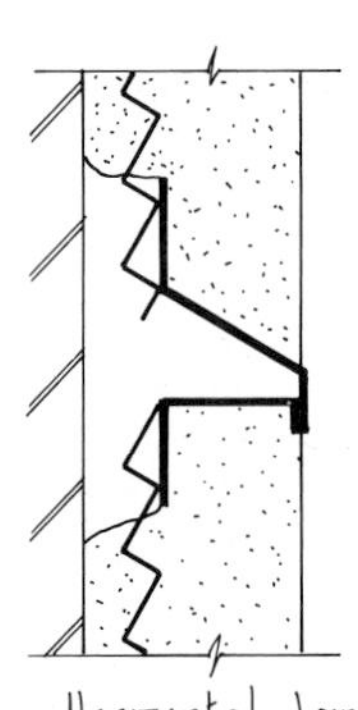

Horizontal Joint

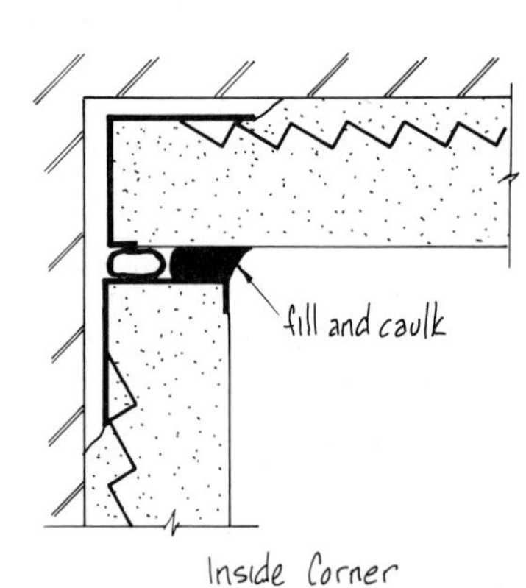

Inside Corner

EXPANSION AND CONTROL JOINTS

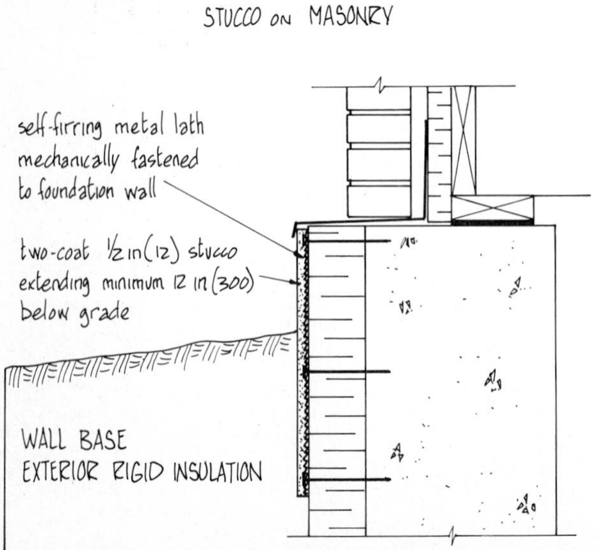

WALL BASE
EXTERIOR RIGID INSULATION

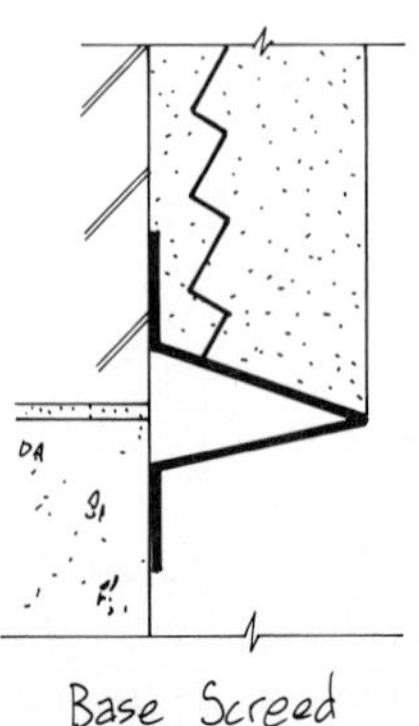

Base Screed

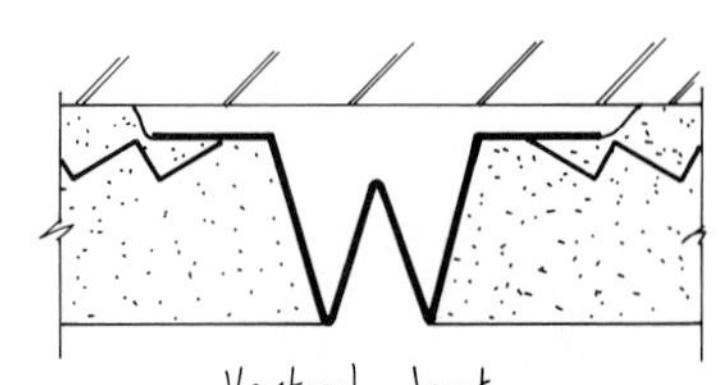

Vertical Joint

In all expansion and control joints, the metal beads must be attached to the structural elements beneath.

STONE VENEER

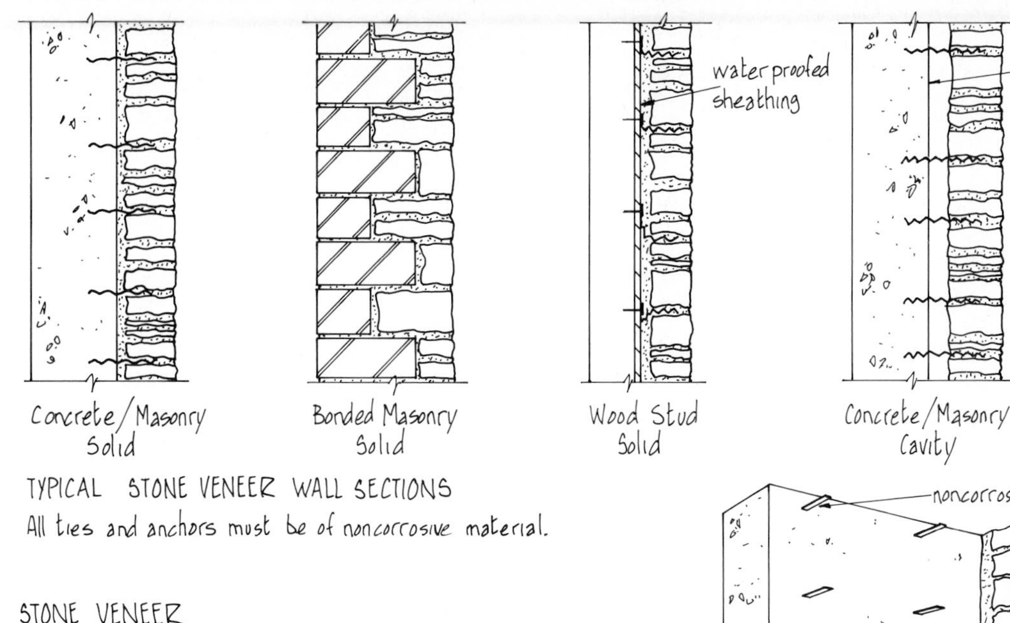

TYPICAL STONE VENEER WALL SECTIONS

All ties and anchors must be of noncorrosive material.

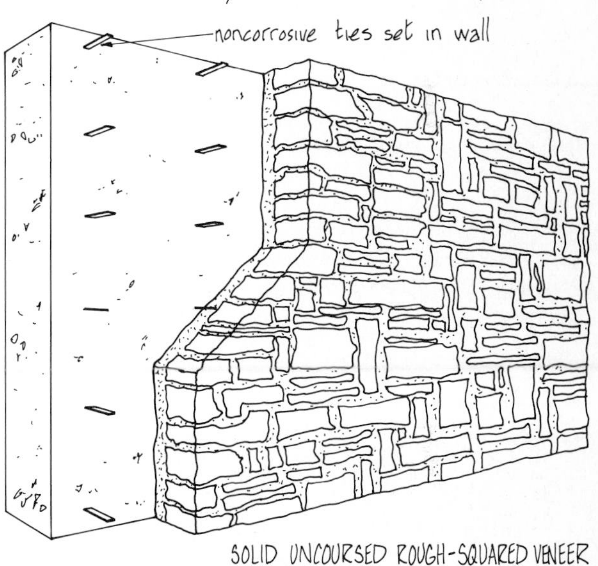

SOLID UNCOURSED ROUGH-SQUARED VENEER

STONE VENEER

Stone veneer can be applied to stud, masonry, or concrete walls in either a solid or cavity veneer as shown in the sections above. A wide range of visual effects can be achieved with a careful choice of stone type and cementing method. Commonly available stone materials include sandstone, granite, marble, limestone, and quartzite, with other stone types available on a regional basis. Stone is also selected on the basis of strength, hardness, or other special qualities and may be applied in a variety of face patterns as shown below.

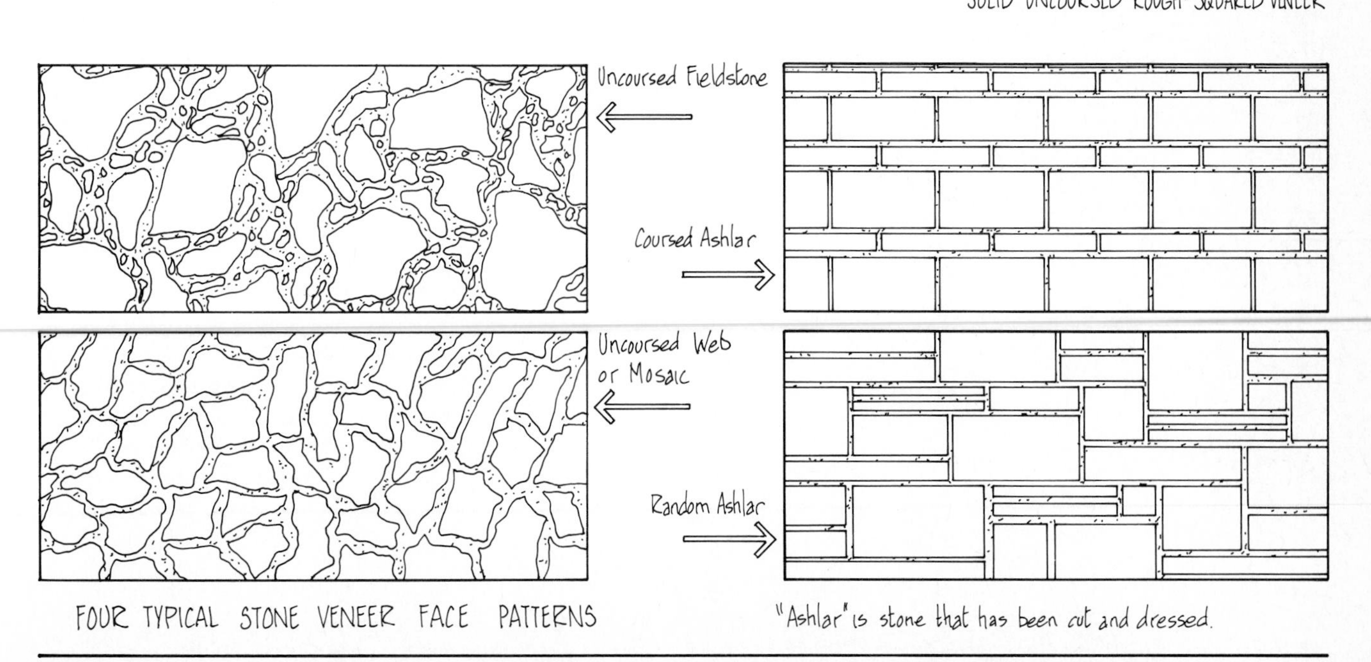

FOUR TYPICAL STONE VENEER FACE PATTERNS

"Ashlar" is stone that has been cut and dressed.

CORNICE (EAVE) CONSTRUCTION

The cornice or eave of a building is the overhang that is formed by the projection of the rafters beyond the wall line. How this projection is finished and trimmed greatly affects the visual aspects of the building and is a major part of the architectural design. On this and the next two pages are examples of typical cornice design in common use.
For most applications, prepainted metal sheet material is used for all exposed surfaces on soffits, fascias, gutters, and downpipes to reduce maintenance costs. Alternatively, plywood or prepainted hardboard can be used for soffits and solid lumber for fascias. Refer to 10-20 and 10-21 for ventilation and insulation information.

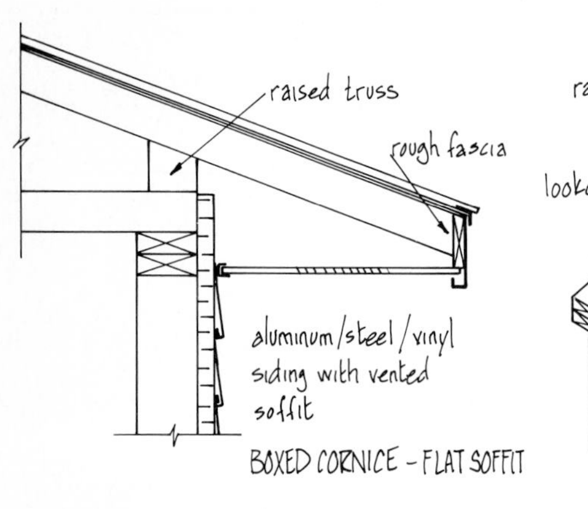

BOXED CORNICE – FLAT SOFFIT

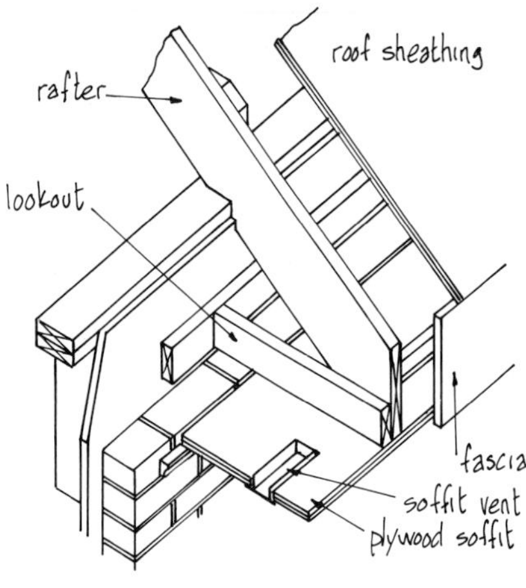

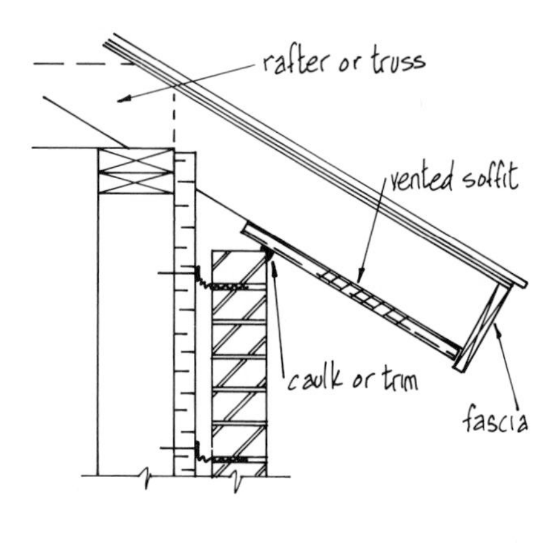

CLOSED (BOX) CORNICE

The closed cornice is the most common type. May be wide with an additional lookout member as shown above center, or narrow with the rafter cut to provide nailing for the trim members. Provides good protection for wall finishes.

SLOPING CLOSED CORNICE

In this case the soffit follows the line of, and is fixed directly to, the rafters, as shown above.

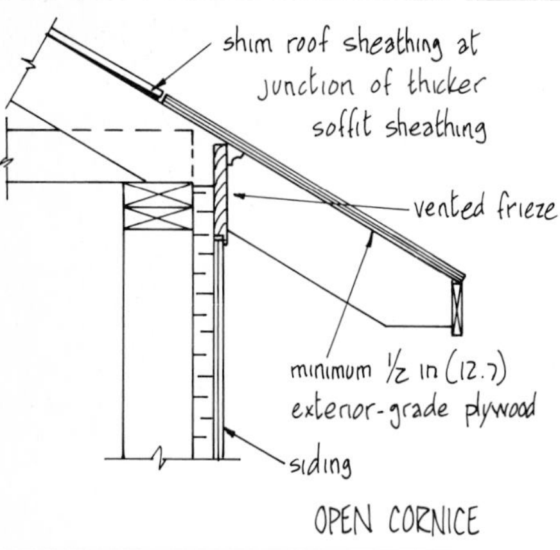

OPEN CORNICE

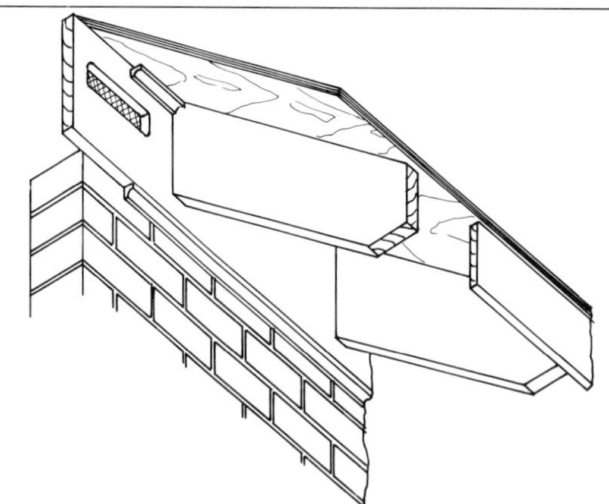

OPEN CORNICE

In this type the rafters are left exposed as a visual feature. Often, only alternate rafters are extended beyond the wall line to reduce visual clutter, provided that the roof sheathing is suitable for the additional span. The exposed roof sheathing must be of exterior-grade material and, to provide a better appearance, tongue-and-grooved V-joint lumber is sometimes used.

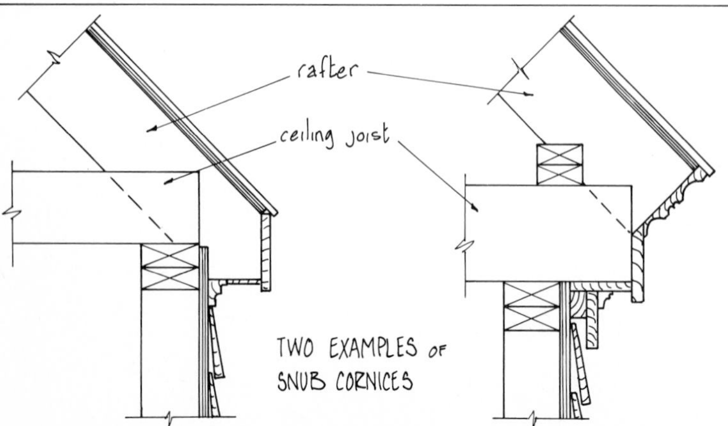

TWO EXAMPLES OF SNUB CORNICES

SNUB CORNICE

This cornice has the fascia at or close to the wall line and all trim is nailed to the cut rafter end. This type saves on material and labor but reduces wall protection and tends to present a less attractive appearance. It also tends to be difficult to provide adequate attic ventilation through the cornice with this type of construction.

CORNICE (EAVE) CONSTRUCTION

RAKES
Sloping rake cornices at the end walls are constructed in much the same way as horizontal rafter-end cornices with the same options in style and materials. Two examples are illustrated here. See 8-14 and 8-20 for rake support structures.

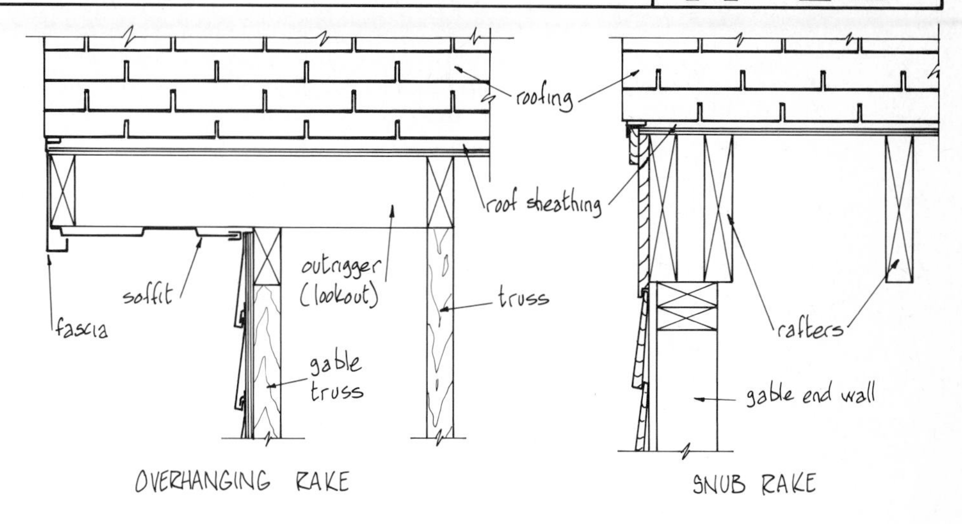

OVERHANGING RAKE

SNUB RAKE

FLAT ROOFS
Raftered or trussed flat roofs can be treated as for sloping rafters with box, open, or snub cornices. Ventilation of the roof space may not be desirable under some conditions – refer to the information on 10-21.

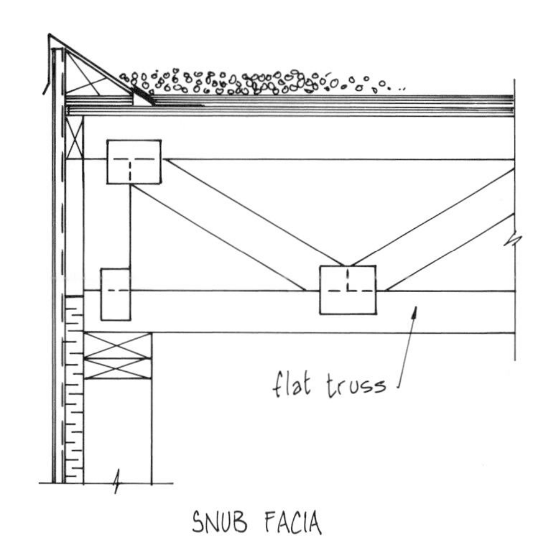

SNUB FACIA

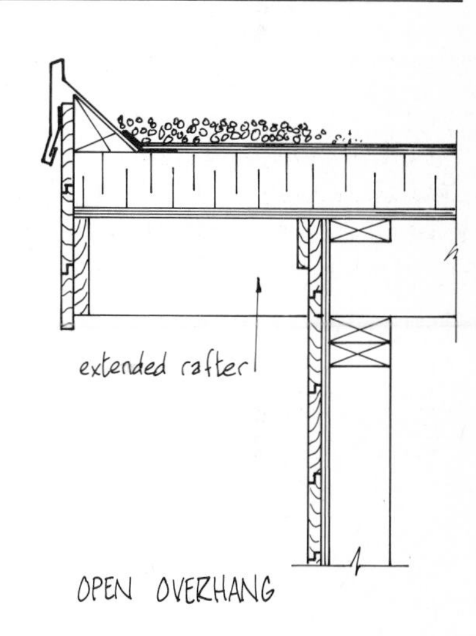

OPEN OVERHANG

PARAPETS
Although a parapet is not generally classified as a cornice, a typical example is shown here for reference (see also 7-32, 7-36, 7-40, and 7-45). The wall is extended upward past the roof line and capped with a flashing. Roof drainage is by gutters at the wall/roof junction or at the roof center with internal roof drains (see 8-30).

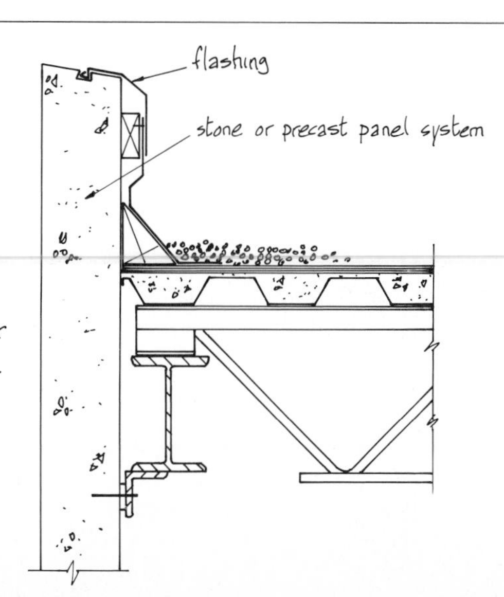

FINISHES

11-24 CORNICE (EAVE) RETURNS

The treatment of the cornice return at the end wall is of importance since it adds greatly to the visual effects of the building. The four examples illustrated on this page are typical of the many possible types of returns.

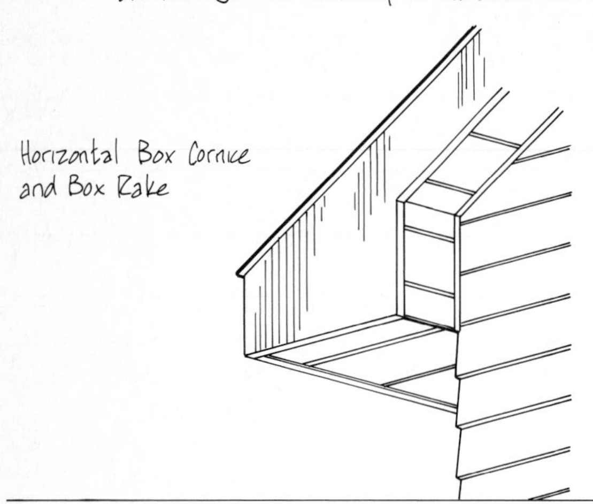

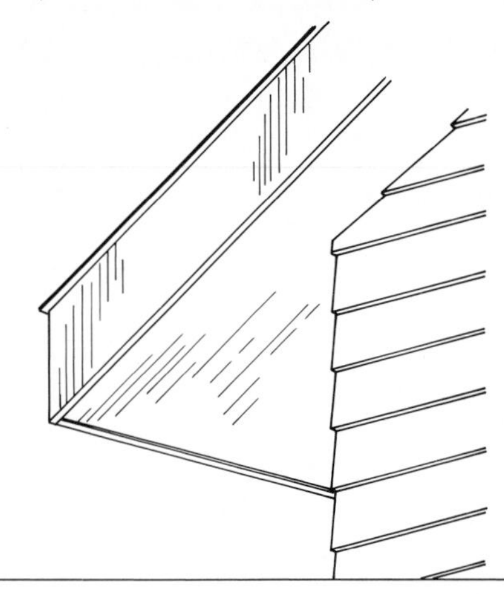

Sloping Box Cornice at Same Angle as Rake

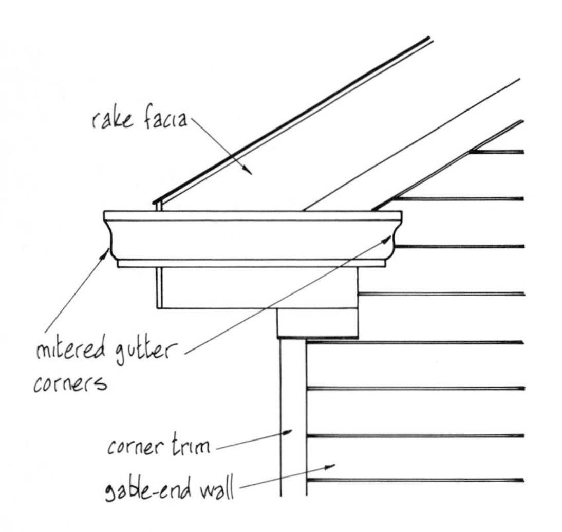

GUTTER RETURN
In this type the gutter is continued around to the end wall for a short distance in line with the rake fascia. A small roof section is constructed to fill the space between gutter and wall.

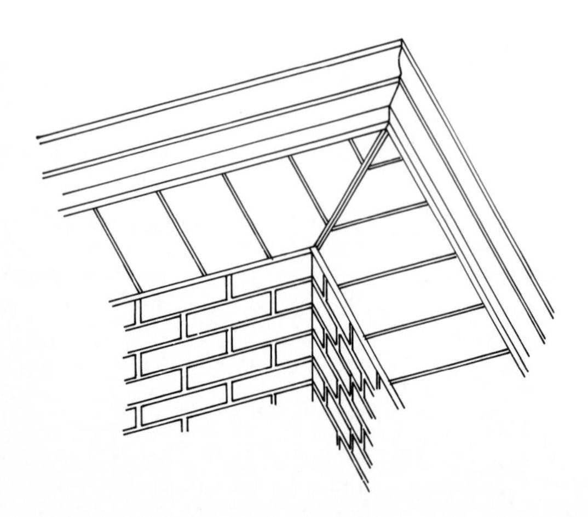

HIP ROOFS
Most hip roof (and flat roof) designs have an equal-dimension horizontal soffit around all sides of the structure. The standard treatment at corners is to miter the soffits as illustrated.

ROOFING - ASPHALT SHINGLES

Roof coverings must be long-lived, waterproof, reasonably fire resistant and should contribute to the esthetic design of the building. For most residences and small commercial structures with sloping roofs, asphalt shingles are the most common form of roofing. The shingles are available in a wide variety of styles and colors and are easy to install and replace. Listed below are the main shingle groups, with additional information on saturated-felt roll-roofing.

SHINGLES

36 in (1000)
12 in (330)

STANDARD
Standard shingles are most used. Wide variety of colors and patterns. Also available with two or four tabs. Dimensions are average and metric shingles are larger than Customary size.

DECORATIVE - DOUBLE LAYER
Double-layer shingles are made with two fiber layers to give a deeper textured look. Limited styles and colors. Generally intended to simulate wood shingles.

DECORATIVE - SINGLE LAYER
Single-layer shingles are available in a wider variety of styles and colors. May have face markings to simulate wood shingles.

LOW-SLOPE
A two-tab wider shingle is designed for low-slope applications. Allows for triple coverage. The tabs must be sealed with adhesive.

INTERLOCKING
Shingles that interlock with each other, virtually eliminating wind uplift.

SHINGLES

- Shingles are made from a layer of asphalt-saturated felt that has a back coating of waterproof asphalt and a top coating of stabilized asphalt with embedded mineral or ceramic granules for protection and coloration. Fiberglass is used as the base layer in some shingles.
- Most shingles have a strip of thermoplastic adhesive on the face, allowing the sun's heat to bond the shingles together.
- Shingles are available in various "weights" based on pounds per 100 square feet of coverage (kg/10 m²). Generally, greater weight means longer life, which, for most types, averages 10-15 years.
- Minimum slope for all standard and most decorative shingles is 4 in 12 (1:3) and 2 in 12 (1:6) for low-slope shingles.
- In high wind regions, use of interlocking or other specialized shingles is necessary to avoid tab uplift.

ROLL-ROOFING

FELT
Felt is available in a roll width of 36 in (1000) and in weights of 15, 20, and 30 pounds per square (7.5, 10, 15 kg/10 m²)* Used over roof sheathing as underlayment for new shingles (see the next page).

ROLL
Roll roofing is available in a roll width of 36 in (1000) and in weights of 50 and 90 pounds per square (25-45 kg/10 m²)*. Used as direct roofing or as flashings.

* Metric values are soft conversions.

SATURATED FELT

- Made from heavy paper or dry felt saturated with tar or asphalt.
- Not classed as a vapor barrier, although vapor diffusion is relatively low.
- Also known as "building paper" or "tar paper."

ROLL ROOFING

- Made in the same way as shingles and with a mineral particle surface coating.

UNDERLAYMENT APPLICATION

UNDERLAYMENT
For new roofing applications a single-layer underlayment of saturated felt or other suitable permeable material is sometimes required (check shingle manufacturer's recommendations). It is applied as shown at the right; in horizontal strips with a 2 in (50) lap over each lower course and a 4 in (100) lap at the ends. Ridges and valleys are lapped 6 in (150). Wide-headed fasteners are used to secure the felt. A metal or plastic drip edge is often recommended at the eaves and rakes. At the eaves it is applied first and the felt lapped over the strip. At the rake the drip is applied over the felt.

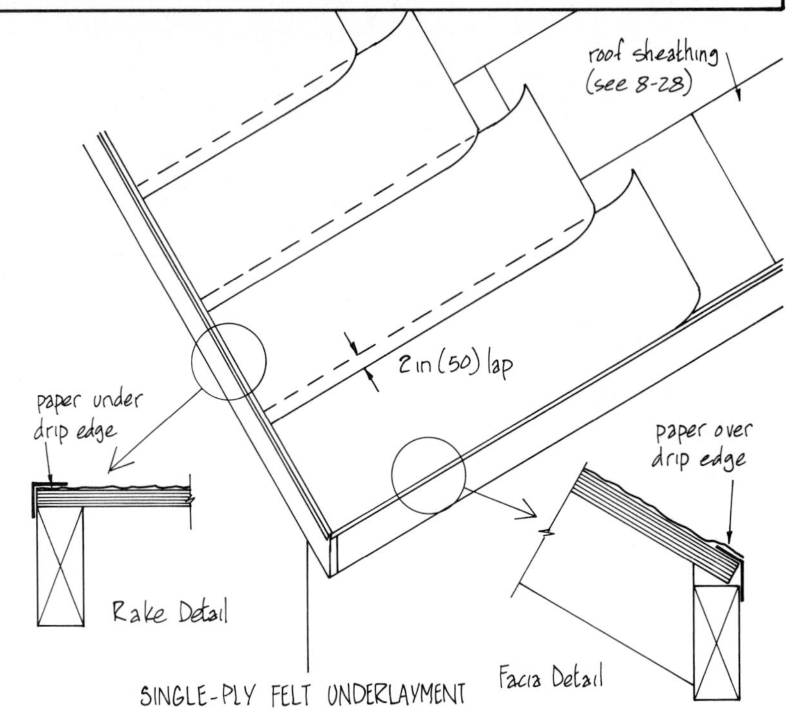

SINGLE-PLY FELT UNDERLAYMENT

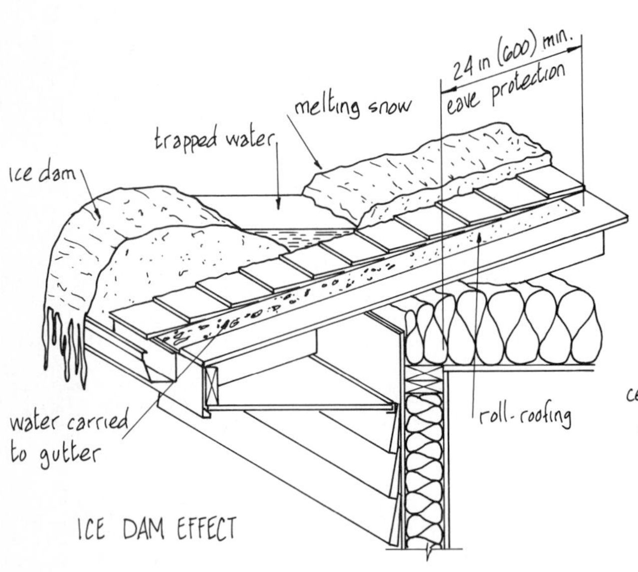

ICE DAM EFFECT

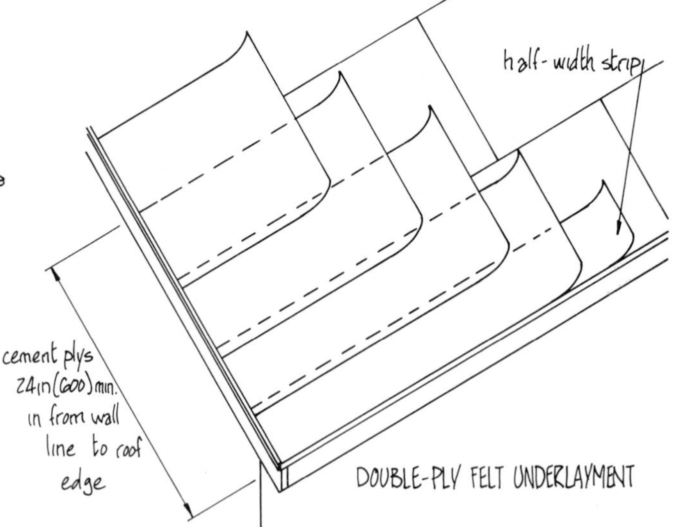

DOUBLE-PLY FELT UNDERLAYMENT

ICE DAMS
Under certain cold weather conditions an ice dam can form on the roof over the wall line. Water trapped by the dam can penetrate to the room below, damaging insulation and finishes. To avoid this potential problem, an additional layer of 50-lb (25-kg*) roll roofing extending from the roof edge to at least 24 in (600) beyond the inner wall line should be applied (polyethylene is not recommended and may not meet some codes).

LOW-SLOPE ROOFS
For low-slope roofs a double layer of felt is usually recommended. In addition, at the eave projection and for a distance of 24 in (600) inside the wall line, the plys are cemented together with a continuous layer of roofing cement.

* Metric value soft converted.

SHINGLE APPLICATION

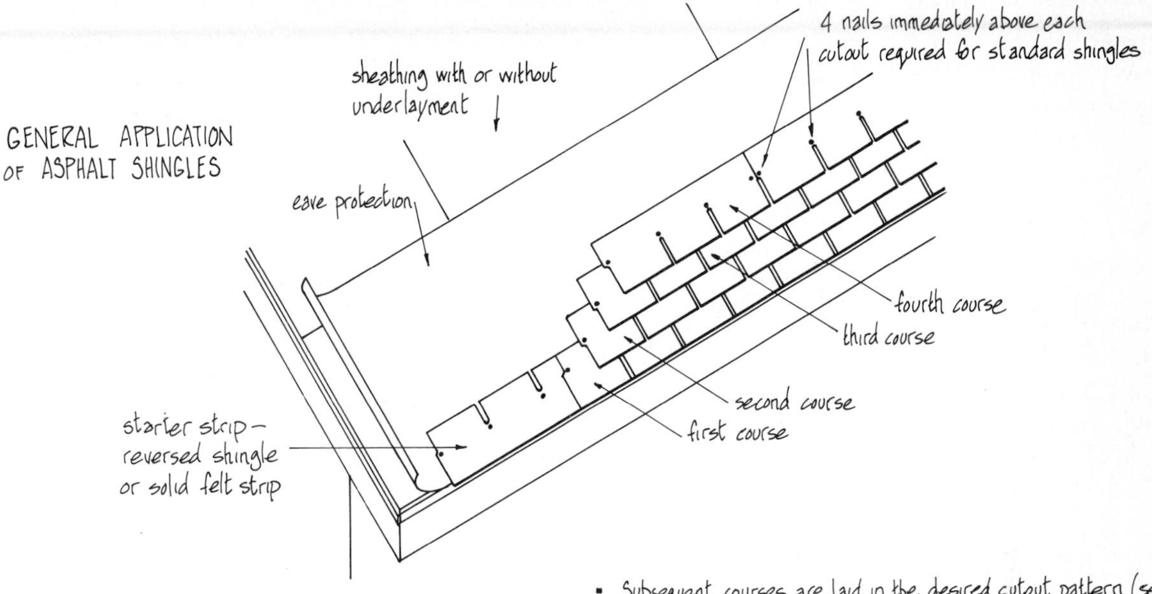

APPLICATION

- For smaller roofs, shingles may be started from either end. Roofs over 30 ft (9 m) should be started from the center.
- A starter strip at least 12 in (300) wide and extending 1/2 in (12) beyond the fascia is first applied. A reversed shingle or roll-roofing is used.
- The first course of shingles is laid in line with the bottom edge of the starter strip. Avoid line-up of joints in starter and first course.
- Subsequent courses are laid in the desired cutout pattern (see below). Chalk lines at suitable intervals up and across the roof should be used to align both cutouts and courses.
- Four roofing nails are required for standard shingles as shown. On very steep roofs (e.g., mansard roofs, see 8-7), double nailing at the cutouts is required.
- Follow manufacturer's recommendations for other types of shingles and low-slope applications.

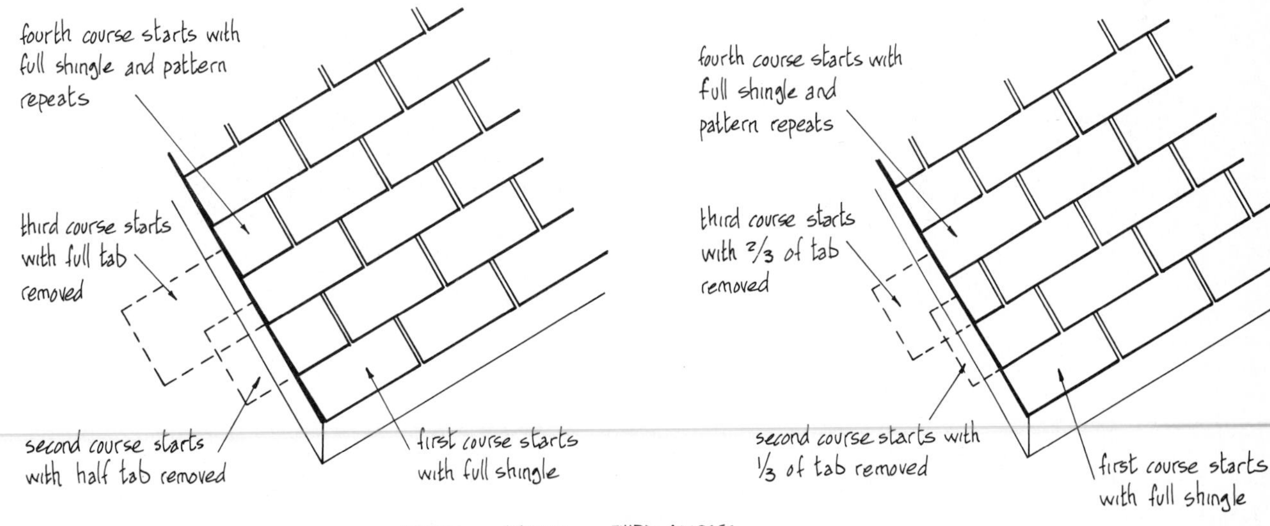

CUTOUTS ON SECOND COURSES

CUTOUTS ON SECOND OR THIRD COURSES
The illustrations above show the starting procedure at the rake for aligning cutouts on every second or every third course. Most shingles have factory-applied alignment marks on the butt face to assist accurate installation.

CUTOUTS ON THIRD COURSES

FINISHES

11-28 VALLEYS, RIDGES, AND HIPS

VALLEYS

The majority of valleys are "open" as illustrated at the right. For a different visual effect, they may also be woven or closed-cut, as shown below.

- In the open method the valley is first flashed with two layers of 90-lb (45-kg*) mineral-surfaced roll-roofing as shown. Metal flashing can also be used. Joints are lapped at least 12 in (300) and sealed with asphalt cement. Two chalk lines are snapped on either side of the valley center to guide shingle cutting. The valley should be 6 in (150) wide at the ridge and increasing in width by 1/8 in per foot (1:100) of downward run.
- For woven valley, only one layer of flashing is required, and alternate layers of shingles are extended up the opposing slopes.
- For a closed-cut valley, all shingles on one side are extended up the opposing slope as shown. Shingles on the second slope are cut on a line 2 in (50) back from the valley center.

* Soft conversion.

18 in (500) wide flashing strip face down

36 in (1000) wide flashing strip face up

clip corner and cement butt to roll roofing

OPEN VALLEY ARRANGEMENT

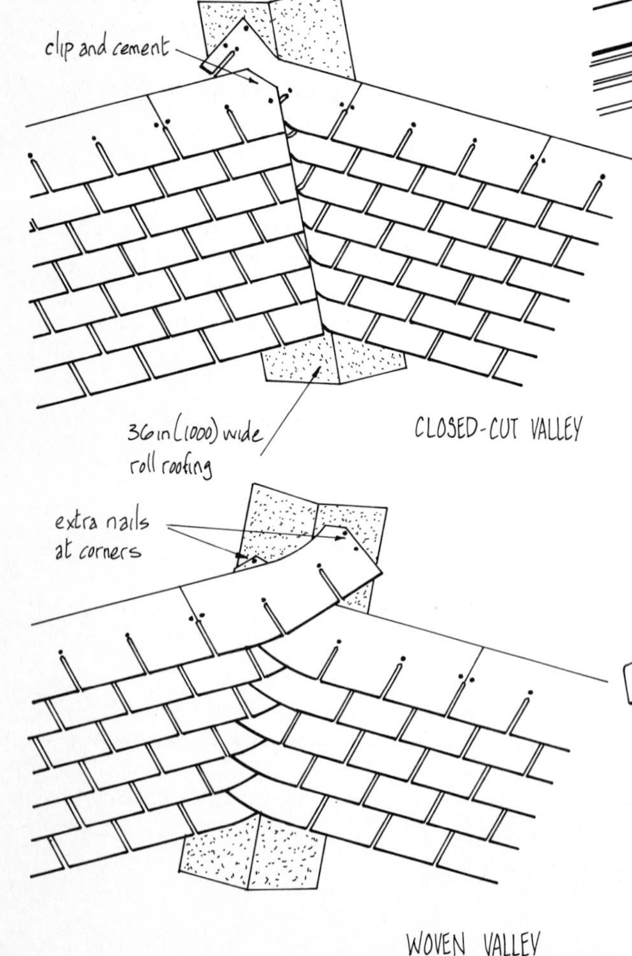

CLOSED-CUT VALLEY

WOVEN VALLEY

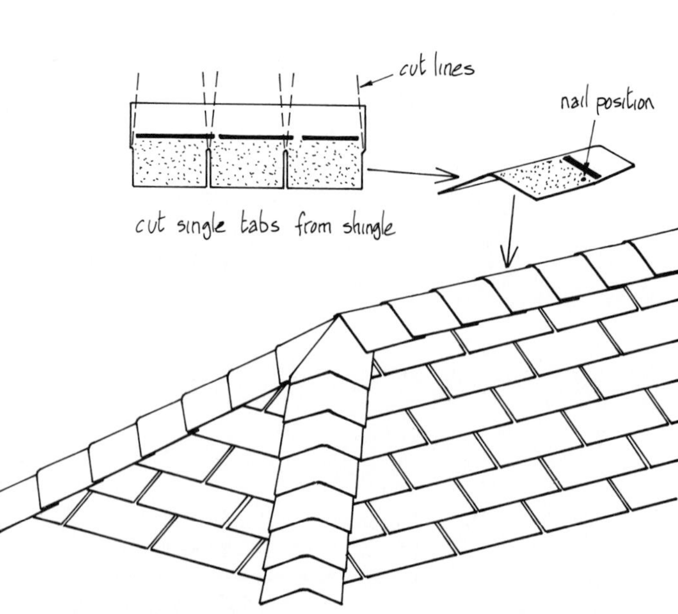

RIDGES AND HIPS

Asphalt shingle squares cut from the tabs of a full shingle are nailed over the ridge or hip and lapped with the same coverage as roof shingles. On ridges, start the shingles at the end away from the prevailing winds. On hips, start at the bottom.

WALLS, VENTS, AND CHIMNEYS

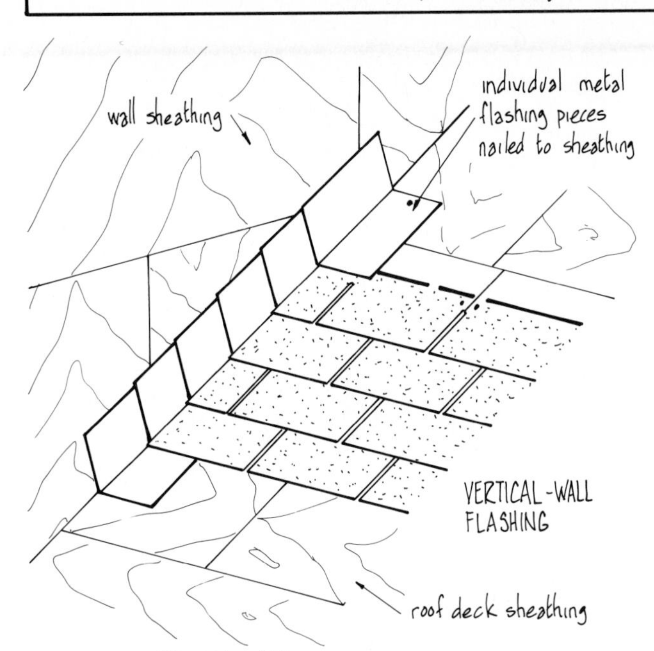

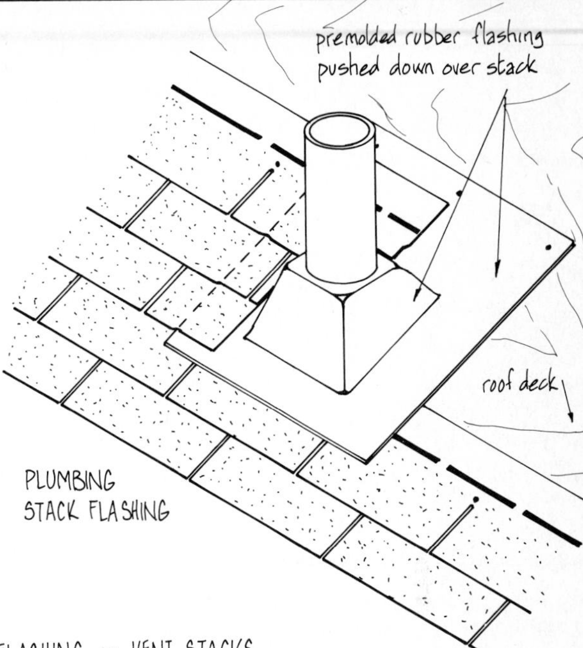

FLASHING AT VERTICAL WALL

Individual metal flashings, sized to suit the type of shingle, are inserted between each course of shingles at the wall line. The flashing is nailed to the roof sheathing (not to the wall) and the overlapping shingle is cemented to the flashing. The flashing should extend 5 in (125) up the wall and across the roof. The wall siding is installed after and over the flashing.

FLASHING AT VENT STACKS

Premolded flashings are available for most standard pipe sizes and roof slopes, or alternatively, roll-roofing can be cut to suit. Shingles are laid up to and cut around the stack, and the flashing is installed and cemented to both shingles and pipe. The upper shingle courses are laid over the flashing and cut around the pipe.

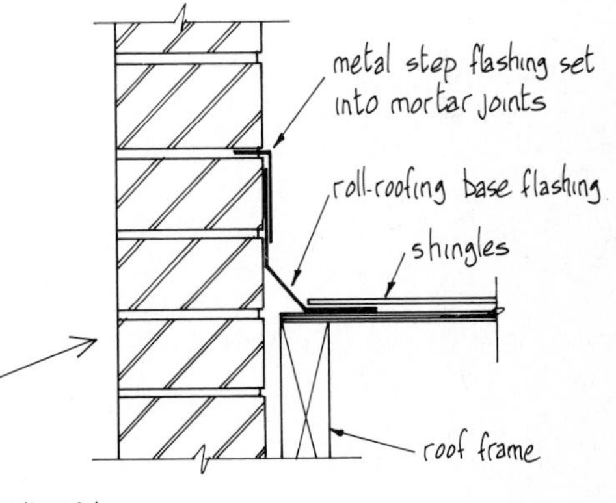

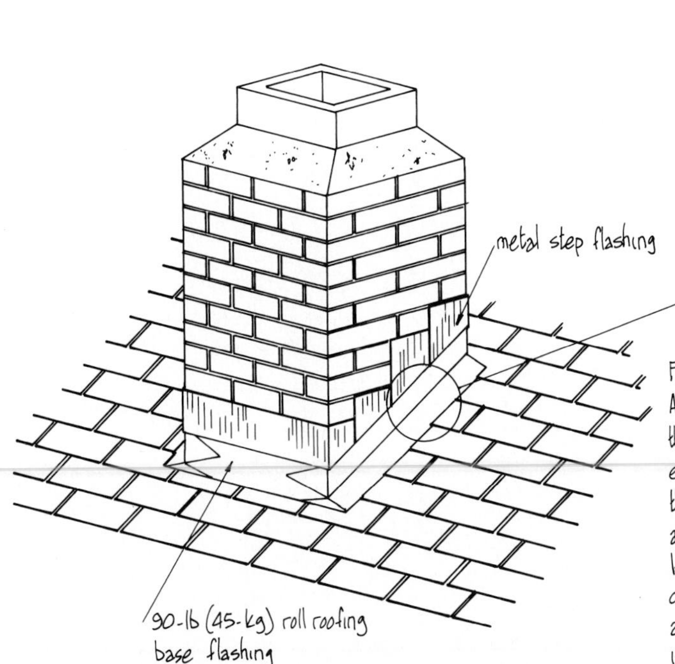

FLASHING AT CHIMNEY

FLASHING AT CHIMNEYS

A two-step process is used to flash around chimneys. The roof is shingled up to the chimney and base flashings of 90-lb (45-kg*) roll roofing are applied to each side of the chimney. The flashing is extended at least 4 in (100) up the chimney and 8 in (200) across the roof. A metal counter flashing is then applied over the base flashing. The top edges of the metal flashing are let into the mortar joints and sealed. The shingle courses are continued over and cemented to the base flashing. To reduce accumulations of ice and snow behind the chimney a saddle or cricket is installed as shown on 8-25.

WOOD SHINGLES AND SHAKES

Wood shingles and shakes are used extensively in residential construction. They provide an interesting visual feature and are very durable, rot resistant, and long-lived. Most are installed without finishes of any kind and will weather to a soft grey-brown color. The material commonly used for shingles and shakes is western red cedar, but other materials, including redwood, white cedar, and cypress are also used.

Although not fireproof (and for this reason may be prohibited by codes in some regions), techniques that include treatment with fire retardant chemicals or interlayering with asbestos-based felts can be used to increase fire resistance. Minimum slope for shingles and shakes is 4 in 12 (1:3) for most applications, and maximum exposure sizes depend on length and quality of the shingles. Flashings at valleys, walls, vent stacks, and chimneys are similar to those already detailed for asphalt shingles.

Shingles are made by taper-sawing a wood blank and have a relatively smooth surface. Shakes are split by hand from a wood block and have a rough textured surface. The three standard types of shakes are illustrated at the right.

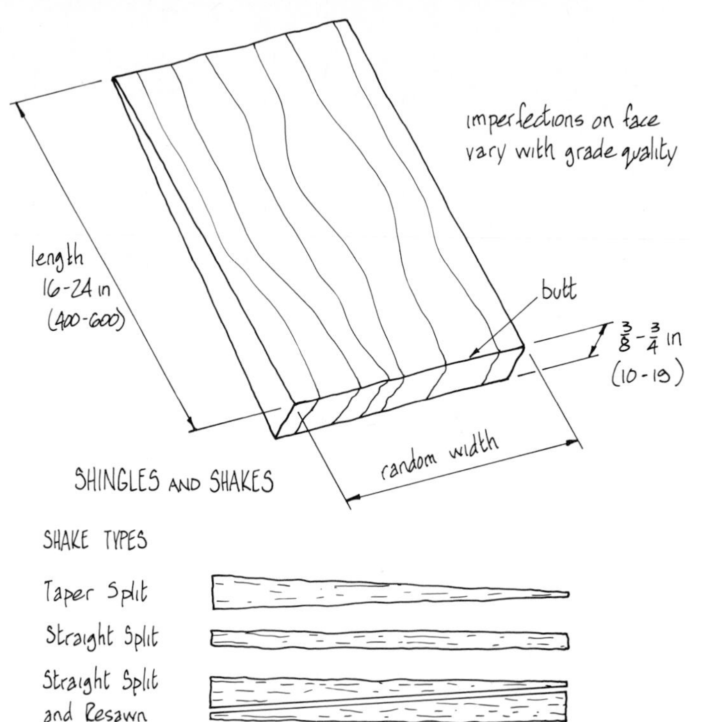

SHINGLE LENGTH AND EXPOSURE – ROOFS

LENGTH	SLOPE – MAXIMUM EXPOSURE ①	
	3 in 12 (1:4)	4 in 12 and up (1:3)
16 in (400)	3 to 3¾ in (75-95)	3½ to 5 in (90-125)
18 in (450)	3½ to 4¼ in (90-105)	4 to 5½ in (100-140)
24 in (600)	5 to 5¾ in (125-145)	5½ to 7½ in (140-190)

① Exposure increases as shingle quality increases.

SHAKES – LENGTH AND EXPOSURE

Based on the minimum recommended slope of 4 in 12 (1:3), exposure is 7½ in (190) for 18 in (450) shakes and 10 in (250) for 24 in (600) shakes.

GRADES

Shingles are available in the following grades:

No. 1 Black Label – premium
No. 2 Red Label – good
No. 3 Black Label – utility
No. 4 Undercoursing – utility
No. 1 & 2 Rebutted and Rejointed – premium grade with machine-trimed edges and butts

Shakes are generally classified by their method of manufacture:

Handsplit & Resawn – split faces and resawn backs
Tapersawn – sawn both sides
Tapersplit – split by hand and tapered
Straight-Split – split by hand and same thickness

WOOD SHINGLES AND SHAKES

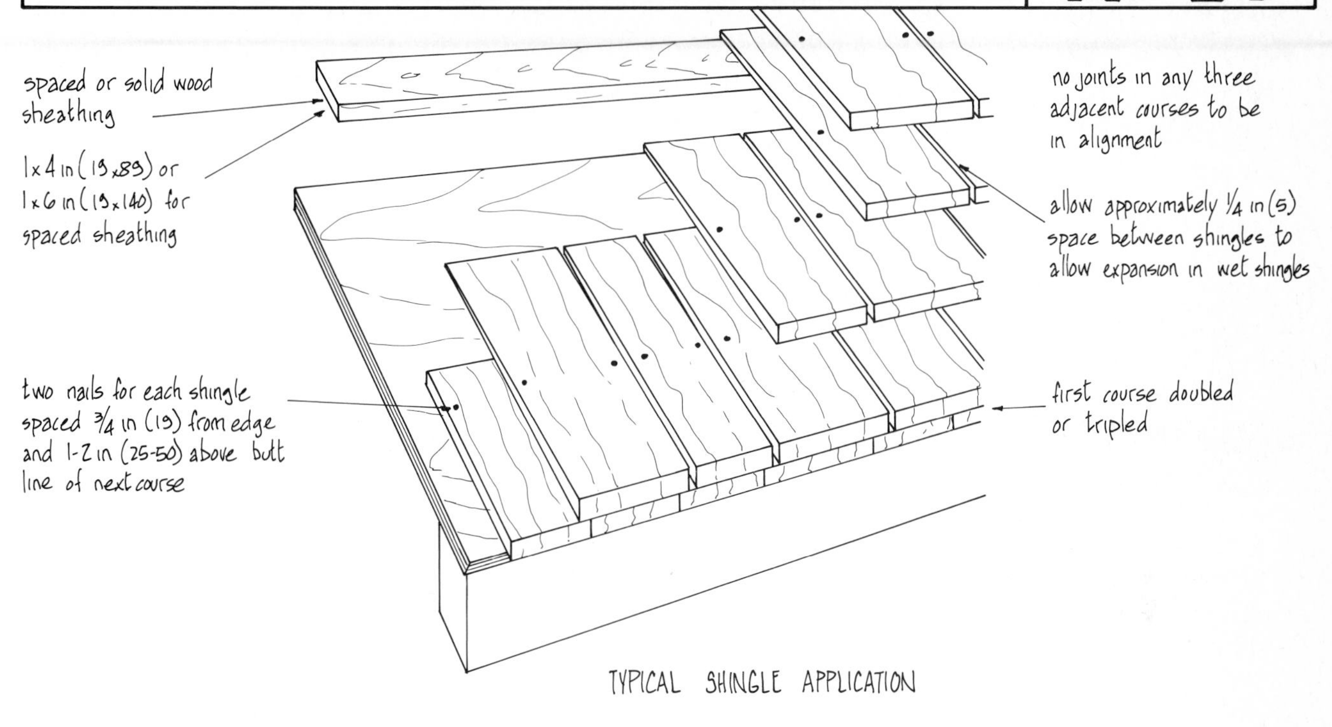

TYPICAL SHINGLE APPLICATION

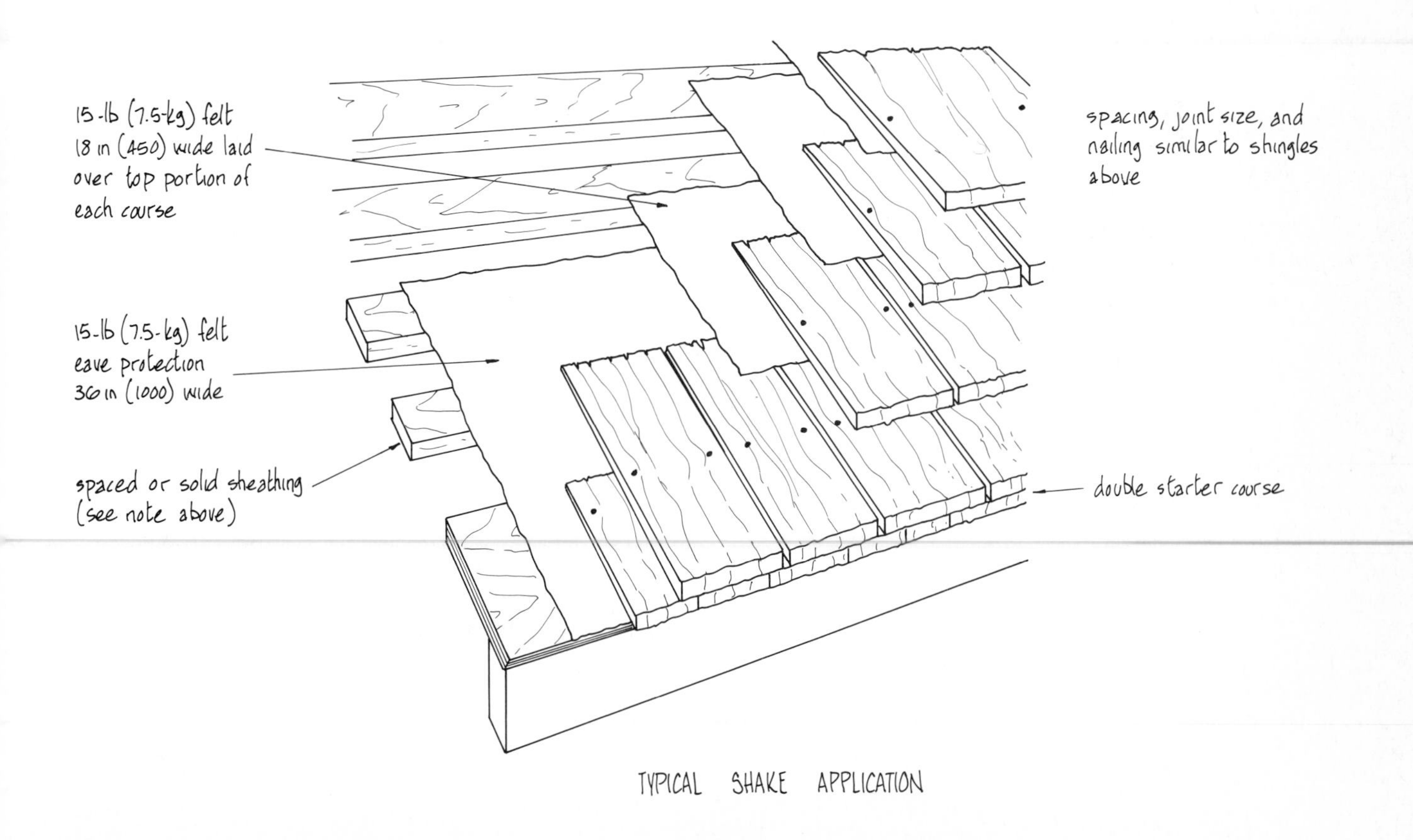

TYPICAL SHAKE APPLICATION

SHINGLE APPLICATION

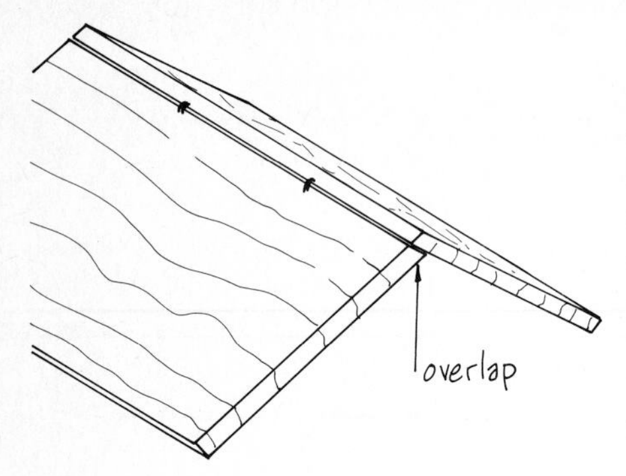

MANUFACTURED RIDGE PIECE

RIDGES AND HIPS

On ridges and hips either factory assembled or site made units are used. Shown here is a factory-made ridge unit and the recommended installation method. Site made units consist of individual shingles or shakes approximately the same width as the weather exposure, blind nailed and edge trimmed in a similar manner to the factory units. A double layer is applied with alternate overlaps.

RIDGE OR HIP CAP SHINGLES

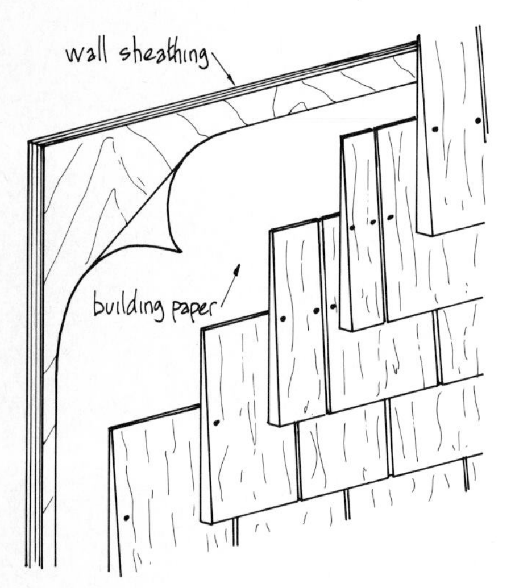

SINGLE-COURSE WALL APPLICATION

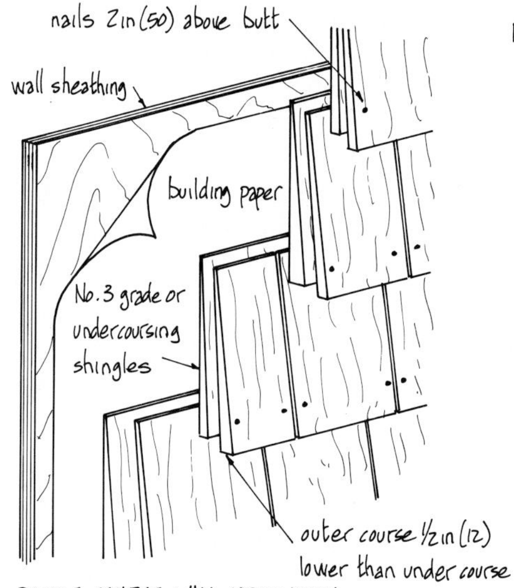

DOUBLE-COURSE WALL APPLICATION

SIDEWALLS

Shingles and shakes can be applied to walls in several ways. Individual units can be applied in single or double courses as shown at the left or panelized and applied in similar manner to wood siding (see below and 11-9)

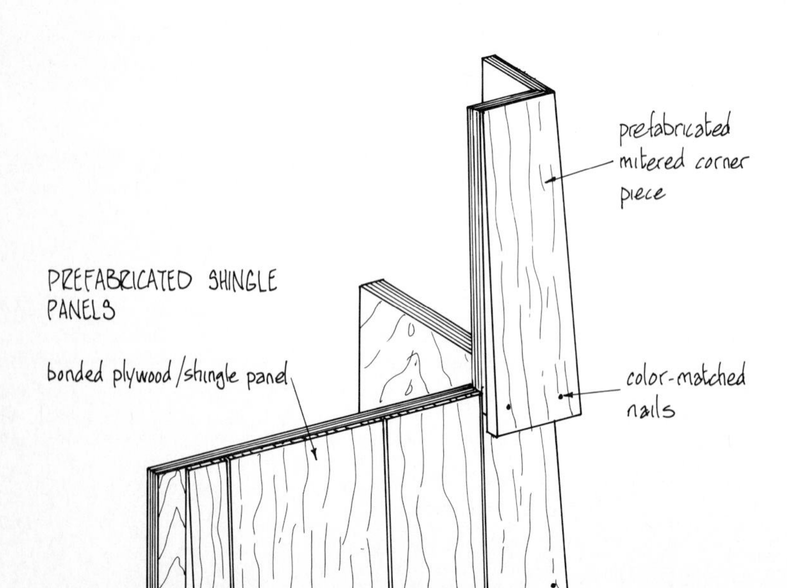

PREFABRICATED SHINGLE PANELS

PANELS

The installation of a prefabricated corner unit for panelized shingles is illustrated at the left. A panelized unit has shingles or shakes bonded to an 8-ft (2400) long plywood backing strip (see 11-18). Weather exposures and laps are designed to suit wall or roof applications and the plywood strip can often act as structural sheathing if allowed by codes. Manufacturers produce a variety of decorative shingle shapes for special architectural effects.

TILE AND METAL ROOFING

CLAY AND CONCRETE TILES

- Clay tiles are generally molded from clay or a mixture of shale and clay and are usually unglazed. Concrete tiles are generally extruded from a mixture of portland cement, aggregate, and color pigments. Both types are extremely durable and have a very long service life.
- Color of natural clay tile varies from orange-yellow to deep red. Glazed tiles can be virtually any color. Concrete tiles are also available in a variety of colors.
- Roof structures must be designed to accommodate the extra loading imposed by these tile systems.
- Depending on type, the tiles are nailed to solid sheathing, to open sheathing, or to horizontal battens.
- Specially molded units are available for ridges, rakes, and fascias.
- Illustrated below are some of the standard tile patterns available in most regions.

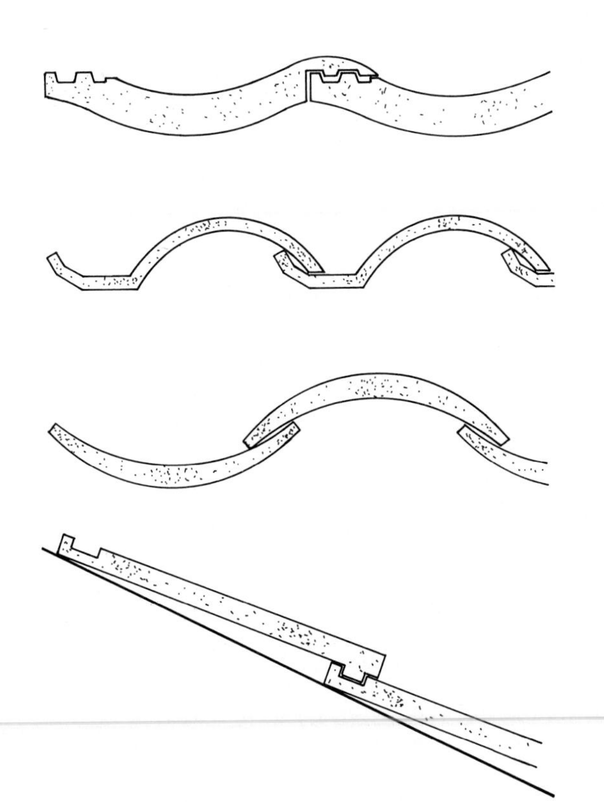

METAL ROOFING

- Metal roofing can be classified into two main groups: flat-seamed and rolled. Examples of flat panels with raised seams are shown below. See 8-38 for examples of rolled sheet roofs.
- Metal used for roofing includes copper, aluminum, stainless steel, and galvanized steel. A wide range of finishes other than plain metal are available, including baked enamel paint, porcelain enamel, and copper, aluminum, or terne metal coatings.
- Most systems, particularly the seamed and roll-formed types, are very specialized and must be erected by the manufacturer. Roof structure and sheathing requirements are subject to roofing system demands.
- Service life is generally long but will vary with the type of material and installation methods.

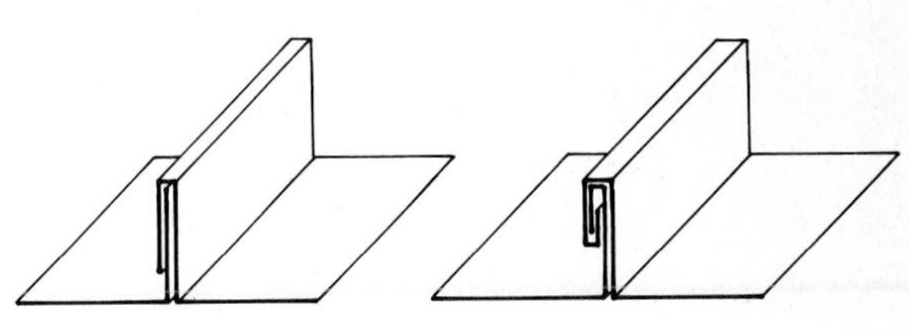

Simple Standing Seam

Double-Backed Standing Seam

SEAM EXAMPLES

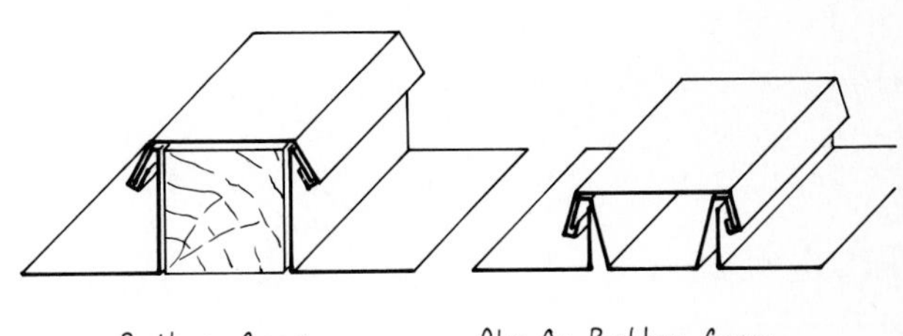

Batten Seam

Clip-On Batten Seam

OTHER FORMED ROOF SYSTEMS

Other materials are used for roofing, although generally only for special light service applications. Materials used include asbestos cement, fiberglass, and clear or opaque plastics.

BUILT-UP AND MEMBRANE ROOFING

For flat or nearly flat roofs, two covering systems are in general use: built-up roofing and single layer membrane roofing. A variety of materials and installation techniques are used for each type. Climatic and service conditions generally determine the system used. Typical examples of each type are illustrated on the next page.
Even though regarded as flat, all such roofs must have some slope for drainage purposes (see 8-30). To prevent damaging condensation some type of warm-side vapor barrier is essential since roofs of this type are not internally vented. Roof insulation can be placed either under or on top of the roof covering, as illustrated below.

BUILT-UP ROOFING

- The roof covering is built up from a series of felt layers set in flood coats of hot tar or asphalt. The felts are topped with a layer of gravel or stone set in tar to provide protection from weather and mechanical damage.
- A variety of felt types are available and the number of layers applied (usually between four and seven) is determined by service requirements.
- Depending on decking and/or insulation type, the base layers may or may not be mechanically fastened to the deck.
- Since built-up roofing is a rigid bonded covering, it is important that the roof structure and decking also be as rigid as possible. Excessive differential movement between structure and roofing may result in failure of the roofing.

SINGLE-PLY MEMBRANE ROOFING

- In this system a single layer of roofing membrane is loosely laid over the entire roof on a prepared deck or insulation base. The membrane seams are bonded chemically or thermally to provide a continuous and sealed roof covering. Fastening to the deck structure generally occurs only at roof edges, openings, or other interruptions, but can also occur at seams with some roofing types.
- The vapor barrier and insulation is also usually laid loosely allowing the entire roof covering to move with the structure although some systems do require mechanical fastening of both base and membrane.
- Several type of membrane materials are available, with PVC and EPDM the most predominant.
- A layer of gravel or stone is applied for protection.

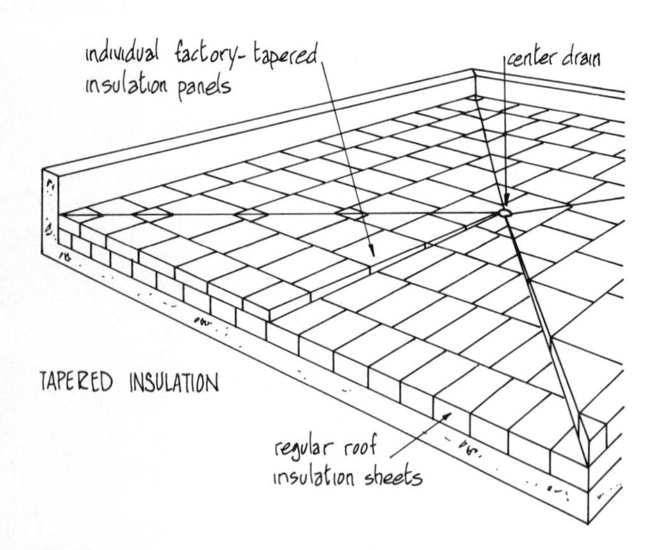

TAPERED INSULATION

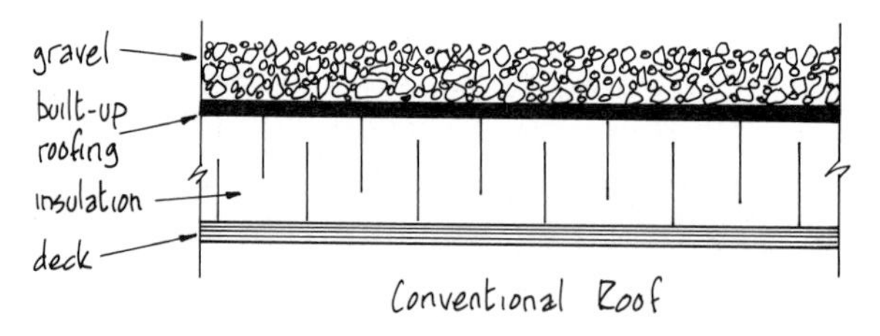

Conventional Roof

INSULATION POSITION

Insulation can be positioned in one of two ways. It can be installed below the roof covering, as shown above, where it acts as part of the roof base. It can also be installed over the roof covering, as shown below. In this case the installation is called an "inverted roof" (also called an upside-down roof) since the traditional application procedure is reversed. The advantage of an inverted roof is that the insulation almost totally protects the roof covering from weather, heat, and mechanical damage.

TAPERED INSULATION

To provide a slope sufficient to drain the roof, the structure can be designed with sloping roof beams or, as illustrated above, a layer of tapered insulation can be applied over the standard base insulation. Tapered insulation sheets are available in a range of slopes and sizes, including mitered 90° corner pieces for center-draining roofs.

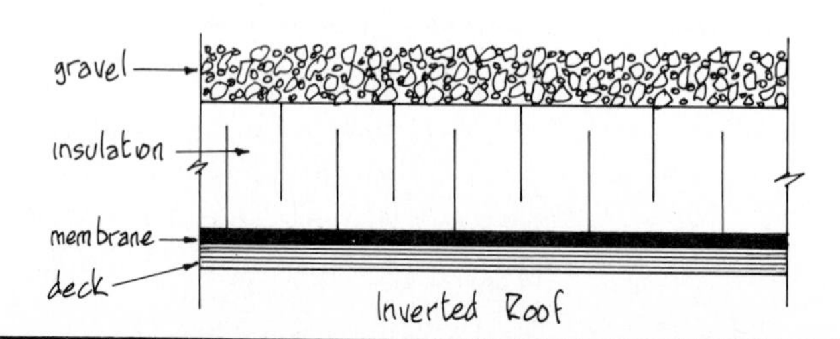

Inverted Roof

BUILT-UP AND MEMBRANE ROOFING

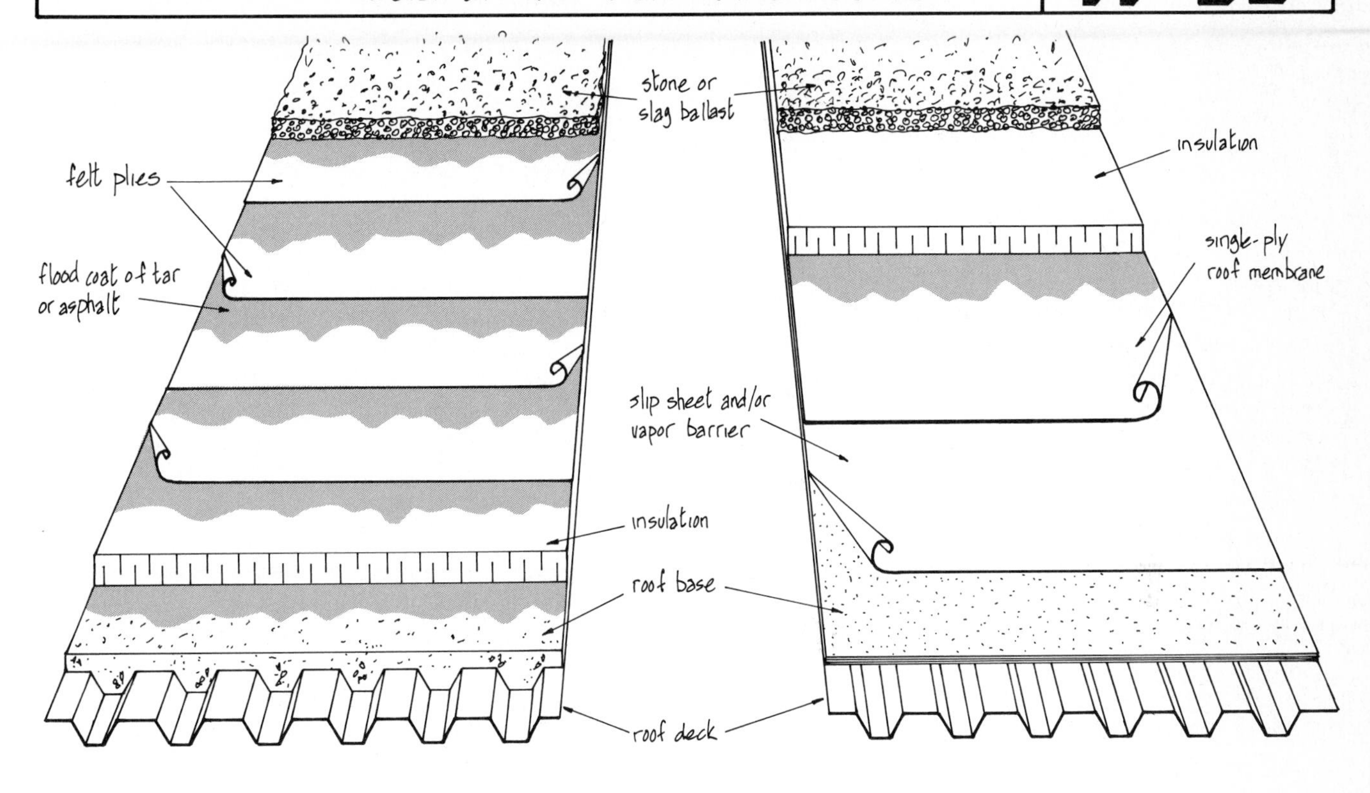

BUILT-UP ROOFING (conventional sequence)

SINGLE-PLY MEMBRANE (inverted sequence)

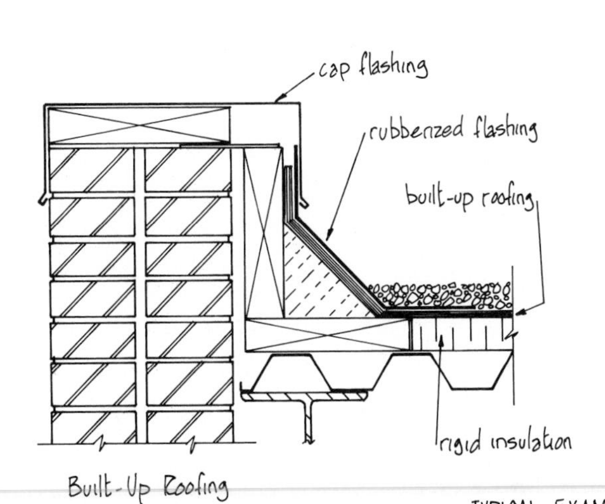

Built-Up Roofing

cap flashing

rigid insulation set in
flood coat of tar

roof membrane

gypsumboard
base

Membrane Roofing

TYPICAL EXAMPLES OF BOTH METHODS

Exact details will vary widely and are dependant upon the type of wall and roof construction and recommended installation methods specified by the roofing manufacturer.

CHAPTER PAGE TITLES

WINDOW TYPES

Windows play an important part in both the functional and aesthetic aspects of a building. They provide natural light and ventilation and contribute greatly to the comfort (or discomfort) of the occupants. Designers must aware of the design challanges inherent in choosing and positioning windows, and contractors must be familiar with modern installation methods that contribute greatly to the building's energy efficiency. There are three main window groupings, each illustrated on this page and the next.

SLIDING WINDOWS

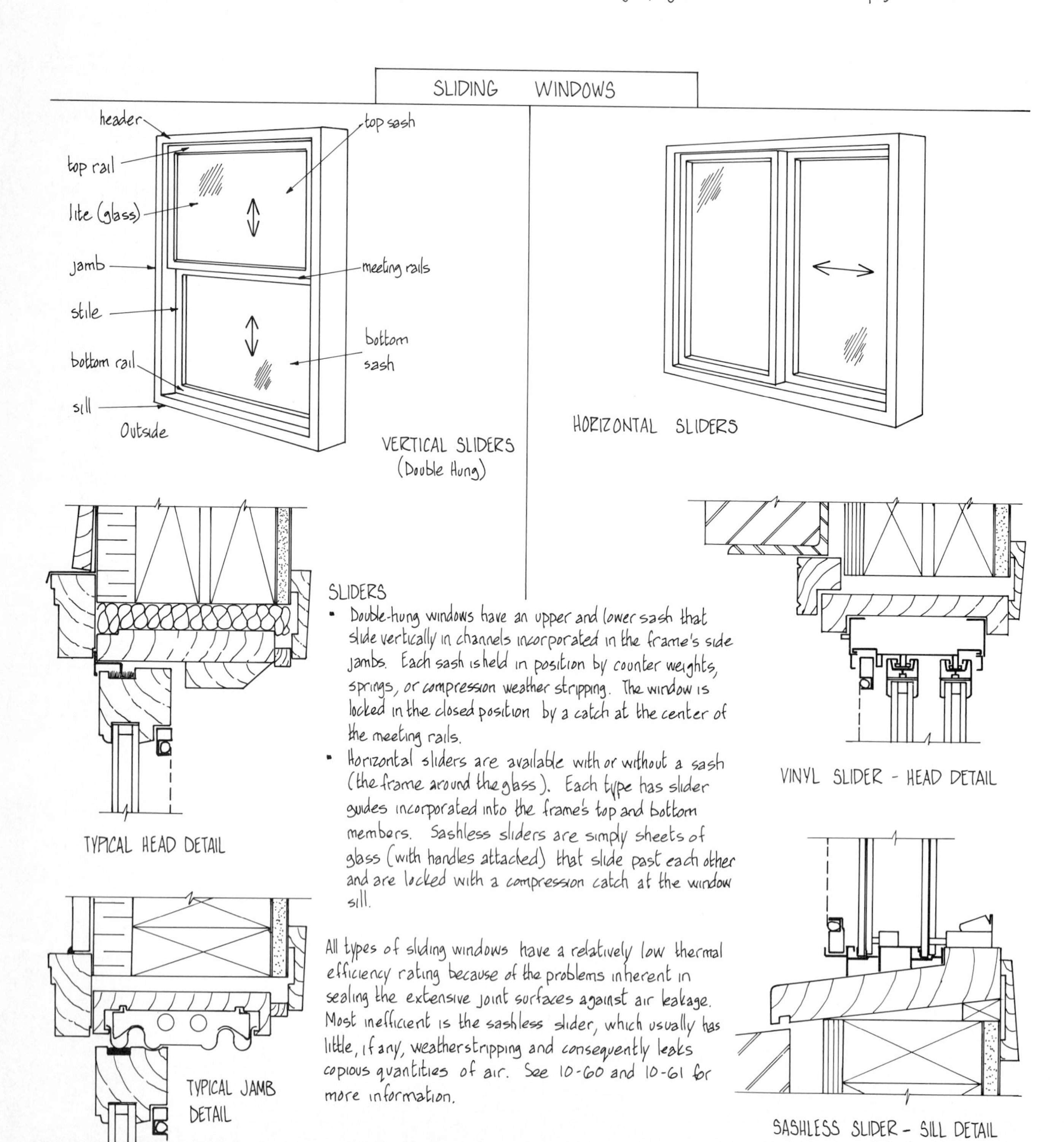

SLIDERS

- Double-hung windows have an upper and lower sash that slide vertically in channels incorporated in the frame's side jambs. Each sash is held in position by counter weights, springs, or compression weather stripping. The window is locked in the closed position by a catch at the center of the meeting rails.
- Horizontal sliders are available with or without a sash (the frame around the glass). Each type has slider guides incorporated into the frame's top and bottom members. Sashless sliders are simply sheets of glass (with handles attached) that slide past each other and are locked with a compression catch at the window sill.

All types of sliding windows have a relatively low thermal efficiency rating because of the problems inherent in sealing the extensive joint surfaces against air leakage. Most inefficient is the sashless slider, which usually has little, if any, weatherstripping and consequently leaks copious quantities of air. See 10-60 and 10-61 for more information.

WINDOW TYPES

HINGED WINDOWS

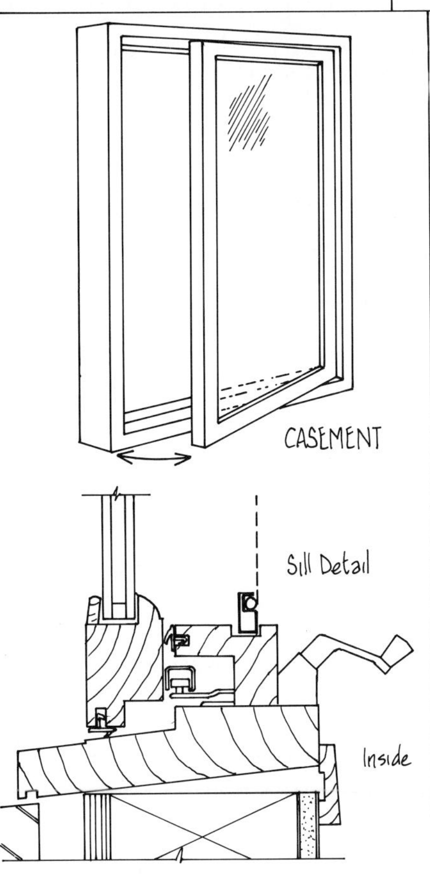

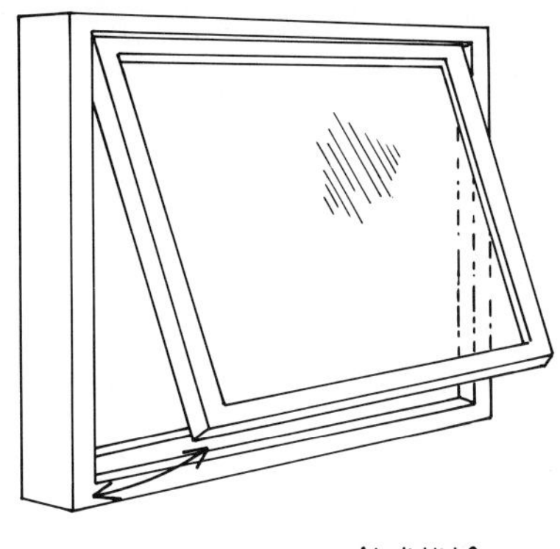

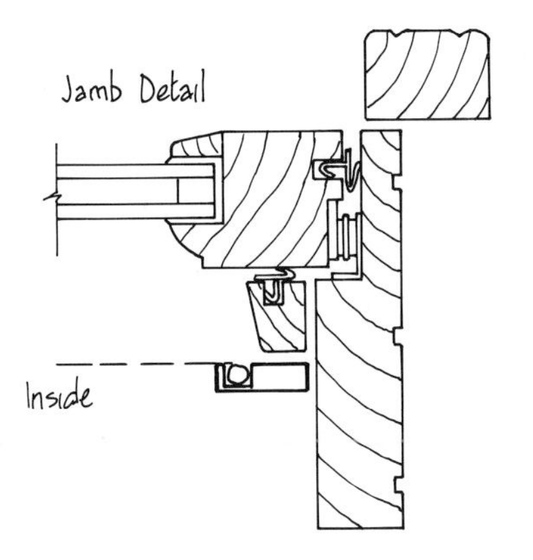

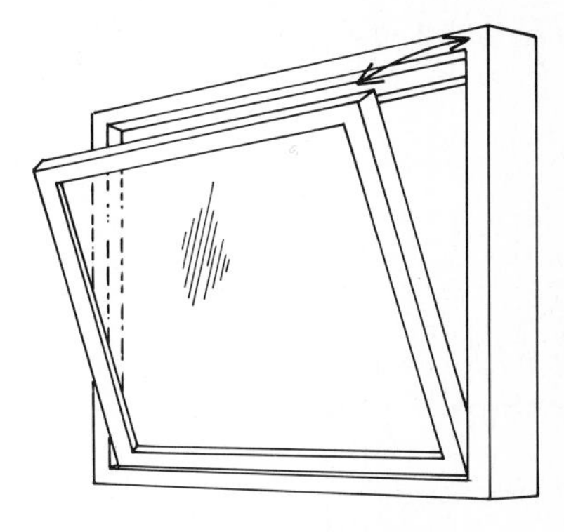

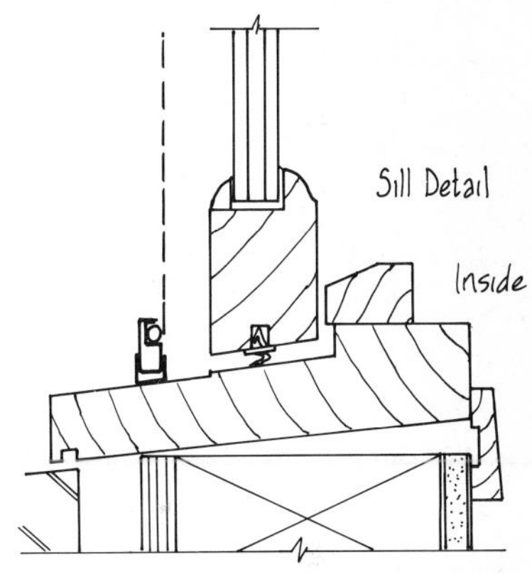

CASEMENT

- Swings outward (in the vertical plane) and is usually operated by a crank or lever mechanism.
- Better-quality casements have sliding type hinges located in the frame head and sill. They allow the sash to be rotated fully outward and away from the jamb, providing access for exterior cleaning. Less sophisticated casements have standard butt hinges (see 12-20) and are opened manually.
- A catch on the jamb opposite the hinges pulls the sash tightly to the weatherstripped frame.

AWNING

- Horizontal version of a casement with the sash swinging upward and outward.
- Sliding hinges, similar to those used on casements but installed on the frame jambs, are standard on good-quality units. A separate locking catch is not usually necessary since the opening mechanism also serves this purpose.

HOPPER

- Sash swings inward and downward and is usually opened by hand.
- Hinges are similar to those of awning windows.
- Catches are installed on the frame head.
- Because hopper windows swing into the room, they may interfere with drapes and reduce usable space near the window.

Hinged windows have a higher energy efficiency rating because they have shorter, simpler joints, while the sliding hinge and locking mechanisms allow the sash to be pulled tightly against the frame. Many hinged windows have weather stripping on both frame and sash, further increasing their airtightness.

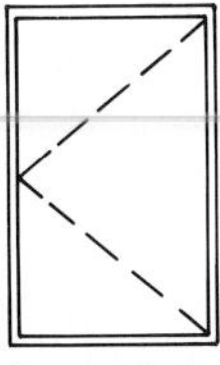

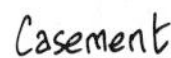

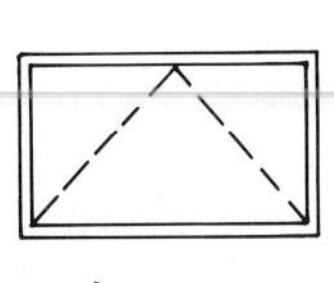

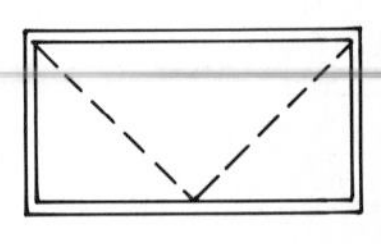

The conventions shown above are used on blueprints to indicate which way a hinged window opens. Broken lines designate an opening window, while the point of the triangle indicates the hinge position.

WINDOW TYPES

FIXED WINDOWS

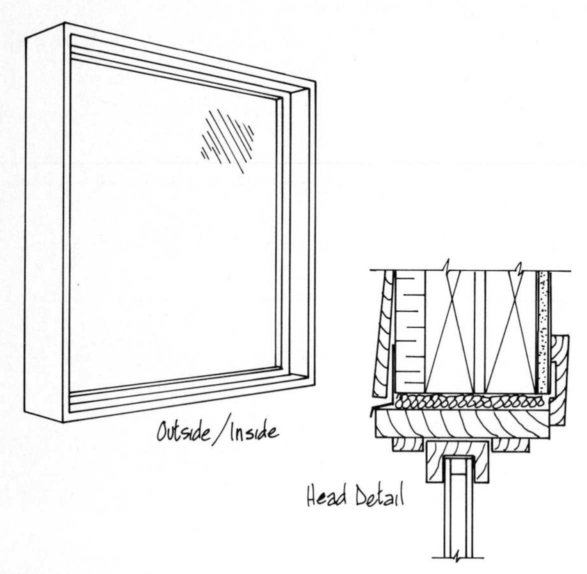

FIXED
- Consists simply of glazing installed directly into the frame.
- Often used when a large uninterrupted glazed area is required or as part of a "picture window" combination with one or more of the opening window types.

JALOUSIE WINDOWS

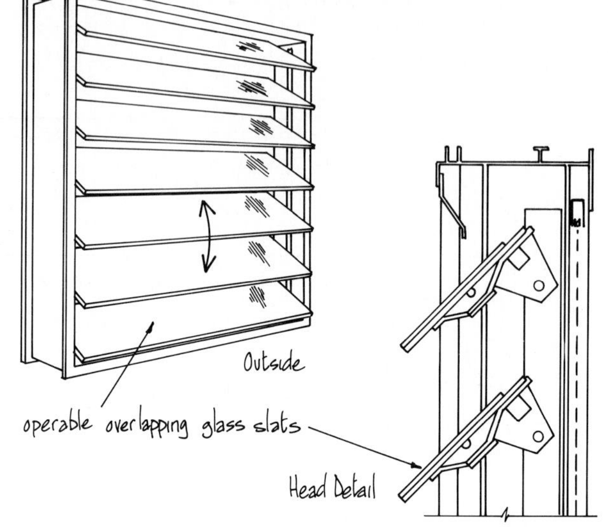

JALOUSIE
- A series of glass slats held in metal frames and attached to each other so that all open or close at the same time. They are essentially the glazed equivalent of a Venetian blind.
- This type of window cannot be made airtight or even totally weather-tight, and it is used extensively only in warm or hot climates.

GLASS BLOCKS

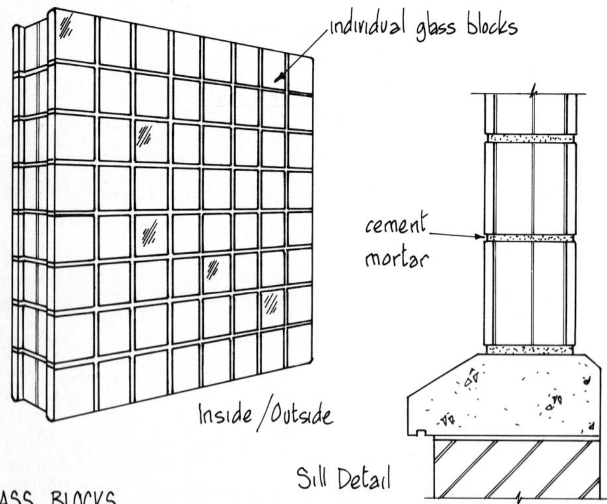

GLASS BLOCKS
- Hollow glass blocks are built into the structure of a building. They provide excellent light levels and are totally airtight and weatherproof.
- They are used often in commercial and industrial applications and enjoy an occasional vogue in residences because of their very low maintenance factor and unique features.
- See 7-21 for more information.

PATIO WINDOWS

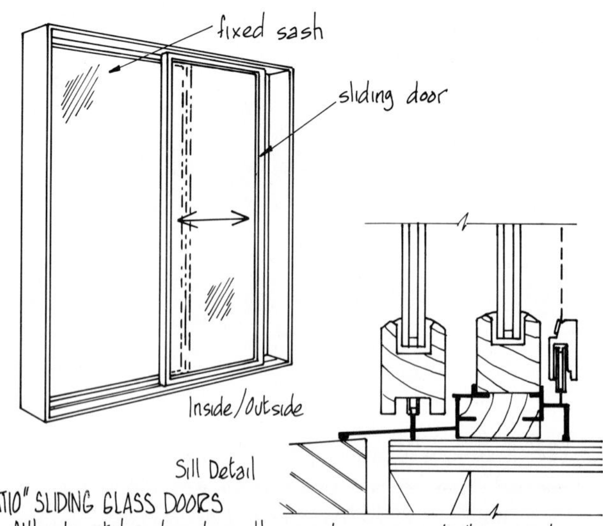

"PATIO" SLIDING GLASS DOORS
- Although catalogued as doors, these units are essentially special-purpose sliding sash windows.
- They are used extensively where large amounts of light, a wide view, and direct and easy access to the outside are required.
- Airtightness problems are typical even with extensive weather stripping. Some manufacturers fabricate patio doors with the slider on the outside so that wind pressure will help press the sash against the frame for a better seal.

FRAME MATERIALS

A wide variety of materials are used in window frame construction, particularly for residential applications. Each material has certain advantages and disadvantages and the factors that will influence a final choice include:

- Ease of maintenance. Prefinished or permanently finished frames are obviously desirable, but a future color change may be difficult or impossible.
- The windows should last the expected life of a house. This is a reasonable expectation for nonferrous metal frames but may not be possible for wood frames without consistent maintenance.
- The finish of the frames should complement the visual aspects of the house. White, brown, or bronze are the standard colors.
- Cost factors are important since windows represent a significant part of a building's total cost.
- Energy conservation factors should also be considered. See the information below and that on 10-61.

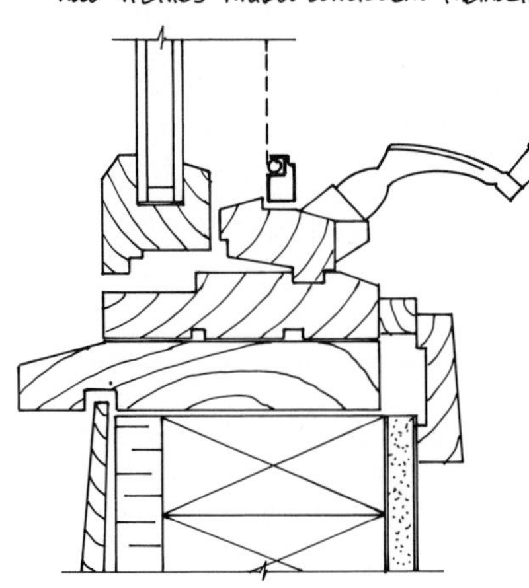

WOOD

Ponderosa pine is the usual material of choice. Redwood and mahogany are also offered by some manufacturers. Frames are factory-treated with a preservative and sometimes primed. Wood frames have distinct thermal advantages over metal and some vinyl frames.

CLADDED WOOD

Wood frames with metal cladding or a plastic coating on the exterior weather surfaces. Combines the advantages of wood with a lifetime protective finish. Interior surfaces are painted or stained.

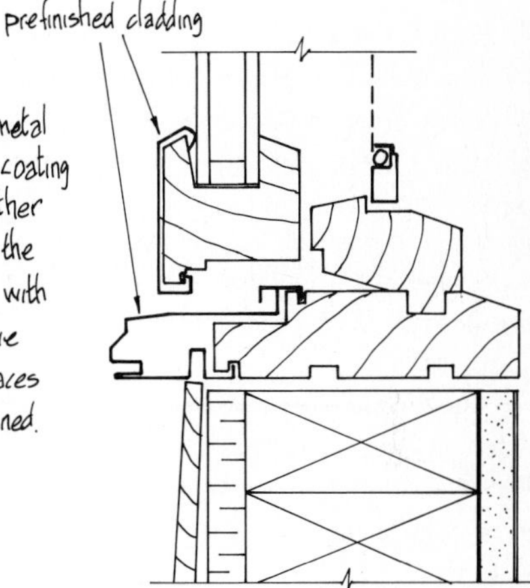

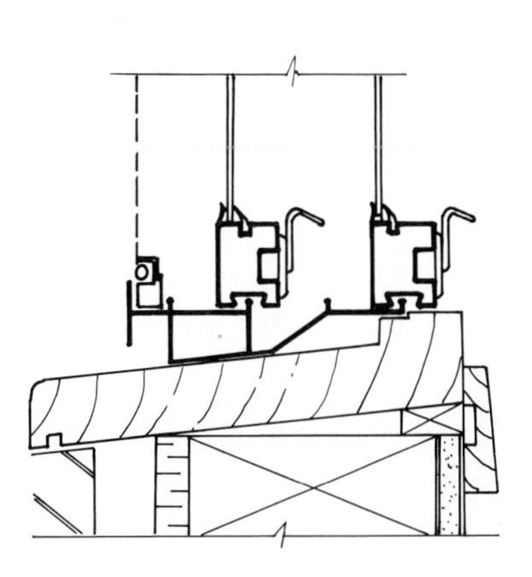

ALUMINUM

Nonrusting low-maintenance frames fabricated from extruded sections. Plain, anodized, or prepainted finishes. Frames must be "thermally broken" to reduce heat loss and surface condensation. May pit and corrode in some acidic atmospheres.

PVC PLASTIC

Similar in appearance to aluminum but with a thermal performance generally equal to that of wood. Color is solid throughout and cannot be lost through wear. Large windows may need strengthening with metal inserts.

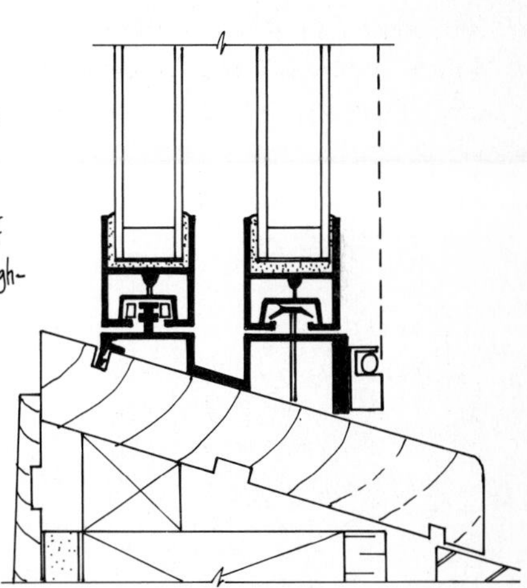

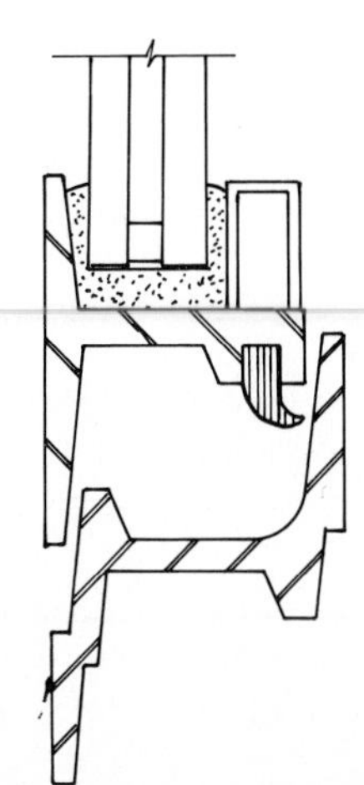

STEEL

This material is rarely used in residential construction but is quite common in commercial high-rise structures (see curtain walls, 7-42) and industrial applications. Aside from stainless steel, finish is usually baked enamel or galvanized.

ENERGY EFFICIENCY AND CONDENSATION

Metal frames conduct heat very rapidly in comparison to wood and plastic frames even though they may be thermally broken between the inner and outer frames. Aside from the heat loss involved, metal frames are subject to "sweating" caused by condensation of warm moist room air on the cold interior frame surfaces. Under certain conditions condensation runoff can be a significant problem. Wood and plastic frames will normally sweat in cold climates only under extreme conditions.

GLASS – DOUBLE GLAZING

GLASS TYPES

There are two generic types of glass in general use for window applications: float glass and tempered glass. Each is manufactured in a range of thicknesses from 3/32 in (2.5) to 7/8 in (22).

- Float glass is made by floating liquid glass over a bed of molten tin. When cooled, the glass has very flat surfaces and little distortion. This type of glass is used for most residential glazing applications with the general exception of patio-type doors, which use tempered glass.
- Tempered glass is produced by reheating and quickly cooling float glass. It is up to five times stronger than regular glass of the same thickness and cannot be cut with standard glass cutters. It is used in patio and swinging doors where strength and impact resistance is important.
- The older "plate glass", made from continuous-cast glass ground and polished to a clear, flat finish on both sides, is now rarely produced in North America, having been superseded by float glass.

For commercial applications a wide variety of specialized glass types are available. Examples include reflective, heat-absorbing, patterned, laminated, wired, vitreous colored, and spandrel glass. Each is chosen to meet operating requirements or to create architectural effects.

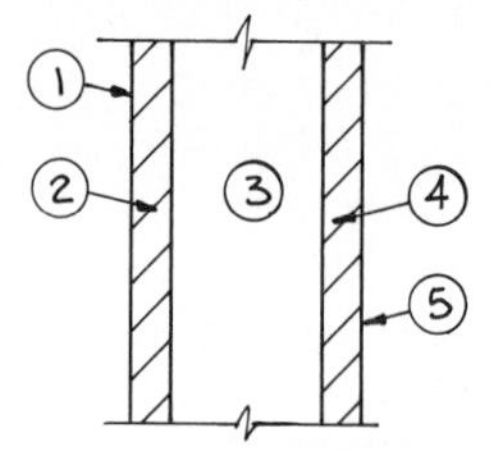

Section - Double Glazing

Element	R (RSI) Value
1. Outside air film	0.17 (0.030)
2. Glass	0.01 (0.002)
3. Air space	0.96 (0.170)
4. Glass	0.01 (0.002)
5. Inside air film	0.68 (0.120)
Total	1.83 (0.324)

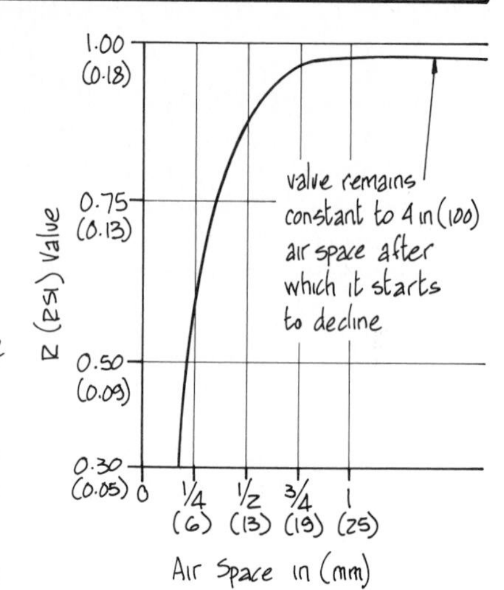

Compared to a typical residential wall structure with a minimum insulation value of R12 (RSI 2.11), windows are a major source of heat loss; single-pane windows have an R of 0.88 (0.15). To reduce heat loss, a second or third pane of glass is added to the window. The trapped air between the panes has a greater insulation value than the glass but the value will vary with the size of the space. The section drawing above shows the elements of the resistance values for sealed double-glazing, and the chart gives R values for various airspaces. Shown below and on the next page are four examples of double-glazing systems.

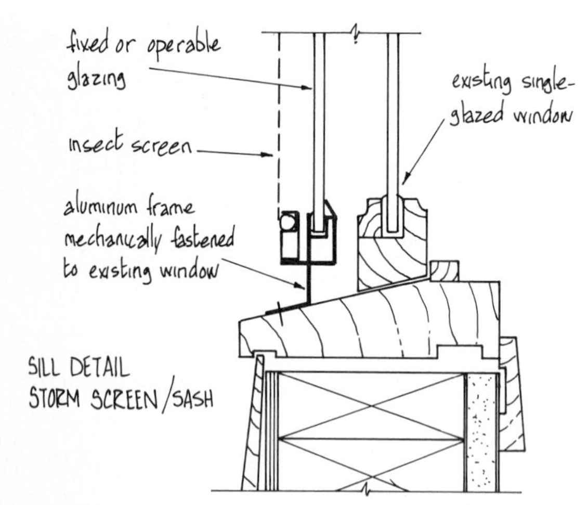

SILL DETAIL
STORM SCREEN/SASH

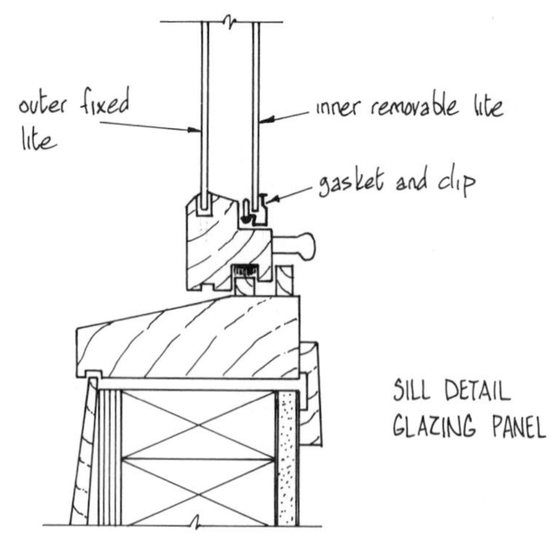

SILL DETAIL
GLAZING PANEL

STORM SASH

The addition of a separate storm sash covering the outside of a single-glazed window has been a traditional solution for window heat loss. Nowadays used almost entirely as an upgrade for existing windows, most storm sash incorporate horizontal or vertical sliding glazing and insect screens. Fixed-glazing storms must be removed to allow summer ventilation. They are also subject to condensation on the inner face if not properly vented.

DOUBLE-GLAZING PANEL

This is a removable glazing panel installed on the interior of each fixed or opening window sash. A large airspace is created and a flexible gasket on the panel provides a relatively tight seal. Insulation values of up to R 2.5 (0.43) are standard for this type of window. Removal is not required for summer ventilation, but condensation between panes can be a problem in very cold climates.

DOUBLE GLAZING / SCREENS / MUNTINS

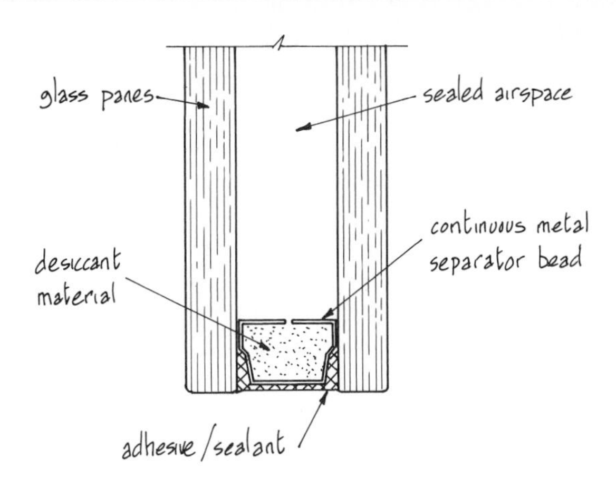

TYPICAL SPACED-EDGE SEALED UNIT

METAL SPACED, SEALED GLAZING UNITS

Also termed "insulating glass", the glazing edges are bonded with special sealants to desiccant-filled metal spacers. This is now the standard sealed glazing panel for most applications. A tough nonmetalic coating may also be applied to the glass edges to reduce installation damage. Provided that the seal remains intact, interior fogging does not occur and because of the relatively wide spacing, resistance values of up to R2.5 (RSI 0.44) are possible. Triple or even quadruple glazing is also available using this method. In addition to standard clear-glass, some manufacturers produce units with special coatings that reduce summer heat gain and maximize winter heat retention. The older type of glazing unit that has fused glass edges is now rarely produced.

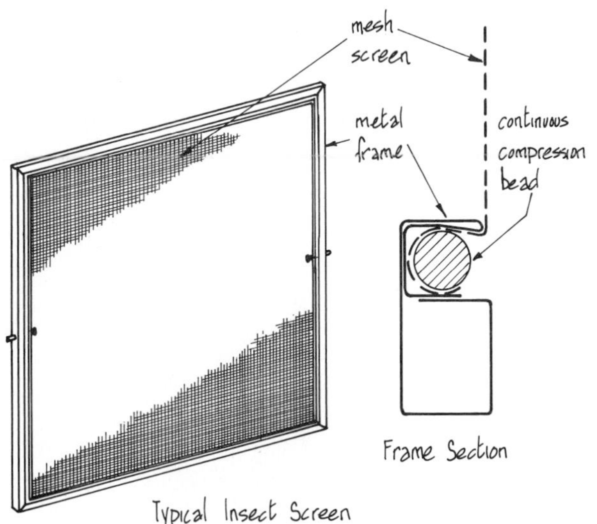

Typical Insect Screen

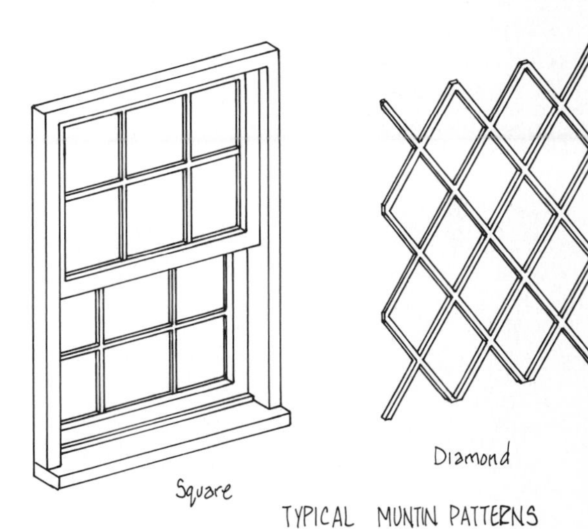

TYPICAL MUNTIN PATTERNS

INSECT SCREENS

- Mesh screens are necessary on all opening windows to keep insects out while maintaining full ventilation.
- Most screens are a mesh of aluminum, fiberglass, or plastic set in a prefinished metal frame. Better-quality meshes are made from stainless steel or bronze.
- The screen is fastened to the window in a way that allows easy removal for cleaning. The type of window determines the position of the screen. Sliders and hoppers have the screen on the outside, while casements and awnings have the screen on the inside.

MUNTINS

- Muntins are window inserts that duplicate the old-fashioned method of using many separate lites to glaze a window.
- Made of plastic or wood, the muntin snaps in and out of the window for easy cleaning. Some manufacturers permanently install muntins in the airspace of sealed double glazing.
- A variety of patterns are available, two of which are shown above.

WINDOW PLACEMENT / FRAME EXTENSIONS

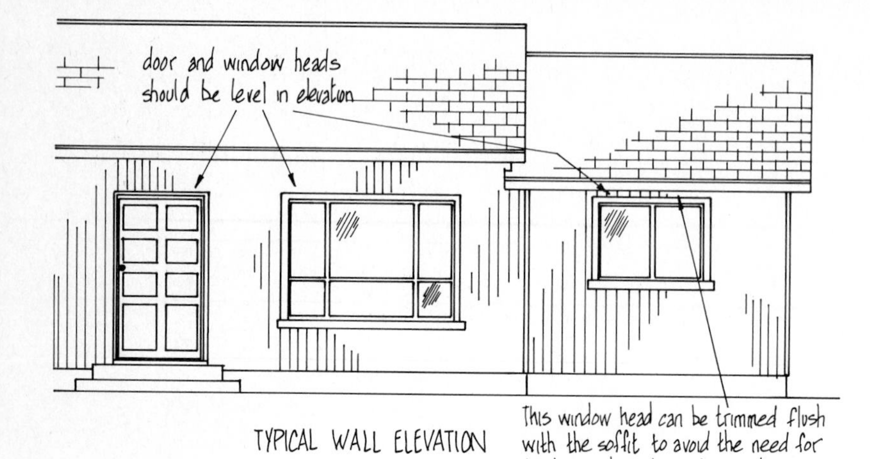

Active consideration should be given to the placement and size of windows in relation to the function of a room:

- Unless an unusual architectural effect is desired, the heads of all windows and doors should be at the same level when viewed from the outside. The wall elevation at the left shows a typical example using different-sized units.
- In addition to architectural effect there is the practical consideration of header sizes. With window and door heads in line, the same-sized header can be used for all units (see 7-6 and 7-7 for header information).

Internally, the window sills should be positioned to suit room usage, and at the right are three such examples. In each example, efforts should be made to avoid windows that have horizontal members at the average eye level. The sill position will determine the overall window depth and also assist in buying decisions when consulting manufacturers' catalogues.

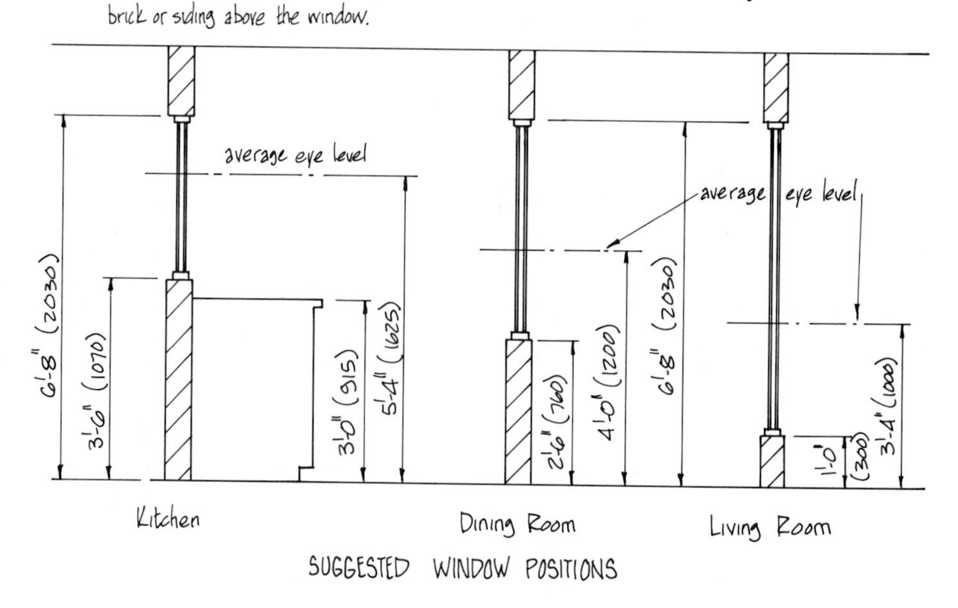

SUGGESTED WINDOW POSITIONS

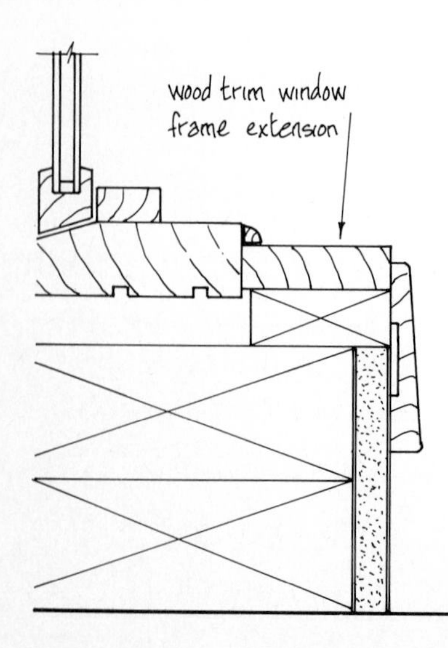

FRAME EXTENSIONS

Most windows and doors are fabricated for installation in 4-in (89)- thick stud walls. This creates a problem in upgraded energy-efficient structures with 6-in (140) or deeper walls (see 10-46 to 10-51 for details). To extend the frame to meet the interior wall line, two basic methods are used:

- A standard 4-in (89) window can be used and either wood or drywall/plaster trim fitted between window and wall line. The wood trim has a better appearance and is less subject to damage, while the drywall/plaster trim allows the wall finish to continue across to the frame.
- Some manufacturers supply frame extensions that are attached to the frame either in the factory or on the job site, while others fabricate frames that directly fit the wider walls.

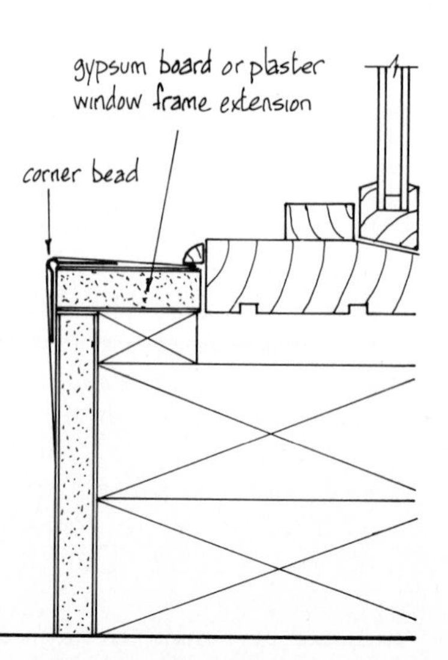

WINDOW SIZES

The window size chart below is typical of most manufacturers' catalogs. Each window type and style is sized separately and dimensions are given for rough frame and masonry openings, frame, and sometimes, glass sizes.

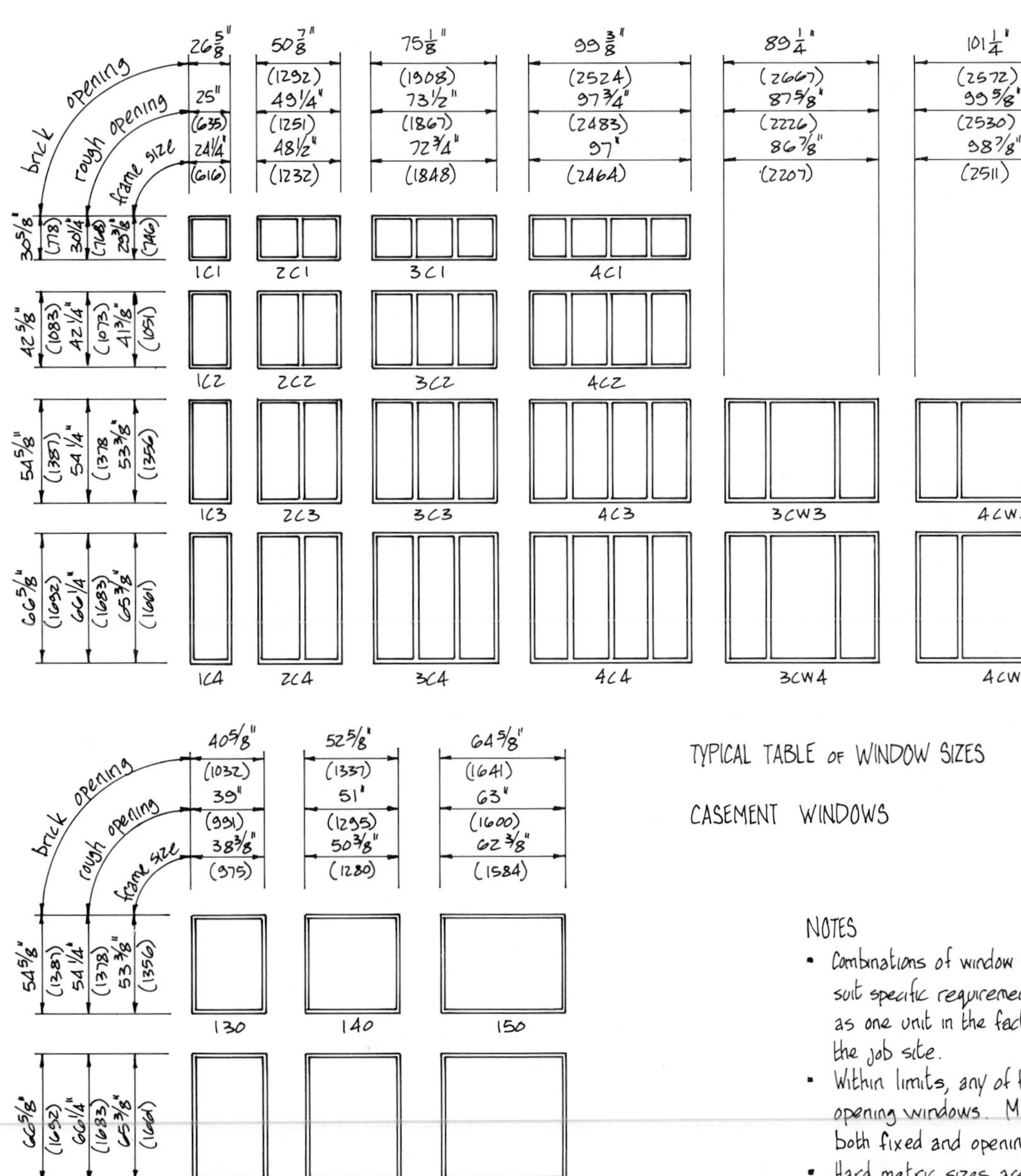

TYPICAL TABLE OF WINDOW SIZES

CASEMENT WINDOWS

All metric dimensions are soft conversions

NOTES

- Combinations of window units can be ordered to suit specific requirements. These will be assembled as one unit in the factory for direct installation at the job site.
- Within limits, any of the units can be ordered as opening windows. Most window combinations have both fixed and opening units.
- Hard metric sizes are also available from some manufacturers.
- Exact frame details vary with the manufacturer, and the designer must specify rough-opening sizes to suit. Some typical frame details are illustrated on 12-10.
- A rough opening larger than that specified by the manufacturer may be used if desired.

WINDOW INSTALLATION

Below are typical frame sections for two types of windows under different structural conditions. Manufacturers' catalogs usually show recommended installation details such as these for each specific type of window they produce.

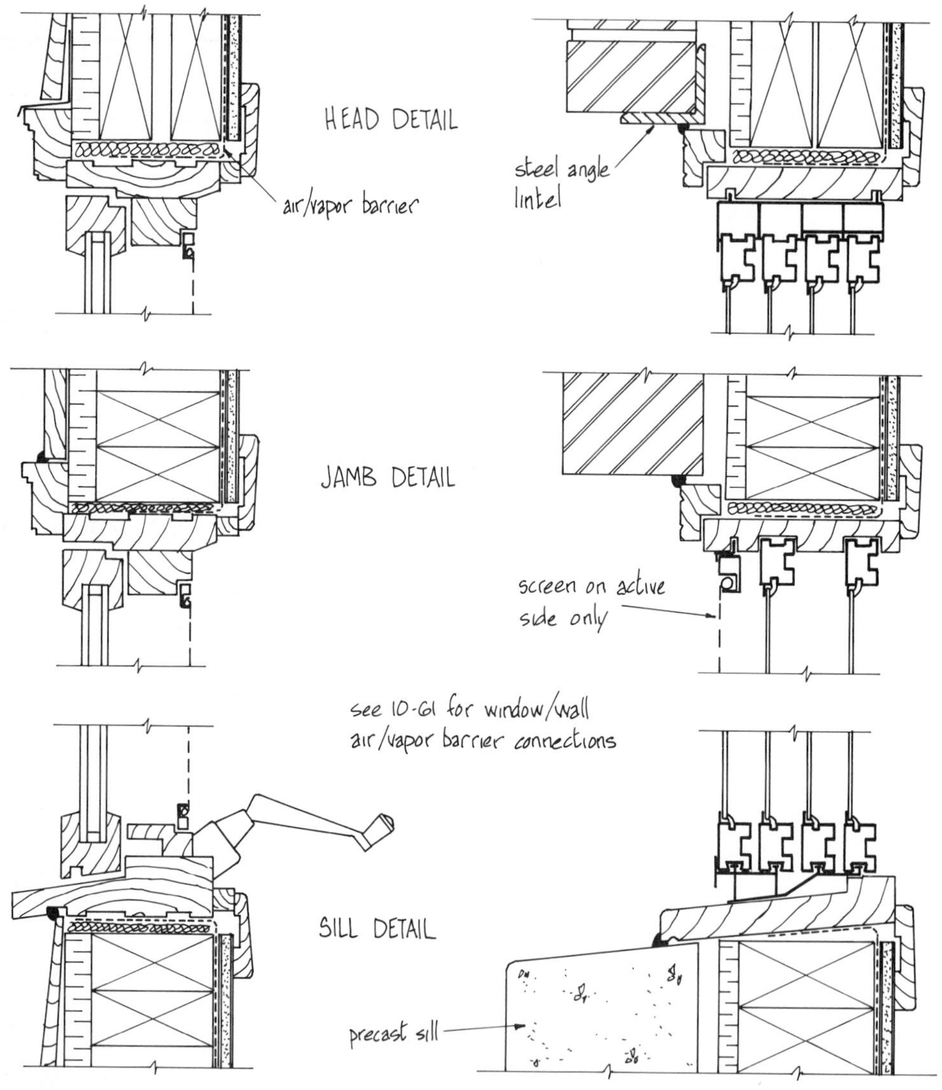

GENERAL INSTALLATION INSTRUCTIONS

Most wood, aluminum, and vinyl prefabricated windows are installed using the same basic method:

- Windows are generally designed to be installed from the outside, fitting into the rough framing or brick opening with a 1/2- to 3/4-in (12-19) clearance on all sides, although a larger rough opening may be used is desired.
- The window is leveled and plumbed using shims placed between frame and opening. The number and positions of shims depends on window size and manufacturer's instructions.
- When correctly set, the window is nailed or screwed to the structural framework through the window's sill, jamb, and head. Where possible, additional nailing is done through the exterior trim to the rough-opening members.
- If this is to be an airtight, energy-efficient installation, insulation is inserted into the cavity between the frame and rough opening and the attached window air/vapor barrier is sealed to the wall a/vb (see 10-60).

FIXED WINDOWS

While the glass or sealed glazing units for most small and medium-sized fixed windows are generally factory installed in the window frames, the larger fixed units are too heavy for this method. For ease of handling they must be separately installed in the prepared sash or frame on the job site. The procedure for wood windows is as follows:

- Both the rough opening and frame construction must be strong enough to support the weight of the window without sag or distortion. Unit seals will break if overstressed.
- Neoprene setting blocks, two per side and at quarter points, are attached to the glazing unit. The blocks isolate the unit from the frame and allow sufficient movement for expansion and contraction.
- A thick bed of glazing sealant is applied to the glazing stop and frame on all sides.
- The unit is lifted onto the sill and carefully pushed up and into position against the inner glazing stops.
- The outer glazing stop is attached and the sealant again applied to ensure a filled joint.

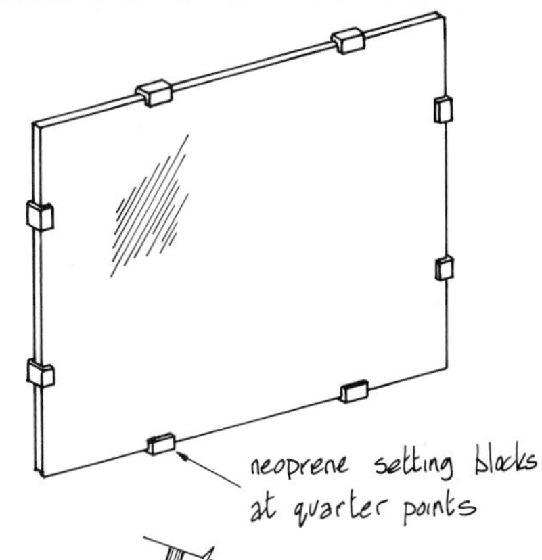

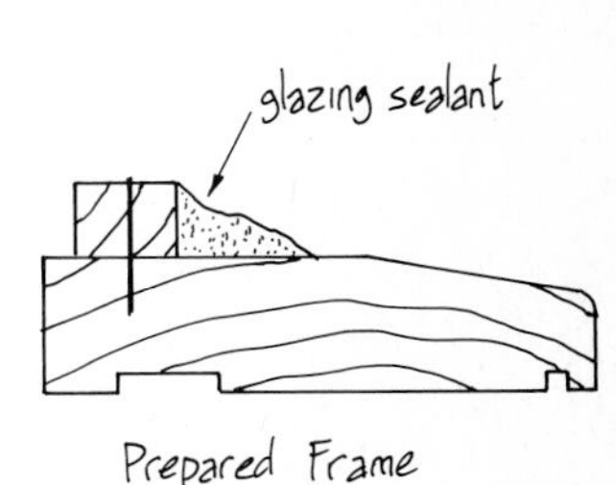

Prepared Frame

neoprene setting blocks

Glazing Unit Lifted Into Position in Bed of Glazing Sealant

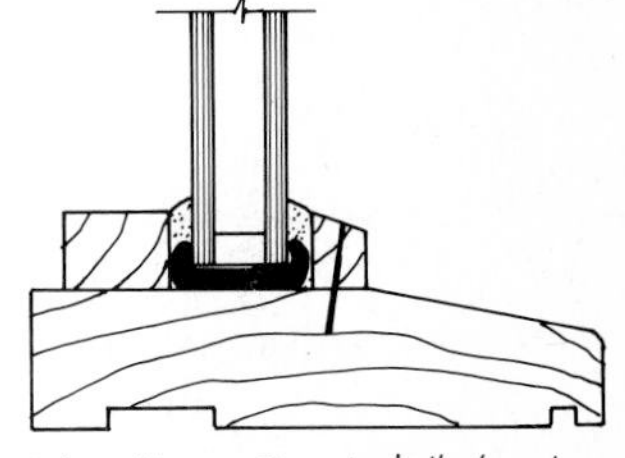
Outer Glazing Stop Installed and Excess Sealant Removed

GLASS/FRAME CLEARANCES

To allow for expansion and contraction of the glazing unit and movement of the frame, there should be no direct contact between glass and frame. Suggested minimum clearances are shown at the right

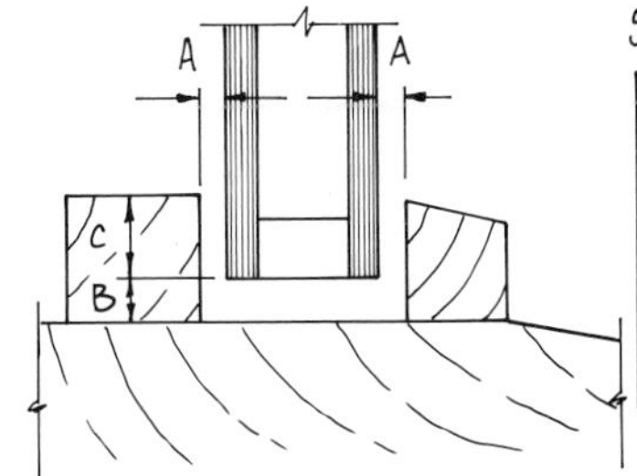

SUGGESTED GLASS UNIT CLEARANCES

GLASS SIZE	A	B	C
1/2" to 3/4" (13 to 19)	1/8" (3)	1/8" (3)	1/2" (13)
7/8" and above (22 -)	3/16" (5)	1/4" (6)	1/2" (13)

OTHER APPLICATIONS

There are many methods of supporting and containing glazing in window frames. Examples at the right include clip-supported double glazing in a metal frame, continuous-bead glazing, and structural gaskets.

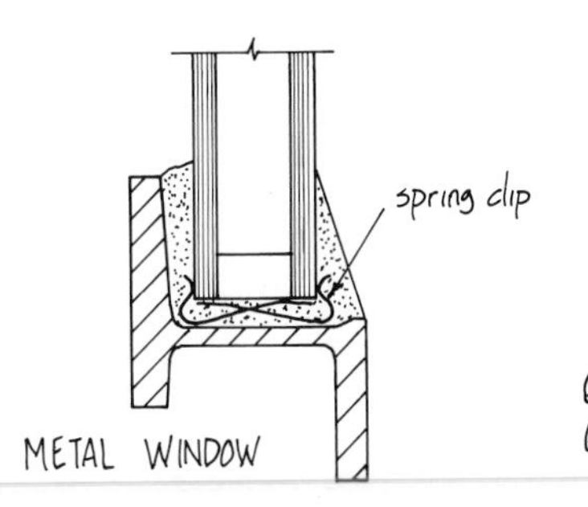

METAL WINDOW

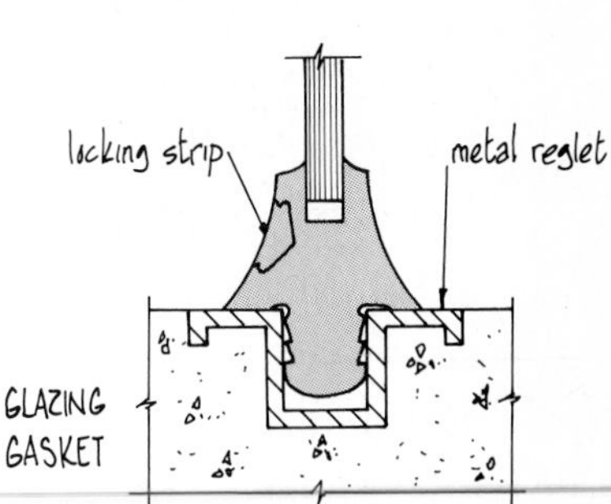

GLAZING GASKET

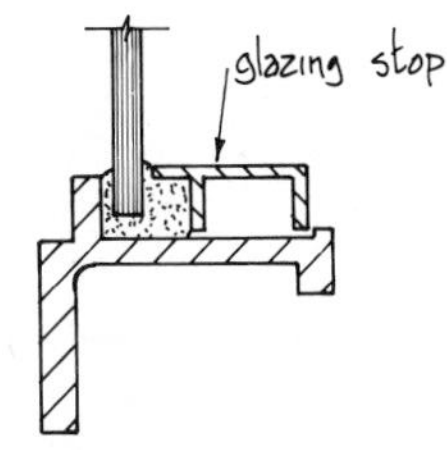

METAL WINDOW

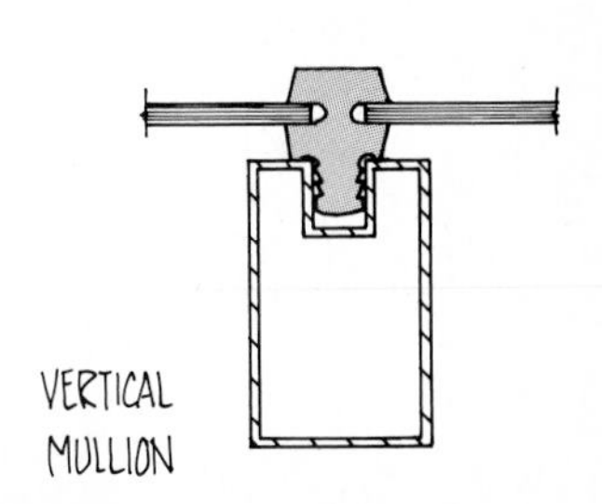
VERTICAL MULLION

SKYLIGHTS

Skylights are ideal for bringing natural light to interior spaces while creating interesting architectural effects.
A well-placed skylight will provide up to five times the light of a similar-sized wall window and ensure excellent privacy as well.

There are two basic types of skylight:

- A low-curb self-flashing frame for use on sloped roofs. This type is installed with the roof shingles overlapping the frame.
- A frame designed to sit on a high, separately flashed curb for use on flat roofs.

The glazing is also generally of two types:

- Flat glass sheets (almost always tempered safety glass).
- Acrylic sheet molded into a variety of shapes (see below right).
- Both glass and acrylic types are often double glazed, especially in cold regions, and are available in clear, tinted, or translucent finishes.

Frame materials are similar to those for standard windows and are subject to the same condensation problems in cold regions. Some units have a condensate drain built into the frame system.

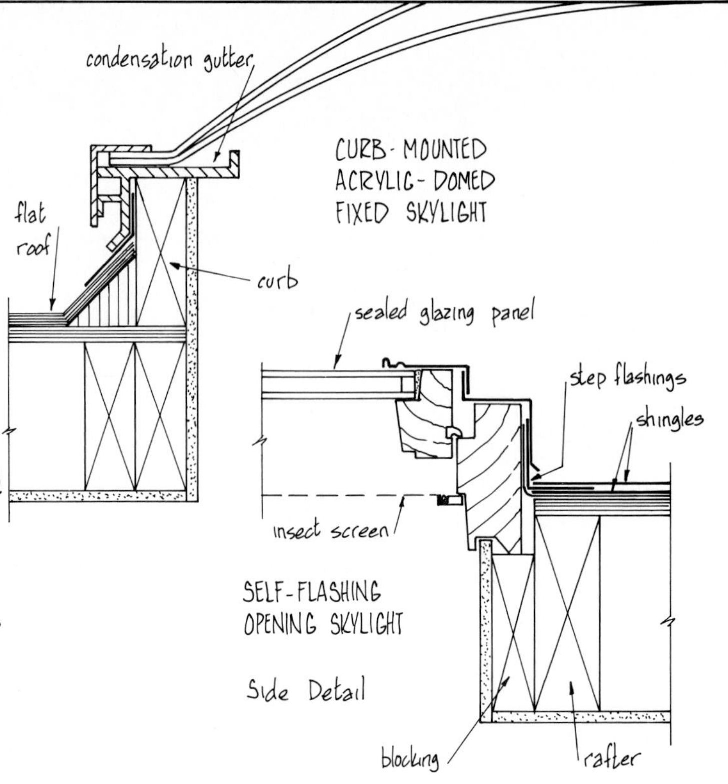

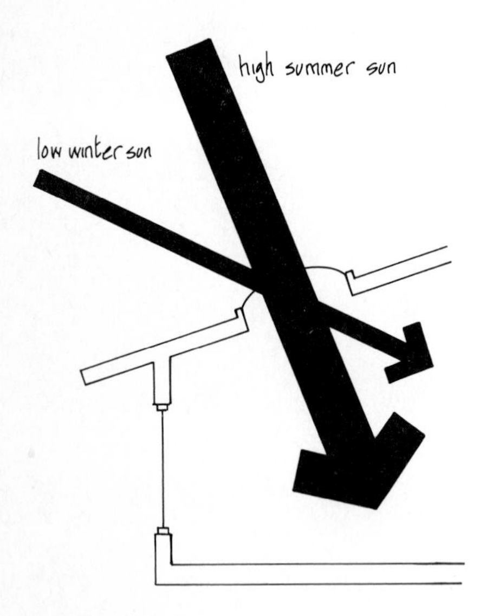

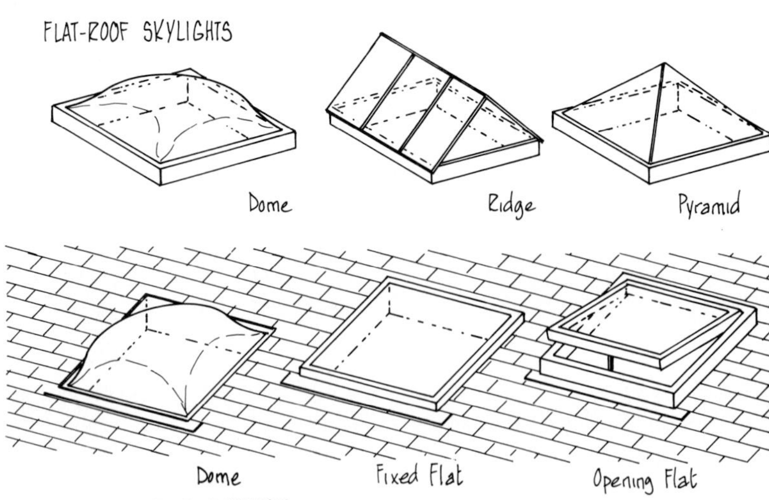

SOLAR ASPECTS

Although skylights are often advocated as collectors for passive solar heating, they tend to function in exactly the opposite way. The illustration above shows the effects of the low roof angle on solar impact. Note that the low winter sun is excluded while high summer sun is unimpeded – the opposite of the desired effect. For more information see 3-16.

Although skylights are a source of heat loss, are subject to condensation problems, and are not good winter solar collectors, they do provide very usefull amounts of light. Ignoring designs that create deliberate architectural effects, skylights should therefore be used in limited sizes, especially in cold climates. In hot climates some form of shading may be necessary to avoid overheating internal spaces.

SKYLIGHTS

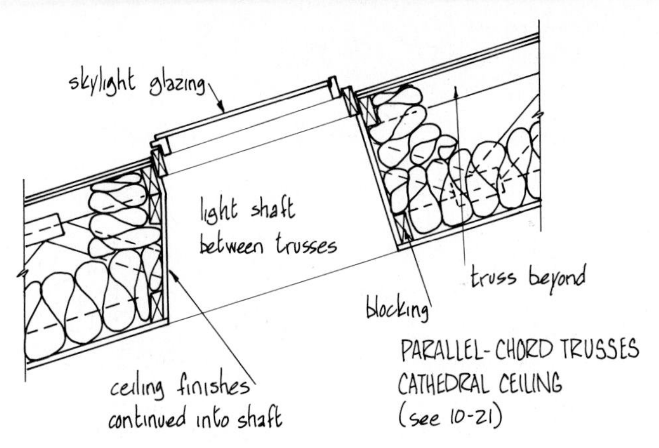

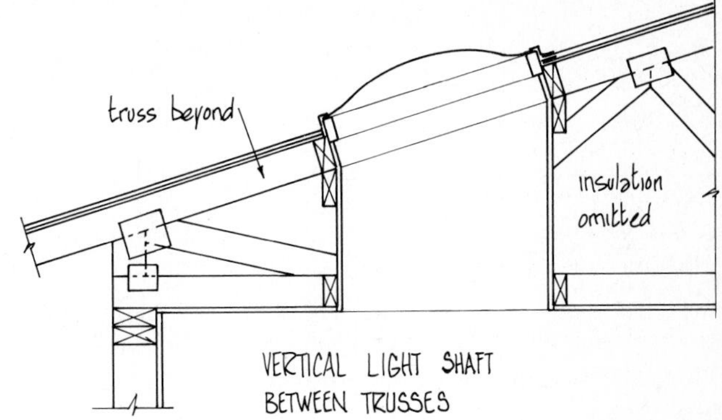

For most roof types, a light shaft will be required to channel illumination to the room below. This can be constructed in several ways, as shown in the three illustrations. To maintain energy efficiency the shaft must be treated as part of the ceiling; fully insulate up to roof sheathing level and extend the air/vapor barrier to the skylight frame. Equip the skylight with an a/vb skirt as for standard windows (see 10-60) and seal to the shaft a/vb.

Roof framing for skylights is similar to that required for other roof openings, e.g., chimneys (see 8-25). Rafters may be cut and headers inserted as shown. Trusses must not be cut, however, and if the skylight is to be wider than one truss space, special box trusses must be ordered to provide structural framing (see 8-13).

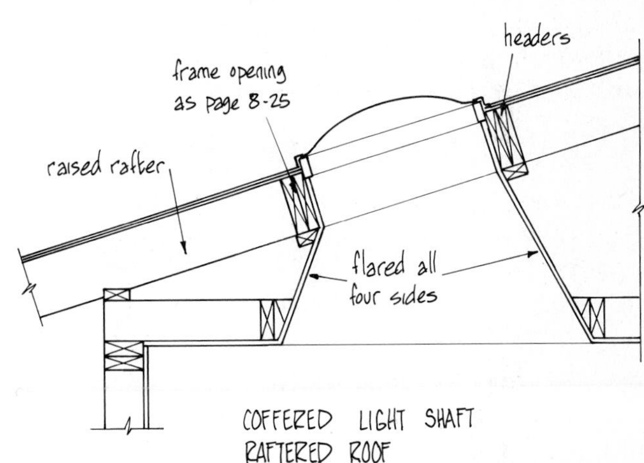

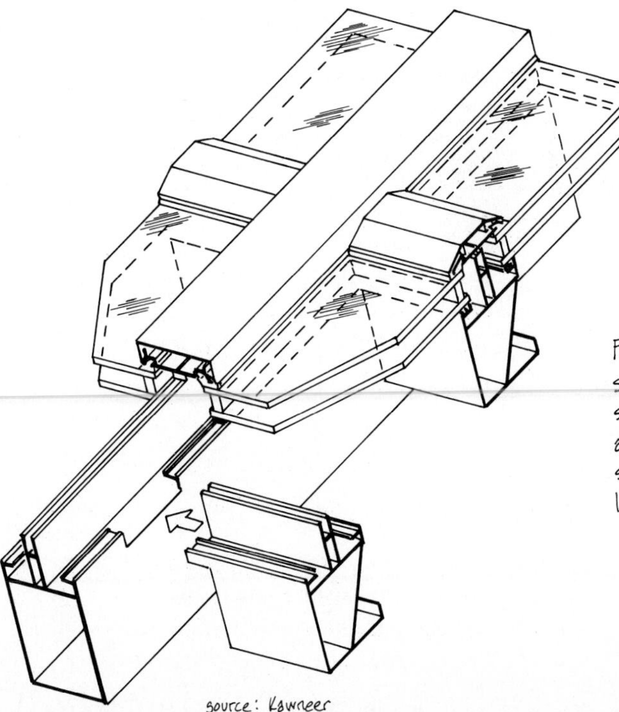

For roof areas where more extensive glazing is desired, a structural glazed roof system is necessary. Most such systems employ an aluminum frame and tempered glass arrangement similar to that illustrated at the left. These systems must be engineered to meet slope, span, and loading requirements.

DOORS - AS MANUFACTURED

Doors for interior and exterior use are fabricated in a vast variety of styles, materials, and sizes. Illustrated on this page are the major groups of physical door types used for most residential and commercial buildings.

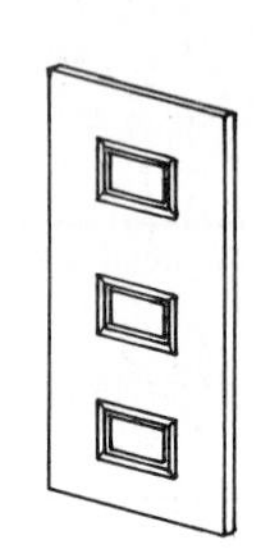

PANEL DOORS

- Styles and rails form the outer framework (see 12-16), with insert panels of solid stock, plywood, or composition material.
- A wide variety of visual effects can be created with raised, recessed, carved, and plain panels and by varing the number, size, and position of panels.
- This type is used extensively for entrance doors.

FLUSH DOORS

- Made by covering both sides of a wood frame with thin layers of sheet material. The core can be either hollow or filled (see 12-16).
- Insulated steel doors are of this type, although most have stamped faces designed to look like a panel door.
- Wood hollow-core flush doors are commonly used for interior residential doors.

GLASSED

- Sash doors are basically panel doors with one or more glazed panels.
- French doors are generally swinging, fully glazed doors that function equally as a window. Used as an exterior door almost entirely in cold climates.
- Sliding glass (patio) doors are generally heavily built with tempered double or triple glazing. See 12-4 for more information.

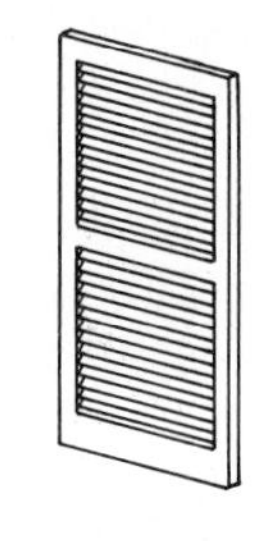

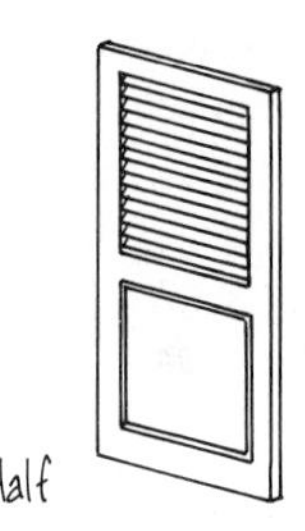

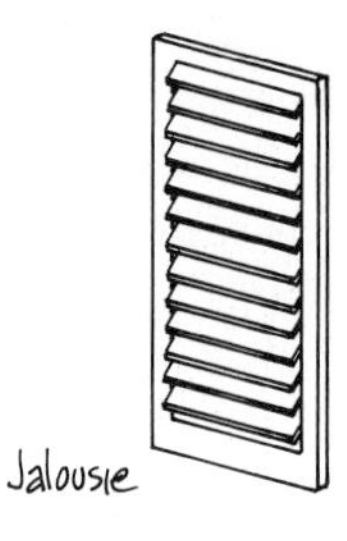

SHUTTER

- Individual wood strips or molded composition material set in a frame. Doors are either fully or half shuttered and may be operable or fixed. Used extensively for closets and cupboards.
- Jalousie doors have individual glass strips set in a frame, all opened or closed with a mechanical operator. Impossible to seal or weatherproof, they are used only in warm climates (see 12-4).

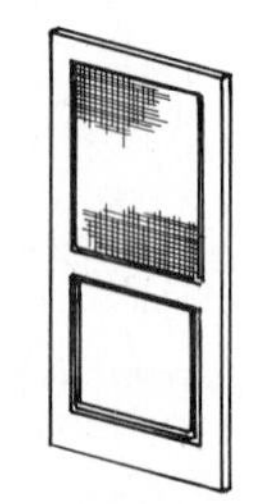

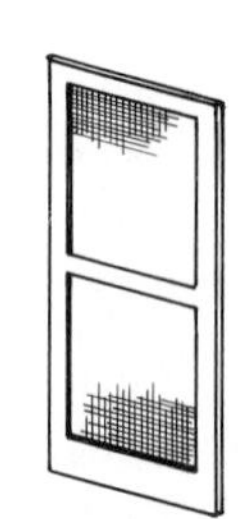

SCREEN

- Commonly seen in storm/screen combination with fixed insect screening and sliding glass storm panels for weather protection, they are installed on the outside of wood exterior doors (not required for steel doors).
- Some types provide only insect protection.

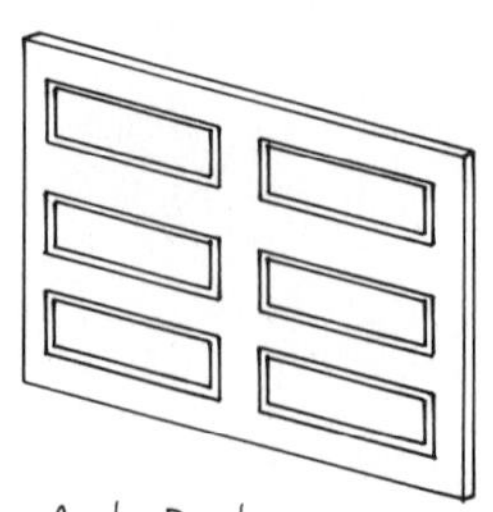

GARAGE

- There are two basic types: single-panel lift-ups and multipanel hinged roll-ups (see the next page).
- Materials include steel, wood, aluminum, and fiberglass. Powered-opening garage doors are common.

DOORS - AS INSTALLED

SWINGING DOORS

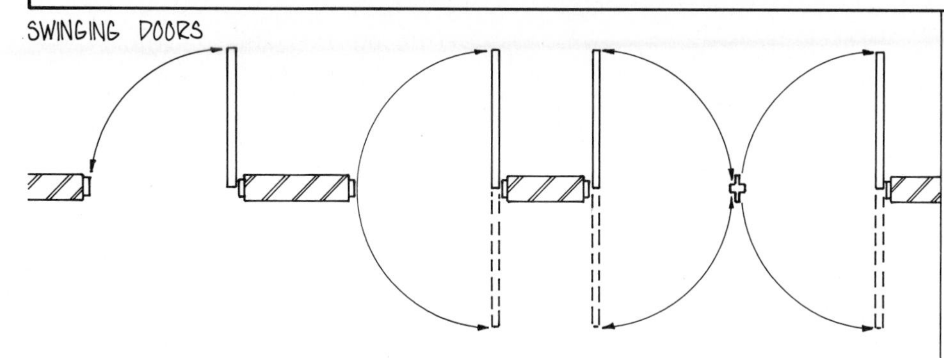

Single-acting with door swinging to one side only. Door frame requres doorstops on jamb and head.

Double-acting with door swinging on either side of wall. No doorstops on frame.

Double doors with a single- or double-acting swing. Optional center mullion for locking purposes on a single swing.

Illustrated here are the five major door operation modes common to most applications. Door operation is chosen to complement the type, position, and access clearances of the door itself. For example, a sliding pocket door, although structurally more complex, may be more practical if the sweep of a swinging door would block access or interfere with furniture or fixtures in a room.

SLIDING DOORS

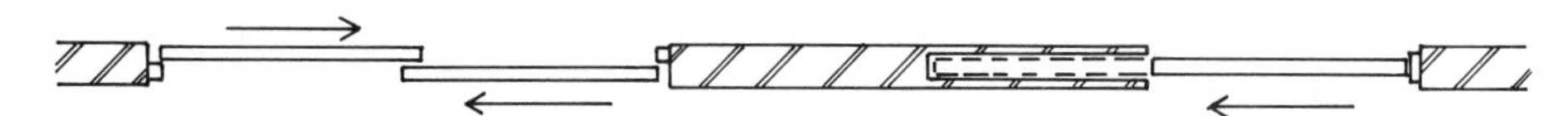

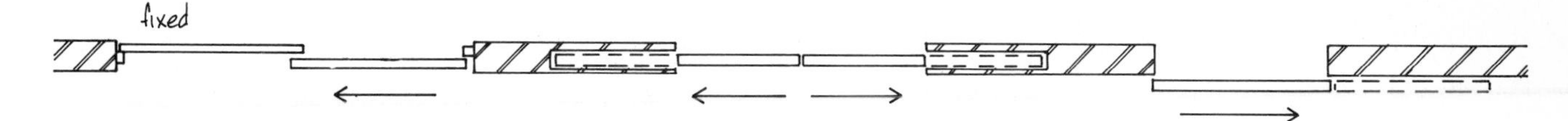

Bipass sliders with doors sliding between frame jambs. Both doors can slide (closet type) or one door can be fixed (patio type).

Pocket slider with a single door sliding into one wall or two opposing sliders sliding into both walls. Track hardware must be built into the wall during the rough-framing stage, making track maintenance or replacement difficult.

Surface slider with door mounted on outer wall face. Requires track hardware mounted on wall face and floor.

FOLDING DOORS

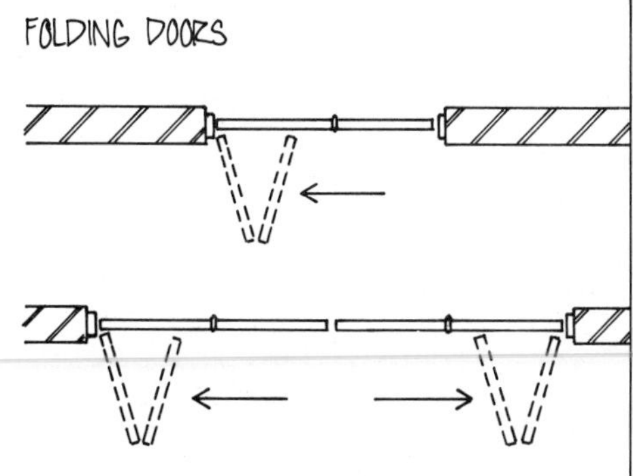

Bifold doors have two or more hinged doors that fold back against the frame jambs but project into the room. Installed in single or double sets depending on opening width.

GARAGE DOORS

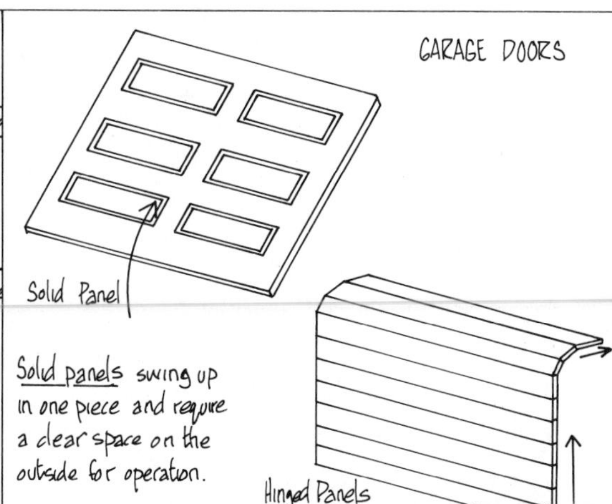

Solid panels swing up in one piece and require a clear space on the outside for operation.

Hinged panels roll up along the door track and do not require outside clearance.

DUTCH DOORS

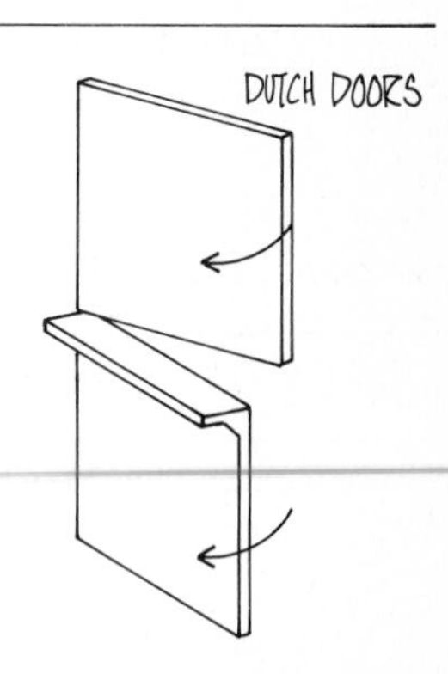

Dutch doors have independently opening top and bottom halves. A locking mechanism also allows both halves to be opened together. The bottom half often has an attached shelf or counter.

12-16 DOOR MATERIALS

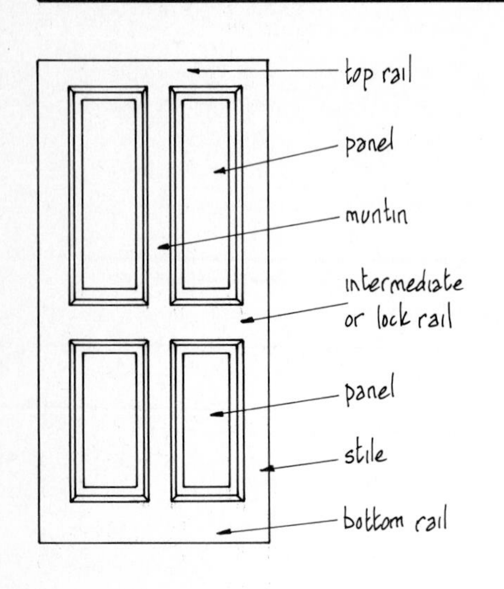

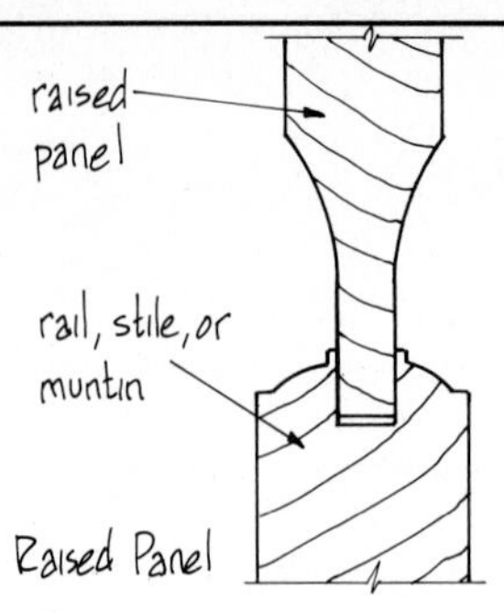

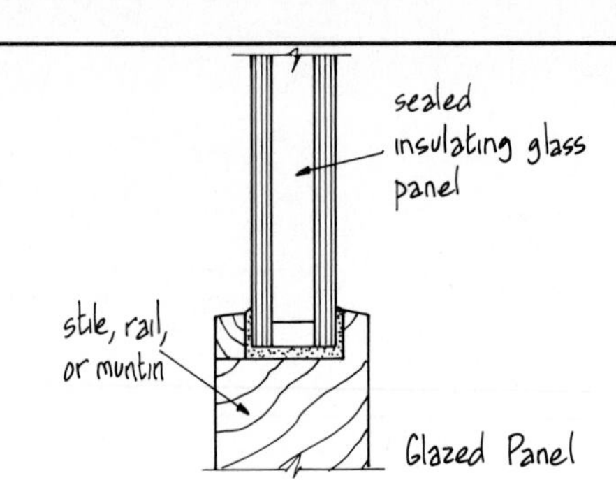

PANEL DOORS

Wood panel doors are assembled using the frame members and infill panels or glazing as shown. Frame materials include pine, mahogany, redwood, and other hardwoods. Panels are solid stock, ply, or veneered composition. In better quality doors the glazing is insulating glass. A wide range of styles and visual effects can be created.

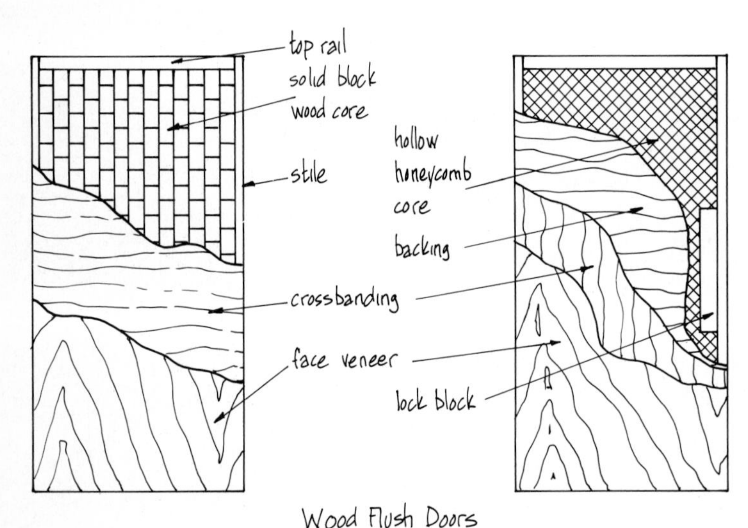

Wood Flush Doors

FLUSH DOORS - WOOD

Flush doors have a solid lumber frame that is covered on both sides with a thin sheet of plywood, composition board, or plastic laminate. There are two general types:

- Solid core, with some type of solid structural material filling the core. The example at the left shows a wood block core, but many other materials are also used. The solid core increases structural strength and improves soundproofing.
- Hollow core, with some type of open spacer material to maintain face sheet separation. The example shown has a preformed honeycomb core, but other materials are also used. This is the standard interior door type for most residential applications.

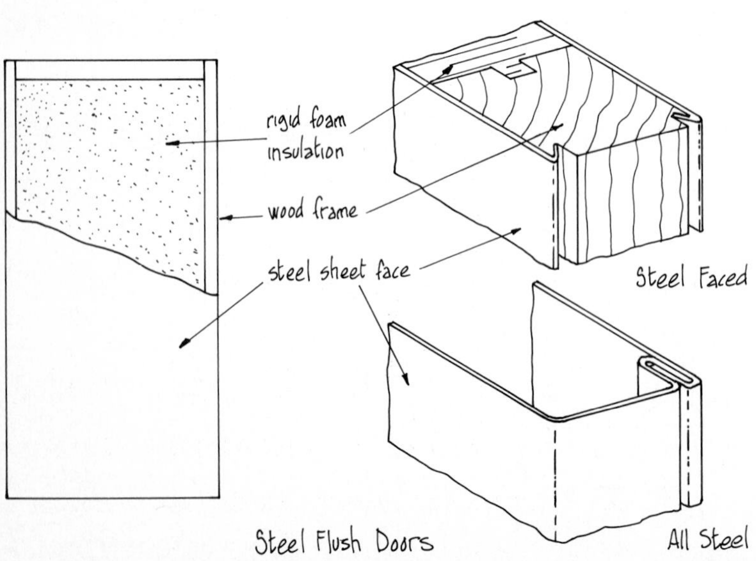

Steel Flush Doors

FLUSH DOORS - STEEL

There are two basic types as illustrated at the left:

- Wood framed with attached sheet steel faces, this type is commonly used for residences. The wood frame provides a thermal break between the faces, and the core is filled with foamed insulation to give an R 15 (RSI 2.6) or higher thermal rating. Usually combined with a wood frame and magnetic weather stripping for energy efficiency (see 12-21). Embossed patterns or attached trim usually applied to face sheets.
- All-steel doors, for commercial and industrial use. Welded construction with reinforcing at hinges and locks and with spacer cores for regular or fire-rated duty.

FRAME MATERIALS - WOOD

INTERIOR DOORS

- Frames for interior doors consist of two jambs and a head with casing trim both sides, all installed to both support the door and cover the partition edges. A sill is not normally required, and door stops are attached after the door is installed. One-inch (19) stock is standard.
- Frame width is cut to equal finished wall thickness. Optional kerfs on the back face of the frame reduce warping and cupping. Jamb edges are beveled slightly to allow casing to fit tightly.
- When frame cannot be precut, an adjustable frame is used. Door stops cover the frame joint.
- Frames are often factory machined and fitted for easy site assembly.

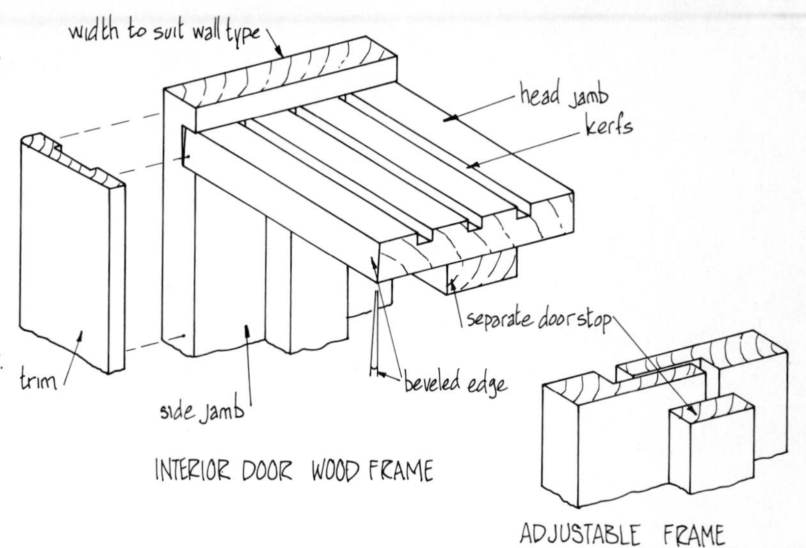

INTERIOR DOOR WOOD FRAME

ADJUSTABLE FRAME

EXTERIOR DOOR FRAMES

- Exterior doors are similar to windows in that they have a head, side jambs, and a sill. Stock size is usually 1¼ in (26) or greater to support the heavier doors and screens.
- A single or double rabbet ½ in (12) deep is cut for the doorstop, although some have separate stops.
- Frame widths should match wall thickness, but frame extensions for deep walls can be factory applied.
- A variety of sill types are available in several different materials. See 12-21 for additional types.
- Many exterior frames are shipped to the site assembled and with doors prehung for accurate and fast installation.
- Air/vapor barrier connections are similar to those for windows (see 10-61).

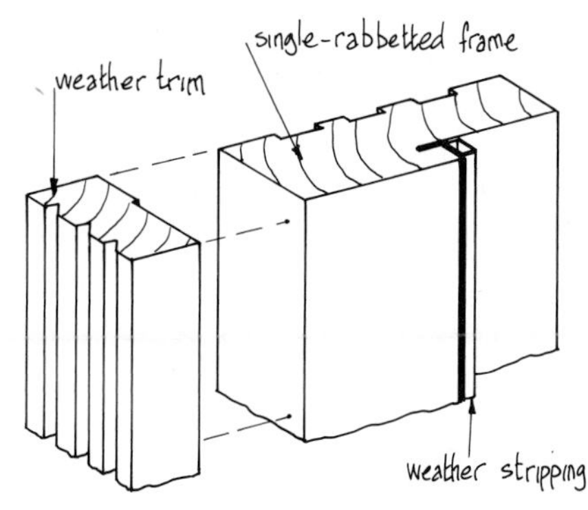

EXTERIOR DOOR FRAME

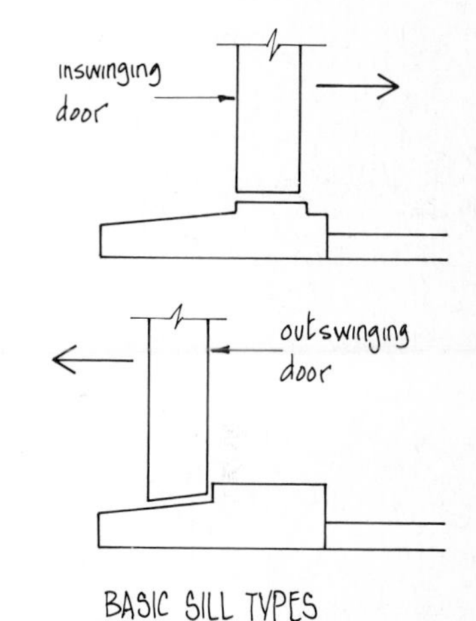

BASIC SILL TYPES

SIDE LIGHTS

- Adding glazed side lights to an exterior door is a popular option. The fixed light is treated as an attached window with matching frame and trim and can be added on one or both sides of a door. Manufacturers offer a variety of glass types in clear, translucent, or tinted colors.

FRAME MATERIALS - METAL

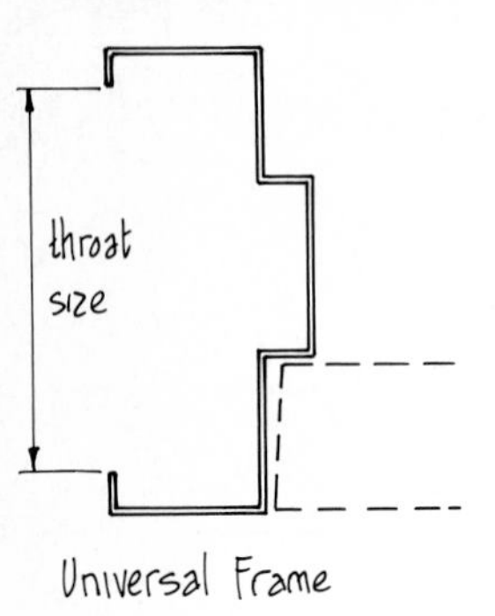

Universal Frame

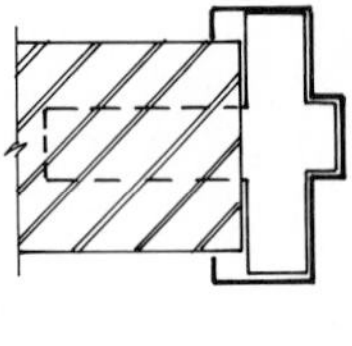

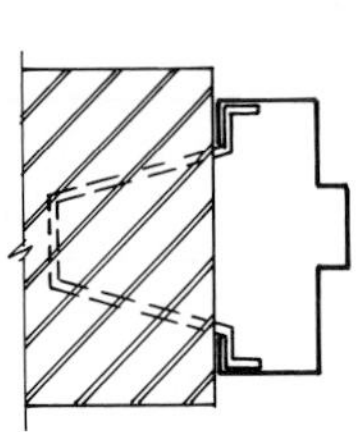

Masonry Walls

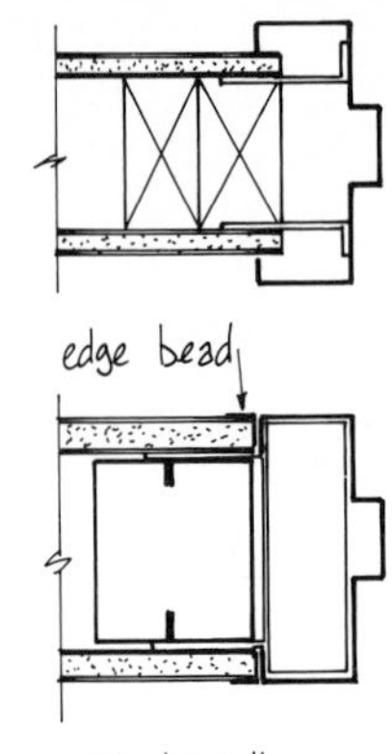

Stud Walls

HOLLOW METAL FRAMES

- This type of frame is standard for many commercial and industrial applications and is rolled-formed from wipe-coat galvanized sheet steel. Any type of door can be hung from these frames.
- The universal profile shown at the left is applicable to almost any type of structural condition. Four typical applications are illustrated.
- The special drywall profile is designed to fit over a finished wall without damaging the drywall.

Three options are generally available for frame corners:

- "Knocked-down", where the mitered corners are fitted with reinforcing clips and tabs and assembled on the site.
- Factory assembled and tack welded corners where looks are not important.
- Factory arc-welded miters with the welds ground smooth.

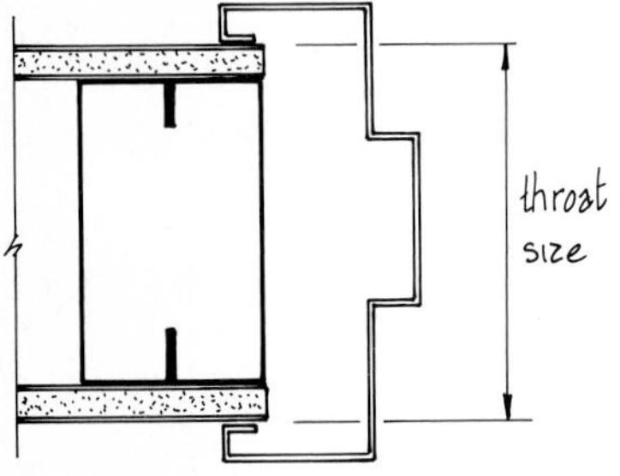

Drywall Frame

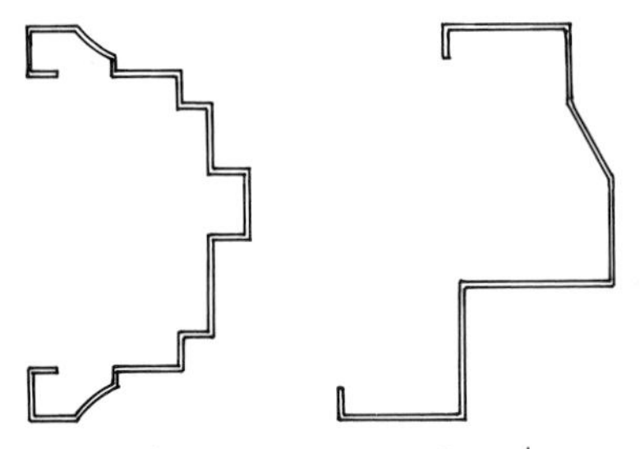

Special Frames – Two Examples
a wide variety of custom frame shapes are available from some manufacturers

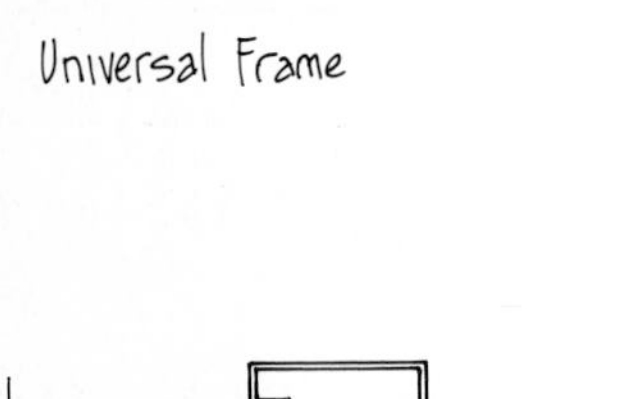

Masonry Wall

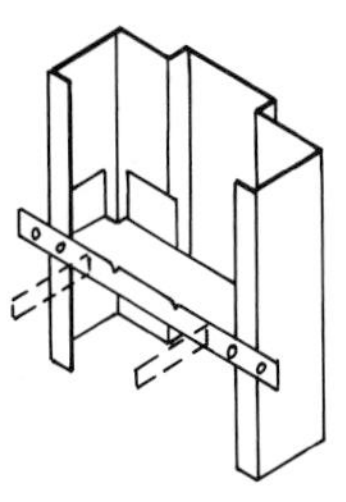

Stud Wall

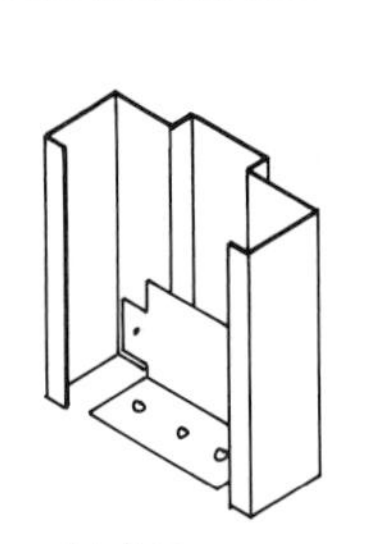

Wall Base

TYPICAL FRAME ANCHORS

Wall anchor options:

- The three illustrations at the left are typical of the many specialized wall anchors available from manufacturers. Each is designed to be attached to or built into any combination of wall construction.

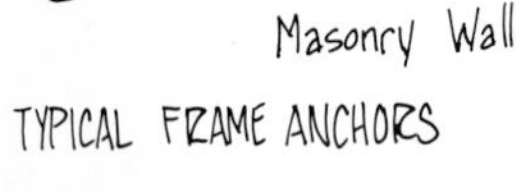

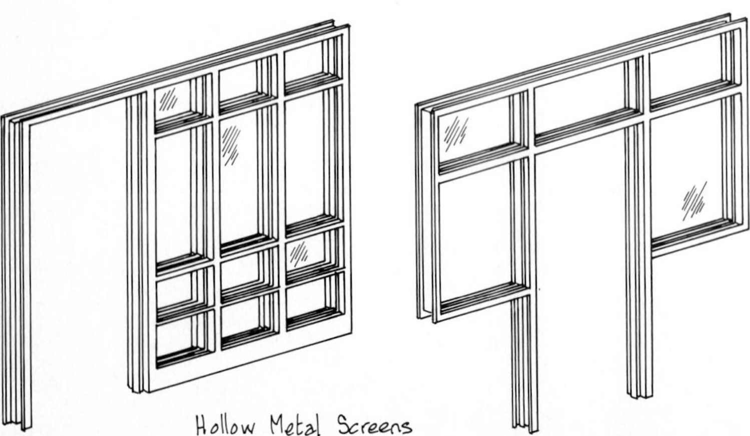

Hollow Metal Screens

HOLLOW METAL SCREENS

As with wood frames, a variety of door and window combinations are available in hollow metal frames. Two typical examples are shown at the left. Since this type of frame is assembled using the "stick method" where standard frame profiles are cut to length and welded, manufacturers can produce frame sets in virtually any size or design.

FRAME MATERIALS - METAL

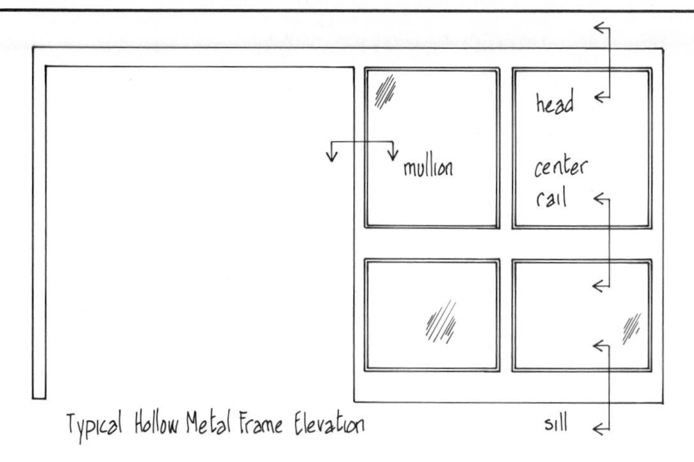

Typical Hollow Metal Frame Elevation

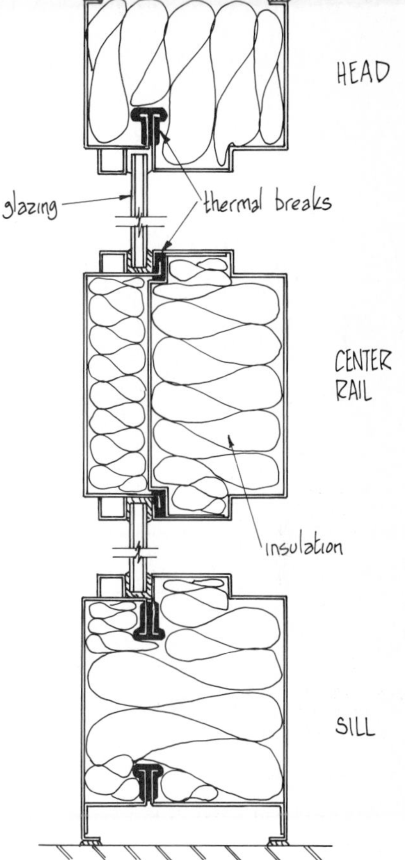

INSULATED HOLLOW METAL FRAMES

- For exterior use in cold regions hollow metal frames must be equipped with a thermal break and be filled with insulation to reduce condensation on room interior frame surfaces.
- Vinyl or rubber inserts generally provide the thermal break, as shown in the sections at the right.

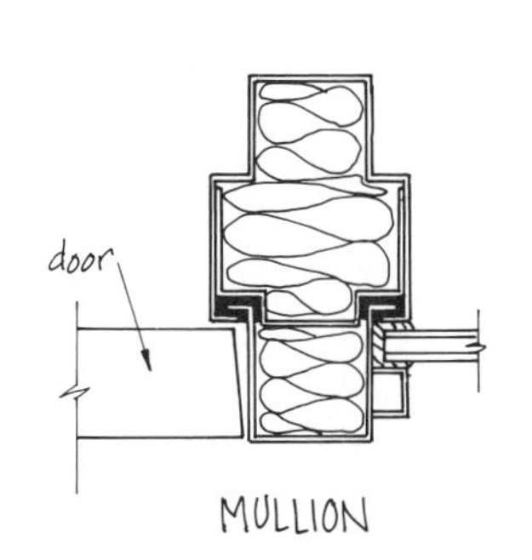

MULLION

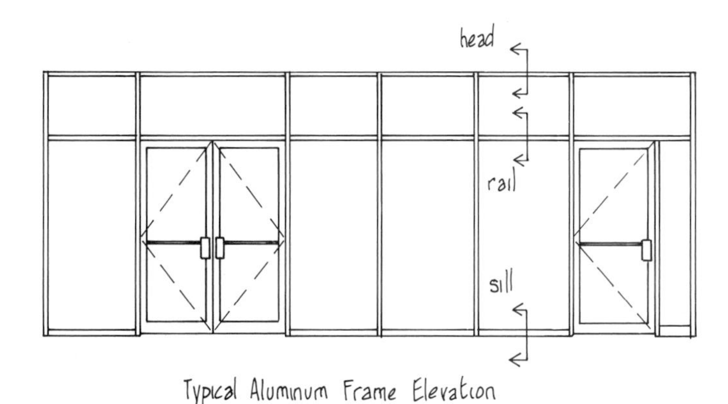

Typical Aluminum Frame Elevation

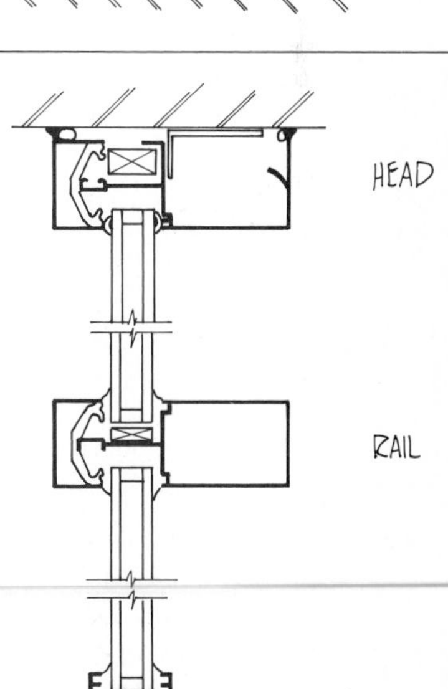

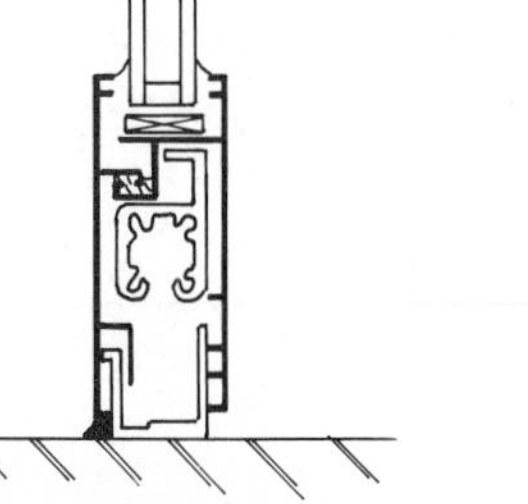

ALUMINUM FRAMES

- Aluminum frames, usually with aluminum framed doors, are used extensively for exterior and interior entrances.
- Extruded aluminum sections in a variety of shapes and styles are commonly available. Frames are usually clear or color anodized to provide weather protection.
- A typical example of an aluminum frame is shown in section at the right.

DOORS

12-20 CLEARANCES / HINGES / SWINGS

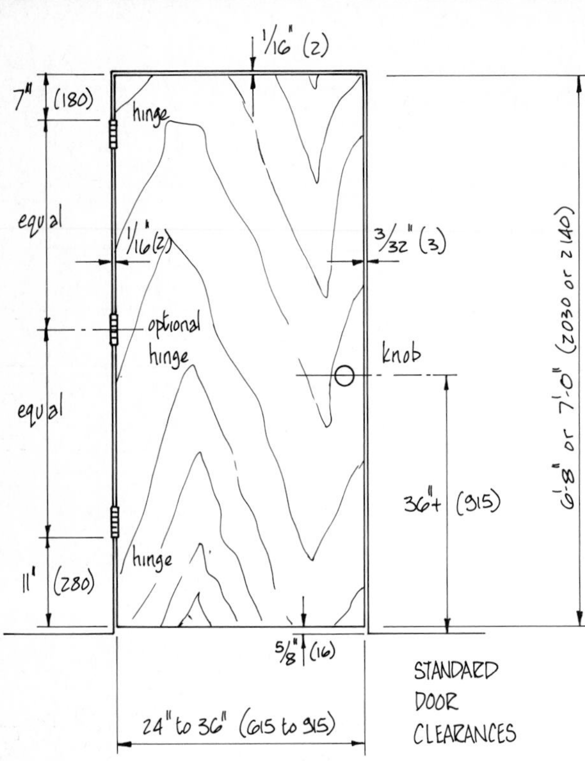

STANDARD DOOR CLEARANCES

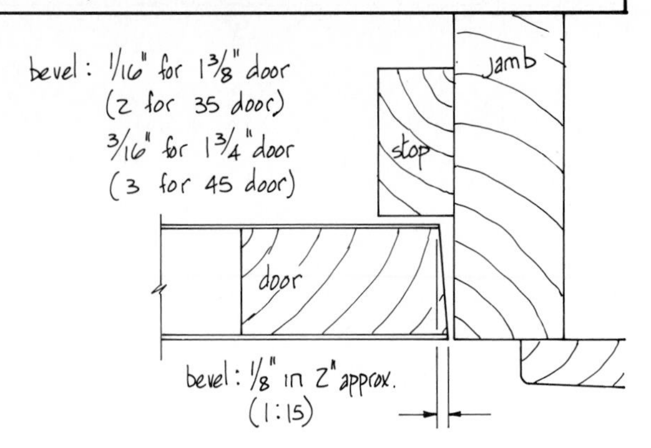

The lock side of the door is bevel-trimmed at the angle shown, to provide clearance when the door is opened.

DOOR CLEARANCES

- The front view above shows the dimensional clearances recommended for standard interior wood doors.
- Clearances for exterior doors are similar except for door bottoms, which must be adjusted to suit the type of sill used.
- The rough-opening dimensions will be about 2½ in (65) higher and wider than the door size.
- Standard door thickness is 1¾ in (45) for exterior doors and 1⅜ in (35) for interior doors. Speciality or commercial doors may vary but will be supplied with frames to suit.
- Interior doors generally have two hinges, but some better-quality interior doors and most exterior doors are supplied with the optional third hinge.

On most swinging doors the loose-pin butt hinge is used. The locking pin can be removed and the hinge leaves separated for initial installation or later door removal. The hinge leaves are cut in flush on door edge and frame jamb. There are several other types of hinges, each designed to serve a specific door movement.

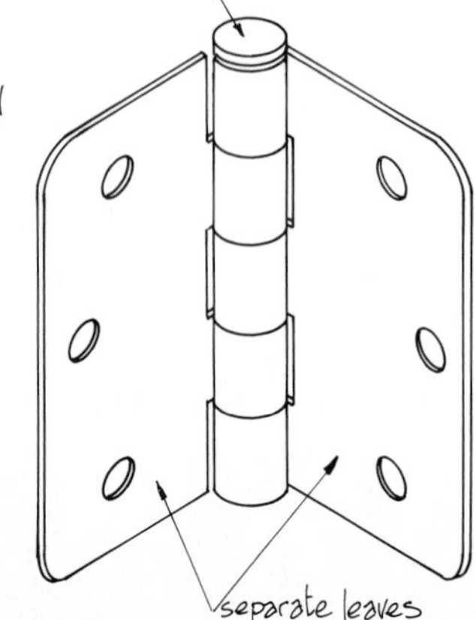

LOOSE-PIN BUTT HINGE

It is necessary to specify the "hand" of the door when ordering so that the door swings in the correct direction. The four standard door hands are as viewed from the outside of an exterior door and from the corridor side of an interior door.

SILLS AND WEATHER STRIPPING

SILLS AND WEATHER STRIPPING

The four examples of sill types and weather stripping illustrated here are typical of those available for residential and commercial use. Sills must be able to withstand foot traffic without excessive wear, and weather stripping must provide a good air seal without hindering or slowing door swing.

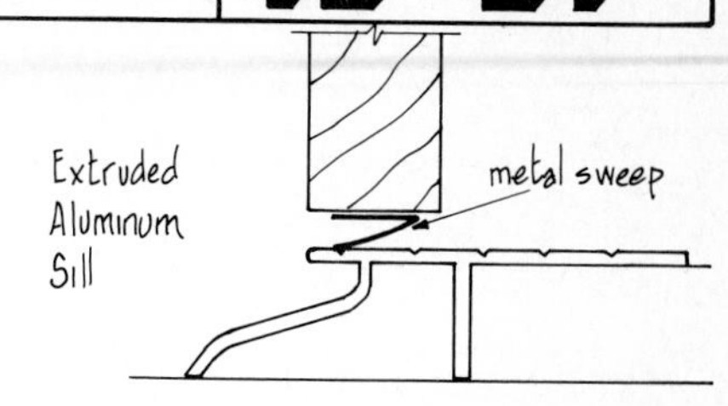

Solid Wood Sill

Plastic-Capped Wood Sill

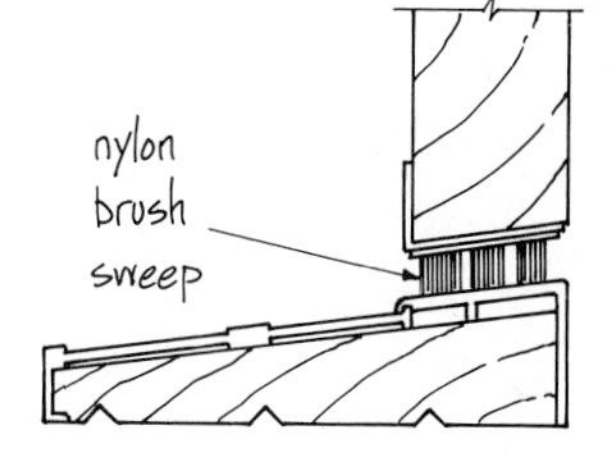

Aluminum Cased Wood Sill

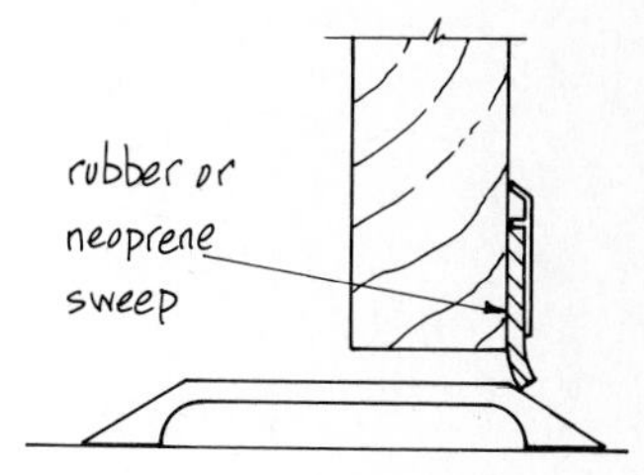

Aluminum Threshold commonly used as a junction between two floor types

TYPICAL SILL TYPES

WEATHER STRIPPING TYPES

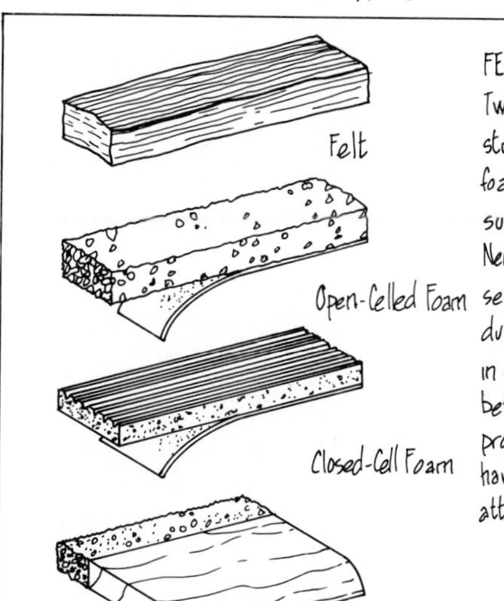

FELT AND FOAM STRIPS

Two types of weather stripping, felt and open-celled foam, are, in general, not suitable for most applications. Neither provides an airtight seal nor is particularly durable. Closed-cell strips in either foam or rubber perform better, but tend to "set" under prolonged compression. Most types have an adhesive strip for attachment.

PILE

Commonly used where extensive movement will occur, pile weather stripping provides a reasonably good seal. Material is usually polyester or polypropylene on a "T" base.

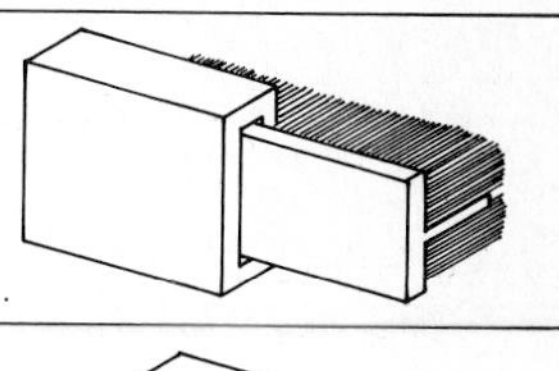

"V" TYPE

Available in many shapes and sizes in vinyl and metal. An effective weather strip under light compression. Will deform under heavy compression and tear with misuse.

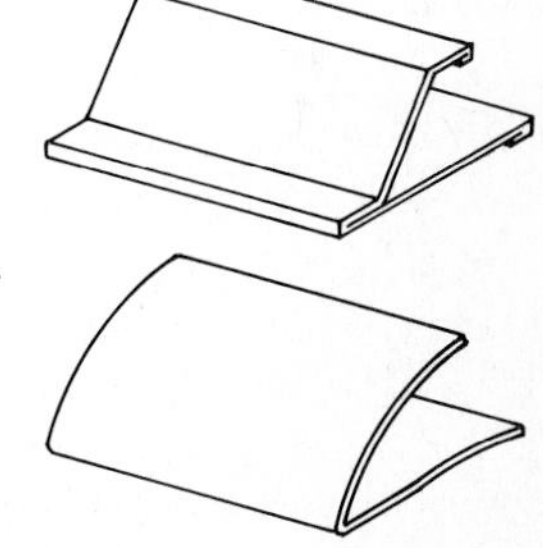

TUBULAR AND SELF ADJUSTING

Usually made from soft plastic or rubber, this type seals well under light compression and allows a reasonable amount of movement. Will also adjust to changing crack widths.

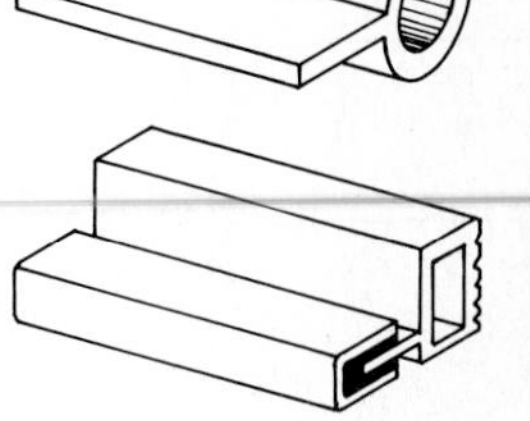

MAGNETIC

Designed for use with steel doors or windows, this type has a magnetic strip in the self adjusting head. Provides an excellent seal. A separate metal strip can be installed on wood or plastic windows or doors.

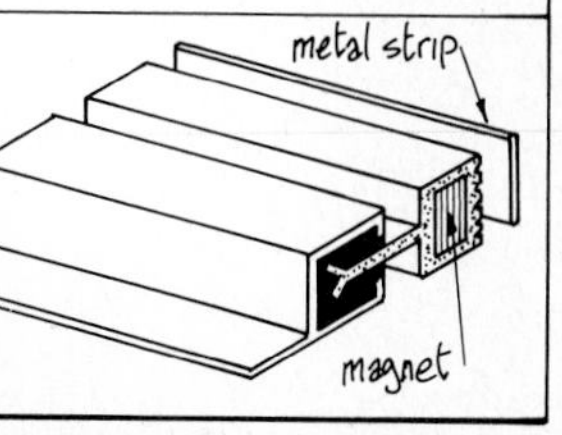

DOOR AND WINDOW WEATHER STRIPPING TYPES

- The major weather stripping types are illustrated together with their principal features. Manufacturers produce several variations of each to suit specific applications.
- The materials used must be capable of performing in all weather extremes without undue deformation or "set" while providing a good air seal and ease of operation.
- Doors and windows are normally delivered to the site with stripping in place, and little if any adjustment is required. Replacement stripping can readily be purchased in all regions.
- The choice of a particular weather stripping will be influenced mainly by the service requirements, particularly the extent of movement and amount of abrasion expected.

DOOR HARDWARE

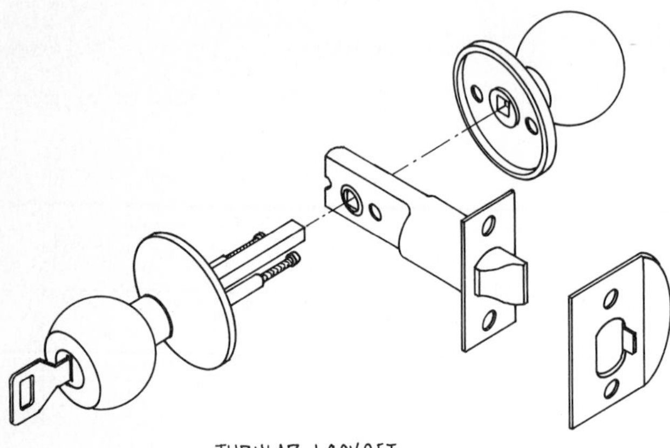

TUBULAR LOCKSET

There is a large variety of hardware available to suit each door type and its intended usage. The items illustrated on this page are a small example of those commonly specified for residential and/or commercial doors.

LOCKS

A typical example of a tubular lock is shown at the left. This is one of the most common types for interior and exterior door use and combines both door handle and lock in one unit. Other types, including unit, cylindrical, and mortise locks, are also in general use depending on the type of operation and level of security desired and may have an incorporated or separate locking mechanism. Detailed installation instructions for all units are provided by the manufacturer.

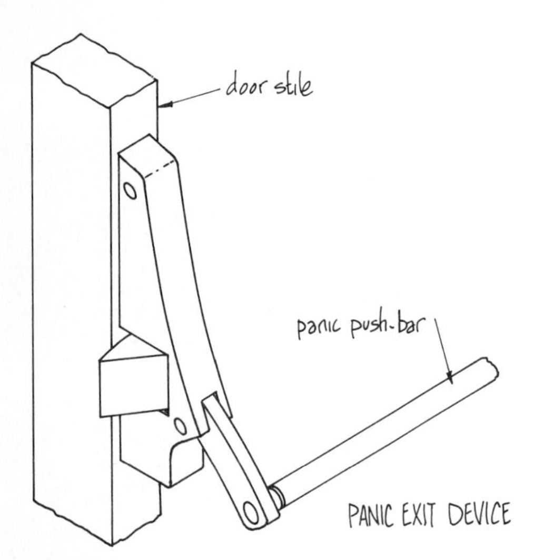

PANIC EXIT DEVICE

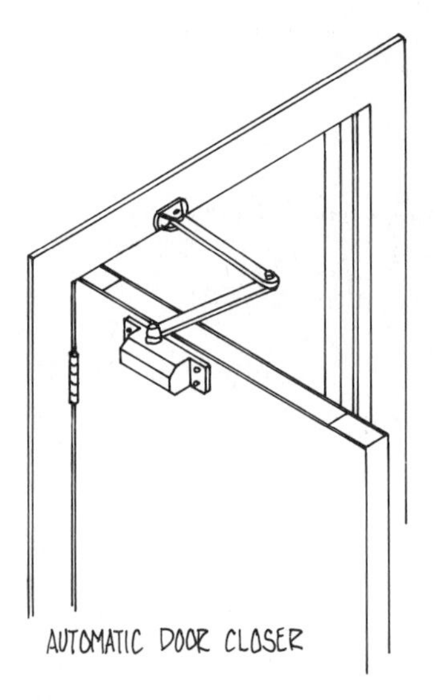

AUTOMATIC DOOR CLOSER

PANIC EXIT DEVICES

This type of hardware is designed to allow quick door opening in an emergency situation when people are likely to crowd the doorway. Simply pushing on the horizontal bar opens the door which must open outwards to a safety area.

CLOSERS

Rarely used in residences but standard in commercial and industrial buildings, these units automatically close exit or other doors after usage. Closing and latching speeds are usually adjustable and hold-open arms are available for most models.

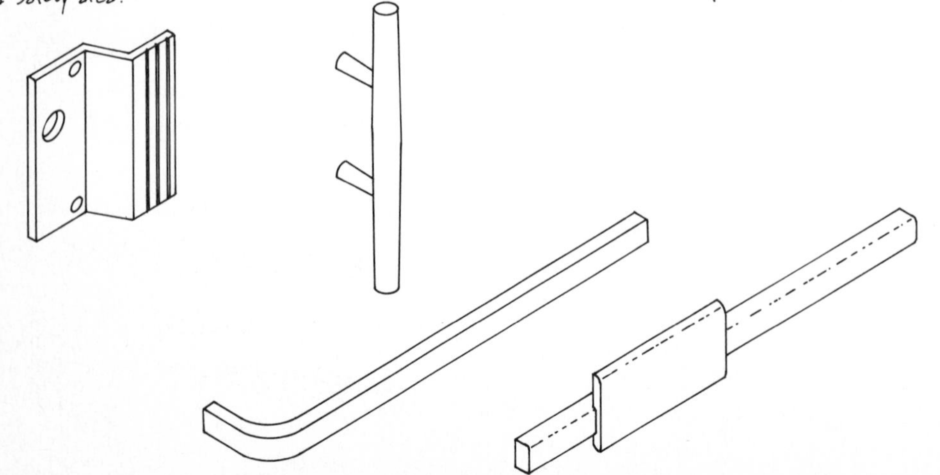

HANDLES

A large variety of door handles are available for architectural effect or simple utility.

GARAGE DOORS

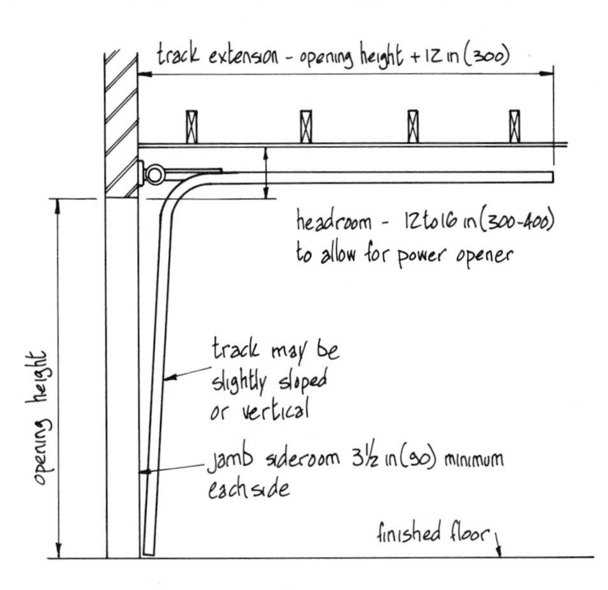

Hinged Rolling Door - Typical Minimum Clearances

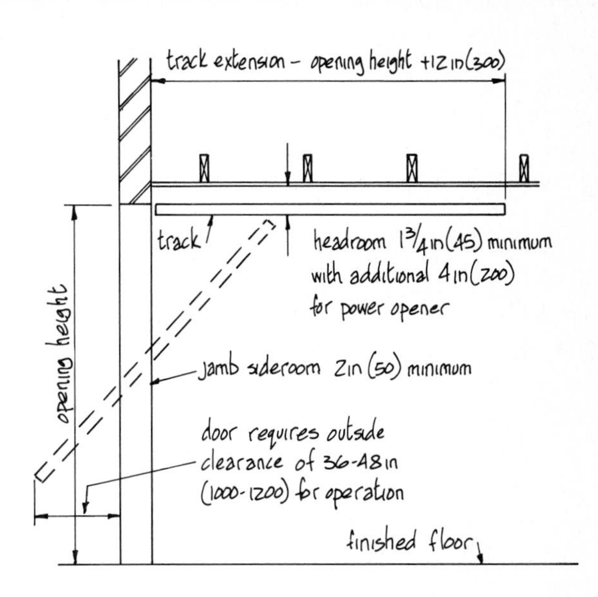

Single Panel Door - Typical Minimum Clearances

Garage and overhead doors employ a track and hinge system that allows the door to be pushed or pulled upward into a horizontal position above the door head. The two general door types shown on 12-15 use essentially the same tracks but different hinges. Certain clearances are required for the door to function properly, and the sections above show typical dimensions for each type. Consult the manufacturer's instructions for exact information.

In contrast to regular house doors the stated size of garage doors is also the exact finished opening size as shown above. The rough-opening size is adjusted only to allow for trim members.
Door materials include wood frames with plywood or composition board infill panels, pressed metal panels and for lightweight doors, fiberglass panels.

STAIRS AND CABINETS

CHAPTER PAGE TITLES

STAIR TYPES

While stairs serve the basic function of providing access between two or more floor levels, they are also used in many cases to add style and character to a building provided that care is taken in their design and construction. Although most stairs can be fabricated by competent carpenters, the highest quality stairs require the same level of skill as that found in fine cabinetmaking.

- Main or finish stairs connect living area floors and are generally installed after walls and floors are finished. Most are prefabricated in millwork shops and shipped as assembled units or "knocked down" for site assembly. Skilled work is still necessary for proper site assembly. Main stairs should be at least 36 in (915) wide and be designed for comfort and safety (see 13-4).
- Service or basement stairs serve nonliving area floors and are designed for function rather than style. This type is often constructed on site and may be steeper than main stairs although no less wide.
- Stairs may be open, closed, or a combination of both. Examples are given at the right.
- There are four basic stair layouts as shown below, each made from combinations of stair flights and landings. The choice of layout depends on access, floor height, space available, and visual effects. Stair platforms provide a change of stair direction to reduce length of stair space, allow a temporary resting place, and provide a safety area in case of falls.

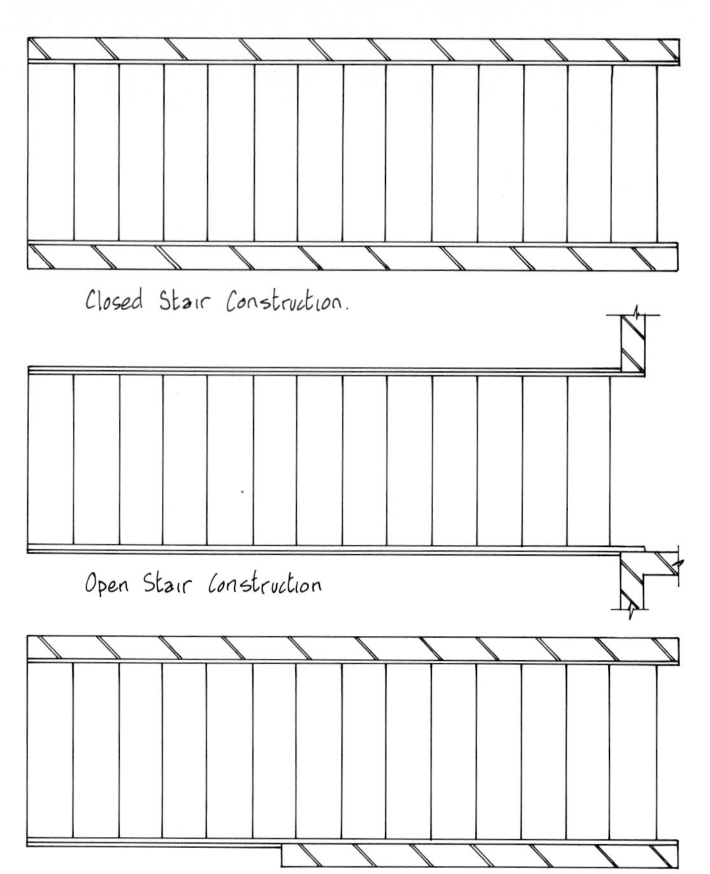

Closed Stair Construction.

Open Stair Construction

Combination Open and Closed Stair Construction

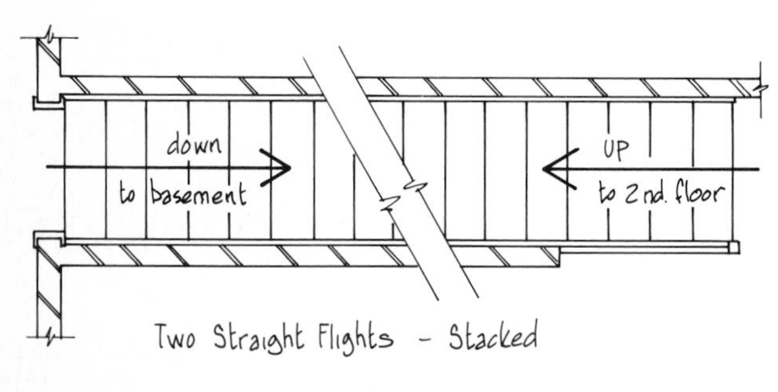

Two Straight Flights - Stacked

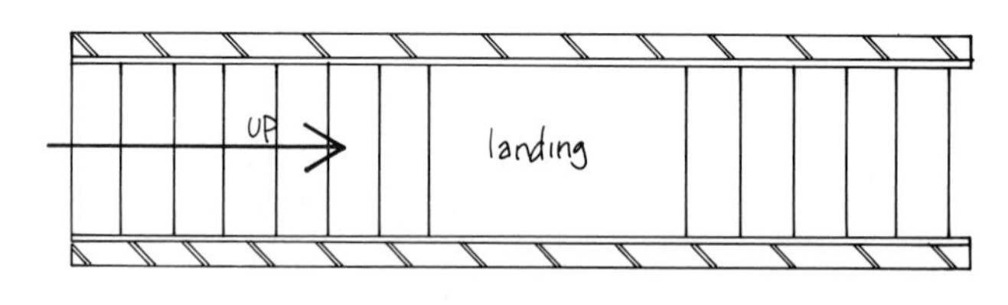

Straight Flight - One Landing

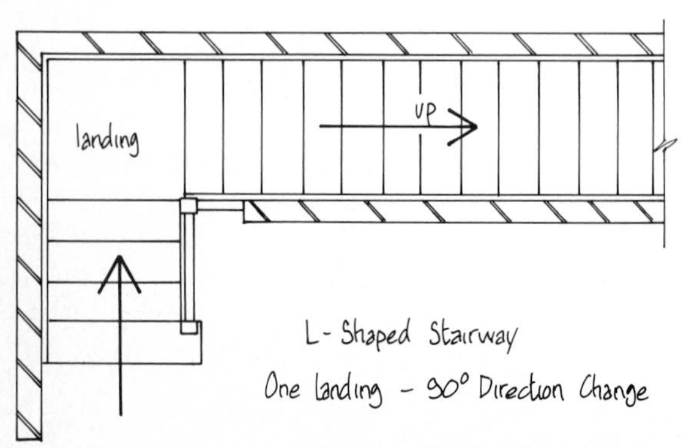

L-Shaped Stairway
One landing - 90° Direction Change

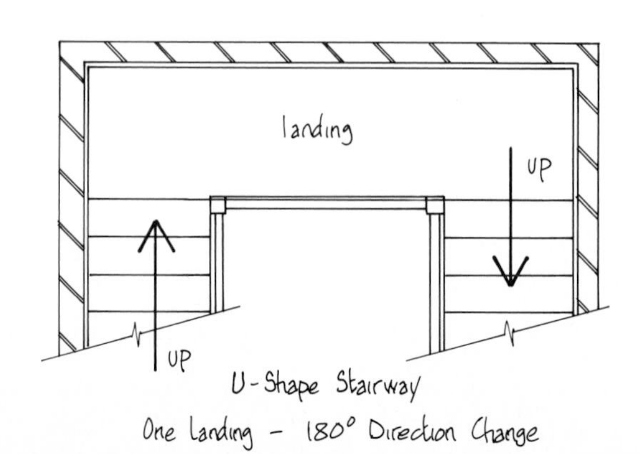

U-Shape Stairway
One Landing - 180° Direction Change

PLATFORMS AND WINDERS

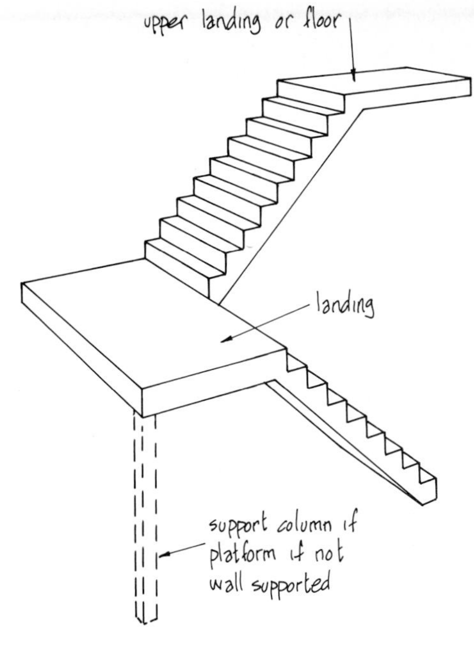

STAIRS WITH INTERMEDIATE PLATFORM

PLATFORMS

There are two basic types of platforms (or landings):

- Flat platforms, which are treated as small floors for construction purposes, must be supported either by attaching the platform joists to the wall framing on each side or by extending posts to the floor below.
- Winders, although not true platforms, do allow a change in stair direction by continuing the stair treads around a corner. They also reduce the length of a stair by adding risers where a platform would normally be located (see 13-4 for more information).

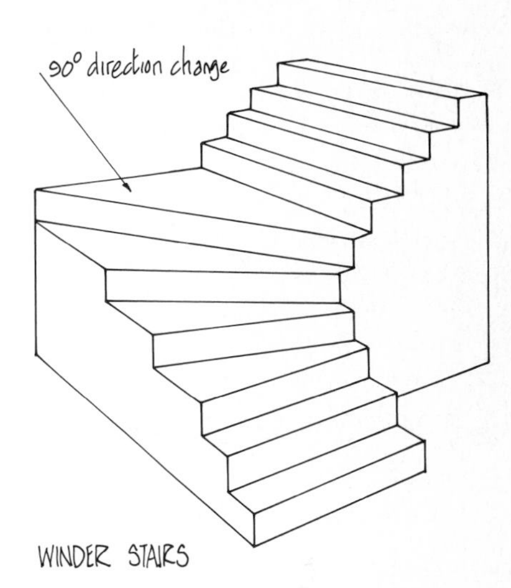

WINDER STAIRS

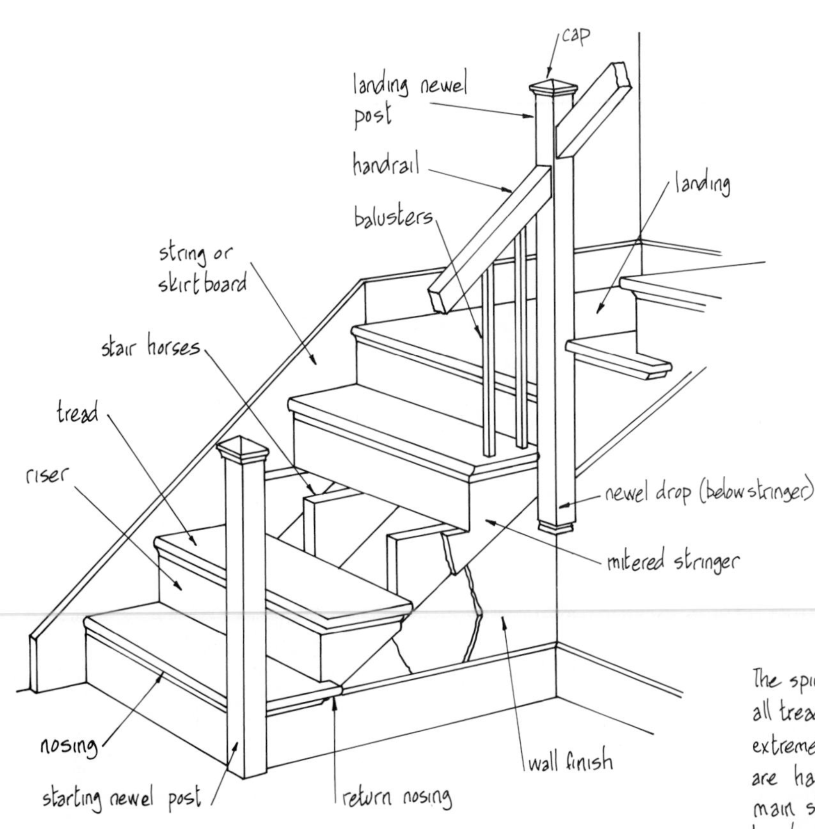

STAIR PART NAMES

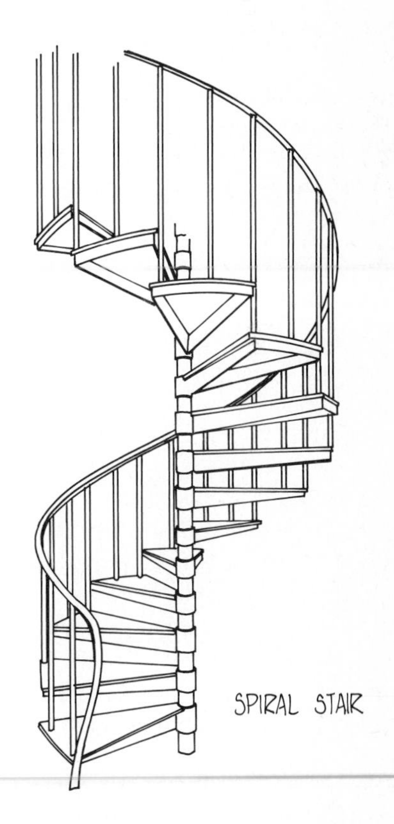

SPIRAL STAIR

The spiral stair above is a special type of winder stair where all treads are winders. Spirals are only used when space is extremely limited or when special effects are required. They are hard to negotiate and should not be used to replace main stairs. The type shown above is made from prefabricated treads and separate spacer rings all assembled on a central structural pole.

STAIR LAYOUT

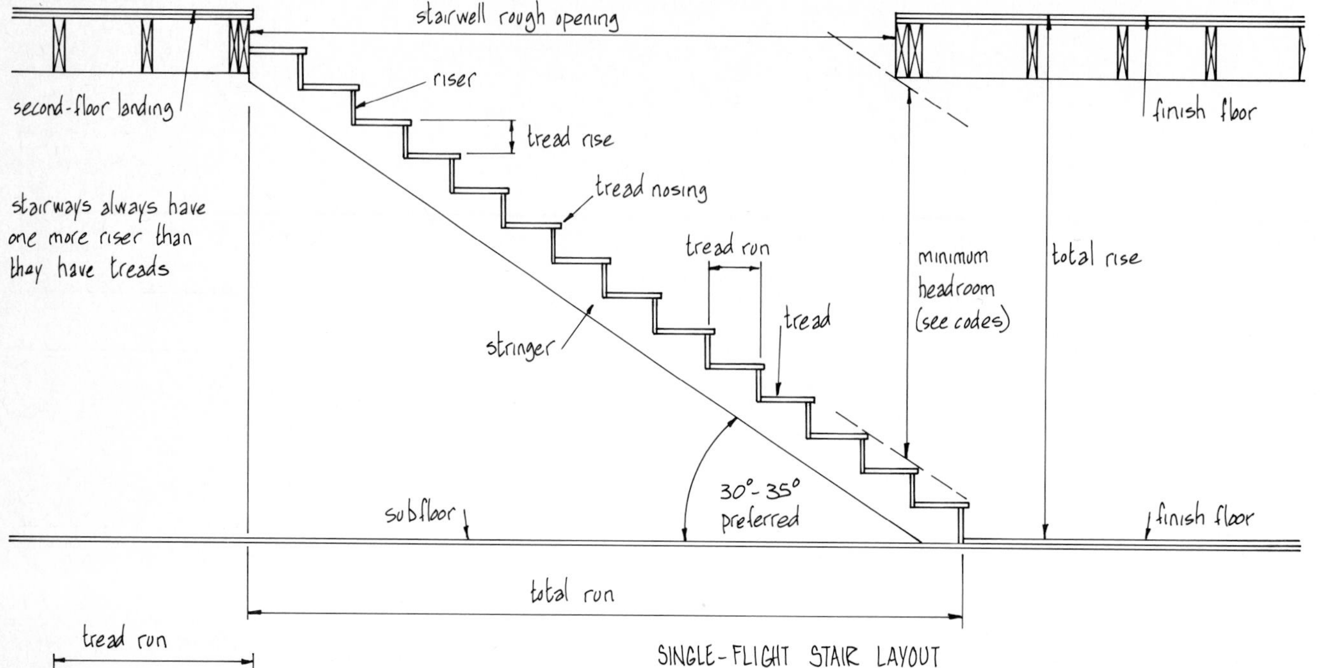

SINGLE-FLIGHT STAIR LAYOUT

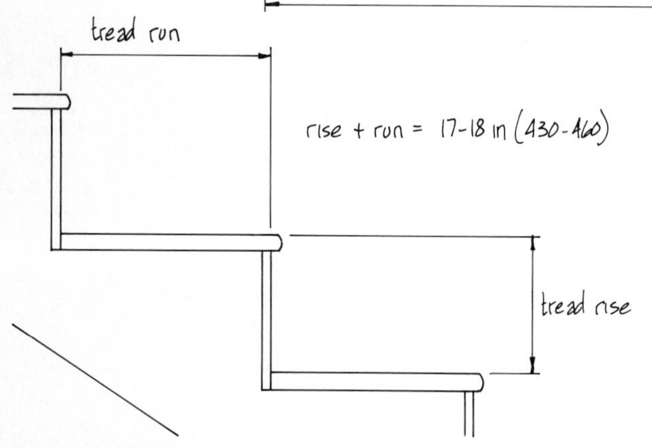

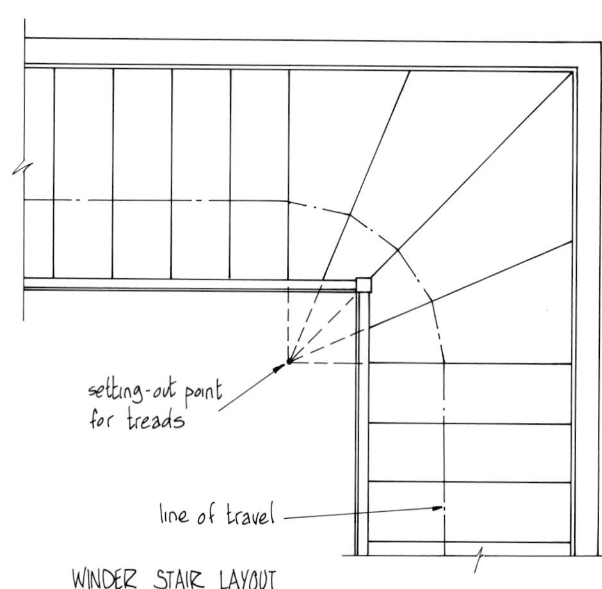

WINDER STAIR LAYOUT

RISE AND RUN

The correct relationship between the vertical risers and the horizontal treads is very important if the stair is to be easy to climb and safe to descend. The following rules should be observed:

- The stair angle should be between 30° and 35°, which translates into a rise of between 7 and 7¾ in (175-195), this being the most comfortable range for main stairs. Some service stairs may be steeper, but even they should not have risers exceeding 8 in (200).
- The sum of one riser and one tread should equal 17 to 18 in (430-460). With a riser of 7½ in (190) the tread will therefore be between 9½ and 10½ in (240 and 270).
- Some codes use a different criterion and require that the sum of two risers and one tread must not be less than 24 in (610) and not greater than 25 in (635).
- All risers in a stair must be the same height and all treads the same width.
- Note that as riser height decreases, tread width increases, as does the opening required for the stair.
- Check local codes for maximum riser heights and minimum tread widths.

WINDER TREADS

The plan above shows the general layout for a set of winders. Note that the tread width, at the line of travel, is equal to the standard tread width. The rise of each winder must also equal the standard risers.

STRINGERS AND CARRIAGES

STAIR STRINGERS

There are two principal methods of stair construction: housed stringer and stair carriage. The variations of each method use different types of stair supports (stringers or carriages) to achieve the required level of finished appearance. Illustrated here are the most common stringer and carriage types for both closed and open-riser stairs. Pages 13-6 and 13-7 show tread and riser connections.

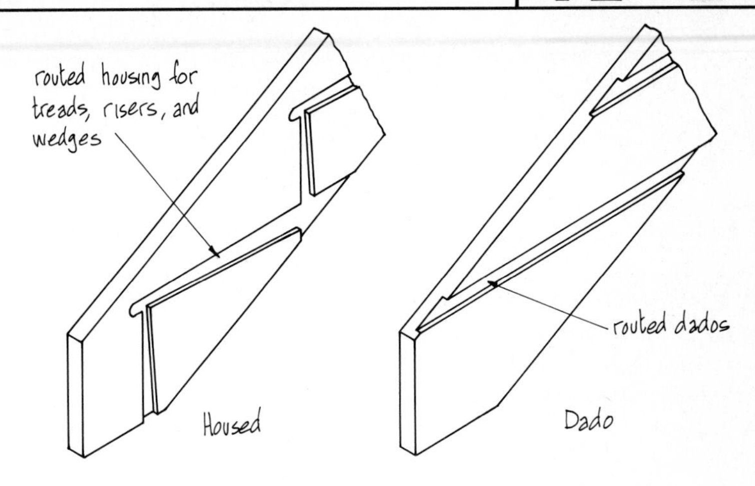

HOUSED STRINGERS Used when treads and risers are fitted into the stringers.

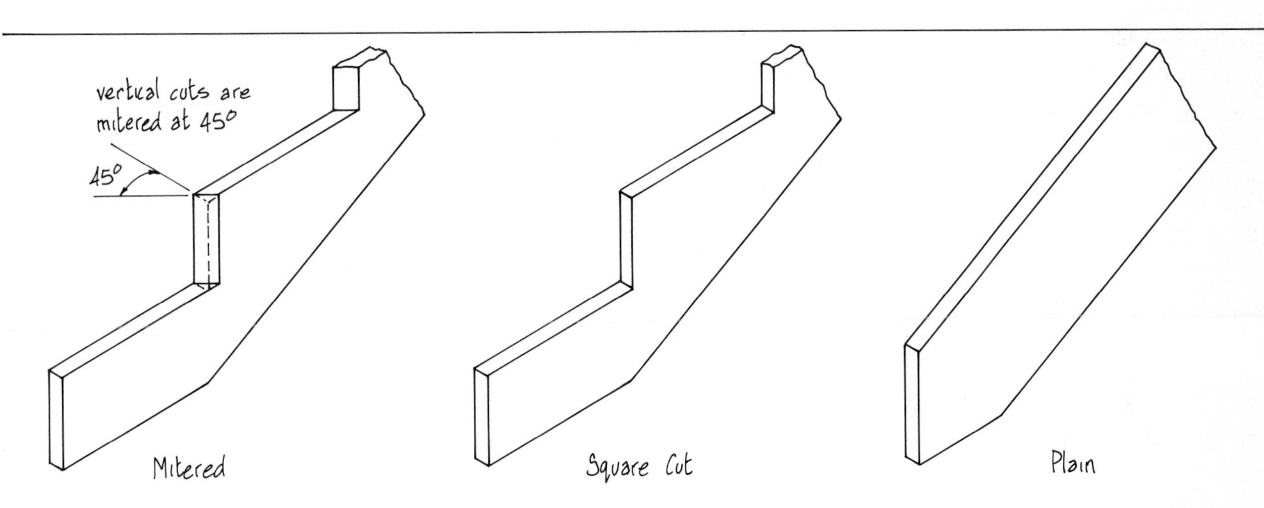

CARRIAGE STRINGERS Used when treads and risers are to sit on the stringer.

STAIR HORSES

These are intermediate stair support members used on almost all stairs except housed stringer types. Depending on stair type and width, up to four horses may be required to properly support the stair.

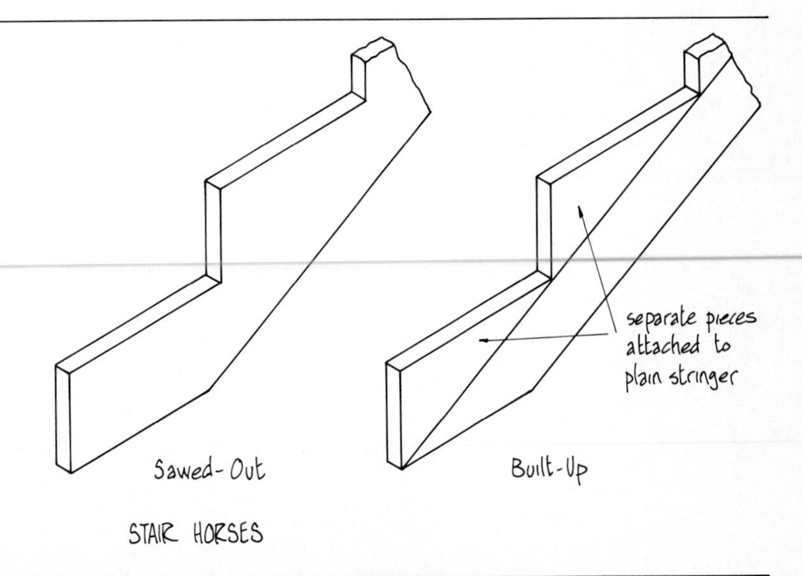

STAIR HORSES

CLOSED-RISER CONSTRUCTION

Closed-riser stairs have solid risers that fill in the vertical space between the treads. There are several methods of construction and finishing, as illustrated here.

HOUSED STRINGERS

Both treads and risers are let-in and nailed to the routed dado housing, hiding the cut ends and providing the necessary support. Glued wedges lock both into position and reduce tread squeak. Glue blocks strengthen the joint between tread and riser.

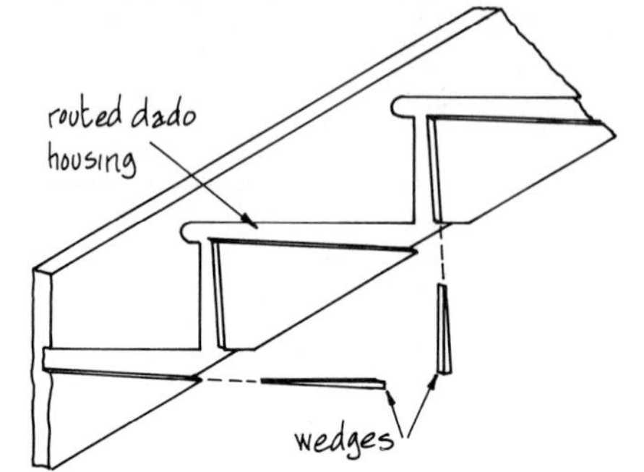

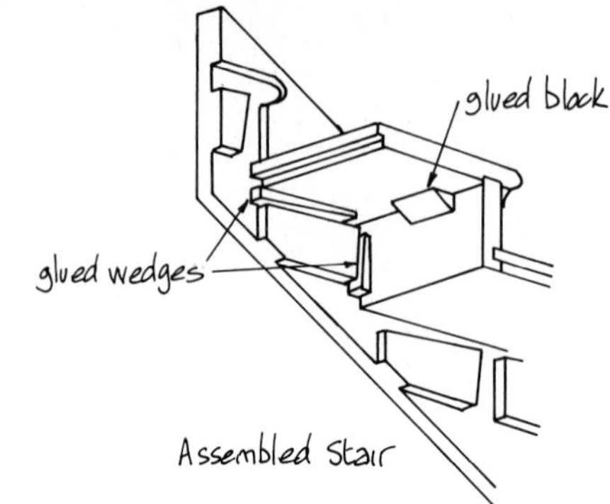

MITERED STRINGERS

Used when the end grain of the risers must not be visible. Both riser and stringer are mitered and carefully fitted. Treads usually overhang stringers.

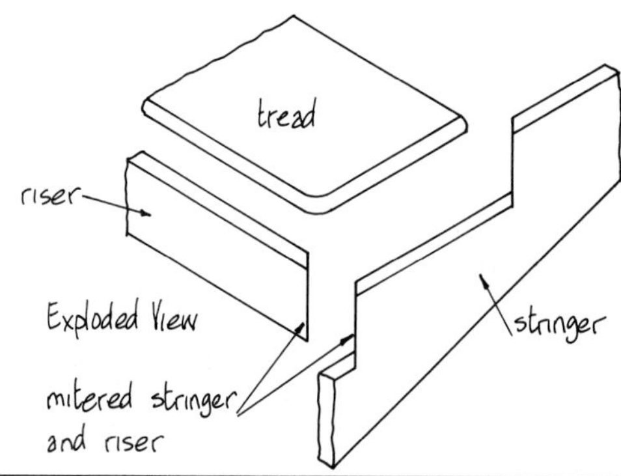

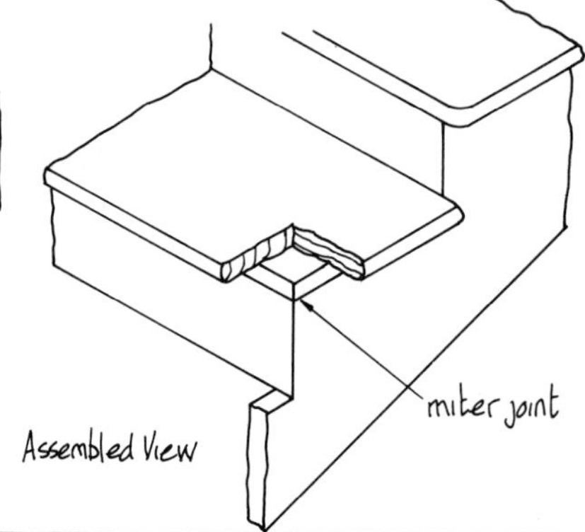

SQUARE CUT

Square ended treads and risers are nailed directly to the stringer. Risers extend slightly beyond the stringer face to simplify finishing.

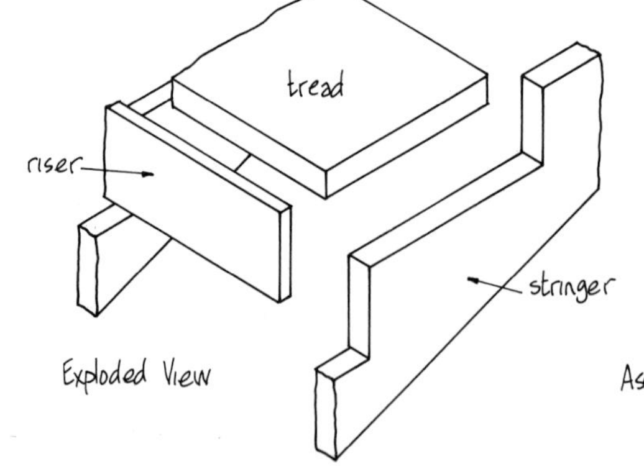

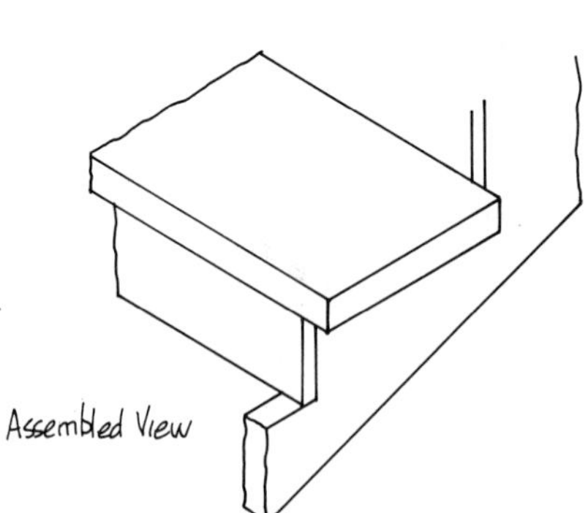

PLAIN STRINGERS

The outer stringer is plain cut and acts as a trim. Treads and risers are supported on a stair horse nailed to the stringer.

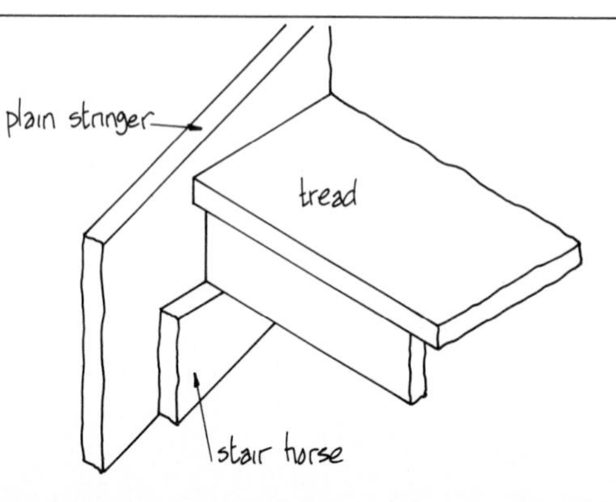

OPEN - RISER CONSTRUCTION

SQUARE CUT STRINGERS
Similar to closed-riser stairs but without a solid riser. In some applications the vertical cut is made at an angle to improve appearance.

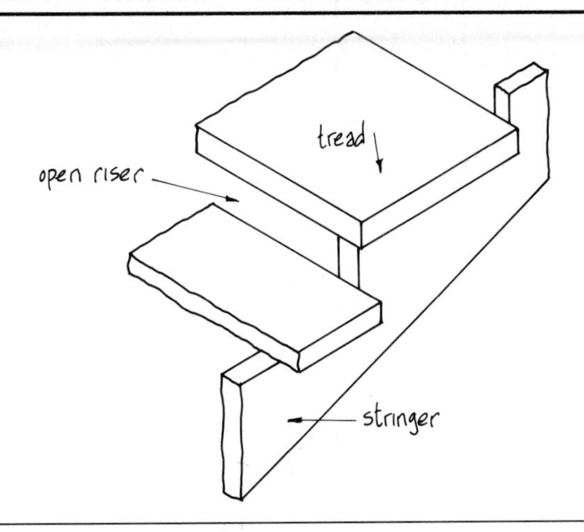

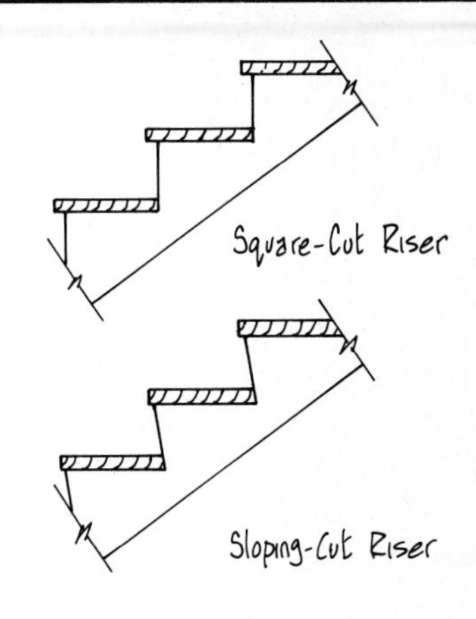

DADO STRINGERS
Individual treads are positioned in the stringer dado cuts. This type of stair is used extensively for basement stairs or other nonliving areas because of its simplicity.

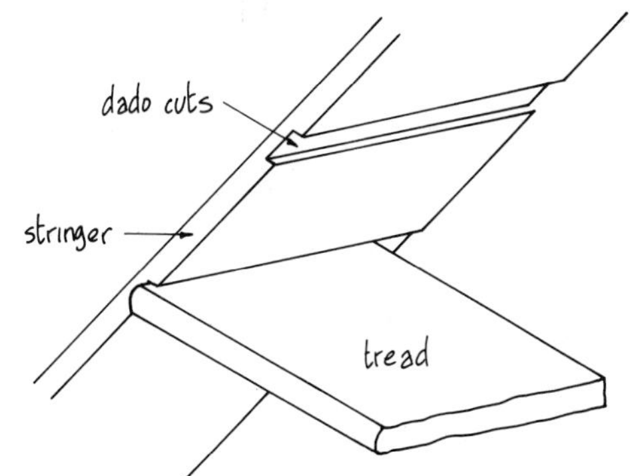

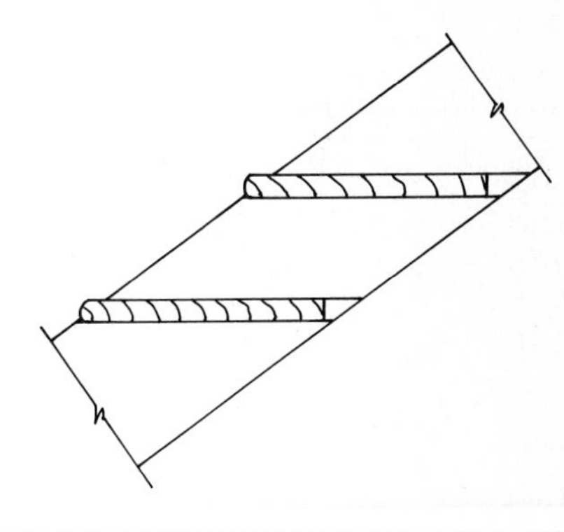

CARRIAGE (STRINGER-LESS)
Stringers are omitted in this type of stair and the treads are carried on one or more stair carriages to create a totally open stair. Treads are made from 2-in (38) or thicker stock to give the extra strength required. Custom steel or concrete stairs designed for special architectural effects are often made this way.

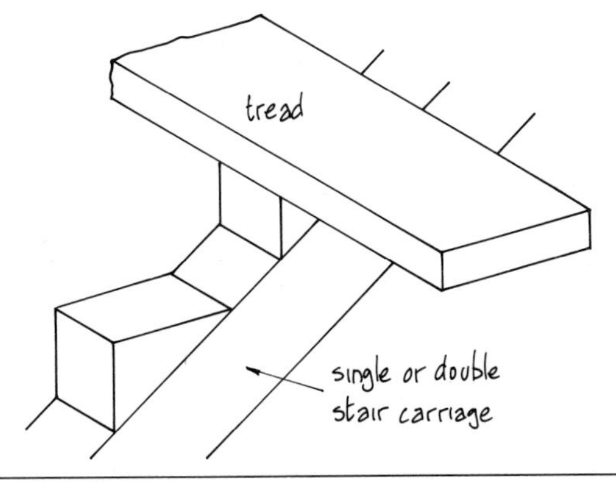

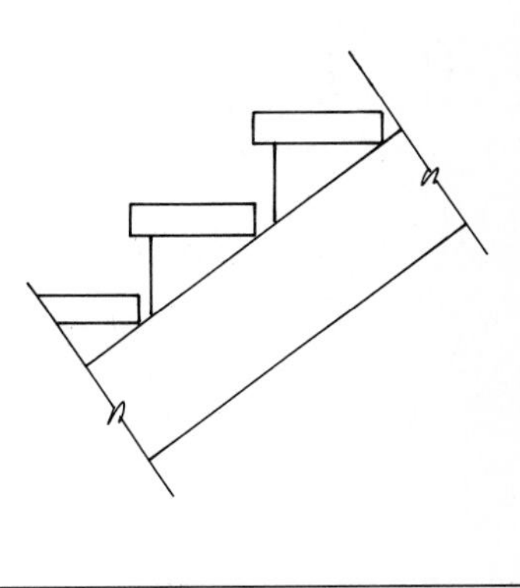

Open-riser stairs have no solid riser material and are simply treads supported on a stringer or carriage. This type of stair is used as a main finished stair for architectural effect or as a simple utility stair for basements or other unoccupied areas.

RISER DETAILS AND HEADROOM

Since it is important that all riser dimensions be exactly the same on a stair, care must be taken when detailing or laying out the stringers or carriages. Particular attention is paid to the first and last risers, as they relate to the floor finishes. Illustrated here are typical first- and last-riser details plus a milled-nosing detail that completes the tread sequence at the floor edge.

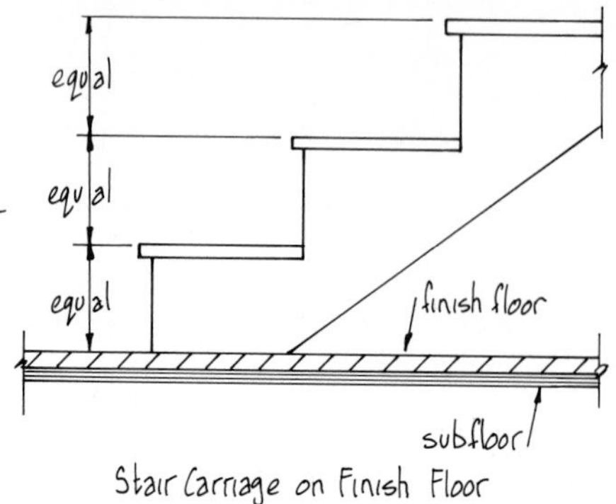

Stair Carriage on Finish Floor

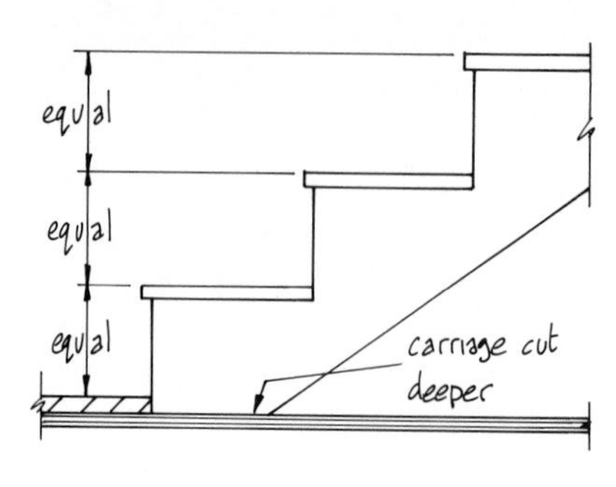

Stair Carriage on Subfloor

CARRIAGE AT FLOOR

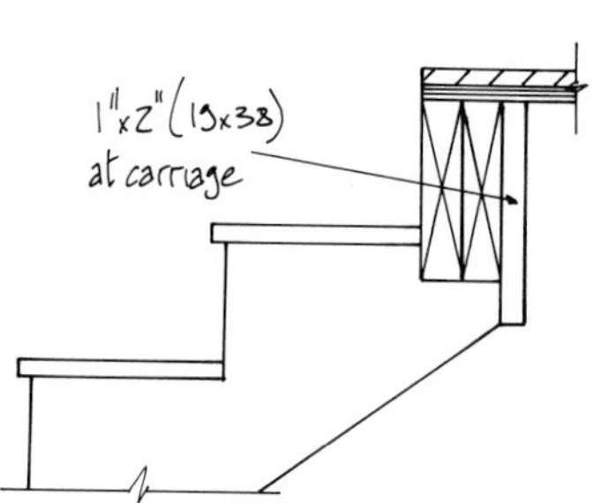

CARRIAGE AT UPPER-FLOOR STAIRWELL

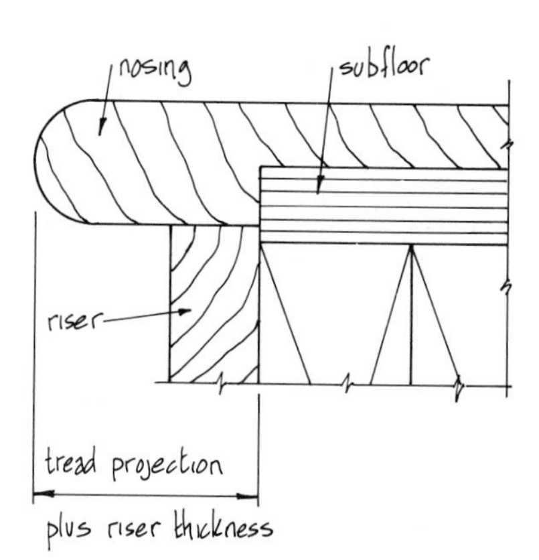

TYPICAL NOSING DETAIL

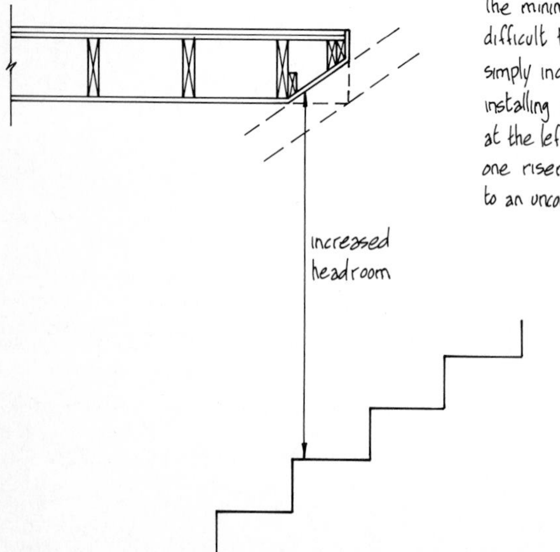

INCREASING HEADROOM

The minimum code headroom is generally 6'-8" (2030). This may be difficult to achieve in some tight design situations where it difficult to simply increase the floor opening length. Additional headroom can be gained by installing a shallower header joist and sloping the finishes as shown at the left. Other options are to slightly shorten each tread or to remove one riser, thereby reducing the total stair run, although this may lead to an uncomfortable stair angle and may not meet code requirements.

STAIRWELL OPENINGS

The opening for a stair is framed as for any other large floor opening. The opening width depends on the type of stair and whether it is open or closed. The opening length must be calculated to give sufficient headroom (see 13-4) and depends on the size and number of treads and risers as well as the ceiling height and floor construction. See 6-11 for typical framing methods for stair openings.

TREAD AND TRIM DETAILS

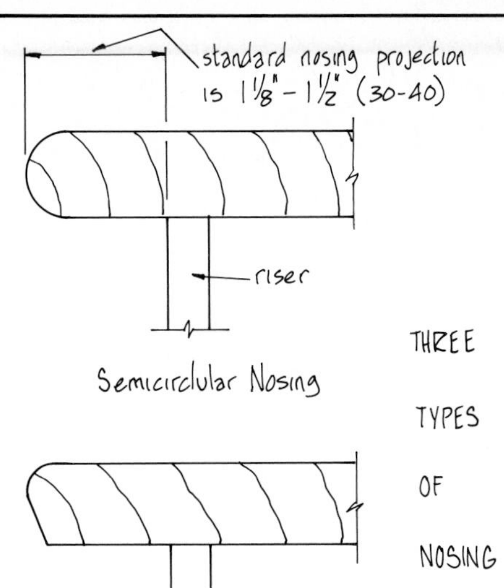

Semicircular Nosing

Rounded and Angled

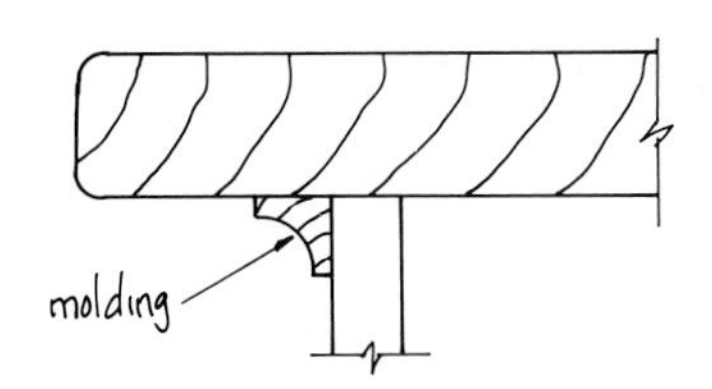

Rounded with Molding

THREE TYPES OF NOSING DESIGNS

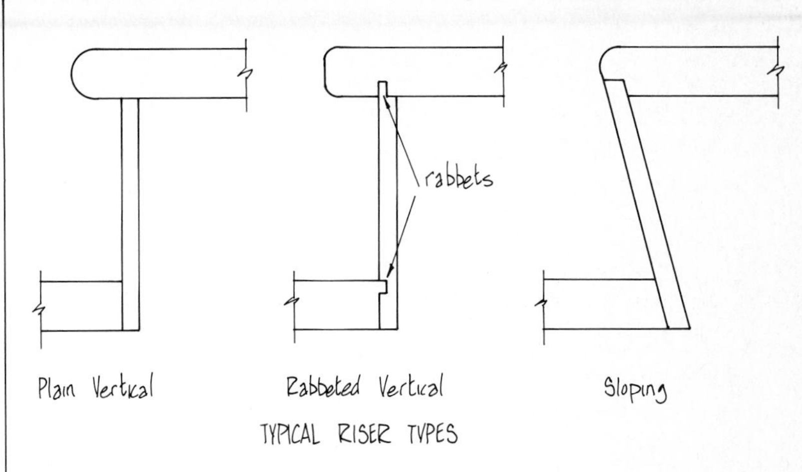

Plain Vertical Rabbeted Vertical Sloping

TYPICAL RISER TYPES

There are many ways that a stair can be "customized" to merge with the surrounding decor or to stand out as a design feature. Millwork shops offer a wide selection of stair options, with emphasis on tread, riser, and trim detail. Illustrated here are a selection of nosings, risers, and moldings, all designed to achieve variations in style.

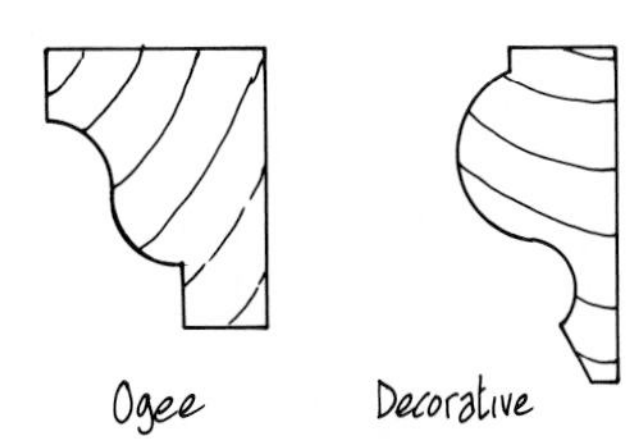

Ogee Decorative Cove

TYPICAL MOLDING SHAPES

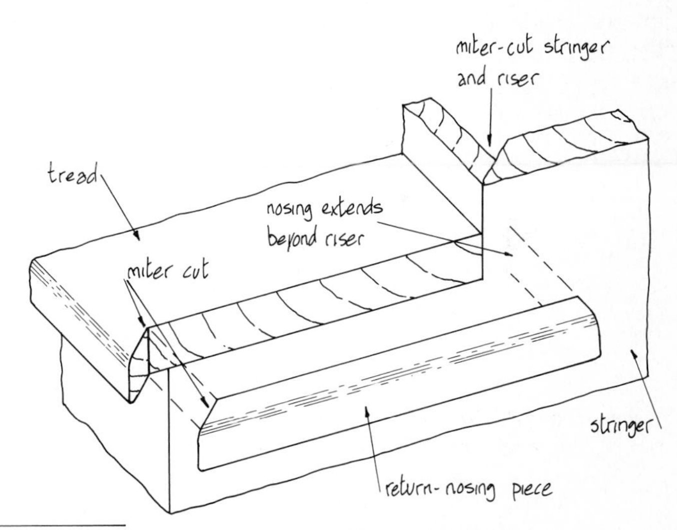

RETURN-NOSING ON OPEN TREAD-END

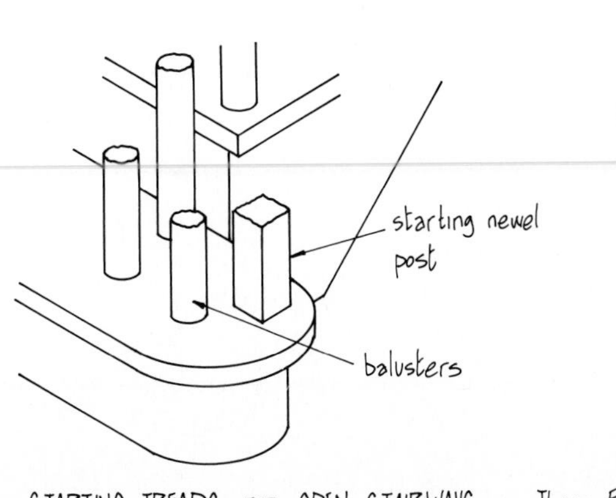

STARTING TREADS FOR OPEN STAIRWAYS — Three Examples

STARTING TREADS

When the handrail and balustrade are designed to curve outward at the bottom, the first riser and tread are made longer to accommodate the newel post. Millwork shops produce a variety of tread shapes for this application, three of which are shown at the left.

BALUSTRADES

A balustrade is the railing that protects the open sides of a stair or stairwell and includes the handrail, balusters, newel posts, and related trim. The balustrade is also a major part of the overall stair design and adds significantly to the visual aspects.

- Handrail height is generally between 32 and 36 in (815-915) above the landing level.
- Newel posts are the main balustrade support members and must be securely fastened to floor joists, wall studs, or stair stringers and carriages.
- Methods of attaching balusters to the stair are shown at the right. Other methods are also used, such as direct attachment to the stringer's outer sides.
- The baluster's top end is bevel-cut to meet the handrail and attached with plugged screws or by being let into slots or rabbets in the rail.
- Details at starting and landing newels are subject to wide style variations. The example at upper right is of a "traditional" stair. The example below is of a more modern stair with open risers and exposed stair carriages.

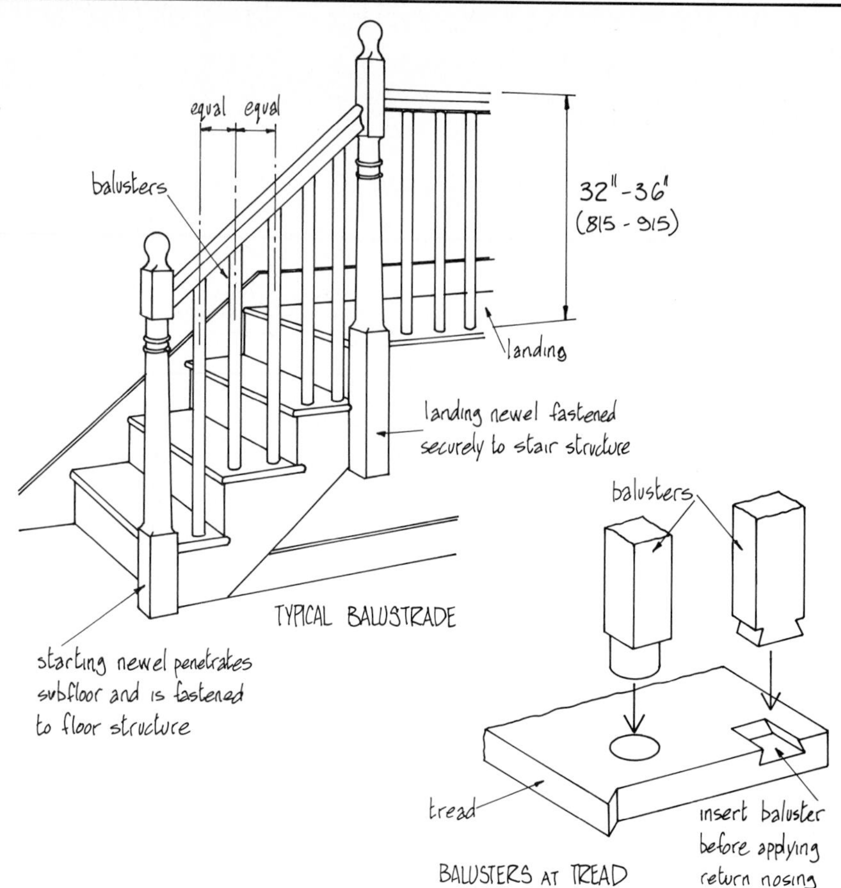

TYPICAL BALUSTRADE

BALUSTERS AT TREAD

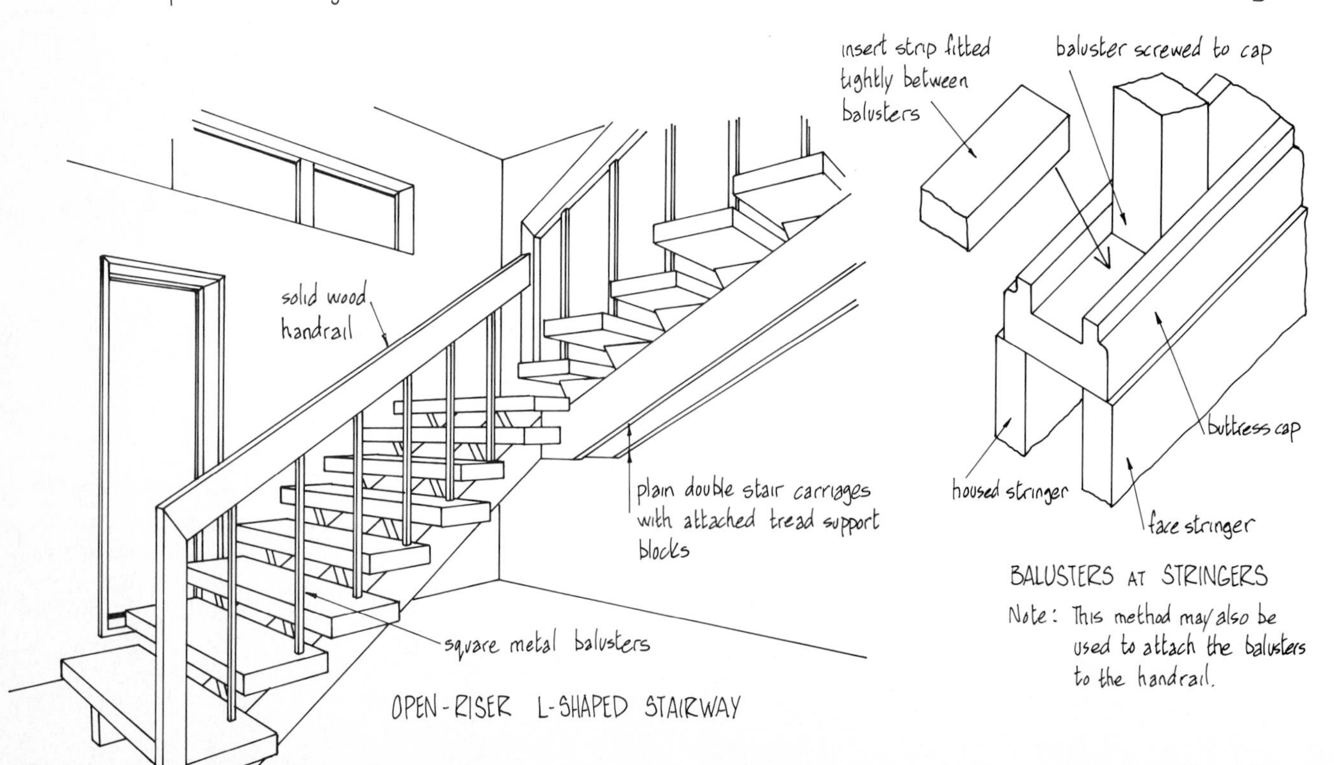

OPEN-RISER L-SHAPED STAIRWAY

BALUSTERS AT STRINGERS

Note: This method may also be used to attach the balusters to the handrail.

BALUSTRADE COMPONENTS

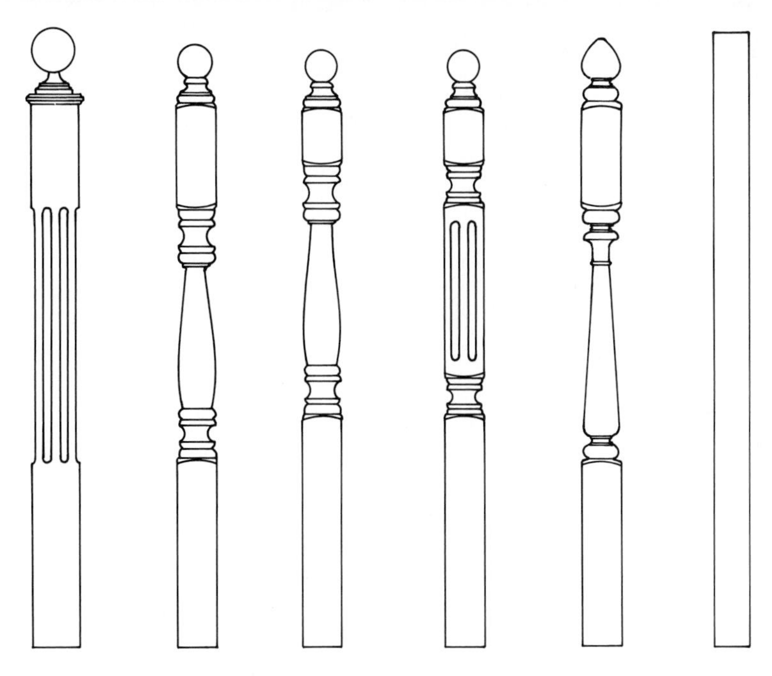

TYPICAL NEWEL POSTS
Each type is usually available in two or three lengths to suit structural conditions and balustrade height.

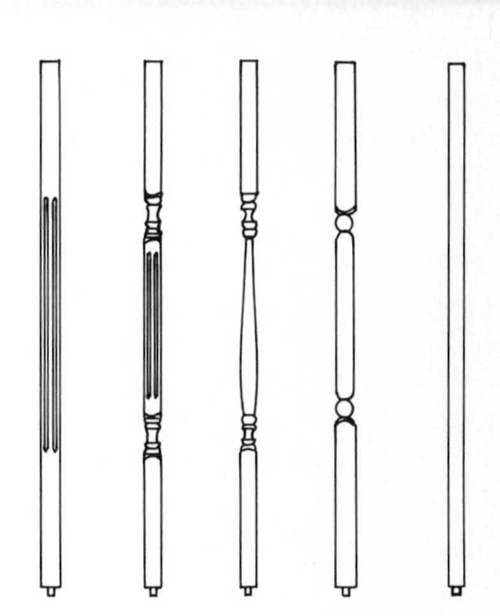

TYPICAL BALUSTERS
Many styles available in varied lengths to suit handrail height.

The balustrade components illustrated on this page are typical of those produced by most stair millwork shops. Materials are generally oak or pine but other materials can be obtained, usually on special order.

Components can be ordered "off the shelf" as standard items for site assembly by contractors or carpenters. When a higher-quality, individualized installation is required, many millwork companies offer custom design, fabrication, and erection services.

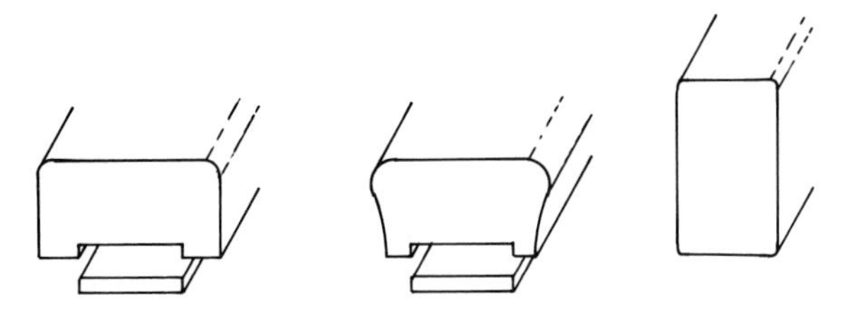

TYPICAL HANDRAIL SECTIONS

Volute

Turnout

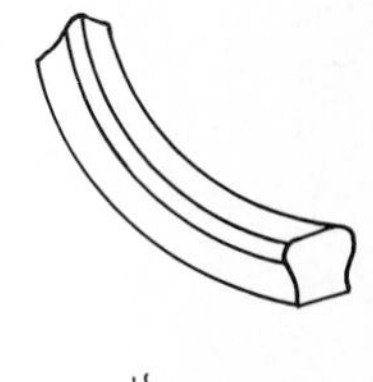

Upeasing

TYPICAL PREFABRICATED HANDRAIL COMPONENTS

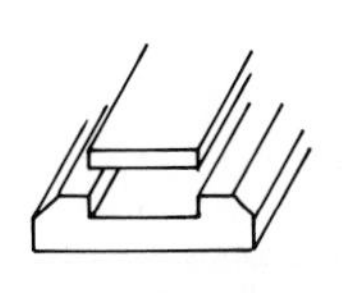

Buttress Cap

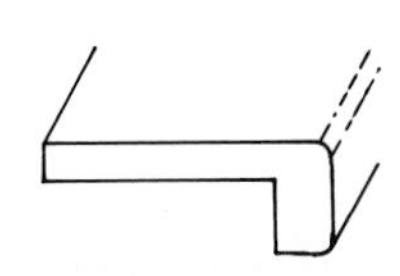

Tread Nosing

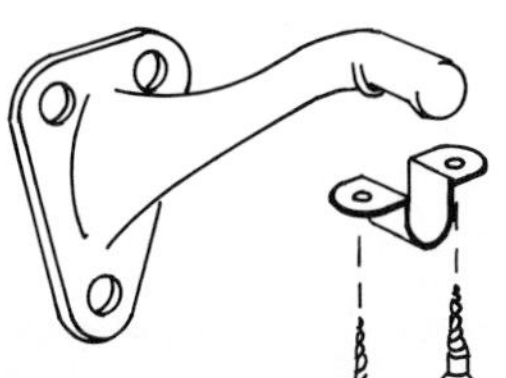

Handrail Bracket

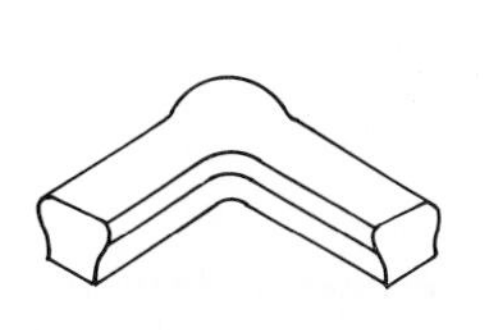

Left Turn

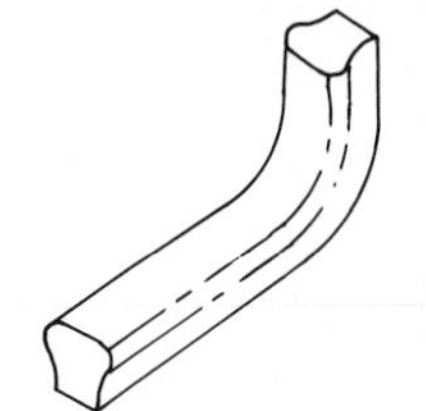

Gooseneck

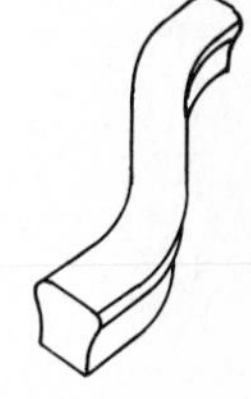

Offset

STEEL STAIRS

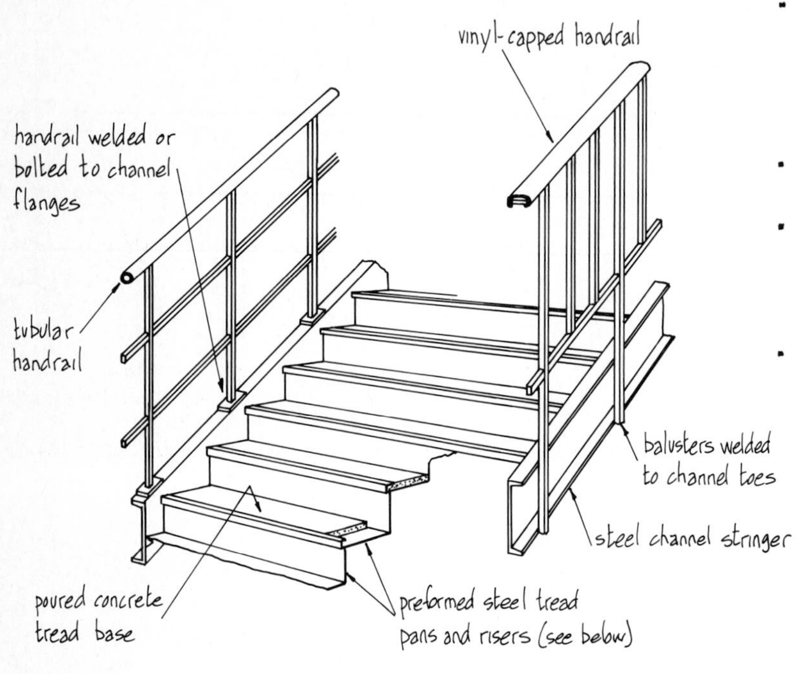

TYPICAL STEEL STAIR

- Steel stairs for general commercial applications are fabricated from open channel section stringers supporting a variety of tread and riser types. Closed box stringers have a plate across the toes or two channels welded toe to toe.
- Stringers can span between landings or be attached to the side walls.
- Steel pans filled with lightweight concrete are commonly used for closed treads and risers, as shown below right. Open riser types include filled single pans, solid plates, and open grilles, as shown below.
- Balustrades can be welded or bolted to the stringers as at the left. Standard handrails include plastic-covered steel rails, metal tubing, and solid or laminated wood.

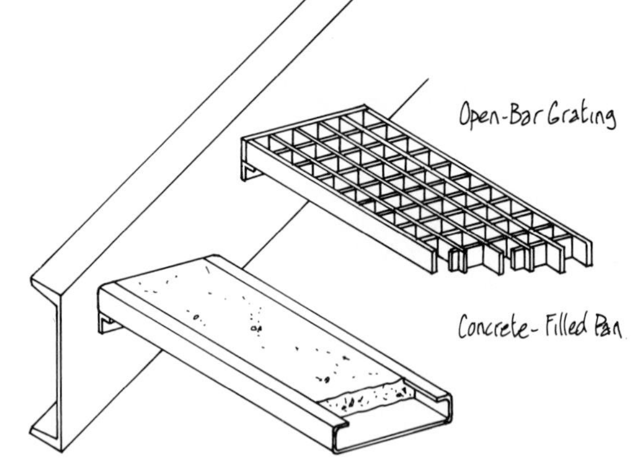

TWO TYPES OF OPEN-RISER TREADS

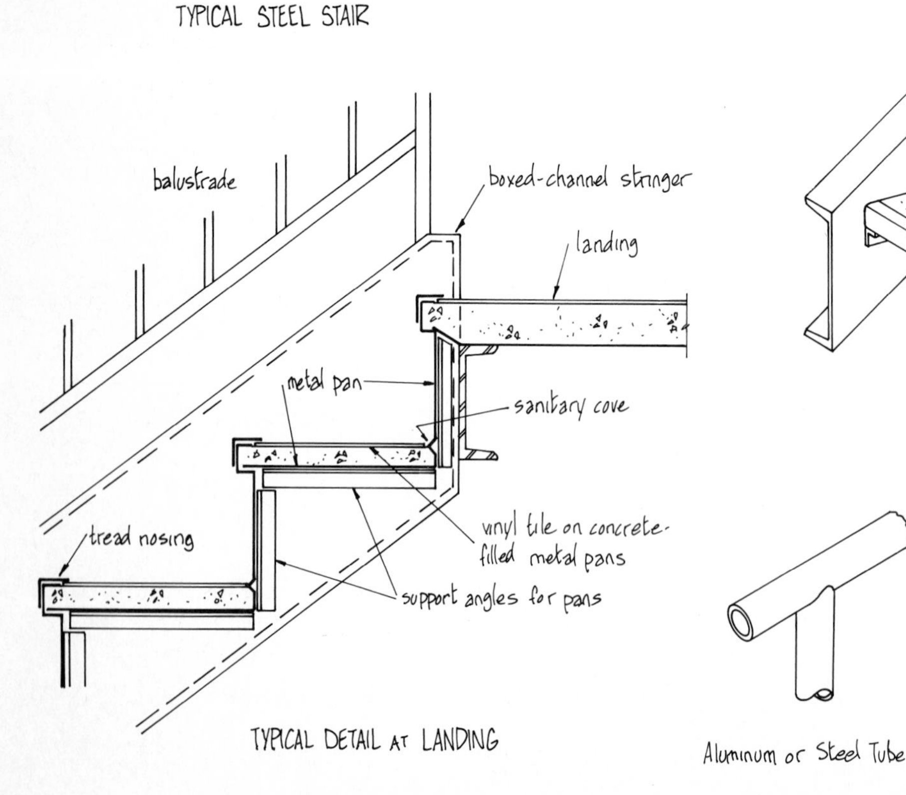

TYPICAL DETAIL AT LANDING

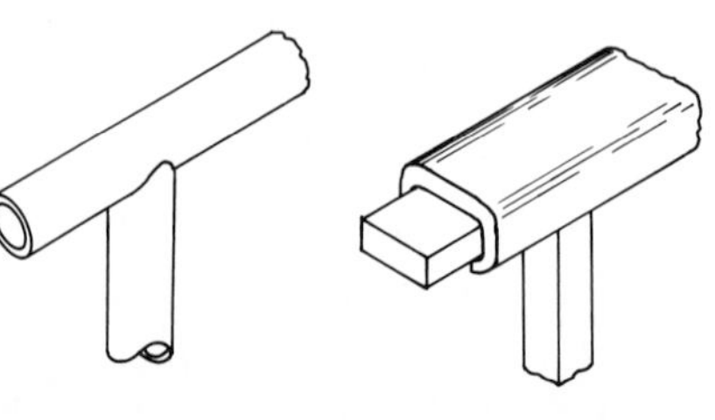
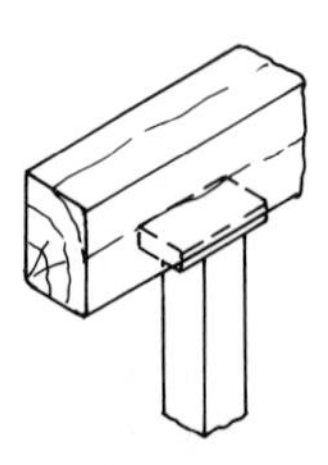

Aluminum or Steel Tube — Plastic Covered — Solid Wood

THREE TYPICAL HANDRAIL TYPES

CONCRETE STAIRS

- Concrete stairs may be formed on site or shipped as precast units. Site forming provides great flexibility in stair design and allows interesting visual effects. However, the stair's weight must be considered in all situations.
- Concrete stairs are used where fireproofing is an important consideration.
- Stairs may be formed as a complete unit or as a plain slab with the treads and risers added later. Each stair must be engineered to meet span and load conditions.
- To increase tread safety and reduce wear, a variety of insertable nosings or covers are available, as shown below.
- With proper design, stairs may be supported by landings or between walls, cantilever from a wall, or be freestanding.

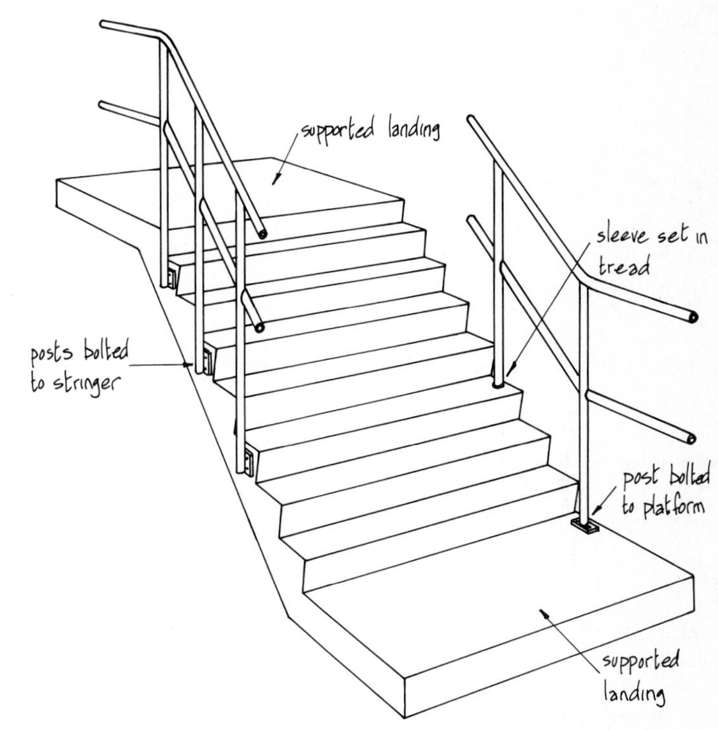

TYPICAL CAST CONCRETE STAIR

open sides, spanning between landings

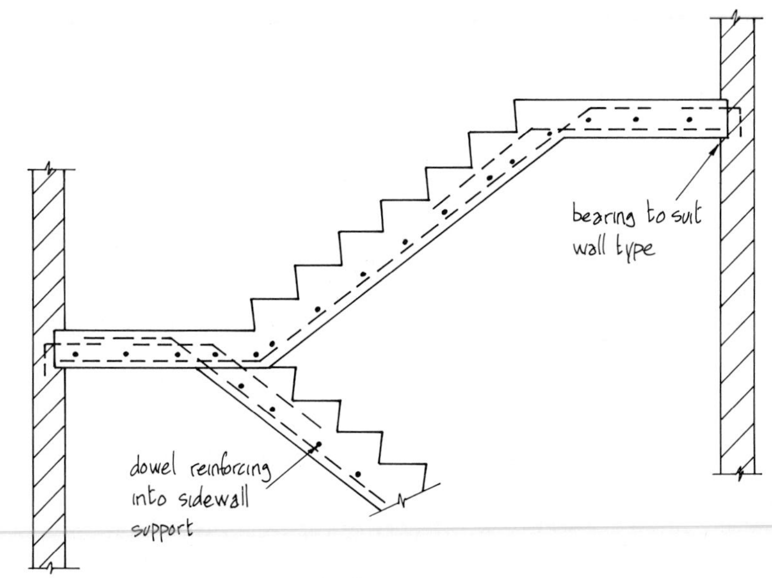

TYPICAL SECTION THROUGH CAST-IN-PLACE CONCRETE STAIR

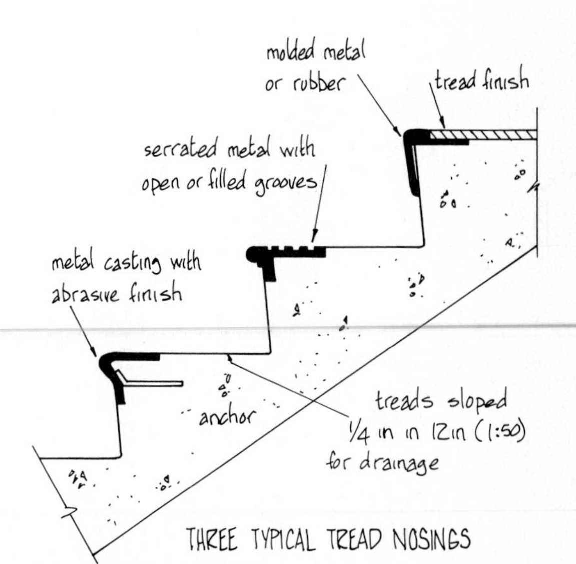

THREE TYPICAL TREAD NOSINGS

LADDERS AND RAMPS

The ladders illustrated here are typical for low-rise commercial and some residential buildings and are generally found in service areas where usage is light and space is restricted. Longer ladders require safety cages and intermediate landings — refer to local codes for details. Materials used are generally combinations of welded steel or aluminum angles, channels, tubes, and flat or round bars. Vertical rung spacing or riser heights of up to 12 in (300) are permitted by most codes.

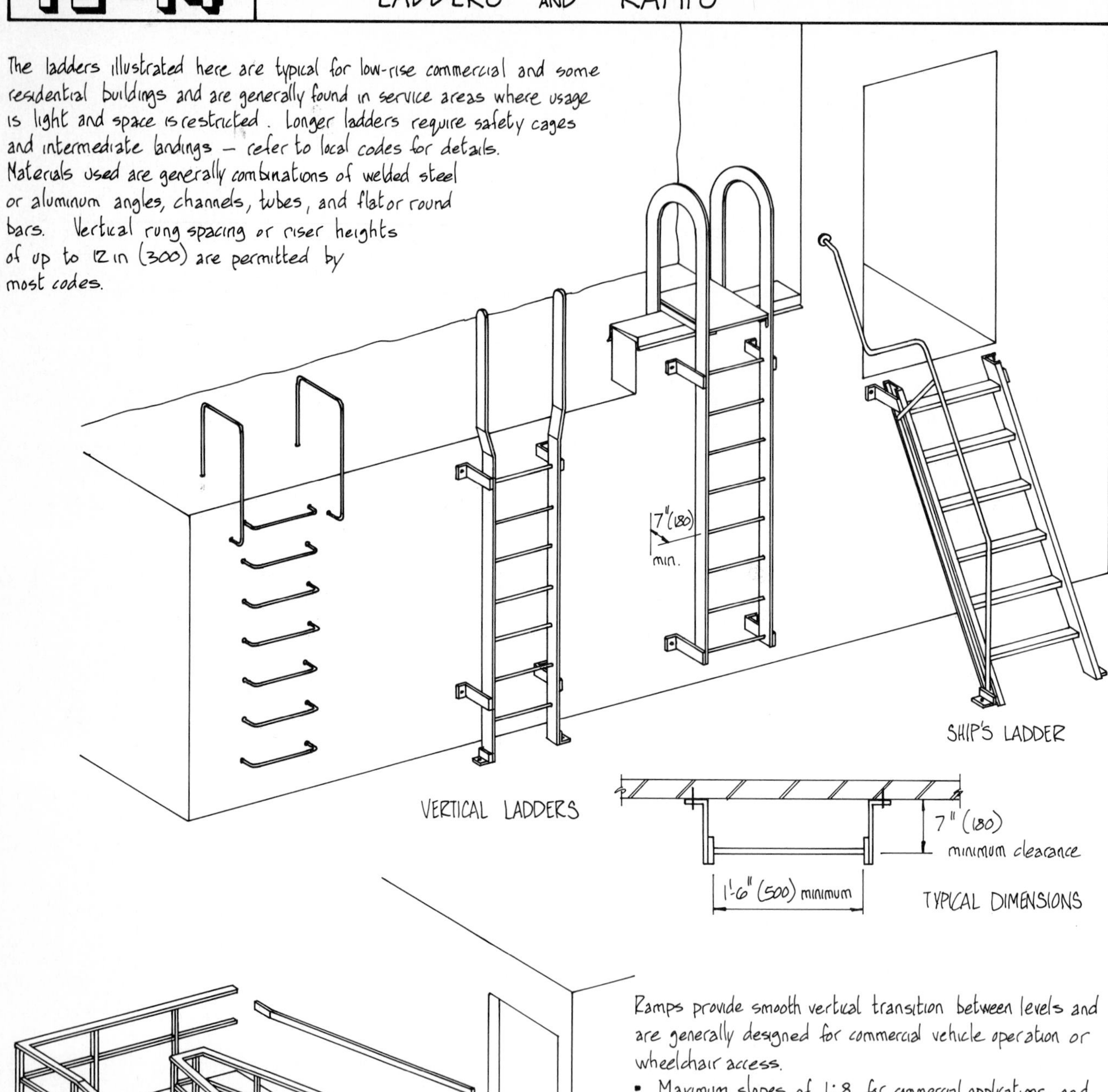

Ramps provide smooth vertical transition between levels and are generally designed for commercial vehicle operation or wheelchair access.

- Maximum slopes of 1:8 for commercial applications and 1:12 for wheelchairs are generally specified by codes.
- Depending on location and usage, codes will also demand a nonslip surface finish on ramp decks, especially for outdoor applications.
- Careful design is essential since the low slope results in long horizontal runs that require large ground areas.
- Landing dimensions and handrail heights must be carefully sized to suit the intended traffic usage.
- Wood, concrete, and steel are the materials of choice.

TYPICAL PEDESTRIAN or WHEELCHAIR RAMP

KITCHEN WORK TRIANGLES

Kitchen cabinets and appliances can be located in a variety of positions, depending on structural necessity and individual preference. The five basic layouts illustrated on this page show the most significant feature of each, the "work triangle".

- The work triangle connects the sink, refrigerator, and stove since most kitchen work traffic is between these points. Through traffic should be routed around the work triangle where possible.
- For maximum efficiency the total length of the triangle's sides should be between 13 and 22 ft (4–6.7 m), although this may be difficult to achieve in some situations.
- The U-shaped kitchen is the most efficient, effectively combining all work elements and excluding through-traffic.
- In large or open kitchen layouts the addition of a peninsula or island unit will greatly increase efficiency. Traffic will be routed properly and the unit can be used to separate kitchen and dining areas.

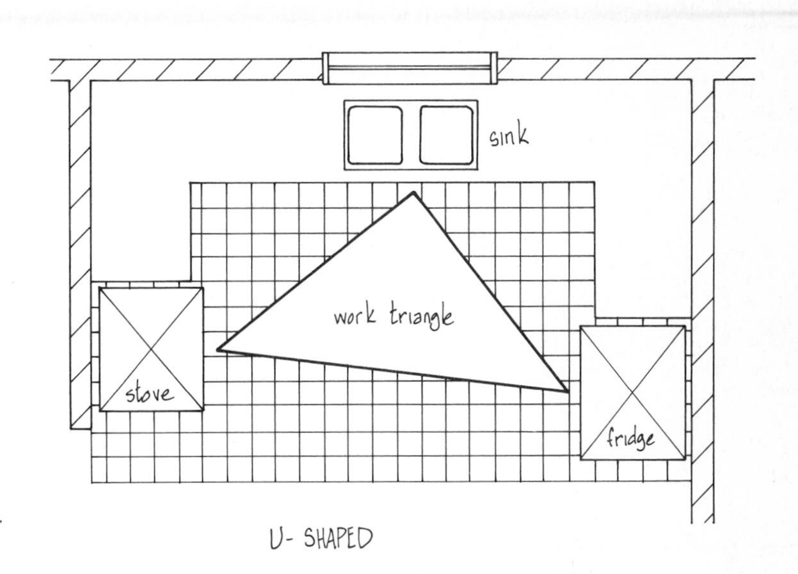

U-SHAPED

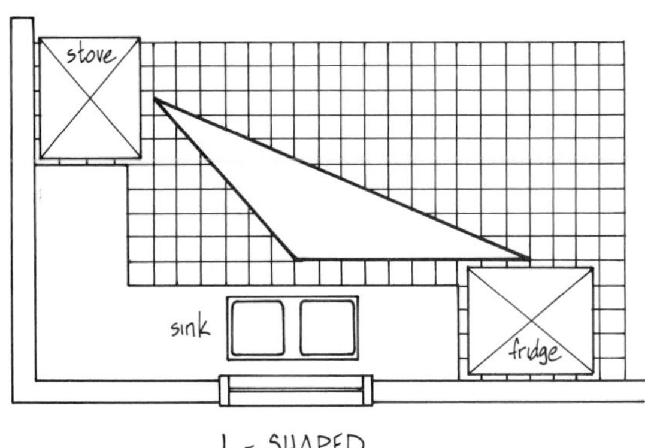

L-SHAPED

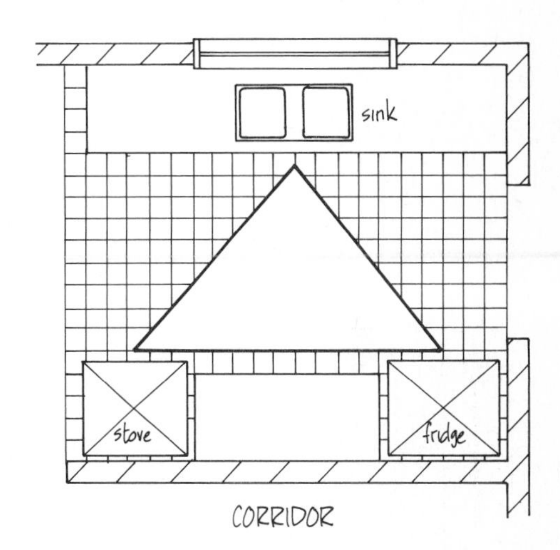

CORRIDOR

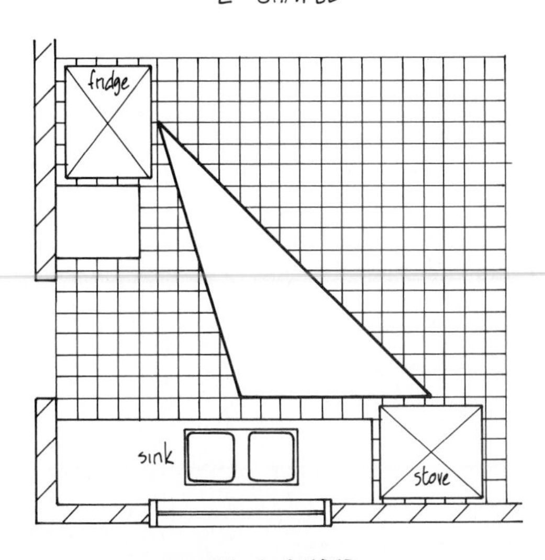

BROKEN L-SHAPED

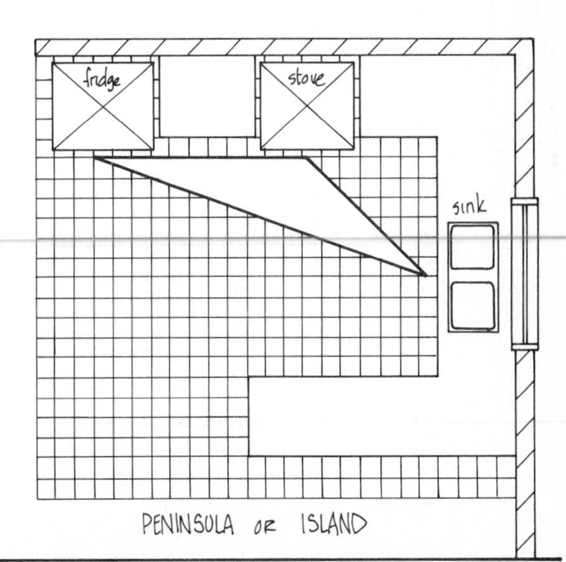

PENINSULA OR ISLAND

STANDARD HEIGHTS AND CLEARANCES

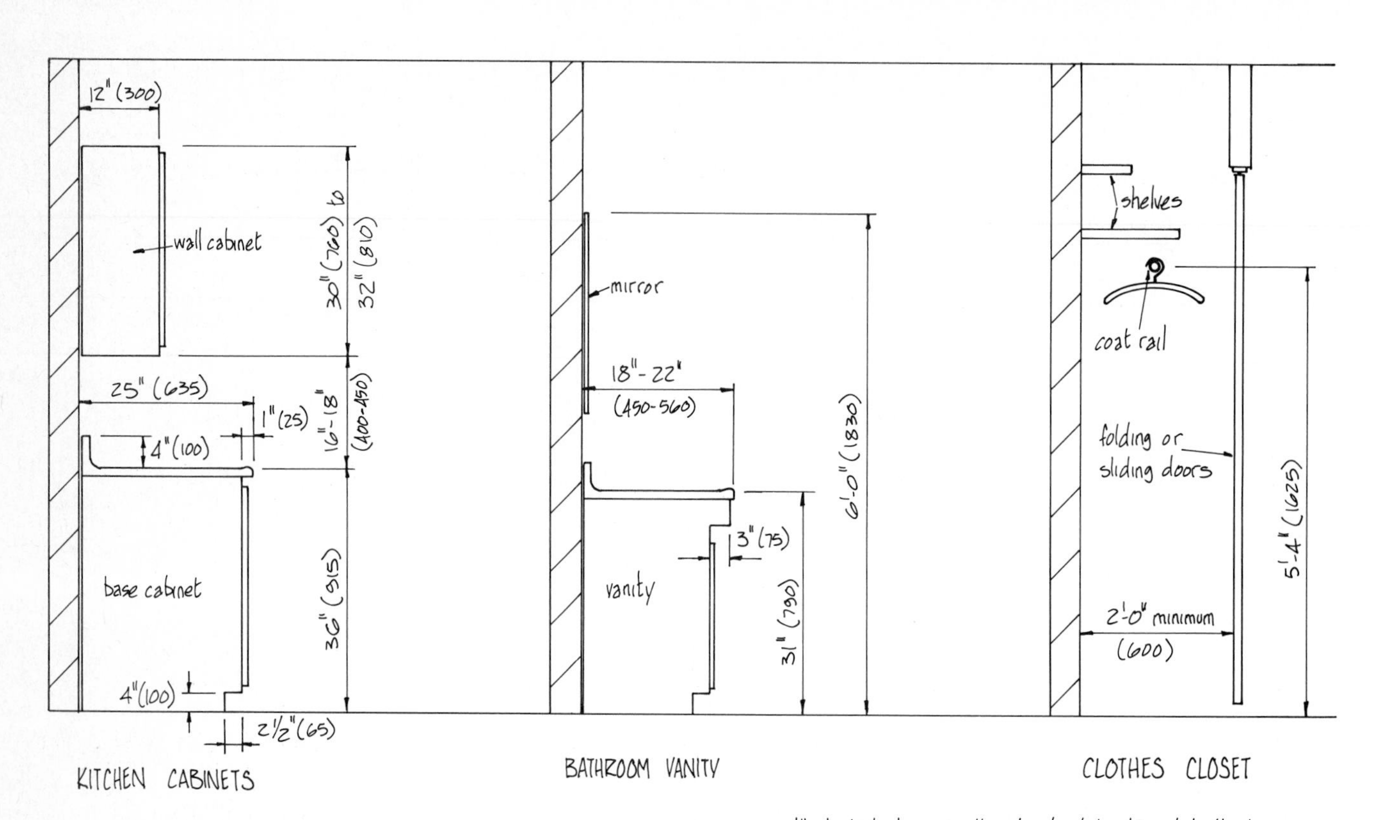

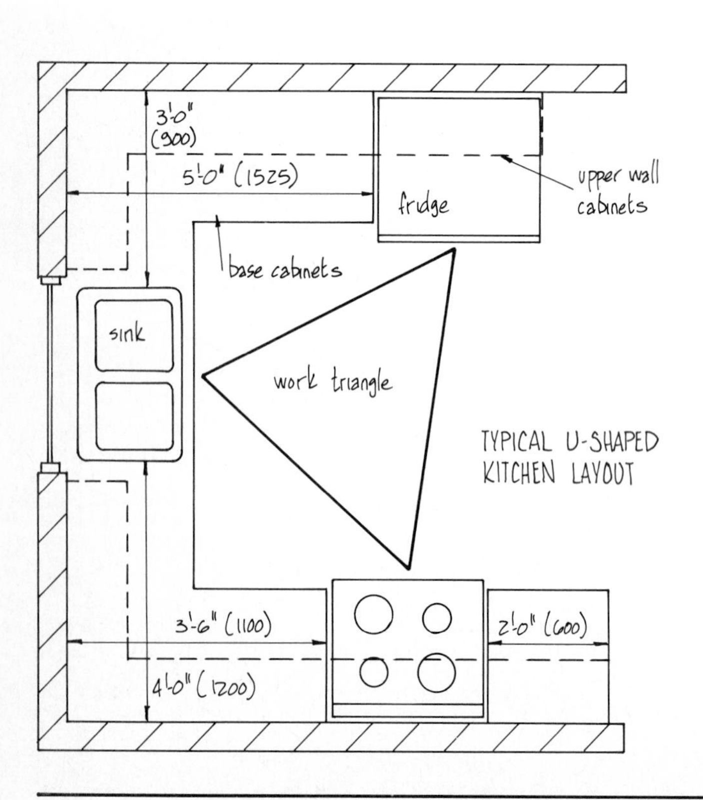

Illustrated above are the standard height and depth dimensions in general use by most manufacturers. All dimensions are based on comfortable usage for average-sized people and installation of standard-sized appliances. Although custom sizing can be obtained for persons shorter or taller than average, to do so may preclude the use of standard appliances and make the house more difficult to sell in the future.

Bathroom vanities may be designed with straight fronts as for kitchen cabinets or, as shown above, with a knee space for a seated person. Broom cupboards are similar to the clothes closet shown above but are generally only as deep as the upper wall cabinets (12 in / 300 mm).

See 13-23 for a selection of standard cabinets.

A typical U-shaped kitchen layout is shown at the left. The dimensions given will allow very efficient usage of the kitchen's facilities but will obviously vary in practice depending on room proportions.

APPLIANCE LOCATION

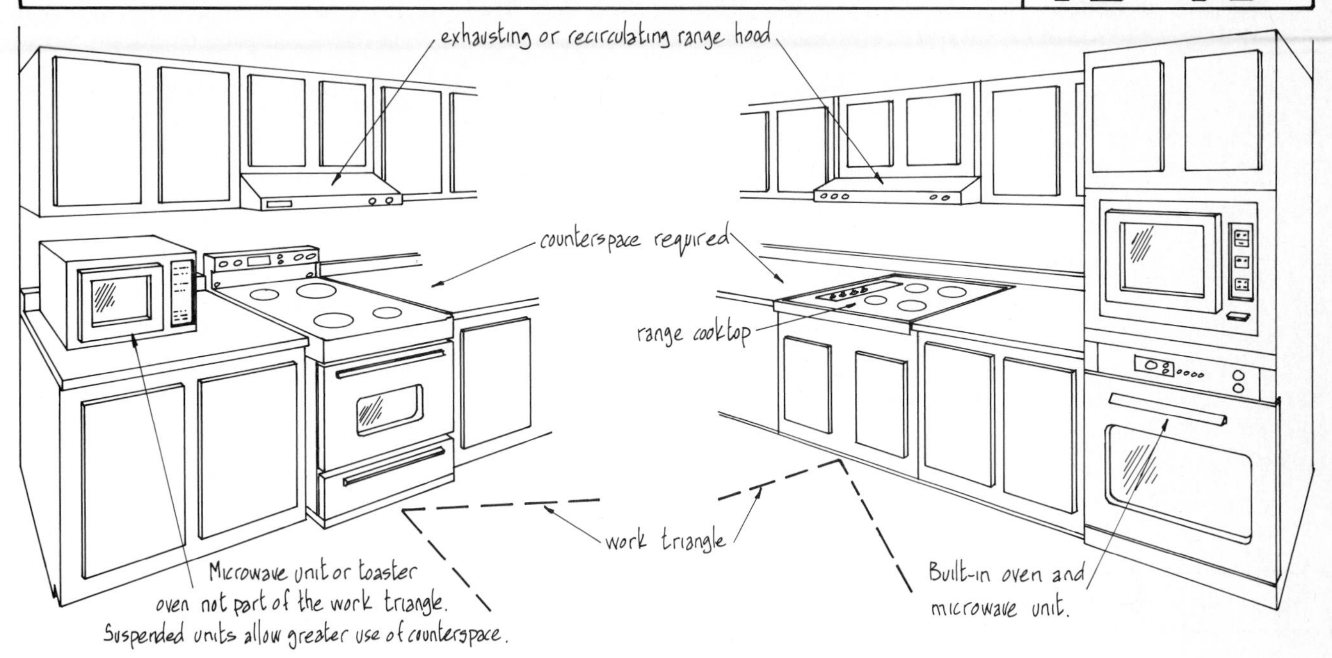

COOKING APPLIANCES

- These must be located with both safety and convenience in mind. It is important to have sufficient counter space on at least one side of the unit for a work or serving area.
- Built-in ovens or microwave units should be located to the side of a countertop cooking surface or range and need not be part of the work triangle since they are used less frequently.
- If separate microwave or toaster ovens are planned, countertop space for their location and a reasonable work area must be allowed.
- Range hoods are commonly used to exhaust cooking fumes from the house. In cold climates the duct must be heavily insulated and exhaust preferably through the soffit. For comments regarding range hoods in energy-efficient houses, see 14-18 (exhaust systems) and 14-21 (gas ranges).

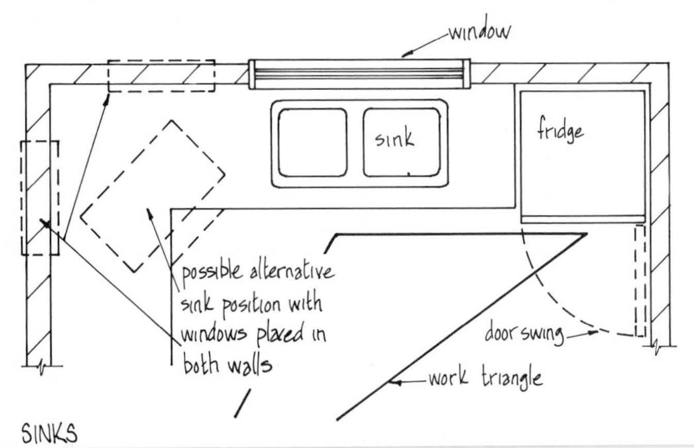

SINKS

- Sinks are best located at a window to provide good working light and for something interesting to look at when using the sink. Washing-up is especially boring when facing a blank wall.
- A double-bowl sink should be used whenever possible. Single bowls are less efficient and should be used only when available space is absolutely minimal.
- Sink units are installed in a hole cut in the countertop and fastened with bolts or clamps from beneath. More expensive units incorporate a draining board that extends to one or both sides of the bowls.

REFRIGERATORS

- The refrigerator door should be hinged to open onto the work triangle for greatest convenience.
- Counter space next to the refrigerator on the opening side is also desirable.
- If the unit is to be placed in a corner, sufficient clearance is required for door operation and shelf removal.

DISHWASHER

- Even though a dishwasher may not be installed initially, space should be made available in the cabinet layout.
- The unit is generally located next to the sink but on the opposite side from the garbage disposal when this unit is installed in a double-bowl sink.
- Where possible, adjacent counter space and storage cupboards should be included.

DOORS

- All work areas should be kept away from the path of a door swing.
- Injury to persons or damage to cabinets or appliances will result if this problem is not recognized and corrected in the planning stages.
- Remember that a dishwasher will require full plumbing services for its operation for which provision must be made before the unit's installation.

MATERIALS

There are a wide variety of materials used in cabinet fabrication ranging through solid lumber, plywood, composition boards, and plastic laminates. In general, the quality and cost of a particular cabinet style determine the ratio of solid lumber to composition materials. Lower quality cabinets will have a greater ratio of man-made materials.

Composition products do, however, have some distinct advantages over solid lumber. The composition boards are especially useful since they are available in large sheets and are not as subject to humidity expansion and contraction as solid material. They are also cheaper than lumber, readily available, and of consistent uniform composition. Depending on quality and style, solid lumber will generally be used only for some parts of the framework and for doors or trim.

SOLID LUMBER

Commonly used material is pine for frames and/or faces, plus oak for door faces and trim. Large fronts are generally made from smaller boards tongued-and-grooved or splined together.

Solid Lumber

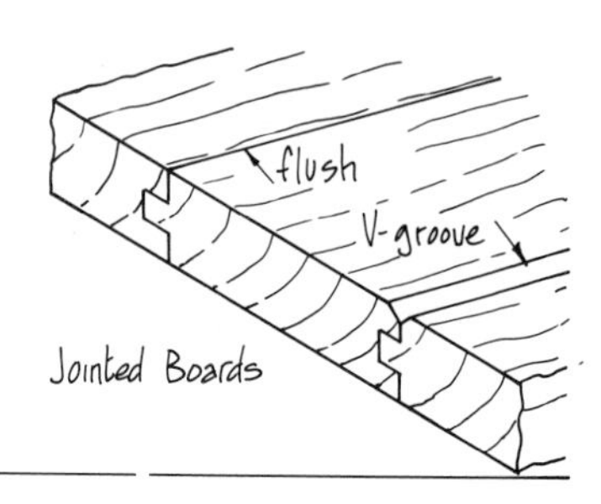

Jointed Boards

COMPOSITION BOARDS

There are several types, each having different properties and uses:

Particleboard, made from finely graded wood chips and sawdust in an adhesive base, is used extensively.

Hardboard, made from glued softwood pulp, is quite strong and durable.

Chipboard is made with larger-sized wood shavings in a glue base. Splits relatively easily and has limited use in cabinetmaking.

TYPICAL EDGE TRIMS

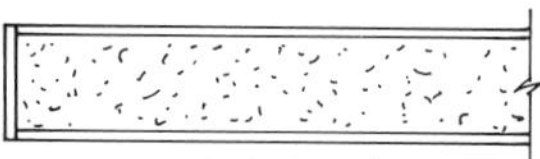

Glue- or Stick-On Veneer

Plastic Insert

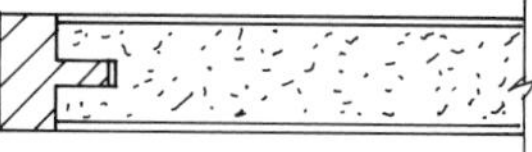

Hardwood Insert

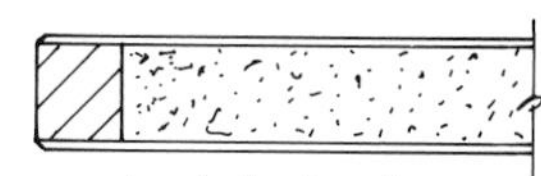

Inset Hardwood

PLYWOOD

A wide variety of plywoods are available. There are two basic groups:

General purpose, for interior panels and some outer panels in cheaper units.

Special purpose, for high-quality face veneers. Veneers may be softwood or hardwood and grain-matched in specific patterns as shown.

Book-Match

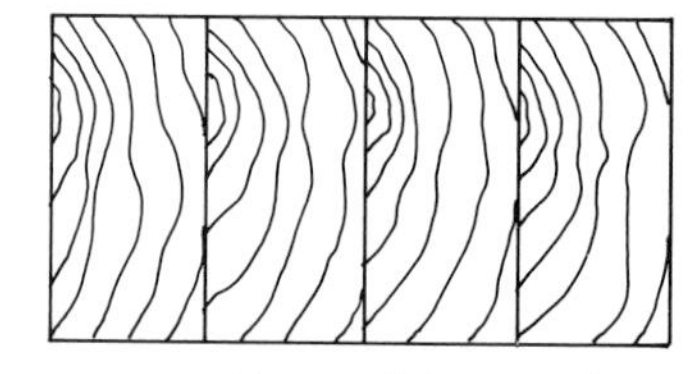

Slip-Match

BLOCKBOARD (LUMBER CORE PLYWOOD)

This product has a core of edge-glued solid lumber with one or two outer veneer plies each side. Its main advantage is that it holds nails and screws well and presents a more finished edge.

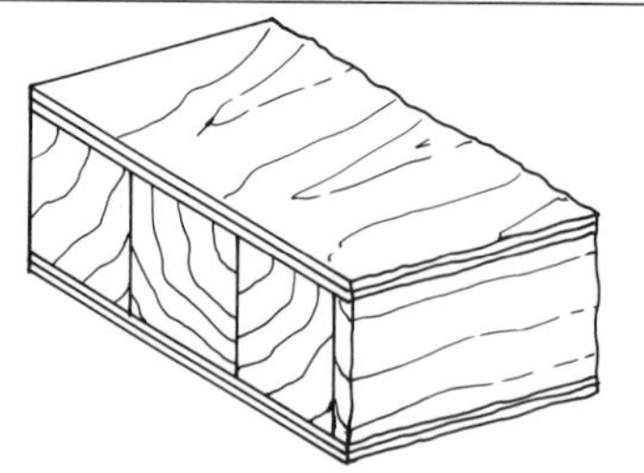

In "blockboard" the solid lumber core is usually assembled without glue.

In "laminboard" the lumber core is in thinner strips which are glued face to face. This produces a better-quality product less subject to surface cracks.

PLASTIC LAMINATE

This is an extremely important product in cabinet fabrication. A wide variety of finishes are available in an immense range of colors and patterns. Laminates are resistant to heat and most household chemicals and are easily cleaned. Laminate sheets are glued to wood cores and edge trimmed as shown.

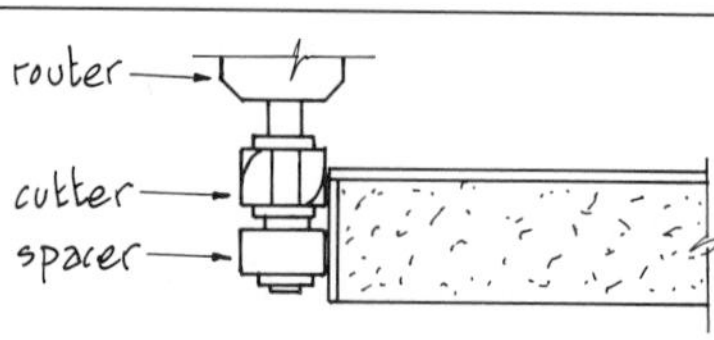

Square Edge

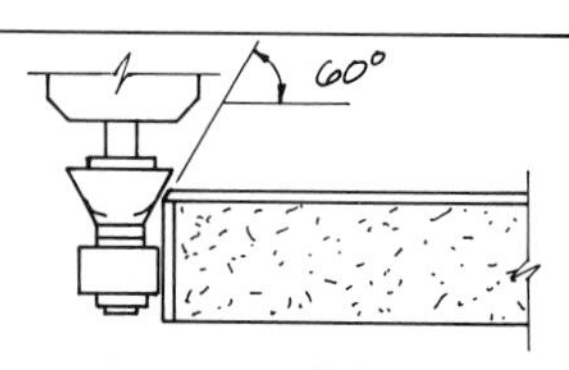

Bevelled Edge

In addition to the edge finishes shown, trim pieces and inserts similar to those for composition material are available.

TYPICAL CABINET ELEMENTS

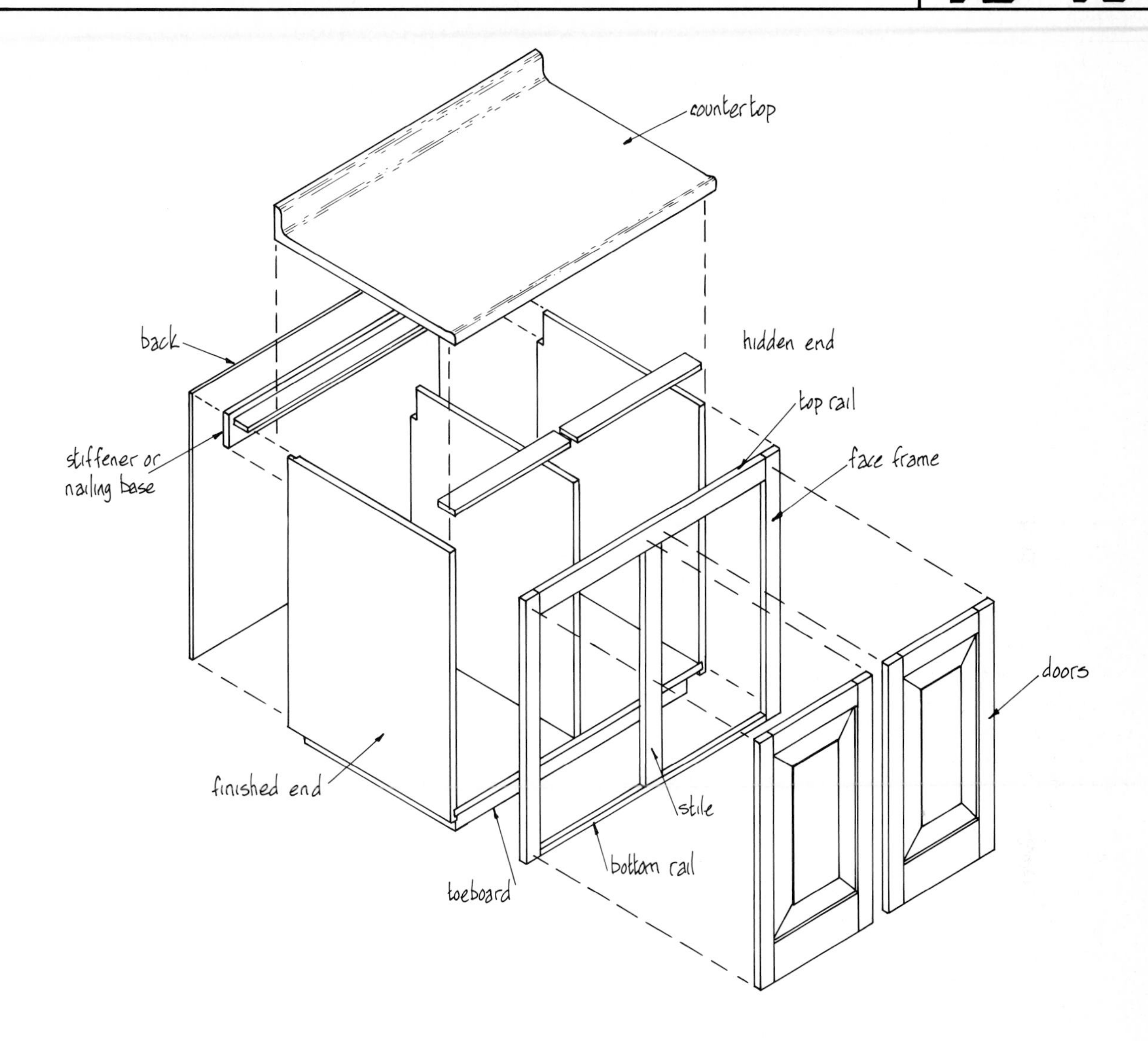

Illustrated above are the elements of a typical kitchen base cabinet. There are many other ways of designing and building cabinets but all follow the same fundamental principles:

- A cabinet is essentially a box with a back, top, bottom, and front, all assembled in a way that ensures structural integrity.
- The inner cabinet space can be divided by partitions, shelves, or drawers to create specific storage options. Specialty options such as "Lazy Susans" (see 13-21) allow maximum use of corner cabinets.
- A full kitchen cabinet set is assembled from a series of connected individual cabinets. The appearance of continuity is achieved by fabricating a single countertop and base that spans the complete set length.
- A cabinet may be assembled from prefabricated standard units or totally custom designed and fabricated for a specific application.
- There are many options available for drawers, doors, and countertops, some of which are detailed on the following pages.

FACE STYLES AND DOOR TYPES

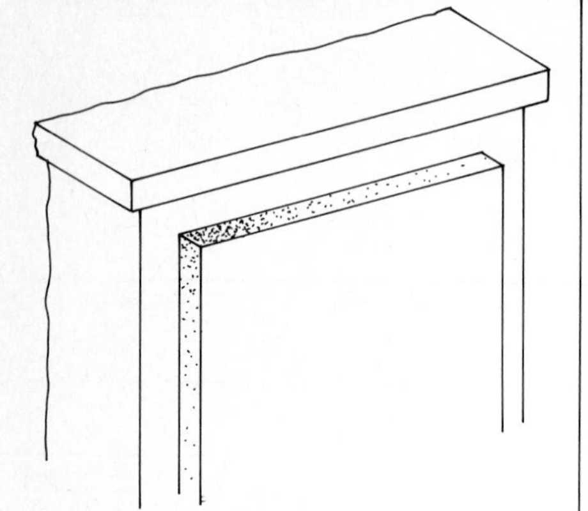

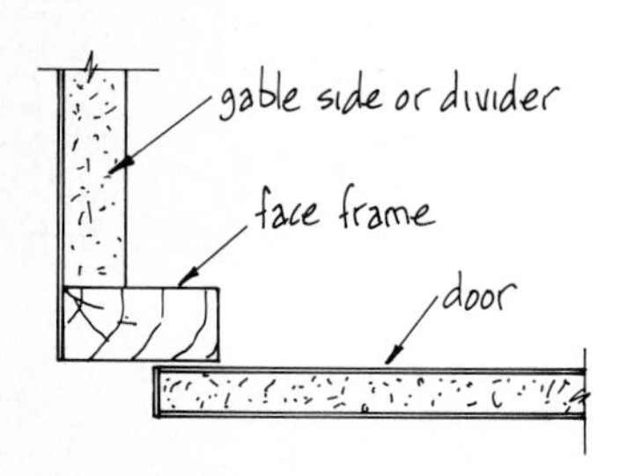

OVERLAY FACE

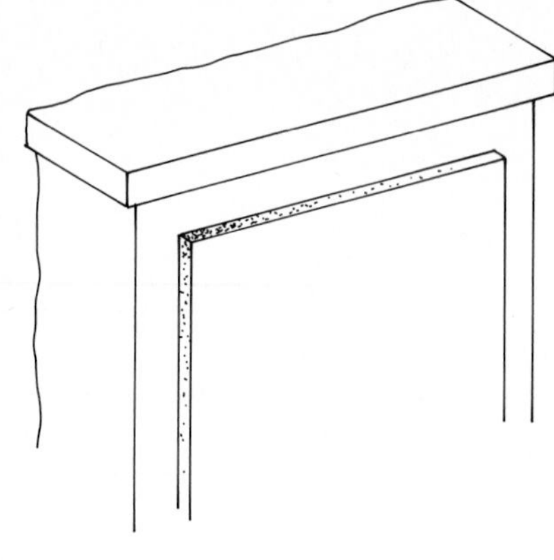

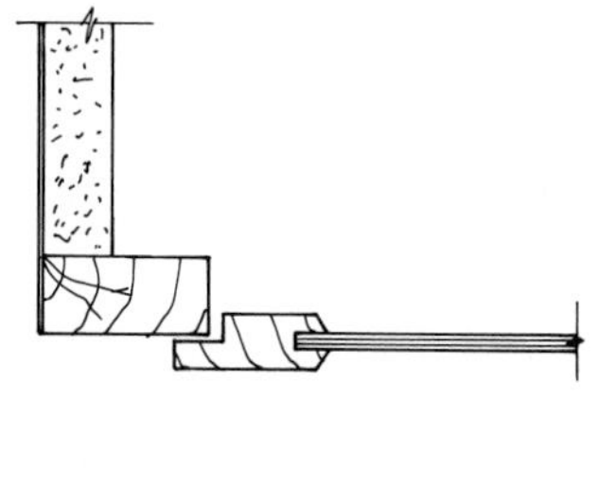

LIP FACE

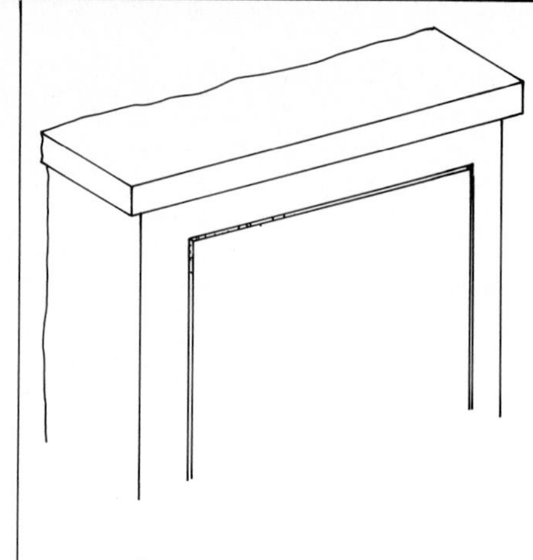

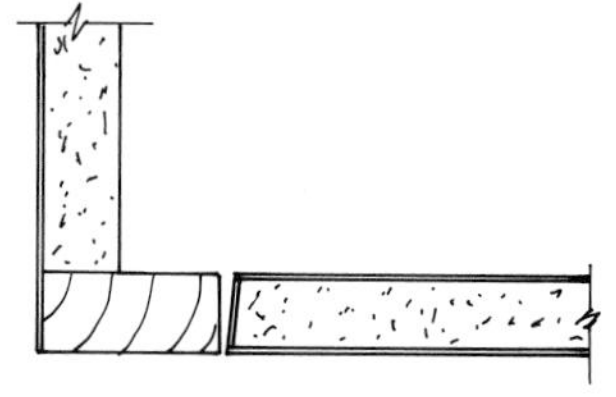

FLUSH FACE

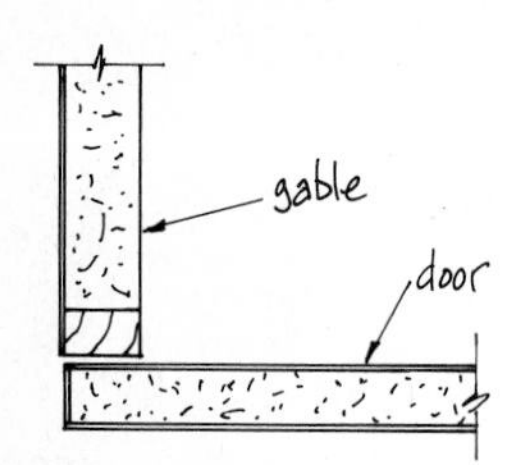

OVERLAY FACE without a face frame

The appearance of a cabinet depends on the type of door and drawer face style used. Shown above are the three main styles.

- Overlay. The total thickness of the face is outside the face frame and is larger than the opening. The face may partially or completely cover the cabinet front. If completely covered, the faces must be carefully made and fitted to reduce visible inaccuracies.
- Lipped. The outer face is larger than the opening, but the edges are rabbeted to fit partially into the opening. Inaccuracies are not as noticeable with this type of face.
- Flush. The face is set flush with the face frame requiring great accuracy in fabrication and fitting. An even and square clearance must be visible on all sides.

DOORS

Cabinet style is also determined by the way the doors are fabricated and the materials that are used. Illustrated at the right are some of the standard door types. Drawer fronts may also be made in the same way to provide continuity or in a different style for contrast.

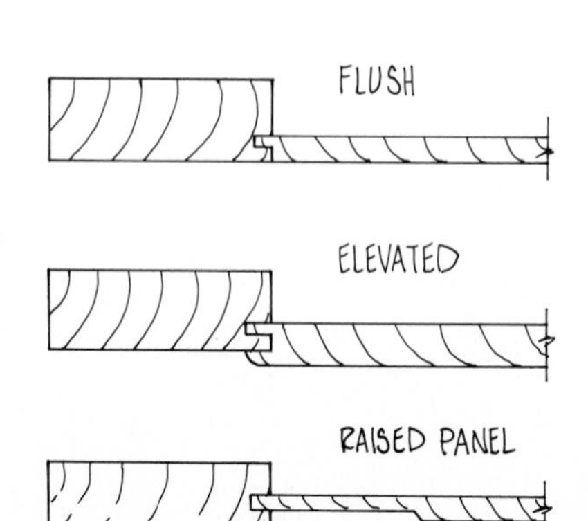

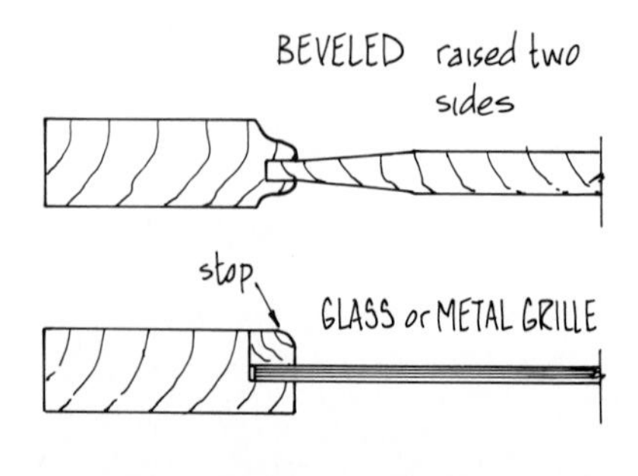

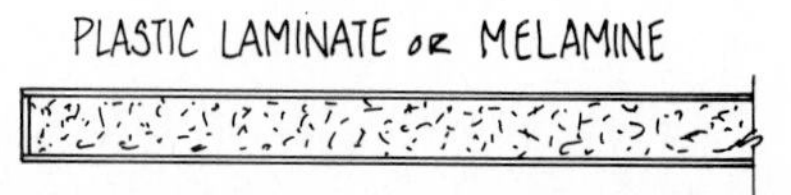

CORNER TREATMENTS

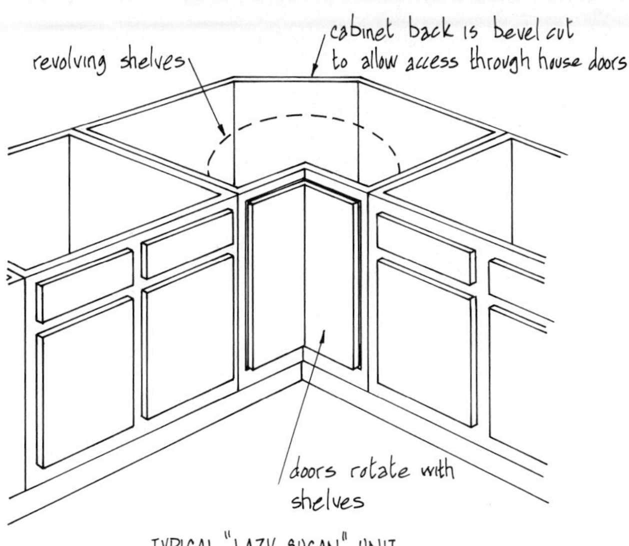

TYPICAL "LAZY SUSAN" UNIT

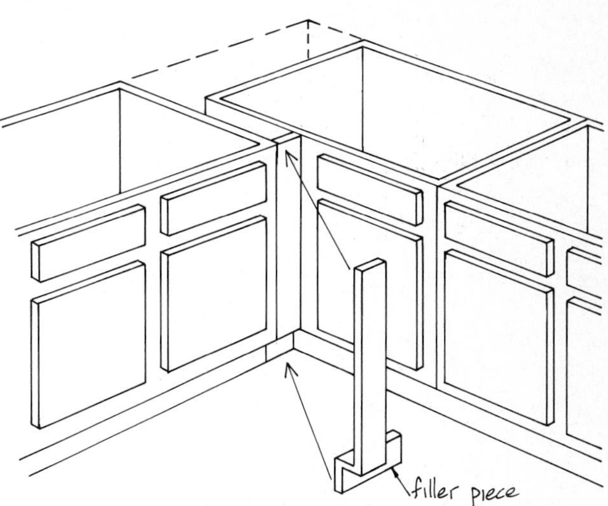

BY-PASS CORNER ARRANGEMENT

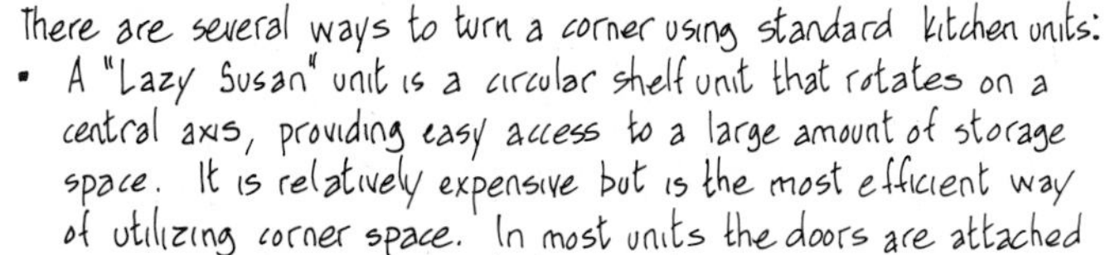

There are several ways to turn a corner using standard kitchen units:

- A "Lazy Susan" unit is a circular shelf unit that rotates on a central axis, providing easy access to a large amount of storage space. It is relatively expensive but is the most efficient way of utilizing corner space. In most units the doors are attached to, and rotate with, the shelves.
- A by-pass cabinet unit is designed to partially overlap the opposing standard cabinet, providing a reasonable amount of storage space in the hidden portion. With this layout a filler piece must be inserted between units to allow space for the doors to open. The filler may also be used to take up odd wall dimensions.
- A third alternative is to use standard units and a corner filler as shown at the right, which is the most economical but wastes all corner storage space. To partially solve this problem the cornering end gables of each cabinet are removed and a floor section is added to the empty corner.
- A similar selection of upper wall cabinets is available, as shown below.

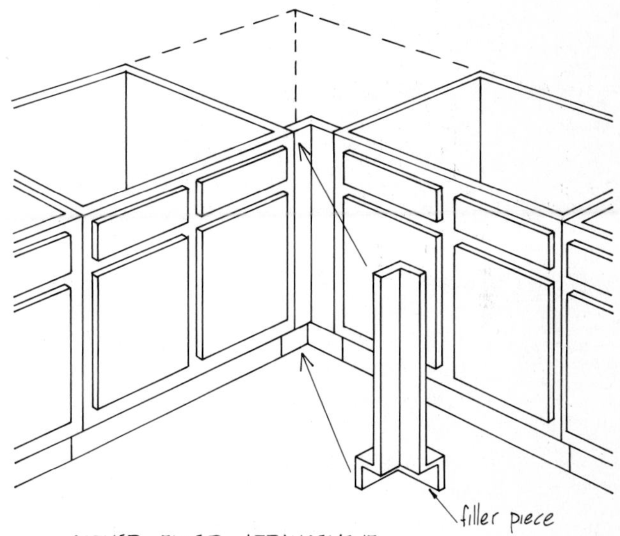

CORNER FILLER ARRANGEMENT

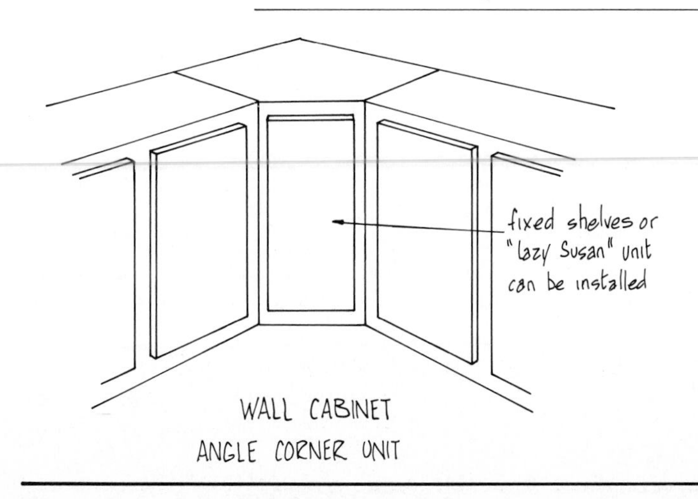

WALL CABINET
ANGLE CORNER UNIT

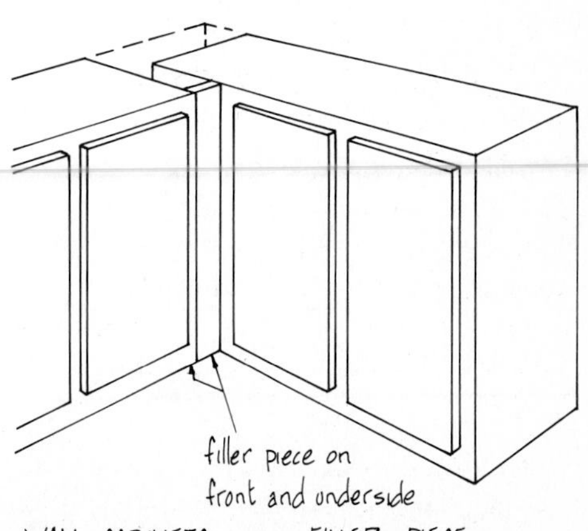

WALL CABINETS WITH FILLER PIECE

DRAWERS AND COUNTERTOPS

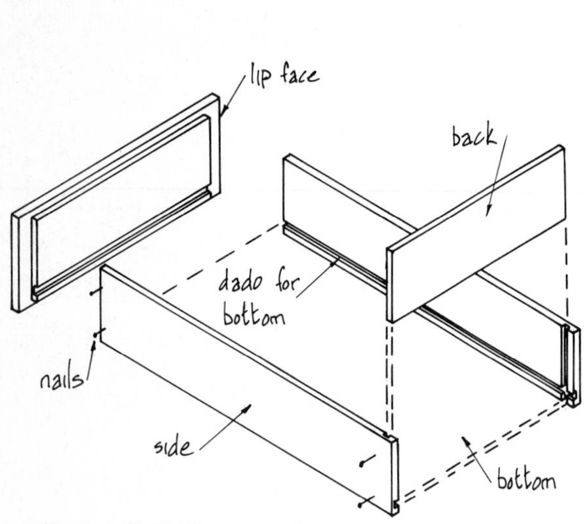

TYPICAL LIP-FACED DRAWER

EXAMPLE OF DRAWER SLIDE MECHANISM

DRAWERS

- Drawer faces will generally conform to the style and type of door face used (see 13-20).
- Methods of drawer fabrication will depend largely on the quality of the cabinet set and the style of the face. The parts drawing above is typical of an average-quality lip-faced drawer, but many other methods are also used.
- The level of quality will also determine the type of drawer guide mechanism. A typical example is shown above right, together with an example of a cabinet door hinge.

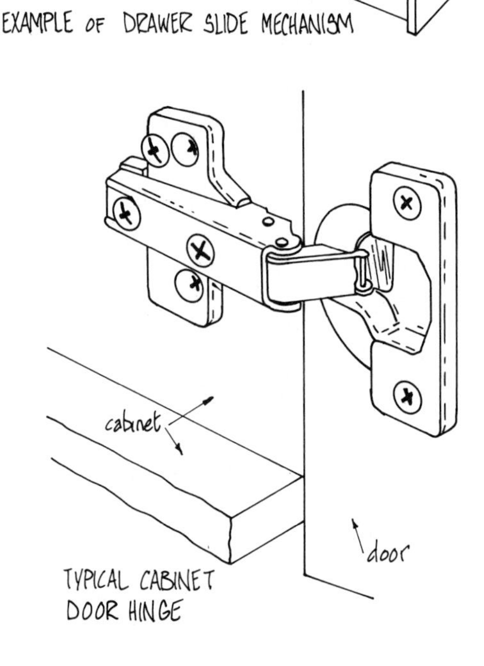

TYPICAL CABINET DOOR HINGE

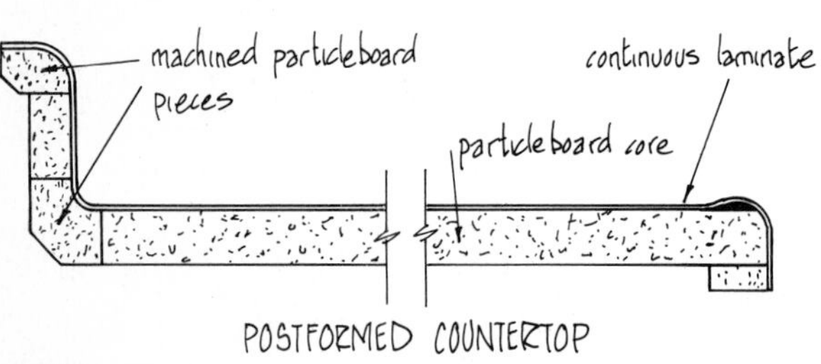

POSTFORMED COUNTERTOP

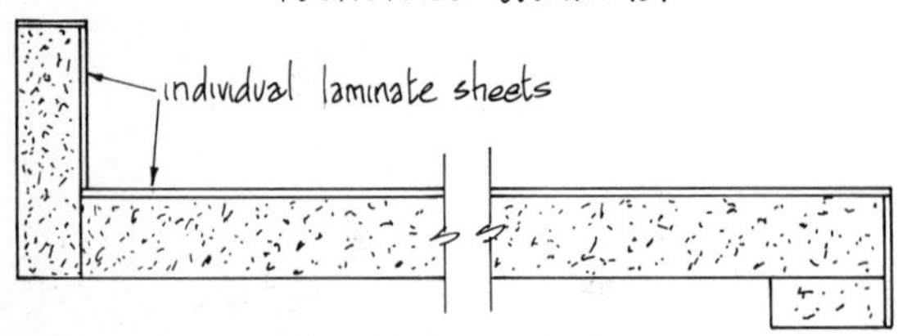

SELF-EDGED COUNTERTOP

Other materials can be used as a countertop finish: ceramic tile, hardwood strips in "butcher-block" style, or natural or synthetic marble are common.

COUNTERTOPS

For most kitchen sets, countertops are made with a particle-board or plywood base that is overlaid with plastic laminate. There are two fabrication methods:

- Postformed, where the laminate is factory applied under heat and pressure using thermosetting glue. Tight rolls and bends can be formed, eliminating all joints between laminate faces and providing a smooth, flowing surface.
- Self-edged, which can be factory or job fabricated and has individual laminate sheets glued to each flat base surface. This type has a "square" look, with many joints and visible edges.
- Countertops of either type are generally fabricated as continuous, single units complete with right-angle mitered corners and end pieces.

STANDARD CABINET UNITS

Illustrated here is a small selection of the standard cabinet units produced by most millwork manufacturers. Refer to suppliers catalogs for the full range of available units.

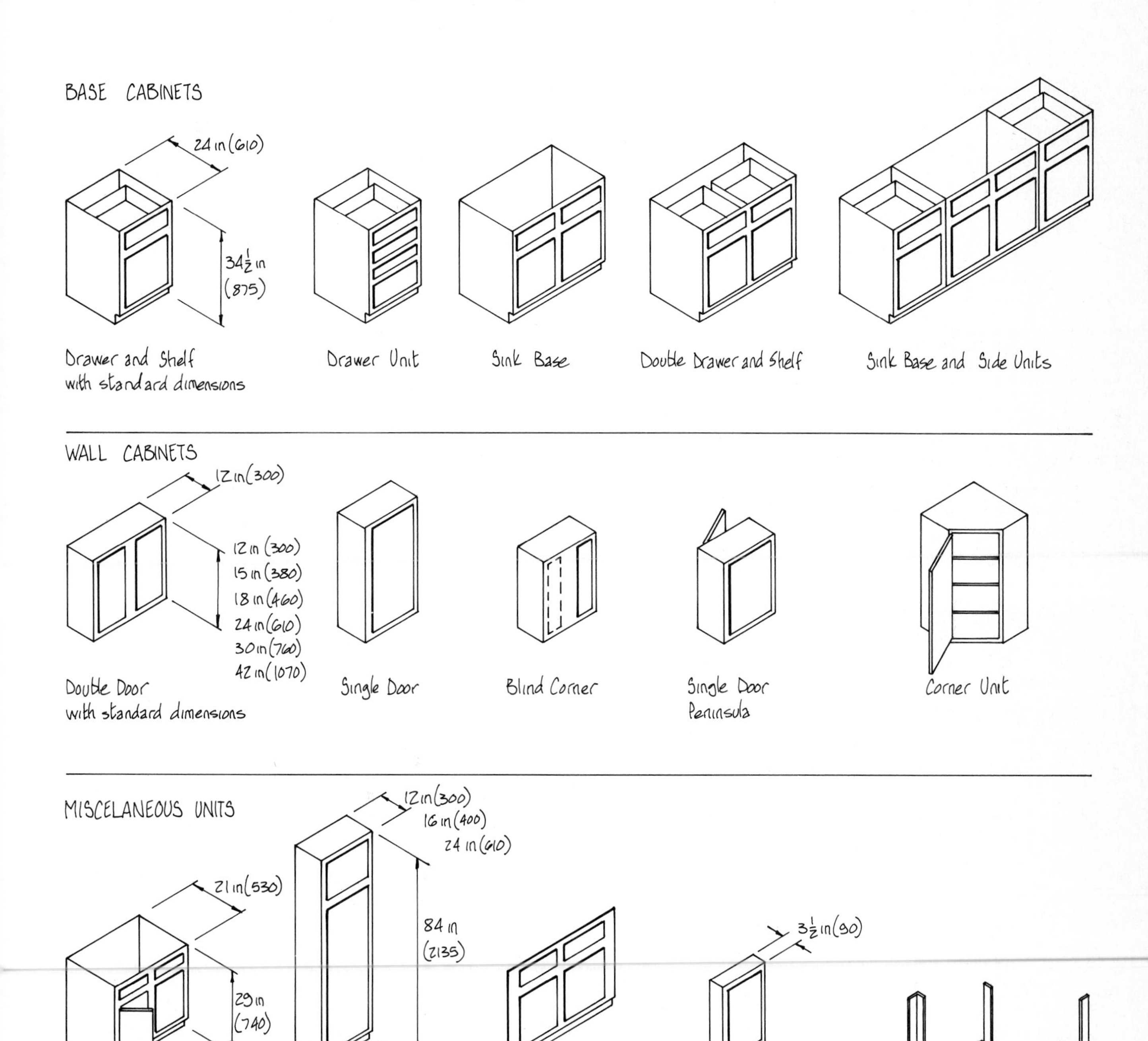

Note: All metric dimensions are rounded soft conversions and are not necessarily metric standard dimensions.

CHAPTER PAGE TITLES

14-2 GENERAL INFORMATION

SYSTEM REQUIREMENTS

- A heating or cooling system must be able to maintain acceptable temperatures in the dwelling space during the coldest part of the winter or the hottest part of the summer.
- In doing so, the system must distribute the heat or cool evenly and without discomfort to the occupants.

SYSTEM SELECTION

There are two main factors in the proper selection of heating or cooling equipment:

- Available fuels. The range of fuels that are economically available at a specific location can vary considerably. Urban sites usually have access to the full range of major energy sources, while rural sites may be limited to two or three options.
- Required system capacity. A reasonably accurate estimate of the heat loss or heat gain of a building is essential. A system that is oversized is uneconomical in terms of both initial capital cost and running cost and can also lead to occupant discomfort. Ideally, the system chosen should have an output that is equal to or slightly greater than the calculated load.

VENTILATION AND MAKEUP AIR

An additional factor in equipment selection is the need for adequate ventilation of the conditioned dwelling and a possible need for "makeup air". Makeup air replaces air that is lost to the outside when used by some combustion-type heating equipment (see oil and gas furnaces, 14-4 and 14-15).
Provision for ventilation and makeup air is especially important in airtight energy-efficient dwellings where air leakage through the structure is carefully controlled. This problem is discussed from 14-16 onward.

ENERGY-EFFICIENT BUILDINGS — EQUIPMENT SIZING

There are two significant factors that must be considered when sizing and choosing heating equipment for energy-efficient housing:

- The method of heat loss calculation must recognize that the structure is extremely airtight and will therefore lose very little heat through air leakage. This is an important factor since most older calculation methods assumed large amounts of air leakage and consequent high heat losses.
- In a well-designed and highly insulated house, the total heat loss will be very small. It may prove quite difficult to find standard combustion-type heating equipment with a capacity that does not exceed the required demand. One approach is to adjust and downrate the smallest available combustion furnace but this results in large inefficiencies. The designer may find that the only recourse is to choose a system such as a heat pump and/or electric furnace that has both low capacity and high efficiency. If this option is chosen, makeup air is not required and results in even greater system efficiency.

FUELS AND EFFICIENCY

HEAT CONTENT OF FUELS

FUEL	CUSTOMARY	METRIC
natural gas	1000 Btu / cu.ft	100 kJ / m^3
electricity	3413 Btu / kWh	3.6 MJ / kWh
oil (domestic)	138,000 Btu / US gal	38 MJ/L
propane	108,000 Btu / US gal	30 MJ/kg
wood (dry)	8000 Btu/lb	18 MJ/kg
coal	13,500 Btu/lb	31 MJ/kg
kerosene	124,000 Btu / US gal	34 MJ/L

FUEL HEAT CONTENT

Each fuel type has a certain heat content. The major fuels, such as oil and gas, have energy values that vary little throughout North America. Other fuels, such as wood and coal, have an energy value that varies according to type and condition. Electricity, of course, has a fixed value per kilowatt.

HEATING EQUIPMENT EFFICIENCY

The efficiency of a heating unit is based on the amount of heat actually delivered to a unit compared to the total heat available from the fuel consumed. E.g., a furnace with a 70% efficiency delivers 70 units of heat to the house for every 100 units available in the fuel. The other 30 units are lost up the chimney as waste heat.

- Combustion units must vent flue gasses (and heat) to the outside. Efficiency increases in proportion to the amount of waste heat recovered.
- Electric heating has a 100% efficiency since all energy consumed is converted directly to usable heat. Heat pumps, which do not create but only move heat, have an efficiency greater than 100% (see 14-7).
- Wood-burning units have a wide range of efficiencies. Open fireplaces without any type of heat-capturing device can actually have a negative efficiency, while well-designed airtight wood stoves can approach 60% efficiency.
- Efficiency is measured in two ways: steady-state, measured only when the furnace is running, and seasonal, measured over the whole heating season. The seasonal figure is the lower but more accurate of the two. Always use this figure when choosing a unit.

SEASONAL HEATING EFFICIENCIES

conventional gas furnace	60 %
condensing gas furnace	85-95%
pulse gas furnace	90-98%
conventional oil furnace	50-60%
gas furnace with heat pump	120-130%
electric furnace or baseboards	100%
electric furnace with heat pump	150-250%
airtight wood stove or insert	20-50%
kerosene heater	98%

Actual efficiency will vary with type of equipment, efficiency options employed, and climate.

Actual cost of fuel consumed compared to delivered cost of fuel

$$\text{actual fuel cost} = \text{delivered fuel cost} \times \frac{100}{\text{seasonal efficiency of equipment}}$$

ACTUAL FUEL COST

In heating systems with a seasonal efficiency of less than 100%, the actual fuel cost to heat the dwelling is found by dividing the fuel's base cost by the equipment's efficiency rating. For example, to supply one unit of heat to a dwelling, a furnace with a 60% rating must actually burn 1.66 units because of heat lost in the combustion process. This additional cost must be taken into account when comparing heating systems.

14-4 GAS / OIL FURNACES

CONVENTIONAL GAS FURNACES

- There are two types of gas burners used in conventional furnaces : atmospheric and forced draft.
- In the atmospheric system ("conventional" furnace) the gas is supplied to the burner at low pressure and mixed with the required amount of air for combustion. The hot gases pass through a heat exchanger, where house air (or water) is heated. The waste gases are then vented through a chimney to the outside. Flue gas temperature is 300-500°F (150-260°C). Note, in the illustration opposite, that the draft hood and combustion chamber are open to the flue, which allows large quantities of warm house air to escape up the chimney even when the chimney is not operating.
- In the forced-draft system a small electric fan exhausts combustion gases to the outside without the need for a chimney or draft hood. A small flue that exits through a wall is normally used.
- Both types of furnace can be made more efficient by replacing the continuously lit pilot light with a spark ignition system and installing a flue damper to stop warm-air escape after burner shutdown.

CONVENTIONAL OIL FURNACES

Except for the type of burner used, oil furnaces operate in much the same way as gas furnaces.

- A spray of oil droplets is forced into the combustion chamber by a pump and ignited by spark ignition. In less efficient types there is a period of up to 25 seconds where combustion is incomplete because of low fuel line pressure and low draft from the burner fan.
- Choose a furnace that overcomes this problem by means of a solenoid valve on the fuel line that keeps fuel from the burner nozzle until adequate pressure and a good draft are established.
- Conventional furnaces are now generally equipped with a flame retention head that gives a hotter, more efficient flame and a flue damper that reduces warm-air losses up the chimney after burner shutdown.

CONTINUOUS CONDENSING FURNACES

- These are very efficient furnaces that capture most of the heat normally lost up the flue. Efficiencies of up to 95% are standard.
- Each has two or more heat exchangers. The first functions in much the same way as a conventional furnace, while the second (and third) heat exchanger, which is usually made of stainless steel, condenses the water vapor in the flue gases to extract the latent heat contained therein. The condensate is drained to a house sewer.
- With this process the flue gas, exhausted through a small flue by an induced draft fan, has a temperature generally between 100 and 180°F (40-80°C) and a PVC flue pipe is often used.
- Spark ignition and a direct outside air feed to the combustion chamber are also standard.
- In a condensing gas furnace, either a conventional burner or a pulse combustion burner is used. A pulse burner has a specialized combustion chamber where gas burns in a rapid series of pulses that push the hot gases through the heat exchangers. Neither pilot light or conventional burner is required for this system.
- A condensing oil furnace is very similar to a gas condensing furnace and has the same energy-saving features.

GAS / OIL FURNACES

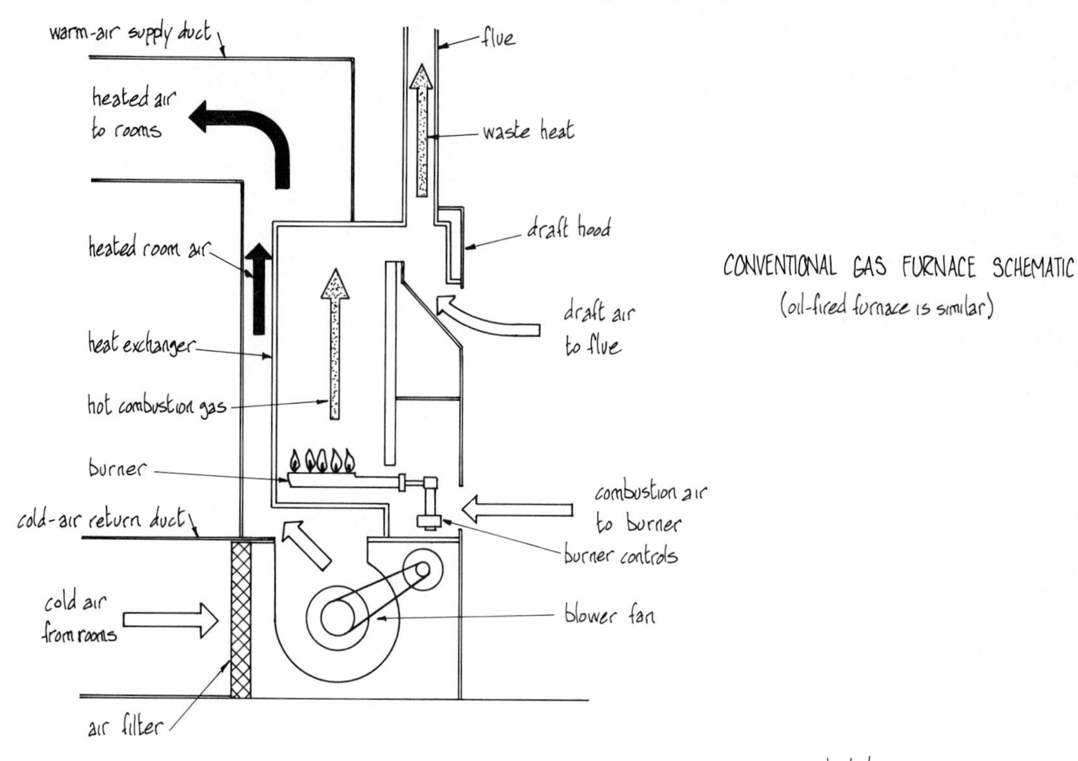

CONVENTIONAL GAS FURNACE SCHEMATIC
(oil-fired furnace is similar)

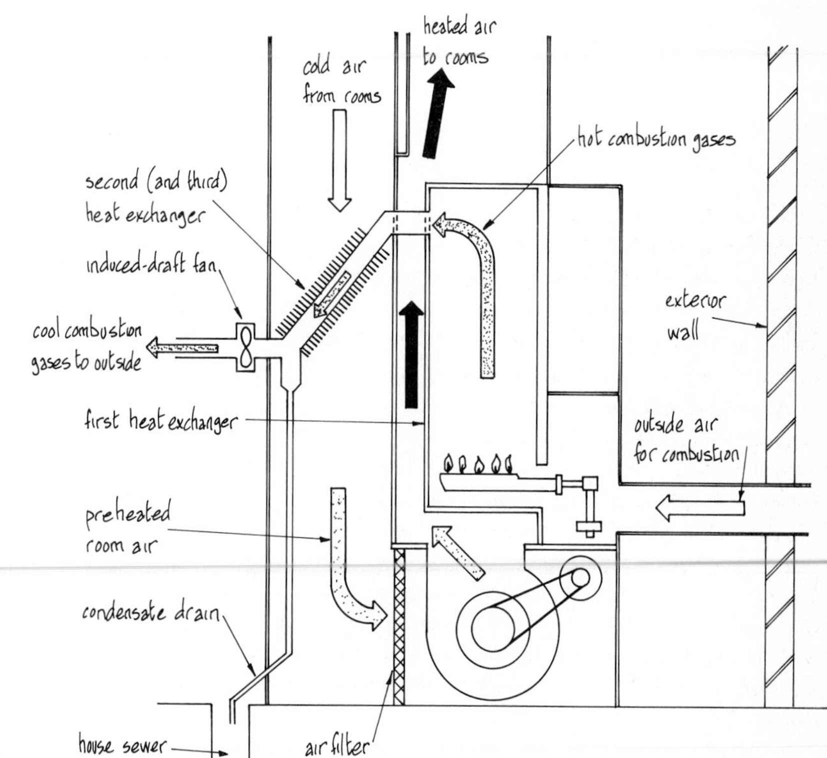

HIGH - EFFICIENCY CONTINUOUSLY - CONDENSING GAS FURNACE SCHEMATIC
(oil-fired furnace is similar)

Note:
For the sake of simplicity, the second (and third) heat exchanger is shown in the cold-air return duct in the schematic drawing. In practice, however, it is actually located inside the furnace where it performs its function of preheating the returning cold air.

ELECTRIC HEATING

ELECTRIC FURNACE

An electric furnace is a relatively simple device compared to gas or oil furnaces.

- There is no burner or flue, and outside air is not required for the heating process. Efficiency is essentially 100%.
- The furnace consists of a series of resistance heating coils, a fan, and appropriate control circuits.
- The number of heating coils in a furnace is determined by the rated capacity of the unit. Coils tend to be rated at 4-5 kW each but vary with the manufacturer.
- Because an electric furnace can be rated at any desired capacity simply by adding or subtracting coils, they are ideal for energy-efficient houses where the low heat loss may rule out any other type of higher-fixed-capacity furnace.
- Choose a furnace that turns on the coils in stages when demanded by the thermostat.

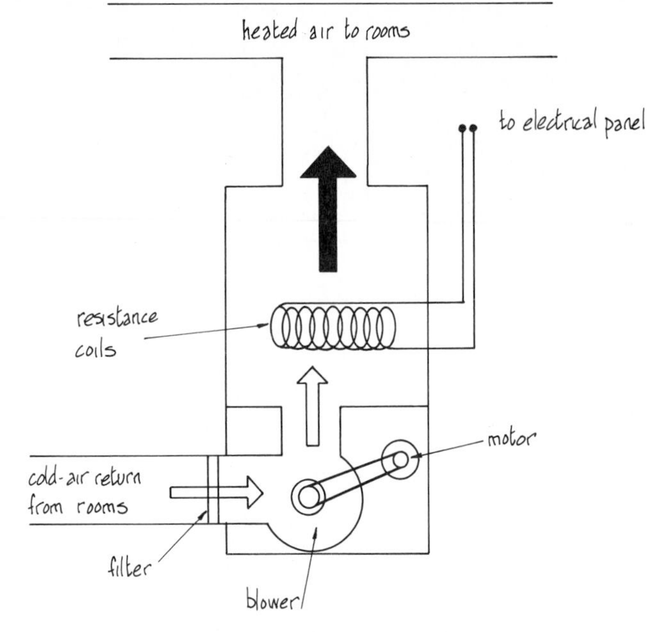

ELECTRIC FURNACE SCHEMATIC

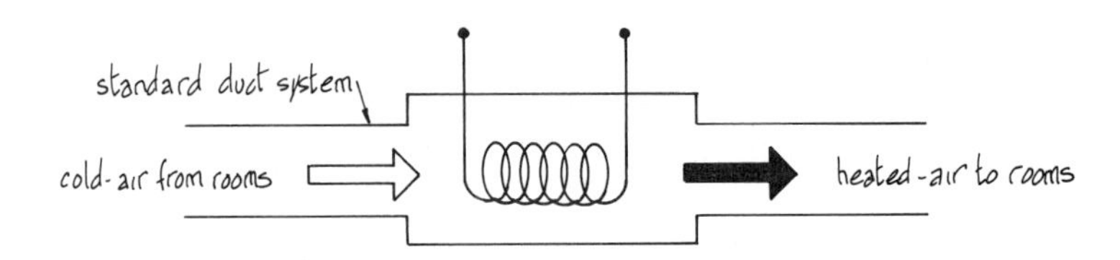

IN-DUCT ELECTRIC HEATER SCHEMATIC

ELECTRIC CONVECTOR HEATING

- Usually found in the form of baseboard heaters, although single-unit heaters set in floors or walls are also available.
- Most systems have the advantage of individual room temperature control through thermostats located in each room. Selected rooms can therefore be closed off or kept at lower temperatures if desired.
- Resistance wires or rods are used to produce heat. Some units have built-in fans to distribute the heat more evenly.
- There are some disadvantages to this type of heating system. Although initial installation costs are quite low, operating costs tend to be high in relation to other systems. There is very little air circulation, leading to temperature stratification and problems with humidity control. In addition, it is very difficult to add features such as heat pumps, air-conditioners, filters, or heat exchangers to the system.

IN-DUCT HEATER

An in-duct heater can be used as an "add-on" to an existing fan-forced furnace system, or as the sole source of heat in a new or renovated system (a blower will also be required in this case).

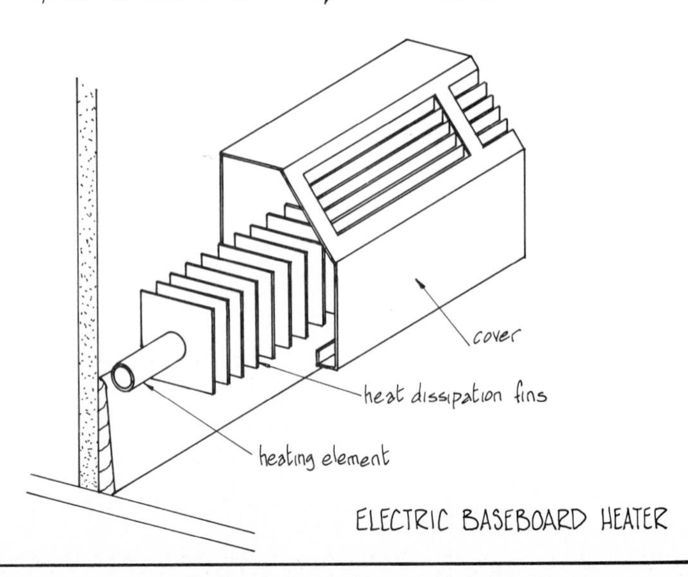

ELECTRIC BASEBOARD HEATER

HEAT PUMPS

GENERAL DESCRIPTION

A heat pump is a form of electrical heating but is considerably more complex mechanically than the heating systems seen so far.

- A heat pump does not create or produce heat itself – it merely moves heat from one place to another using the same mechanism that a refrigerator uses to move heat from the cold interior to the outside.
- In winter the pump extracts heat from the outside air and transfers it inside the house. In summer the process is reversed and interior heat is pumped to the outside as in a conventional air conditioner.
- Heat pumps utilize the "refrigeration cycle" and since it takes less energy to simply "move" heat than to generate it, heat pumps can have energy efficiencies of well over 100%.

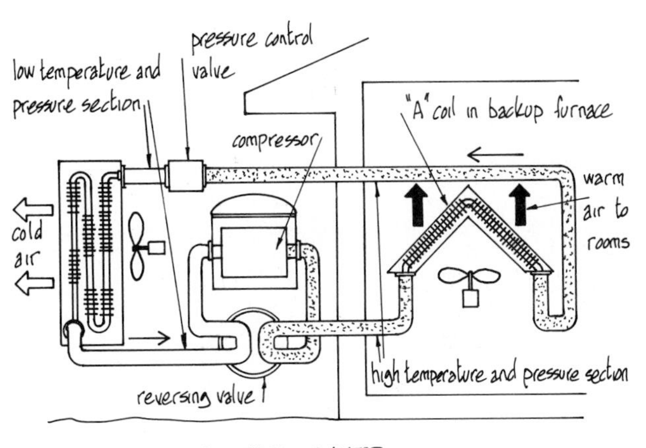

HEATING CYCLE – WINTER

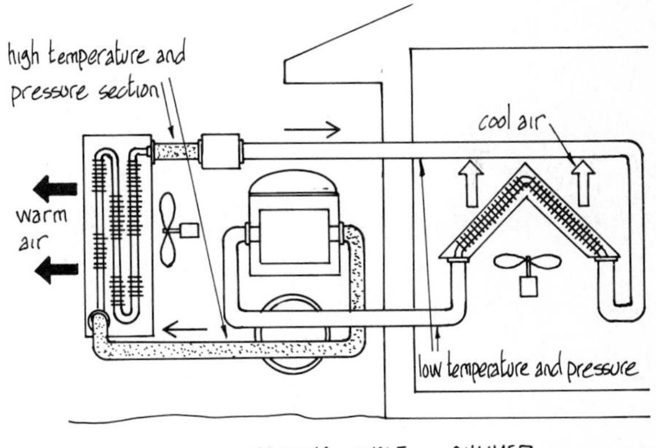

COOLING CYCLE – SUMMER

THE REFRIGERATION CYCLE – AIR-TO-AIR SYSTEMS

Heat pumps work on an evaporation-condensation cycle and transfer heat by circulating a refrigerant such as Freon through the system.

- On the heating cycle, refrigerant is evaporated at low pressure in the outside coil, where it absorbs heat from the outside air. The compressor pumps the refrigerant vapor to the inside coil, where it condenses under high pressure and gives up its heat to the house air being blown through the furnace. The liquid refrigerant then returns to the outside coil for evaporation again. This process is continuous while the heat pump is in operation.
- In the summer the system is switched to air-conditioning mode and the reversing valve reverses the refrigerant flow so that heat is pumped from the house to the outside.

EFFICIENCY MEASUREMENT

Heat pump efficiency is measured in two ways:

- Coefficient of performance (COP). This is a measure of continuous performance and is dependent on exterior temperature. Most heat pumps have a COP of 3 at 50°F (10°C), meaning that the pump will supply 3 kW of heat for 1 kW of power used. Performance declines as temperature declines. At 32°F (0°C) COP is about 2.3 and at 0°F (-18°C) about 1.4. Even at this level the heat pump is still more efficient than electric resistance heating (COP 1) or gas/oil furnaces (COP 0.5 - 0.9).
- Seasonal performance factor (SPF). This is a measure of overall seasonal performance and varies with local climate. In very cold regions the SPF may average 1.3, while in mild regions an SPF of 2.3 or higher is usual. This must again be compared with the much lower SPFs of electric heating and gas/oil furnaces.

HEAT PUMPS

PRICE CONSIDERATIONS

Although heat pumps are very efficient, they are also quite expensive in terms of capital cost and all cost/benefit factors must be considered.

- An analysis of anticipated fuel savings versus the higher initial investment cost is important. There must be a reasonable pay-back period over which the extra costs are recovered.
- Best economy is achieved in regions with high fuel costs and/or a temperate-to-cool climate.
- Best economy is also achieved when the pump is used for heating and cooling. Heating use only will reduce investment return.

CONVENTIONAL BACKUP

In colder regions where a heat pump alone will not normally supply enough heat on the coldest days, a conventional heat source must also be added to the system.

- The cheapest and most practical solution is to add an electric hot air furnace to the heat pump. With this system the heat pump will run continuously on extremely cold days while the electric furnace will cycle on and off as determined by the room thermostat.
- If a gas or oil furnace is added, an "economizer switch" will probably be added to the control system. This turns off the heat pump and turns on the furnace at a preset outside temperature.

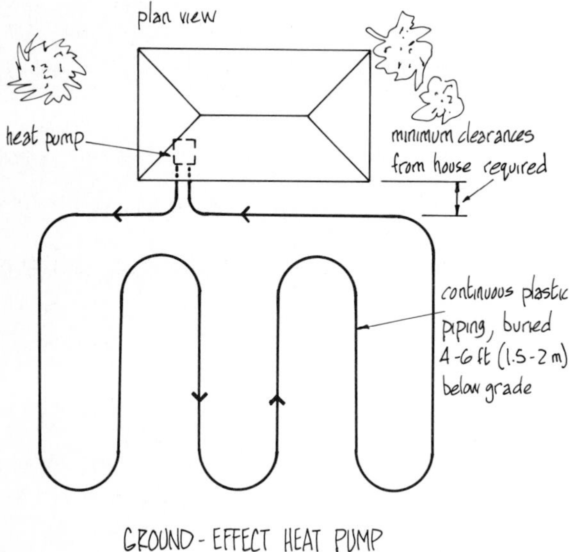

GROUND-EFFECT HEAT PUMP

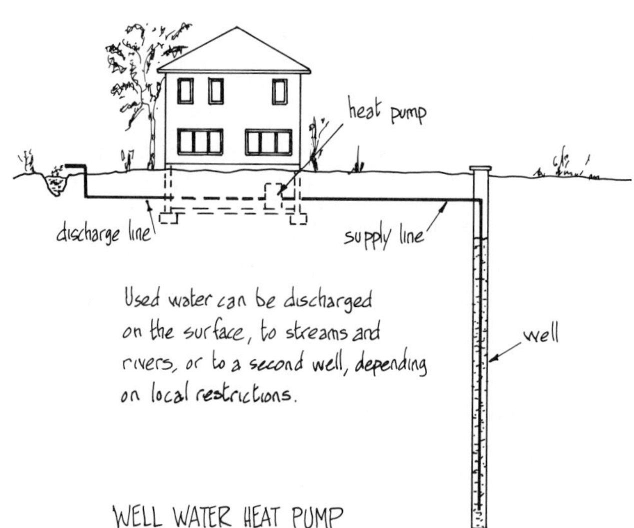

WELL WATER HEAT PUMP

OTHER HEAT PUMP TYPES

The heat pump system described on 14-7 was an air-to-air system that extracted heat from the outside air. Other heat sources (as shown above) can be utilized, such as:

- Ground-to-air, where either an antifreeze solution or Freon is circulated through a coil buried in the ground. Heat is extracted from the earth, which provides a large thermal mass at reasonable temperatures. Generally, coils are buried 4-6 ft (1.5-2 m) below grade and are 2000-5000 ft (600-1500 m) long.
- Water-to-air, where a large volume of water provides the heat source. Lakes, ponds, or streams are most effective, while wells can be used provided that sufficient water volume is available.

Other systems are available, and although most tend to be more efficient than the simpler air-to-air pumps, they also tend to be considerably more expensive.

FIREPLACES AND WOOD STOVES

TRADITIONAL FIREPLACES

A traditional fireplace produces very little in the way of usable heat to a dwelling. Its only value may be an aesthetic one, especially if the home owner does not have a source of wood available at little or no cost.

- Most open fireplaces actually have a negative efficiency, since the low pressure created by the chimney draft forces cold, outside air, to infiltrate through holes and cracks in the structure. The heat required to warm this infiltrating air is often more than that recovered from the fireplace itself.
- A hot fire can create so much draft that the dangerous flue gases of other combustion equipment can be forced back down their flues and into the house.
- Conversely, with a low, smoldering fire, the chimney draft is quite weak. Other functioning combustion equipment elsewhere in the house can create a greater draft and force the fireplace gases to flow backwards into the room.
- Chimneys built into the exterior wall of a dwelling will radiate heat to the outside as well as the inside. Where possible, locate chimneys as freestanding interior structures or as part of an interior wall.

EFFICIENCY OPTIONS

- There are many types of "zero clearance" fireplaces or fireplace "inserts" that provide insulated fireboxes and chimneys and have some means of controlling air supply to the fire (airtight glass doors on the firebox front are common). An example of a typical unit is illustrated at the right.
- Many of these inserts have air blowers that increase efficiency by circulating cool room air behind the firebox and extracting heat by forced convection.
- For all types of fireplaces install a separate outside air feed to the front of the firebox. The ducting should be insulated and dampered.

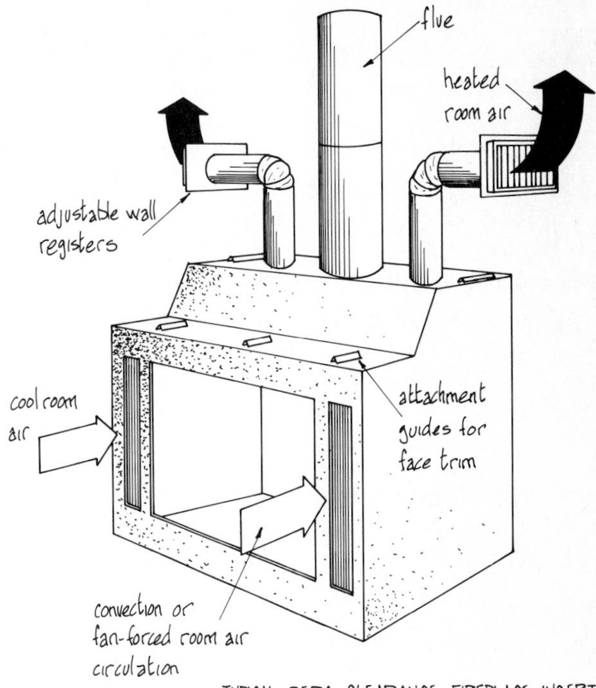

TYPICAL ZERO-CLEARANCE FIREPLACE INSERT

WOOD STOVES

Wood stoves are freestanding units that function much the same as fireplaces but with some significant advantages:

- Most units are "airtight" in that combustion airflow can be closely controlled. The size and hotness of the fire can therefore be adjusted to reflect heating demands.
- Seasonal efficiencies of up to 65% are possible with some units when an adequate combustion fresh-air supply is provided.
- Safety precautions must be rigidly observed, especially with regard to installation details. All codes have strict requirements in this regard.
- Special attention must be paid to creosote buildup in chimneys caused by cool, slow-burning fires. Easy access to the chimney for cleaning is essential.

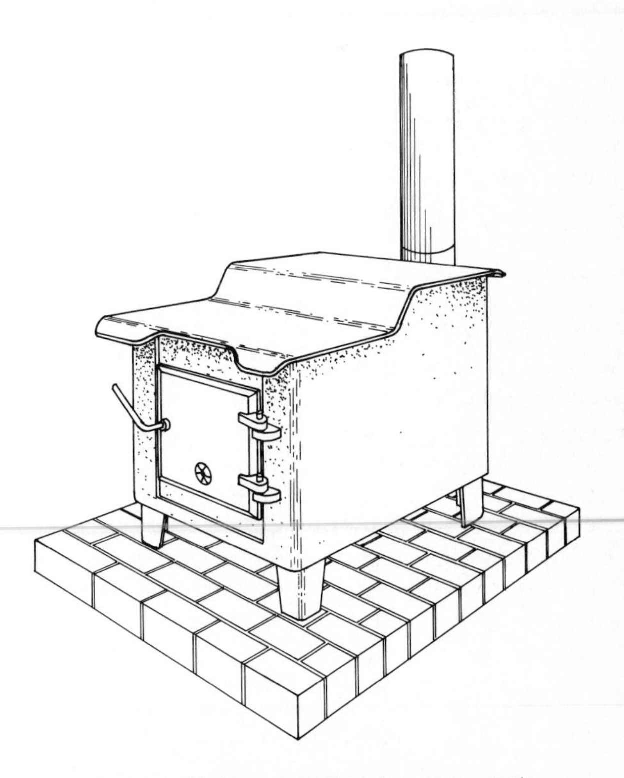

TYPICAL FREESTANDING WOOD STOVE

FORCED-AIR SYSTEMS

FORCED WARM-AIR SYSTEMS – GENERAL

The use of a forced-air system to deliver warm or cool air to the rooms of a dwelling is usually the preferred method for several reasons:

- Any type of combustion furnace, electric furnace, or heat pump can be used as the primary heat source. In most such equipment the electric fan that drives the forced air is part of the furnace itself.
- The furnace can be located in almost any part of the house, although a central location is best.
- With the fan system, ducts can be relatively small and can generally be located to allow the best positioning of diffusers and return air grilles.
- The fan system is ideal for ventilation and humidity control. In addition, accessories such as electronic air cleaners, humidifiers, dehumidifiers, and separate air-conditioning can be added to the system with ease.
- The forced-air movement reduces heat stratification in a room, leading to greater comfort levels.

SYSTEM DESIGN

Although heating and cooling systems should be designed by a competant contractor, some of the basic points are as follows:

- The exact type of delivery system depends on the type of structure. Five typical systems are shown on the next five pages.
- Where possible, warm-air supply registers should be located against outside walls, preferably beneath windows so that the rising warm air counteracts the falling cold air.
- To complete the circulation pattern, cold-air return grilles should be located in or near inside partition walls.
- Registers or grilles can be set into either floors or walls with equal efficiency.
- Ducts should not be run in outside walls or other cold spaces. Where this situation cannot be avoided the ducts should be sealed and insulated to the same R value as that of the wall or floor itself.

FAN OPERATION

The way in which the furnace fan distributes air to the dwelling is subject to the type of equipment installed and the preference of the homeowner. There are two modes of operation: cyclical and continuous.

- In the cyclical mode the fan is turned on and off as determined by the heating equipment. When the room thermostat calls for heat from a combustion furnace, the burner will run for a preset period to preheat the heat exchanger before the fan is started. After the burner is off the fan will continue running for another preset period to capture residual heat before being turned off. With an electric furnace or heat pump the fan turns on and off at the same time as the equipment.
- Alternatively, the fan can be set to run continuously without regard to heating or cooling demand by the thermostat.
- In energy-efficient houses the cyclical method can result in long periods between cycles, making it difficult to control air quality. A continuously running fan helps control air quality but may give rise to complaints of cool drafts. One answer is to install a two-speed fan that runs continuously at low speed but switches to high speed on the heating or cooling cycle.
- Whichever mode is chosen, the air delivery at each register must be "balanced" over the whole system to ensure the best comfort conditions in each room.

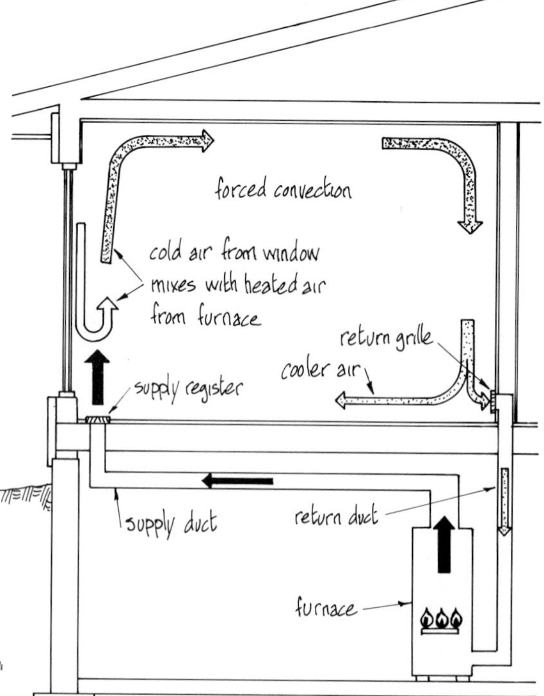

This illustration shows the recommended positioning of supply registers and cold air returns in a room so that heat is distributed with optimum efficiency.

BASEMENT FORCED-AIR SYSTEMS

EXTENDED PLENUM DUCT SYSTEM

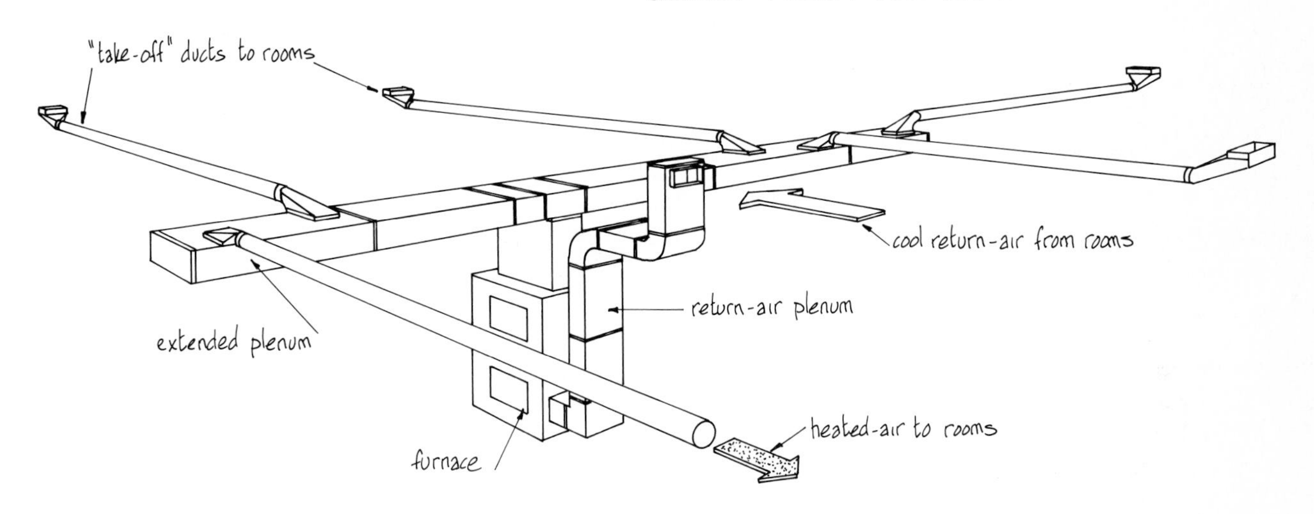

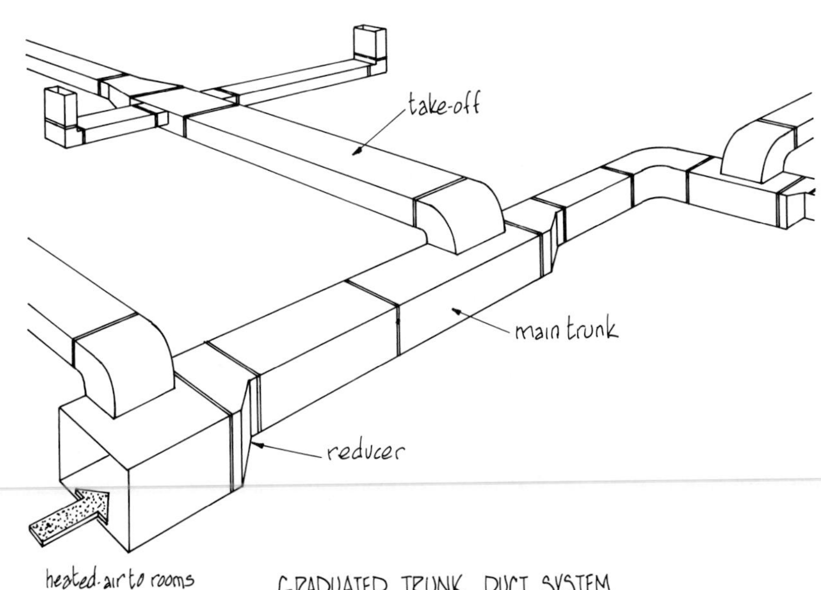

GRADUATED TRUNK DUCT SYSTEM

EXTENDED PLENUM SYSTEM

- This system is typically installed in basements where headroom is important, and is one of the commonly used systems.
- It is a relatively simple but flexible system, although generally limited to small-and medium-sized installations.
- A rectangular plenum, usually located alongside the main supporting beam, is extended from one or both sides of the furnace.
- Individual room "takeoff ducts" extend at right angles to the plenum and can often be hidden in the space between floor joists for an unrestricted headroom.
- Either a single floor-level central cold-air return or a matching extended plenum return air duct with several pickups can be used.

GRADUATED TRUNK SYSTEM

- This is similar to the extended plenum system but has a main supply trunk that is graduated in size to help balance delivery pressure after each takeoff.
- This an ideal but expensive system and is generally used only for larger or more complex installations.

CRAWL SPACE FORCED-AIR SYSTEMS

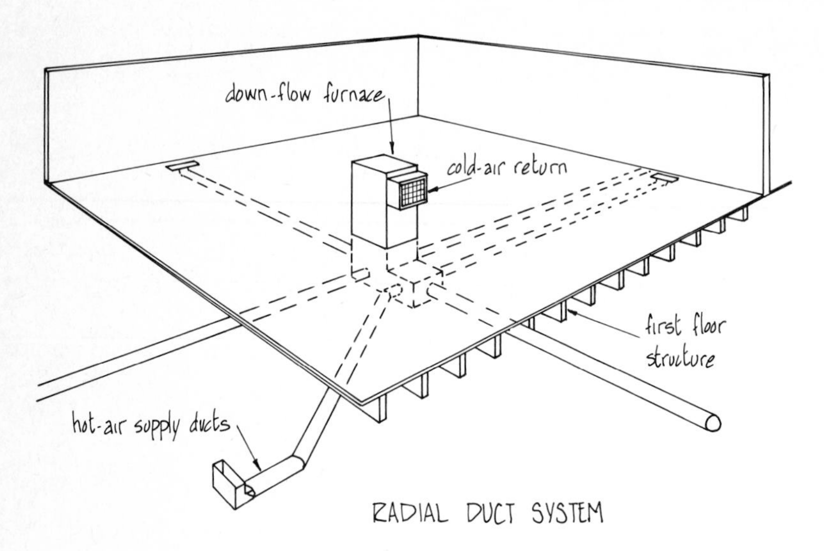

RADIAL DUCT SYSTEM

RADIAL SYSTEM

- This system requires a "down-flow" type of furnace that blows warm air downward to a distribution plenum beneath the floor.
- Individual room supply ducts radiate out from the supply plenum, as shown at the left, and are positioned below the floor joists.
- This system is generally used only in crawl space or slab-on-grade construction, where headroom below the supply ducts is of no concern.
- The cold-air return is usually through a single grille on or near the furnace head.
- In an unheated crawl space the supply ducts must be sealed and well insulated.

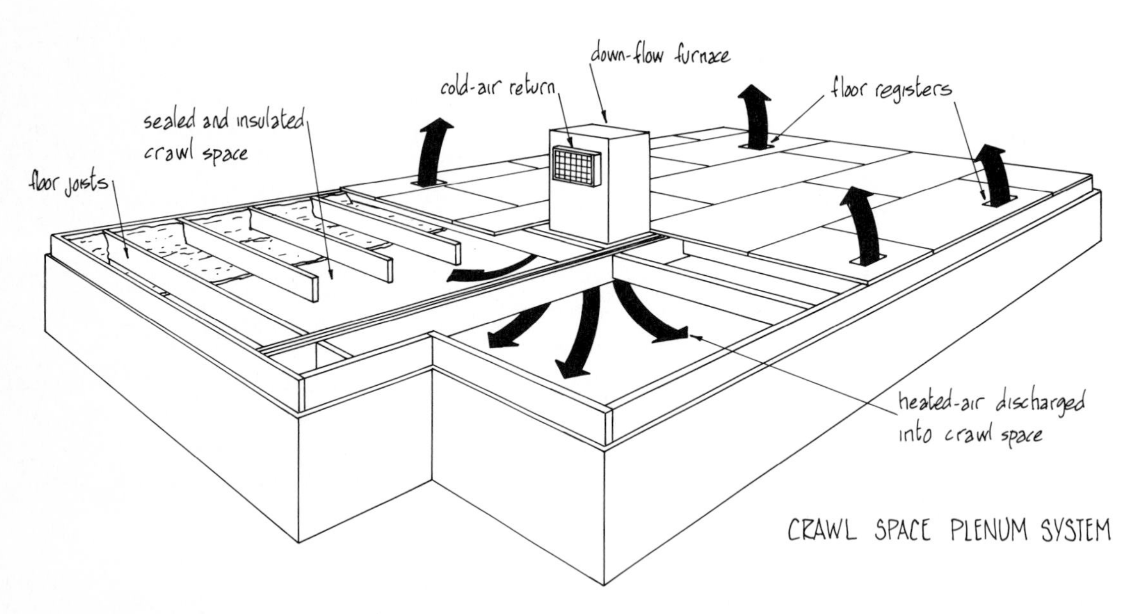

CRAWL SPACE PLENUM SYSTEM

CRAWL SPACE PLENUM SYSTEM

- In this system the crawl space itself is used as a plenum to distribute the warm air to the individual rooms.
- A down-flow furnace is located on the first floor and discharges warm air into the crawl space, where a positive pressure is created. The warmed air then flows upward through the open floor registers into the first-floor rooms. In addition to the convection effect, the floor is warm and radiates heat. This increases the mean radiant temperature (see 3-6) and raises comfort levels.
- Variations in crawl space depth have little effect on the system's overall efficiency, and any type of foundation construction can be used.
- The foundations and crawl space floor must be treated as a fully heated basement and insulated to suit. The crawl space must be made airtight using a continuous and sealed air/vapor barrier (see 10-45).
- System costs are relatively low since no supply ducts are required and overall system efficiency tends to be quite high.

source: Southern Forest Products Assoc.

SLAB-ON-GRADE FORCED-AIR SYSTEM

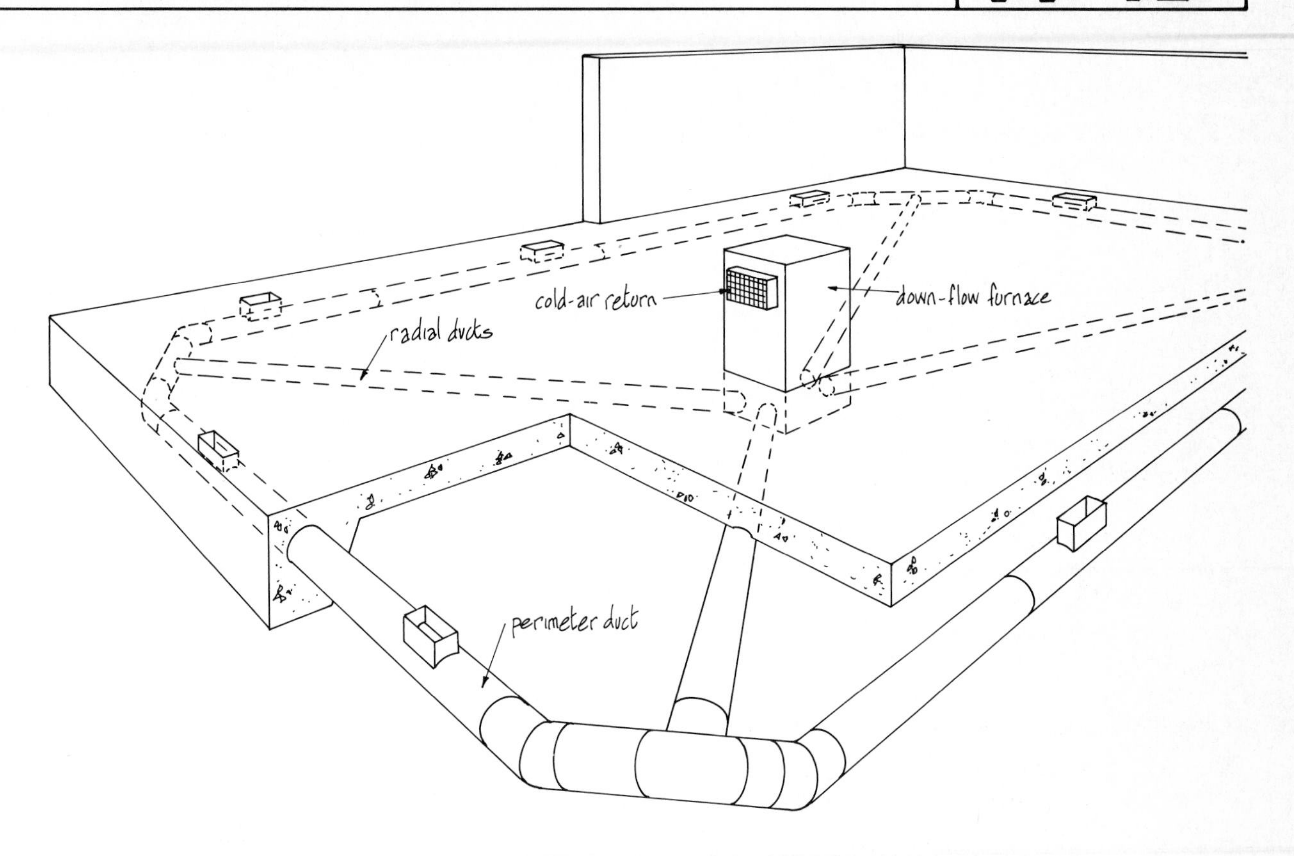

PERIMETER LOOP SYSTEM

This is similar to the radial system shown opposite with the addition of a perimeter loop connected to the radial supply ducts.

- Supply registers are located in the outer loop at suitable room positions.
- Fewer radial feeder ducts are needed, and the slab perimeter is kept relatively warm provided that the perimeter loop is not insulated.
- The feeders and loop are set in place before the concrete floor slab is poured. Ensure that the ducts are sloped downward toward the supply plenum for the collection of any water that may accumulate in the system.

The radial system can also be used in slab-on-grade construction but tends to result in cold floor areas at the slab perimeter.

HYDRONIC HEATING SYSTEMS

GENERAL INFORMATION

- Hydronic systems use either steam or hot water as the heat transfer medium, although a steam system is rarely used in dwellings.
- The central boiler is usually a gas, oil, or electric unit and pumps water through the system at temperatures between 120 and 212°F (50-100°C). The boiler usually has a small capacity, 1-3 gal (4-12 L), allowing a fast response to heating demands and reduced standby losses.
- In the two most common systems shown here, the radiators can be adjusted individually. Heat to the rooms is supplied through a combination of convection and radiation. In large systems the supply loop can be divided into series of separate zones for even greater control.
- The major disadvantage of any system where forced air is not the heat transfer medium is that adequate ventilation and humidity control is very difficult to achieve (see 14-16 for more information). It is equally hard to add air conditioning to the dwelling other than relatively inefficient and noisy individual window units.

SINGLE-PIPE HOT-WATER SYSTEM

- This is the simplest and most common type of pumped hot-water system.
- The radiators are in parallel with the single main supply line. This allows water to be diverted to each radiator with only minimal effect on the others in the circuit.

DOUBLE-PIPE HOT-WATER SYSTEM

- In this system there is both a hot supply and a cold return line.
- This allows the greatest flexibility and assures equalization of heat at each radiator.
- Naturally, this method is the most expensive and is normally used only on large and complex structures.

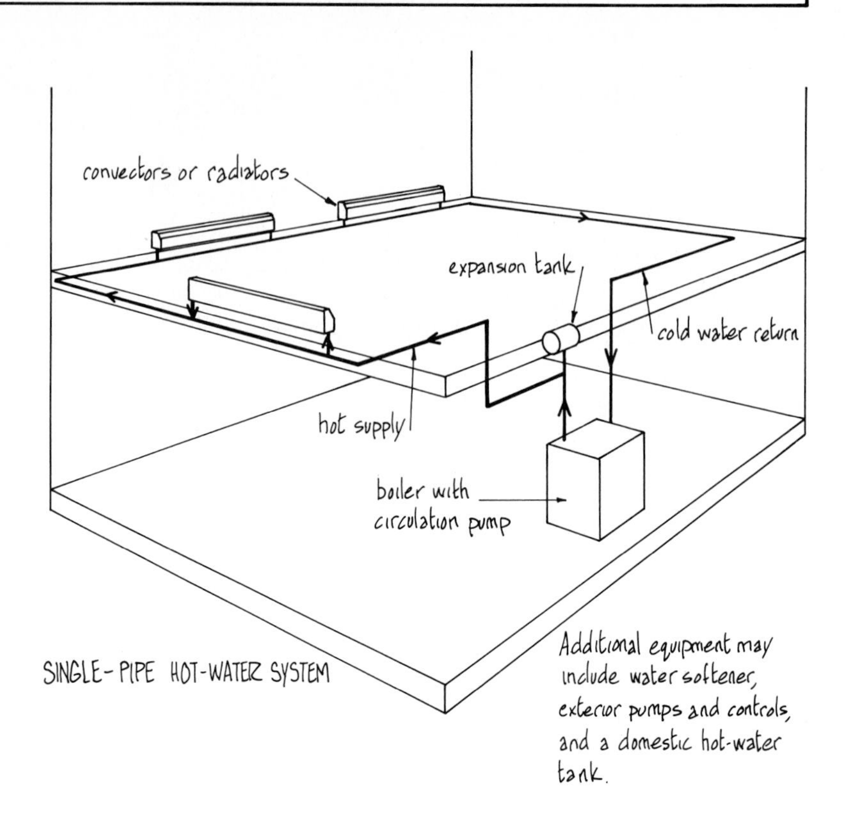

SINGLE-PIPE HOT-WATER SYSTEM

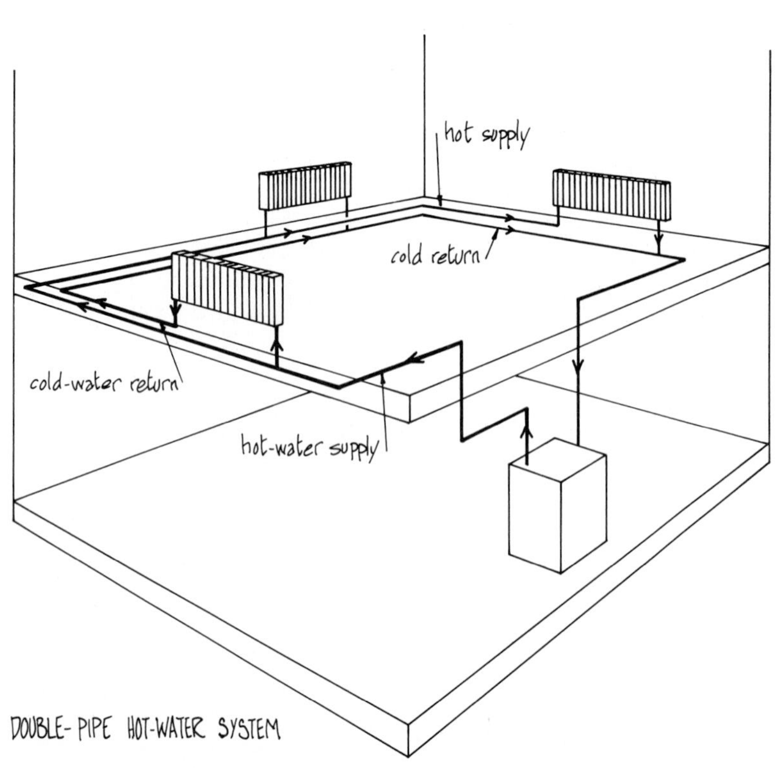

DOUBLE-PIPE HOT-WATER SYSTEM

RADIANT HEATING

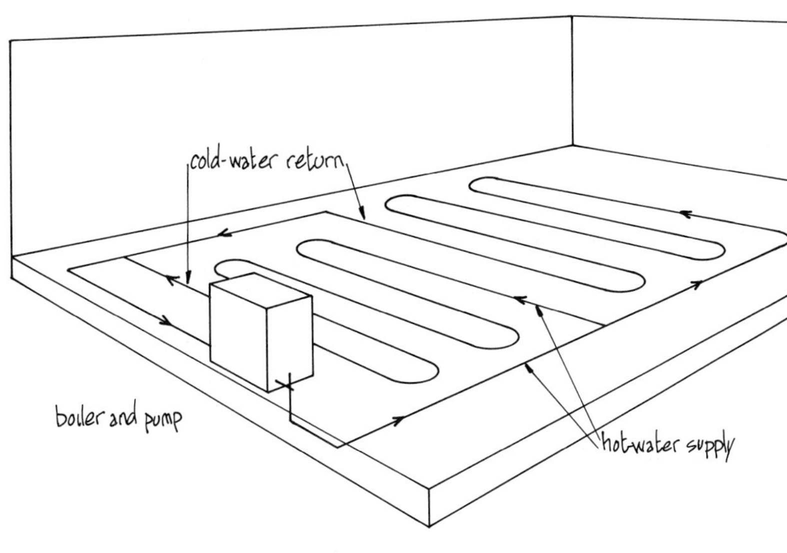

UNDER-SLAB HOT-WATER SYSTEM

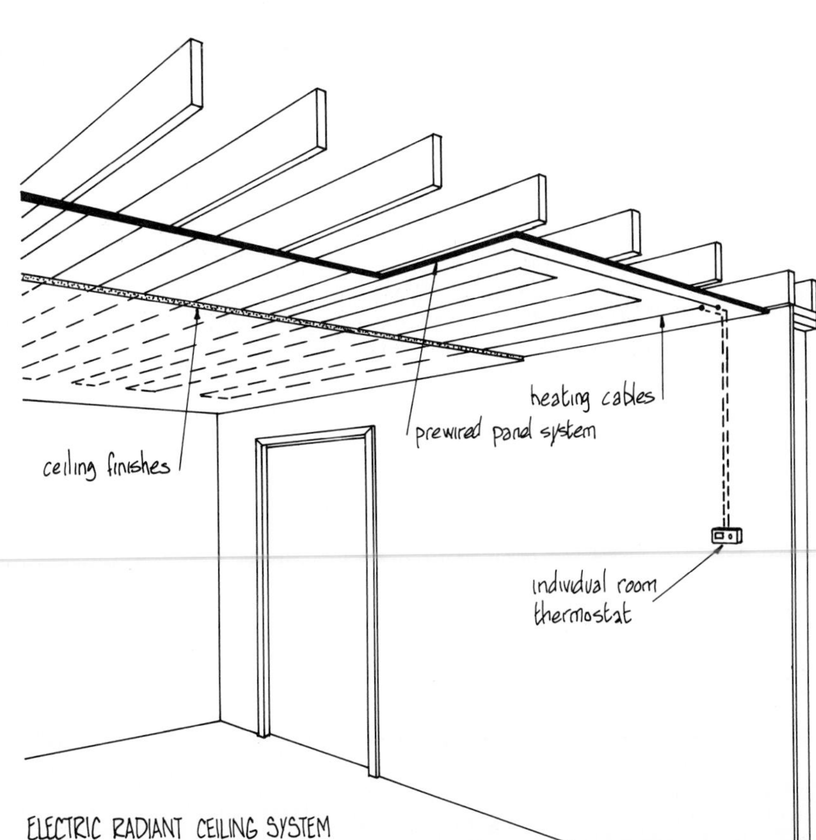

ELECTRIC RADIANT CEILING SYSTEM

GENERAL INFORMATION

- Radiant systems employ heating coils installed in concrete floor slabs or behind ceiling finishes.
- Heat transfer is by convection and radiation from floors but almost entirely by radiation from a ceiling.
- A major disadvantage is the extended temperature lag caused by the high thermal capacity of the systems. This makes it hard to control room temperature under rapidly changing weather conditions.
- As with the hot-water radiator systems opposite, adding air-conditioning and controlling ventilation and humidity are difficult (see 14-16 for more information).
- Since these systems become part of the structure, leaks in the slab piping or failure of the electrical coils can result in major repair expenses.

SLAB HOT-WATER LOOP

- The system is similar in nature to the radiator system opposite except that the hot supply lines are buried in the slab before pouring.
- Floor temperatures are limited to a maximum of 85°F (29°C) or occupant discomfort will result.
- Furnishings, particularly carpets, have an effect on the heat response time to varying degrees.
- Individual room temperature control is difficult, and most systems divide the structure into small zones of one to three rooms each to retain a measure of control.

ELECTRIC RADIANT CEILING SYSTEMS

- Electric resistance heating coils either above or incorporated into the ceiling finishes are used.
- Surface temperatures up to 120°F (50°C) are acceptable, and response time is slightly quicker than floor systems.
- It may be necessary to install insulation in the ceilings of lower stories to avoid excessive heat transfer to the floor above. Ceilings of the top story will also lose large quantities of heat to the attic unless very well insulated.
- Individual room thermostats can be used with this system.

VENTILATION AND MAKEUP AIR

GENERAL INFORMATION

In any type of building where people live or work, control of the quality and movement of air is extremely important. If fresh air is not supplied in sufficient quantities, a range of serious and even dangerous situations can occur that affect the health of the occupants and the structural integrity of the building.

> These problems are especially important in an energy-efficient airtight structure where infiltration is kept to an absolute minimum. In this type of structure the designer <u>must</u> include a means of supplying adequate quantities of fresh air if potential problems are to be avoided.

There are two overall requirements for air in any dwelling:
ventilation air and makeup air

- Ventilation air is required to meet the health needs of the occupants. There must be some direct and controllable means of bringing this air into the dwelling while exhausting a similar quantity of stale air.
- Makeup air has two separate components. First is the supply side of the ventilation/exhaust system — exhausted air must be replaced with fresh air. Second is air supplied directly to combustion heating equipment for use in the combustion process.

The way in which each air requirement affects the design and operation of a dwelling is discussed on the following pages. The chart below gives an overall visual illustration of those air requirements.

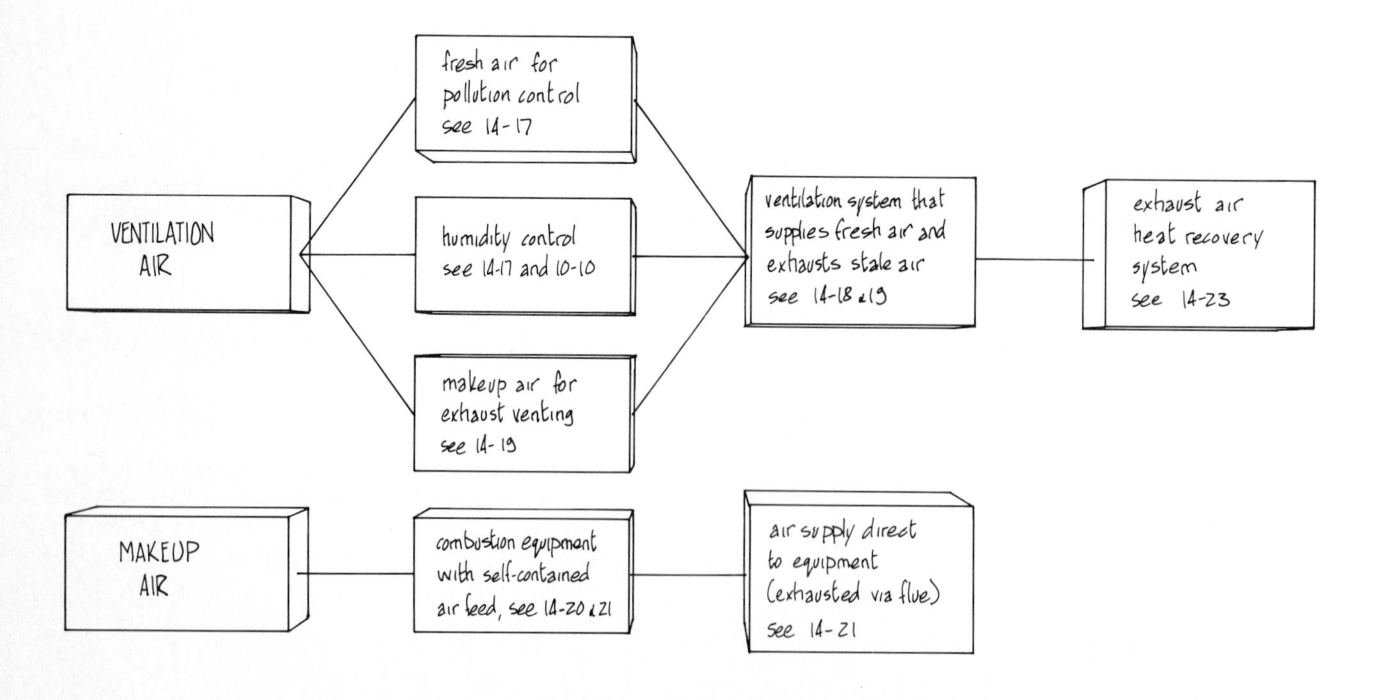

VENTILATION REQUIREMENTS

INDOOR POLLUTANTS

In any closed environment such as a well-sealed house, pollution and humidity from various sources can build up to dangerous levels if not properly vented. Aside from standard pollutants such as cooking fumes, cigarette smoke, and odors, there are other, more exotic, pollution sources, such as radon gas, chemical fumes and outgassing:

- Radon is a naturally formed radioactive gas produced by certain rocks and mineral materials. It tends to be prevalent in basement areas and is impossible to detect without sophisticated equipment.
- Chemical pollution is caused by deliberately introduced materials, such as cleaning fluids, hair sprays, insecticides, paints and varnishes, and other common household items.
- Outgassing is the leakage of gaseous chemicals from items that are part of the dwelling's structure or furnishings. Especially prevalent is formaldehyde, which outgases from plywood glue and from improperly installed urea-formaldehyde foam insulation. (U-F insulation is permanently banned in Canada for this reason.)

The other major air quality problem is that of excess humidity. This subject was discussed in detail on 10-10. Please refer to the information on that page and integrate with the above.

CONVENTIONAL vs ENERGY-EFFICIENT HOUSING

In conventional houses with standard combustion furnaces and considerable air leakage, indoor pollution is rarely a problem. Air infiltration caused by stack, wind, and combustion effects (see 10-8) provided ventilation in excessive, and expensive amounts. In energy-efficient houses where infiltration is reduced to an absolute minimum, natural ventilation does not occur and suitable systems must be installed to take its place.

VENTILATION BY ROOM

each habitable room	10 cfm (5 L/s)
large open areas	2 cfm/20 sf (1 L/s per $2m^2$)

Habitable rooms include bedrooms, living rooms, dining rooms, family rooms, kitchens, bathrooms, and other main-floor rooms. Habitable rooms in a basement should also be included.

VENTILATION QUANTITIES

Ventilation quantities are best expressed by a method that takes into consideration the size of the dwelling, the number of people therein, and the type of combustion equipment used. One of two methods is generally used for the occupant load:

- Cubic feet per minute (cfm) per room. Metric – liters per sec (L/s).
- Cfm (L/s) per person.

Added to each calculation are the equipment load makeup air requirements, discussed on 14-20.

The ventilation air requirements shown at the right are <u>minimum continuous</u> quantities and should be distributed to all parts of the dwelling. Note also that equal quantities of stale air must be extracted through a suitable exhaust system (see 14-18) that takes air from selected rooms within the dwelling.

Check local codes for exact requirements – some codes may require greater air quantities than indicated on this page.

VENTILATION BY PERSON

each person	10 cfm (5 L/s)

The obvious difference between the per person and per room air quantities exist because while the room numbers are fixed, the occupancy load will vary greatly. Use the highest of the two quantities in each situation.

The minimum quantities shown above will not be adequate for certain overload situations. To cater for such occasions, it is recommended that the ventilation and exhaust systems (see 14-18) be equipped with two-speed fans that allow the doubling of ventilation rates on demand. The lower speed should maintain the continuous ventilation rate, while the high speed can be switched either manually or automatically through a humidistat.

14-18 CENTRAL AIR SYSTEMS - EXHAUST

SYSTEM MECHANICS

A central air-change system consists of two basic parts:

- An exhaust system that extracts stale air from certain rooms in the dwelling and expels it to the outside.
- An intake system that supplies fresh, conditioned air to all rooms of the dwelling. This system may also supply makeup air to certain appliances.

The exhaust system is basically the same whatever type of heat distribution system is used. The intake system, however, should be designed to integrate with the heating system and usually takes one of two forms. Whatever design combinations are used, exhaust and intake air volumes must be closely balanced.

EXHAUST SYSTEMS

The illustration at the right shows a central exhaust system for a two-story house. Note the following points:

- Air is exhausted directly from kitchens and bathrooms. This ensures that stale and humid air is removed from the rooms where it is most likely generated.
- Design the system so that twice the amount of air is exhausted from a kitchen as from each bathroom or powder room. (E.g., in a house with one kitchen and two bathrooms, half the air would come from the kitchen and a quarter would come from each bathroom.)
- In the kitchen the range should not be vented to the outside but should be equipped with a recirculating hood and grease trap. Position the exhaust grille on an interior wall and equip it with a grease filter.
- All interior doors should be undercut to leave a 1/2-to 1-in (12-25) gap between the underside of the door and the top of the finished flooring so that air can move freely through the house.

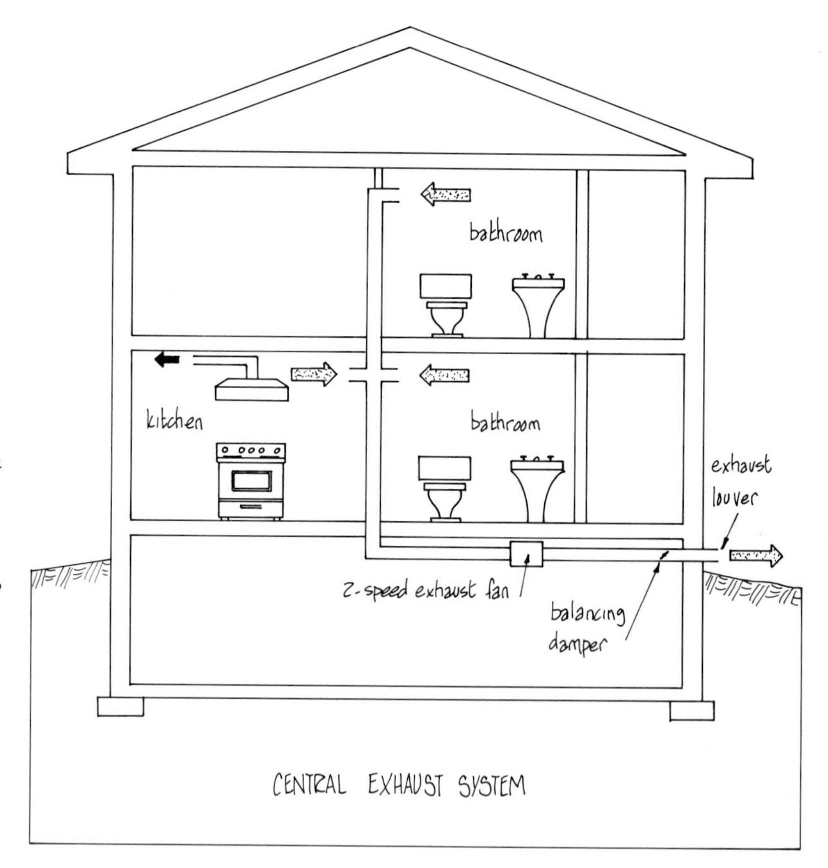

CENTRAL EXHAUST SYSTEM

Note that this exhaust system must not be used on its own without a balancing supply system, especially if a standard combustion furnace and/or an open fireplace is installed in the house. To do so would induce dangerous backdrafts and lower the neutral plane.

source: Canadian Homebuilding Assoc.

CENTRAL AIR SYSTEMS - INTAKE

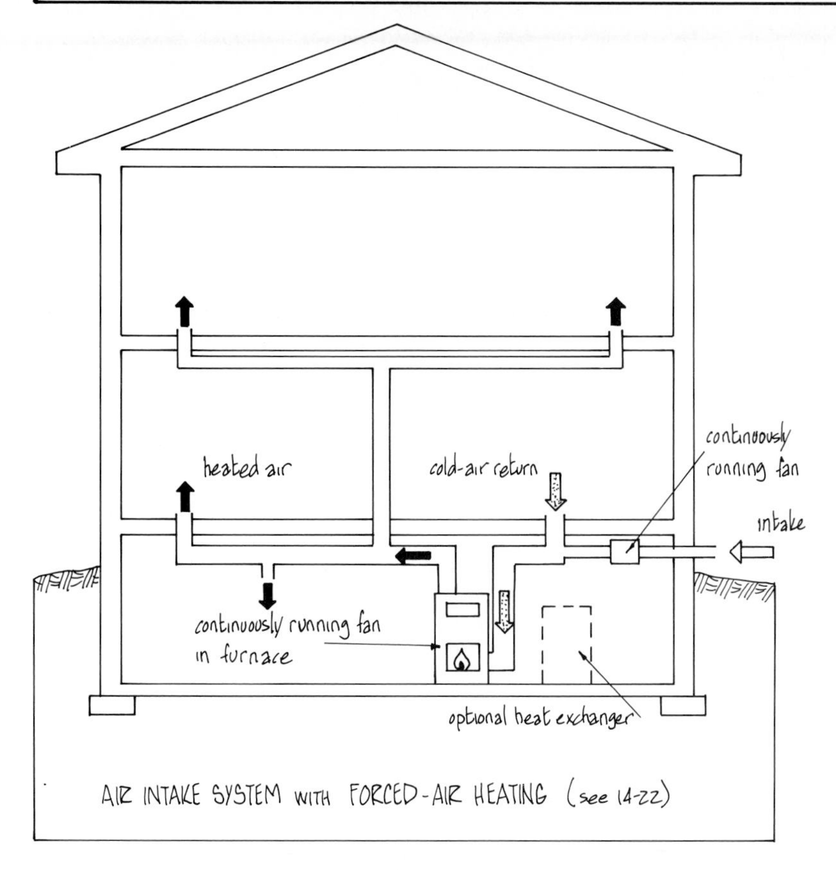

AIR INTAKE SYSTEM WITH FORCED-AIR HEATING (see 14-22)

INTAKE SYSTEMS

The setup of a fresh-air supply system depends on the type of heating equipment installed in the house. The ideal system is forced-air heating, which has advantages beyond those of a simple air distribution system.

FORCED-AIR HEATING SYSTEMS

- The fresh air intake is connected directly to the cold-air return of the heating system but before the furnace return plenum. This allows the fresh air to mix and be tempered by the house return air before being filtered and conditioned by the furnace.
- The intake must be fully insulated over its full length to avoid frosting in cold weather. The opening should be protected from rain, snow, and wind and be equipped with insect and rodent screens.
- The furnace fan should run continuously at low speed (except when on the heating cycle) to maintain even distribution.
- Installation of a heat exchanger is recommended to recover the heat normally lost with the exhausted air. See 14-22 for information on this equipment.

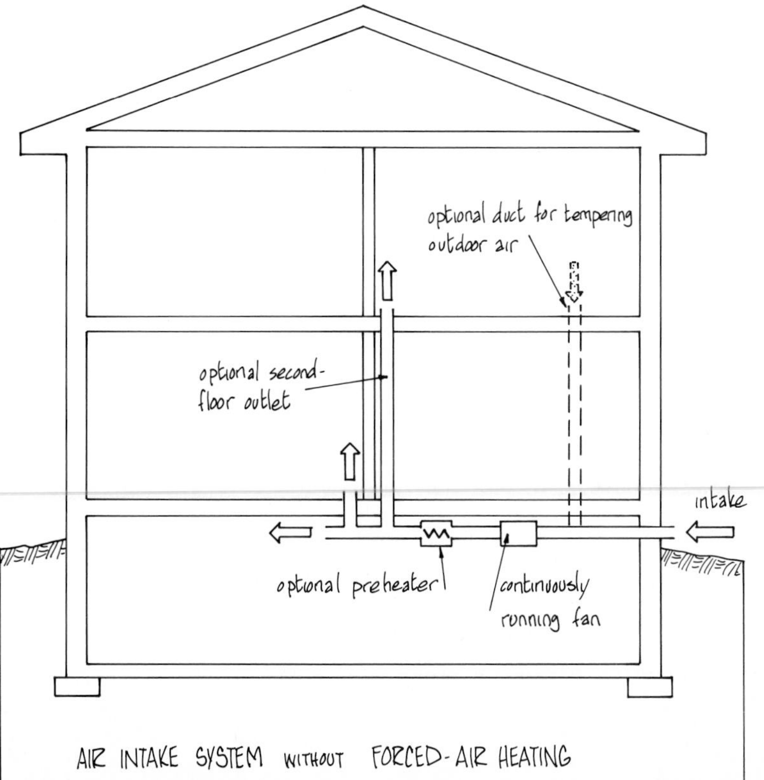

AIR INTAKE SYSTEM WITHOUT FORCED-AIR HEATING

ELECTRIC BASEBOARD, RADIANT, AND HYDRONIC HEATING SYSTEMS

- With no warm-air duct system to circulate fresh air, some other simple form of air distribution should be included in the house design.
- The simplest method is to discharge fresh air directly into the basement (or other suitable room in dwellings without basements).
- As an option, supply ducts can be extended to upper floors for better distribution.
- In cold climates, the incoming air must be preheated before distribution. Ideally, a heat exchanger should be interfaced between the supply and exhaust systems, or failing that, an electric in-duct plenum heater should be installed.

Whatever fresh air supply system is installed, it should be fan-driven and balanced with the exhaust system already described.

source: Canadian Homebuilding Assoc.

14-20 MAKEUP AIR REQUIREMENTS

The air supply requirements for heating equipment varies widely depending on the type of fuel consumed and the specific heat generation process. More significant, however, is whether or not the heating equipment takes combustion and draft control air from inside the house. If so, provision must be made to supply outside air directly to the equipment if backdrafting is to be avoided and efficiency is to be maintained. Shown below are the three main equipment groups and their air requirements (refer to 14-4 to 14-9 for specific equipment details).

TYPICAL INDOOR AIR REQUIREMENTS - HEATING EQUIPMENT

TYPE OF EQUIPMENT	ENERGY SOURCE FOR HEATING	TYPICAL AIR REQUIREMENTS CFM (L/S)	AIR SUPPLY SOURCE
TYPE 1 Not Requiring Indoor Air Supply			
electric resistance baseboard or plenum	electric	none	—
heat pump	electric	none	—
condensing gas heat pump	gas	none	unit located outside the house
sealed combustion unit	gas/oil	none	all air requirements from outside

EQUIPMENT NOT REQUIRING INDOOR AIR

There are two main groups in this category:

- Electric systems, such as heat pumps, furnaces, baseboard units, etc., that do not use combustion for heat generation.
- Combustion units that have direct outside air feeds to the combustion chamber and direct, usually low-temperature exhaust flues to the outside. Some units in this group have the combustion chamber outside the house, thus solving the air-supply problem completely.

TYPE 2 Requiring Combustion Air Only From Inside The House			
furnace with induced draft	gas/oil	24 (12)	motorized damper linked to furnace
condensing furnace	gas/oil	16 (8)	motorized damper linked to furnace recommended
high efficiency oil furnace	oil	20 (10)	room the furnace is located in
stove (used as an airtight unit)	wood/coal	10 (5)	room the stove is located in

EQUIPMENT REQUIRING ONLY COMBUSTION AIR FROM INSIDE THE HOUSE

The two main groups in this category are:

- High-efficiency condensing gas or oil furnaces that are equipped with a forced-draft fan. The fan eliminates the need for a standard flue and allows the unit to operate safely in negative pressure conditions.
- Wood- or coal-burning airtight stoves that require no air for draft control.

Equipment in this category should have outside air supplied near or directly to the combustion chamber (see the next page for details).

TYPE 3 Requiring and Draft Control Air From Inside the House			
conventional furnace	gas oil	110 (55) 140 (70)	room the furnace is located in
conventional furnace fitted with flame retention head	oil	130 (65)	room the furnace is located in
stove with open doors	wood/coal	110 (55)	room the stove is located in
fireplace	wood/coal	400 (190)	room the fireplace is located in

EQUIPMENT REQUIRING COMBUSTION AND DRAFT AIR FROM INSIDE THE HOUSE

There are again two groups in this category:

- Conventional gas or oil furnaces with open combustion chambers and draft hoods.
- Open wood stoves or standard fireplaces.

Units of this type are very susceptible to backdrafting under negative air-pressure conditions. Ideally, they should not be used in airtight energy-efficient houses, but if they are, a direct outside air feed to the unit should be supplied (see the next page).

source: Canadian Homebuilders Assoc.

DIRECT AIR FEEDS

FLUE BACKDRAFTING

Flue backdrafting can occur in any open-type flue where negative interior house pressure is present and results in dangerous combustion gases being drawn back down the flue and into the house. The negative pressure condition is caused by other combustion equipment or air extraction systems, such as dryers and range hoods exhausting air from the house faster that it can be replaced by infiltration or ventilation. To combat this problem, makeup air should be supplied directly to the equipment in question, and two methods of doing so are given below.

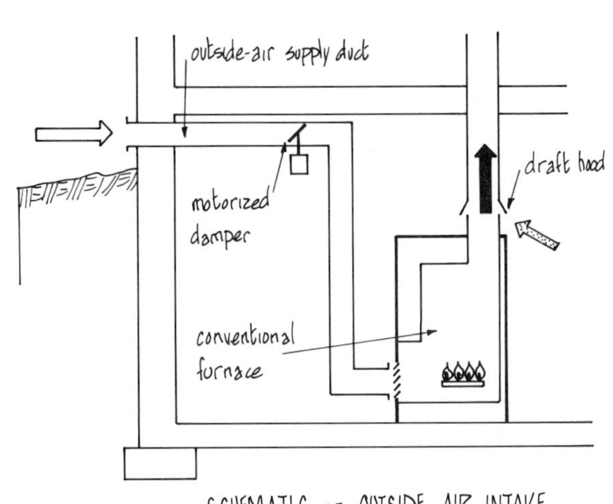

SCHEMATIC OF OUTSIDE-AIR INTAKE

ensures a positive pressure near the combustion chamber

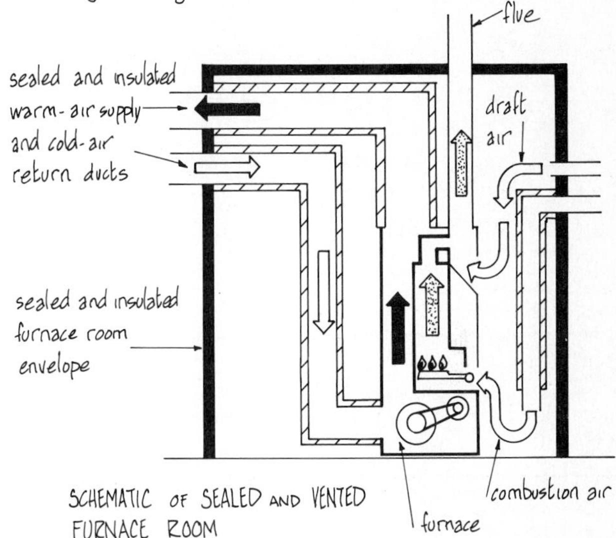

SCHEMATIC OF SEALED AND VENTED FURNACE ROOM

MAKEUP AIR FEEDS

- For furnaces, the diagram above applies. Note that the motorized damper is tied to furnace operation and will only open when the furnace is on. Conversely, the furnace cannot operate if the damper is closed.
- The furnace room concept is designed to make the furnace and gas water heater independent of house air by isolating them in an insulated, airtight room that is open to the outside air. Although this approach is effective in providing makeup air, it also creates many other operating problems, caused mostly by the low room temperature. This method is still being evaluated and is not recommended unless thoroughly researched by the designer.
- Gas clothes dryers have their own forced draft fan and are not normally subject to backdrafting. Gas dryers must be vented directly to the outside. Electric dryers can be vented to the inside to retain their heat, but this is not recommended because of the large water vapor load that this places on the house.
- Gas ranges must also be vented to the outside through a range hood equipped with a grease filter and vent pipe run over the shortest practical route. However, gas ranges are <u>not</u> recommended for use in airtight houses because of the pollution problems they create.

DESIGN CONSIDERATIONS

It is obvious from the information presented so far that indoor pollution and the need for makeup air can pose serious problems in airtight energy-efficient houses. These problems can, however, be reduced to manageable levels or be eliminated entirely with careful planning. Designers and contractors should therefore be aware of the following points:

- In designing an energy-efficient house (or any other efficient structure for that matter), all factors must be carefully integrated to create a fully functional whole. Energy-efficient structures are very sensitive to the internal environment, and greater control over causative factors is important.
- There is very little point in using heating equipment or appliances that function well in conventional houses but cause pollution or make-up air problems in energy-efficient houses. As discussed previously, there are a number of proven and readily available systems that eliminate these problems.
- Although these advanced systems tend to have a higher initial cost, it must be kept firmly in mind that there will be considerable long-term fuel cost savings, compared to conventional housing, because of the low energy design. In addition, the low energy requirements reduce the size of equipment needed, thereby allowing further cost reductions.

Source: Energy, Mines, and Resources Canada.

AIR-TO-AIR HEAT EXCHANGERS

WASTE HEAT RECOVERY

The central exhaust system detailed on 14-18 is very effective in reducing indoor pollution and humidity problems. However, along with the stale air the system is also exhausting large amounts of heat to the outside, and it makes sense to recover some of this heat where practical.

The best approach for houses (and small commercial projects) is to install a whole-house air-to-air heat exchanger. There are several types of heat exchanger, but all operate on the same basic principle. Outgoing warm stale air is exchanged for incoming cold fresh air, and in the process some of the heat from the stale air is transferred to the fresh air. The heat is transferred through a thin metal or plastic membrane or by mixing the airstreams with a high-speed heat wheel. Both types are described below.

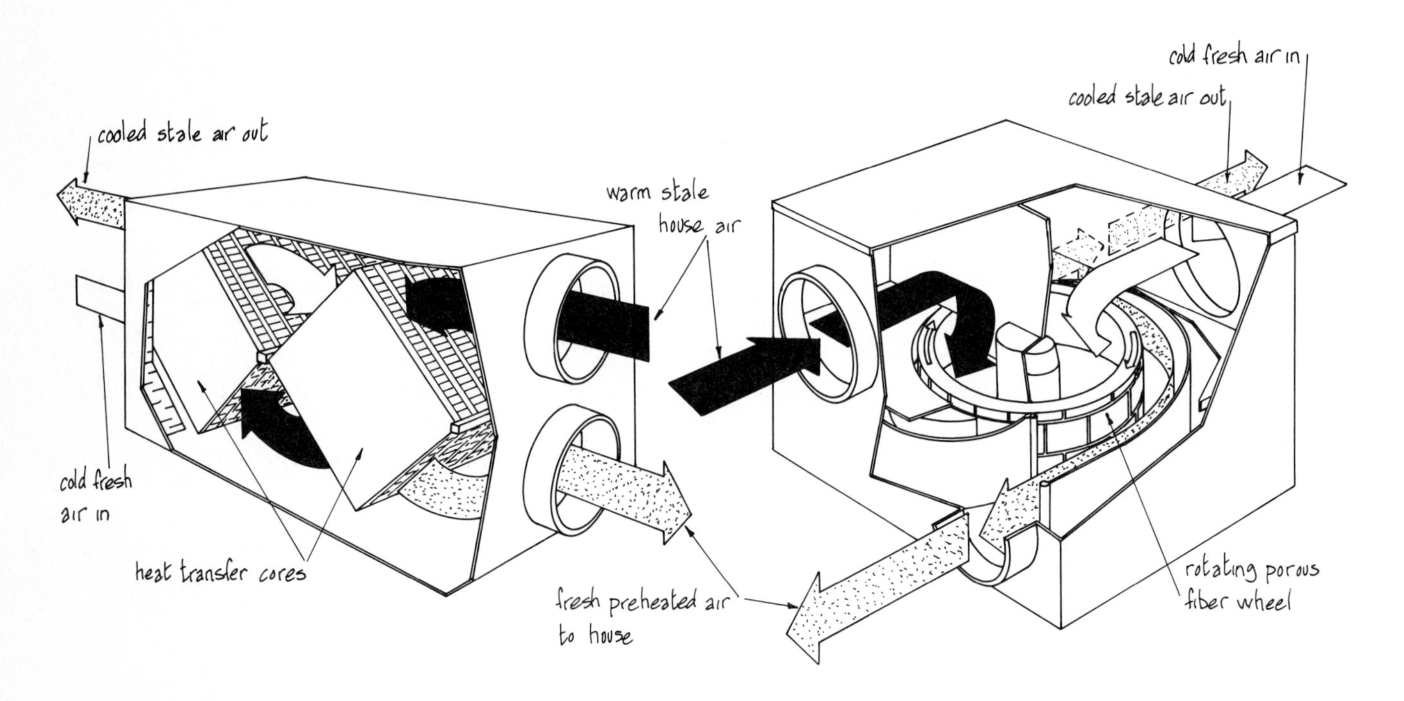

PLATE-TYPE HEAT EXCHANGERS

- Opposing streams of stale and fresh air are passed through thin metal or plastic cores without mixing. Simple conduction transfers heat from the stale air to the cold fresh air.
- Two fans are necessary to provide the correct balance between airstreams.
- Condensation will occur on the exhaust side of the system and a condensate drain is required. Most units also incorporate some type of defrost mechanism since the condensate is prone to frosting in very cold climates.
- Ideally, the exchanger cores should be removable for periodic cleaning.
- Heat recovery efficiencies of 50-80% can be expected with this type of exchanger, depending on airflow speed.

source: Conservation Energy Systems Inc.

SINGLE HEAT WHEEL EXCHANGERS

- These units use a single, rapidly rotating porous fiber wheel as the heat transfer medium. A single variable-speed electric motor drives the wheel and provides the necessary air movement without the need for additional fans.
- Both warm stale air and cold fresh air streams are ducted separately to the center of the wheel. The centrifugal action of the rotating wheel forces each airstream through the porous wheel, where heat exchange takes place by a combination of conduction and mixing. The warmed fresh air and cooled stale air are ducted to the house interior or exterior, respectively.
- Some mixing of stale and fresh air does occur, but this is not generally considered a significant problem.
- No defrost or condensate drain systems are required.
- Sensible heat recovery varies with airflow speed, but latent heat recovery remains constant at around 45%.

source: P.M. Wright Ltd

AIR-TO-AIR HEAT EXCHANGERS

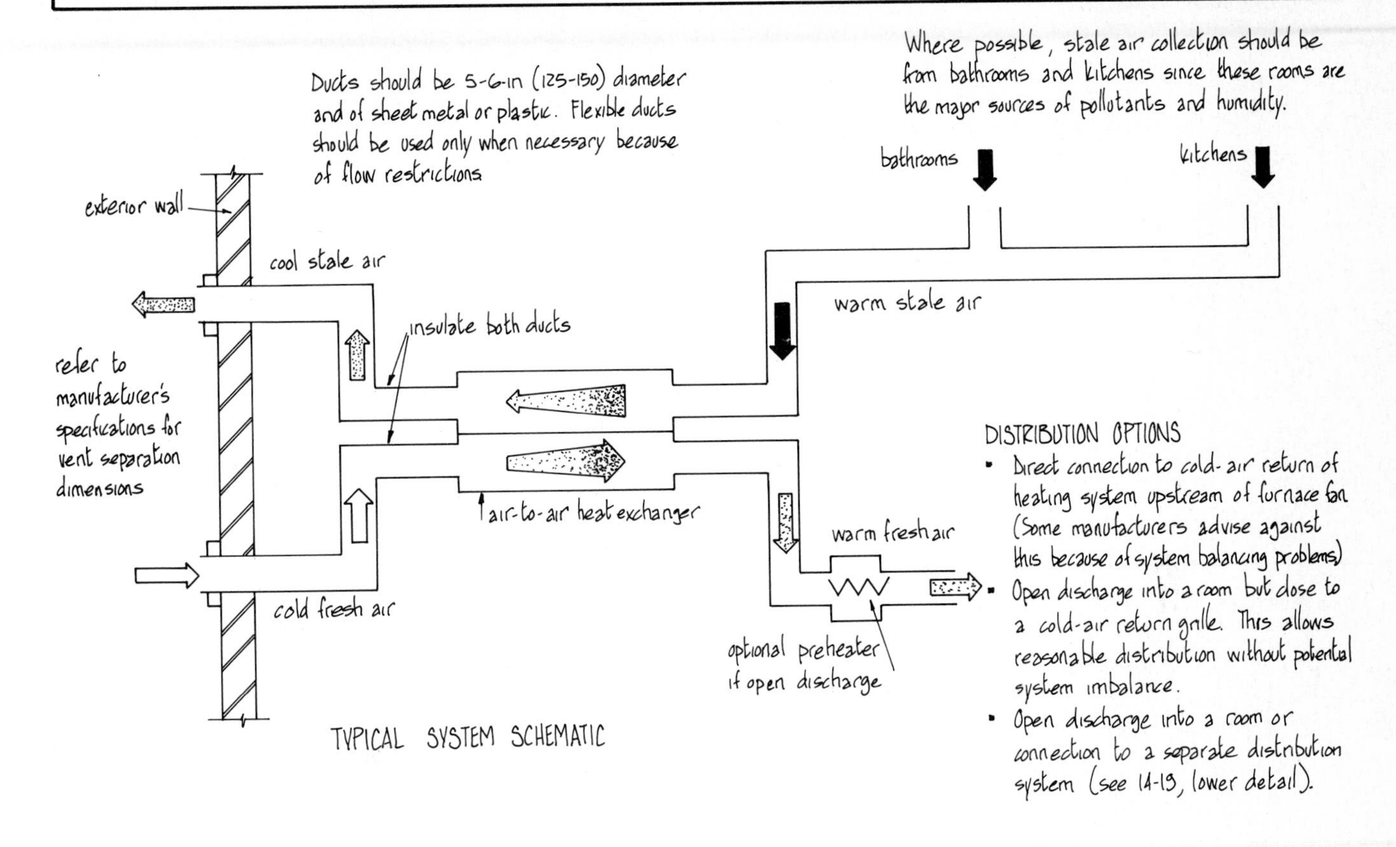

TYPICAL SYSTEM SCHEMATIC

SYSTEM SETUP

The schematic above is typical of most air-to-air heat exchanger systems. Exact air collection and distribution details depend on house design and type of heating/cooling equipment installed.

General information:

- Some exchanger units, particularly plate types, are quite large and reach sizes equivalent to those of a standard domestic hot water heater. Unit location is important since access for regular maintenance is required. Ideally, a conditioned space such as a basement is best, with utility or laundry rooms as a second choice. Crawl spaces and attics should be used only as a last resort.
- The unit should also be isolated from structural members to stop fan vibration resonating through the house.
- Maximum heat transfer is normally achieved when the airflow volumes through the exchanger are in balance. Maintaining that balance is often difficult under very windy conditions, but careful positioning of the exterior intake and exhaust ducts will help control this problem.
- Heat exchangers cannot be used to supply makeup air for combustion equipment. Separate feeds must be provided for this type of equipment (see 14-21).
- Neither cooking ranges nor clothes dryers should be directly connected to the heat exchanger. If cooking fumes are to be taken directly from a kitchen range hood, the exchanger core may become contaminated and clogged. Clothes dryers will induce lint blockage, and the high temperature may actually damage some exchanger cores (a blocked dryer can also result in a hazardous overheating situation).
- After installation, the exhaust and intake airflows must be balanced to reduce the possibility of creating positive or negative pressures within the building.

ELECTRICAL

CHAPTER PAGE TITLES

ELECTRICAL SERVICE SCHEMATIC

The illustration on this page shows in schematic form the electrical power supply sequence from power company transformer to individual house circuits as it applies to most residential and small commercial buildings. The individual components with their typical wiring details are described on the following pages.

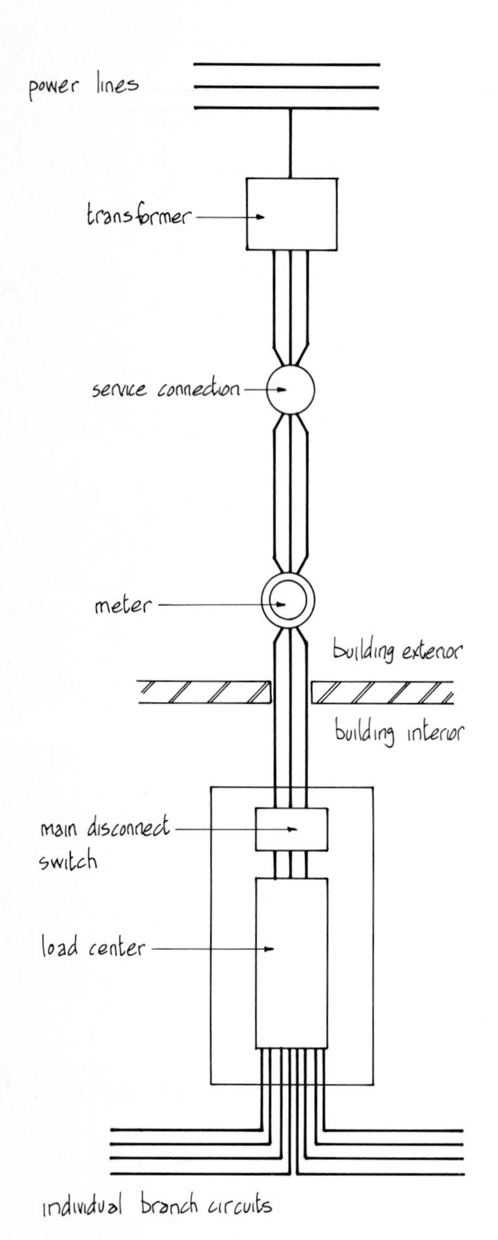

POWER COMPANY TRANSFORMER
Positioned at central locations in dense urban areas or on individual power poles at more remote rural areas, the transformer reduces high long-distance transmission voltages down to standard 120- to 240-volt (V) house service requirements. See the next page for details.

SERVICE CONNECTION
This is the point at which the power company lines are connected to the house wiring system. The type of service connection used depends on whether the power cables from the transformer are run above or below ground. In new urban areas they are likely to be underground while in older urban and most rural areas they are likely above ground (see 15-4 and 15-5 for details).

METER
Usually installed on the house exterior for easy inspection by power company personnel, the meter records total power consumption in watt-hours. The meter plugs into a meter base and its removal isolates the house from the power supply.

MAIN DISCONNECT SWITCH or MAIN BREAKER
In most residences this is a single switch capable of disconnecting the entire house wiring system from the incoming power service. The switch may be an integral part of the load center or installed at a separate location depending on local codes and the load center panel's proximity to the service entrance. See 15-6 for details.

LOAD CENTER, PANELBOARD, or DISTRIBUTION PANEL
Normally located inside the house, this panel contains the power connection points for the individual house electrical circuits. (Depending on load requirements there may be more than one load center; see 15-7.) First-line overcurrent protection devices (fuses or breakers) for each circuit are in this panel (see 15-7 for details).

BRANCH CIRCUITS
These are individual circuits extending to all parts of the house. In each circuit the wire size, number of wires, and connection procedures are determined by the type of appliance or equipment the circuit is to serve (see 15-8 for details).

TRANSFORMERS

TRANSFORMER

The schematic at the right shows the voltage step-down procedure for reducing high-voltage transmission power to low-voltage house power.

- The power company transmits electrical energy in high-voltage three-phase power. To meet house requirements this must be reduced to 120/240 V single-phase power.
- To do this, one high-voltage phase conductor is fed to a center-tapped single-phase transformer which reduces the voltage to 240 V on the house side. To provide the required 120/240 V service the single 240 V circuit is divided into two separate circuits and grounded in the middle.
- There are now two "hot" conductors and one grounded wire (neutral). This gives 240 V across the two hot conductors and 120 V between each hot conductor and the grounded neutral. The neutral conductor is grounded at the transformer and at the house service load center. It is important that the applied loads be balanced as closely as possible across both hot conductors to reduce current through the neutral. Balancing is one of the load center functions (see 15-6).
- The three-wire system is fed through the service connection to the load center and extends throughout the house in a variety of circuit types, depending on appliance and equipment load and voltage requirements.

Note: Residences rarely require the higher load and voltage capacity of three-phase power normally used by commercial and industrial companies. Consequently, this type of power service may not be readily available in residential locations, and homeowners requiring three-phase power should check with their power company to establish service requirements.

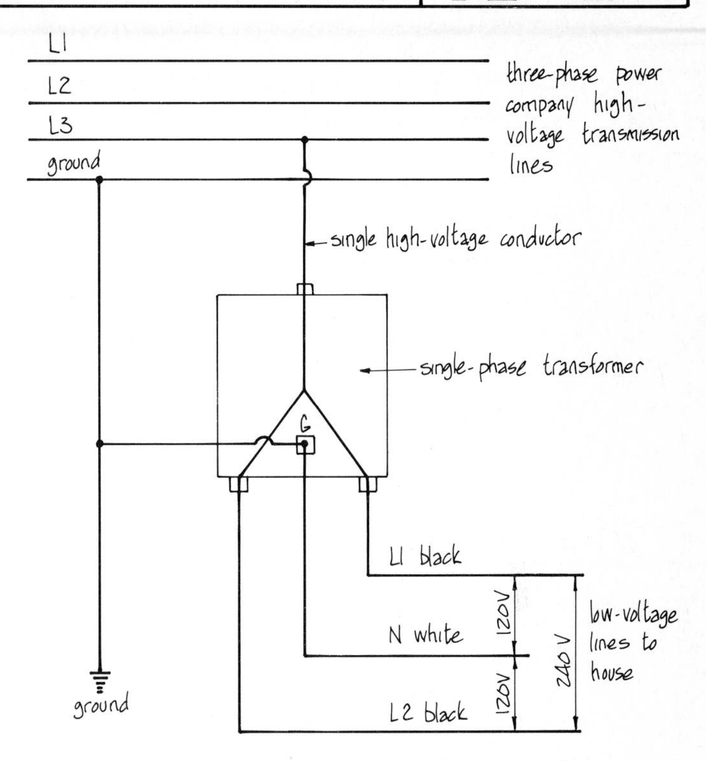

Schematic showing three-phase high-voltage stepdown to single-phase low-voltage house supply.

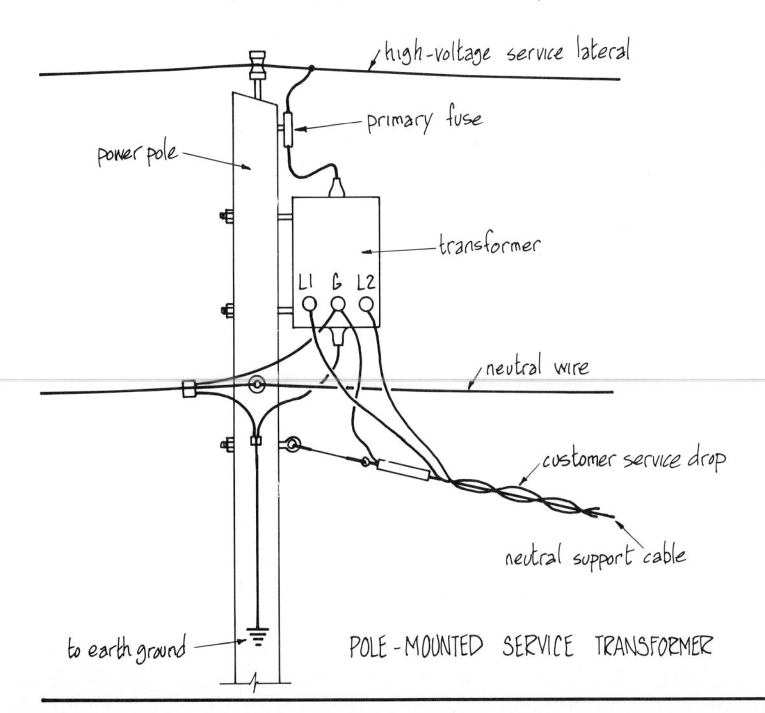

POLE-MOUNTED SERVICE TRANSFORMER

TRANSFORMER LOCATION

- Where full underground service is provided (as in most newly developed subdivisions) a pad-mounted transformer at ground level (or underground) is used. This eliminates the need for unsightly power poles and greatly reduces weather damage to power lines.
- For aboveground service, the transformer is usually pole-mounted.
- In some locations the transformer is pole-mounted but the house conductors are routed down the pole and fed underground.
- Note that each transformer can service the power requirements of several houses.

SERVICE ENTRANCE - UNDERGROUND

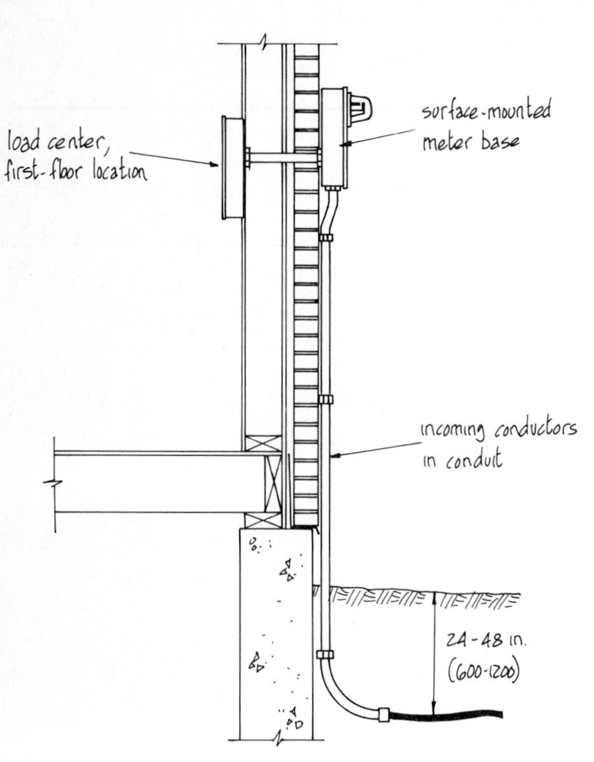

SURFACE-MOUNTED METER
UNDERGROUND SERVICE AND FIRST-FLOOR LOAD CENTER

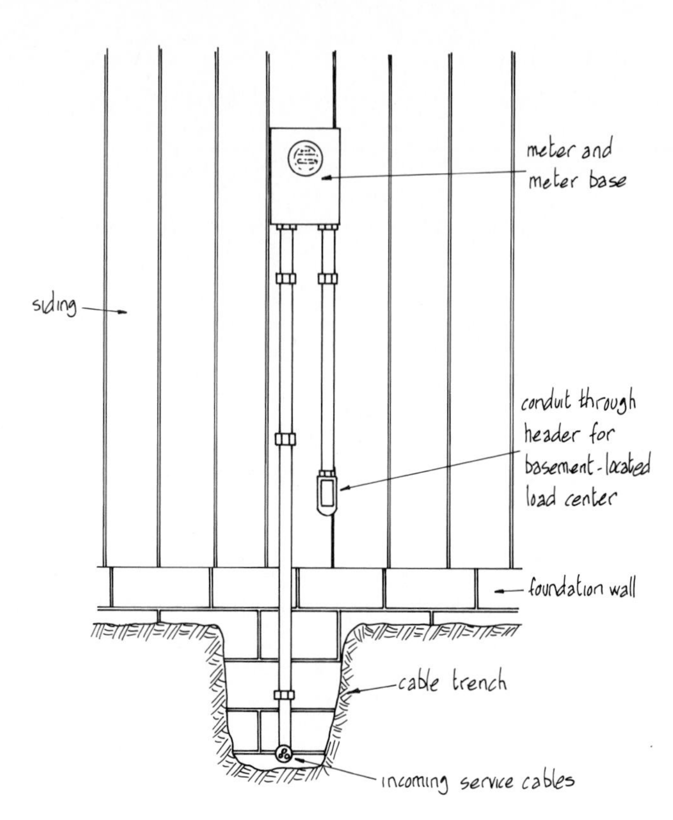

FRONT VIEW OF BASEMENT ENTRY SERVICE CONNECTION

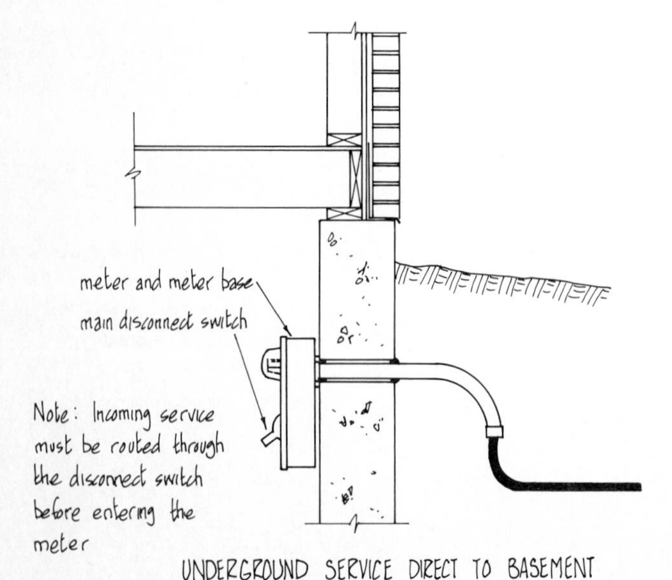

UNDERGROUND SERVICE DIRECT TO BASEMENT

UNDERGROUND SERVICE ENTRANCES

- Refer to local or national codes for exact installation requirements.
- The incoming service conductors are usually one of two types: a multiconductor sheathed cable designed for direct burial, or single conductors individually insulated.
- Buried conductors are not normally run in conduit unless subject to possible subsurface damage.
- Meter base and vertical conduit may be either recessed in the wall structure or surface-mounted (depending on local codes). Load center location determines routing of conductors from the meter base (two examples are shown above).
- Where conductors enter the house the conduit must be carefully sealed against weather and air infiltration.

SERVICE ENTRANCE - OVERHEAD

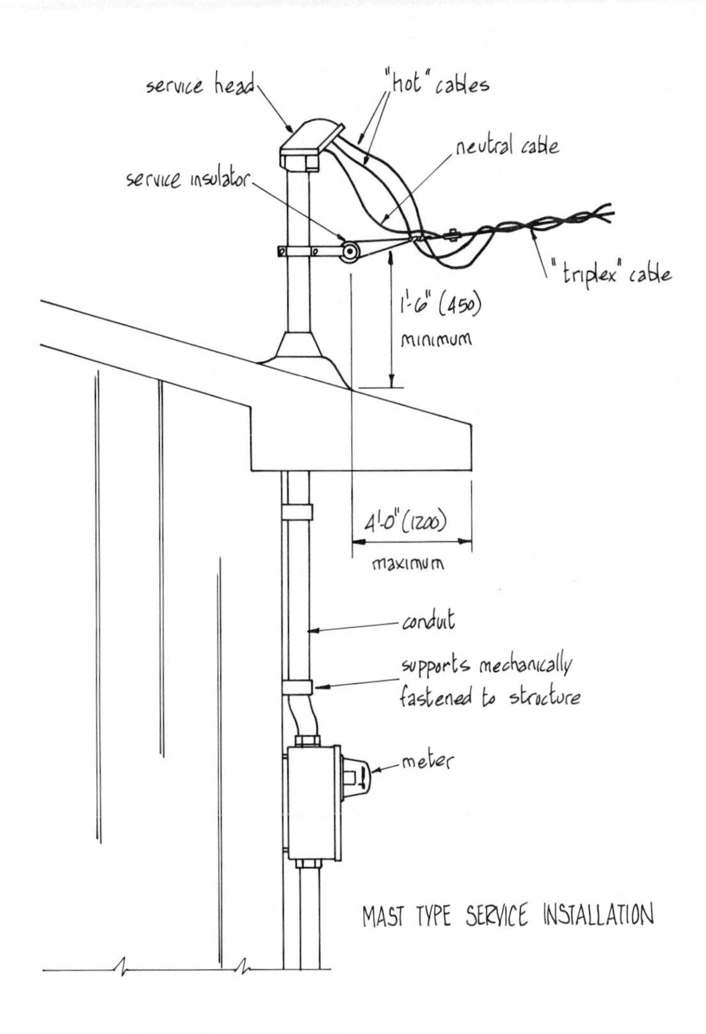

MAST TYPE SERVICE INSTALLATION

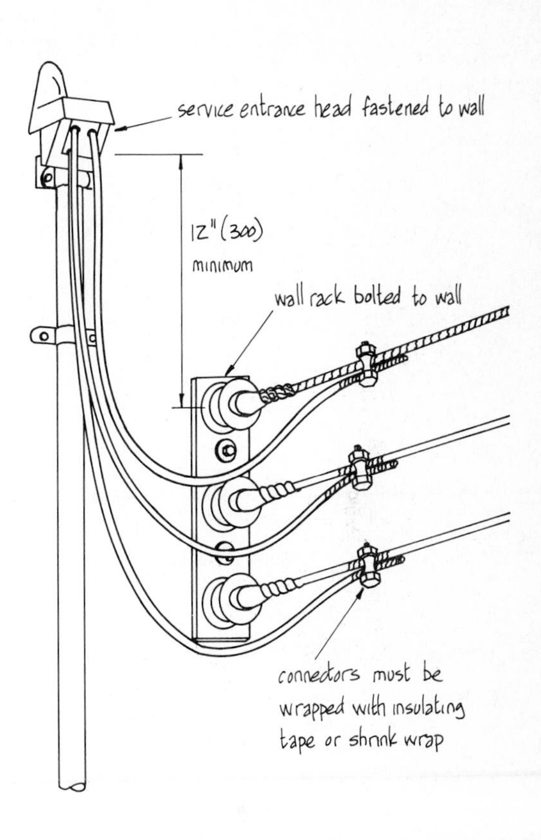

WALL MOUNTED SERVICE HEAD WITH WALL RACK AND THREE SEPARATE CONDUCTORS

OVERHEAD SERVICE ENTRANCES

- Refer to local codes for exact installation requirements.
- The location of the service drop, and to some extent the load panel, is usually determined by the power company. In most regions the electrician is responsible for both the service drop equipment and the conductors from the house to the entrance head. The power company will supply and install the service drop conductors from the entrance head to the power pole.
- Triplex cable (a bare neutral supporting two insulated "hot" conductors) is generally used for the service drop (see above, left). In some locations three separate conductors are supplied and an insulator rack is required (see above, right)
- There are minimum height clearances that must be observed for the service conductors above sidewalks and roadways. Check local codes.
- Meter bases and service entrances to the house are generally the same as for underground installations.

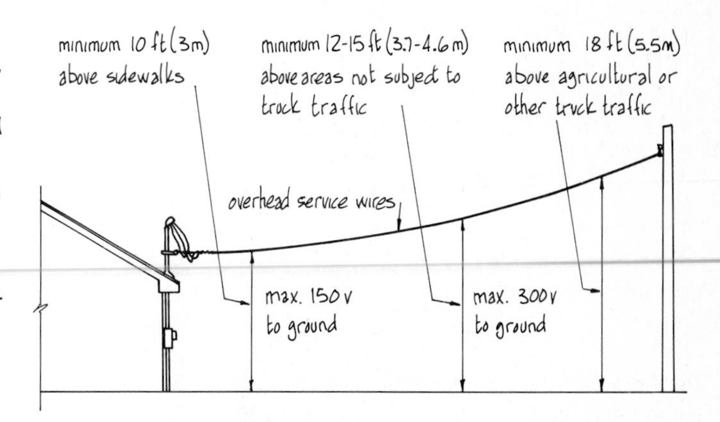

GENERAL MINIMUM CLEARANCES FOR SERVICE DROPS

see local codes for exact clearances

ELECTRICAL

15-6 LOAD CENTER - PANEL BOARD

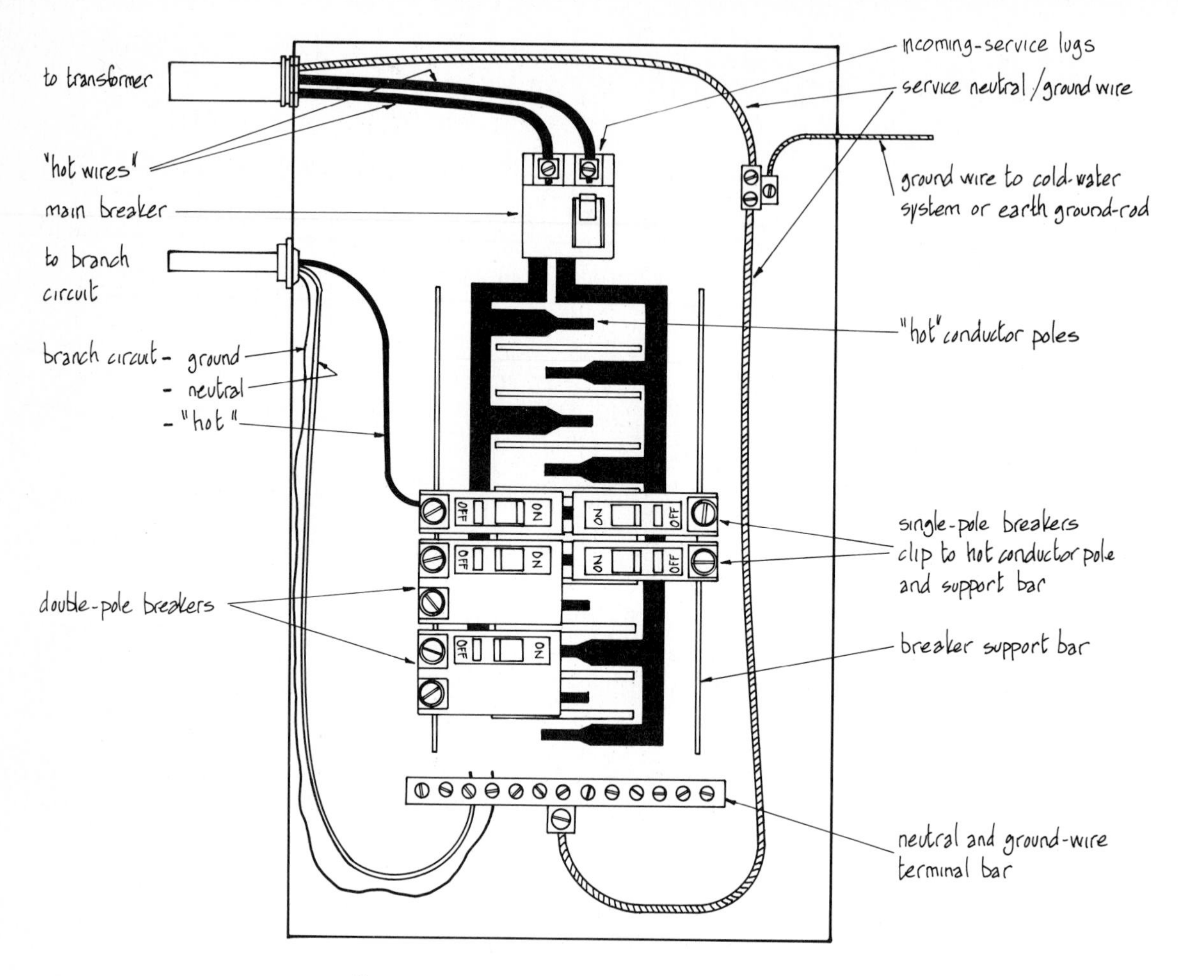

SCHEMATIC OF 100 AMP LOAD CENTER

The load center above is typical of the circuit panels in general use for most residences and small commercial establishments.

- The panel shown uses circuit breakers for overcurrent protection but panels with fuses are also available (see the next page). The individual breakers or fuses are rated to match the type of circuit they protect.
- The main breaker, which disconnects the entire panel from the power supply, is rated for the total potential panel capacity.
- An important function of the load center is the balancing of the circuits so that the two hot conductors from the transformer are loaded as equally as possible. To achieve this the two hot connection bars extending downward from the incoming service lugs are extended into the panel center as single poles on an alternate basis. Single-pole breakers down each side then connect to alternate poles for 120V service while double-pole breakers for 240V equipment automatically connect to both sides of the service.
- The panel size determines the number of circuits available; a 100-ampere (A) panel averages 12 circuits, a 200A panel averages 30 circuits.
- The grounded neutral and ground-wire terminal bar serves as the connection point for the insulated white neutrals and bare ground wires of the house circuits. It is grounded by the incoming service neutral from the transformer and the ground wire to the cold-water system or ground-rod. Some panels combine the neutral and ground connections as in the panel above, while in some the neutral bar is separate from the ground connections. Ideally, the service neutral will carry no current to ground if all loads are properly balanced. In reality, loads are rarely totally balanced and some current will flow to ground. See 15-12 for important grounding information.
- All connections should be as tight as possible. Keep wire lengths reasonably short for easy access to components and to avoid excessive heat buildup.

CIRCUIT BREAKERS AND FUSES

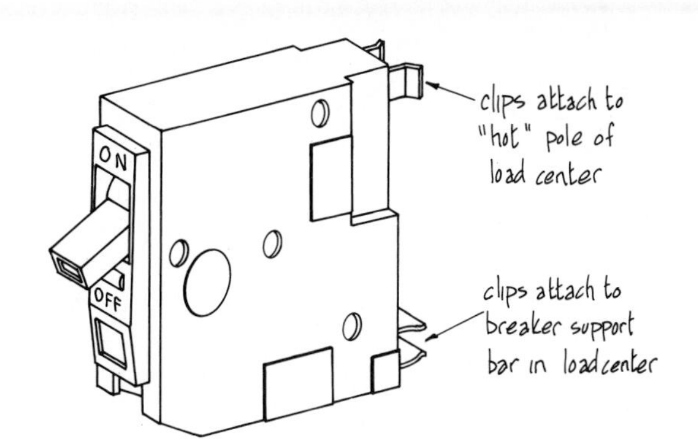

SINGLE-POLE CIRCUIT BREAKER

A circuit breaker acts as both a switch and a fuse. A red flag is visible in the front panel when the breaker has been tripped. Double-pole breakers are also available (see the load center opposite) for split circuits or 240 V appliances or equipment.

FUSE TYPES

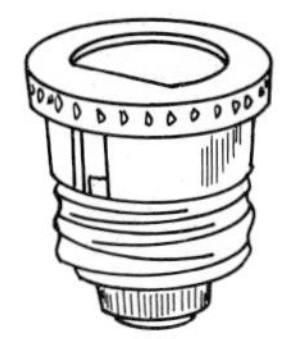

Edison Base Fuse
Now permitted as replacement fuses in 120 V circuits only. A flat, plastic, fuse base insert is available that prevents installation of a higher-rated fuse.

Type "S" Fuse and Adapter
Used for new installations. The adapter, which must be installed first, is constructed to accept only one size of fuse, making it impossible to install a higher-rated fuse.

CARTRIDGE FUSES

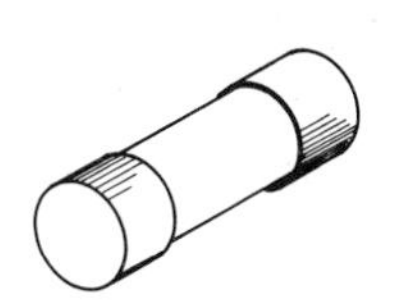

Ferrule
Rated from 10 to 60 A and generally used in individual 120/240 V appliances.

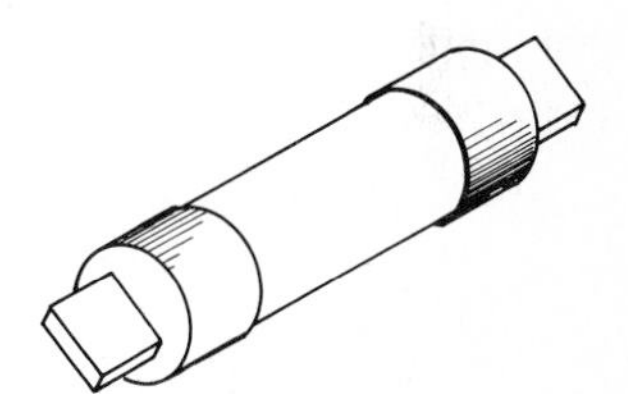

Knife-Blade
Rated at 70 A and above and generally used in the main disconnect switch of a load center.

CIRCUIT BREAKERS AND FUSES

- Breakers and fuses are usually the first line of defense against circuit overloading. The size and type of each must match the current and loading characteristics of the equipment or appliances they protect.
- Breakers are usually installed in the panel by being clipped to each hot pole and to a side support bar. The hot (black) wire of each circuit is attached to the breaker's terminal, while the neutral (white) and bare ground are attached to the grounded neutral bar. The neutral and ground connections are on separate bars in some panel designs (see the preceding page).
- Fuses are screwed into a fuse-holder/terminal connection block to which the hot wire is attached.
- After a current overload has occurred, a breaker is reset by turning its switch first to "off" then to "on". A fuse that has "blown" cannot be used again and must be replaced.

MULTIPLE LOAD CENTERS

In many larger residences and most commercial establishments, total power demand will exceed the capacity of the largest allowable load center (check local codes for maximum sizes). In this case two or more load centers must be installed and a typical schematic example is shown at the right.

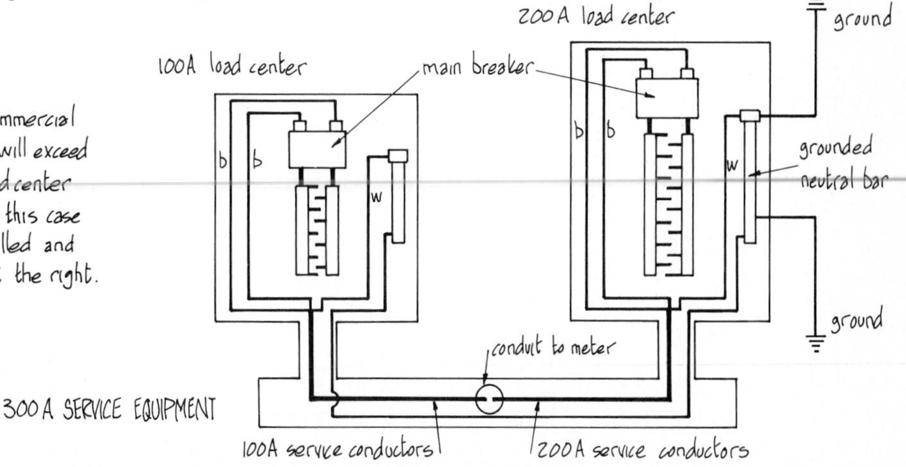

BRANCH CIRCUITS

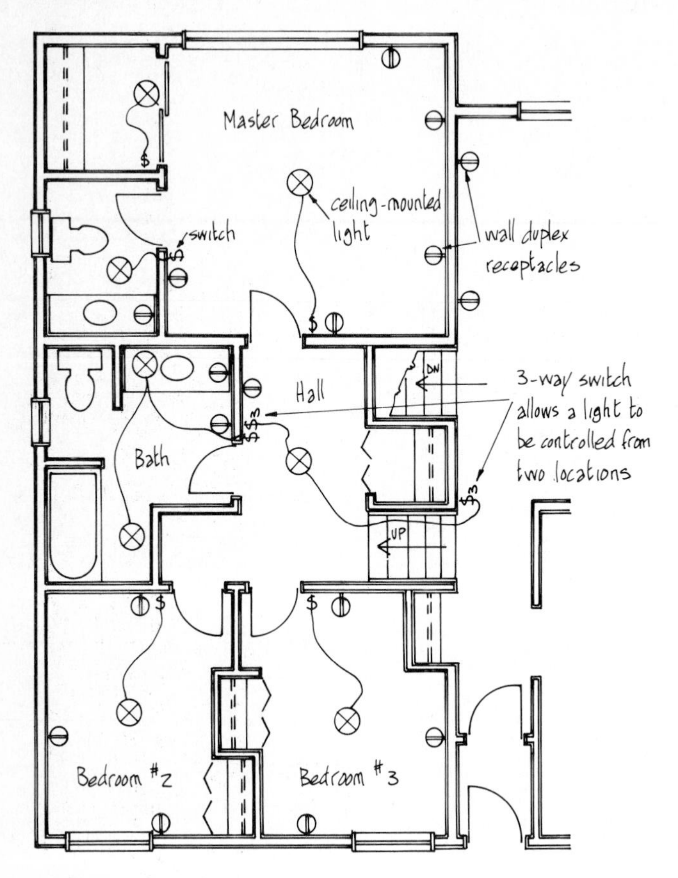

TYPICAL FLOOR PLAN
Showing approximate locations of receptacles, light fixtures and switches.
Note that the symbols are not universal - they will vary between drawing sets.

BRANCH CIRCUITS
The floor plan at the left shows a typical layout of receptacles, light fixtures, and switches in symbol form. On most residential plans, only the connections between lights and switches are shown, with the branch circuit runs being left to the electrician's discretion. On commercial projects, where an electrical engineer has designed the system, all branch circuits are clearly shown.

- There are code restrictions on the number and types of electrical devices that may be included in any circuit. There are also minimum requirements for the placement of receptacles and other devices in any particular type of room. Check local codes for exact information.
- The size and type of overcurrent protection required depend on the appliances and fixtures on a circuit. The illustration below indicates typical breaker or fuse sizes for selected circuits.

TYPICAL RESIDENTIAL CIRCUIT BREAKER OR FUSE AMPERAGE SIZES

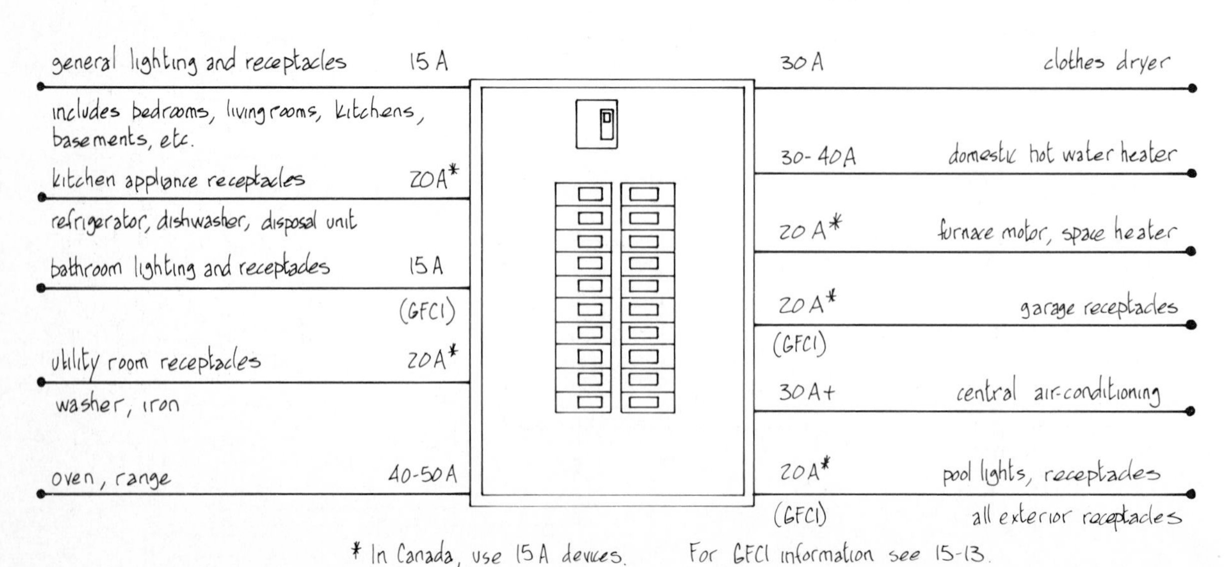

* In Canada, use 15 A devices. For GFCI information see 15-13.

BRANCH CIRCUITS

Shown below are several typical circuits in schematic form. There are many other circuit types and there is often more than one way of wiring a circuit. Because of the distinct possibility of incorrect (and dangerous) circuit construction, it is strongly recommended that all but the simplest circuits be installed by a competent and licenced electrician.

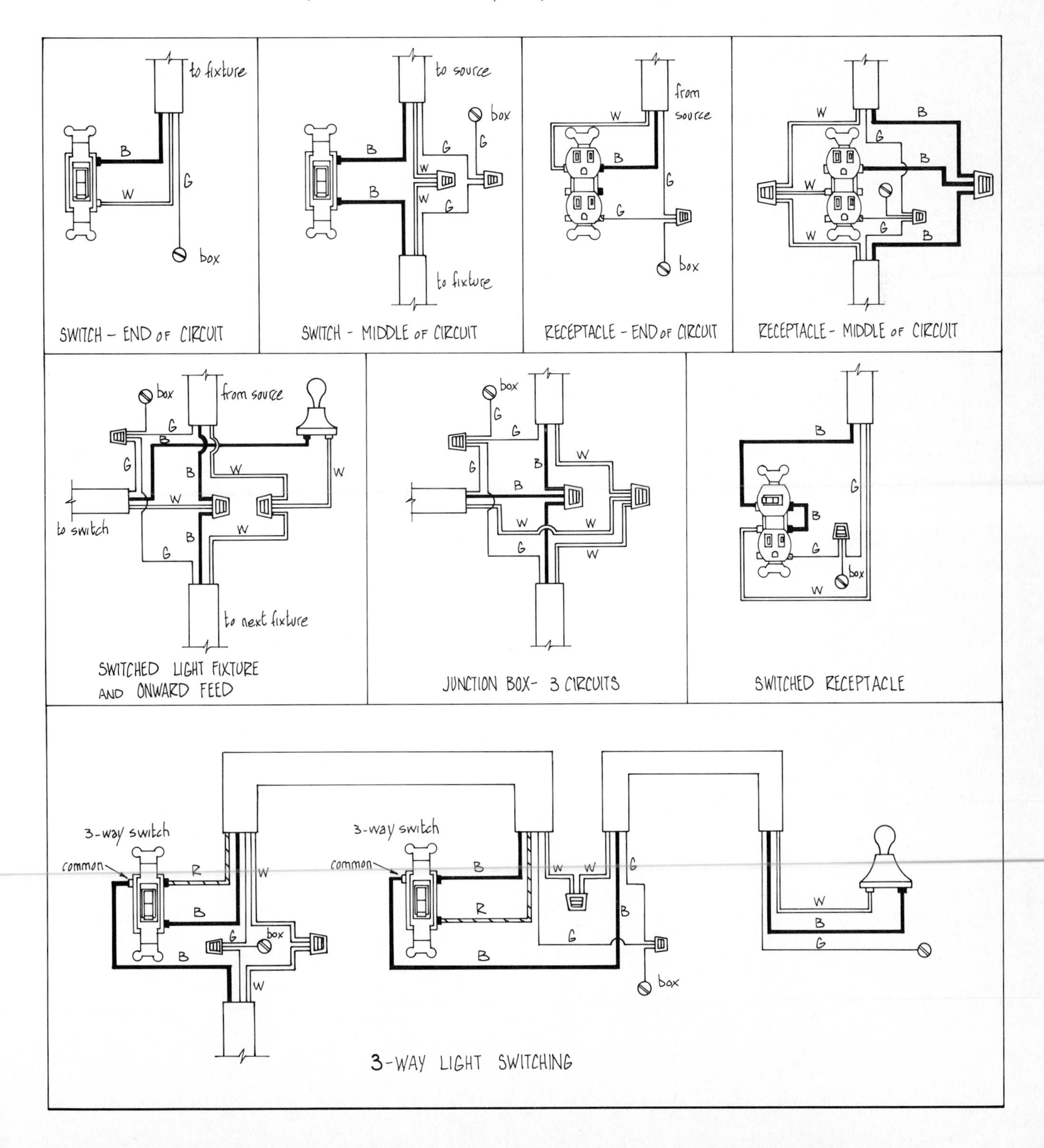

ELECTRICAL

15-10 SWITCHES, RECEPTACLES, CABLES, AND CONDUITS

Shown at the right are the face views of the most common types of switches and receptacles. Each is rated for a specific maximum amperage and voltage, and for use with copper or aluminum wire (see 15-11). This information is clearly indicated on the device.

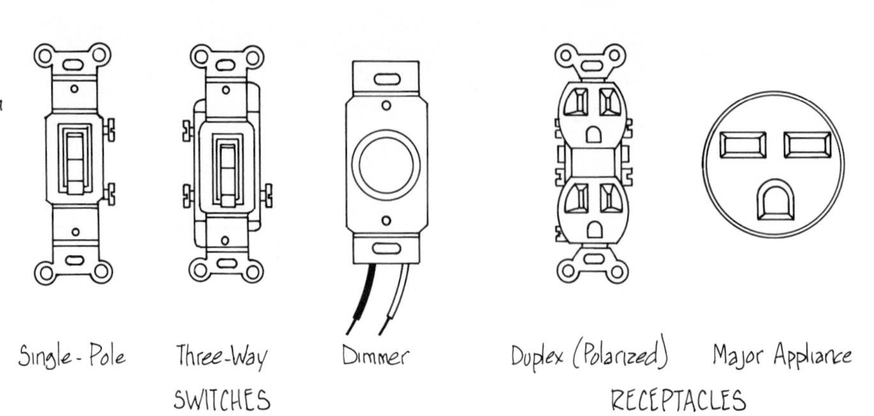

Shown below are the standard wire and conduit types used in most residential and commercial buildings. It is important to note that the wire size required for a particular circuit is dependent on the load carried by that circuit and the conductor material (copper or aluminum). Many types, sizes, and classifications of wire are available; check manufacturer's specifications for conductor applications and local codes for applicability.

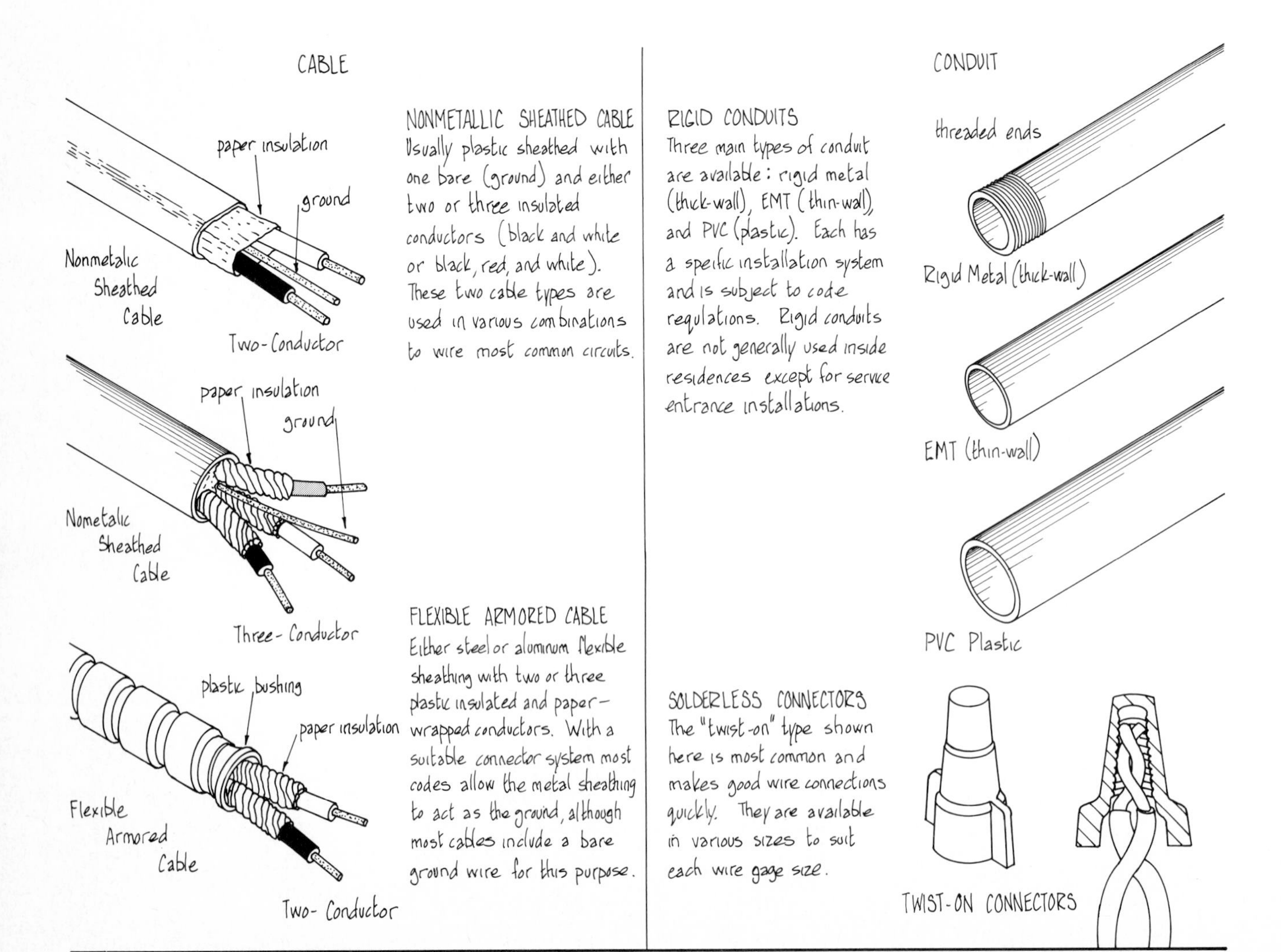

NONMETALLIC SHEATHED CABLE

Usually plastic sheathed with one bare (ground) and either two or three insulated conductors (black and white or black, red, and white). These two cable types are used in various combinations to wire most common circuits.

FLEXIBLE ARMORED CABLE

Either steel or aluminum flexible sheathing with two or three plastic insulated and paper-wrapped conductors. With a suitable connector system most codes allow the metal sheathing to act as the ground, although most cables include a bare ground wire for this purpose.

RIGID CONDUITS

Three main types of conduit are available: rigid metal (thick-wall), EMT (thin-wall), and PVC (plastic). Each has a specific installation system and is subject to code regulations. Rigid conduits are not generally used inside residences except for service entrance installations.

SOLDERLESS CONNECTORS

The "twist-on" type shown here is most common and makes good wire connections quickly. They are available in various sizes to suit each wire gage size.

WIRE SIZES AND MOUNTING HEIGHTS

WIRE SIZES

The maximum designed amperage load of a particular circuit determines the minimum gage size of the wires used for that circuit. Wire sizes are indicated in American Wire Gage (AWG), and the smaller the number, the larger the wire and the greater its current-carrying capacity. The chart below gives standard wire sizes for three types of residential circuits, together with matching breaker or fuse size (see also 15-8).

TYPE OF CIRCUIT	AWG SIZE COPPER	AWG SIZE ALUMINUM	MAXIMUM BREAKER/ FUSE SIZE
general purpose lights and receptacles	14 or 12	12 or 10	15A or 20A (15A in Canada)
small appliances refrigerator, toaster, iron, hair-dryers, etc.	12 (14 in Canada)	10	20A (15A in Canada)
large appliances heating/cooling equipment	10+	8+	30A+

Note that aluminum wire sizes are one size larger than copper sizes for the same current capacity because of aluminum's greater electrical resistance. Also, for operation in circuits where aluminum wire is used, all switches and receptacles must be marked "CO/ALR" and all equipment rated at 30A and over, marked "AL-CU".

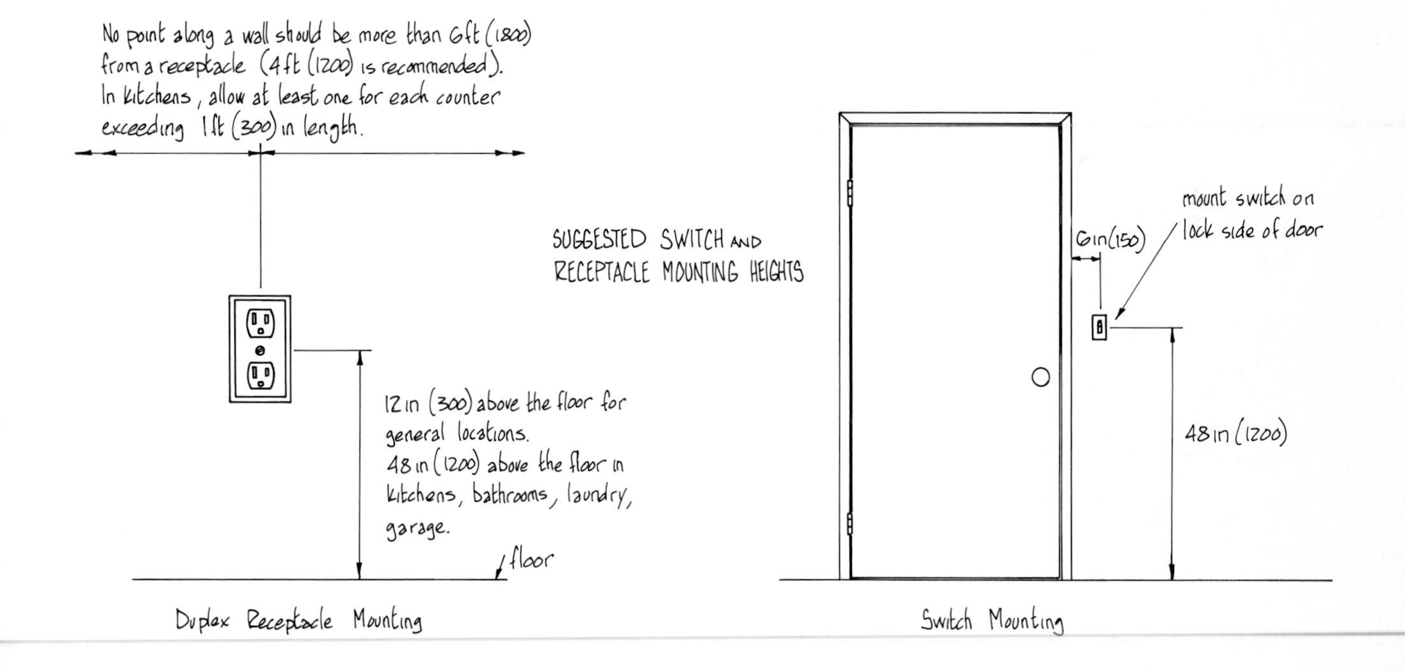

SWITCH AND RECEPTACLE MOUNTING HEIGHTS

Shown above are typical mounting heights that meet most standard residential requirements. The spacing of receptacles should at least meet the minimum standards, but it is strongly recommended that extras be included and consideration given to the probable placement of furniture, fixtures, and equipment, particularly in kitchens or other work areas.

GROUNDING

Grounding is the deliberate and direct connection of specific parts of the electrical system to the earth. Correct grounding procedures are a significant part of an electrical installation and are extremely important for occupant safety and the protection of appliances and equipment.

The electrical installation is grounded in two ways: system grounding (discussed below) and equipment grounding (discussed on the next page).

SYSTEM GROUNDING

- As was seen on page 15-6, all white neutral wires from each branch circuit, plus the incoming neutral from the transformer, are connected to the neutral bar in the load center. The neutral bar, in turn, is connected to ground with a bare ground wire as illustrated at the right.
- This procedure protects the circuits from high voltage situations that originate outside the building from sources such as lightning strikes or transformer malfunctions.
- The white neutral wire must be continuous throughout each circuit and must not be interrupted by switches, breakers, or fuses (there are some exceptions to this noninterruption rule — refer to codes).
- The neutral is always white, but in some of the more complex circuits, a white wire may not always be a neutral.
- One way of ensuring white-wire continuity is shown below.

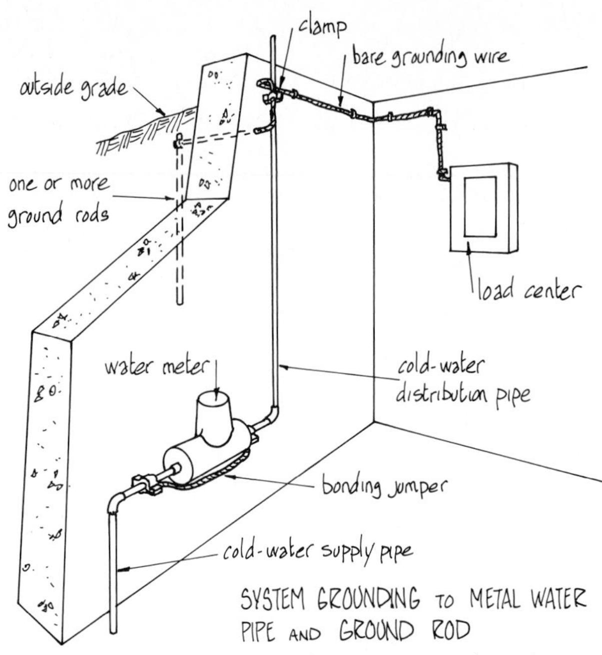

SYSTEM GROUNDING TO METAL WATER PIPE AND GROUND ROD

Note: This grounding connection is not legal in Canada. The grounding wire must make direct contact with the supply side of the cold-water line.

The illustration above shows a typical residential grounding system.

- The bare copper ground conductor from the grounding bar in the load center is connected to the incoming cold-water supply pipe and to one or more exterior grounding electrodes (exterior electrodes are not required in Canada). The water pipe must be in direct contact with the earth for at least 10 ft (3 m).
- Ground conductor size is determined by the ampacity of the largest service entrance conductor. E.g., with No. 1 or 0 copper service conductors, a No. 6 copper ground conductor is used. Check codes for other size requirements.
- Other types of connection to ground are also permissible — refer to codes for details.
- Note the use of a bonding jumper across the water meter, installed to ensure continuity of the ground circuit. Jumper wires should also be installed across any other equipment or insulated sections of pipe where continuity may be broken.

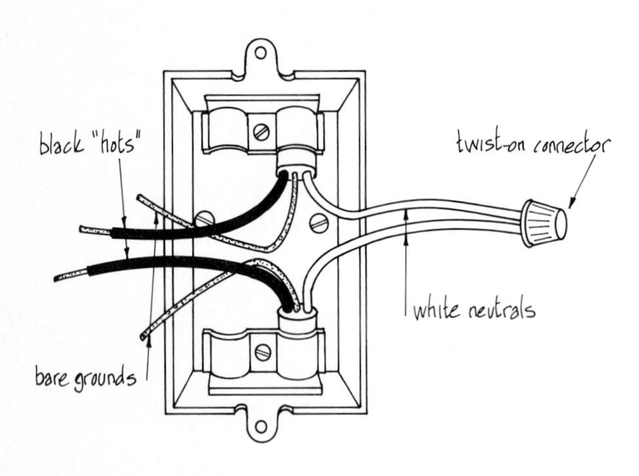

The illustration above shows a typical solderless connection of the neutral wire at an electrical box prior to installation of a switch. For neutral connection at a receptacle, see 15-9.

GROUNDING

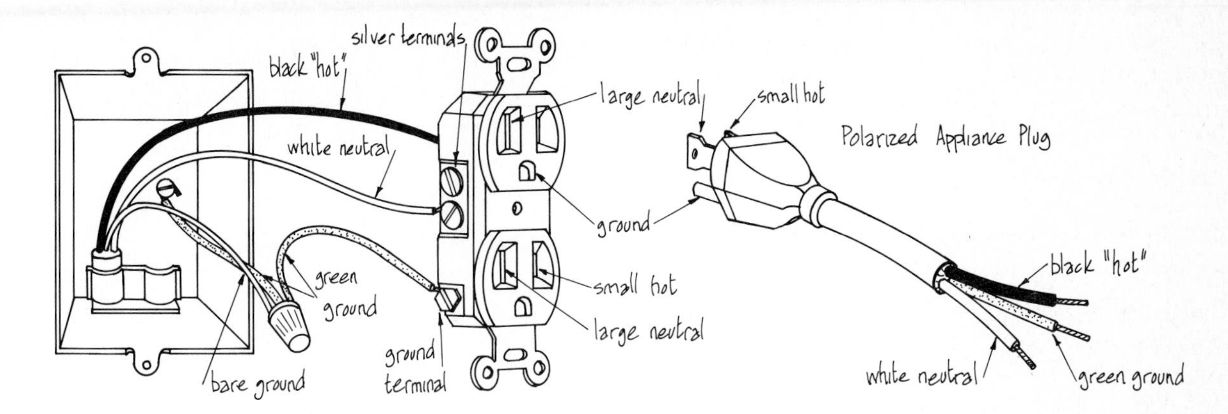

EQUIPMENT GROUNDING

- All metallic, non-current carrying parts of the electrical system must be grounded. This includes all metal boxes containing switches, receptacles, or other electrical devices and the frames of equipment, appliances, motors, and conduit.
- These items are grounded to reduce the shock hazard from boxes or equipment that may become "hot" because of a conductor insulation failure or loose connection. If the short circuit is large enough, the circuit breaker or fuse will be activated to protect the wiring and equipment. For human protection see GFCIs below.
- The bare grounding conductor in a typical residential circuit is connected to the neutral bar in the load center and to each box or electrical item in turn. (In some wiring systems the ground wire is insulated. If so, it must be green or green with yellow stripes.)

The illustration above shows typical wire connections for a box, receptacle, and appliance chord.

- The incoming bare ground is connected to the box and the receptacle by green insulated jumper wires as shown. (With plastic boxes grounding may not be necessary, and the bare ground wire is connected directly to the receptacle ground screw.)
- The green wire of the appliance chord is connected to the ground screw on the appliance (this may be a direct connection or a plug-in type connection). When the appliance chord is plugged into the receptacle, the ground circuit is established from the appliance to the neutral bar in the load center

POLARIZATION: Care must be taken to ensure the correct connection of wire colors to the appropriate terminal screw colors on devices such as the receptacle above. Note also the differing size of receptacle holes and plug prongs that ensure a one-way-only insertion. These procedures are to ensure that the continuity of the black, white, and ground wires is maintained throughout each circuit.

GROUND FAULT CIRCUIT INTERRUPTERS (GFCIs)

GFCIs are special-purpose units that detect potentially dangerous ground fault current in a circuit. They are extremely sensitive and will turn off a circuit when as little as 0.005 A is flowing to ground. Codes require GFCI protection for all outside receptacles, around swimming pools, in bathrooms and for some garage outlets. GFCIs also provide standard 15 A or 20 A overcurrent circuit protection.

There are three basic types of GFCI:

- A combination circuit breaker-GFCI installed in the load panel. This protects all receptacles on that circuit.
- A unit that replaces a standard receptacle in an electrical box. This also protects all other outlets beyond it on the same circuit.
- A portable unit that plugs into an existing receptacle and protects only appliances, which are in turn plugged into the GFCI.

white pigtail is attached to the neutral bar in the load center

The branch circuit wires are each attached to identified terminals on the GFCI.

If the unit is working, the test button will trip the circuit. Monthly testing is recommended.

ON
15
OFF
PUSH TO TEST

GFCI TYPE CIRCUIT BREAKER

PLUMBING

CHAPTER PAGE TITLES

WATER SUPPLY - MUNICIPAL

WATER SUPPLY SOURCES

There are two major sources of water supply:

- A municipal or city system supplying water under pressure to a large number of individual users (see below).
- A private well system serving an individual user or a small group of users (see the next page).

Whatever the source, the water supplied must be safe for drinking or other household uses (called "potable water") and be free from unpleasant taste, odors, or impurities. In a public system, large volumes of water must be supplied and raw water is generally obtained from lakes, rivers, or deep wells. Such raw water is unlikely to be acceptable for human consumption and may need to be heavily treated. Private well water, however, tends to require little or no treatment provided that contamination sources can be avoided.

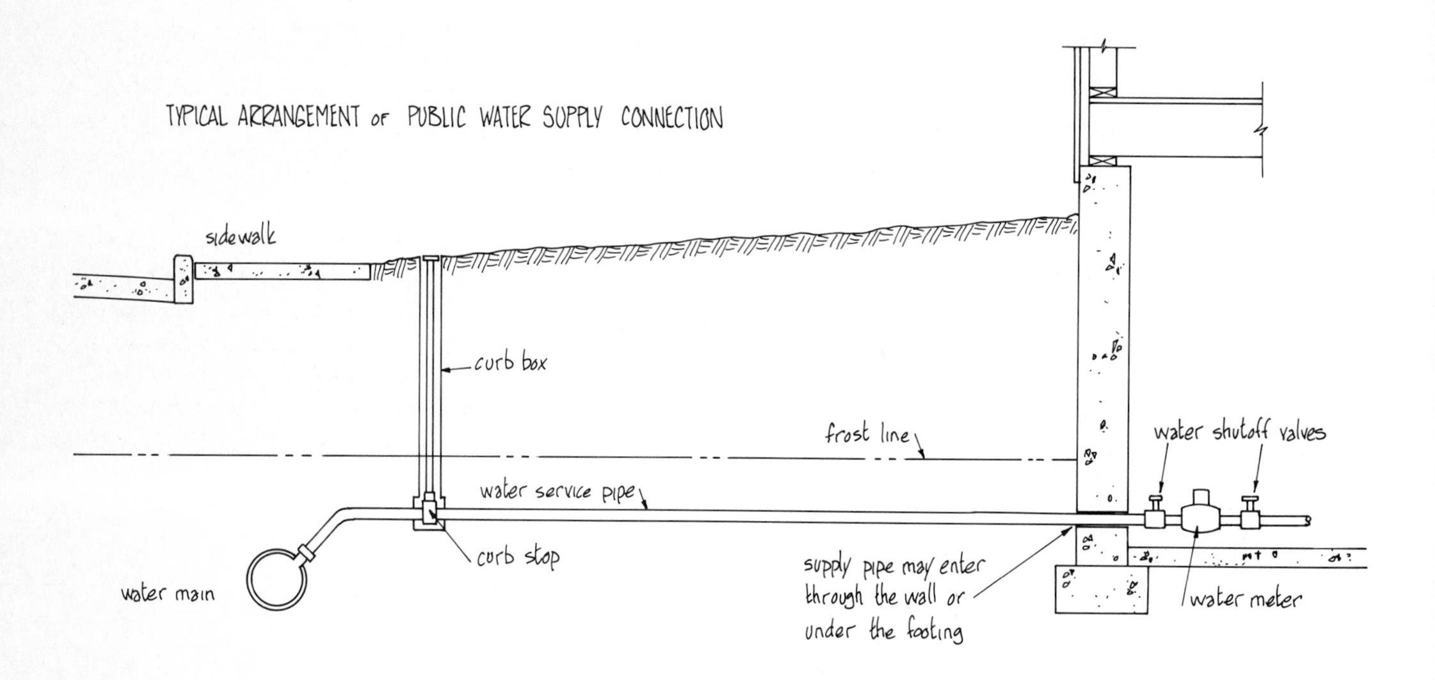

WATER MAIN

Carries public water for community use and is part of the overall distribution system. Water pressure varies with location but averages 50 psi (345 kPa). Individual users may be connected to the system on payment of a fee and issuance of appropriate permit.

WATER SERVICE PIPE

Installation and maintenance of this pipe from the curb stop onward is generally the responsibility of the property owner. Pipe must be installed below the frost line and deep enough to avoid mechanical damage and may need protection from acidic or other soil conditions.

CURB STOP VALVE

Installed by the water company, this valve is used to shut off the water supply to the entire building for service or emergency procedures. It is operated by a key that extends to grade level in a protected curb box.

WATER METER

Meters the amount of water used and is installed by the water company. Shutoff valves are installed on both sides of the meter to allow easy repair or replacement. Meter may be installed outside in frost-free regions.

WATER SUPPLY - PRIVATE WELLS

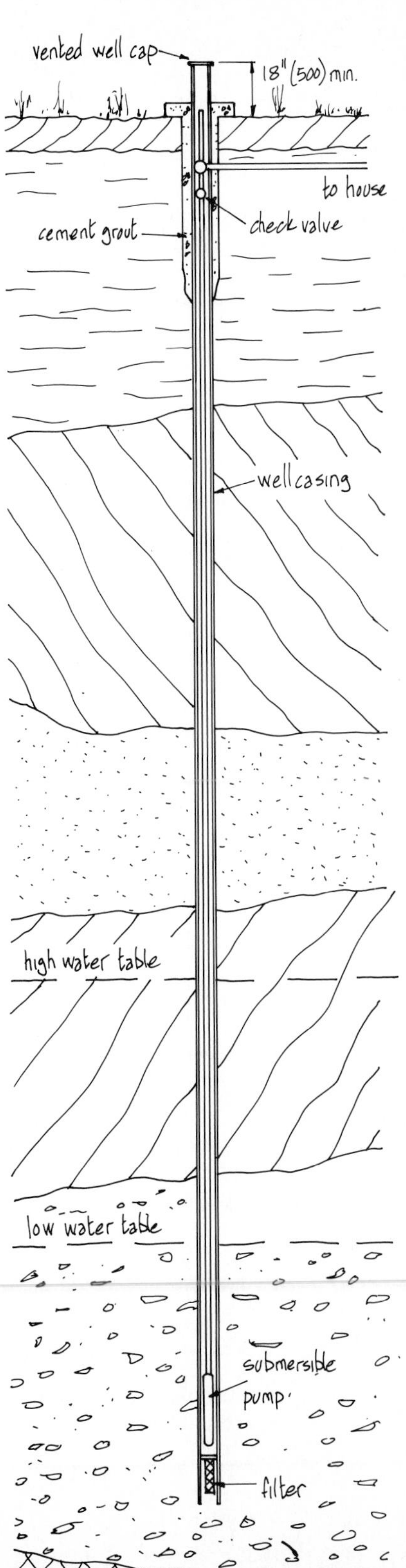

WELL HEAD

The well must be capped at the casing head with a watertight seal around the suction pipe to prevent groundwater from entering the well. A concrete collar at the well head and grout around the casing to a minimum depth of 10 ft (3 m) also keeps groundwater from the well. The pump may or may not be located at the well head, depending on pump type and local climate.

WELL TYPES

Water depth and subsoil conditions dictate the type of well:

- Bored holes are made with an earth auger. Medium depths in reasonably soft subsoil. Casing required.
- Driven wells are made by forcing a hardened well-point into the subsoil by repeated hammer blows. Medium depth only in rock-free subsoil, but casing not required.
- Drilled well using rotary or percussion drilling rig. Well can be any depth in any subsoil or rock type and needs casing.

WATER TABLE

The water table is the upper surface of a water-bearing subsurface layer. Its position will vary from just below to hundreds of feet (m) beneath ground level depending on geology and seasonal changes. The well must extend below the dry season water-table level for a dependable water supply.

CASING END

In good rock formations an open-ended casing can be used. Where loose sand or gravel is present, some form of filter is necessary.

GENERAL INFORMATION

- The type of pump used for a particular well is generally determined by well depth, water demand volume, and initial cost. Pump types include centrifical, reciprocating, jet, and rotary. Also, depending on type, the pump may be installed at ground level, in the well casing above the water level, or be submerged below the water level.
- Well location is important. To avoid contamination a well must be at least 100 ft (30 m) from pollution sources such as a septic tank leach field, discharge well, or feedlot. In locations where other toxic wastes may be present, a careful analysis of the site must be undertaken to determine potential pollution problems.
- Besides pollution problems some well water may need treatment to reduce high mineral or bacterial content. Impurities include iron, manganese, calcium, acids, and hydrogen sulfide, with treatment methods being specific to the type of pollutant present. Untreated water will be unpleasant at least and at worst may result in damage to supply systems and fixtures.

WATER PRESSURE

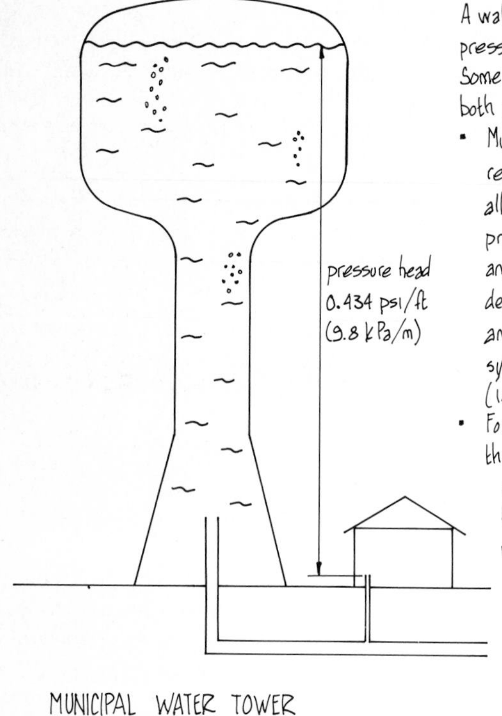

MUNICIPAL WATER TOWER

A water supply system must provide adequate pressure and water volume at the point of usage. Some form of water storage tank that meets both pressure and volume requirements is ideal:

- Municipal water systems use large-scale reservoirs or water towers to store water, allowing gravity to create the necessary pressure in the supply system. Water pressure and flow rates at any particular usage point depends on the elevation below the reservoir and the friction losses from the distribution system. Pressures range from 20 to 65 psi (140-450 kPa) in most systems.
- For private well systems hydropneumatic tanks that contain air and water under pressure are the economical solution. Air in the tank head is pressurized as water is pumped in from the well on a cycle controlled by the high/low pressure switch. Tank sizes for most residences range from 20 to 120 gal (90-350 L) and provide high/low-pressure cycles ranging between 20 and 60 psi (140-415 kPa).

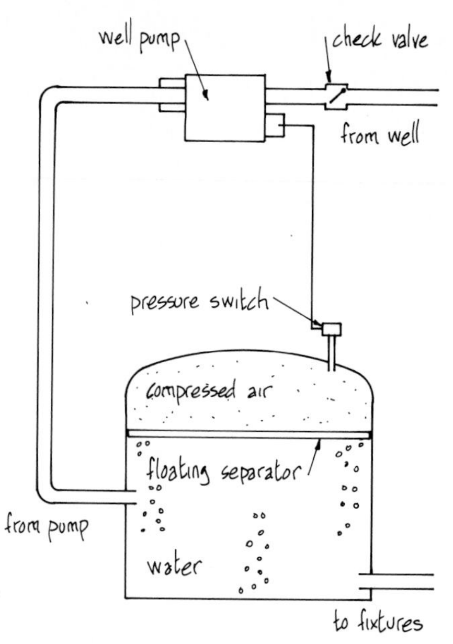

PRIVATE WELL PRESSURE TANK

FRICTION LOSS

When water flows through a pipe, pressure is lost to friction from the pipe walls and turbulence caused by the pipe fittings. The size of the pressure loss is the sum of the size, length, and type of pipe, the number and type of fittings, and the rate of water flow. The pressure loss for a fitting is expressed as an equivalent length of the same diameter pipe. The tables at the right give pressure losses for typical pipes and assorted fittings.

Plumbing fixtures require specific pressures and flow rates to function efficiently (see the next page). Care must be taken to correctly size a plumbing system to avoid excessive pressure loss between the service connection and each fixture.

FRICTION LOSS BASED ON 5 GPM (0.4 L/s) WATER FLOW

PRESSURE LOSSES	TYPE L COPPER			SCHEDULE 40 PLASTIC		
	PIPE DIAMETER			PIPE DIAMETER		
	$\frac{3}{8}$" (10)	$\frac{1}{2}$" (13)	$\frac{3}{4}$" (19)	$\frac{3}{8}$" (10)	$\frac{1}{2}$" (13)	$\frac{3}{4}$" (19)
psi/100ft	53	18	3	—	10	2.5
kPa/30m	360	120	20	—	68	17

FRICTION LOSS ALLOWANCE FOR COPPER FITTINGS
Expressed As Equivalent Lengths Of Pipe

FITTING SIZE	FRICTION LOSS FT (M)*				
	90° ELL	45° ELL	90° T SIDE BRANCH	GLOBE VALVE	GATE VALVE
$\frac{3}{8}$" (10)	0.5 (0.15)	0.3 (0.10)	0.75 (0.23)	4 (1.20)	0.1 (0.03)
$\frac{1}{2}$" (13)	1 (0.30)	0.6 (0.20)	1.5 (0.45)	7.5 (2.30	0.2 (0.06)
$\frac{3}{4}$" (19)	1.25 (0.40)	0.75 (0.23)	2 (0.60)	10 (3.0)	0.25 (0.08)

* Note: All metric pipe sizes are soft conversions.

SUPPLY SYSTEM DESIGN

For low-rise residential and small commercial installations, designing the water supply system is relatively simple and is based on the known demand factors of the different fixtures and the ability of specific pipe sizes to supply that demand. The table at the right lists standard fixtures and their recommended supply pipe sizes, plus minimum pressure and flow requirements.

In addition to the recommended sizes for branch pipes there are three rules of thumb that apply to the main supply lines:

- Up to three 3/8-in (10) branch lines may be supplied by a 1/2-in (13) pipe.
- Up to three 1/2-in (13) branch lines may be supplied by a 3/4-in (19) pipe.
- Up to three 3/4-in (19) branch lines may be supplied by a 1-in (25) pipe.

For more complex installations, two other considerations apply:

- Length of supply and branch lines and the number and type of fittings. This was discussed on the previous page.
- Height of the fixtures above the point of water entry into the building. This is a major factor in water supply design for high-rise buildings, where pressure loss on upper floors may be significant. As indicated on the preceding page, water pressure decreases by 0.434 psi per vertical foot (9.8 kPa/m).

PIPE SIZE, MINIMUM PRESSURE, AND FLOW RATES

FIXTURE TYPE	PIPE SIZE in. (mm)*	MINIMUM PRESSURE psi (kPa)	FLOW RATE gpm (L/s)
bathtub faucet	1/2 (13)	5 (35)	6 (0.45)
basin faucet	3/8 (10)	8 (55)	3 (0.20)
kitchen sink	1/2 (13)	5 (35)	4.5 (0.35)
shower	1/2 (13)	12 (85)	5 (0.40)
water closet - flush tank	3/8 (10)	15 (100)	3 (0.20)
water closet - flush valve	1 (25)	10-20 (70-140)	15-40 (1.15-3.0)
hose bib	1/2 (13)	15 (100)	5 (0.40)
laundry tub	1/2 (13)	5 (35)	5 (0.40)
urinal - flush valve	3/4 (19)	15 (100)	15 (1.15)

* Note: metric pipe sizes are soft conversions.

Shown below is a typical residential installation indicating fixture types and pipe sizes. Only the cold-water system is shown; the hot-water system is presumed to be parallel and of the same size.

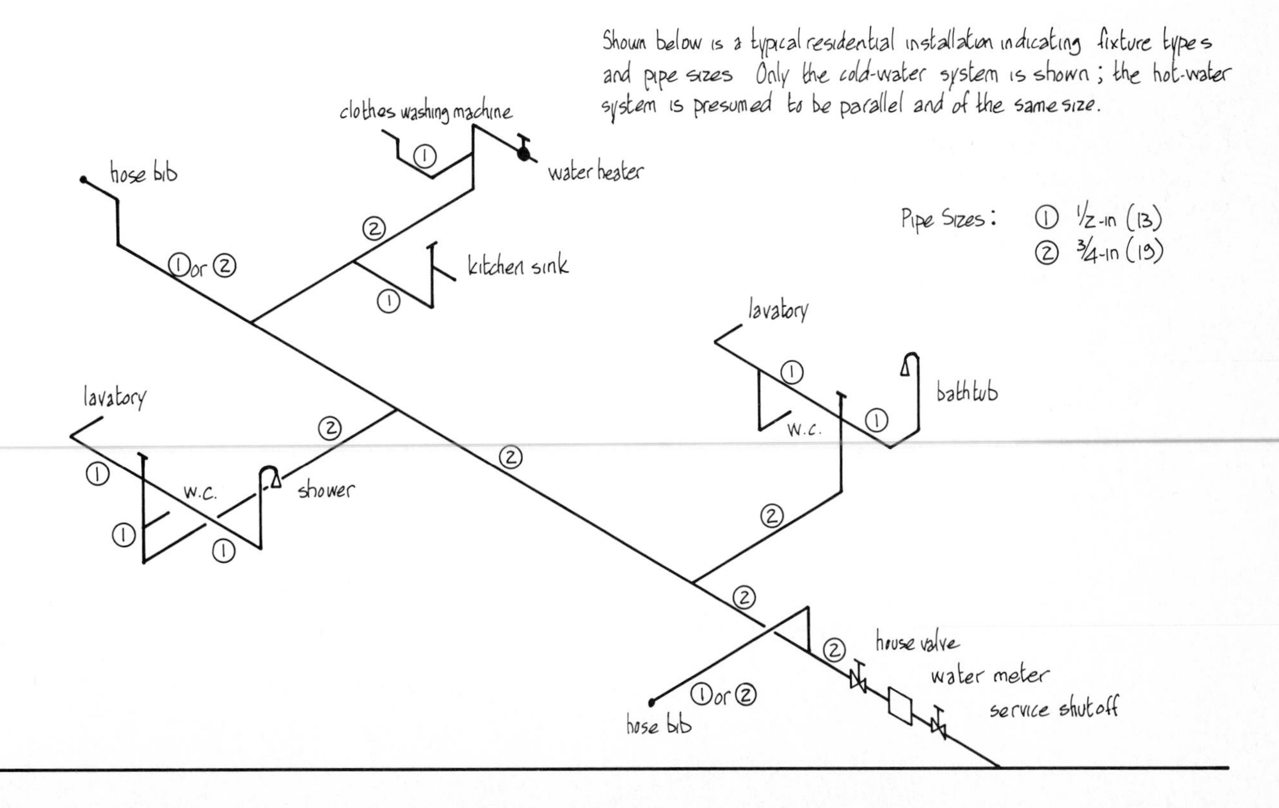

HOT WATER SYSTEMS

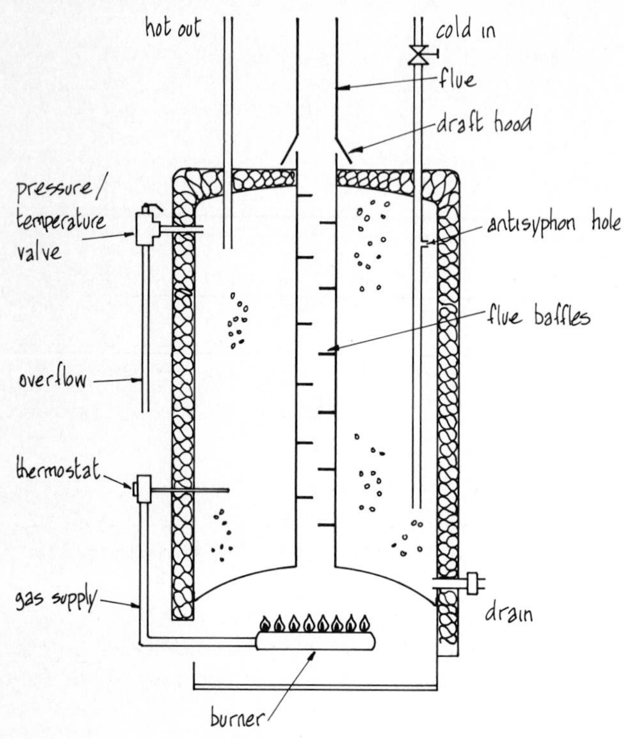

GAS HOT-WATER HEATER

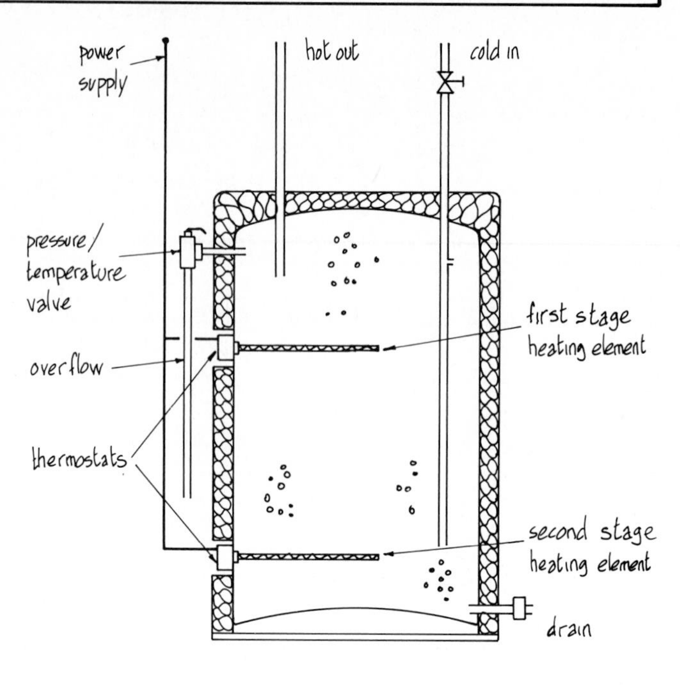

ELECTRIC HOT-WATER HEATER

STANDARD HEATER TYPES

Two types of hot-water heaters are in general use for residential and small commercial buildings: natural gas fired and electric. The schematics above show the general method of heating for each type. Note the connection to the main cold-water supply pipe and the hot-water takeoff pipe that parallels the cold supply.

Choice of heater is generally determined by fuel availability, required recovery times, and intended installation location:

- Gas heaters have a quick recovery time (the time needed to heat a tank of cold water) and tend to be more economical but require a flue pipe and a reasonable supply of combustion air. The flue also contributes to house heat loss, which may be undesirable (see 10-8 and 14-16). A high-efficiency gas heater, similar to a high-efficiency gas furnace (see 14-5), partially solves these problems.
- Electric heaters have a slower recovery time and are generally more expensive to operate but do not need a flue and can be located almost anywhere in a building.

SIZING THE TANK

Choice of tank size depends on anticipated hot water demand. The table above indicates minimum recommended sizes based on average recovery times. Consideration should be given to increasing tank size if an automatic washer or other high-demand appliance is likely to be used extensively.

SUGGESTED TANK SIZES

TANK SIZE gal (L)	NO. OF BATHROOMS	NO. OF BEDROOMS
30 (115)	1 to 1½	2 to 4
40 (150)	2 to 2½	3 to 5
50 (190)	3 to 3½	4 to 6

WATER TEMPERATURE

All heaters are equipped with a thermostat to control maximum water temperature by turning the burner or elements on or off on demand. For domestic use the recommended temperature is 130-140°F (54-60°C). This is cool enough to reduce risk of scalding but hot enough for domestic use.

SAFETY FEATURES

All modern hot-water tanks are factory equipped with a temperature/ pressure relief valve installed at the hottest part of the tank. This valve will open on excess temperature and/or pressure, which is usually caused by a faulty thermostat. It is extremely important that these valves be installed on all hot-water storage tanks; failure to do so may result in tank failure and a powerful explosion.

Other safety features include a gas shutoff valve on gas heaters that closes if the pilot light goes out and circuit connections on electric heaters that allow only one element to heat at any one time.

WATER SUPPLY SYSTEMS - MATERIALS

GRADES OF COPPER PIPE

TYPE	COLOR CODE	LENGTHS	COILS	APPLICATIONS
K	green	12 ft ① (3.6 m) *	60† and 100 ft (18 and 30 m)	underground and interior service
L	blue	12 ft ① (3.6 m) *	③	aboveground service
M	red	12 ft ② (3.6 m) *	not available	aboveground supply, drain, waste and vent

① Hard and soft temper.
② Hard temper only.
③ Soft temper only.
† Some manufacturers produce 66 ft (20 m) coils.
* Some manufacturers produce 20 ft (6 m) lengths.

PLASTIC PIPE

TYPE	JOINTING	APPLICATIONS
ABS acrylonitrile-butadiene-styrene	solvent cement	drain, waste, and vent sewer lines and water supply
PVC polyvinyl-chloride	solvent welding	water supply
	threaded/flanged	DWV, sewers, process
CPVC chlorinated-polyvinyl chloride	solvent welding	hot and cold water supply
	threaded/flanged	process
PE polyethylene	compression/flange fusion welding	water supply natural gas/oil field
SR styrene-rubber	solvent welding	storm, agriculture drains

10 ft (3 m)
sewer
water
Double Trench
1 ft (300)
water
sewer
Single Trench

SEWER AND WATER PIPE INSTALLATION

Water supply piping must meet certain requirements:

- All pipe must have a minimum pressure rating of 160 psi (1100 kPa). All metallic pipes exceed this limit, but some plastic products do not.
- Some water supplies corrode or leave mineral deposits in the supply lines. A pipe type that avoids these problems should be chosen.
- Buried pipes must be protected from mechanical damage and freezing. In some regions the pipe exterior may need to be protected from corrosive soil or other contaminants.
- In all regions, local codes will determine which materials may or may not be installed.

MATERIAL TYPES

- Copper pipe is available in four grades, three of which are suitable for water supply piping (see the table at the left). In addition, two of the grades may be ordered in either soft or hard temper. Soft-tempered pipes are flexible and are usually supplied in 60- or 100-ft (18 m or 30 m) coils. Hard-tempered pipes are rigid and available only in lengths of up to 20 ft (6 m). All joints between pipe and fittings are soldered.
- Plastic pipe for hot- and cold-water service is rapidly gaining acceptance in most regions. There are several types of pipe available for water supply systems (see the table at the left). Fitting connections are made either by mechanical couplings or are cemented or welded with a solvent. It is important to note that the correct solvent must be used for each material type to ensure a watertight joint.
- Steel pipe is also available for water supply systems. Only galvanized pipe is allowed by most codes, in order to reduce rusting and deposit problems. Standard-weight pipe is used for most plumbing installations, and all connections are made with taper-threaded pipe and fittings. This type of pipe is rarely used inside residences but may be used outside where rough service or mechanical damage is likely.

BELOW GRADE INSTALLATION

The two illustrations at the left show single- and double-trench installations and the clearances required between water supply pipes and sewers. The dimensions given are typical for most codes, but must be confirmed locally.

PLUMBING

16-8

SANITARY DRAINAGE AND VENT SYSTEMS

A sanitary gravity-driven drainage system must meet certain criteria:

- It must meet all code requirements and be a safe and sanitary installation.
- The system piping and fittings must remain water- and gastight for the life of the building.
- Pipe sizes must be adequate for the intended design load at each fixture and for the whole installation.
- Correct installation of traps and vents to ensure an airtight seal at fixtures.

One of the most important parts of DWV (drainage, waste, and venting) system design is the correct identification of each pipe as it relates to the codes. The illustration below shows a typical isometric drawing (see 1-7) of a residential DWV system with the name and a brief description of each pipe.

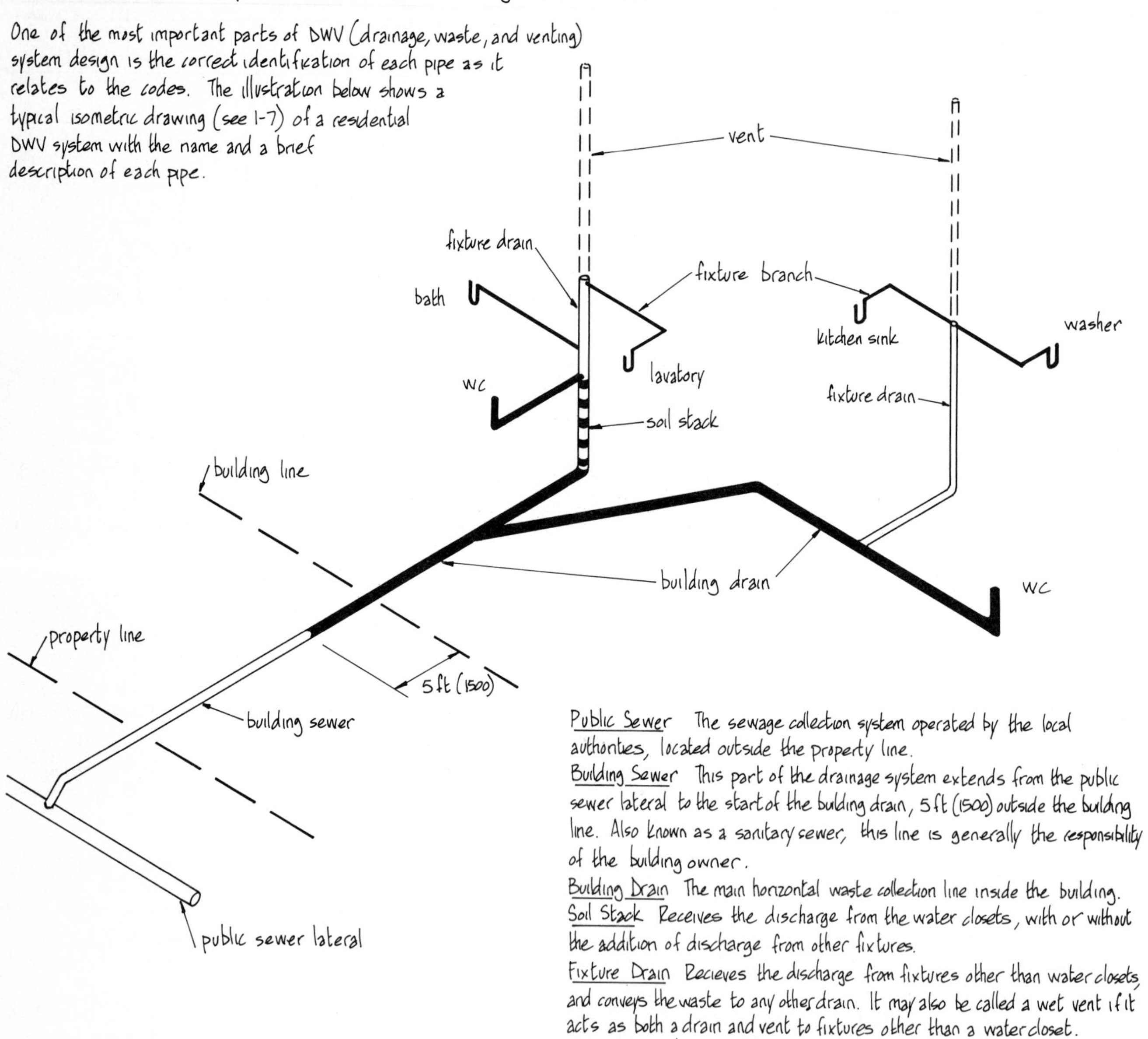

<u>Public Sewer</u> The sewage collection system operated by the local authorities, located outside the property line.

<u>Building Sewer</u> This part of the drainage system extends from the public sewer lateral to the start of the building drain, 5 ft (1500) outside the building line. Also known as a sanitary sewer, this line is generally the responsibility of the building owner.

<u>Building Drain</u> The main horizontal waste collection line inside the building.

<u>Soil Stack</u> Receives the discharge from the water closets, with or without the addition of discharge from other fixtures.

<u>Fixture Drain</u> Recieves the discharge from fixtures other than water closets, and conveys the waste to any other drain. It may also be called a wet vent if it acts as both a drain and vent to fixtures other than a water closet.

<u>Fixture Branch</u> Connects the fixture trap (see 16-12) to the vent serving that fixture, either directly to a vertical vent stack or to a horizontal wet vent below the floor.

<u>Stack Vent</u> An extension of the soil or waste stack up through the roof of the building that provides atmospheric venting to the drainage system.

<u>Horizontal Branch</u> (not shown). A pipe extending laterally from a waste or soil stack that receives waste discharge from one or more fixture branches.

FLOW RATES AND PIPE SIZES

DESIGN BASIS – FIXTURE UNITS

In a sanitary drainage system, pipe sizes are based on the total discharge rate of each fixture measured in gallons per minute. To simplify sizing, each fixture type is given a fixture unit rating (also called the load factor) that expresses the discharge rate. The fixture unit is based on the average flow from a lavatory (a hand-wash sink) of 7.5 gpm or approximately 1 cfm. This particular fixture therefore has a 1 F.U. rating. The table at the right gives F.U. ratings for a selection of common fixtures.

FIXTURE UNITS PER FIXTURE TYPES

FIXTURE TYPE	FIXTURE UNITS
water closet - tank	4
water closet - flush valve	8
bath tub	2
shower	2
lavatory	2
bathroom set (w.c., b, lav.)	6
bathroom set (w.c. with flush valve)	8
dishwasher	2
kitchen sink	2

Note: Metric fixture unit specifications are not yet available.

RECOMMENDED SIZES AND SLOPES — FIXTURE AND BUILDING DRAINS

PIPE DIAMETER in (mm)	MAXIMUM NUMBER OF FIXTURE UNITS THAT MAY BE CONNECTED TO ANY PORTION OF THE BUILDING DRAIN OR SEWER			ONE VERTICAL STACK OF THREE STORIES OR LESS
	1/8" (1:100) FALL	1/4" (1:50) FALL	1/2" (1:25) FALL	
1¼ (30)	1	1	3	2
1½ (40)	4	6	10	4
2 (50)	10	21	26	10
2½ (65)	14	24	31	20
3 (75)	28	36	40	30
4 (100)	180	216	250	240
5 (125)	390	480	575	540
6 (150)	700	840	1000	960

PIPE SIZES

The various pipes leading from each fixture to the public sewer are sized according to the F.U. load they must carry. The table at the left gives pipe sizes and their maximum F.U. rating for horizontal branches and drains and vertical soil and waste stacks.
Note the effect that slope (fall per foot) has on horizontal pipe capacity. A fall of 1/8 in per foot (1:100) is preferred for pipes of 4 in (100) diameter or greater, and a fall of 1/4 in per foot (1:50) is preferred for pipes of 3 in (150) or less. In addition, pipes should not be deliberately oversized because the reduced scouring action of the water allows solids to settle and cause blockages. Larger pipes are also more expensive and difficult to install.

HOW CODES AFFECT DESIGN

It would seem from the information above that selecting pipe sizes is only a matter of calculating the total F.U. ratings for each section of the drainage system and extracting pipe sizes from the tables. Such is not always the case, however, since the plumbing codes specify minimum pipe sizes for specific applications. In most such cases the size specified will exceed that called for by tables. There is also a bewildering array of exceptions, limitations, and restrictions that affect the design of each drainage system, and designers must be totally familiar with national and local codes.

ROOM STACKING

ROOM STACKING

A primary consideration in the design phase of a building is to locate rooms having extensive plumbing requirements as close together as possible, both horizontally and vertically. The vertical alignment of plumbed rooms in buildings of two stories or more is especially important:

- The vertical "stacking" of bathrooms, kitchens, utility rooms, etc., allows a single common soil stack to service all the drainage requirements.
- Since the soil stack is the most expensive part of the system, especially in terms of labor, significant savings can be realized.
- With closely grouped rooms, pipe runs are simplified and can be located together in special plumbing walls or furred spaces.

It is the designer's responsibility to take advantage of stacking and the plumber's responsibility to read the drawings and understand the design intent. The drawings at the right show a typical example of room stacking in a residence.

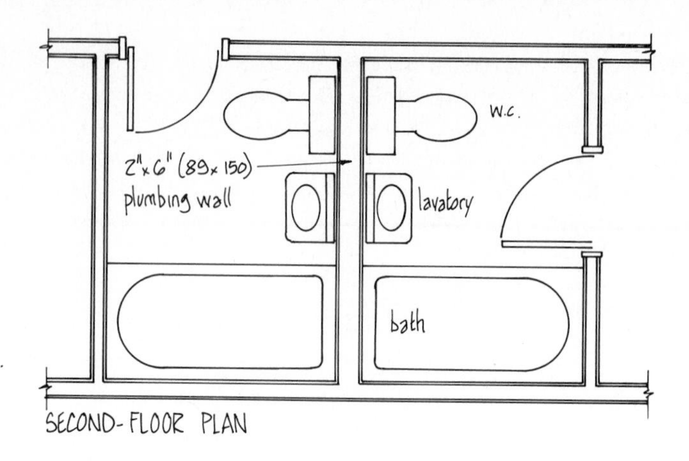

SECOND-FLOOR PLAN

The residential first- and second-floor plans shown above and below are designed for the vertical stacking of upper bathroom and lower kitchen and laundry facilities. The common soil stack and associated drain piping is shown on the next page.

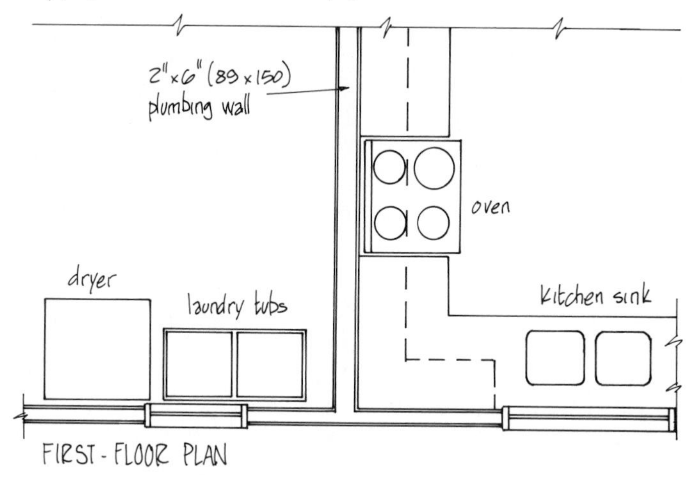

FIRST-FLOOR PLAN

THE KITCHEN WORK TRIANGLE

For maximum work efficiency, the workstations and appliances in a kitchen should be located at the points of a triangle, the total side length of which should be as short as possible (see 13-15). Plumbing pipe location is therefore a major consideration, especially if lower-floor ceilings are to be finished.

BATHROOM DESIGN

Careful location of fixtures in bathrooms is equally important. Privacy, access to fixtures, and location of plumbing are prime considerations. Where possible, plumbing should be installed in a single wall or arranged to run between floor joists (not across or through joists). Two typical examples of bathrooms layouts are illustrated below.

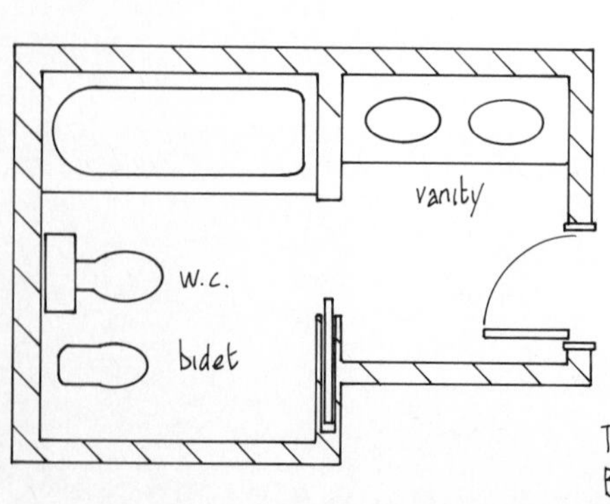

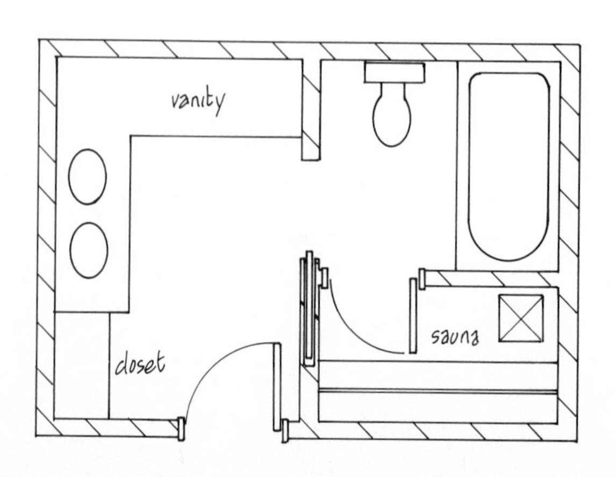

TWO EXAMPLES OF BATHROOM PLANS

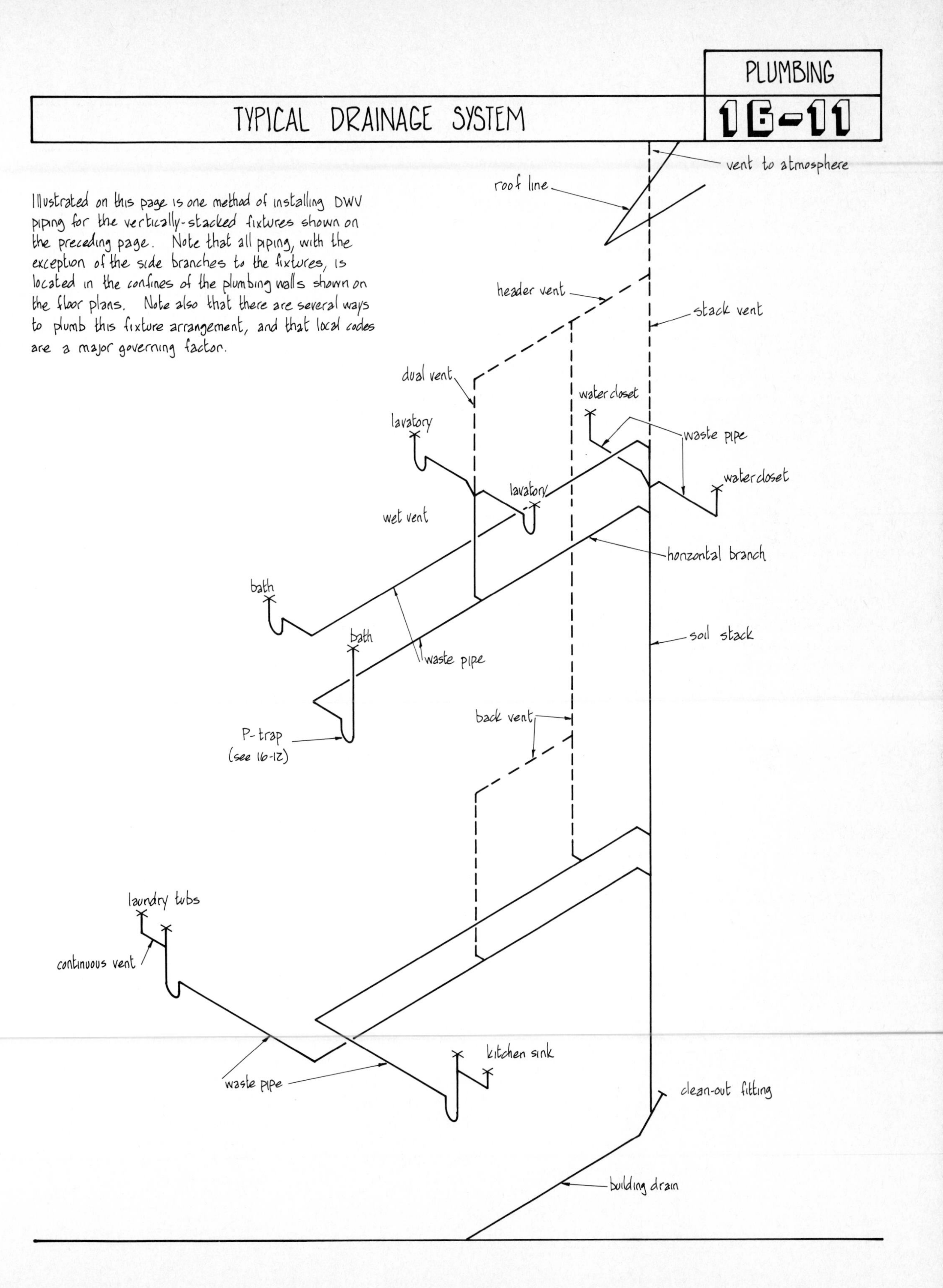
PLUMBING
16-11
TYPICAL DRAINAGE SYSTEM
Illustrated on this page is one method of installing DWV piping for the vertically-stacked fixtures shown on the preceding page. Note that all piping, with the exception of the side branches to the fixtures, is located in the confines of the plumbing walls shown on the floor plans. Note also that there are several ways to plumb this fixture arrangement, and that local codes are a major governing factor.
vent to atmosphere
roof line
header vent
stack vent
dual vent
water closet
lavatory
waste pipe
water closet
lavatory
wet vent
horizontal branch
bath
bath
waste pipe
soil stack
back vent
P-trap
(see 16-12)
laundry tubs
continuous vent
kitchen sink
waste pipe
clean-out fitting
building drain

TRAP SEALS AND VENTING

Proper installation of trap seals and vents is a major part of a drainage system. Traps provide an air seal at each fixture and prevent the entry of sewer gases into the building. Vents are open to the outside, equalizing atmospheric pressure in the piping system, allowing the traps to maintain their seal.

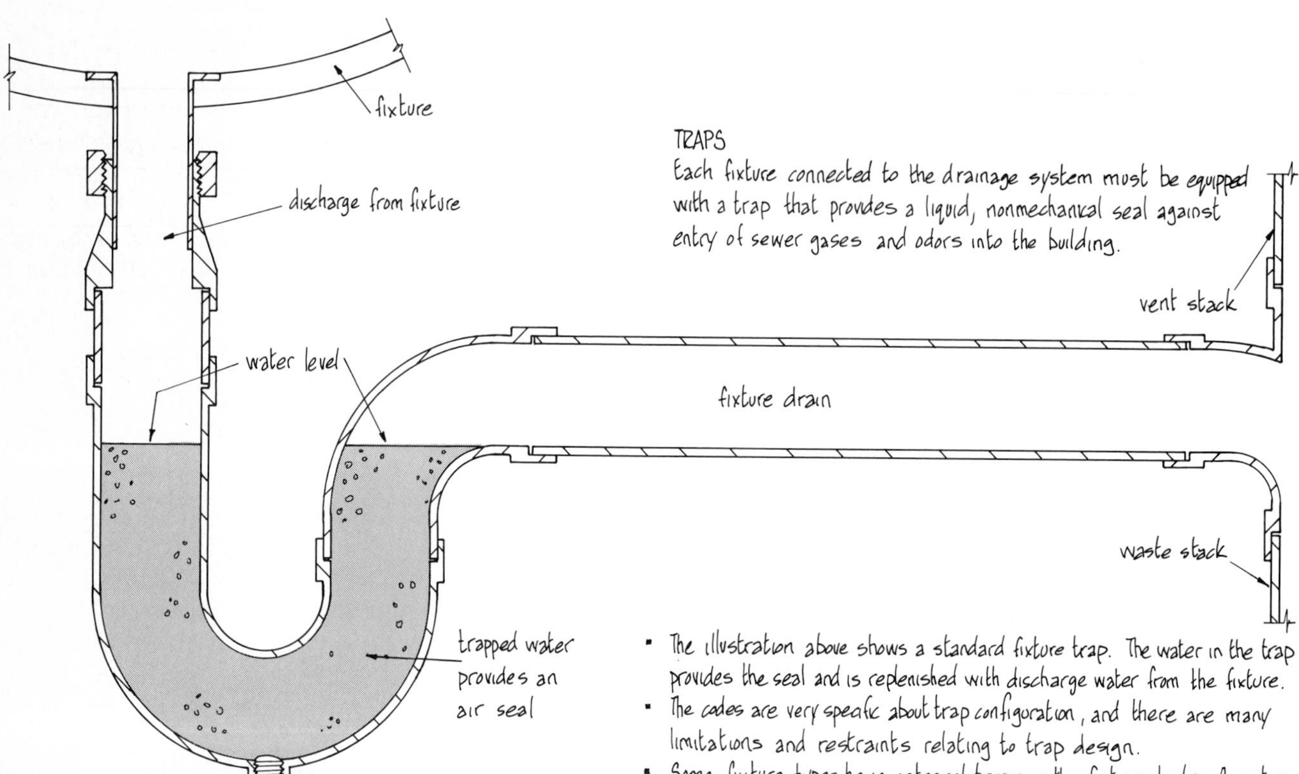

TYPICAL P-TRAP CONFIGURATION

TRAPS

Each fixture connected to the drainage system must be equipped with a trap that provides a liquid, nonmechanical seal against entry of sewer gases and odors into the building.

- The illustration above shows a standard fixture trap. The water in the trap provides the seal and is replenished with discharge water from the fixture.
- The codes are very specific about trap configuration, and there are many limitations and restraints relating to trap design.
- Some fixture types have integral traps in the fixture body. A water closet is a good example (see below, left). These fixtures must not be separately trapped (double trapping).
- Some specific trap types are prohibited by most codes since they encourage loss of seal. Examples are bell traps, S-traps, pot traps, and crown vented traps.
- Seal can also be lost through evaporation, capillary action (a fibrous material in the trap siphoning water away), and excessive wind pressure.

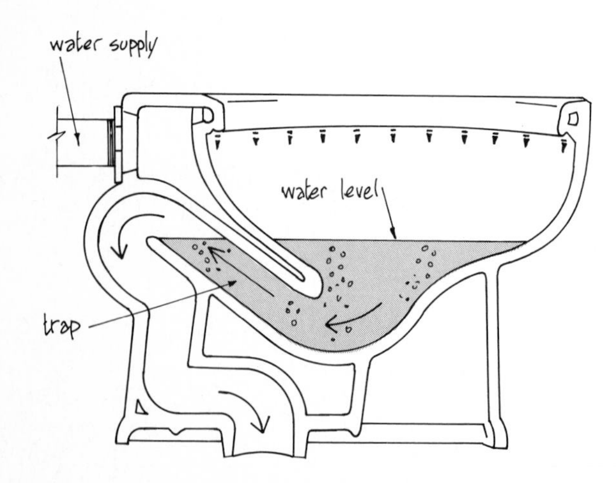

SECTION THROUGH A REVERSE-TRAP WATER CLOSET

VENTS

Vents are designed to circulate air through the waste pipe system, maintaining atmospheric pressure on both sides of the trap seals and venting sewer gases to the outside.

- Slugs of water, flushing through the waste system, can create high negative or positive pressures (in relation to atmospheric pressure) at the traps. If nearly equal pressure is not maintained, trap seals can be lost through these pressure forces.
- Venting system design is also fully controlled by code requirements. Some typical vent systems are illustrated on the next page.
- Besides loss of trap seal, problems occuring from improper venting include slow fixture drainage, noisy "gurgling" drainage, and back pressure strong enough to blow sewer gases through the trap seal into the building.

VENTING SYSTEMS

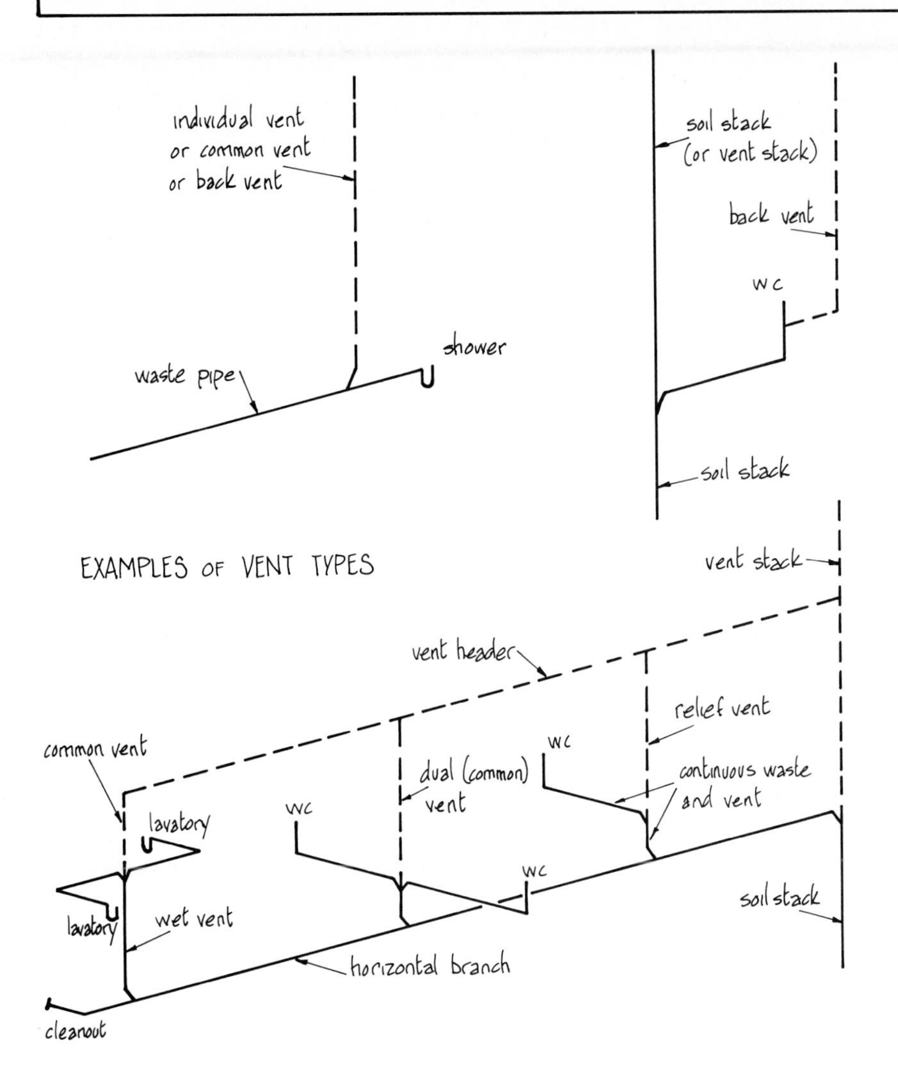

Shown here are three of the several types of venting systems commonly used on residential and commercial buildings. A thorough knowledge of code requirements is essential for correct installation. Vent sizes are determined in much the same way as for waste pipes. Fixture unit loading and minimum vent sizes for specific fixtures applies, but as a general rule vents must not be smaller than the largest branch entering them.

The service requirements for DWV pipes differ from water supply pipe in that minimum pressure requirements are much lower (usually a 5 to 10 ft (1.5-3 m) head of water) and the fittings used must ensure a smooth and uninterrupted flow of waste.

MATERIAL TYPES

- Plastic pipe, particularly ABS (see plastic materials table on 16-7), is extensively used for DWV piping, both above and below grade. There are a wide variety of fittings readily available to suit virtually every plumbing design situation. Pipe lengths are easily cut with a hand saw, and fitting joints are solvent welded.
- Rigid copper pipe and copper fittings are also available, but are used mostly on commercial and industrial projects above grade. Fitting joints are soldered or braised.
- Vitrified clay pipe, used below grade with plastic or rubber compression fittings, has largely been replaced by plastic pipe in other than special situations.
- Cast iron or steel pipe and fittings are also available, but are generally used only on commercial or industrial projects.
- Glass, stainless steel, or aluminum pipes are used in special installations where corrosive chemical waste is discharged.

SEPTIC SYSTEMS

Cities and towns usually have a sanitary sewer system that collects waste from public and private buildings for disposal at a centralized sewage treatment plant. In regions where this service is not available, other forms of private disposal systems must be used and a septic system is the most common method.
Liquid and solid waste is collected in a septic tank for treatment and then discharged to a leach field, where the clarified liquid is allowed to drain into the ground.
Assuming that all parts of the system operate within design limits and local codes are strictly observed, the process is considered to be relatively safe. The illustration below shows the arrangement of a typical system.

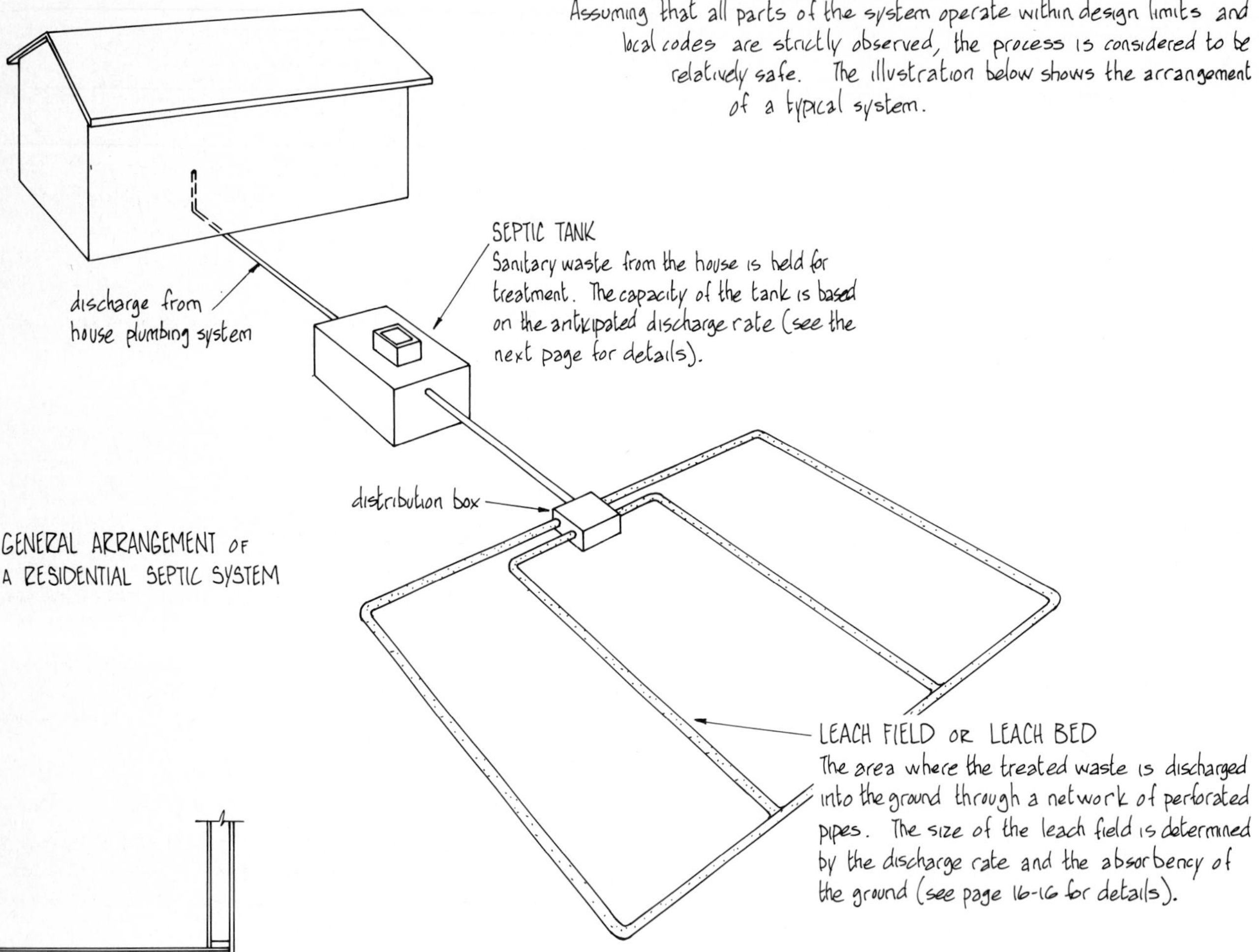

SEPTIC TANK
Sanitary waste from the house is held for treatment. The capacity of the tank is based on the anticipated discharge rate (see the next page for details).

discharge from house plumbing system

distribution box

GENERAL ARRANGEMENT OF A RESIDENTIAL SEPTIC SYSTEM

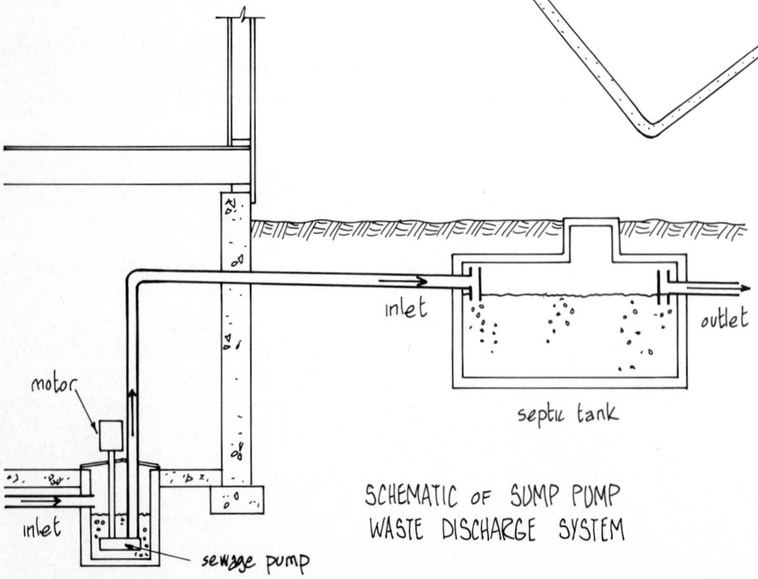

LEACH FIELD OR LEACH BED
The area where the treated waste is discharged into the ground through a network of perforated pipes. The size of the leach field is determined by the discharge rate and the absorbency of the ground (see page 16-16 for details).

inlet

outlet

septic tank

motor

inlet

sewage pump

SCHEMATIC OF SUMP PUMP WASTE DISCHARGE SYSTEM

BUILDING AND SEPTIC TANK CONNECTION
In buildings where the lowest drain level is below that of the septic tank (usually buildings with basements), the the plumbing system will discharge its waste into a sump at or below the lowest floor level. The collected waste is then pumped up and out of the building to the septic tank, as shown at the left. Buildings with the lowest drain level above the septic tank will normally not require a sump and can discharge directly into the septic tank.

SEPTIC TANKS

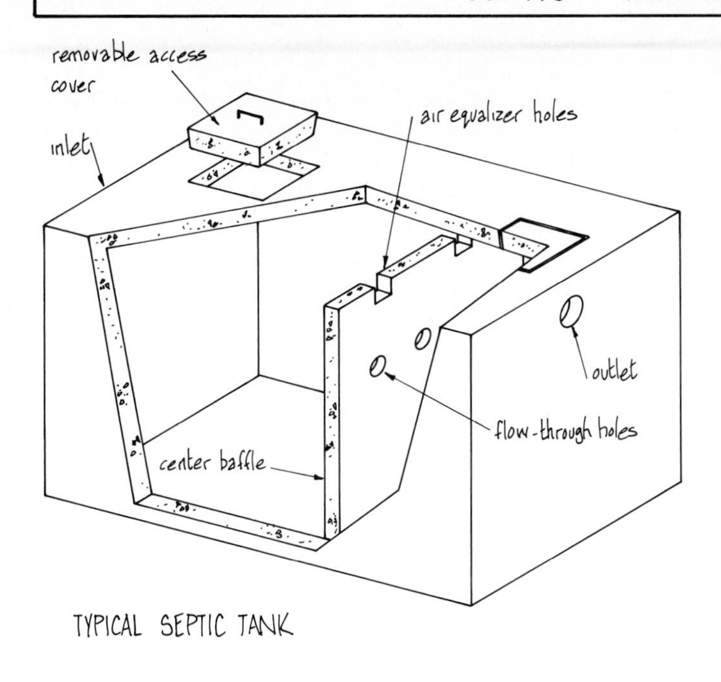

TYPICAL SEPTIC TANK

A septic tank is a watertight container wherein the collected sanitary liquid and solid waste is treated before disposal in the leach field.

- Sanitary waste contains on average up to 1 lb of solids per 100 gal of liquid (1 kg/1000 L). The heavier parts of the waste fall to the tank bottom while lighter waste and grease rises to the surface.
- Natural bacterial action digests the solids, yelding gases and relatively harmless liquids. As new sewage enters the tank from the building, an equal amount of clarified liquid is displaced into the leach field disposal piping while the gases are expelled through the building's vent system.
- After a number of years, there will be a residue of undigestable sludge in the tank. This must be pumped out and disposed of by a licenced contractor specializing in this type of work.
- Tank construction ranges from the more usual precast concrete to other materials, notably steel or fiberglass.

SEPTIC TANK SIZES

NUMBER OF BEDROOMS	MINIMUM LIQUID CAPACITY gal (L)
2 or less	750 (2900)
3	900 (3400)
4	1050 (4000)
5	1200 (4500)

Note: In all cases, minimum liquid depth is 4 ft (1200) with minimum 8 in (200) air space above the liquid.

TANK SIZING

- The table at the left gives minimum sizes for residential applications using the number of bedrooms as the determining factor. For commercial buildings, refer to local codes. Capacities are based on expected flow rates and the retention time needed for proper digestion.
- Also refer to codes for tank location restrictions. These will probably specify minimums of 5 ft (1.5 m) from the outer building lines, 100 ft (30 m) from water supply wells, and 10 ft (3 m) from water supply lines.

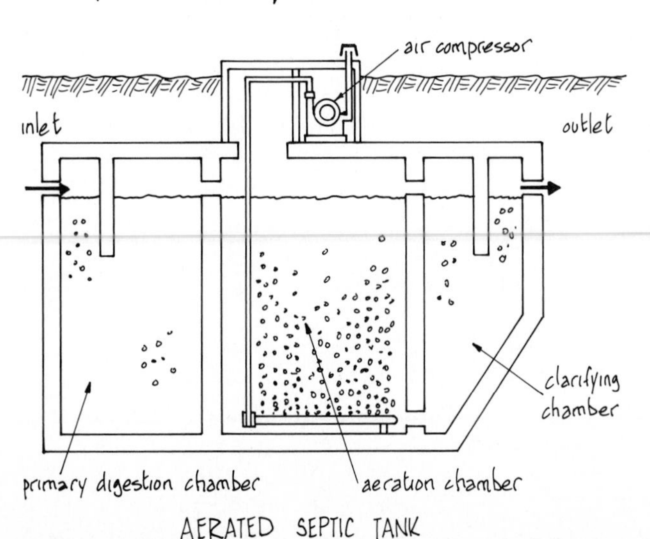

AERATED SEPTIC TANK

PRESSURIZED SYSTEMS

In regions with poor soil conditions (and where codes allow), a low-pressure pipe system (LPP) may be specified. The basic septic system is the same except that the effluent is pumped to the leach field under pressure to ensure a more even distribution. Field pipes are smaller and have fewer perforations, while pump operation is timed to provide a preset on/off cycle.

UPGRADED PROCESSING

In some regions it may be necessary to install an aeration-type tank. As shown at the left, it processes sewage in the same way as in a standard septic tank but with the addition of forced-air aeration and a separate clarification chamber. Air, blown through the liquid that has been partially digested in the primary chamber, promotes increased digestion of the solids and results in an effluant that is relatively pure.

LEACH FIELDS

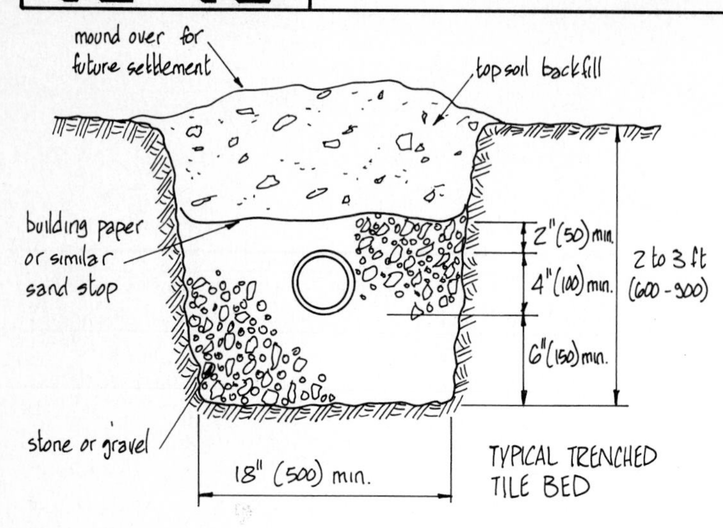

TYPICAL TRENCHED TILE BED

INSTALLATION

The section at the left shows the basic method of leach field pipe installation in a trench. An open excavation over the whole tile bed may also be used. Trenching is done when the existing grade is to be relatively undisturbed, while an open excavation is useful when the site must receive considerable backfill. In either case, the pipe must be located 16 to 24 in (400-600) below the finished grade, even in cold regions.

MATERIALS

Perforated plastic pipe, supplied in continuous rolls or rigid lengths, are in general use. Clay pipe, similar to footing drain tile (see 5-9), is now rarely used because of cost factors. All pipes to the leach field must be nonperforated.

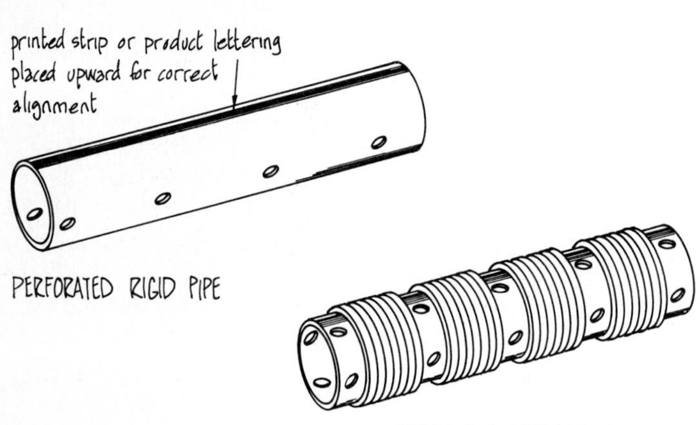

PERFORATED RIGID PIPE

FLEXIBLE PERFORATED PIPE

SIZING THE FIELD

Field size is determined by the soil percolation rate and the number of bedrooms:

- Local health authorities will test the porosity of the soil at the tile bed depth to determine the soil percolation rate. The faster the rate, the smaller the tile bed required.
- The number of bedrooms in relation to the percolation rate dictates the total area of the tile bed necessary to safely and efficiently absorb the expected effluent flow.
- Special allowances must be made for low-absorbency soil and fields for commercial establishments. Check local codes for details.

TWO EXAMPLES OF LEACH BED PATTERNS

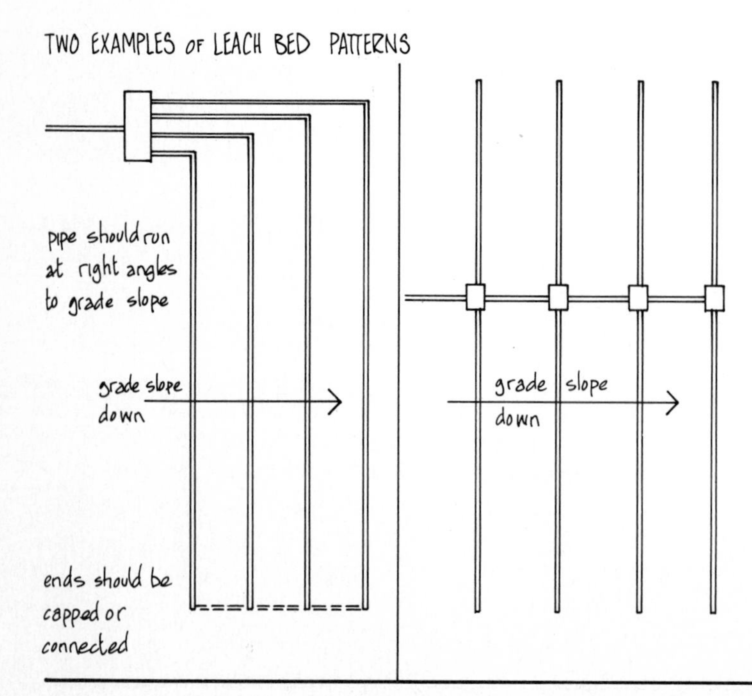

GENERAL INFORMATION

- Pipe in the tile bed must be laid with slope not exceeding ½ in in 10 ft (1:240) to achieve even distribution of effluent over the entire field. Particular care must be taken with field layout on excessively sloped building lots. See the typical layouts at the left.
- In general, only grass should be grown over the leach field since the roots of shrubs or trees can easily clog the tiles. Vehicular traffic over the field should also be avoided.
- The minimum clearances between field and water supply wells, shore lines, property lines, buildings, etc., must be carefully observed. Check local codes for a list of clearance requirements.

ALTERNATE WASTE DISPOSAL SYSTEMS

In regions where soil conditions preclude leach fields, other waste disposal means must be employed. Alternative methods fall into three general classes:

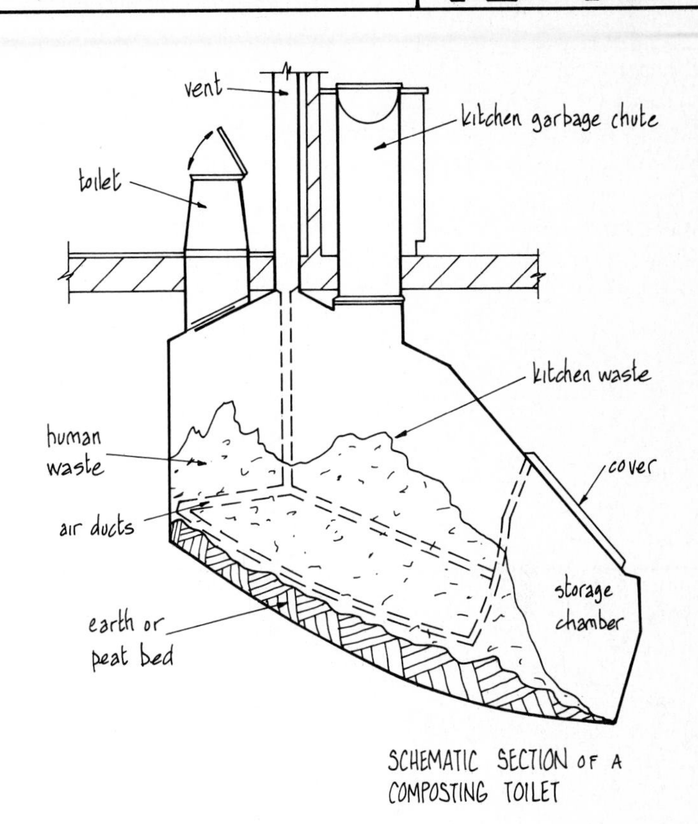

SCHEMATIC SECTION OF A COMPOSTING TOILET

HOLDING TANKS

Simple nontreated storage of waste in a large holding tank. The tank is pumped dry at suitable intervals and the waste transported to a central disposal site (usually a municipal sewage treatment plant).

CHEMICAL WASTE TREATMENT

These are closed systems where waste is collected in a holding tank and chemically treated to remove odors and purify the water. The more sophisticated systems filter and reuse the treated water for flushing the water closet. The holding tank must be pumped dry and the chemicals recharged after a certain number of uses. A typical example of this system is illustrated below.

ORGANIC WASTE TREATMENT SYSTEMS

These are essentially holding tanks wherein organic processes transform the solid and liquid waste into humus that can be safely used as a garden fertilizer. Vapor and other gases are vented to atmosphere. These units will accept both bathroom and kitchen wastes, are sized to meet most residential demands and can be installed in indoor or outdoor locations. See above for a typical system schematic.

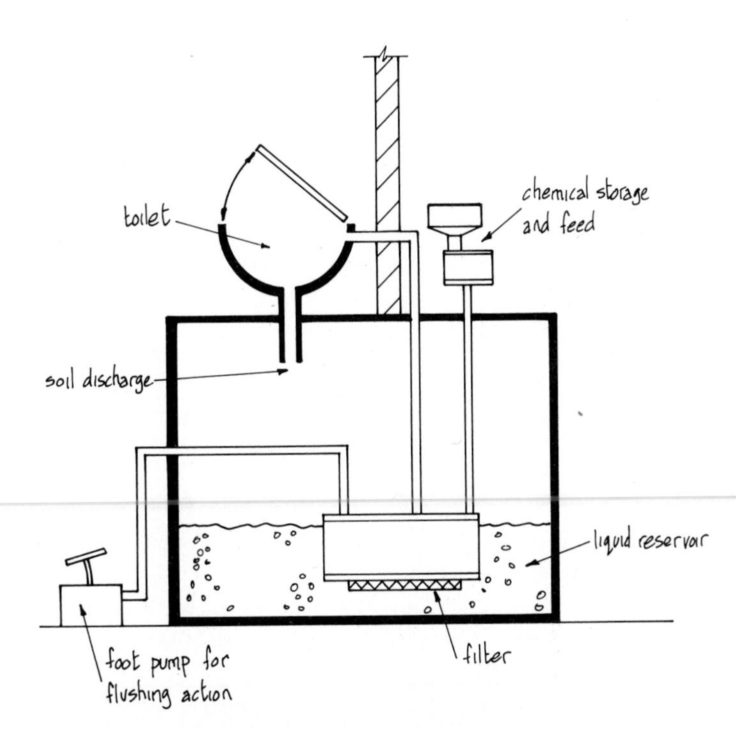

SCHEMATIC OF A CHEMICAL TOILET

RAINWATER DISPOSAL

The collection and legal disposal of rainwater is considered an important part of the design process. In many regions storm water volume is considerable and when improperly collected or channeled for disposal can be a danger to the structure as well as being a potential health hazard.

Typical collection systems for residential and commercial applications are illustrated on this page and the next. Actual disposal methods vary throughout the country but generally fall into three groups:

- Storm sewers, separate from sanitary sewers, are common in most population centers. Where they are in place, local codes will require connection of rainwater collection systems to the sewer. Cross-connections between storm and sanitary sewers at any point in the plumbing system are strictly forbidden.
- Surface discharge, where collected water is discharged directly onto the outside grade but away from the building. With this method, care must be taken to correctly slope the finished grade away from building lines.
- Discharge into soaking pits, street gutters, open storm drains, or bodies of water are subject to local approval and regulation.

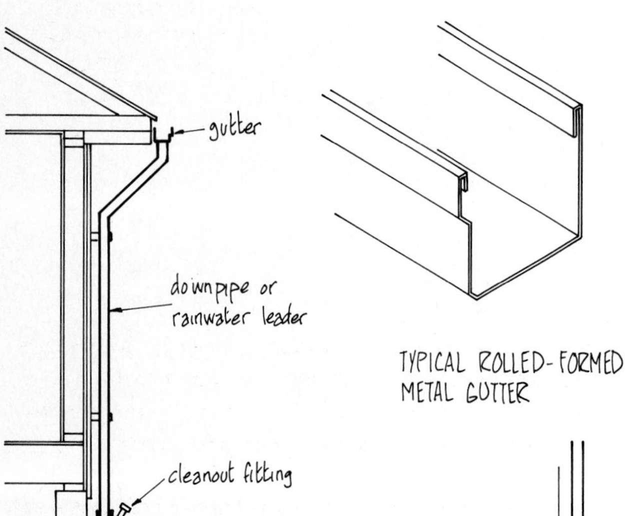

TYPICAL ROLLED-FORMED METAL GUTTER

ALUMINUM/STEEL PREPAINTED MATERIAL
- Gutters can be roll-formed at the job-site to virtually any continuous length, thus eliminating all gutter joints.
- Variety of colors are available.

PLASTIC MOLDED MATERIAL
- Short lengths with watertight joints and fittings.
- Solid color throughout material, but usually only white or brown.

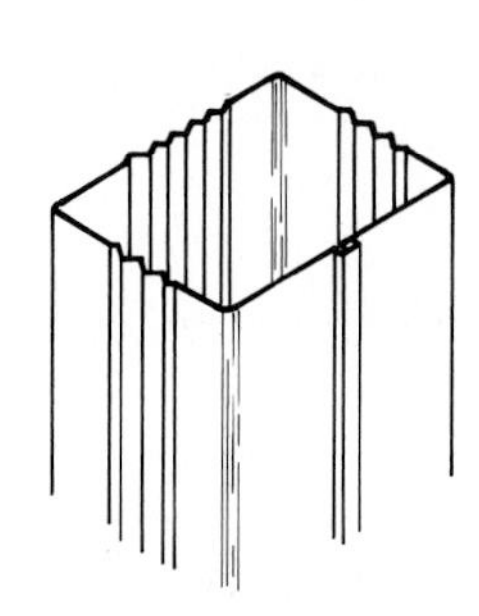

ROLLED-FORMED METAL DOWNPIPE

storm sewer collector
foundation weeping tile

STORM SEWER COLLECTION

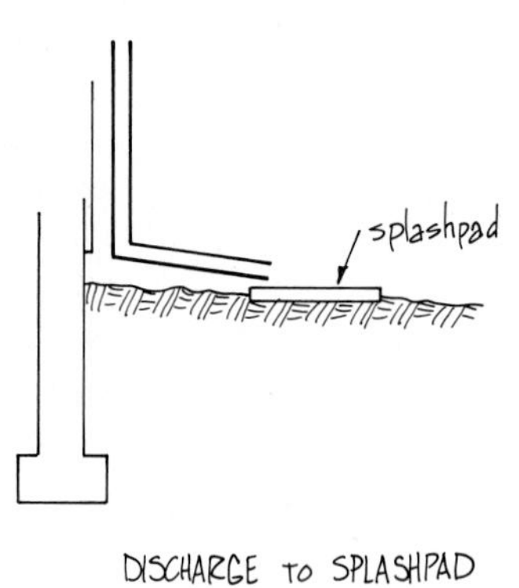

DISCHARGE TO SPLASHPAD

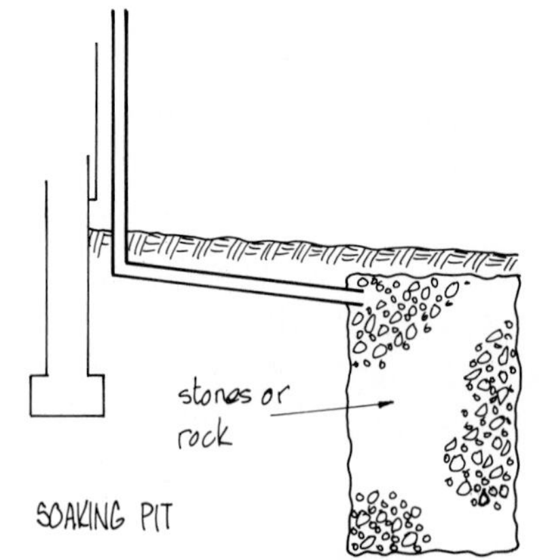

SOAKING PIT

RESIDENTIAL SYSTEMS

The illustrations above show three typical rainwater collection and disposal methods. Gutters on the outer edges of sloping roofs collect rainwater runoff and direct it to a downpipe (or rainwater leader).

- Connection of the downpipe to the storm sewer may be made inside or outside the building (local codes must be consulted). A cleanout fitting is required at the top of the below-grade portion of the rainwater leader.
- For surface discharge, the downpipe is extended outward from the wall line at or close to grade level. A splashpad to reduce erosion is recommended.
- A soaking pit (or drywell) is essentially a large below-grade hole filled with gravel, rocks, or concrete blocks that acts as a holding tank until the storm water can percolate into the surrounding earth. An overflow fitting should be installed above grade level on the downpipe. Exact size, construction, and location of the pit are governed by local codes.

RAINWATER DISPOSAL

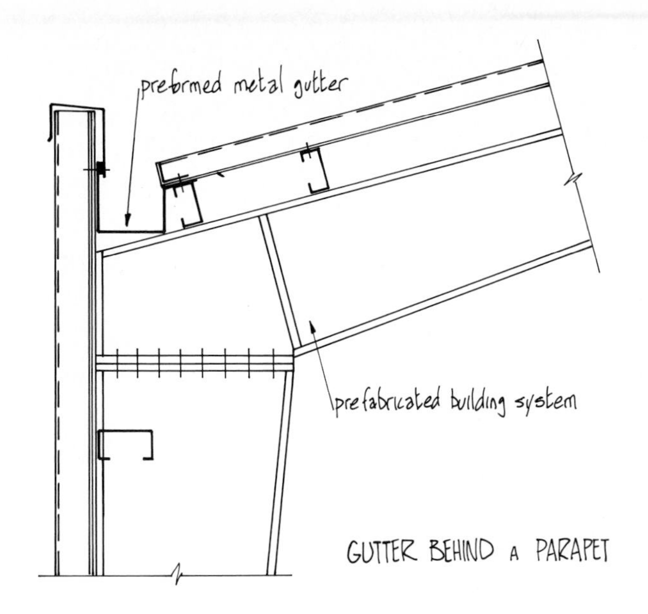

GUTTER BEHIND A PARAPET

COMMERCIAL AND INDUSTRIAL BUILDINGS

Although there is obviously a wide variation in roof shapes, buildings in this category tend to be designed with flat roof systems. Flat roofs, however, are never actually flat – there is always a slope for drainage purposes (see 8-30 for typical examples). Depending on roof type, storm water will be collected at roof-edge gutters or at central roof drains.

- Gutters can overhang the exterior walls as in residential construction or can be part of a parapet wall design as shown at the left. Interior or exterior downpipes, located along the gutter system, will direct the runoff water for disposal.
- Roof drains, located at the lowest levels of the roof, directly collect runoff water for disposal. In some roof systems the roof drain is elevated to create pooling of storm water for summer cooling purposes. An example of a roof drain is shown at the center left.

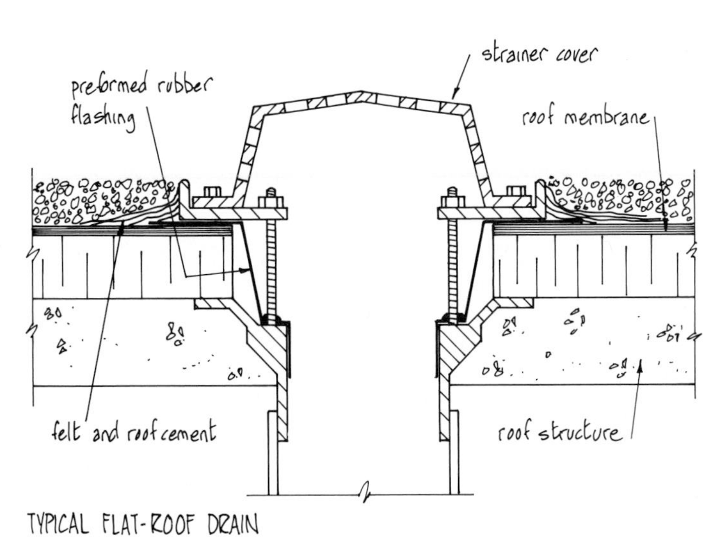

TYPICAL FLAT-ROOF DRAIN

STORM SYSTEM SIZING

For most residential buildings standard-sized gutters and downpipes are more than adequate in all regions. For commercial and industrial buildings the storm systems are specifically engineered to meet code and regional rainfall requirements. Systems are designed on the basis of:

- The roof area served by each gutter, downpipe, and storm sewer. This will include not only the actual roof but other surfaces, such as vertical walls, parapets, etc.
- Maximum expected rainfall per hour. This figure varies with locality and will be specified in local codes. Values range from 1.7 in/hour (43) at Juneau, Alaska, to 9.4 in/hour (239) at Pensacola, Florida.

The table below gives typical sizes for storm system components based on 5 in/hour (127) maximum rainfall. Refer to codes for other storm values.

ROOF AREA / PIPE SIZE FOR 5 in (127) RAINFALL PER HOUR

	MAXIMUM ROOF AREA FT² (M²)				
PIPE SLOPE	PIPE SIZE in (mm)				
in/ft (ratio)	2 (50)	3 (75)	4 (100)	5 (125)	6 (150)
1/8 (1:100)	270 (83)	747 (228)	1560 (475)	2810 (850)	4450 (1360)
1/4 (1:50)	380 (115)	1080 (330)	2210 (675)	4000 (1220)	6290 (1920)
1/2 (1:25)	460 (140)	1270 (390)	3080 (940)	5620 (1715)	8880 (2700)
gutters	—	635 (195)	1540 (470)	2810 (850)	4440 (1350)
leaders	460 (140)	1270 (390)	3080 (940)	5620 (1715)	8880 (2700)

A

B

C

SERVICE ENTRANCE - STUCCO